Grundlehren der mathematischen Wissenschaften 251

A Series of Comprehensive Studies in Mathematics

Springer
New York
Berlin
Heidelberg
Barcelona
Budapest
Hong Kong
London
Milan
Paris
Santa Clara
Singapore
Tokyo

Shui-Nee Chow   Jack K. Hale

Methods of Bifurcation Theory

Corrected Second Printing

With 97 Illustrations

Springer

Shui-Nee Chow
Department of Mathematics
Michigan State University
East Lansing, Michigan   48824
U.S.A.

Jack K. Hale
Division of Applied Mathematics
Brown University
Providence, Rhode Island   02912
U.S.A.

AMS Subject Classifications 34Bxx, 34Cxx, 34Kxx, 35Bxx,
47A55, 47H15, 47H17, 58Cxx, 58Exx

Library of Congress Cataloging in Publication Data

Chow, Shui-Nee.
  Methods of bifurcation theory.
  (Grundlehren der mathematischen
Wissenschaften; 251)
  Bibliography: p.
  Includes index.
  1. Functional differential equations.
2. Bifurcation theory.   3. Manifolds
(Mathematics)   I. Hale, Jack K.   II. Title.
III. Series.
QA372.C544   515.3′5   81-23337   AACR2

9 8 7 6 5 4 3 2 (Corrected second printing, 1996)

ISBN 0-387-90664-9   Springer-Verlag New York Heidelberg Berlin
ISBN 3-540-90664-9   Springer-Verlag Berlin Heidelberg New York

To Marie and Hazel

Preface

An alternative title for this book would perhaps be Nonlinear Analysis, Bifurcation Theory and Differential Equations. Our primary objective is to discuss those aspects of bifurcation theory which are particularly meaningful to differential equations.

To accomplish this objective and to make the book accessible to a wider audience, we have presented in detail much of the relevant background material from nonlinear functional analysis and the qualitative theory of differential equations. Since there is no good reference for some of the material, its inclusion seemed necessary.

Two distinct aspects of bifurcation theory are discussed—static and dynamic. Static bifurcation theory is concerned with the changes that occur in the structure of the set of zeros of a function as parameters in the function are varied. If the function is a gradient, then variational techniques play an important role and can be employed effectively even for global problems. If the function is not a gradient or if more detailed information is desired, the general theory is usually local. At the same time, the theory is constructive and valid when several independent parameters appear in the function. In differential equations, the equilibrium solutions are the zeros of the vector field. Therefore, methods in static bifurcation theory are directly applicable.

Dynamic bifurcation theory is concerned with the changes that occur in the structure of the limit sets of solutions of differential equations as parameters in the vector field are varied. For example, in addition to discussing the way that the set of zeros of the vector field (the equilibrium solutions) change through the static theory, the stability properties of these solutions must be considered. In fact, there is an intimate relationship between changes of stability and bifurcation. The dynamics in a differential equation can also introduce other types of bifurcations; for example, periodic orbits, homoclinic orbits, invariant tori. This introduces several difficulties which require rather advanced topics from differential equations for their resolution.

The introductory chapter is designed to acquaint the reader with some of the types of problems that occur in bifurcation theory. The tools from nonlinear functional analysis are presented in Chapter 2. Some of this material is used more extensively in the text than others, but all topics are a necessary part of the vocabulary of persons working in bifurcation theory.

Some of the presentations and details of proofs are different from standard ones.

Chapter 3 gives applications of the Implicit Function Theorem. These are not bifurcation problems. Some of the applications were chosen because the material is needed in later chapters. Others give good illustrations of some of the tools in Chapter 2.

Chapters 4–8 deal with static bifurcation theory. Chapter 4 contains the fundamental elements of variational theory together with serious applications to Hamiltonian systems, elliptic and hyperbolic problems.

Chapters 5–8 deal almost entirely with analytic methods in local static bifurcation theory. In Chapter 5, for functions depending on a scalar parameter, conditions are given to ensure that there is always a bifurcation near equilibrium. These conditions are based on the linear approximation and are independent of the nonlinearities. Some global results are also included.

In Chapter 6, the case of a one-dimensional null space for the linear approximation is analyzed in detail under generic conditions on the quadratic and cubic terms. The effects of symmetry are also discussed. Chapter 7 is concerned with the case where the linear approximation has a two-dimensional null space with the quadratic and cubic terms satisfying some nondegeneracy conditions. Both of these chapters contain constructive procedures in the analysis. Chapter 8 contains applications to the buckling of plates, chemical reactions and Duffing's equation.

Chapters 9–13 are devoted to dynamic bifurcation theory. Chapter 9 is concerned with the bifurcation from an equilibrium point in the case when the linear approximation has either one zero eigenvalue or a pair of purely imaginary eigenvalues. It is shown that all relevant information on existence and stability is contained in the bifurcation function obtained via the alternative method or the method Liapunov–Schmidt. The hypotheses on the linear part are the typical situation for one parameter families of vector fields. Chapter 10 is devoted to the other bifurcation phenomena that occur in the plane for typical one parameter families of autonomous vector fields.

In Chapter 11, we discuss periodic planar vector fields and especially Hamiltonian systems with a small damping and small periodic forcing term. Emphasis is placed on the existence of subharmonic solutions and the role of successive bifurcations through subharmonics in the creation of homoclinic points and a type of random behavior.

In Chapter 12, averaging, the theory of normal forms and the theory of integral manifolds for ordinary differential equations are presented. This material is relevant to the discussion of bifurcation to tori considered in that chapter as well as the problems in Chapter 13, which is devoted to the behavior of the solutions of a differential equation near an equilibrium point when the linear part of the vector field is typical of two parameter problems.

The topics in Chapter 14 on perturbation of the spectrum of linear operators is distinct from the ones in the previous chapters. It is included because the same methods can be applied to yield elementary proofs of some results in this field.

The material in this book can be easily adapted to several types of one semester courses. For example, four possible reasonable arrangements could be:

    I.   Chapters 1, 5, 6, 7, 8 with Sections 2.3–2.8 from Chapter 2.
   II.   Chapter 2, 3, 4.
 III.   Chapters 1, 9, 10 with Sections 2.3, 2.4, 2.5 from Chapter 2.
 IV.   Chapters 11, 12, 13.

Examples I, II, III are independent and require minimal knowledge of differential equations. Example IV can only be taught after III and requires more sophisticated concepts from differential equations.

The authors are indebted to numerous colleagues and students for their assistance in this work. We especially thank John Mallet–Paret with whom we had so many stimulating conversations about technique and method of presentation. Luis Magalhães also was of great assistance, especially in the presentation in Chapter 11 and the examples in Chapter 12. We have also been assisted by many persons in the preparation of the final manuscript. We are indebted especially to Eleanor Addison, Dorothy Libutti, Sandra Spinacci, Kate MacDougall, Mary Reynolds and Diane Norton. The second author is also indebted to the Guggenheim Foundation for a Fellowship during 1979–80.

Contents

Chapter 1

Introduction and Examples

1.1. Definition of Bifurcation Surface

Every problem in applications contains several physical parameters which may vary over certain specified sets. Thus, it is important to understand the qualitative behavior of the system as the parameters vary. A good design for a system will always be such that the qualitative behavior does not change when the parameters are varied a small amount about the value for which the original design was made. However, the behavior may change when the system is subjected to large variations in the parameters. A change in the qualitative properties could mean a change in stability of the original system and thus the system must assume a state different from the original design. In vague terms, the values of the parameters where this change takes place are called bifurcation values. Knowledge of the bifurcation values is absolutely necessary for the complete understanding of the system. Our objective is to give an overall view of methods and results in this area. To facilitate the introductory discussion, we first define the basic problem and give a precise definition of a bifurcation value.

Suppose X, Z are Banach spaces, Λ is an open set in a Banach space, $M: \Lambda \times X \to Z$ is continuous together with its first Frechet derivative. The set Λ will be called the *parameter set*. Sometimes, more derivatives on M are required and it will always be assumed that M has as many derivatives as necessary if it is not always explicitly stated. Consider the equation

$$(1.1) \qquad\qquad M(\lambda, x) = 0$$

for $\lambda \in \Lambda$, $x \in X$.

A *solution* of Equation (1.1) is a point $(\lambda, x) \in \Lambda \times X$ such that Equation (1.1) is satisfied. Let $S \subset \Lambda \times X$ denote the set of solutions of Equation (1.1) and, for any $\lambda \in \Lambda$, let

$$S_\lambda = \{x \in X : (\lambda, x) \in S\}.$$

In a physical system, Equation (1.1) generally represents the equilibrium positions of the system or, more generally, equations for the state of the system which satisfy certain boundary conditions. The dynamics of the system are not included in Equation (1.1). Stability of a solution of Equation (1.1) often requires a discussion of a differential equation $du/dt = \bar{M}(u, \lambda)$ for u near $x \in S_\lambda$ and $\bar{M}$ related to M in some way.

If S is any closed set in $\Lambda \times X$, one can always construct an M such that S is the solution set for Equation (1.1). As a consequence, it is impossible to give a complete description of S_λ as λ varies. In the applications, the most typical solution sets have components consisting of various pieces of the solution set depicted in Figure 1.1.

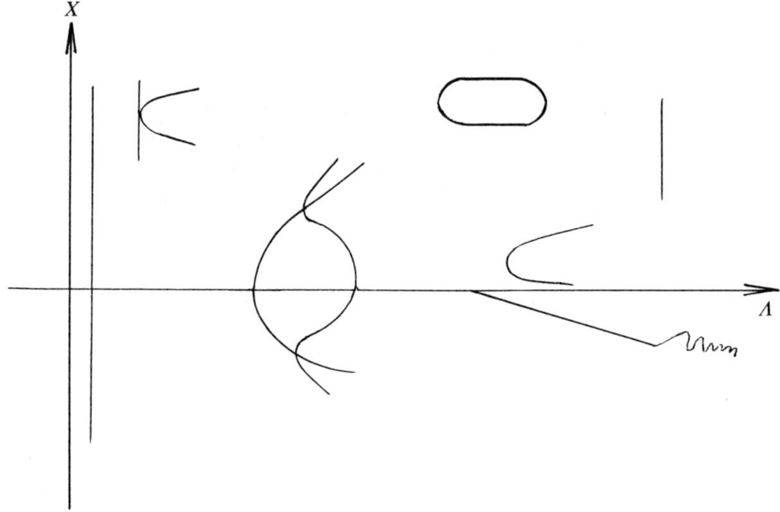

Figure 1.1

The basic problem is to discuss the dependence of the set S_λ on λ. Suppose U is an open set in X. We say S_λ is *equivalent to* S_μ, $S_\lambda \sim S_\mu$, in U if $S_\lambda \cap U$ is homeomorphic to $S_\mu \cap U$. We say λ_0 is a *bifurcation point for* $(S, \sim)$ if, for any neighborhood V of λ_0, there is an $x_0 \in S_{\lambda_0}$, a neighborhood U of x_0 and λ_1, λ_2 in V such that $S_{\lambda_1} \not\sim S_{\lambda_2}$ in U. In particular, a point λ_0 is a bifurcation point if $S_{\lambda_0} \neq \varnothing$, the empty set, and there is an $x_0 \in S_{\lambda_0}$ such that, for any neighborhood U of (λ_0, x_0), there are two distinct solutions (λ, x_1), $(\lambda, x_2) \in U$; that is, there is a $\lambda \in \Lambda$, $x_1, x_2 \in X$, $x_1 \neq x_2$, such that (λ, x_1), $(\lambda, x_2) \in U$ and (λ, x_1), (λ, x_2) satisfy Equation (1.1). Whenever possible, it is desirable to have the complete characterization of the solution set S in a neighborhood U of a solution (λ_0, x_0) for which λ_0 is a bifurcation point.

The Implicit Function Theorem shows that the characterization is trivial near some solutions (λ_0, x_0).

Lemma 1.1. *If* $(\lambda_0, x_0) \in S$ *and* $D_x M(\lambda_0, x_0)$ *has a bounded inverse, then there is a neighborhood* Ω *of* (λ_0, x_0) *such that* $S \cap \Omega$ *is a diffeomorphic image of a* *neighborhood of λ_0; more precisely, there is a neighborhood Λ_0 of λ_0 and a continuously differentiable function $x^*: \Lambda_0 \to X$ such that $S \cap \Omega = \{(\lambda, x^*(\lambda)), \lambda \in \Lambda_0\}$.*

As a consequence of Lemma 1.1, the only points $(\lambda_0, x_0) \in S$ that require further discussion are those for which $D_x M(\lambda_0, x_0)$ is singular.

1.2. Examples with One Parameter

If $X = Z$, $M(\lambda, x) = Bx - \lambda x$, where $\lambda \in \mathbb{R}$ and $B: X \to X$ is a bounded linear operator, then any eigenvalue λ_0 of B is a bifurcation point. In fact, if λ_0 is an eigenvalue of B, then, for any $\varepsilon > 0$, there is an $x_0 \in X$, $|x_0| = \varepsilon$ such that $Bx_0 = \lambda_0 x_0$. On the other hand, if

$$(2.1) \qquad M(\lambda, x) = Bx - \lambda x + 0(|x|^2 + |\lambda - \lambda_0|^2 |x|)$$

as $x \to 0$, $\lambda \to \lambda_0$ where λ_0 is an eigenvalue of B, the point λ_0 may not be a bifurcation point. Notice that $M(\lambda, 0) = 0$ for all $\lambda \in \Lambda$ so that $(\lambda, 0)$ is a solution. In fact, suppose $x = (x_1, x_2) \in \mathbb{R}^2$ and consider the equations

$$(2.2) \qquad \begin{aligned} x_2(x_1^2 + x_2^2) + \lambda x_1 &= 0 \\ -x_1(x_1^2 + x_2^2) + \lambda x_2 &= 0. \end{aligned}$$

Any solution of Equation (2.2) must satisfy $(x_1^2 + x_2^2)^2 = 0$ for all λ; that is, $x_1 = x_2 = 0$. Thus, $\lambda = 0$ is not a bifurcation point and it is a double eigenvalue of the linear part of Equation (2.2).

To emphasize the role of the nonlinearities in Equation (2.2), we make only a sign change and bifurcation will occur at $\lambda = 0$. In fact, the equations

$$(2.3) \qquad \begin{aligned} x_2(x_1^2 + x_2^2) + \lambda x_1 &= 0 \\ x_1(x_1^2 + x_2^2) + \lambda x_2 &= 0 \end{aligned}$$

have the solutions $x_2 = x_1$, $x_1(2x_1^2 + \lambda) = 0$. Thus, $\lambda = 0$ is a bifurcation point.

It is also possible to give a similar example in $\mathbb{R}^2$ for the case in which the null space $\mathcal{N}(B)$ has dimension one. In fact, consider the equation

$$(2.4) \qquad \begin{aligned} x_2 + \lambda x_1 &= 0 \\ \lambda x_2 - x_1^3 &= 0 \end{aligned}$$

near $(\lambda_0, x_{10}, x_{20}) = (0, 0, 0)$. The matrix B for this case is

$$B = \begin{pmatrix} 0 & 1 \\ 0 & 0 \end{pmatrix}$$

which has $\lambda_0 = 0$ as an eigenvalue of multiplicity two but $\mathcal{N}(B)$ has dimension one. The above equation is equivalent to $x_2 + \lambda x_1 = 0$, $x_1(\lambda^2 + x_1^2) = 0$ which has only the solution $x_1 = x_2 = 0$ for all $\lambda \in \mathbb{R}$. Therefore, $\lambda_0 = 0$ is not a bifurcation point.

If $\lambda_0 \in \mathbb{R}$ is a simple eigenvalue of B and M is given as in Relation (2.1), then λ_0 is always a bifurcation point. Although an even more general result will be given in Chapter 5, it is instructive to do the simplest case here. Suppose $X = Z = \mathbb{R}^n$, $\Lambda = \mathbb{R}$ and M is given in Relation (2.1). If λ_0 is a simple eigenvalue of B, then we may assume that $B = \mathrm{diag}(0, B_0)$ where $B_0 - \lambda_0 I$ is an $(n-1) \times (n-1)$ nonsingular matrix. If $x = (y, z)$, $y \in \mathbb{R}$, $z \in \mathbb{R}^{n-1}$, then $M(\lambda, x) = 0$ is equivalent to:

$$(2.5) \quad \begin{aligned} &\text{(a)} & \mu y &= g(\mu, y, z) \\ &\text{(b)} & (B_0 - \lambda_0 I)z &= h(\mu, y, z) \end{aligned}$$

where $\mu = \lambda - \lambda_0$, and both g, h are $0(y^2 + |z|^2 + \mu^2(|y| + |z|))$ as $y, z, \mu \to 0$. Since $B_0 - \lambda_0 I$ is nonsingular, one can invoke the Implicit Function Theorem to solve Equation (2.5b) for $z = z^*(\mu, y)$ near $\mu = 0$, $y = 0$, $z^*(\mu, 0) = 0$. Therefore, Equations (2.5) are equivalent to the scalar equation

$$(2.6) \quad \mu y = g(\mu, y, z^*(\mu, y)).$$

Since $z^*(\mu, 0) = 0$, $g(\mu, 0, 0) = 0$, one can divide Equation (2.6) by y to obtain the equivalent equation for $y \neq 0$

$$(2.7) \quad \mu = \bar{g}(\mu, y)$$

where $\bar{g}(\mu, y) = g(\mu, y, z^*(\mu, y))/y$ satisfies $\bar{g}(0, 0) = 0$, $D_\mu \bar{g}(0, 0) \neq 0$. Thus, the Implicit Function Theorem implies there is a solution $\mu = \mu^*(y)$ for y near zero and $\mu^*(0) = 0$. This proves $\mu = 0$ (or $\lambda = \lambda_0$) is a bifurcation point. In Figure 2.1, we have depicted a possible solution set. Without further restrictions on the nonlinearity, we can only assert that the curve $\mu = \mu^*(y)$ is a smooth curve passing through $(0, 0)$.

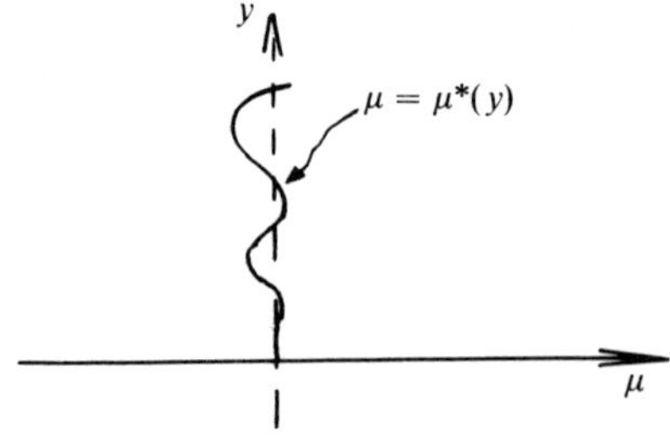

Figure 2.1

The procedure used in the above example is known as the method of Liapunov-Schmidt or the method of alternative problems. The method in abstract form is fundamental in bifurcation theory.

1.3. The Euler–Bernoulli Rod

To illustrate some other ideas that occur in bifurcation theory, let us consider the Euler–Bernoulli problem of the buckling of a rod. In equilibrium position, suppose the rod of length l is represented in the (x, y)-plane by the set $\{(x,0), 0 \leq x \leq l\}$, the rod is fixed at $(0,0)$ and the other end $(l,0)$ is allowed to vary along the x-axis when it is subjected to a constant horizontal force P at the right end. If s represents arc length along the displaced rod, $\phi(s)$ represents the angle which the unit tangent vector to the rod makes with the x-axis (see Figure 3.1) and it is assumed that the change in curvature is proportional to the moment of force, then the equations describing the displacement of the rod are

(3.1)
$$Py = -k \frac{d\phi}{ds}$$

$$\frac{dy}{ds} = \sin \phi$$

where k is a constant and the boundary conditions are

(3.2)
$$y(0) = y(l) = 0, \qquad x(0) = 0.$$

If $P \neq 0$, Equations (3.1) and Boundary Conditions (3.2) are equivalent to

(3.3)
$$\frac{d^2\phi}{ds^2} + \lambda \sin \phi = 0$$

$$\phi'(0) = \phi'(l) = 0.$$

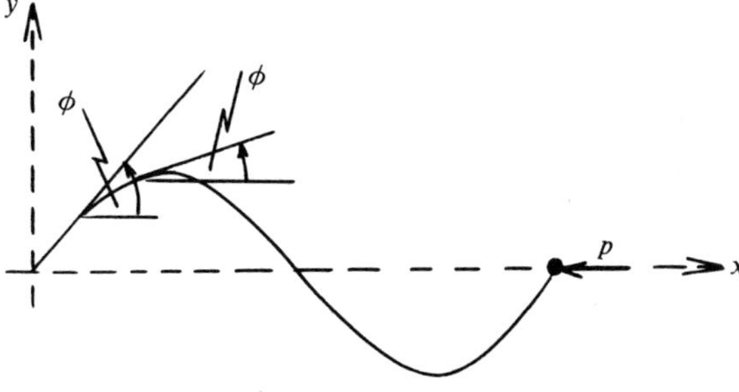

Figure 3.1

Let $C^k[0, l] = \{f : [0, l] \to \mathbb{R} : f$ is continuous together with derivatives up through order $k\}$. For any $f \in C^k[0, l]$, let

$$|f|_k = \max\{\sup[|f^{(j)}(s)| : 0 \le s \le l], j = 0, 1, \ldots, k\}.$$

Let $X = \{f \in C^2[0, l] : f'(0) = f'(l) = 0\}$, $Z = C^0[0, l]$ and define

$$(3.4) \qquad F : \mathbb{R} \times X \to Z, \qquad F(\lambda, \phi) = \frac{d^2\phi}{ds^2} + \lambda \sin \phi.$$

Our objective is to determine all solutions of the equation

$$(3.5) \qquad F(\lambda, \phi) = 0, \qquad (\lambda, \phi) \in \mathbb{R} \times X$$

which have ϕ "close" to the trivial solution $\phi = 0$.

Let us first consider the linear problem

$$(3.6) \qquad D_\phi F(\lambda, 0)\psi = 0$$

which is equivalent to

$$\frac{d^2\psi}{ds^2} + \lambda\psi = 0$$

$$(3.7)$$

$$\psi'(0) = \psi'(l) = 0.$$

It is easy to verify that this equation has a nontrivial solution if and only if $\lambda = \lambda_m = m^2\pi^2/l^2$, $m = 0, 1, 2, \ldots$ and, for $\lambda = \lambda_m$, every solution is a constant multiple of $\psi_m(s) = \cos\sqrt{\lambda_m}s$. For the original problem of the bar, $m = 0$ is not possible since $m = 0$ and $P \ne 0$ implies that $y(s) = 0$, $0 \le s \le l$. Figure 3.2 shows the displaced rod corresponding to λ_m for $m = 1, 2, 3$.

If $\phi(s)$ is a solution of Equation (3.3) and $p(s) = d\phi(s)/ds$, then

$$G(\lambda, p(s), \phi(s)) = G(\lambda, p(0), \phi(0)), \qquad 0 \le s \le l,$$

$$(3.8)$$

$$G(\lambda, p, \phi) = \frac{p^2}{2} + \lambda(1 - \cos \phi).$$

In the (p, ϕ)-plane, the orbits are shown in Figure 3.3. The Boundary Condition (3.2) imply that we are only interested in those solutions satisfying Relation (3.8) for which ϕ is small and which begin at $s = 0$ on the ϕ-axis and end at $s = l$ on the ϕ-axis; that is, the only candidates for solutions are those which are periodic in s and encircle $(p, \phi) = (0, 0)$. Of course, the solution may encircle the origin any number of times before it returns to the ϕ-axis at $s = l$; that is the period may be very small compared with l. In terms of the original rod, this means more zeros (nodes) as shown in Figure 3.2.

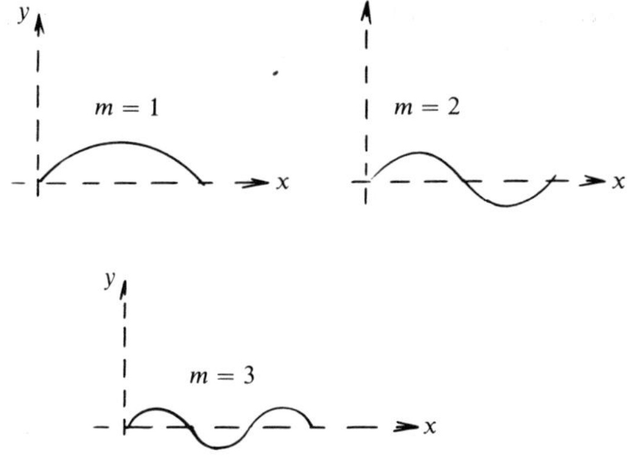

Figure 3.2

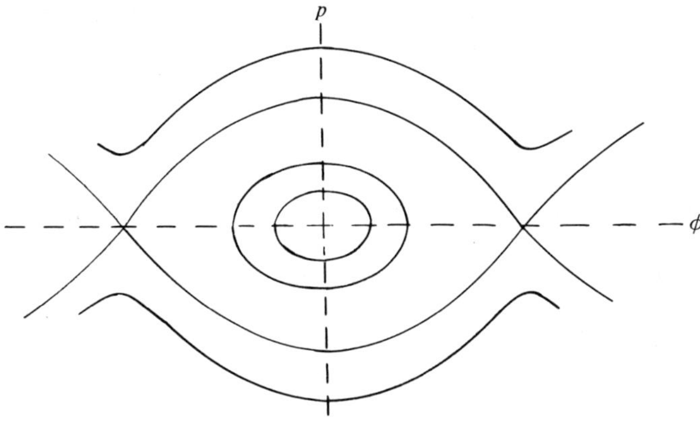

Figure 3.3

To determine the solutions mentioned above, we must solve the equation

$$(3.9) \qquad G(\lambda, 0, \phi(l)) = G(\lambda, 0, \phi_0)$$

for (λ, ϕ_0). Using the fact that every solution of Equation (3.1) must satisfy Relation (3.8), we first determine the minimum positive value s_0 as a function of (λ, ϕ_0), $\phi_0 > 0$ such that a solution of Equation (3.1) that begins at $(0, -\phi_0)$ at $s = 0$ takes on the value $(0, \phi_0)$ at $s = s_0$. From Relation (3.8)

$$\frac{d\phi}{ds} = \sqrt{2\lambda}[\cos \phi - \cos \phi_0]^{1/2}$$

$$= 2\sqrt{\lambda}\left[\sin^2 \frac{\phi_0}{2} - \sin^2 \frac{\phi}{2} \right]^{1/2}$$

From this expression, the symmetry in the equation and the change of variable $\sin \phi/2 = \sin \phi_0/2 \sin \psi$, one obtains

$$
s_0 = \frac{1}{2\sqrt{\lambda}} \int_{-\phi_0}^{\phi_0} \frac{d\phi}{\left[\sin^2 \dfrac{\phi_0}{2} - \sin^2 \dfrac{\phi}{2} \right]^{1/2}}
$$

$$
\text{(3.10)} \qquad = \frac{1}{\sqrt{\lambda}} \int_0^{\phi_0} \frac{d\phi}{\left[\sin^2 \dfrac{\phi_0}{2} - \sin^2 \dfrac{\phi}{2} \right]^{1/2}}
$$

$$
= \frac{2}{\sqrt{\lambda}} \int_0^{\pi/2} \frac{d\psi}{\left[1 - \sin^2 \dfrac{\phi_0}{2} \sin^2 \psi \right]^{1/2}} \overset{\text{def}}{=} \frac{\gamma(\phi_0)}{\sqrt{\lambda}}
$$

With this notation, it follows that (λ, ϕ_0) corresponds to a solution of the original Equation (3.1) with Boundary Conditions (3.2) if and only if there is an integer $m \geq 1$ such that

$$
\text{(3.11)} \qquad\qquad \lambda = m^2 \gamma^2(\phi_0)/l^2, \qquad \gamma(0) = \pi,
$$

It is easy to verify that $\gamma'(\phi_0) > 0$ for $0 < \phi_0 < \pi$ and, thus, all the solutions are depicted as in Figure 3.4 even the solutions which are far away from zero. Because of the specific nonlinearity $\sin \phi$ in Equation (3.1), the solution set has the simple form shown in contrast to the example of a simple eigenvalue discussed in Section 2.

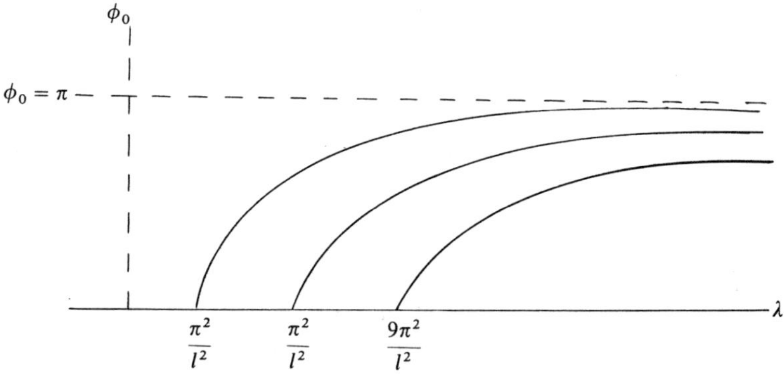

Figure 3.4

For the linear problem (3.7), the corresponding solution diagram is shown in Figure 3.5. Qualitatively, the Figures 3.4 and 3.5 are the same with the effect of the nonlinearity being to bend the bifurcation curves. In general, one cannot expect the global behavior of the solutions to be so similar. It

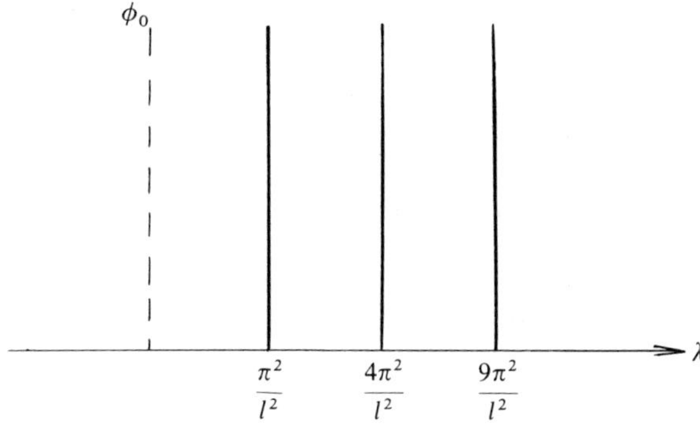

Figure 3.5

will only be so when the nonlinearities have a very special form. However, the local behavior near $\phi_0 = 0$ can be preserved in a very general setting. The basic reason for preserving this behavior locally in this example is that the null space of $D_\phi F(\lambda_m, 0)$ is one dimensional; that is, the boundary value problem (3.7) has only one linearly independent solution. Recall the corresponding remarks in Section 2 for equations in $\mathbb{R}^n$. As we shall see that the local structure of solutions is always preserved near simple eigenvalues in an abstract setting.

Another interesting aspect of this example is that, on any sufficiently small sphere S_ε in the space X, there exist $\phi_m \in S_\varepsilon$ and distinct values $\lambda_m \in \mathbb{R}$, $m = 1, 2, \ldots$ such that $F(\lambda_m, \phi_m) = 0$ where $F(\lambda, \phi)$ is defined in Equation (3.4). That is, there exist infinitely many distinct eigenvalues of the nonlinear operator equation $F(\lambda, \phi) = 0$. In some situations, one can also prove similar results for more general nonlinear equations and this is one of the applications of Ljusternik-Schnirelman theory of critical points of mappings. Some aspects of this theory will be discussed below.

In this example, we were also able to determine the global structure of all solutions. In more complicated problems one will not be able to obtain such complete information. However, for a simple eigenvalue (and more generally), one can discuss the extension of the branch of solutions that begins at $(\lambda_m, 0)$ and discuss whether or not it is unbounded in solution space. The answer to the question is closely related to the number of zeros of the eigenfunctions in the example and to degree theory in the general setting. This theory is also discussed below.

1.4. The Hopf Bifurcation

As another illustration of a bifurcation problem which at first glance appears to be of a different nature than the previous ones but actually is of the same type, consider the bifurcation of a periodic orbit from the equilibrium position of an autonomous second order differential equation—the so called *Hopf bifurcation*. This type of bifurcation is one of the most elementary ones in the theory of nonlinear oscillations.

Consider the second order equation

$$(4.1) \qquad \dot{x} = A(\alpha)x + f(\alpha, x) \qquad (\cdot = d/dt)$$

where $x \in \mathbb{R}^2$, $\alpha \in \mathbb{R}$, $A(\alpha)$ is a 2×2 matrix, $A(\alpha)$, $f(\alpha, x)$ have continuous derivatives up through order one, $D_{\alpha x}f(\alpha, x)$ is continuous for $|\alpha| < \alpha_0$, $x \in \mathbb{R}^2$, $f(\alpha, 0) = 0$, $D_x f(\alpha, 0) = 0$ for $|\alpha| < \alpha_0$. Also, suppose that the eigenvalues of $A(\alpha)$ are given by $\mu(\alpha) \pm iv(\alpha)$, where $v(0) > 0$, $\mu(0) = 0$, $D\mu(0) \neq 0$; that is, at $\alpha = 0$, the linear part of Equation (4.1) has two purely imaginary roots which cross the imaginary axis as α increases through $\alpha = 0$. Without loss in generality, we may choose a coordinate system, rescale the time t and redefine α in such a way that

$$A(\alpha) = \begin{bmatrix} \alpha & \beta(\alpha) \\ -\beta(\alpha) & \alpha \end{bmatrix}, \qquad \beta(0) = 1.$$

The linear equation

$$(4.2) \qquad \dot{y} = A(\alpha)y$$

has all solutions approaching zero exponentially as $t \to \infty$ for $\alpha < 0$, all solutions periodic of period 2π for $\alpha = 0$ and all solutions becoming unbounded exponentially as $t \to \infty$ for $\alpha > 0$. The phase portraits in the (x_1, x_2)-plane for Equation (4.2) are shown in Figure 4.1. If the nonlinear equation were such that one could show that an orbit beginning near the origin for $\alpha > 0$ were bounded and the Equation (4.1) had no equilibrium point except $(0,0)$, then the Poincaré–Bendixson Theorem would imply the existence of a periodic orbit. Even without this hypothesis, one can prove the following result.

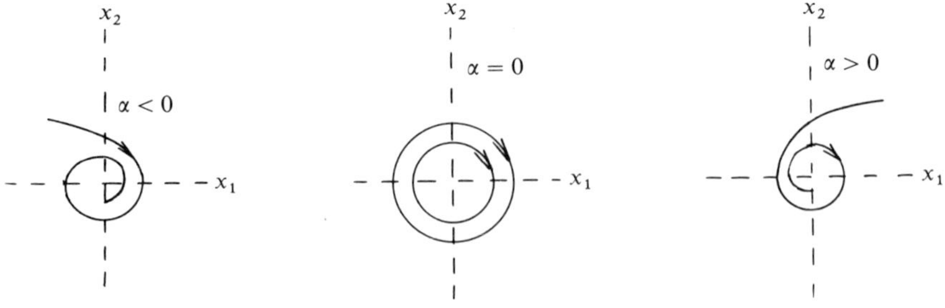

Figure 4.1

Theorem 4.1. *Under the above hypotheses, there are constants $a_0 > 0$, $\alpha_0 > 0$, $\delta_0 > 0$, functions $\alpha(a) \in \mathbb{R}$, $\omega(a) \in \mathbb{R}$, $\alpha(0) = 0$, $\omega(0) = 2\pi$, and an $\omega(a)$-periodic function $x^*(a)$ with all functions having continuous derivatives up through order one for $|a| < a_0$ such that $x^*(a)$ is a solution of Equation (4.1) with $\alpha = \alpha(a)$ and*

$$x_1^*(a) = a \cos \omega(a)t + o(|a|)$$
$$x_2^*(a) = -a \sin \omega(a)t + o(|a|)$$

as $|a| \to 0$. Furthermore, for $|\alpha| < \alpha_0$, $|\omega - 2\pi| < \delta_0$, every ω-periodic solution x of Equation (4.1) with $|x(t)| < \delta_0$ must be given by $x^*(a)$ except for a translation in phase.

Proof. All periodic solutions of Equation (4.1) near $x = 0$ can be obtained by using polar coordinates. If $x_1 = \rho \cos \theta$, $x_2 = -\rho \sin \theta$ in Equation (4.1), then one obtains

$$
\begin{array}{lll}
(4.3) & \text{(a)} & \dot{\theta} = \beta(\alpha) + \Theta(\alpha, \theta, \rho) \\
 & \text{(b)} & \dot{\rho} = \alpha\rho + \mathcal{R}(\alpha, \theta, \rho)
\end{array}
$$

where $\Theta(\alpha, \theta, 0) = 0$, $\mathcal{R}(\alpha, \theta, 0) = 0$, $D_\rho\mathcal{R}(\alpha, \theta, 0) = 0$ and these functions are 2π-periodic in θ. Eliminating t in Equations (4.3), one obtains

$$(4.4) \qquad \frac{d\rho}{d\theta} = \frac{\alpha}{\beta(\alpha)}\rho + \mathcal{R}^*(\alpha, \theta, \rho)$$

where $\mathcal{R}^*$ is 2π-periodic in θ, $\mathcal{R}^*(\alpha, \theta, 0) = 0$, $D_\rho\mathcal{R}^*(\alpha, \theta, 0) = 0$. The 2π-periodic solutions of (4.4) near $\rho = 0$ yield periodic solutions of Equation (4.1) near $x = 0$ and conversely.

If $X = \{f : \mathbb{R} \to \mathbb{R}; f \text{ is } 2\pi\text{-periodic with } f, Df \text{ continuous}\}$, $Z = \{f : \mathbb{R} \to \mathbb{R}; f \text{ is } 2\pi\text{-periodic and continuous}\}$, then the linear operator $d/d\theta : X \to Z$ has a one dimensional null space and the complement of the range is one dimensional. It is like a simple eigenvalue and a proof of the theorem could be constructed from this idea. We choose another proof rather than repeat the argument analogous to the one in Section 2 for a simple eigenvalue.

If $\rho(\theta)$ is a solution of Equation (4.4) with $\rho(0) = a$, then $\rho(\theta + 2\pi) = \rho(\theta)$ if and only if $\rho(2\pi) = \rho(0) = a$. Applying the variation of constants formula to Equation (4.4), this latter condition is easily seen to be equivalent to

$$(4.5) \qquad (1 - e^{-2\pi\alpha/\beta(\alpha)})a = h(\alpha, a)$$

where $h(\alpha, 0) = 0$ for all α. Dividing by a, one has an equivalent equation

$$(4.6) \qquad (1 - e^{-2\pi\alpha/\beta(\alpha)}) - g(\alpha, a) = 0$$

where $g(0,0) = 0$, $D_\alpha g(0,0) \neq 0$. One may now apply the Implicit Function Theorem to determine $\alpha = \alpha^*(a)$ satisfying Equation (4.6) and $\alpha^*(0) = 0$.

To determine the period $\omega^*(a)$, one must use equation (a) and this development is left to the reader. This will complete the proof of the theorem. $\quad\square$

Equation (4.4) can also be effectively used to compute the function $\alpha^*(a)$. In fact, if we were so lucky to have Equation (4.4) in the form

$$\frac{d\rho}{d\theta} = \rho(\alpha - \rho^2)$$

then $\alpha^*(a) = a^2$. Actually, this is the typical situation in the sense that most $f(\alpha, x)$ in Equation (4.1) lead to an $\alpha^*(a) = a^2 + 0(|a|^3)$ as $|a| \to 0$. To prove this, one usually employs computational procedures based on averaging which will be discussed later.

One final remark on this example is that $x \in \mathbb{R}^2$ is not essential. An analogue of Theorem 4.1 will be given for $x \in \mathbb{R}^n$ and even in certain infinite dimensional spaces.

1.5. Some Generic Examples

For the examples in the previous sections containing a scalar parameter λ, the existence of a bifurcation point λ_0 was proved by making assumptions only about the linear terms in x in the Taylor series of $M(\lambda, x)$. For example, when M has the special form in Relation (2.1), the value λ_0 is a simple eigenvalue of B. On the other hand, we have seen by example in Section 2 that an eigenvalue of multiplicity two may or may not correspond to a bifurcation point. The specific form of the nonlinearities plays an essential role. It is extremely important to have procedures which are more analytic in nature and which will clarify the role of the nonlinearities. One would hope to be able to specify some generic conditions on the nonlinearities which will permit a complete classification of the solutions near a point (λ_0, x_0). This must be done even when λ is a vector parameter. Much of the discussion in later chapters is devoted to a clarification of these remarks. In this section, we give a few elementary examples which illustrate the ideas.

As a first example, suppose $x \in \mathbb{R}$, $\lambda \in \mathbb{R}$ and consider the equation

$$(5.1) \qquad M(\lambda, x) \stackrel{\text{def}}{=} x^2 - \lambda + 0(|x|^3 + |\lambda x| + |\lambda|^2) = 0$$

Our objective is to compare the solutions of Equation (5.1) with the solution of the equation obtained by considering only the lowest order terms in the Taylor series of $M(x, \lambda)$; namely,

$$(5.2) \qquad x^2 - \lambda = 0.$$

In particular, we wish to know if the qualitative behavior of the solutions of Equation (5.1) is the same as for Equation (5.2)? The point $\lambda = 0$ is a bifurcation point for Equation (5.2) and there are no solutions for $\lambda < 0$ and two solutions for $\lambda > 0$.

The following discussion shows that the point $\lambda = 0$ is also a bifurcation point for Equation (5.1). In a neighborhood of $(x, \lambda) = (0, 0)$, there is a unique function $x^*(\lambda)$ near $\lambda = 0$, $x^*(0) = 0$, such that $D_x M(\lambda, x^*(\lambda)) = 0$ since $D_x^2 M(0, 0) = 2 \neq 0$. The point $x^*(\lambda)$ corresponds to minimum of $M(\lambda, x)$. If $\alpha(\lambda) = M(\lambda, x^*(\lambda))$, then $\alpha(\lambda) > 0$ implies no solutions near $(0, 0)$ and $\alpha(\lambda) < 0$ implies two solutions near $(0, 0)$. Thus, those λ for which $\alpha(\lambda) = 0$ correspond to bifurcation points. On the other hand, it is easy to verify that $\alpha(\lambda) = -\lambda + 0(\lambda^2)$ as $\lambda \to 0$ and so $\alpha(\lambda)$ has the sign of $-\lambda$ near $\lambda = 0$. Thus, $\lambda = 0$ is a bifurcation point. In Figure 5.1 we have depicted the bifurcation set in parameter space with the number of solutions as well as the solution set in $\mathbb{R}^2$.

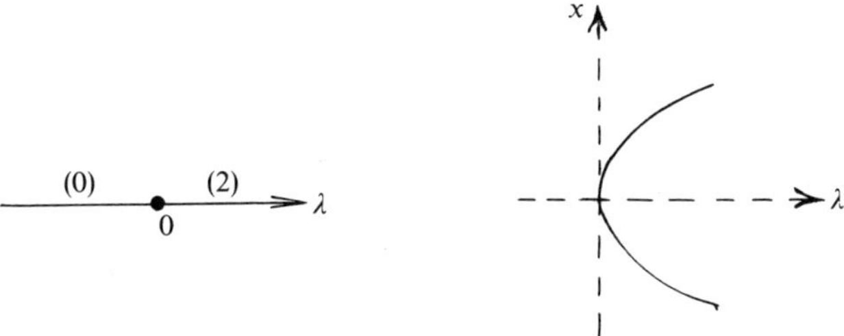

Figure 5.1

As a slightly more complicated example, suppose $x \in \mathbb{R}$, $\lambda \in \mathbb{R}^2$, $\lambda = (\lambda_1, \lambda_2)$ and consider the equation

$$(5.3) \qquad M(\lambda, x) = x^3 - \lambda_1 x - \lambda_2 + 0(|x|^4 + |\lambda_1 x^2| + |\lambda_1^2 x| + |\lambda_2 x| + \lambda_2^2) = 0$$

in a small neighborhood of $(\lambda_1, \lambda_2, x) = (0, 0, 0)$. As before, consider the polynomial equation

$$(5.4) \qquad\qquad\qquad x^3 - \lambda_1 x - \lambda_2 = 0$$

and try to compare Equation (5.3) with Equation (5.4).

For Equation (5.4), it is easy to determine the bifurcation curves in (λ_1, λ_2)-space. In fact, these curves are precisely the curves across which the number of solutions changes; that is, multiple solutions occur on these curves. It is well known that this corresponds to the discriminant of the cubic polynomial being equal to zero. The explicit computation is easy to perform. In fact, any multiple solution of Equation (5.4) must satisfy Equation (5.4) as well as the equation

$$(5.5) \qquad\qquad\qquad 3x^2 - \lambda_1 = 0$$

Equations (5.4) and (5.5) uniquely define λ_1, λ_2 as functions of x; namely, $\lambda_1 = 3x^2$, $\lambda_2 = 2x^3$. This is the parametric representation of the cusp $4\lambda_1^3 - 27\lambda_2^2 = 0$ (the discriminant equal to zero). In Figure 5.2, we have depicted the bifurcation set in parameter space, typical cross sections of the solution set for a fixed λ_2 and the complete solution set in $\mathbb{R}^3$.

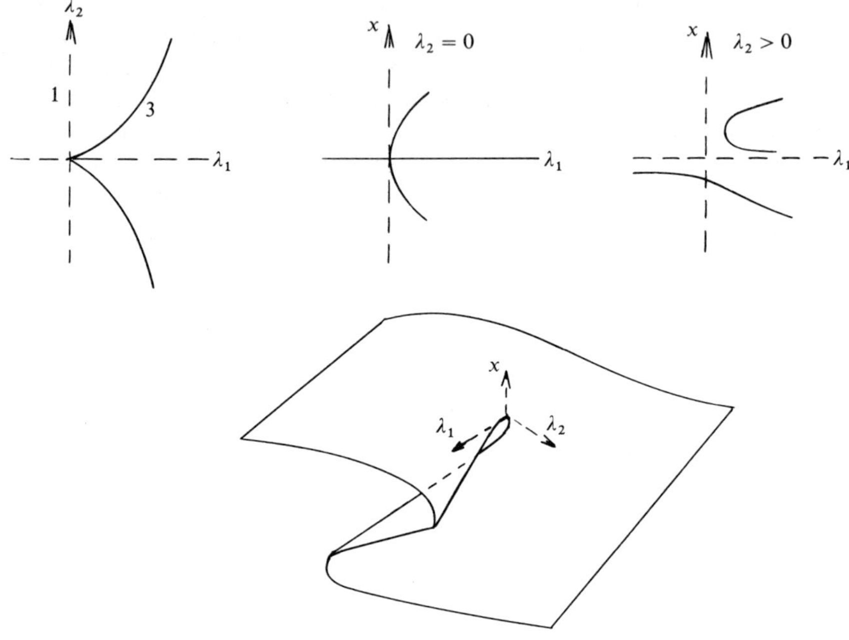

Figure 5.2

The basic question is the following: do the solutions of Equation (5.3) have the same qualitative behavior as the solutions of Equation (5.4)? The answer is obtained by following the same procedure as for the polynomial equation. The bifurcation curves must correspond to multiple solutions of Equation (5.3); that is, $M(\lambda, x) = 0$, $D_x M(\lambda, x) = 0$. Near $\lambda = 0$, $x = 0$, these equations uniquely determine λ_1, λ_2 as functions of x. It is easy to check that $\lambda_1 = 3x^2 + 0(|x|^3)$, $\lambda_2 = 2x^3 + 0(|x|^4)$ as $x \to 0$. Again, these are the parametric representation for a cusp which has essentially the same form as in Figure 5.2. One must now show that the number of solutions of Equation (5.3) changes by exactly two as this cusp is crossed. This is left to the reader and involves discussing $D_x^2(\lambda, x)$, $D_{\lambda_2}(\lambda, x)$ along this curve. In this way one obtains the complete structure of the bifurcations of Equation (5.3) as shown in Figure 5.2.

An elementary but important remark about these two examples is the following. By imposing some conditions on the nonlinearities in the function (the square term in x is present in the first example and the cubic term in the second), we were able to determine the complete behavior of the solution set even when several parameters are involved.

In the applications, Equation (5.1) is obtained through an application of the method of Liapunov-Schmidt used in Section 2. To illustrate how interesting information is obtained in this way, let us consider the nth order equation

$$(5.6) \quad \begin{array}{ll} \text{(a)} & y^2 - \lambda - f(\lambda, y, z) = 0 \\ \text{(b)} & B_0 z - g(\lambda, y, z) = 0 \end{array}$$

where $\lambda \in \mathbb{R}$, $y \in \mathbb{R}$, $z \in \mathbb{R}^{n-1}$, B_0 is a nonsingular matrix, $f = 0(|y|^3 + |yz| + |z|^2)$, $g = 0(|y|^2 + |z|^2)$ as $|y|, |z| \to 0$, $|\lambda| < \lambda_0$. For λ, y, z small, Equation (5.6b) has a unique solution $z = z^*(\lambda, y)$, $z^*(\lambda, y) = 0(|y|^2)$ as $y \to 0$, and Equations (5.6) are equivalent to the equation

$$(5.7) \quad y^2 - \lambda - f(\lambda, y, z^*(\lambda, y)) = 0$$

and $f(\lambda, y, z^*(\lambda, y)) = 0(|y|^3)$ as $y \to 0$, $|\lambda| < \lambda_0$. Equation (5.7) has the form of Equation (5.1) and, thus, $\lambda = 0$ is a bifurcation point.

Equation (5.3) may also arise in a similar manner.

In spite of the simplicity of the methods used in these two examples, we shall see that they can be systematically generalized to equations for which x as well as λ are multidimensional parameters.

1.6. Dynamic Bifurcation

As remarked in Section 1.1, the solutions of Equation (1.1) often represent equilibrium positions of some physical system. The actual behavior of the system in a neighborhood of this equilibrium is often determined by the stability properties of the solutions of an evolutionary equation. The bifurcation of equilibrium solutions is intimately connected with exchanges of stability in the system. Also, when the dynamics of the system is taken into consideration, other types of bifurcations can occur. The purpose of this section is to illustrate these remarks with some elementary examples. The general subject of dynamic bifurcation will be treated in Chapters 9–13.

The definition of bifurcation for evolutionary equations usually proceeds as follows. An equivalence relation is given for vector fields. For this equivalence relation, $\sim$, a vector field f is said to be *structurally stable* if there is a neighborhood V of f such that $g \sim f$ for every $g \in V$. The vector field f is a *bifurcation point* if f is not structurally stable. Let us say $f \sim g$ if there is a homeomorphism of the base space which maps the orbits defined by f onto the orbits defined by g and the homeomorphism preserves the sense of direction of the orbits in time. We will not be more precise, but simply illustrate the ideas with examples.

Consider the scalar equation

(6.1) $$\dot{x} = -x^2 + \lambda$$

where λ is a real parameter. The point $\lambda = 0$ is a bifurcation point for the function $-x^2 + \lambda$ and also a bifurcation point for the vector field $-x^2 + \lambda$. The orbits are depicted in Figure 6.1 with the arrows representing the direction of the flow.

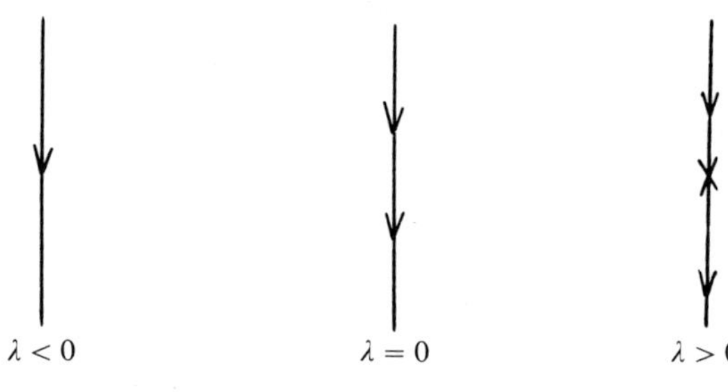

Figure 6.1

For the equation

(6.2) $$\dot{x} = -x^3 + \lambda x$$

the point $\lambda = 0$ is a bifurcation point with the orbits depicted in Figure 6.2. This is a good illustration of the exchange of stability at the bifurcation point. The solution $x = 0$ gives up its stability to the solutions $\pm \lambda^{1/2}$.

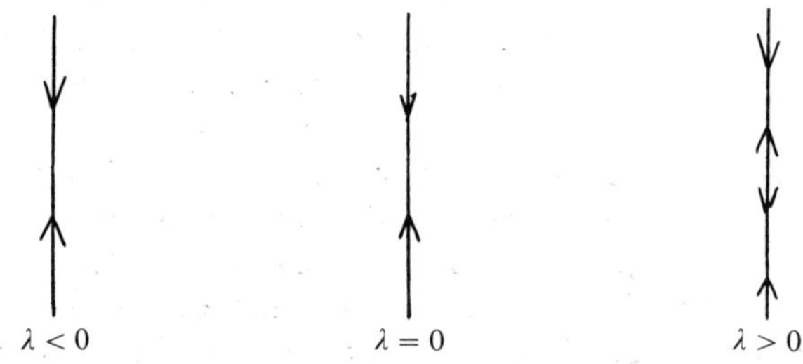

Figure 6.2

For the equation

(6.3) $$\begin{aligned} \dot{x} &= y - x(x^2 + y^2 - \varepsilon) \\ \dot{y} &= -x - y(x^2 + y^2 - \varepsilon) \end{aligned}$$

the point $\varepsilon = 0$ is a bifurcation point with the flow depicted in Figure 6.3. The periodic orbit $x^2 + y^2 = \varepsilon$ bifurcates from the equilibrium solution $(0, 0)$

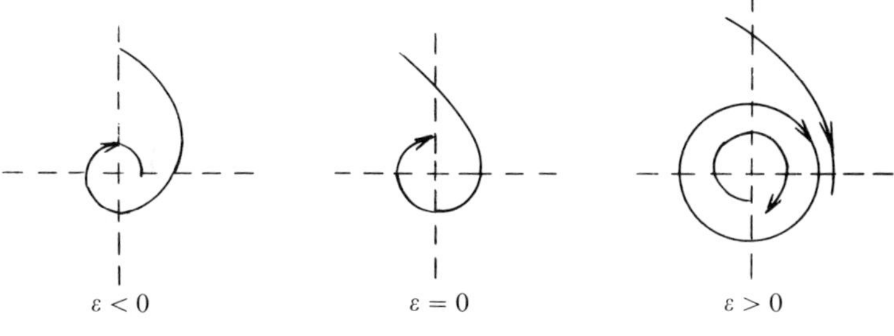

Figure 6.3

and there is a similar exchange of stability as in example (6.2). This type of bifurcation (the Hopf bifurcation) is a consequence of the dynamics.

For the equation

$$(6.4) \qquad \begin{aligned} \dot{x} &= y \\ \dot{y} &= x - x^2 + \varepsilon y \end{aligned}$$

the point $\varepsilon = 0$ is a bifurcation point with the orbits depicted in Figure 6.4. The equilibrium point $(1,0)$ exchanged its stability properties as ε passed from negative to positive values.

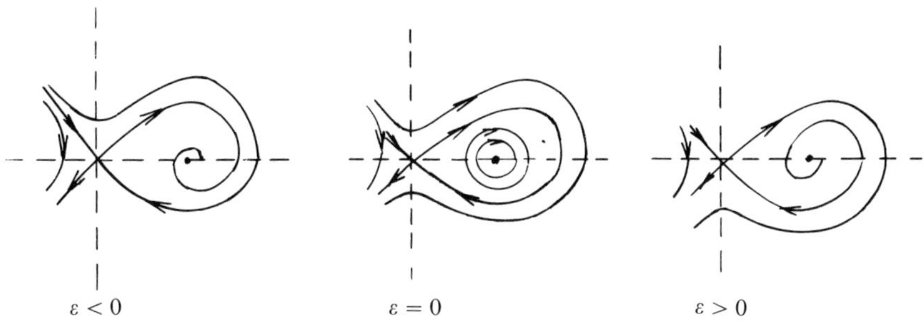

Figure 6.4

System (6.4) with $\varepsilon = 0$ is Hamiltonian and is very degenerate as far as bifurcation theory is concerned. One can make other perturbations of this equation which will create other types of bifurcations. For example, it is possible to find an equation

$$(6.5) \qquad \begin{aligned} \dot{x} &= y + g(x, y, \varepsilon) \\ \dot{y} &= x - x^2 + h(x, y, \varepsilon) \end{aligned}$$

where g, h vanish for $\varepsilon = 0$ so that the point $\varepsilon = 0$ is a bifurcation point and the orbits are as depicted in Figure 6.5. This is a much more complicated

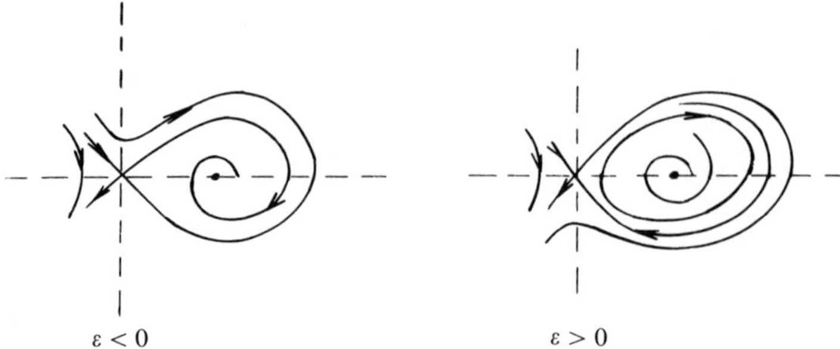

Figure 6.5

type of bifurcation where a periodic orbit grows to touch a saddle point, creating the homoclinic orbit through the saddle point.

In the plane, the above examples are indicative of the types of bifurcation that can occur. In higher dimensions, the situation is very difficult to understand. Some parts of this subject will be treated in later chapters.

Chapter 2

Elements of Nonlinear Analysis

2.1. Calculus

This section is devoted to the fundamental concepts of differentiability and analyticity of mappings from a Banach space X to a Banach space Y. We let $L(X, Y)$ denote the Banach space of bounded linear operators from X to Y with the operator topology. We also assume the reader is familiar with integration of functions with values in a Banach space. For h in a Banach space, the symbol $0(|h|)$ as $|h| \to 0$ denotes a function such that $0(|h|)/|h|$ is bounded for h in a neighborhood of zero, $o(|h|)$ denotes a function such that $o(|h|)/|h| \to 0$ as $|h| \to 0$.

Definition 1.1. Let U be an open set in X. A function $f : U \to Y$ is said to be *Gateaux differentiable* at $x_0 \in U$ if for every $h \in X$, there is an element $df(x_0, h) \in Y$ such that

$$\lim_{t \to 0} \left| \frac{f(x_0 + th) - f(x_0)}{t} - df(x_0, h) \right| = 0$$

The element $df(x_0, h)$ is called the *Gateaux derivative* of f at x_0 and we write

$$\frac{d}{dt} f(x_0 + th) \bigg|_{t=0} = df(x_0, h).$$

Definition 1.2. Let U be an open subset of X. A function $f : U \to Y$ is said to *Fréchet differentiable* at x_0 if there is a bounded linear operator $Df(x_0) : X \to Y$ such that

$$|f(x_0 + h) - f(x_0) - Df(x_0)h| = o(|h|) \quad \text{as } |h| \to 0$$

for every $h \in X$ such that $x_0 + h \in U$. The operator $Df(x_0)$ is called the *Fréchet derivative* of f at x_0.

It is not difficult to see that the Fréchet and Gateaux derivatives are unique. The relationship between the Fréchet and Gateaux derivatives is contained in the following result.

Theorem 1.3. *If U is an open set in X, $f: U \to Y$ is Fréchet differentiable at x_0, then it is Gateaux differentiable at x_0 and $Df(x_0)h = df(x_0, h)$ for every $h \in X$. Conversely, if the Gateaux derivative $df(x_0, h)$ at x_0 is such that $df(x_0, \cdot) \in L(X, Y)$ and $df(x, h)$ is defined and continuous in x as a map from X to $L(X, Y)$ for x in a neighborhood of x_0, then f is Fréchet differentiable at x_0 and $df(x_0, h) = Df(x_0)h$ for every $h \in X$.*

Proof. The proof that Fréchet differentiability implies Gateaux differentiability is immediate from the definition. To prove the converse, note that the hypothesis implies $df(x, h) = df(x)h$ where $df(x) \in L(X, Y)$ and x belongs to a neighborhood of x_0. Since

$$\frac{d}{dt} f(x + ty) = df(x + ty)y$$

it follows that

$$|f(x + h) - f(x) - df(x)h| = \left| \int_0^1 [df(x + th)h - df(x)h] \, dt \right|$$

$$\leq \int_0^1 |df(x + th) - df(x)| \, |h| \, dt$$

$$= o(|h|) \quad \text{as } |h| \to 0.$$

Thus, the Fréchet derivative exists and is equal to $df(x)$. $\quad\square$

The Fréchet derivative has properties very similar to the derivative in finite dimensions. For example, if $f: X \to Y$, $g: Y \to Z$ are Fréchet differentiable, then the chain rule holds for the function $(g \circ f)(x) = g(f(x))$:

$$Dg \circ f(x) = Dg(f(x))Df(x).$$

If $f: X \to Y$, $Y = \prod_{i=1}^n Y_i$, $f = (f_1, \ldots, f_n)$ with each $f_i: X \to Y_i$ being Fréchet differentiable, then f is Fréchet differentiable and $Df(x) = (Df_1(x), \ldots, Df_n(x))$. If $X = \prod_{i=1}^N X_i$, $f: X \to Y$, $f = f(x_1, \ldots, x_N)$, $x_i \in X_i$, then one can also define partial Fréchet derivatives $D_i f(x) \in L(X_i, Y)$ with respect to x_i in the usual way:

$$f(x_1, \ldots, x_i + h, \ldots, x_N) - f(x_1, \ldots, x_i, \ldots, x_N) - D_i f(x)h = o(|h|)$$

as $|h| \to 0$. If $Df(x)$, $D_i f(x)$ exist and $h = (h_1, \ldots, h_N) \in X$ with $h_i \in X_i$, then

$$Df(x)h = \sum_{i=1}^N D_i f(x)h_i.$$

It is also easy to prove that the following mean value theorem is valid.

Theorem 1.4. *If $f:[a,b] \to X$ is Fréchet differentiable and there is a continuously differentiable function $\alpha:[a,b] \to \mathbb{R}$ such that $|Df(t)| \le \alpha'(t)$ for $t \in [a,b]$, then*

$$|f(b) - f(a)| \le \alpha(b) - \alpha(a)$$

$$|f(b) - f(a)| \le \left(\sup_{t \in [a,b]} |Df(t)| \right) |b - a|$$

To define higher order derivatives, we need the concept of multilinear operator. An operator $f:X \to Y$ is called *multilinear* if $X = \prod_{j=1}^{n} X_j$ and for each $x = (x_1, \ldots, x_n) \in X$ and each integer $k = 1, 2, \ldots, n$, $f(x_1, \ldots, x_n)$ is linear in x_k for all other variables fixed.

Lemma 1.5. *Suppose $X = \prod_{i=1}^{n} X_i$ and Y are Banach spaces. If $f:X \to Y$ is multilinear, then the following statements are equivalent:*

(i)   *f is continuous at each point $x \in X$;*
(ii)   *f is continuous at $x = 0$;*
(iii) *f takes bounded sets to bounded sets and there is a constant $K > 0$ such that*

$$|f(x_1, \ldots, x_n)| \le K \prod_{i=1}^{n} |x_i|_{X_i}.$$

(iv) *f is Fréchet differentiable.*

Proof. Obviously, (iv) implies (i) implies (ii). We now show (ii) implies (iii). Since f is continuous at $x = 0$, $|f(x)| \le 1$ on a ball of sufficiently small radius. Since $f(\alpha x) = \alpha^n f(x)$ for any real α, it follows that f is bounded on every ball. In particular there is a constant K such that $|f(x)| \le K$ for $|x| \le 1$. Thus, $|f(x)| \le K|x|^n$ for any $x \in X$. Since there is no loss in generality in assuming the $\sup_i |x_i|_{X_i}$ is the norm in X, we obtain (iii).

To prove (iii) implies (iv), observe first that the Gateaux derivative of f always exists and is given by

$$df(x, h) = \sum_{i=1}^{n} f(x_1, \ldots, x_{i-1}, h_i, x_{i+1}, \ldots, x_n)$$

where $h = (h_1, \ldots, h_n)$. The function $df(x, h)$ is linear in h. Also, for any $\alpha > 0$, (iii) implies there is a $\beta > 0$, such that $|df(x, h)| \le \beta |h|$ for $|x| \le \alpha$ and all $h \in X$. Using (iii) again, we have

$$|f(x + h) - f(x) - df(x, h)| = o(|h|) \quad \text{as } |h| \to 0.$$

Thus, $df(x, h) = Df(x)h$ and f is Fréchet differentiable. $\quad\square$

The *norm* $|f|$ *of a multilinear map* from X to Y is the infimum of those K satisfying (iii). The set of bounded linear maps from $X = \prod_{j=1}^{n} X_j$ to Y will be denoted by $L(X_1, \ldots, X_n, Y)$. This is a Banach space with the usual rules for addition and scalar multiplication and the above norm. If $X_1 = \cdots = X_n = X$, we let $L(X, \ldots, X, Y) = L^n(X, Y)$.

The proof of the following lemma is left as an exercise for the reader.

Lemma 1.6. *There is an isometric isomorphism between* $L(X_1, \ldots, X_n, Y)$ *and* $L(X_1, L(X_2, L(X_3, \ldots, L(X_n, Y))))$.

A multilinear form $f \in L^n(X, Y)$ is *symmetric* if $f(x_1, \ldots, x_n)$ is invariant under any permutation $\sigma(1, 2, \ldots, n)$ of the integers $1, 2, \ldots, n$. To any multilinear form $f \in L^n(X, Y)$, we can associate the symmetric multilinear form

$$\operatorname{Sym} f(x_1, \ldots, x_n) = \frac{1}{n!} \sum_{\sigma(i_1, \ldots, i_n)} f(x_{i_1}, \ldots, x_{i_n})$$

and the polar form $f(x) = f(x, \ldots, x)$. If f is symmetric then it is not difficult to show that

$$f(x_1, \ldots, x_n) = \frac{1}{n!} \frac{\partial^n}{\partial t_1 \cdots \partial t_n} f\left(\sum_{i=1}^{n} t_i x_i\right)\bigg|_{t_1 = \cdots = t_n = 0}.$$

This relation implies that if the polar forms of two symmetric multilinear forms are equal, then the multilinear forms are equal.

If U is an open set in X and $f: U \to Y$ has a Fréchet derivative $Df(x)$ for $x \in U$, then we say f has a second Fréchet derivative $D^2f(x_0)$ at x_0 if $Df(\cdot): U \to L(X, Y)$ has a Fréchet derivative at x_0. In this case, $D^2f(x_0) \in L(X, L(X, Y)) = L^2(X, Y)$ by Lemma 1.6. It is possible to show that $D^2f(x_0)$ is a symmetric bilinear form; that is $D^2f(x_0)(h_1, h_2) = D^2f(x_0)(h_2, h_1)$ for all $(h_1, h_2) \in X \times X$. By induction, one can define the Fréchet derivatives of all orders with the Nth derivative $D^N f(x_0) \in L^N(X, Y)$ and symmetric.

Definition 1.7. If X, Y are Banach spaces and U is an open subset of X, the space $C^N(U, Y)$, $N \geq 0$, is the space of functions $f: U \to Y$ such that the jth Fréchet derivative $D^j f(x)$ exists for each $x \in U$, $0 \leq j \leq N$, and the mapping $x \mapsto D^j f(x)$ of U into $L^N(X, Y)$ is continuous for each x in U. If $f \in C^N(U, Y)$ for all integers $N \geq 0$, we say $f \in C^\infty(U, Y)$.

For notational purposes, let x^k be the k-tuple $(x, x, \ldots, x)$.

Theorem 1.8 (Taylor's Theorem). *If* $f \in C^n(U, Y)$, *then*

$$(1.1) \quad f(x + h) = f(x) + Df(x)h + \cdots + \frac{1}{(n-1)!} Df^{n-1}(x)h^{n-1} + R_n(x, h)$$

for $x \in U$, $x + sh \in U$, $0 \le s \le 1$, where

$$(1.2) \qquad R_n(x, h) = \frac{1}{(n-1)!} \int_0^1 (1 - s)^{n-1} Df^n(x + sh)h^n \, ds.$$

Also,

$$(1.3) \qquad f(x + h) = f(x) + Df(x)h + \cdots + \frac{1}{n!} D^n f(x)h^n + g(x, h)$$

where $g(x, h) = o(|h|^n)$ as $|h| \to 0$.

Proof. Since the map $s \mapsto D^n f(x + sh)h^n$ is continuous for $0 \le s \le 1$, the right hand side of (1.1) is well defined. For a continuous linear functional y^* on Y, let $g(t) = \langle y^*, f(x + th) \rangle$, $0 \le t \le 1$. Since g is a C^n scalar–valued function on $[0, 1]$, one can apply Taylor's theorem. We obtain

$$\langle y^*, f(x + h) \rangle = \left\langle y^*, f(x) + \sum_{j=1}^{n-1} D^j f(x)h^j \right\rangle$$

$$+ \frac{1}{(n-1)!} \int_0^1 (1 - s)^{n-1} \langle y^*, Df^n(x + sh)h^n \, ds \rangle.$$

Since y^* commutes with the integral and this relation is true for all y^*, Relation (1.1) follows.

To prove (1.3), observe that $g(x, h)$ is given by

$$g(x, h) = \frac{1}{(n-1)!} \int_0^1 (1 - s)^{n-1} [D^n f(x + sh)h^n - D^n f(x)h^n] \, ds$$

and satisfies the desired property. $\square$

Definition 1.9. Let X, Y be Banach spaces over the complex numbers and let U be a connected open set of X. A function $f : X \to Y$ is *complex analytic* in U if, for each $x \in U$, $h \in X$, there is a $\delta(x, h) > 0$ such that, for each $y^* \in Y^*$, $f(x)$ is single valued and $\langle y^*, f(x + th) \rangle$ is an analytic function of t for $|t| < \delta(x, h)$.

There are several equivalent concepts of analyticity which are stated below. The proofs are omitted.

Theorem 1.10. *If U is an open connected set of X, $f : U \to Y$ is single valued and locally bounded, then the following statements are equivalent:*

(i)   *f is complex analytic in U;*
(ii) *f is Gateaux differentiable in U;*
(iii) *f is Fréchet differentiable in U;*
(iv) *f has infinitely many Gateaux derivatives;*
(v)   *f has infinitely many Fréchet derivatives;*

(vi)   *For any $x \in U$, there is a $\delta = \delta(x) > 0$ such that, whenever $|h| < \delta$,*

$$f(x + h) = \sum_{k=0}^{\infty} \frac{1}{k!} D^k f(x) h^k$$

with the series converging uniformly in x, h.

EXAMPLE 1.11. Let $f(x_1, \ldots, x_n)$ be a multilinear form and define $g(x) = f(x, \ldots, x)$. Prove that $D^{n+1} g(x) = 0$ for all x and so $g(x)$ is analytic.

EXAMPLE 1.12. Let $X = Y = C([0,1], \mathbb{C})$ and define $f(x)(t) = \sin x(t)$, $0 \le t \le 1$. Prove that $f(x)$ is analytic. If $X = Y = L_2([0,1], \mathbb{C})$, show that $f(x)$ is Lipschitz continuous, but not differentiable. Thus, it is not analytic.

2.2. Local Implicit Function Theorem

A fixed point of a transformation T on some space S is an element $x \in S$ such that $Tx = x$. If S is a metric space with metric d, the mapping T is said to be a *contraction on S* if there is a $\theta \in [0, 1)$ called the contraction constant such that

$$d(Tx, Ty) \le \theta d(x, y) \quad \text{for all } x, y \in S.$$

Theorem 2.1 (Banach–Cacciopoli). *If (S, d) is a complete metric space and $T: S \to S$ is a contraction, then there is a unique fixed point $\bar{x}$ of T in S. Also, for any $x_0 \in S$, if $T^n x_0 = T(T^{n-1} x_0)$, $n \ge 1$, $T^0 x_0 = x_0$, then $T^n x_0 \to \bar{x}$ and*

$$d(T^n x_0, \bar{x}) \le \theta^n d(x_0, \bar{x})/(1 - \theta)$$

where θ is the contraction constant for T.

Proof. If $x = Tx$, $y = Ty$, $x, y \in S$, then

$$d(x, y) = d(Tx, Ty) \le \theta d(x, y)$$

implies $x = y$. Thus, the fixed point is unique if it exists.

We now prove existence. For any $x_0 \in S$, $T^n x_0 \in S$, $n \ge 0$ and

$$d(T^{n+1} x_0, T^n x_0) \le \theta d(T^n x_0, T^{n-1} x_0) \le \cdots \le \theta^n d(Tx_0, x_0).$$

Thus, for any $m > n$,

$$d(T^m x_0, T^n x_0) \le d(T^m x_0, T^{m-1} x_0) + \cdots + d(T^{n+1} x_0, T^n x_0)$$

$$\le (\theta^{m-1} + \cdots + \theta^n) d(Tx_0, x_0)$$

$$\le \frac{\theta^n}{1 - \theta} d(Tx_0, x_0)$$

The sequence $\{T^n x_0\}$ therefore forms a Cauchy sequence. Since S is complete, there is an $\bar{x} \in S$ such that $T^n x_0 \to \bar{x}$ as $n \to \infty$. Also,

$$d(T\bar{x}, \bar{x}) = \lim_{m \to \infty} d(T^{m+1} x_0, T^m x_0) = 0,$$

$T\bar{x} = \bar{x}$ and $\bar{x}$ is a fixed point. This proves existence. In the previous estimate, taking the limit as $m \to \infty$ shows that $d(T^n x_0, \bar{x}) \le \theta^n d(x_0, \bar{x})/(1 - \theta)$. $\square$

In the following, we need more information on the dependence of fixed points on various parameters. If (S, d) is a metric space, Λ is a set, we say $T: S \times \Lambda \to S$ is a *uniform contraction* on S if there is a $\theta \in [0, 1)$ such that

$$d(T(x, \lambda), T(y, \lambda)) \le \theta d(x, y)$$

for all $x, y \in S, \lambda \in \Lambda$.

Theorem 2.2 (Uniform Contraction Principle). *Let U, V be open sets in Banach spaces X, Y, let $\bar{U}$ be the closure of U, $T: \bar{U} \times V \to \bar{U}$ a uniform contraction on $\bar{U}$ and let $g(y)$ be the unique fixed point of $T(\cdot, y)$ in $\bar{U}$. If $T \in C^k(\bar{U} \times V, X), 0 \le k < \infty$, then $g(\cdot) \in C^k(V, X)$. If there is a neighborhood U_1 of $\bar{U}$ such that T is analytic from $U_1 \times V$ to X, then the mapping $g(\cdot)$ is analytic from V to X.*

Proof. If $k = 0$, then

$$\begin{aligned}
|g(y + h) - g(y)| &= |T(g(y + h), y + h) - T(g(y), y)| \\
&\le \theta|g(y + h) - g(y)| + |T(g(y), y + h) - T(g(y), y)|
\end{aligned}$$

implies

$$|g(y + h) - g(y)| \le (1 - \theta)^{-1}|T(g(y), y + h) - T(g(y), y)|.$$

This proves $g(y)$ is continuous in y.

Suppose $k = 1$. Since $|T(x_1, y) - T(x_2, y)| \le \theta|x_1 - x_2|$ for all $x_1, x_2 \in U$, $y \in V$, it follows that $|D_x T(x, y)| \le \theta < 1$ for $(x, y) \in U \times V$. Formally, differentiating $g(y) = T(g(y), y)$, the operator $Dg(y)$ should be a solution of the operator equation in M

$$(2.1) \qquad\qquad M - D_x T(g(y), y)M = D_y T(g(y), y)$$

Since $|D_x T(g(y), y)| \le \theta < 1$, this equation has a unique solution $M(y)$. We must show that

$$|g(y + \eta) - g(y) - M(y)\eta| = o(|\eta|) \quad \text{as } \eta \to 0.$$

If $\gamma = \gamma(\eta) = g(y + \eta) - g(y)$ then

$$
\begin{aligned}
\gamma &= T(g(y) + \gamma, y + \eta) - T(g(y), y) \\
&= D_x T(g(y), y)\gamma + D_y T(g(y), y)\eta + \Delta \\
\Delta &= T(g(y) + \gamma, y + \eta) - T(g(y), y) - D_x T(g(y), y)\gamma \\
&\quad - D_y T(g(y), y)\eta.
\end{aligned}
\tag{2.2}
$$

Since T is C^1, for any $0 < \varepsilon < 1 - \theta$, there is a $\delta > 0$ such that $|\Delta(\gamma, \eta)| < \varepsilon(|\gamma| + |\eta|)$ if $|\gamma| < \delta$, $|\eta| < \delta$. Since $\gamma = \gamma(\eta)$ is continuous in η, $\gamma(\eta) \to 0$ as $\eta \to 0$, we may further restrict δ so that $|\Delta(\gamma(\eta), \eta)| < \varepsilon(|\gamma(\eta)| + |\eta|)$ for $|\eta| < \delta$. From Equation (2.2),

$$
|\gamma(\eta)| \le \frac{1}{1 - \theta - \varepsilon}\,(|D_y T(g(y), y)| + \varepsilon)|\eta| \stackrel{\text{def}}{=} k|\eta|
$$

if $|\eta| < \delta$. Therefore, $|\Delta(\gamma, \eta)| < \varepsilon(1 + k)|\eta|$ for any $0 < \varepsilon < 1 - \theta$, $|\eta| < \delta$. From Equation (2.1), (2.2),

$$
[I - D_x T(g(y), y)][\gamma(\eta) - M(y)\eta] = \Delta(\gamma(\eta), \eta).
$$

Therefore, $|\gamma(\eta) - M(y)\eta| < \varepsilon(1 + k)(1 - \theta)^{-1}|\eta|$ for any $0 < \varepsilon < 1 - \theta$, $|\eta| < \delta$. This implies $Dg(y)$ exists and is equal to $M(y)$.

Suppose the result holds for $k - 1$ and $k > 1$. Thus, if T is C^k, then g is C^{k-1} at least and the fact that $Dg(y)$ satisfies (2.1) implies that g is C^k.

In the analytic case, there is a complex neighborhood of $(g(y), y)$ on which T is analytic and a uniform contraction. The above argument proves differentiability in this complex neighborhood and thus analyticity of g. $\square$

An application of Theorem 2.2 gives the following version of the Implicit Function Theorem.

Theorem 2.3 (Implicit Function Theorem). *Suppose X, Y, Z are Banach spaces, $U \subset X$, $V \subset Y$ are open sets, $F : U \times V \to Z$ is continuously differentiable, $(x_0, y_0) \in U \times V$, $F(x_0, y_0) = 0$ and $D_x F(x_0, y_0)$ has a bounded inverse. Then there is a neighborhood $U_1 \times V_1 \in U \times V$ of (x_0, y_0) and a function $f : V_1 \to U_1$, $f(y_0) = x_0$ such that $F(x, y) = 0$ for $(x, y) \in U_1 \times V_1$ if and only if $x = f(y)$. If $F \in C^k(U \times V, Z)$, $k \ge 1$ or analytic in a neighborhood of (x_0, y_0), then $f \in C^k(V_1, X)$ or is analytic in a neighborhood of y_0.*

Proof. If $L = [D_x F(x_0, y_0)]^{-1}$, $G(x, y) = x - LF(x, y)$, then the fixed points of G are the solutions of $F = 0$. The function G has the same smoothness properties as F, $G(x_0, y_0) = x_0$, $D_x G(x_0, y_0) = 0$. Therefore, $|D_x G(x, y)| \le \theta < 1$ for all x, y in a neighborhood $U_1 \times V_1$ of (x_0, y_0). It is easy to choose this neighborhood so that $G: U_1 \times V_1 \to U_1$. The result now follows from Theorem 2.2. $\quad\square$

In some applications, Theorem 2.3 cannot be used because $F(x, y)$ is not differentiable in y for all x. Using the same method of proof as above, one can give the following generalization of Theorem 2.3. The proof is omitted since it follows the same ideas as before.

Theorem 2.4. *Suppose F is a closed subset of a Banach space X, Int $F \neq \phi$, Λ is an open subset of a Banach space Y, where Int F denotes the interior of F. Assume that $T: F \times \Lambda \to F$, $(x, \lambda) \mapsto T(x, \lambda)$, satisfies the following set of hypotheses:*

- (i)   *$T(x, \cdot): \Lambda \to F$ is continuous.*
- (ii)   *$T(\cdot, \lambda): F \to F$ is continuous and has, for each Λ, a unique fixed point $x(\lambda)$ which depends continuously on Λ.*
- (iii) *If $x(\Lambda) = F_1$, then $T(x, \lambda)$ is continuously differentiable in λ for $(x, \lambda) \in F_1 \times \Lambda$.*
- (iv)   *There is an open set $F_2 \subset X$, $F \subset F_2$, and a $\delta \in [0, 1)$ such that the derivative of $T(x, \lambda)$ with respect to x is continuous and has norm $< \delta$ for all $(x, \lambda) \in F_2 \times \Lambda$.*

Then the fixed point $x(\lambda)$, $\lambda \in \Lambda$, of $T(x, \lambda)$ is continuously differentiable in λ.

2.3. Global Implicit Function Theorem

In this section, we prove a global implicit function theorem without differentiability hypotheses.

Definition 3.1. If X, Y are topological spaces, $\phi: X \to Y$ is continuous, then ϕ is called *locally invertible* at a point $x_0 \in X$ if there is a neighborhood U of x and a neighborhood V of $y_0 = \phi(x_0)$ such that ϕ is a homeomorphism from U onto V.

Definition 3.2. If X, Y are topological spaces, a mapping $\phi: X \to Y$ is called *proper* if the inverse image of a compact set is compact.

Definition 3.3. If X, Y are topological spaces, $\phi: X \to Y$ is continuous, then a point $x \in X$ is a *critical point* of ϕ if ϕ is not locally invertible at x. If x is a critical point of ϕ then $y = \phi(x)$ is a *critical value* of ϕ. The set of critical points of ϕ will be denoted by W and the set of critical values by $\phi(W)$.

Lemma 3.4. *Suppose X, Y are metric spaces, $\phi: X \to Y$ is continuous and proper, and, for each $y \in Y$, let $N(y)$ be the cardinal number of $\phi^{-1}(y)$. Then $N(y)$ is finite and constant on each connected component of $Y \backslash \phi(W)$.*

Proof. Consider the restriction of ϕ to $X\backslash\phi^{-1}(\phi(W)) \stackrel{\text{def}}{=} X_1$. Then $\phi: X_1 \to Y\backslash\phi(W)$ is locally invertible at each point of X_1. The conclusion of the lemma clearly will follow if we prove $N(y)$ is locally constant and finite for each $y \in Y\backslash\phi(W)$.

Suppose $y \in Y\backslash\phi(W)$. Then $\phi^{-1}(y)$ consists of isolated points since ϕ is locally invertible and it is also compact since ϕ is proper. Therefore, $\phi^{-1}(y)$ is finite. Let $N(y) = k$, $\phi^{-1}(y) = \{x_1, \ldots, x_k\}$ and let U_j be a neighborhood of x_j, V a neighborhood of y such that ϕ is a homeomorphism from U_j onto V for $j = 1, 2, \ldots, k$. For any $\bar{y} \in V$, it thus follows that $N(\bar{y}) \geq N(y)$. We claim there must be a neighborhood $W \subset V$ of y such that $N(\bar{y}) = k$ for each $\bar{y} \in W$. If this is not the case, then there is a sequence $y_n \to y$ such that $N(y_n) > k$ for each $n = 1, 2, \ldots$. Also, there is a sequence $\xi_n \notin \bigcup_j U_j$ such that $\phi(\xi_n) = y_n$. But, $\phi^{-1}(\{y_n\} \cup y)$ is compact and we may assume without loss of generality that $\xi_n \to \xi \notin \bigcup_j U_j$. Obviously, $\phi(\xi) = y$ and this contradicts the fact that $\phi^{-1}(y) = \{x_1, \ldots, x_k\}$. $\square$

The fundamental result is

Theorem 3.5. *Suppose X, Y are metric spaces, $\phi: X \to Y$ is continuous and proper. If $X\backslash\phi^{-1}(\phi(W))$ is nonempty, arcwise connected and $Y\backslash\phi(W)$ is simply connected, then ϕ is a homeomorphism from $X\backslash\phi^{-1}(\phi(W))$ onto $Y\backslash\phi(W)$.*

Corollary 3.6 (Global Implicit Function Theorem). *If the hypotheses of Theorem 3.5 are satisfied, $W = \phi$ and Y is simply connected, then ϕ is a homeomorphism from X onto Y.*

The proof of Theorem 3.5 will be broken down into several lemmas.

Lemma 3.7. *If the hypotheses of Theorem 3.5 are satisfied, $\phi(x_0) = y_0 \notin \phi(W)$, $F: [0, 1] \to Y\backslash\phi(W)$ is continuous, $F(0) = y_0$, then there is one and only one continuous function $G: [0, 1] \to X$ such that $G(0) = x_0$, $\phi(G(t)) = F(t)$, $0 \leq t \leq 1$.*

Proof. To prove uniqueness, suppose G_1, G_2 are continuous functions satisfying the properties stated. Since ϕ is locally invertible at x_0 and G_1, G_2 are continuous, there is a maximal closed interval $[0, \xi]$ in $[0, 1]$ such that $G_1(t) = G_2(t)$, $t \in [0, \xi]$. If $x^* = G_1(\xi) = G_2(\xi)$, then $y^* = \phi(x^*) = F(\xi) \in Y\backslash\phi(W)$ implies $x^* \notin W$ and ϕ is locally invertible at x^*. Therefore, the interval $[0, \xi]$ is open in $[0, 1]$, which implies $\xi = 1$.

To prove existence, let ξ be the supremum of $\tau \in [0, 1]$ such that there is a curve $\tilde{G}: [0, \tau] \to X$, $\tilde{G}(0) = x_0$, $\phi(\tilde{G}(t)) = F(t)$, $t \in [0, \tau]$. Then there is a curve $G: [0, \xi) \to X$ such that $G(0) = x_0$, $\phi(G(t)) = F(t)$, $t \in [0, \xi)$. Let ξ_n be an increasing sequence of real numbers approaching ξ. Since ϕ is proper, there is a subsequence which we label the same so that $G(\xi_n) \to x^* \in X\backslash W$ as

$n \to \infty$. If $y^* = \phi(x^*)$, then $y^* = F(\xi)$. Since ϕ is locally invertible at x^*, this implies one can extend G to a continuous function on the closed interval $[0, \xi]$ in $[0, 1]$, and $\phi(G(t)) = F(t)$, $t \in [0, \xi]$. Since local invertibility implies $[0, \xi]$ is open in $[0, 1]$, it follows that $\xi = 1$ and the lemma is proved. $\square$

Lemma 3.8. *Suppose the hypotheses of Theorem 3.5 are satisfied, $\phi(x_0) = y_0 \in Y \backslash \phi(W)$ and let $Q = \{(s, t) : 0 \leq s, t \leq 1\}$. If $F : Q \to Y \backslash \phi(W)$ is continuous, $F(0, 0) = y_0$, then there is a unique continuous function $G : Q \to X$ such that $G(0, 0) = x_0$, $\phi(G(s, t)) = F(s, t)$, for $(s, t) \in Q$.*

Proof. If q is any point of Q, consider the line segment γ connecting q to $(0, 0)$. If there exist two functions G_1, G_2 satisfying the conditions stated in the lemma, then $\phi \circ G_i \circ \gamma = F \circ \gamma$. By Lemma 3.7, $G_1 \circ \gamma = G_2 \circ \gamma$ and $G_1(q) = G_2(q)$ for every $q \in Q$. This proves uniqueness.

We now prove existence. By Lemma 3.7, there is a continuous function $\widetilde{G}(t)$ such that $\widetilde{G}(0) = x_0$, $\phi(\widetilde{G}(t)) = F(0, t)$, $t \in [0, 1]$. For any fixed $t \in [0, 1]$, one can determine a function $G(s, t)$, continuous in $s \in [0, 1]$, $G(t, t) = \widetilde{G}(t)$, such that $\phi(G(s, t)) = F(t, s)$ by Lemma 3.7. It remains only to show that G is continuous. We first show that G must be continuous at any point $(0, \bar{t})$. In fact, for any $\bar{t} \in [0, 1]$, the fact that ϕ is locally invertible implies there is a rectangle $S = \{(s, t) : 0 \leq s \leq s_0, |t - \bar{t}| \leq t_0\}$ and a unique continuous function $\Lambda : S \to X$ such that $\phi(\Lambda(s, t)) = F(s, t)$ for $(s, t) \in S$. But, then $\Lambda(s, t) = G(s, t)$ for $(s, t) \in S$ and G is continuous at $(0, \bar{t})$. A similar argument shows that G is continuous at each point of Q. $\square$

Proof of Theorem 3.5. The surjectivity follows immediately from Lemma 3.4, since $Y \backslash \phi(W)$ is connected and $X \backslash \phi^{-1}(\phi(W)) \neq \varnothing$ by hypothesis. Suppose now there are points $x_0 \neq x_1$ in $x \backslash \phi^{-1}(\phi(W))$ such that $\phi(x_0) = \phi(x_1) = y$. Choose $\bar{x} \in X \backslash \phi^{-1}(\phi(W))$ distinct from x_0 and x_1. Since $X \backslash \phi^{-1}(\phi(W))$ is arcwise connected, there is a continuous curve $x_0(t)$ joining $\bar{x}$ to x_0 and a continuous curve $x_1(t)$ joining $\bar{x}$ to x_1. Then both image curves join $\bar{y} = \phi(\bar{x})$ to y. Since Y is simply connected, there is a continuous function $F : Q \to Y$ such that $F(0, t) = y_0(t)$, $F(1, t) = y_1(t)$, $F(s, 0) = \bar{y}$, $F(s, 1) = y$, $0 \leq s, t \leq 1$. From Lemma 3.8, there is a continuous $G : Q \to X$ such that $\phi \circ G = F$, $G(0, t) = x_0(t)$, $G(1, t) = x_1(t)$. But then $\phi(G(s, 1)) = y$ for $0 \leq s \leq 1$. This contradicts the fact that ϕ is locally invertible. $\square$

When ϕ is continuously differentiable, the following result can sometimes be easier to check in the applications.

Theorem 3.9. *Suppose X, Y are Banach spaces, $\phi \in C^1(X, Y)$, $[D\phi(x)]^{-1}$ exists and there is a constant K such that $|[D\phi(x)]^{-1}| \leq K$ for all $x \in X$. Then ϕ is a homeomorphism of X onto Y.*

To prove this result, we need the following lemma which is the analogue of Lemma 3.8.

Lemma 3.10. *Under the hypotheses of Theorem 3.9, if $Q = \{(s, t), 0 \leq s, t \leq 1\}$, and $F(s, t)$ satisfies*

(i) $F : Q \to Y$,
(ii) *$F(s, t)$ is continuous in (s, t) and, for each s, $0 \leq s \leq 1$, is continuously differentiable in t,*
(iii) *$F(s, 0) = y_0 = \phi(x_0)$, $F(s, 1) = y_1$ $0 \leq s \leq 1$,*

then there exists a function $G : Q \to X$ satisfying (ii) *and $\phi \circ G = F$.*

Proof. By the Implicit Function Theorem, there exist neighborhoods V of y_0, U of x_0 such that ϕ is a homeomorphism of U onto Y. There exist an $\varepsilon > 0$ and a function $G(s, t) = \phi^{-1}(F(s, t))$, $0 \leq t \leq \varepsilon$, $0 \leq s \leq 1$, $G(s, t)$ is continuous in t, s and for each fixed s, is continuously differentiable in t with

$$D_t G(s, t) = [D\phi(G(s, t))]^{-1} D_t F(s, t)$$
$$|D_t G(s, t)| \leq K |D_t F(s, t)|.$$

This estimate holds on the set of existence of $G(s, t)$ and shows that there is a function A_s such that

$$|G(s, t_1) - G(s, t_0)| \leq A_s |t_1 - t_0|$$

for $0 < s \leq 1$. This equicontinuity in t implies the function $G(s, t)$ can be defined continuously on a maximal square $0 \leq t \leq a$, $0 \leq s \leq 1$ in Q. The local Implicit Function Theorem shows that $a = 1$ and the lemma is proved.

Proof of Theorem 3.9. For $y_0 = \phi(0)$ and any point $y \in Y$, let $y(t)$, $0 \leq t \leq 1$, be the straight line segment joining y_0 to y. Lemma 3.10 implies there is a continuous curve $x(t)$, $0 \leq t \leq 1$, such that $\phi(x(t)) = y(t)$. Then $\phi(x(1)) = y$ and ϕ is onto.

  The remainder of the proof is supplied exactly as in the proof of Theorem 3.5. $\square$

2.4. Alternative Methods

Many problems in analysis and applied mathematics can be reduced to the determination of the zeros of a function in a Banach space. There may be several different ways to obtain such a function. A satisfactory solution of the problem often depends upon the ingenuity of the investigator to determine a function for which the nature of the zeros can be analyzed effectively by classical mathematical methods. Of course, no general procedure can be prescribed which has universal application. On the other hand, for a particular class of problems, there is a method which has been applied on several occasions which goes under the general title of the *alternative method*. The purpose of this section is to describe this method.

We begin with the simplest case and use it to motivate the more general method. For notation, if X is a Banach space, the symbol X_P will denote there is a continuous projection $P: X \to X$ such that $X_P = PX$, the range $\mathscr{R}(P)$ of P. The null space of a linear operator A on X is denoted by $\mathscr{N}(A)$.

Suppose X, Z are Banach spaces, $A: X \to Z$ is a continuous linear operator, $N: X \to Z$ is a continuous operator. The problem is to determine the solutions of the equation

$$(4.1) \qquad\qquad Ax - Nx = 0$$

for $x \in X$.

The following simple lemma is fundamental.

Lemma 4.1. *If*

$$\mathscr{N}(A) = X_U, \qquad \mathscr{R}(A) = Z_E,$$

then there is a bounded linear operator $K: Z_E \to X_{I-U}$, called the right inverse of A, such that $AK = I$ on Z_E, $KA = I - U$ on X and Eq. (4.1) is equivalent to the equation

$$x = y + z, \qquad y \in X_U, \qquad z \in X_{I-U}$$

$$(4.2) \quad \begin{array}{l} \text{(a)} \\ \text{(b)} \end{array} \qquad \begin{array}{l} z - KEN(y + z) = 0 \\ (I - E)N(y + z) = 0 \end{array}$$

Finally, the bound on K can be obtained from A since, for every $v > 0$, $|K| \le v$ if and only if

$$(4.3) \qquad\qquad |(I - U)x| \le v|Ax| \quad \text{for } x \in X.$$

Proof. The map A is one-to-one from X_{I-U} onto Z_E. Thus, the existence of the right inverse $K: Z_E \to X_{I-U}$ satisfying $AK = I$ on Z_E, $KA = I - U$ on X is clear. The fact that K is bounded follows from the closed graph theorem. If $x = y + z$ as in (4.2), then $KAx = KAy + KAz = z$ for all $x \in X$. Since Eq. (4.1) is equivalent to

$$(I - E)(A - N)(y + z) = 0, \qquad E(A - N)(y + z) = 0$$

and $EA = A$, $KA(y + z) = z$, we obtain (4.2). To prove the last assertion of the lemma, suppose $w \in Z_E$ and $Ax = w$. Then

$$Kw = (I - U)Kw = (I - U)KAx = (I - U)x.$$

This completes the proof of the lemma. $\square$

Remark 4.2. In the applications, the operators A, N may arise as maps from domains $D(A) \subseteq D(N) \subseteq Z$ into Z. If A is a closed linear operator, that is, the graph $\{(x, Ax): x \in D(A)\}$ in $Z \times Z$ is closed, then the linear space $X = D(A)$ with $|x|_X = |x|_Z + |Ax|_Z$ is a Banach space, which can be considered continuously embedded in Z. The solution of the equation $Ax - Nx = 0$ for $x \in D(A) \subseteq Z$ is then equivalent to Eq. (4.1) for $x \in X$.

We now consider the solution of Eq. (4.2) considered as two equations in the two unknowns $y \in X_U$, $z \in X_{I-U}$. If it is possible to find a solution $z^*(y)$ of Eq. (4.2a) for z as a function of y, then Eq. (4.1) becomes equivalent to the equation

$$(4.4) \qquad\qquad (I - E)N(y + z^*(y)) = 0.$$

When this can be done, we refer to Eq. (4.4) as the *determining equations* or *bifurcation equations*. The function in (4.4) is called the *determining* (or *bifurcation*) *function*. These equations involve only the unknown parameter y from the null space of A. The problem concerning the nature of the zeros of the function $A - N$ in a Banach space X has been reduced to an alternative problem concerning the nature of the zeros of a function from $\mathcal{N}(A)$ to Z_{I-E}. Generally, this latter problem is in a space of dimension much smaller than the dimension of X. We refer to this general procedure of reducing Eq. (4.1) to (4.4) for y in $\mathcal{N}(A)$ as the *alternative method applied to the null space of A.*

Let us now discuss a particular case of Eq. (4.1) which is very important in the theory of bifurcation. Suppose X, Z, Λ are Banach spaces, $A: X \to Z$ is a bounded, linear operator, $N: X \times \Lambda \to Z$ is continuous together with its Fréchet derivative in X,

$$(4.5) \qquad\qquad N(0, 0) = 0, \qquad \partial N(0, 0)/\partial x = 0$$

and consider the equation

$$(4.6) \qquad\qquad Ax - N(x, \lambda) = 0$$

for (x, λ) in a neighborhood of $(0, 0)$. With the notation as before, this equation is equivalent to

$$x = y + z, \qquad y \in X_U, \qquad z \in X_{I-U}$$
$$(4.7) \quad \text{(a)} \qquad z - KEN(y + z, \lambda) = 0$$
$$ \quad \text{(b)} \qquad (I - E)N(y + z, \lambda) = 0$$

Applying the Implicit Function Theorem to Eq. (4.7a), there is a neighborhood V of $(0, 0) \in X_U \times \Lambda$ and a function $z^*: V \to X_{I-U}$, $z^*(y, \lambda)$ continuous together with its Fréchet derivative in y, $z^*(0, 0) = 0$, such that $z^*(y, \lambda)$ satisfies

(4.7a) and is the only solution in a neighborhood of zero. If the function $N(x, \lambda)$ has k continuous derivatives in x, λ (or is analytic in x, λ) then $z^*(y, \lambda)$ has k continuous derivatives in y, λ (or is analytic in x, λ). Thus, Eq. (4.6) has a solution in a sufficiently small neighborhood of $(x, \lambda) = (0,0)$ if and only if $(y, \lambda) \in V$ satisfy the determining equations (bifurcation equations)

$$(4.8) \qquad (I - E)N(y + z^*(y, \lambda), \lambda) = 0.$$

When the alternative method is applied to $\mathcal{N}(A)$ for special equations of the form (4.6) for the solutions in a neighborhood of $(x, \lambda) = (0,0)$, we shall refer to this as the *method of Liapunov–Schmidt* for Eq. (4.6).

There are several reasons for extending the above procedure to situations where the determining function is a map from X_P into Z and the projection P is not a projection onto the null space of A. This more general procedure will be referred to as the *alternative method* or the *method of alternative problems*.

We make the following hypotheses:

(H_1) There exist continuous projections P on X, Q on Z such that $(I - Q)A = AP$

(H_2) There is a continuous linear map $K : Z_Q \to X_{I-P}$ satisfying
    (i) $KQA = I - P$
    (ii) $AKQN = QN.$

We can now prove the following result.

Lemma 4.3. *Eq.* (4.1) *has a solution if and only if* $x = y + z,\ y \in X_P,\ z \in X_{I-P}$

$$(4.9) \quad \begin{array}{ll} \text{(a)} & z - KQN(y + z) = 0 \\[1mm] \text{(b)} & (I - Q)(A - N)(y + z) = 0 \end{array}$$

Proof. If $Ax - Nx = 0$, then $x = y + z$ satisfies (4.9b) and $Q(Ax - Nx) = 0$. Applying K to this last equation and using (i) of (H_2), we obtain (4.9a). Conversely, if $x = y + z$ satisfies (4.9), then operating on (4.9a) with A and using (ii) of (H_2), we obtain $Az - QN(y + z) = 0$. Using (H_1), we have $(I - Q)A(y + z) = Ay$ so that (4.9b) is equivalent to $Ay - (I - Q)N(y + z) = 0$. Thus, $A(y + z) - N(y + z) = 0$, which is Eq. (4.1). The lemma is proved. $\quad\square$

Eq. (4.9) has the same form as Eq. (4.2) except with the projections U, E replaced respectively by the projections P, Q. It is interesting to note that hypotheses (H_2) implies that X_P contains the null space of A. In fact, if $Ax = 0$, then (i) of (H_2) implies $(I - P)x = 0$ or $Px = x$ and $x \in X_P$. On the other hand, the hypotheses do not imply that $Z_Q \subseteq Z_E$; that is, Z_Q is in the range of A. If it is supposed that $Z_Q \subseteq Z_E$, then it is possible to estimate the norm of KQ in terms of A and P. One can show that

$$(4.10) \qquad |KQ| \le v \quad \text{if and only if} \quad |(I - P)x| \le v|Ax| \quad \text{for } x \in X.$$

In fact, if $w \in Z_Q$ and $Ax = w$, then (i) of (H_2) implies

$$KQw = KQAx = (I - P)x$$

and relation (4.10) follows immediately.

If it is possible to solve Eq. (4.9a) for a function $z^*(y)$, then Eq. (4.1) is equivalent to the equation

$$(4.11) \qquad\qquad (I - Q)(A - N)(y + z^*(y)) = 0.$$

The equations are called the *determining* (or *bifurcation*) *equations* and the general procedure of reducing (4.1) to the alternative problem (4.11) for $y \in X_P$ is referred to as the *alternative method applied to* X_P.

If $N(x, \lambda)$ satisfies (4.5), and there are operators P, Q, K satisfying (H_1), (H_2), then we can use the Implicit Function Theorem to solve Eq. (4.9a) for a unique continuous function $z^*(y, \lambda)$ defined on a neighborhood V of $(0, 0)$ in $X_P \times \Lambda$, $z^*(0, 0) = 0$. The smoothness properties of $z^*(y, \lambda)$ are the same as the smoothness properties of $N(x, \lambda)$. Thus, Eq. (4.6) has a solution in a sufficiently small neighborhood of $(0, 0)$ if and only if $(y, \lambda) \in V$ satisfy

$$(I - Q)\left[Ax^*(y, \lambda) + N(x^*(y, \lambda), \lambda)\right] = 0$$
$$x^*(y, \lambda) = y + z^*(y, \lambda).$$

This procedure will be referred to as the *alternative method applied to* X_P *for solutions in a neighborhood of zero.*

2.5. Embedding Theorems

Let Ω be an arbitrary bounded open set of $\mathbb{R}^n$. The symbol $L^p(\Omega)$, $p \geq 1$, denotes the Banach space of Lebesgue measurable functions ϕ defined on Ω such that $\int_\Omega |\phi(\theta)|^p \, d\theta < \infty$, with norm $|\cdot|$ defined by

$$|\phi|^p = \int_\Omega |\phi(\theta)|^p \, d\theta$$

and the usual identification through the a.e. equivalence relation.

Definition 5.1. A function $f : \Omega \times \mathbb{R} \to \mathbb{R}$ is said to satisfy the *Carathéodory conditions* if $f(x, y)$ is continuous in y for almost all x and is measurable in x for all $y \in \mathbb{R}$.

If f satisfies the Carathéodory conditions, then we define an operator N_f, called the *Nemitskii operator*, on the set of functions $\phi : \Omega \to \mathbb{R}$ by $N_f(\phi)(x) = f(x, \phi(x))$ for $x \in \Omega$. If ϕ is measurable, then $N_f(\phi)(x)$ is measurable. The $N_f(\phi)$ is also continuous with respect to convergence in measure.

A more important property of this operator is contained in the following result which is stated without proof.

Theorem 5.2. *If $N_f : L^{p_1}(\Omega) \to L^{p_2}(\Omega)$, $p_1, p_2 \geq 1$, then N_f is continuous, takes bounded sets into bounded sets and, if $p_2 < \infty$, there is a constant $b > 0$ and a function $a \in L^{p_2}(\Omega)$ such that*

$$(5.1) \qquad |f(x, y)| \leq a(x) + b|y|^{p_1/p_2} \quad \text{a.e. in } x, \text{ for all } y \in \mathbb{R}.$$

If $p_2 = \infty$, there is a constant $M > 0$ such that

$$(5.2) \qquad |f(x, y)| \leq M \quad \text{a.e. in } x, \text{ for all } y \in \mathbb{R}.$$

If either (5.1) is satisfied for $p_2 < \infty$ or (5.2) is satisfied for $p_2 = \infty$, it is clear that N_f is defined from $L^{p_1}(\Omega)$ to $L^{p_2}(\Omega)$. Thus, these inequalities are sufficient for the existence, continuity and boundedness of N_f.

Even if it is only necessary to define N_f on some open set $U \subseteq L^{p_1}(\Omega)$, the inequality (5.1) (respectively (5.2)) still must be satisfied. In fact, it is possible to prove that, if N_f is defined from any open set U in $L^{p_1}(\Omega)$ to $L^{p_2}(\Omega)$, then N_f can be defined on all of $L^{p_1}(\Omega)$ into $L^{p_2}(\Omega)$.

We now examine the differentiability of N_f. If $f(x, y)$ has a derivative $D_2 f(x, y)$ with respect to y which satisfies the Caratheodory conditions, then one can define the Nemitskii operator $N_{D_2 f}$. If $N_{D_2 f} : L^{p_1}(\Omega) \to L^{p_1 p_2/(p_1 - p_2)}(\Omega)$, $p_1 > p_2 \geq 1$, then an application of Hölder's inequality implies that N_f has a continuous Fréchet derivative at $\phi \in L^{p_1}(\Omega)$ given by

$$(5.3) \qquad [DN_f(\phi)h](x) = D_2 f(x, \phi(x))h(x), \qquad x \in \Omega, h \in L^{p_1}(\Omega)$$

From Theorem 5.2, this implies that there is a constant $d > 0$ and a function $c \in L^{p_1 p_2/(p_1 - p_2)}$ such that f satisfies

$$(5.4) \qquad |D_2 f(x, y)| \leq c(x) + d|y|^{(p_1 - p_2)/p_2}$$

a.e. in x in Ω for every $y \in \mathbb{R}$.

The above shows that one can obtain Fréchet differentiability of N_f if $p_1 > p_2 \geq 1$. This restriction is important. In fact, if $p_1 = p_2 = 2$ and $N_f : L^2(\Omega) \to L^2(\Omega)$, then Theorem 5.2 implies that there is an $a \in L^2(\Omega)$ and constant $b > 0$ such that

$$|f(x, y)| \leq a(x) + b|y|$$

a.e. in x for all y. If, in addition, there is a constant $M > 0$ such that $|f(x, y)| \leq M$ a.e. in x for all y, then it can be shown that N_f has a Gateaux derivative. On the other hand, this function has a Fréchet derivative if and only if $f(x, y) = \alpha(x) + \beta y$ for some $\alpha \in L^2(\Omega)$ and constant β.

In the following we need some Sobolev imbedding theorems. Suppose Ω is an open set in $\mathbb{R}^n$ with a smooth boundary; for example, $\Omega = \mathbb{R}^n$ or Ω a bounded open set in $\mathbb{R}^n$ whose boundary is C^1. For $1 \le p < \infty$, let $L^p(\Omega, X)$ be the pth power integrable functions from Ω into a Banach space X with

$$|f|_{L^p(\Omega,X)} = \left[\int_\Omega |f(x)|_X^p \, dx \right]^{1/p}.$$

Let $W^{k,p}(\Omega, X)$ be the Sobolev space of $f \in L^p(\Omega, X)$ which have generalized derivatives of order $\le k$ which are p^{th} power integrable and let

$$|f|_{W^{k,p}(\Omega,X)} = \left[\int_\Omega \sum_{j=0}^{k} |f^{(j)}(x)|_X^p \, dx \right]^{1/p}.$$

Let $W_0^{k,p}(\Omega, X)$ be the closure in $W^{k,p}(\Omega, X)$ of the C^∞ functions with compact support in Ω.

The following imbeddings hold and are compact:

$$W^{k,p}(\Omega, \mathbb{R}) \subset L^q(\Omega, \mathbb{R}) \quad \text{if } \frac{1}{p} \ge \frac{1}{q} > \frac{1}{p} - \frac{k}{n} > 0,$$

$$W^{k,p}(\Omega, \mathbb{R}) \subset C(\Omega, \mathbb{R}) \quad \text{if } kp > n.$$

The first imbedding is continuous if $1/q = 1/p - k/n$.

2.6. Weierstrass Preparation Theorem

Suppose $f : \mathbb{C} \times \mathbb{C}^n \to \mathbb{C}$ is an analytic function of the $(n + 1)$-complex variables (w, z), $w \in \mathbb{C}$, $z \in \mathbb{C}^n$. If $f(0,0) = 0$ and $D_w f(0,0) \ne 0$, then the Implicit Function Theorem implies there is a unique solution $w = w(z)$ of

$$(6.1) \qquad\qquad\qquad f(w, z) = 0$$

in a neighborhood of $(0,0)$. Furthermore, there is a neighborhood of $z = 0$ such that, the equation $D_w f(w(z), z)h + D_z f(w(z), z) = 0$ has a unique solution $h = h(z)$ which is continuous in z. One easily shows that this implies $w(z)$ is continuously differentiable in z and $Dw(z) = h(z)$ in a neighborhood of $z = 0$. Thus, $w(z)$ is an analytic function of z in a neighborhood of $z = 0$ and we have proved:

Theorem 6.1. *If $f : \mathbb{C} \times \mathbb{C}^n \to \mathbb{C}$ is analytic in a neighborhood of $(0,0)$ and $f(0,0) = 0$, $D_w f(0,0) \ne 0$, then there exist $\delta > 0$, $\varepsilon > 0$ such that for every z, $|z| < \delta$, the Equation (6.1) has a unique solution $w(z)$ which is analytic in a neighborhood of zero.*

The Weierstrass Preparation Theorem is a generalization of this result.

Theorem 6.2 (Weierstrass Preparation Theorem). *Suppose* $f: \mathbb{C} \times \mathbb{C}^n \to \mathbb{C}$ *is an analytic function in a neighborhood of* $(0,0)$ *satisfying*

(6.2)
$$f(w, 0) = w^k g(w), \quad g \text{ analytic in a neighborhood of } 0 \in \mathbb{C},$$
$$g(0) \neq 0$$

Then there exists a function q *analytic in a neighborhood* V *of zero of* $\mathbb{C} \times \mathbb{C}^n$ *and functions* $a_0(z), \ldots, a_{k-1}(z)$ *analytic in a neighborhood of zero in* $\mathbb{C}^n$ *such that* $a_0(0) = \cdots = a_{k-1}(0) = 0$, $q(0,0) \neq 0$ *and*

$$(6.3) \qquad q(w, z)f(w, z) = w^k + \sum_{i=0}^{k-1} a_i(z)w^i, \qquad (w, z) \in V.$$

The detailed proof given below brings out the relationship between Theorem 6.2 and the division theorems for analytic functions. This method of proof can be generalized to C^∞-functions. Another proof using the Implicit Function Theorem will be indicated, but it does not generalize to the C^∞-case. Theorem 6.2 is a special case of the following result.

Theorem 6.3 (Weierstrass Division Theorem). *Let* f, g, k *be as in Theorem* 6.2 *and let* $G: \mathbb{C} \times \mathbb{C}^n \to \mathbb{C}$ *be any function analytic in a neighborhood of zero. Then there exist functions* $q, r: \mathbb{C} \times \mathbb{C}^n \to \mathbb{C}$ *analytic in a neighborhood of zero such that*

$$G = qf + r$$
$$r(w, z) = \sum_{i=0}^{k-1} r_i(z)w^i$$

where each r_i *is analytic in a neighborhood of zero and* q, r *are unique on some neighborhood of zero.*

Theorem 6.3 implies Theorem 6.2. In fact, choose $G(w, z) = w^k$, $a_i(z) = -r_i(z)$. To show $q(0, 0) \neq 0$, note that

$$w^k = q(w, 0)f(w, 0) + r(w, 0)$$
$$= q(w, 0)w^k g(w) + \sum_{i=0}^{k-1} r_i(0)w^i.$$

This clearly implies $q(0,0)g(0) = 1$ and thus $q(0,0) \neq 0$. This proves that Theorem 6.3 implies Theorem 6.2. $\square$

Proof of uniqueness in Theorem 6.3. If

$$G = qf + r = q_1 f + r_1$$

then

$$(q - q_1)f = r - r_1.$$

The right hand side of this equation is a polynomial in w of degree at most $k - 1$. Therefore, it has at most $k - 1$ zeros for any fixed z. On the other hand, if $q \neq q_1$, then a simple application of Rouche's theorem implies the left hand side has k zeros in a neighborhood of $w = 0$ for each z in a neighborhood of zero. This contradiction implies $q = q_1$ and thus $r = r_1$. Uniqueness is proved. $\square$

Basic to the proof of existence in the Weierstrass Division Theorem is the Polynomial Division Theorem.

Theorem 6.4 (Polynomial Division Theorem). *Let* $\lambda \in \mathbb{C}^k$, $P(w, \lambda) = w^k + \sum_{i=0}^{k-1} \lambda_i w^i$ *and suppose* $G(w, z)$ *is analytic in a neighborhood of* $(0, 0) \in \mathbb{C} \times \mathbb{C}^n$. *Then there exist functions* $q(w, z, \lambda)$, $r(w, z, \lambda)$ *analytic in a neighborhood of zero in* $\mathbb{C} \times \mathbb{C}^n \times \mathbb{C}^k$ *satisfying*

$$G(w, z) = q(w, z, \lambda)P(w, \lambda) + r(w, z, \lambda)$$

$$r(w, z, \lambda) = \sum_{i=0}^{k-1} r_i(z, \lambda)w^i$$

and each $r_i(z, \lambda)$ *is analytic in a neighborhood of zero.*

Moreover, if there exists a function $g : \mathbb{C} \to \mathbb{C}$ *analytic in a neighborhood of the origin such that* $g(0) \neq 0$ *and* $G(w, 0) = w^k g(w)$ *for* $|w|$ *sufficiently small, then* $q(0, 0, 0) \neq 0$.

Theorem 6.4 implies Theorem 6.3. If f, G satisfy the conditions of Theorem 6.3, then Theorem 6.4 implies the existence of analytic functions q_f, r_f, q_G, r_G with r_f a polynomial of degree $k - 1$, $q_f(0, 0) \neq 0$ since $f(w, 0) = g(w)w^k$, $g(0) \neq 0$, r_G a polynomial in w of degree at most k such that

$$f = q_f P + r_f, \qquad G = q_G P + r_G.$$

If we can choose $\lambda = \lambda(z)$ in such a way that $r_f = 0$, then $G = q_G(q_f)^{-1}f + r_G = q(w, z)f(w, z) + r(w, z)$ where $r(w, z)$ is a polynomial in w of at most degree $k - 1$ since $f(w, 0) = g(w)w^k$, $g(0) \neq 0$. To show such a $\lambda(z)$ exists, note first that

$$w^k g(w) = f(w, 0) = q_f(w, 0, 0)P(w, 0) + r_f(w, 0, 0)$$
$$= q_f(w, 0, 0)w^k + r_f(w, 0, 0).$$

Since $g(0) \neq 0$, this obviously implies $q_f(0, 0, 0) \neq 0$ and $r_f(w, 0, 0) = 0$ for all w in a neighborhood of zero. If

$$r_f(w, z, \lambda) = \sum_{i=0}^{k-1} r_i(z, \lambda)w^i$$

the above remark implies $r_i(0,0) = 0$, $i = 0, 1, \ldots, k - 1$. Next, we compute $\partial r_i(0,0)/\partial \lambda_j$. For $z = 0$, we have

$$w^k g(w) = f(w,0) = q_f(w,0,\lambda)\left(w^k + \sum_{i=0}^{k-1} \lambda_i w^i\right) + \sum_{i=0}^{k-1} r_i(0,\lambda)w^i$$

Differentiating with respect to λ_j and evaluating at $\lambda = 0$, one obtains

$$0 = \frac{\partial q_f}{\partial \lambda_j}(w,0,0)w^k + q_f(w,0,0)w^j + \sum_{i=0}^{k-1} \frac{\partial r_i}{\partial \lambda_j}(0,0)w^i$$

for all w in a neighborhood of zero. This implies $\partial r_i(0,0)/\partial \lambda_j = 0$ for $i < j$ and $\partial r_j(0,0)/\partial \lambda_j = q_f(0,0,0) \neq 0$, $j = 1, 2, \ldots, n$. Thus, $\det(\partial r_i(0,0)/\partial \lambda_j) \neq 0$. Theorem 6.1 implies there exists a function $\lambda(z)$ analytic in a neighborhood of zero such that $\lambda(0) = 0$ and $r_i(z, \lambda(z)) = 0$, $i = 0, 1, \ldots, k - 1$; that is, $r_f(z, \lambda(z)) = 0$. This completes the proof that Theorem 6.4 implies Theorem 6.3 since $q_f(w, z, \lambda(z)) \neq 0$ in a neighborhood of zero. $\qquad \square$

Proof of Theorem 6.4. Suppose γ is a simple closed curve in $\mathbb{C}$ containing zero. If w is in the interior of γ, then

$$G(w, z) = \frac{1}{2\pi i} \int_\gamma \frac{G(\eta, z)}{\eta - w}\, d\eta.$$

Since $P(w, 0) = w^k$ and $P(w, \lambda)$ is a polynomial in w of degree k, one can be sure that all the zeros of $P(w, \lambda)$ lie inside γ if λ is sufficiently small. The form of P implies

$$P(\eta, \lambda) - P(w, \lambda) = (\eta - w) \sum_{i=0}^{k-1} s_i(\eta, \lambda)w^i$$

where each $s_i(\eta, \lambda)$ is analytic in a neighborhood of zero. Therefore,

$$\frac{P(\eta, \lambda)}{\eta - w} = \frac{P(w, \lambda)}{\eta - w} + \sum_{i=0}^{k-1} s_i(\eta, \lambda)w^i$$

$$G(w, z) = \frac{1}{2\pi i} \int_\gamma \left(\frac{G(\eta, z)}{(\eta - w)}\right) \frac{P(\eta, \lambda)}{P(\eta, \lambda)}\, d\eta$$

$$= \left(\frac{1}{2\pi i} \int_\gamma \frac{G(\eta, z)}{(\eta - w)P(\eta, \lambda)}\, d\eta\right) P(w, \lambda)$$

$$\quad + \sum_{i=0}^{k-1} \left(\frac{1}{2\pi i} \int_\gamma \frac{G(\eta, z)}{P(\eta, z)} s_i(\eta, \lambda)\, d\eta\right) w^i$$

$$\stackrel{\text{def}}{=} q(w, z, \lambda)P(w, \lambda) + r(z, \lambda).$$

Also, if G satisfies the additional hypothesis stated in the last part of the theorem, then

$$q(0,0,0) = \frac{1}{2\pi i} \int_\gamma \frac{G(\eta,0)}{\eta P(\eta,0)}\, d\eta = \frac{1}{2\pi i} \int_\gamma \frac{g(\eta)}{\eta}\, d\eta = g(0) \neq 0.$$

This proves the theorem. $\square$

Remark 6.5. It is possible to prove the Weierstrass Preparation Theorem using the Implicit Function Theorem. Fix $\delta > 0$ and let $A = \{a:\{|z| \leq \delta\} \to \mathbb{C}$ analytic$\}$. If $a(z) = \sum_{k=0}^{\infty} a_k z^k$, define $|a| = \sum_{k=0}^{\infty} |a_k|\, \delta^k$. Let $Q = \{q:\{(w,z): |w| \leq \delta, |z| \leq \delta\} \to \mathbb{C}\}$ with a similar norm. If $a_j \in A$, $j = 0, 1, \ldots, q \in Q$, define

$$F(q, a_0, a_1, \ldots, a_k, w, z) = f(w,z) - \left(\sum_{j=0}^{h} a_j(z)w^j\right) q(w,z)$$

Then $F: Q \times A \times \cdots \times A \times \mathbb{C} \times \mathbb{C}^n \to \mathbb{C}$, $F(g(w), 0, \ldots, 0, 1, 0, 0) = 0$. Now show that $D_{q, a_0, a_1, \ldots, a_k} F$ is an isomorphism at $(g(w), 0, \ldots, 0, 1, 0, 0)$.

Corollary 6.6. *If $f: C \times C^n \to C$ is analytic in a neighborhood of $(0,0)$ and satisfies Hypothesis (6.2), then there are $\delta > 0$, $\varepsilon > 0$ such that, for every z, $|z| < \delta$, there are exactly k solutions $w_1(z), \ldots, w_k(z)$ of Equation (6.1) satisfying $|w_j(z)| < \varepsilon$, $j = 1, 2, \ldots, k$. Furthermore, each $w_j(z)$ is continuous in z, $|z| < \delta$, and $w_j(0) = 0$.*

Proof. By the Weierstrass Preparation Theorem, $f(w,z)$ satisfies Relation (6.3) with $q(w,z) \neq 0$, $|w| < \varepsilon$, $|z| < \delta$. Thus, $f(w,z) = 0$ if and only if (w,z) satisfy the polynomial equation

$$(6.4) \qquad\qquad w^k + a_{k-1}(z)w^{k-1} + \cdots + a_0(z) = 0.$$

Since each $a_j(z)$ satisfies $a_j(0) = 0$, it follows from Rouche's theorem that there are $\delta > 0$, $\varepsilon > 0$ and k zeros $w_j(z)$ of Equation (6.4) satisfying the properties stated in the corollary. $\square$

From Corollary 6.6, it is therefore only necessary to discuss the behavior of the zeros of a polynomial

$$(6.5) \qquad\qquad w^k + a_{k-1}(z)w^{k-1} + \cdots + a_0(z)$$

with coefficients analytic in z, vanishing at $z = 0$.

The next question is the following: How can the zeros of a Polynomial (6.5) depend upon the parameter z? How complicated is the branch point at $z = 0$?

At any point $z_0 \neq 0$ for which all solutions of Equation (6.4) are distinct, it follows from Theorem 6.1 that the solutions of Equation (6.4) are analytic and single valued in a neighborhood of $z = z_0$. However, this is not the case at $z = 0$ and the solutions are generally complicated functions of z near $z = 0$.

For $z \in \mathbb{C}$, the situation is completely understood. Suppose all solutions of Equation (6.4) are distinct for every $z \neq 0$, $|z| < \delta$. Let $w_1(z), \ldots, w_k(z)$ be the single valued branches of solutions of Equation (6.4) given by the Implicit Function Theorem 6.1. These functions are not single valued in a neighborhood of $z = 0$. In fact, it is known that if z moves along a Jordan curve Γ around 0, then $w_j(z)$ generally will have changed when one returns to the initial point on Γ. Therefore, $w_j(z)$ after this rotation along Γ must coincide with some other solution $w_k(z)$. This divides the set $\{w_1(z), \ldots, w_k(z)\}$ into cyclic sets $A_r(z) = \{w_{j_1}(z), \ldots, w_{j_r}(z)\}$, $r = 1, 2, \ldots$ such that the sets $A_r(z)$ undergo a cyclic permutation when z traverses Γ one time. In a later section, we give a specific way to determine these cyclic sets by means of the Newton polygon. To apply the Newton polygon, it is necessary to know something about the coefficients in the power series expansions of the functions $a_k(z)$. One can prove the following result for $n = 1$.

Lemma 6.7. *If $f: \mathbb{C} \times \mathbb{C} \to \mathbb{C}$ is an analytic function in a neighborhood of $(0,0)$ and Hypothesis (6.2) is satisfied,*

$$f(w, z) = \sum_{i=0}^{k} b_j(z)w^j + 0(|w|^{k+1}) \quad \text{as } |w| \to 0$$

$$b_j(z) = c_{jl_j} z^{l_j} + 0(|z|^{l_j+1}) \qquad \text{as } |z| \to 0$$

$$c_{jl_j} \neq 0, \qquad j = 0, 1, 2, \ldots, k,$$

then $l_0 = 0$ and the coefficients $a_j(z)$, $j = 0, 1, \ldots, k - 1$ in Relation (6.3) satisfy

$$a_j(z) = \frac{c_{jl_j}}{c_{00}} z^{l_j} + 0(|z|^{l_j+1}) \quad \text{as } |z| \to 0,$$

$$j = 0, 1, \ldots, k - 1.$$

The above classification of the behavior of the solutions of Equation (6.4) near zero for $z \in \mathbb{C}$ is interesting but is not sufficient for the applications. In fact, for bifurcation problems, the function $f(w, z)$ is real for w, z real and one is interested in the structure of the real solutions as a function of real parameters z. Even when z is a one dimensional parameter, this makes the problem much more difficult. As remarked before, the Newton polygon will permit one to obtain complete information in this case.

When $z \in \mathbb{C}^n$, $n > 1$, the available knowledge of the analytic structure of the solutions of Equation (6.4) for w as a function of z is not as specific as for $n = 1$. It has not led to a constructive way of determining the solutions. On the other hand, the Weierstrass Preparation Theorem asserts that it is sufficient to consider polynomials. Furthermore, by the transformation

$$w = \eta - \frac{1}{k} a_{k-1}$$

in the polynomial equation

$$w^k - \sum_{i=0}^{k-1} a_i w^i = 0$$

it is sufficient to consider the equation

(6.6)
$$P(\eta, b) \overset{\text{def}}{=} \eta^k - \sum_{i=0}^{k-2} b_i \eta^i = 0$$
$$b = (b_0, \ldots, b_{k-2})$$

where the b_i vary in a neighborhood of zero. A discussion of the roots of $P(\eta, b) = 0$ as a function of the $(k-1)$-vector b leads to a complete answer to the question of the zeros of $f(w, z) = 0$.

Let us discuss a few cases for real roots of $P(\eta, b) = 0$ for $b \in \mathbb{R}^{k-1}$. The cases $k = 2, 3$ were considered in Section 1.5.

If $k = 4$, then

$$P(\eta, b_0, b_1, b_2) = \eta^4 + b_2 \eta^2 + b_1 \eta + b_0 = 0$$

can have 0, 2, or 4 simple solutions. The values of (b_0, b_1, b_2) in $\mathbb{R}^3$ at which the number of solutions changes are the simultaneous solutions of the equations

$$P(\eta, b_0, b_1, b_2) = \eta^4 + b_2 \eta^2 + b_1 \eta + b_0 = 0$$
$$\frac{\partial P}{\partial \eta}(\eta, b_0, b_1, b_2) = 4\eta^3 + 2b_2 \eta + b_1 = 0.$$

These equations define a surface in (b_0, b_1, b_2)-space called the *swallow tail*. The cross sections of this surface for various values of b_2 are shown in Figure 6.1 with the number of real zeros of $P(\eta, b)$ as labeled.

For arbitrary k, the simultaneous solution of the equations

$$P(\eta, b) = 0$$
$$\frac{\partial P}{\partial \eta}(\eta, b) = 0$$

determines a surface $\mathscr{B}_k$ in the parameter space where the number of solutions changes. This surface is called the *bifurcation surface*.

One final remark on the generality of the Weierstrass Preparation Theorem. The theorem was stated for the parameter $z \in \mathbb{C}^k$. However, the finite dimensionality of z was not used in a significant manner—only analyticity was important. This remark will be used in some applications.

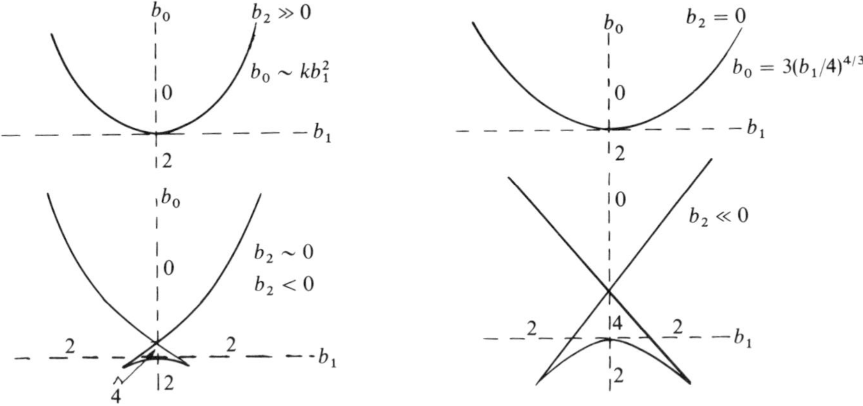

Figure 6.1

2.7. The Malgrange Preparation Theorem

A function $f: \mathbb{R}^k \times \mathbb{R}$ is said to be *smooth* in an open set $U \in \mathbb{R}^k$ if f has continuous derivatives of all orders in U; that is, $f \in C^\infty(U, \mathbb{R})$. The Malgrange Preparation Theorem is a generalization to smooth functions of the Weierstrass Preparation Theorem.

Theorem 7.1 (Malgrange Preparation Theorem). *Suppose $U \subset \mathbb{R} \times \mathbb{R}^n$ is an open set, $(0,0) \in U$, and $f \in C^\infty(U, \mathbb{R})$ satisfies*

$$f(t, 0) = t^k g(t)$$

where g is smooth in a neighborhood of zero, $g(0) \neq 0$. Then there exists a function q smooth in a neighborhood V of zero in $\mathbb{R} \times \mathbb{R}^n$, and functions $a_0(x), \ldots, a_{k-1}(x)$, smooth in a neighborhood of zero in $\mathbb{R}^n$, such that $q(0,0) \neq 0$;

$$q(t, x) f(t, x) = t^k + \sum_{i=0}^{k-1} a_i(x) t^i, \qquad (t, x) \in V.$$

The proof of this theorem follows the spirit of the above proof of the Weierstrass Preparation Theorem, but there are a number of technical difficulties which must be overcome.

Exactly as in the proof that Theorem 6.3 implies Theorem 6.2, one shows that Theorem 7.1 is a consequence of the following Division Theorem.

Theorem 7.2 (Division Theorem). *With f as in Theorem 7.1, and G any smooth real valued function defined in a neighborhood of zero in $\mathbb{R} \times \mathbb{R}^n$, there are* *functions q and r smooth in a neighborhood V of zero in $\mathbb{R} \times \mathbb{R}^n$ such that*

$$G = qf + r$$

$$r(t, x) = \sum_{i=0}^{k-1} r_i(x)t^k, \qquad (t, x) \in V.$$

Exactly as in the analytic case, one shows that Theorem 7.2 is a consequence of the following result.

Theorem 7.3 (Polynomial Division Theorem). *If G is a complex valued function defined and smooth on a neighborhood of zero in $\mathbb{R} \times \mathbb{R}^n$, $P(t, \lambda) = t^k + \sum_{i=0}^{k-1} \lambda_i t^i$, then there are smooth, complex valued functions $q(t, x, \lambda)$, $r_i(x, \lambda)$, $i = 0, 1, \ldots, k - 1$, defined in a neighborhood V of zero in $\mathbb{R} \times \mathbb{R}^n \times \mathbb{R}^k$ such that*

$$G(t, x) = q(t, x, \lambda)P(t, \lambda) + r(t, x, \lambda)$$

$$r(t, x, \lambda) = t^k + \sum_{i=0}^{k-1} r_i(x, \lambda)t^i \quad \text{in } V.$$

Moreover, if G is real valued, then q, r can be chosen to be real valued.

For the analytic case, the proof of the Polynomial Division Theorem was fairly routine because one could take advantage of the Cauchy integral formula. For functions which are only C^∞, this formula is no longer valid. However, there is an analogue of this formula which can be obtained from Green's function. From this point, only the ideas of the proof of Theorem 7.3 will be given.

If x, y are real numbers, $z = x + iy$, $\bar{z} = x - iy$, then, for any $f : \mathbb{C} \to \mathbb{R}$, we can define

$$\frac{\partial f}{\partial \bar{z}} = \frac{1}{2}\left(\frac{\partial f}{\partial x} + i\frac{\partial f}{\partial y}\right).$$

If f is complex valued, $f = \alpha + i\beta$,

$$\frac{\partial f}{\partial \bar{z}} = \frac{\partial \alpha}{\partial \bar{z}} + i\frac{\partial \beta}{\partial \bar{z}} = \frac{1}{2}\left[\left(\frac{\partial \alpha}{\partial x} - \frac{\partial \beta}{\partial y}\right) + i\left(\frac{\partial \alpha}{\partial y} + \frac{\partial \beta}{\partial x}\right)\right],$$

If $f : \mathbb{C} \to \mathbb{C}$ is a smooth function (when viewed as a mapping from $\mathbb{R}^2$ to $\mathbb{R}^2$), γ is a simple closed curve in $\mathbb{R}^2$ with interior D, then one can show that, for any $w \in D$,

$$F(w) = \frac{1}{2\pi i}\int_\gamma \frac{F(z)}{z - w}\,dz + \frac{1}{2\pi i}\iint_D \frac{\partial F}{\partial \bar{z}}(z)\frac{dz\,d\bar{z}}{z - w}$$

where, by definition, $dz\,d\bar{z} = -2i\,dx\,dy$.

Now suppose $G(t, x)$ satisfies the properties stated in Theorem 7.3 and suppose $\tilde{G}(w, x, \lambda)$ is a smooth function defined in a neighborhood of zero in $\mathbb{C} \times \mathbb{R}^n \times \mathbb{C}^k$ with $\tilde{G}$ an extension of G; that is, $\tilde{G}(t, x, \lambda) = G(t, x)$ for all real t. Formally, following the same steps as in the proof of the analytic version of the Polynomial Division Theorem, one obtains

$$\tilde{G}(w, x, \lambda) = q(w, x, \lambda)P(w, \lambda) + r(w, x, \lambda)$$

$$q(w, x, \lambda) = \frac{1}{2\pi i} \int_\gamma \frac{\tilde{G}(\eta, x, \lambda)}{P(\eta, \lambda)} \frac{d\eta}{(\eta - w)} + \frac{1}{2\pi i} \iint_D \frac{(\partial \tilde{G}/\partial \bar{z})(\eta, x, \lambda)}{P(\eta, \lambda)} \frac{d\eta \, d\bar{\eta}}{(\eta - w)}$$

$$r_i(x, \lambda) = \frac{1}{2\pi i} \int_\gamma \frac{\tilde{G}(\eta, x, \lambda)}{P(\eta, \lambda)} s_i(\eta, \lambda) \, d\eta$$

$$+ \frac{1}{2\pi i} \iint_D \frac{(\partial \tilde{G}/\partial \bar{z})(\eta, x, \lambda)}{P(\eta, \lambda)} s_i(\eta, \lambda) \, d\eta \, d\bar{\eta}.$$

One must show these formulas have meaning to complete the proof. The only troublesome terms are the integrals over D since some zeros of P may be in D. We must also have q, r smooth and this means that $\partial \tilde{G}/\partial \bar{z}$ must vanish to infinite order on the zero of P for w real. To show such an extension $\tilde{G}$ of G exists is the difficult step and is the substance of the next result which is stated without even an indication of the proof.

Theorem 7.4 (Nirenberg Extension Lemma). *Let $G(t, x)$ be a smooth complex valued function defined on a neighborhood of zero in $\mathbb{R} \times \mathbb{R}^n$. Then there is a smooth complex valued function $\tilde{G}(t, x, \lambda)$ defined in a neighborhood of zero in $\mathbb{R} \times \mathbb{R}^n \times C^k$ such that $\tilde{G}(t, x, \lambda)$ is an extension of $G(t, x)$, $\partial \tilde{G}/\partial \bar{z}$ vanishes to infinite order on $\mathrm{Im}\, z = 0$ and $\{P(z, \lambda) = 0\}$.*

It should be remarked that q and r are not unique as in the analytic case. The basic reason for this is that the Taylor series for a smooth function can have all coefficients zero and yet the function is not zero. For example e^{-1/t^2} in a neighborhood of $t = 0$.

The Malgrange Preparation Theorem is also valid for the case where the parameter x belongs to a Banach space.

2.8. Newton Polygon

In this section, we suppose $f : \mathbb{C} \times \mathbb{C} \to \mathbb{C}$ is analytic in a neighborhood of $(0, 0)$ and satisfies the relations

$$(8.1) \quad f(w, z) = w^k + a_{k-1}(z)w^{k-1} + \cdots + a_0(z) + a_{k+1}(z)w^{k+1} + \cdots$$

$$(8.2) \quad a_j(z) = a_j^{(p_j)}z^{p_j} + a_j^{(p_j+1)}z^{p_j+1} + \cdots$$

$$(8.3) \quad a_0^{(p_0)} \neq 0$$

where each $a_j(z)$ is analytic in a neighborhood of $z = 0$.

Our objective is to give a constructive method for finding the solutions of

$$(8.4) \qquad\qquad f(w, z) = 0.$$

The difficulty consists in recognizing in these complicated power series which terms are important and which ones may be ignored. To get an idea of what happens, let us consider some examples.

For the equation

$$(8.5) \qquad\qquad w^2 + a_0^{(1)}z + a_0^{(3)}z^3 = 0, \qquad a_0^{(1)} \neq 0$$

the linear term in z should be more important than the cubic. Therefore, the solutions should be close to the solutions of $w^2 + a_0^{(1)}z = 0$ which are functions of $z^{1/2}$. If $w = z^{1/2}v$, then v satisfies the equation

$$v^2 + a_0^{(1)} + a_0^{(3)}z^2 = 0.$$

Since $a_0^{(1)} \neq 0$, this equation has two simple solutions $\pm(-a_0^{(1)})^{1/2}$ for $z = 0$. Thus, Theorem 2.3 implies there are two solutions $v_j(z)$, analytic in z for $|z| < \delta$, sufficiently small. This implies the solutions w of Equation (8.5) are given by $w_j(z) = z^{1/2}v_j(z)$, $j = 1, 2$. By Corollary 6.6, for δ sufficiently small, there are only two solutions and thus all solutions are obtained in this way. Of course, one could have solved Equation (8.5) explicitly to obtain the same result. However, the simple procedure of scaling can be used in more general situations.

As another example, consider the equation

$$(8.6) \qquad\qquad w^3 + a_2^{(1)}zw^2 + a_1^{(3)}z^3w + a_0^{(4)}z^4 = 0$$

with all coefficients not zero. It is no longer clear how w should depend upon z. Let us, therefore, assume that $w = z^\alpha v$ and try to determine $\alpha \in (0, \infty)$ and a function $v(z)$ continuous in z, $v(0) \neq 0$, so that Equation (8.6) is satisfied. One obtains

$$z^{3\alpha}v^3 + a_2^{(1)}z^{1+2\alpha}v^2 + a_1^{(3)}z^{3+\alpha}v + a_0^{(4)}z^4 = 0.$$

Since $v(0) \neq 0$ and we wish to obtain a solution $v(z)$ for $|z| < \delta$, the constant α must certainly be chosen so that at least two of the exponents of z are equal. For such an α, one should be able to divide the equation by this common exponent and obtain an equation with coefficients well defined at $z = 0$. This puts a minimality requirement on this common exponent. For this example, $3\alpha = 1 + 2\alpha$ implies $\alpha = 1$ and the exponents of z are $(3, 3, 4, 4)$, satisfying the requirements specified. Dividing by z^3, one obtains

$$v^3 + a_2^{(1)}v^2 + a_1^{(3)}zv + a_0^{(4)}z = 0.$$

This equation for $z = 0$ has the simple zero $v = -a_2^{(1)}$. Thus, Theorem 2.3 implies there is an analytic solution $v(z)$ near $z = 0$ with $v(0) = -a_2^{(1)}$. One solution of the original Equation (8.6) is $w(z) = zv(z)$.

Checking the other possible values of α, one obtains $1 + 2\alpha = 4$, $\alpha = \frac{3}{2}$ with the corresponding exponents for z being $(\frac{9}{2}, 4, \frac{9}{2}, 4)$. Our requirements on α are satisfied. Dividing by z^4, one obtains

$$(8.7) \qquad z^{1/2}v^3 + a_2^{(1)}v^2 + a_1^{(3)}z^{1/2}v + a_0^{(4)} = 0.$$

For $z = 0$, there are two distinct solutions $v = \pm(-a_0^{(4)}/a_2^{(1)})^{1/2}$. Theorem 2.3 implies there are two functions $v_j(\zeta)$, $j = 1, 2$, analytic for ζ near zero such that Equation (8.7) has the solution $v_j(z^{1/2})$, $j = 1, 2$, near $z = 0$. Thus, the original Equation (8.6) has the two distinct solutions $w_j(z) = z^{3/2}v_j(z^{1/2})$, $j = 1, 2$. These two solutions plus the one found before give three solutions of Equation (8.6), which are all of the solutions by Corollary 6.6.

The Newton polygon systematizes the procedure used to determine the exponents α in the above examples and applies to the general Equation (8.4) where f satisfies Relations (8.1), (8.2), (8.3).

Introduce rectangular coordinates (i, j) in the plane and assign to every term in Relation (8.1) a point A_i with coordinates (i, p_i), $i = 0, 1, \ldots, k$, where p_i is the exponent of z in the first nonzero coefficient of the power series expansion of $a_i(z)$. If $a_i(z) = 0$ for all z, do not assign any point in the (i, j)-plane. Consider the convex polygonal line L passing through points of the set $\{A_i\}$ joining $(0, p_0)$ to $(k, 0)$ such that each A_i lies above or on L. This convex polygonal line is known as Newton's polygon or Newton's diagram (see Figure 8.1).

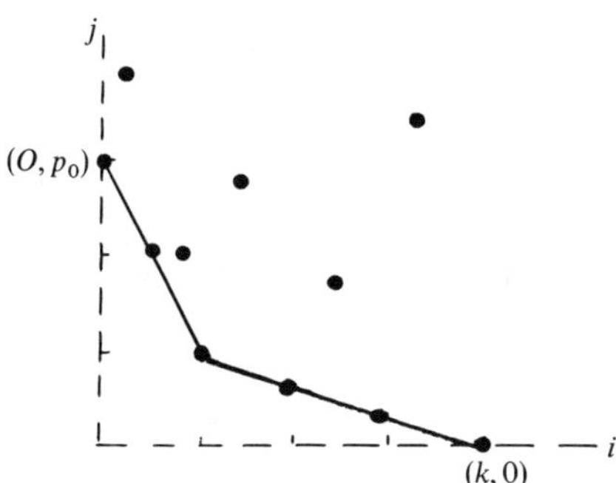

Figure 8.1

In Example (8.6), the A_i are $(3, 0), (2, 1), (1, 3), (0, 4)$ and the Newton polygon is shown in Figure 8.2.

The Newton polygon consists of a finite number of line segments L_j with end points $(A_{i_{j-1}}, A_{i_j})$, slopes $-\alpha_j$ and the distance along the abscissa between A_{i_j} and $A_{i_{j-1}}$ is $n_j = i_j - i_{j-1}$, $j = 1, 2, \ldots, r$. Also, $A_{i_0} = (0, p_0)$, $A_{i_r} = (k, 0)$.

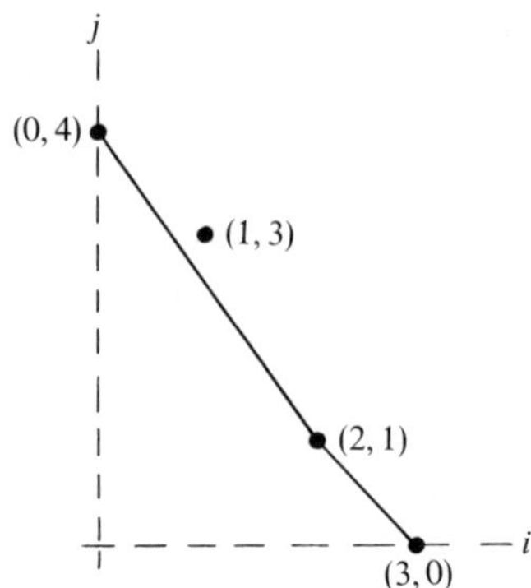

Figure 8.2

We are going to prove that corresponding to each L_j there are n_j solutions of Equation (8.4) of the form $z^{\alpha_j} v_{jl}(z)$, $l = 1, 2, \ldots, n_j$, where each $v_{jl}(z)$ is analytic in some fractional power of z in a neighborhood of $z = 0$. Since $\sum_{j=1}^{r} n_j = k$, it follows from Corollary 6.6 that this process will yield all solutions of Equation (8.4).

If α_j is one of the numbers defined above corresponding to the line segment L_j and points $A_{i_{j-1}} = (i_{j-1}, p_{i_{j-1}})$, $A_{i_j} = (i_j, p_{i_j})$, then $p_{i_{j-1}} + \alpha_j i_{j-1} = p_{i_j} + \alpha_j i_j \le p_i + \alpha_j i$ for all $i = 0, 1, \ldots, k$. Equality may hold here for some other points A_i but, for simplicity in notation, we suppose it does not. From this inequality, it follows that one can let

$$(8.8) \qquad\qquad w = z^{\alpha_j} v$$

in Equation (8.4) divide the resulting equation by $z^{p_{i_j} + \alpha_j i_j}$ to obtain a "scaled" equation for v of the form

$$(8.9) \qquad\qquad a_{i_j}^{p_{i_j}} v^{i_j} + a_{i_{j-1}}^{p_{i_j-1}} v^{i_j - 1} + g_{i_j}(z, v) = 0$$

where $g_{i_j}(z, v)$ is analytic in v and some fractional power of z near $(0, 0)$, $g_{i_j}(0, v) = 0$.

For $z = 0$, the possible candidates for approximate solutions of Equation (8.9) are the solutions of the equation

$$(8.10) \qquad\qquad a_{i_j}^{p_{i_j}} v^{n_j} + a_{i_{j-1}}^{p_{i_j-1}} = 0$$

since $n_j = i_j - i_{j-1}$. This equation has n_j nonzero distinct roots and Theorem 2.3 implies there is a unique solution of Equation (8.9) corresponding to each of these roots which is analytic in some fractional power of z. Returning to the original variable w given by Relation (8.8), one obtains n_j solutions of Equation (8.4). Repeating this process for each j, one obtains all solutions of Equation (8.4).

If more than two of the numbers $p_i + \alpha_j i$ are equal, then the resulting Equation (8.10) involves more powers of v and there is the possibility of a multiple root of the equation for $z = 0$. Near this multiple root, one must repeat the same process as above. Rather than carry out this analysis in detail, we illustrate it below with an example.

For Example (8.6), the Newton polygon is given in Figure 8.2. From the above results, it follows immediately that $\alpha_1 = \frac{3}{2}$, $n_1 = 2$, $\alpha_2 = 1$, $n_2 = 1$ and so there is one solution analytic in z and two solutions of the form $z^{3/2}v(z)$ where $v(z)$ is analytic in some fractional power of z. This fractional power is determined from the "scaled" Equation (8.9), which is $z^{1/2}$ for this example.

As another example, consider the equation

$$(8.11) \qquad f(w, z) \stackrel{\text{def}}{=} w^5 + 2zw^4 - zw^2 - 2z^2w - z^3 - z^4 = 0.$$

The Newton diagram is shown in Figure 8.3. From this Figure, we obtain $\alpha_1 = 1$, $n_1 = 2$, $\alpha_2 = \frac{1}{3}$, $n_2 = 3$.

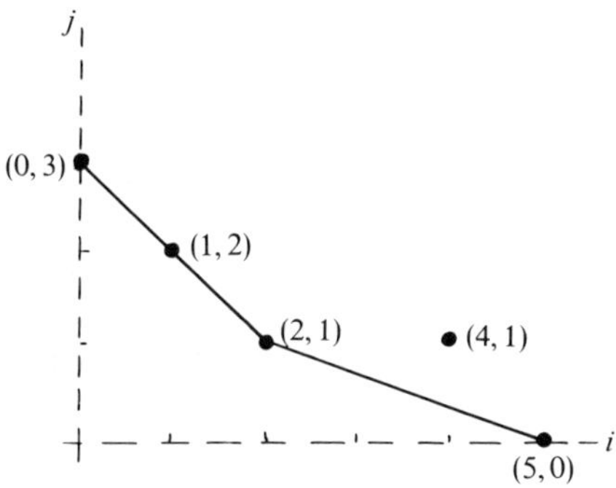

Figure 8.3

If $w = z^{1/3}v$, then v satisfies the "scaled" equation

$$(8.12) \qquad v^5 + 2z^{2/3}v^4 - v^2 - 2z^{2/3} - z^{4/3} - z^{7/3} = 0.$$

For $z = 0$, this equation has three distinct roots given by $v^3 = 1$. Thus, Theorem 2.3 implies Equation (8.12) has three roots $v_j(z^{1/3})$ analytic in $z^{1/3}$ near $z = 0$ and $v_j^3(0) = 1$. The functions $w_j(t) = z^{1/3}v_j(z^{1/3})$ yield three solutions of Equation (8.11).

If $w = zv$, then v satisfies the "scaled" equation

$$z^2v^5 + 2z^2v^4 - v^2 - 2v - 1 - z = 0.$$

For $z = 0$, this equation has a double zero at $v = -1$. If $v + 1 = \bar{v}$, then $\bar{v}$ satisfies the equation

$$(1 - 2z^2)\bar{v}^2 + 3z^2\bar{v} + z - z^2 + 0(z\bar{v}^3) = 0.$$

The Newton polygon for this equation is shown in Figure 8.4. Thus, $\alpha_1 = \frac{1}{2}$, $n_1 = 2$. If $\bar{v} = z^{i/2}\bar{\bar{v}}$, then $\bar{\bar{v}}$ satisfies the "scaled" equation

$$\bar{\bar{v}}^2 + 1 + 0(z) = 0.$$

This equation has two solutions $\bar{\bar{v}}_j(z)$ analytic in z which implies Equation (8.11) has two solutions of the form $w = z^{3/2}\bar{\bar{v}}_j(z)$. These two solutions and the previous three solutions yield all solutions of Equation (8.11) near $(0,0)$.

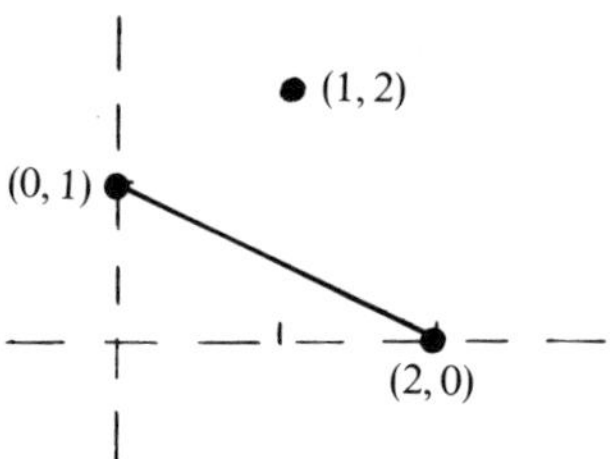

Figure 8.4

In the application of Newton's polygon to bifurcation problems in later sections, the function $f(w, z)$ is real if w, z are real and one is primarily interested in real solutions. There need not be k real solutions of Equation (8.4) and this puts restrictions on the region of the parameter z in which there are real solutions. Let us illustrate this with the Example (8.6) with $a_1^{(3)} = 0$; that is,

$$f(w, z) = w^3 + a_2^{(1)}zw^2 + a_0^{(4)}z^4, \qquad a_2^{(1)} \neq 0, \qquad a_0^{(4)} \neq 0.$$

The Newton diagram is given in Figure 8.2 without the point $(1,3)$. The solution analytic in z is real if z is real and given by

$$w = -a_2^{(1)}z + 0(|z|^2).$$

As observed before, the other two solutions of Equation (8.6) have the form $w(z) = z^{3/2}v(z)$ where $v(z)$ is analytic in $z^{1/2}$. If one wants only real solutions of Equation (8.6), then the scaled equations should be obtained from the transformation

$$w = |z|^{3/2}v, \quad z \text{ real},$$
$$|z|^{1/2}v^3 \pm a_2^{(1)}v^2 + a_0^{(4)} = 0$$

where the $+$ corresponds to $z > 0$ and the $-$ corresponds to $z < 0$. For $z = 0$, the solutions are $v^2 = \mp a_0^{(4)}/a_2^{(1)}$ which are real if $a_0^{(4)}a_2^{(1)} \operatorname{sgn} z > 0$. The diagram below (Figure 8.5) illustrates the solutions for $a_2^{(1)} < 0$, $a_0^{(4)}a_2^{(1)} > 0$.

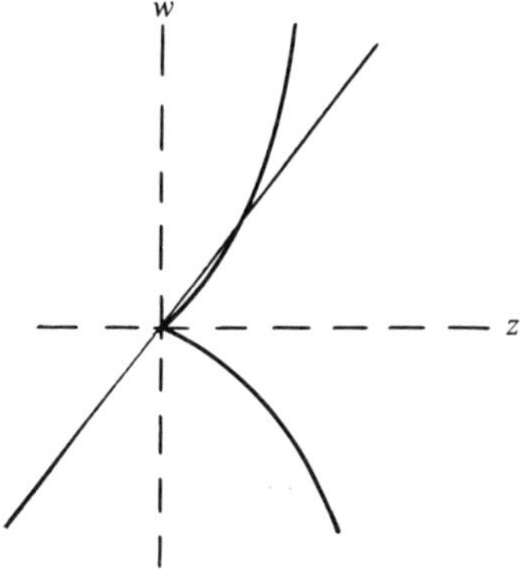

Figure 8.5

One final remark on the application of Newton's polygon. In general, one is not discussing a polynomial in w but a power series in w with coefficients analytic in z. The Weierstrass Preparation Theorem implies the problem can be reduced to a polynomial equation in w with coefficients analytic in z. Lemma 6.7 specifies the manner in which the leading coefficients in Relation (8.2) are determined and, therefore, determines the Newton polygon.

2.9. Manifolds and Transversality

In this section, we collect a few facts from the theory of manifolds. In order to avoid excessive notation, we sometimes give details only for finite dimensional spaces and indicate how everything can be generalized to Banach spaces.

Definition 9.1. Let X, Z be Banach spaces, $U \subset X$ be an open set and $f \in C^1(U, Z)$. A point $x \in U$ is called a *regular point* for f if $Df(x): X \to Z$ is surjective (onto). A point $x \in U$ is a *critical point* for f if x is not a regular point for f. A point $y \in Z$ is a *regular value* of f if each point of $f^{-1}(y)$ is a regular point for f. A point $y \in Z$ is a *critical value* of f if it is not a regular value of f; that is, $f^{-1}(y)$ contains a critical point for f. If $X = \mathbb{R}^n$, $Z = \mathbb{R}^k$, then x is a critical point for f if and only if rank $Df(x) < k$.

Definition 9.2. Suppose $r \geq 1$ is an integer. A set $M \subseteq \mathbb{R}^n$ is said to be a C^r-*manifold of dimension m* if, for each $x \in M$, there is an open set U contain x and a function $F \in C^r(U, \mathbb{R}^{n-m})$ which is regular at each point of U and $M \cap U = \{x : F(x) = 0\}$.

EXAMPLE 9.3. If U is an open set in $\mathbb{R}^m$ and $f \in C^r(U, \mathbb{R}^k)$, then the graph of f, $Gr(f) = \{(x, f(x)) : x \in U\} \subset \mathbb{R}^m \times \mathbb{R}^k = \mathbb{R}^n$, is a C^r-manifold of dimension

m in $\mathbb{R}^n$, $n = m + k$. In fact, we let $x = (u, v)$, $u \in \mathbb{R}^m$, $v \in \mathbb{R}^k$, then

$$Gr(f) = \{(u, v) \in \mathbb{R}^n : v = f(u),\ u \in U\}.$$

If $F : U \times \mathbb{R}^k \to \mathbb{R}^k$ is defined by $F(u, v) = v - f(u)$, then $Gr(f) = \{(u, v) : F(u, v) = 0,\ u \in U\}$. Thus, it remains only to show that F is regular at each point of $U \in R$. Since $\partial F(u, v)/\partial(u, v) = (-\partial f/\partial u, I)$, where I is the $k \times k$ identity matrix, $Df(u, v)$ is surjective and $Gr(f)$ is a C^r-manifold of dimension m in $\mathbb{R}^n$.

By using the Implicit Function Theorem, it is not difficult to prove that this example represents the general situation as stated in

Theorem 9.4. *Every C^r-manifold $M \subseteq \mathbb{R}^n$ of dimension m is locally the graph of a C^r-function. More precisely, one can express $\mathbb{R}^n$ as the product $\mathbb{R}^m \times \mathbb{R}^k$ so that, if $x = (u, v)$, $u \in \mathbb{R}^m$, $v \in \mathbb{R}^k$, $x_0 = (u_0, v_0)$ then there is a neighborhood U of x_0, a neighborhood V of u_0 and $f \in C^r(V, \mathbb{R}^k)$ such that $M \cap U = \{(u, v) : v = f(u),\ u \in V\}$.*

As a consequence of Theorem 9.4, one can obtain a local parametrization of a C^r-manifold.

Theorem 9.5. *Suppose M is a C^r-manifold of dimension m in $\mathbb{R}^n$. For any $x_0 \in M$, there is a neighborhood U of x_0 and a function $\phi \in C^r(U \cap M, \mathbb{R}^m)$ with $D\phi(x)$ an isomorphism for each $x \in U \cap M$.*

Proof. From Theorem 9.4, the set M is representable as the graph of a function in a neighborhood U of $x_0 : M \cap U = \{(u, v) : v = f(u),\ u \in V\}$. If $\phi(u, v) = u$, then the assertions of the theorem are true. $\square$

The ideas expressed in Theorem 9.5 lead to an alternative definition of a C^r-manifold of dimension m. Suppose M is a Hausdorff space. A C^r-*chart on M modeled on* $\mathbb{R}^m$ is a pair (U, ϕ) such that U is an open subset of M and $\phi \in C^r(U, \mathbb{R}^m)$ is a homeomorphism. The pair $(U \cap M, \phi)$ in Theorem 9.5 is a C^r-chart modeled on $\mathbb{R}^m$. Two C^r-charts (U, ϕ), (V, ψ) on M modeled on $\mathbb{R}^m$ are *compatible* if the map

$$\phi \circ \psi^{-1} : \psi(U \cap V) \to \phi(U \cap V)$$

is a C^r-isomorphism. A C^r-*atlas on M modeled on* R^m is a collection $\{(U, \phi)\}$ of C^r-charts on M modeled on R^m any two of which are compatible. An atlas is *maximal* if it contains each chart which is compatible with all of its members. A C^r-*manifold M modeled on* $\mathbb{R}^m$ is a Hausdorff space M together with a maximal C^r-atlas on M modeled on $\mathbb{R}^m$. It can be shown that this definition coincides with the previous one in Definition 9.2. On the other hand, the same definition can be used to define a *C^r-manifold M modeled on a Banach space E* if M is a Hausdorff space. We do not repeat the details.

If M is a C^r-manifold and N is a subset of M, then it is possible to define N as a submanifold in the following way. The set $N \subset M$ is a *C^r-submanifold* of M if at every $y \in N$, there is an admissible chart (U, ϕ) such that $\phi(U) = V_1 \times V_2$ where $V_1 V_2$ are open neighborhoods of the origin in Banach spaces F_1, F_2, $\phi(y) = (0,0)$ and $\phi(N \cap V) = V_1 \times \{0\}$.

The submanifolds of codimension k will play an especially important role for us. Thus, we give an alternative definition which is easy to verify analytically.

Definition 9.6. Suppose $r \geq 1$ is an integer and X is a Banach space. A set $M \subset X$ is said to be a *C^r-submanifold of codimension k*, if for each $x \in M$, there is an open set U containing x and a function $F \in C^r(U, \mathbb{R}^k)$ which is regular at each point of U and $M \cap U = \{x : F(x) = 0\}$.

If $X = \mathbb{R}^n$ and $M \subset X$ is a submanifold of codimension k, then Definition 9.2 implies that M is a manifold of dimension $m = n - k$.

It is not very difficult to show that $M \subseteq X$ being a C^r-manifold of codimension k implies that M is a C^r-manifold modeled on E for any Banach space E for which $X = Y \oplus E$ where Y is k-dimensional.

EXAMPLE 9.7. If X is a real Banach space, U is an open set in X and $f \in C^r(U, \mathbb{R}^k)$, then the graph $Gr(f)$ of f is a C^r-submanifold of codimension k in $Z = X \times \mathbb{R}^k$. The proof is the same as the one in Example 9.3.

In a later section, we need the following result on submanifolds of codimension one.

Proposition 9.8. *If $r \geq 1$ and M is a closed, connected C^r-submanifold of codimension one in a Banach space X, then $X \backslash M$ can have at most two components.*

Proof. Suppose, on the contrary, that there are three disjoint, nonempty, open sets A_1, A_2, A_3 such that $X \backslash M = A_1 \cup A_2 \cup A_3$. Each A_j is open in X since M is closed. If F_j is the boundary of A_j, then F_j is nonempty and belongs to M. For any $x_0 \in M$, the fact that M has codimension one implies there is a neighborhood U of x_0 such that $U \cap (X \backslash M)$ has only two components. Thus, $U \cap M$ intersects only two of the sets F_i. Each of these sets F_i must be open and closed in M. Since M is connected, we have that $M = F_i$ which contradicts the fact that each point of $U \cap M$ can appear in only two of the F_i. $\square$

Our next objective is to define the tangent space at a point x in a manifold M. We begin with the simplest case of a manifold in $\mathbb{R}^n$.

Definition 9.9. Suppose M is a C^r-manifold in $\mathbb{R}^n$ of dimension $m = n - k$ and $x_0 \in M$, $M \cap U = \{x : F(x) = 0\}$, $F \in C^r(U, \mathbb{R}^k)$ for a neighborhood U of x_0. If x_0 is a regular point for F, then the *tangent space*, $T_{x_0}(M)$, to M at x_0 is defined by

$$T_{x_0}(M) = \{x \in \mathbb{R}^n : DF(x_0)x = 0\}.$$

For any $x_0 \in M$, there is a decomposition of X as $X = E_{x_0} \oplus Y_{x_0}$ where Y_{x_0} is k-dimensional, E_{x_0} is m-dimensional, each subspace is invariant under $DF(x_0)$ and $T_{x_0}(M) = E_{x_0}$. Thus, $T_{x_0}(M)$ is isomorphic to $\mathbb{R}^m$.

It is possible to show that $T_{x_0}(M)$ is determined by M and not by the particular function F that is used to define the manifold M. Also, one can show that $T_{x_0}(M)$ has a very simple geometric characterization. Let $\phi : [-1, 1] \to M$ be a C^1-curve through x_0, $\phi'(t) = d\phi(t)/dt \neq 0$ for $-1 \le t \le 1$, $\phi(0) = x_0$. The curve defined by ϕ is a manifold in $\mathbb{R}^n$, and the tangent space to this curve at x_0 is $\phi'(0)$. But it can be shown that $\phi'(0)$ also belongs to $T_{x_0}(M)$. One can actually prove the converse.

Theorem 9.10. *If M is a C^r-manifold of R^n of dimension m, then the tangent space to M at x_0 consists of all tangent vectors at x_0 of all C^1-curves on M that pass through x_0.*

The geometric characterization of $T_{x_0}(M)$ given in Theorem 9.10 can be generalized to a C^r-manifold of a Banach space X modeled on a Banach space E. We do not give the detailed definition.

Another important concept is transversality.

Definition 9.11. Let M, N be C^1-manifolds, $f \in C^1(M, N)$, and $W \subset N$ a submanifold of N. The function f is *transversal* to W at a point $x \in M$ if either $f(x) \notin W$ or $f(x) \in W$ and

(i) $Df(x) : T_x(M) \to T_{f(x)}(N) \backslash T_{f(x)}(W)$ is onto,
(ii) If $X_1 = \{y \in T_x(M) : Df(x)y = 0\}$, then there is a subspace X_2 of $T_x(M)$ such that $T_x(M) = X_1 \oplus X_2$.

Condition (ii) is always satisfied if the null space of $Df(x)$ is finite dimensional. Condition (i) simply states that the span of the set of vectors consisting of those from the range of $Df(x)$ and $T_{f(x)}W$ span $T_{f(x)}(N)$.

2.10. Sard's Theorem

Let X and Z be Banach spaces and $U \subset X$ be an open subset. Let $f : U \to Z$ be C^1. Recall that $x \in U$ is called a *critical point* of f if and only if $Df(x)$ is not surjective (onto); x is a *regular point* if and only if it is not a critical point. A point $y \in Z$ is a *critical value* of f if and only if $f^{-1}(y)$ contains a critical point; y is a *regular value* of f if and only if it is not a critical value. If $X = \mathbb{R}^n$ and $Z = \mathbb{R}^p$, then x is a critical point of f if and only if

$$\text{rank } Df(x) < p.$$

More generally, consider a C^1 map $f: M^n \to N^p$, from a smooth manifold of dimension n to another of dimension p. A point $x \in M^n$ is critical if and only if the differential at x

$$df(x): T_x M^n \to T_{f(x)} N^p$$

has rank less than p (i.e., is not surjective), where TM^n and TN^p denote the tangent bundles over M^n and N^p.

We will present first Sard's Theorem on the density of the set of regular values of a smooth map. This result is fundamental for differential topology and generic theory in differential equations. Surprisingly, it is also applicable to numerical computations of fixed points.

Theorem 10.1 (Sard's Theorem). *Let* $f: U \to \mathbb{R}^p$ *be* C^r, *where* $U \subset \mathbb{R}^n$ *is open and*

$$r > \max(0, n - p).$$

If $W \subset U$ *is the set of critical points of* f,

$$W = \{x \in U : \text{rank } Df(x) < p\},$$

then $f(W)$ *has Lebesgue measure zero.*

Since M^n can be covered by a countable collection of neighborhoods, each diffeomorphic to $\mathbb{R}^n$, we have the following

Corollary 10.2. *If* $f: M^n \to N^p$ *is* C^r, $r > \max(0, n - p)$, *then the set of regular values of* f *is dense in* N^p.

Proof of Theorem 10.1. If $i \geq 1$ and

$$W_i = \{x \in U : D^k f(x) = 0 \text{ for all } 1 \leq k \leq i\},$$

then $W \supset W_1 \supset W_2 \supset \cdots$ Since

$$f(W) = f(W - W_1) \cup f(W_1 - W_2) \cup \cdots \cup f(W_{r-1} - W_r) \cup f(W_r)$$

we will complete the proof provided the following lemmas are true.

Lemma A $f(W - W_1)$ *has measure zero.*
Lemma B $f(W_i - W_{i+1})$ *has measure zero for* $1 \leq i \leq r$.
Lemma C $f(W_r)$ *has measure zero.*

Proof of Lemma A. Since $W = W_1$ when $p = 1$, we may assume that $p \geq 2$. Let $\bar{x} \in W - W_1$. Then we may assume

$$\frac{\partial f_1}{\partial x_1}(\bar{x}) \neq 0.$$

Consider the map

$$h(x) = (f_1(x), x_2, \ldots, x_n)$$

Since $Dh(\bar{x})$ is nonsingular, h is locally a diffeomorphism. Let $g = f \circ h^{-1}$. Note that g is only defined locally in an open neighborhood V of $\bar{x}$ and the set of critical values of g is precisely $h(V \cap W)$. Hence, if W' is the set of critical points of g, then $g(W') = f(V \cap W)$. Moreover

$$g(x_1, x_2, \ldots, x_n) = \begin{pmatrix} x_1 \\ \tilde{g}(x_2, \ldots, x_n) \end{pmatrix}$$

If $n = 1$, then for some $c \in \mathbb{R}^{p-1}$

$$g(x^1) = \begin{pmatrix} x_1 \\ c \end{pmatrix}$$

It is trivial that $g(W')$ has measure zero. If $n = 2$, then

$$g(x_1, x_2) = \begin{pmatrix} x_1 \\ \tilde{g}(x_2) \end{pmatrix}$$

Hence, (x_1, x_2) is a critical point of g if and only if x_2 is a critical point of $\tilde{g}$. Since the Lemma A is true for $n = 1$, the intersection of $g(W')$ and $x_1 = $ constant has (1-dimensional) measure zero. By Fubini's Theorem, $g(W')$ has measure zero. An induction will now show that $g(W')$ has measure zero for any $n \geq 1$. Since W is compact, finitely many such V's will cover W. Thus, $f(W - W_1)$ has measure zero.

Proof of Lemma B. Let $\bar{x} \in W_i - W_{i+1}$. By definition, all i^{th} derivatives vanish but not all $(i + 1)^{\text{th}}$ partial derivatives vanish. Let α be an appropriate i^{th} partial derivative of f with (without loss of generality)

$$\frac{\partial \alpha}{\partial x_1}(\bar{x}) \neq 0$$

Let

$$h(x) = \begin{pmatrix} \alpha(x) \\ x_2 \\ \vdots \\ x_n \end{pmatrix} \quad \text{and} \quad g = f \circ h^{-1}.$$

Again, g is only defined locally in an open neighborhood V of $\bar{x}$. By definition, $\alpha(W_i) = 0$, so $h(W_i \cap V) \subset \{0\} \times \mathbb{R}^{p-1}$. Let $\tilde{g}$ be the restriction of g to $(\{0\} \times \mathbb{R}^{p-1}) \cap V$. So

$$\tilde{g}(h(W_i \cap V)) = f(W_i \cap V)$$

Since each point in $h(W_i \cap V)$ is a critical point of $\tilde{g}$, $\tilde{g}(h(W_i \cap V))$ has measure zero by induction on n. Since W_i is compact, $f(W_i - W_{i+1})$ has has measure zero.

Proof of Lemma C. Let $I^n \subset U$ be a cube of length $a > 0$. By Taylor's theorem,

$$|f(x) - f(y)| \le K|x - y|^{r+1}, \quad \text{for all } x, y \in I^n$$

where $K > 0$ is a constant independent of x and y. Let $k \ge 1$ be an integer. Subdivide the cube I^n into k^n cubes, $J_1, \ldots, J_{k^n}$, each of which has length a/k. By the above estimate, for each $i = 1, \ldots, k^n$,

$$|f(x) - f(y)| \le 2(K\sqrt{na})^{r+1}k^{-r-1} \quad \text{for } x \in J_i \cup W, \ y \in J_i.$$

This says that $f(J_i \cup W_k)$ lies in a cube of length $K(a/k)^{r+1}$. Hence, $f(W_k)$ lies in the union of (k^n) cubes whose volume is

$$[2(K\sqrt{na})^{r+1}]^p k^{n-p(r+1)},$$

If k is sufficiently large, then the above number will tend to zero since $r > \max(0, n - p)$.

This completes the proof of Sard's Theorem. $\square$

Remark. Note that if $n = p = 1$, then the proof of Lemma C actually yields the complete proof of Sard's Theorem. In fact, in this case we only have to assume $f \in C^1$ so that

$$|f(x) - f(y)| \le K\delta(|x - y|)$$

where $K \ge 0$ and δ is the modulus of continuity of f.

For mappings with parameters, we have the following version of Sard's Theorem, also known as the *transversality density theorem*. In fact, this is a special case of a more general theorem (Theorem 10.10).

Theorem 10.3. *Let $f : \mathbb{R}^m \times \mathbb{R}^n \to \mathbb{R}^p$ be C^r, $r > \max(m - p, 0)$. If $b \in \mathbb{R}^p$ is a regular value of f, then the set $\{\lambda \in \mathbb{R}^n : b$ is critical value of $f(\cdot, \lambda)\}$ has measure zero. In other words, for almost every $\lambda \in \mathbb{R}^n$, b is a regular value of $f(\cdot, \lambda)$.*

Proof. Since b is a regular value of f,

$$S = f^{-1}(b)$$

is a C^r-manifold of dimension $(m + n - p)$ (see Section 9). Consider the projection

$$\pi : \mathbb{R}^m \times \mathbb{R}^n \to \mathbb{R}^n.$$

Since S is C^r, the restriction $\pi^1 = \pi | S$ is C^r. By Corollary 10.2, almost every $\lambda \in \mathbb{R}^n$ is a regular value of π^1. Let $\bar{\lambda}$ be a regular value of π^1. Thus, the linear map

$$D\pi^1(\bar{x}, \bar{\lambda}) : T_{(\bar{x},\bar{\lambda})}S \to \mathbb{R}^n$$

is surjective for every $(\bar{x}, \bar{\lambda}) \in S$. Since the tangent space $T_{(\bar{x},\bar{\lambda})}S$ consists of precisely the elements $(y, \mu) \in \mathbb{R}^m \times \mathbb{R}^n$ satisfying

$$D_x f(\bar{x}, \bar{\lambda})y + D_\lambda f(\bar{x}, \bar{\lambda})\mu = 0,$$

the surjectivity of $D\pi^1(\bar{x}, \bar{\lambda})$ implies that, for every $\mu \in \mathbb{R}^n$, there exists $y \in \mathbb{R}^m$ with $D_x f(\bar{x}, \bar{\lambda})y + D_\lambda f(\bar{x}, \bar{\lambda})\mu = 0$. Hence,

$$\text{Range } D_x f(\bar{x}, \bar{\lambda}) = \text{Range } D_x f(\bar{x}, \bar{\lambda}) + \text{Range } D_\lambda f(\bar{x}, \bar{\lambda})$$
$$= \mathbb{R}^p,$$

b is a regular value for f since this implies that b is a regular value for $f(\cdot, \bar{\lambda})$ and completes the proof. $\square$

Corollary 10.4. *Let M, N and P be smooth manifolds of dimensions m, n and p respectively. Let $f : M \times N \to P$ be C^r, $r > \max(m - p, 0)$. If $b \in P$ is a regular value of f, then for almost all $\lambda \in N$, b is a regular value of $f(\cdot, \lambda)$.*

  As a simple application, we give the following constructive proof of the *Brouwer fixed point theorem*.
  Let $B^n = \{x : x \in \mathbb{R}^n, |x| < 1\}$.
  Let $f : \bar{B}^n \to \bar{B}^n$ be smooth, and define

$$g(a, x) = x - a, \qquad a, x \in \mathbb{R}^n, \quad \text{and}$$
$$\phi(a, \lambda, x) = (1 - \lambda)g(a, x) + \lambda(x - f(x))$$

In seeking fixed points of f, we examine the zeroes of g and ϕ. Let Γ_a be the component of $\phi_a^{-1}(0) \cap (0, 1) \times \bar{B}^n$ which has $(0, a)$ as an endpoint.

Theorem 10.5. *Consider the mapping*

$$\phi : B^n \times (0, 1) \times B^n \to \mathbb{R}^n$$

defined as above. We have

  (a) *$0 \in \mathbb{R}^n$ is a regular value of ϕ*
  (b) *for almost every $a \in B^n$, Γ_a is a smooth curve in $(0,1) \times B^n$ joining $(0,a)$ to a fixed point of f.*

Remark. In practice, for most f, the curve Γ_a forms a smooth arc leading to a fixed point of f. However, possibilities remain that Γ_a will not actually meet the fixed point but will converge to a set of fixed points of f. Hence $\phi_a^{-1}(0)$ is smooth when restricted to $(0,1) \times B^n$ or $[0,1) \times B^n$ but not necessarily in $[0,1] \times B^n$.

Proof. Let $(\bar{a}, \bar{\lambda}, \bar{x}) \in B^n \times (0,1) \times B^n$ and $\phi(\bar{a}, \bar{\lambda}, \bar{x}) = 0$. Then $D_a\phi(\bar{a}, \bar{\lambda}, \bar{x}) = -(1-\lambda)I$, $I = $ identity matrix, and, for $\lambda \neq 1$,

$$\text{Range } D\phi(\bar{a}, \bar{\lambda}, \bar{x}) \supset \text{Range } D_a\phi(\bar{a}, \bar{\lambda}, \bar{x}) = \mathbb{R}^n$$

This proves (a). By Corollary 10.4, for almost every $a \in B^n$, 0 is regular value of ϕ_a. Hence $\phi_a^{-1}(0)$ is a smooth manifold in $(0,1) \times B_n$; each component is a smooth curve and is either diffeomorphic to a circle or an open interval.

Consider the map ϕ_a with domain $(-\infty, 1) \times \mathbb{R}^n$. Since $\phi_a(0,a) = 0$ and $D_x\phi_a(0,x) = I$, the Implicit Function Theorem implies there is a neighborhood V of $(0,a)$ and a smooth function $c(\lambda)$ such that $c(0) = a$ and the only solution of $\phi_a = 0$ in V is $(\lambda, c(\lambda))$; that is,

$$\phi_a(\lambda, c(\lambda)) = 0, \qquad c(0) = a.$$

This implies that $\Gamma_a \subset B^n \times (0,1) \times B^n$ is not diffeomorphic to a circle. It remains to be shown that Γ_a has no limit points on the lateral surface $(0,1) \times S^{n-1}$. Such a limit point would satisfy $\phi_a(\lambda, x) = 0$, $0 < \lambda < 1$; that is, $x = (1-\lambda)a + \lambda f(x)$. Hence x is on the line segment between a and $f(x)$. Since $a \in B^n$, $f(x) \in \bar{B}^n$, we have $x \in B^n$. Hence Γ_a has no limit points on $(0,1) \times S^{n-1}$. Of course, any limit point of Γ_a is in $\phi_a^{-1}(0)$. So the only limit point of Γ_a in $\{0\} \times B^n$ is $(0,a)$. Furthermore, we have seen that Γ_a is homeomorphic to an open interval and $(0,a)$ lies at one end. By compactness of $[0,1] \times \bar{B}^n$, there must be at least one additional limit point of Γ_a and all such limit points must lie in $\{1\} \times B^n$. If $(1, x_1)$ is such a point, then x_1 is a fixed point of f. This completes the proof. $\square$

Notice that, if $I - Df(x)$ is nonsingular for every fixed point x of f, then $\bar{\Gamma}_a$ is a smooth curve in $[0,1] \times \bar{B}^n$ and so has finite arc length.

In order to see that Theorem 10.5 is constructive, we show that Γ_a is the solution of a differential equation which is defined on an open set. Hence the curve may be followed by a computer using any of the available algorithms for solving differential equations.

Let $K(\lambda, x)$ be $\mathcal{N}(D\phi_a(\lambda, x))$, where $D\phi_a$ denotes $D_{(\lambda,x)}\phi_a$ and $\mathcal{N}(D\phi_a(\lambda, x))$ is $\{v \in \mathbb{R}^{n+1} : D\phi_a(\lambda, x)v = 0\}$. Let $\tilde{0}$ be the set of $(\lambda, x) \in [0, 1) \times B^n$ such that the subspace $K(\lambda, x)$ is one dimensional. Let $\{e_1, \ldots, e_n\}$ be a complete set of basis vectors for $\mathbb{R}^n$. Let (λ, x) be in $\tilde{0}$. Let $H(\lambda, x)$ be the n-dimensional subspace in $\mathbb{R}^{n+1}$ which is perpendicular to $K(\lambda, x)$. By construction, for $i = 1, \ldots, n$, there exist unique vectors $\tilde{e}_i \in H(\lambda, x)$ such that $D\phi_a(\lambda, x)\tilde{e}_i = e_i$. For $v \in \mathbb{R}^{n+1}$, let $\mu(v) = \mu(v, \lambda, x)$ be the determinant of the ordered set $(v, \tilde{e}_1, \ldots, \tilde{e}_n)$; that is, we consider these vectors to be columns of an $(n + 1) \times (n + 1)$ matrix and evaluate its determinant. Notice that, if $v \in K(\lambda, x)$, $v \neq 0$, then $\{v, \tilde{e}_1, \ldots, \tilde{e}_n\}$ is a basis for $\mathbb{R}^{n+1}$, and so $\mu(v) \neq 0$.

Let a be chosen so that zero is a regular value of ϕ_a. Then $\tilde{0}$ is a neighborhood of Γ_a. There are two natural vector fields, that is, differential equations on $\tilde{0}$. Write y for (λ, x), and let $v^+(y)$ be the unit vector in $K(y)$ for which $\mu(v^+(y)) > 0$, and let $v^-(y)$ be the other unit vector, $-v^+(y)$. If y is a point of Γ_a, both $v^+(y)$ and $v^-(y)$ are tangent to Γ_a at y. Writing $v(y)$ for either v^+ or v^-, the corresponding differential equation is

$$(10.1) \qquad \frac{dy}{ds} = v(y) \qquad y(0) = (0, a) \in R^{n+1}$$

The solution $y(s)$ has Γ_a as its trajectory and s corresponds to arc length since $|v(y)| = 1$. One of these choices traces out Γ_a with positive s, and the other, negative s. Let v_0 be the unit vector in $K(0, a)$ for which the first coordinate (i.e. the λ coordinate) is positive. If $v_0 = v^+(0, a)$, we choose $v(y) = v^+(y)$, and $v^-(y)$ in the alternative case. Hence we have now chosen a single differential equation on $\tilde{0}$.

It should be noticed that for any $y \in \tilde{0}$, $(d/ds)\phi_a(y(s)) = 0$ since

$$D\phi_a(y)v(y) = 0;$$

that is, ϕ_a is constant along trajectories of (10.1). Of course, for an arbitrary point $y \in \tilde{0}$, $y \notin \Gamma_a$, the solution of (10.1) may trace out a closed loop or it may reach a boundary of $\tilde{0}$ at which $\lambda \neq 1$; that is, a point at which v is not defined.

An alternative approach for defining v (and v^+ and v^-) is available. Given a point $(\lambda, x) \in [0, 1) \times B^n$, we could choose $\tilde{a} = \tilde{a}(\lambda, x)$ so that $\phi(\tilde{a}, \lambda, x) = 0$. If $D_{(\lambda,x)}\phi(\tilde{a}, \lambda, x)$ has rank n, then we can define $v(\lambda, x)$ by choosing it to be tangent to the curve through (λ, x) on which ϕ_a is 0. Of course, we still have to choose the orientation of v by examining an appropriate determinant.

In many applications we will need the regular values of mappings in an infinite dimensional space. In finite dimensional contexts, the set of critical values is described in a measure theoretical sense, while the set of regular values has the density property. Thus, we will give the generalizations in terms of regular values.

Let X and Z be Banach spaces. Recall that a linear map $L: X \to Z$ is *Fredholm* with *Fredholm index* $\alpha(L)$ if and only if

(i)   dim $\mathcal{N}(L) < \infty$, co dim $\mathcal{R}(L) < \infty$; and
(ii) $\alpha(L) = $ dim $\mathcal{N}(L) - $ co dim $\mathcal{R}(L)$.

A nonlinear C^1 map $f: X \to Z$ is *Fredholm* if and only if $Df(x)$ is Fredholm for every $x \in X$. We define similarly Fredholm maps between Banach manifolds.

Theorem 10.6. *Let X be Lindelöf (every open cover of X has a countable subcover) and $f: X \to Z$ be C^r, where*

$$r > \max(0, \alpha(df(x))) \qquad for\ every\ x \in M$$

Then the set of regular values of f is residual, i.e., the countable intersection of open dense sets.

Corollary 10.7. *Let M and N be C^r Banach manifolds modeled respectively on X, Z with X Lindelöf. Let $f: M \to N$ be C^r and Fredholm. If*

$$r > \max(0, \alpha(df(x))) \qquad for\ every\ x \in M$$

where $df(x): T \times M \to T_{f(x)}N$ is the differential, then the set of regular values is residual.

Proof of Theorem 10.6. Essentially, we reduce the problem by the Liapunov-Schmidt method to a finite dimensional problem where Sard's theorem is applicable. Since X is Lindelöf we only have to show that the theorem holds locally. Let $\bar{x} \in X$ be fixed. For simplicity, let $\bar{x} = 0$. By using a local C^r-diffeomorphic change of variables, we assume that

$$f(u, v) = \begin{pmatrix} \eta(u, v) \\ v \end{pmatrix}$$

where $u \in \mathbb{R}^n$, $\eta(u, v) \in \mathbb{R}^p$; $n = $ dim $\mathcal{N}(Df(0))$, $p = $ co dim $\mathcal{R}(Df(0))$; and v is in some Banach space Y which is isomorphic to both the complement of $\mathcal{N}(Df(0))$ and $\mathcal{R}(Df(0))$. Note that $r > \max(0, n - p)$. By Sard's theorem, for each $v \in Y$, the set of regular values of $\eta(\cdot, v)$ is residual. On the other hand,

$$Df(u, v) = \begin{bmatrix} D_u\eta(u, v) & D_v\eta(u, v) \\ 0 & I \end{bmatrix}$$

So $Df(u, v)$ is surjective if and only if $D_u\eta(u, v)$ is surjective. Thus, the set of regular values of f intersects every plane $\mathbb{R}^n \times \{v\}$ in a residual set and is therefore dense in $\mathbb{R}^n \times Y$. If z is a regular value of f, then every z_1 sufficiently close to z is also a regular value. This shows that the set of regular values of f is open and dense. $\quad\square$

EXAMPLE 10.8. Let X be the space of C^2 functions on $[0,1]$ vanishing at $t = 0, 1$ with the sup norm and Z be the space $C[0,1]$. Let $\varphi:\mathbb{R}^2 \to \mathbb{R}$ be C^2. Consider the mapping

$$f:X \to Z$$
$$u \mapsto \ddot{u} + \varphi(u,\dot{u})$$

We note that

$$Df(u):X \to Z$$
$$v \mapsto \ddot{v} + \varphi_u(u,\dot{u})v + \varphi_{\dot{u}}(u,\dot{u})\dot{v}.$$

Hence, f is Fredholm with Fredholm index zero at every $u \in X$. By Theorem 10.6, there exists a residual set $Z^1 \subset Z$ such that for every $p \in Z^1$ the two point boundary value problem

$$\ddot{u} + \varphi(u,\dot{u}) = p(t) \qquad 0 \leq t \leq 1$$
$$u(0) = u(1) = 0$$

has local uniqueness.

The above example may be generalized to elliptic problems with appropriate spaces (e.g., Hölder continuous spaces). We also note that in bifurcation theory, one studies the problem for precisely those p's in the complement of Z^1.

Remark. Theorem 10.6 is false in general if f is not Fredholm.

It is now natural to generalize Theorem 10.6 to one with parameters in infinite dimensional space. We see that the proof of Theorem 10.3 may be carried over word for word except the projection π would have to be Fredholm. In the case that the space of parameters is a space of functions, which is generally the case in applications, the evaluation map is appropriate for the representation of π. We first give a simple example to illustrate the ideas of the evaluation map.

EXAMPLE 10.9. Consider the two point boundary value problem:

$$\ddot{u} - u = p(u,\varepsilon), \qquad 0 \leq t \leq 1$$
$$u(0) = u(1)$$

where $p = 0(|\varepsilon|)$ is a C^2 bounded function. It is well-known that this problem has a unique solution which depends smoothly on ε (linear equation has a unique solution) for $|\varepsilon|$ small. If ε is a vector, then the same holds. Suppose we replace $p(u, \varepsilon)$ by $f(u)$ with f sufficiently "small" and ask: does the same conclusion hold? In fact, it does and the proof is quite easy if we use the evaluation map. Let $a > 0$ and $C^2[-a, a]$ denote the Banach space of real-valued functions defined on $[-a, a]$ with sup norm. Let $X = \{u \in C^2[0, 1] : u(0) = u(1)\}$ and $Z = C[0, 1]$. Consider the mapping (*evaluation map*)

$$ev: X \times C^2[-a, a] \to Z$$
$$(u, f) \mapsto \ddot{u} - u - f(u)$$

We note that ev is C^2. Since $ev(0, 0) = 0$, $D_u ev(0, 0)v = \ddot{v} - v$, the implicit function theorem implies there exists a unique smooth curve $u = u(f)$, $u(0) = 0$, such that $ev(u(f), f) = 0$. This proves our claim.

We note that we did not consider the set of C^∞ functions on $[-a, a]$. The reason is simply that this set of functions may not be made into a Banach space. It is a Fréchet space with the Whitney topology and the implicit function theorem is false in Fréchet spaces. This is also the reason for presenting Sard's Theorem without assuming the functions to be C^∞.

Before we state the parametrized Sard's theorem, the following notations are needed: Let M and N be C^r Banach manifolds, $r \geq 1$, $C^r(M, N)$ denote the set of all C^r functions from M to N. If M and N are open subsets of Banach spaces, then we have a natural C^r-norm on $C^r(M, N)$ defined as follows:

$$|f|_r = \sup_{x \in M} \{|f(x)| + |Df(x)| + \cdots + |D^r f(x)|\}$$

where $|\cdot|$ denotes the norms in space of multilinear linear symmetric maps between Banach spaces. It is easy to see that $C^r(M, N)$ is a Banach space with the norm $|\cdot|_r$. If M and N are manifolds, then by using local coordinates we have a similar C^r-norm which makes again $C^r(M, N)$ into a Banach space. In the future, we will drop the subscript r in the C^r-norm if there is no confusion.

Theorem 10.10. *Let X and Z be Banach spaces and $r \geq 1$. The evaluation map*

$$ev: X \times C^r(X, Z) \to Z$$
$$(u, f) \mapsto ev(u, f) = f(u)$$

is C^r. In fact, for all $v \in X$, $h \in C^r(X, Z)$,

$$Dev(u, f) \cdot (v, h) = h(v) + Df(u) \cdot h$$
$$D^2 ev(u, f) \cdot ((v, h), (v, h)) = 2Dh(u) \cdot v + D^2 f(u) \cdot (v, v)$$
$$\cdots\cdots\cdots\cdots$$
$$D^k ev(u, f) \cdot (v, h)^k = kD^{k-1}h(u) \cdot v^{k-1} + D^k f(u) \cdot v^k.$$

Proof. We will consider only a formal expansion in a small parameter ε. We have

$$ev(u + \varepsilon v, f + \varepsilon h) = ev(u, f) + \varepsilon Dev(u, f) \cdot (v, h)$$
$$+ \tfrac{1}{2}\varepsilon^2 D^2 ev(u, f) \cdot (v, h)^2$$
$$+ \cdots \cdots \cdots$$

By the definition of ev,

$$ev(u + \varepsilon v, f + \varepsilon h) = (f + \varepsilon h)(u + \varepsilon v)$$
$$= f(u + \varepsilon v) + \varepsilon h(u + \varepsilon v)$$
$$= f(u) + \varepsilon Df(u) \cdot v + \tfrac{1}{2}\varepsilon^2 D^2 f(u) \cdot v^2 + \cdots$$
$$+ \varepsilon[h(u) + \varepsilon Dh(u) \cdot v + \cdots]$$

Equating the coefficients, we obtain the desired formulae. $\square$

We note that the evaluation map given by Theorem 10.10 is not exactly the kind we have in Example 10.9. For this reason, we introduce the following notation.

Definition 10.11. Let Λ, M and N be C^r Banach manifolds, $r \geq 1$, and $\rho: \Lambda \to C^r(M, N)$ be a map. For each $\lambda \in \Lambda$, we write ρ_λ instead of $\rho(\lambda)$. We say that ρ is a C^r *representation* if and only if the evaluation map

$$ev_\rho: M \times \Lambda \to N$$
$$(x, \lambda) \mapsto \rho_\lambda(x)$$

is C^r.

EXAMPLE 10.12. Let $\Lambda = C^2[-a, a]$, $M = C^2[0, 1]$ and $N = C[0, 1]$. Let

$$\rho: \Lambda \to C^2(M, N)$$

be defined as follows: if $f \in \Lambda$,

$$\rho_f(u) = \ddot{u} - u + f(u)$$

Then ρ is a C^2 representation. Moreover, we may apply Theorem 10.10 formally to compute the derivative of ev_ρ. In fact,

$$Dev_\rho(u, f) \cdot (v, h) = \ddot{v} - v + h(u) + Df(u) \cdot v$$
$$D^2 ev_\rho(u, f) \cdot (v, h)^2 = 2Dh(u) + D^2 f(u)v^2.$$

We now state the *parametrized Sard's theorem* in infinite dimensional spaces. The proof is omitted.

Theorem 10.13. *Let Λ, M and N be C^r manifolds and $\rho:\Lambda \to C^r(M,N)$ be a C^r representation with evaluation map ev_ρ. Assume that:*

    (i) $\dim M = n < \infty$, $\dim N = p < \infty$
    (ii) *Λ is second countable*
    (iii) *$r > \max(0, n - p)$*
    (iv) *$b \in N$ is a regular value of ev_ρ*

Then the set of $\lambda \in \Lambda$ such that b is a regular value of ρ_λ is residual.

The above theorem has the following generalization which is useful in applications.

Theorem 10.14. *Assume the hypotheses of Theorem 10.13 hold. If $W \subset N$ is a submanifold, not necessary closed, and ev_ρ is transversal to W, then*

$$\{\lambda \in \Lambda : \rho_\lambda \text{ is transversal to } W\}$$

is residual in Λ.

2.11. Topological Degree, Index of a Vector Field and Fixed Point Index

Here we give a brief outline of degree theory. The approach is in some sense nonstandard but is related to bifurcation theory.

Let Ω be a bounded open set in $\mathbb{R}^n$ and

$$f:\bar{\Omega} \to \mathbb{R}^n$$

be C^2. Suppose that $f(x) \neq 0$ for all $x \in \partial\Omega$. One says that f is *generic* if $Df(x)$ is nonsingular for each $x \in \Omega$ for which $f(x) = 0$. For generic f, we define the (*topological*) *degree* of f with respect to 0 and Ω by the following formula:

$$(11.1) \qquad \deg(f,\Omega,0) = \sum_{x \in f^{-1}(0)} \text{sign det } Df(x)$$

We note that, if f is generic, then $f^{-1}(0)$ consists of only a finite number of points in Ω. Thus, (11.1) is well-defined. Similarly, for any $f \in C^2(\bar{\Omega})$ and $p \notin f(\partial\Omega)$, we may define the topological degree of f with respect to p and Ω by a similar formula as (11.1) provided f is generic (i.e., p is a regular value of f).

Our goal is to extend this definition of degree to all continuous functions $h:\bar{\Omega} \to \mathbb{R}^n$ with respect to $p \notin h(\partial\Omega)$. To begin, let $h:\bar{\Omega} \to \mathbb{R}^n$ be continuous and $0 \notin h(\partial\Omega)$. We may approximate h in the $C^0(\bar{\Omega})$-topology by a C^∞ function $f^0:\bar{\Omega} \to \mathbb{R}^n$. Moreover, $0 \notin f^0(\partial\Omega)$ and f^0 is generic. If f^0 is not generic, then, by Sard's theorem, 0 is a regular value of $f^0 - \varepsilon = g$ for almost all ε near 0. Clearly, for ε small, $0 \notin g(\partial\Omega)$. We may thus always assume f^0 is generic.

We may define

$$(11.2) \qquad \deg(h, \bar\Omega, 0) = \deg(f^0, \bar\Omega, 0)$$

provided (11.2) is independent of the choice of f^0. That is, if f^1 is generic, $0 \notin f^1(\partial\Omega)$ and f^1 is also near h in the $C^0(\bar\Omega)$ topology, it must be shown that

$$(11.3) \qquad \deg(f^0, \Omega, 0) = \deg(f^1, \Omega, 0)$$

Indeed, this is the case. We will give a proof of (11.3) via the theory of generic bifurcation. The advantage here is that we will have the homotopy property of topological degree as a simple corollary.

Consider any $f^0, f^1 \in C^\infty(\bar\Omega)$ not necessarily near each other. Assume that $0 \notin f^0(\partial\Omega) \cup f^1(\partial\Omega)$ and both are generic. Suppose f^α is C^∞ in $(x, \alpha) \in \bar\Omega \times [0, 1]$, $0 \le \alpha \le 1$, and is a homotopy from f^0 to f^1 such that $0 \notin f^\alpha(\partial\Omega)$ for any $0 \le \alpha \le 1$. We will show that

$$\deg(f^0, \bar\Omega, 0) = \deg(f^1, \bar\Omega, 0)$$

In particular, this will prove (11.3) for f^0 and f^1 that are close to h since we may take

$$f^\alpha = \alpha f^1 + (1 - \alpha) f^0$$

We note that the above homotopy also shows that, in the definition (11.2), we may take any $f \in C^\infty(\bar\Omega)$ with

$$\sup_{x \in \bar\Omega} |f(x) - h(x)| < \inf_{x \in \partial\Omega} |h(x)|$$

Consider now the homotopy f^α, $0 \le \alpha \le 1$. The result in the appendix to this section shows that we may choose a homotopy g^α, $0 \le \alpha \le 1$, near f^α such that $g^0 = f^0$, $g^1 = f^1$, g^α is near f^α in $C^2(\bar\Omega)$, $0 \notin g^\alpha(\partial\Omega)$, and g^α is generic except for finitely many α's, say $0 < \alpha_1 < \cdots < \alpha_k < 1$. Moreover, at each $\alpha = \alpha_j$, $1 \le j \le k$, there exists precisely one $x_j \in \Omega$ such that $g^{\alpha_j}(x_j) = 0$ and $g^\alpha(x)$ satisfies the conclusions of the theorem in the appendix. We refer to this as a "generic bifurcation" at $x = x_j$ and $\alpha = \alpha_j$. If, for $\varepsilon > 0$ sufficiently small,

$$(11.4) \qquad \deg(g^{\alpha_j - \varepsilon}, \bar\Omega, 0) = \deg(g^{\alpha_j + \varepsilon}, \bar\Omega, 0)$$

$$(11.5) \qquad \deg(g^0, \bar\Omega, 0) = \deg(g^\varepsilon, \bar\Omega, 0)$$

then this will show that $\deg(g^\alpha, \bar\Omega, 0)$ is constant for $0 \le \alpha \le 1$ and hence

(11.3) is true. We will first show (11.5). If $g^0(\bar{x}) = 0$ and $Dg^\circ(\bar{x})$ is nonsingular, then by the implicit function theorem, there exists a unique smooth curve $x(\alpha)$ for $|\alpha|$ sufficiently small such that $x(0) = \bar{x}$, $g^\alpha(x(\alpha)) = 0$ and $Dg^\alpha(x(\alpha))$ is nonsingular. Moreover, the sign of $\det Dg^\alpha(x(\alpha))$ is constant for α small. It is also clear that the number of inverse images of zero in $\bar{\Omega}$ remains constant for α small. Thus, (11.5) is proved. It remains to show (11.4). For simplicity, we let $\beta = \alpha - \alpha_j$ and $x_j = 0$. Since $g^{\alpha_j}(x)$ has a generic bifurcation at $x = 0$, we may assume that after a change of variables

$$Dg^\alpha(0) = \left(\begin{array}{c|c} \beta & 0 \\ \hline 0 & B(\beta) \end{array} \right)$$

where $B(\beta)$ is a nonsingular $(n - 1) \times (n - 1)$ matrix for all β small. Thus, for $|\beta|$ small, sgn $\det B(\beta) = $ constant, and sgn $\det Dg^\alpha(0) = $ sgn $\beta \det B(\beta)$. By the theorem in the appendix, $D_1^2 g^{\alpha_j}(0) \neq 0$, where $D_1^2 g^\alpha$ is the second derivative with respect to the first component of x. An application of the method of Liapunov-Schmidt similar to the one described in Section 1.5 for Equation (1.5.6) for finding solutions to $g^\alpha(x) = 0$ near $\alpha = \alpha^j$, $x = 0$, shows there exists a smooth curve $x(\beta)$ such that $x(0) = 0$, $x(\beta) \neq 0$ for $\beta \neq 0$, $g^\alpha(x(\beta)) = 0$ and $Dg^\alpha(x(\beta))$ is nonsingular. Moreover,

$$\text{sgn } \beta \det Dg^\alpha(x(\beta)) = \text{constant.}$$

This implies that $\deg(g^{\alpha_j - \varepsilon}, \Omega^1, 0) = \deg(g^{\alpha_j + \varepsilon}, \Omega^1, 0)$ where Ω^1 is a sufficiently small neighborhood of x_j. Summarizing the above, we have the following.

Theorem 11.1. (a) *For any continuous $h:\bar{\Omega} \to \mathbb{R}^n$ and $p \notin h(\partial\Omega)$, there exists a C^∞ function $f:\Omega \to \mathbb{R}^n$ such that $p \notin f(\partial\Omega)$, p is a regular value of f and*

$$\sup_{x \in \bar{\Omega}} |f(x) - h(x)| < \inf_{x \in \partial\Omega} |h(x) - p|$$

Moreover,

$$\deg(h, \bar{\Omega}, p) = \deg(f, \bar{\Omega}, p)$$

is well defined.

  (b) *If h^α, $0 \le \alpha \le 1$, is a continuous homotopy of h^0 and h^1, $p \notin h^\alpha(\partial\Omega)$ for all $0 \le \alpha \le 1$, then*

$$\deg(h^\alpha, \bar{\Omega}, p) = constant$$

Now we give some applications of degree theory.

Theorem 11.2 (Borsuk). *Let $\Omega \subset \mathbb{R}^n$ be open, bounded and symmetric with respect to the origin, i.e., if $x \in \Omega$, then $-x \in \Omega$. Suppose $h:\bar{\Omega} \to \mathbb{R}^n$ is contin-*

uous and odd ($h(x) = -h(-x)$) and $0 \notin h(\partial\Omega)$. Then

$$\deg(h, \bar{\Omega}, 0) = odd \quad if\ 0 \in \Omega$$
$$\deg(h, \bar{\Omega}, 0) = even \quad if\ 0 \notin \Omega.$$

Proof. Let $\tilde{f}:\Omega \to \mathbb{R}^n$ be a generic smooth approximation of h. Let

$$f(x) = \tfrac{1}{2}[\tilde{f}(x) - \tilde{f}(-x)]$$

Then f is an odd generic approximation of h and

$$\deg(f, \bar{\Omega}, 0) = \deg(h, \bar{\Omega}, 0)$$

Since the zeros of f occur in pairs except for $x = 0$ and $Df(x) = Df(-x)$ for any $x \in \Omega$, the result follows immediately from the definition. $\square$

Corollary 11.3. *Let $S^{n-1} = \{x \in \mathbb{R}^n : |x| = 1\}$ and $h:S^{n-1} \to \mathbb{R}^n$ be continuous and odd. Then there exists $x \in S^{n-1}$ such that $h(x) = 0$.*

Proof. Suppose not. Then there exists $\varepsilon > 0$ such that $\inf\{|h(x)| : x \in S^{n-1}\} = \varepsilon > 0$. By Tietze's extension theorem, we may continuously extend h to $\bar{B}^n = \{x : |x| \leq 1\}$ and

$$\inf\{|h(x)| : x \in \bar{B}^n\} > 0.$$

Moreover, we may assume that the extension is odd. Since $0 \in \bar{B}^n$,

$$\deg(h, \bar{B}^n, 0) = odd \neq 0.$$

Thus, there exists at least one $x \in \bar{B}^n$ such that $h(x) = 0$. This contradicts the above infimum conditions. $\square$

Theorem 11.4 (Brouwer). *Let $h:\bar{B}^n \to \bar{B}^n$ be continuous. Then there exists a fixed point $x \in \bar{B}^n$ of h.*

Proof. Suppose that $x \neq h(x)$ for all $x \in \partial B^n$. Consider the homotopy $H^\alpha(x) = \alpha x + (1 - \alpha)(x - h(x))$. Thus, $H^\alpha(x) \neq 0$ for all $x \in \partial B^n$. Hence, by Theorem 11.1,

$$1 = \deg(H^1, B^n, 0) = \deg(H^0, B^n, 0) = 1$$

This implies that h has a fixed point. $\square$

The degree of a mapping may often be computed from the local behavior. The notion of index is then convenient to use. Let $h:\bar{\Omega} \to \mathbb{R}^n$ be continuous and $h(x_0) = 0$. If there exists an open neighborhood Ω' of x_0 such that

$\bar{\Omega}' \cap h^{-1}(0) = \{x_0\}$, we define the *index of h at x_0* by:

$$\text{ind}(h, x_0) = \deg(h, \Omega', 0)$$

We note that as long as x_0 is isolated, $\text{ind}(h, x_0)$ is well-defined. Moreover, this enables one to define index of h when h is defined on a manifold.

We now state two theorems omitting the proofs.

Theorem 11.5. *Let A be a nonsingular $n \times n$ matrix and $h(x) = Ax + 0(|x|^2)$ as $x \to 0$. Then*
$$\text{ind}(h, 0) = (-1)^k$$

where

$$k = \textit{number (counting multiplicities) of eigenvalues of } A \textit{ in } (-\infty, 0).$$

Theorem 11.6 (Poincaré formula). *Let $h: \mathbb{R}^2 \to \mathbb{R}^2$ be C^1 and $h(0) = 0$. Suppose that $x = 0$ is an isolated zero of h, then*

$$\text{ind}(h, 0) = 1 + \tfrac{1}{2}(E - H)$$

where E and H are integers associated with the flow $\dot{x} = h(x)$; $E = $ number of elliptic regions, and $H = $ number of hyperbolic regions (see Fig. 11.1).

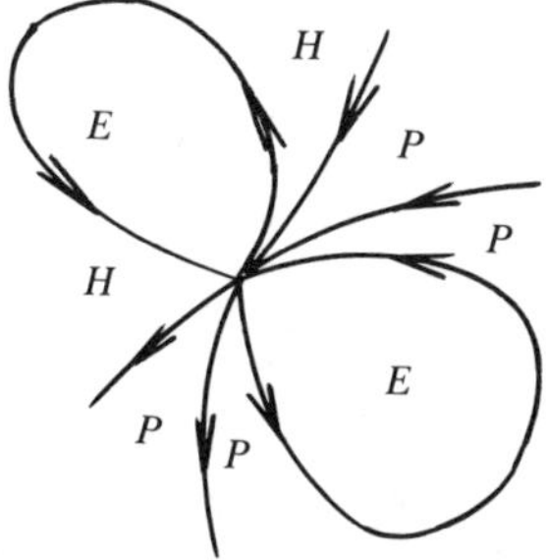

Figure 11.1. Flow of $\dot{x} = h(x)$ about $x = 0$ with elliptic regions (E), hyperbolic regions (H) and parabolic regions (P).

Here is a simple application. Consider the flow (see Fig. 11.2).

$$\begin{aligned} \dot{x}_1 &= x_1^2 - x_2^2 \\ \dot{x}_2 &= 2x_1 x_2 \end{aligned} \qquad h(x) = \begin{pmatrix} x_1^2 - x_2^2 \\ 2x_1 x_2 \end{pmatrix}$$

By the Poincaré formula, the index of h at $x = 0$ is 2. We may also consider the perturbation

$$\begin{aligned} \dot{x}_1 &= -\varepsilon x_1 + x_1^2 - x_2^2 \\ \dot{x}_2 &= -\varepsilon x_2 + 2x_1 x_2 \end{aligned} \qquad \varepsilon > 0$$

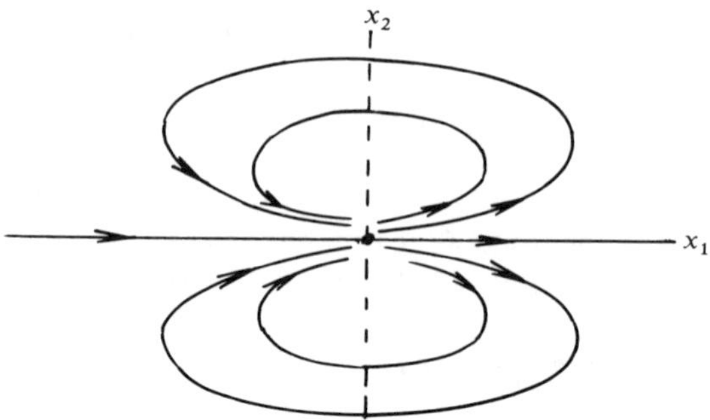

Figure 11.2. Flow of $\dot{x}_1 = x_1^2 - x_2^2$, $\dot{x}_2 = 2x_1 x_2$.

The flow is shown in Fig. 11.3. Note that there are two inverse images of $x = 0$, namely, $(0, 0)$ and $(\varepsilon, 0)$. By Theorem 11.5, the index of h at each point is 1. Thus, the total index is 2.

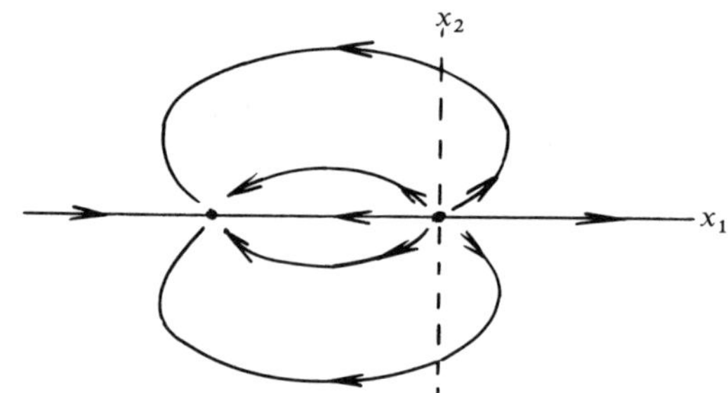

Figure 11.3. Flow of $\dot{x}_1 = -\varepsilon x_1 + x_1^2 - x_2^2$, $\dot{x}_2 = -\varepsilon x_2 + 2x_1 x_2$.

In the previous example, the vector field h consists of the real and imaginary parts of an analytic function $w(z) = z^2$ with the identification

$$z = x_1 + ix_2$$

(11.6)

$$w(z) = h_1(x_1, x_2) + ih_2(x_1, x_2).$$

More generally, given a continuous, not necessary analytic, $w : \mathbb{C} \to \mathbb{C}$ with $w(0) = 0$, if $z = 0$ is an isolated zero of $w(z)$, we define the *index of* w at $z = 0$ by $\mathrm{ind}(w, 0) = \mathrm{ind}(h, 0)$, where $h : \mathbb{R}^2 \to \mathbb{R}^2$ is defined by (11.6). Similarly, we may define the index of a continuous function $w : \mathbb{C}^n \to \mathbb{C}^n$.

Theorem 11.7. *Let* $w : \mathbb{C} \to \mathbb{C}$ *be analytic and* $z = 0$ *be an isolated zero of* w. *Then* $\mathrm{ind}(w, 0) > 0$.

Proof. By making a perturbation of w, we may assume that $w^{-1}(0)$ has only a finite number of points. Moreover, 0 is a regular value of w. Let $w(z_0) = 0$. Since 0 is a regular value, $|w'(z_0)| \neq 0$. By the Cauchy-Riemann equation and (11.6), $\det Dh(z_0) = |w'(z_0)|^2 > 0$. This completes the proof. $\square$

The above theorem is also true for analytic $w : \mathbb{C}^n \to \mathbb{C}^n$. If $f : \mathbb{C}^n \to \mathbb{C}$ is analytic and $z = 0$ is an isolated critical point of f, then the *Milnor number* of f at $z = 0$ is precisely the index of $f' : \mathbb{C}^n \to \mathbb{C}^n$ at $z = 0$.

EXAMPLES.

(a) $\mathrm{ind}(z^n, 0) = n$ for any $n \geq 1$
(b) $\mathrm{ind}(\bar{z}^n, 0) = -n$.

We note that the mapping $z \to \bar{z}^n$ is not analytic.

In many problems, one is interested in the fixed points of a continuous map. We define the *local fixed point index* $\tau(h, x)$ of a map h at a fixed point x by the following

$$\tau(h, x) = \mathrm{ind}(I - h, x)$$

where I is the identity. Note that the number $\tau(h, x)$ is defined locally at each isolated fixed point of h.

EXAMPLES. (a) Let $h(z) = z + z^2$, where $z \in \mathbb{C}$. Then

$$\mathrm{ind}(h, 0) = 1, \qquad \tau(h, 0) = 2.$$

(b) Let

$$w = e^{2\pi i/3}$$

and

$$h(z) = wz + z^n \qquad n \geq 2.$$

Let $h^j(z)$ denote the jth iterate of h. We have

$$h^2(z) = w^2 z + (w + w^n)z^n + \cdots$$
$$h^3(z) = z + (w^2 + w^{n+1} + w^{2n})z^n + \cdots$$

Thus, $\tau(h, 0) = 1$ and $\tau(h^2, 0) = 1$. If $n = 1 \bmod (3)$, then $\tau(h^3, 0) = n$. In fact, we claim that $\tau(h^n, 0) = 1 \bmod (3)$ for any n. To prove the claim, we consider a generic perturbation $f(z)$ of $h(z)$ such that $f(0) = 0$. The fixed point index of f at $z = 0$ is clearly 1. If $z \neq 0$ is a fixed point of f, then

$$z, \quad f(z), \quad f^2(z)$$

are also fixed points of $f(z)$. Since $f(z)$ is close to wz in a neighborhood of $z = 0$, the above three points are distinct. Thus, fixed points of $f(z)$ appear in triplets except for $z = 0$. This proves the claim.

If we label the fixed point index of $h^j(z)$ in a sequence, then we have the following:

$$1, 1, C, 1, 1, C, \ldots.$$

where $C = 1 \bmod (3)$. The periodicity of the above sequence is in fact true for more general mappings.

In the following, we discuss the fixed indices for iterates of a C^1 mapping. The C^1 smoothness is important here.

Theorem 11.8. *Let $f : \Omega \to \mathbb{R}^n$ be C^1 and $x = 0$ be an isolated fixed point for f^k for all $k \geq 1$. If*

$$I + Df(0) + \cdots + Df^{k-1}(0) = \sum_{j=0}^{k-1} Df^j(0)$$

is nonsingular, then

$$\tau(f, 0) = \pm \tau(f^k, 0).$$

Proof. Let Df^k be the map $x \mapsto Df^k(0)x$, $f = Df + \theta_1$ and $f^k = Df^k + \theta_k$. Then, by the rule of differentiation for composition of maps.

$$
\begin{aligned}
I - f^k &= I - Df^k - \theta_k \\
&= (I + Df + \cdots + Df^{k-1})(I - Df) - \theta_k \\
&= (I + Df + \cdots + Df^{k-1})(I - f) \\
&\quad + (I + Df + \cdots + Df^{k-1})\theta_1 - \theta_k
\end{aligned}
$$

We will show by induction that, for any $k \geq 1$ and $\varepsilon > 0$, there exists a neighborhood $U_{k,\varepsilon}$ of $x = 0$ such that

$$\left| \left(\sum_{j=0}^{k-1} Df^j \theta_1 - \theta_k \right)(x) \right| < \varepsilon |x - f(x)| \quad \text{for all } x \in U_{k,\varepsilon}.$$

Thus, if

$$\sum_{j=0}^{k-1} Df^j(0)$$

is nonsingular, then

$$\tau(f^k, 0) = \mathrm{ind}\left(\left(\sum_{j=0}^{k-1} Df^j\right)(I - f), 0\right) = \pm\tau(f, 0)$$

To estimate

$$\sum_{j=0}^{k-1} Df^j\theta_1 - \theta_k,$$

first observe by induction that

$$\theta_k = \sum_{j=0}^{k-1} Df^{k-j-1}\theta_1 \circ f^j.$$

So,

$$\sum_{j=0}^{k-1} Df^j\theta_1 - \theta_k = \sum_{j=0}^{k-1} Df^{k-j-1}\theta_1 - \sum_{j=0}^{k-1} Df^{k-j-1}\theta_1 \circ f^j$$

$$= \sum_{j=1}^{k-1} Df^{k-j-1}(\theta_1 - \theta_1 \circ f^j).$$

By the mean value theorem,

$$\left|\left(\sum_{j=0}^{k-1} Df^j\theta_1 - \theta_k\right)(x)\right| \le \sum_{j=1}^{k-1} |Df^{k-j-1}|\,|D\theta_1|\,|x - f^j(x)|,$$

where $|\cdot|$ denotes either the norm in $\mathbb{R}^n$ or the sup norm in $C(\bar{U}_{k,\varepsilon})$. Since $D\theta_1(0) = 0$, it suffices to prove inductively that, given $j < k$, there is a neighborhood V_j of $x = 0$ and $c_j \ge 0$ such that

$$|x - f^j(x)| \le c_j|x - f(x)|, \quad \text{for all } x \in V_j.$$

Since

$$I - f^j = \left(\sum_{i=0}^{j-1} Df^i\right)(I - f) + \sum_{i=0}^{j-1} Df^i\theta_1 - \theta_j,$$

we can inductively choose $U_{j,\varepsilon}$ so that

$$|x - f^j(x)| \le \sum_{i=0}^{j-1} |Df^i|\,|x - f(x)| + \varepsilon|x - f(x)|$$

and we are done. $\square$

Theorem 11.9. *If $f:\Omega \to \mathbb{R}^n$ is C^1 and $x = 0$ is an isolated fixed point of f^k for all k, then $\{\tau(f^k, 0)\}$ is a periodic sequence in $k \geq 0$.*

Proof. Let f be replaced by f^l. Theorem 11.8 shows that, if

$$A = I + Df^l(0) + \cdots + Df^{l(k-1)}(0)$$

is nonsingular, then $\tau(f^{lk}, 0) = (-1)^\mu \tau(f^l, 0)$, where μ is the number of eigenvalues (counting multiplicities) of A that lie in $(-\infty, 0)$. It follows from the spectral mapping theorem that every eigenvalue of A has the form

$$1 + \lambda^l + \cdots + \lambda^{l(k-1)} = \frac{1 - \lambda^{lk}}{1 - \lambda^l}$$

Hence

$$\tau(f^k, 0) = \tau(f, 0) \qquad \text{if } k \text{ is odd}$$
$$\tau(f^{kl}, 0) = -\tau(f^l, 0) \quad \text{if } l \text{ is even.}$$

This shows that, if $Df(0)$ has no roots of unity as an eigenvalue, then the sequence $\{\tau(f^k, 0)\}$ is periodic in k with period 2.

Let $S = \{p : Df(0)$ has a primitive p^{th} root of unity$\}$ and m be the least common multiple of $p \in S$. Since $1 + \lambda^m + \cdots + \lambda^{m(k-1)} = 0$, for any $k \geq 1$, the sequence $\{\tau(f^{mk}, 0)\}$ has period 2 in $k \geq 1$ by using the arguments above. Now it is not difficult to see that $\{\tau(f^k, 0)\}$ is periodic in $k \geq 0$ with period $2m$. $\quad\square$

We now give an application of index to area-preserving maps.

If $f:\mathbb{R}^2 \to \mathbb{R}^2$ is C^1, then f is said to be *area preserving* if $\det Df(x) = 1$ for all $x \in \mathbb{R}^2$. Let $x = 0$ be an isolated fixed point of f. Let λ and μ be the eigenvalues of $Df(0)$. Since f is area preserving, $\lambda\mu = 1$. There are three possibilities for these eigenvalues:

(a) $0 < |\lambda| < 1 < |\mu|$ (hyperbolic)
(b) $|\lambda| = |\mu| = 1$, but $\lambda \neq 1$ (elliptic)
(c) $\lambda = \mu = 1$ (parabolic)

Since the eigenvalues of $I - Df(0)$ are precisely $1 - \lambda$ and $1 - \mu$, we conclude from Theorem 11.5 that $\tau(f, 0) = \pm 1$ in case (a) and $\tau(f, 0) = +1$ in case (b). It remains to compute $\tau(f, 0)$ in case (c). The basic idea is to find a vector field in $\mathbb{R}^2$ which has a critical point at $x = 0$ with index equal to $\tau(f, 0)$. For vector fields in $\mathbb{R}^2$, the Poincaré index formula is applicable. Since f is area preserving, we suspect that the vector field would be Hamiltonian. This is indeed the case.

Consider the differential 1-form

$$\Omega = (f_1 - x_1)d(f_2 + x_2) - (f_2 - x_2)d(f_1 + x_1).$$

Since $\det Df(x) = 1$,

$$d\Omega = 2(df_1 \wedge df_2 - dx_1 \wedge dx_2)$$
$$= 2(\det Df(x) - 1)\, dx_1 \wedge dx_2$$
$$= 0.$$

This says that Ω is a closed 1-form and, therefore, there must exist a scalar-valued function $H(x)$, called the generating function, such that $dH = \Omega$. Assume that -1 is not an eigenvalue of $Df(0)$. We may introduce the new variables $w = f(x) + x$, so that

$$dH = (f_1 - x_1)\, dw_2 - (f_2 - x_2)\, dw_1$$

and

$$J\nabla H(w) = (I - f) \circ (I + f)^{-1}(w), \qquad J = \begin{pmatrix} 0 & 1 \\ -1 & 0 \end{pmatrix}.$$

Since $(I + f)$ is locally a C^1 diffeomorphism, fixed points of f and critical points of H are in one-to-one correspondence. Next, for any $|x|$ small, $I + f$ is close to the identity and therefore preserves the orientation, i.e., $\text{Det } D(I + f) = 1$. Hence, the index of $J\nabla H(w)$ at $w = 0$ is equal to $\tau(f,0)$. Consider the flow

$$\dot{w} = J\nabla H(w).$$

Since the flow is Hamiltonian, there are no elliptic regions at $w = 0$ (see Fig. 11.1). By the Poincaré index formula (Theorem 11.6),

$$\tau(f,0) = 1 + \frac{-H}{2} \leq 1.$$

In summary, we have the following theorem:

Theorem 11.10. *Let $f: \mathbb{R}^2 \to \mathbb{R}^2$ be C^1 and area preserving. If $x = 0$ is an isolated fixed point of f, then the local fixed point index $\tau(f,0) \leq 1$.*

We now present a constructive proof of the existence of a zero of a vector field on an even dimensional sphere based on transversality theory and degree theory. This theorem will be used in Chapter 4 to prove the Krasnoselskii theorem on bifurcation.

Let TS^n be the tangent bundle of S^n. For any given smooth vector field $F: S^n \to TS^n$, we can write $F(x) = (x, f(x))$, where $x \in S^n$, $x \cdot f(x) = 0$ and x, $f(x) \in \mathbb{R}^{n+1}$. For any fixed $a \in S^n$, define the "trivial" vector field

$$G(x) = (x, g(a, x)), \qquad g(a, x) = a - (a \cdot x)x$$

(G is a vector field whose trajectories lead from the south pole "$-a$" to the north pole "a".) Let ϕ be the homotopy between G and F defined by

$$\phi(a, \lambda, x) = (x, (1 - \lambda)g(a, x) + \lambda f(x)).$$

Theorem 11.11. *For the mapping*

$$\phi: S^n \times (0, 1) \times S^n \to TS^n$$

we have

(a) ϕ *is transversal to* $W = \{(x, y) \in TS^n : y = 0\}$;
(b)   *if n is even, then for almost every a, $\Gamma_a = \phi_a^{-1}(W)$ contains a curve in* $(0, 1) \times S^n$ *such that Γ_a leads from a to a zero of the function f.*

Proof. Suppose that $\phi(\bar{a}, \bar{\lambda}, \bar{x}) \in W$. Hence

$$(1 - \bar{\lambda})g(\bar{a}, \bar{x}) + \bar{\lambda}f(\bar{x}) = 0.$$

Since $\bar{\lambda} \neq 0, 1$, $g(\bar{a}, \bar{x})$ and $f(\bar{x})$ are proportional. Thus,

$$\text{Range } D_a\phi(\bar{a}, \bar{\lambda}, \bar{x}) = \{b - (b \cdot \bar{x})\bar{x} : \bar{a} \cdot b = 0\},$$
$$\text{Range } D_\lambda\phi(\bar{a}, \bar{\lambda}, \bar{x}) = \{(\bar{a} - (\bar{a} \cdot \bar{x})\bar{x})\mu : \mu \in \mathbb{R}\}.$$

Moreover,

$$T_{\phi(\bar{a}, \bar{\lambda}, \bar{x})}W = \{y : y \cdot \bar{x} = 0\}.$$

It is clear that

$$\text{Range } D_a\phi(a, \lambda, x) \oplus \text{Range } D_\lambda\phi(a, \lambda, x) = T_{\phi(a, \lambda, x)}W.$$

This proves (a). Next, by Theorem 10.14, for almost every $a \in S^n$, ϕ_a is transversal to W. Since codim $W = $ codim $\phi_a^{-1}(W) = n$, the set $\phi_a^{-1}(W)$ is diffeomorphic to either a circle or an open interval. It follows from the Implicit Function Theorem that $\Gamma_a = \phi_a^{-1}(W)$ is diffeomorphic to an open interval.

Since the manifold $(0, 1) \times S^n$ has no boundary, Γ_a must have limit points on $\{0\} \times S^n$ or $\{1\} \times S^n$. Suppose that Γ_a has no limit points on $\{1\} \times S^n$. Then, Γ_a has two limit points, namely, $(0, a)$ and $(0, -a)$. Let $\Omega \subset (0, 1) \times S^n$ be an open neighborhood of Γ_a such that the boundary $\partial\Omega$ of Ω contains no points of $\{1\} \times S^n$. Let $\Omega(\lambda) = \{x : (\lambda, x) \in \Omega\}$. Since n is even, the indices of the vector field $\phi_a(0, \cdot)$ at a or $-a$ are both equal to 1. Thus, $\deg(\phi_a(\lambda, \cdot), \Omega(\lambda)) = 2$ for $0 < \lambda$ small, where "deg" denotes the total indices of $\phi_a(\lambda, \cdot)$ on $\Omega(\lambda)$. By the homotopy invariance, $\deg(\phi_a(\lambda, \cdot), \Omega(\lambda)) = 2$ for all $\lambda \in (0, 1)$. But, for λ sufficiently close to 1, $\Omega(\lambda)$ is empty and $\deg(\phi_a(\lambda, \cdot), \Omega(\lambda)) = 0$. This is a contradiction and proves (b). $\square$

APPENDIX

Let M and P be C^r manifolds ($r > 1$) of dimension n and 1 respectively. In applications, P is the parameter manifold. Let $\mathscr{F}$ be the space of C^r maps $F : M \times P \to M$ and

$$Z(F) = \{(x, \alpha) \in M \times P : F(x, \alpha) = x\}$$
$$Z_0(F) = \{(x, \alpha) \in Z(F) : D_x F(x, \alpha) - I \text{ is singular}\},$$

where I is the identity map on M.

Theorem.

    (1)   *There exists an open dense subset $\mathscr{F}_1 \subset \mathscr{F}$ such that for all $F \in \mathscr{F}_1$, $Z(F)$ is a 1-dimensional submanifold of $M \times P$ and for $(x, \alpha) \in Z_0(F)$, 1 is an eigenvalue of $D_x F$ of multiplicity 1.*

    (2)   *There exists an open dense subset $\mathscr{F}_0 \subset \mathscr{F}_1$ such that for each $F \in \mathscr{F}_0$ and each $(x, \alpha) \in Z_0(F)$ we have $\pi_N D_\alpha F \neq 0$, $D_x^2(F\,|\,N) \neq 0$, rank $DF = n$, where N is the eigenspace of 1 and π_N is the projection onto N. Also, $Z_0(F)$ consists of isolated points.*

Proof. For simplicity, we will use local coordinates $(x, \alpha) \in \mathbb{R}^n \times \mathbb{R}$ only. Let $j^1 F(x, \alpha) = (x, \alpha, F(x, \alpha), DF(x, \alpha))$, i.e., the $1 - $ jet of F at (x, α). If we think of the $1 - $ jet bundle $J^1(M \times P, M)$ as the set of $(n^2 + 3n + 1)$-tuples (x, α, y, u, v), then $j^1 F : M \times P \to J^1(M \times P, M)$. Let

$$E = \{(x, \alpha, y, u, v) \in J^1(M \times P, M) : x = y\} \quad \text{and}$$
$$E_0 = \{(x, \alpha, y, u, v) \in E : \det(u - I) = 0\}.$$

We note that E is a submanifold of codimension n and E_0 is a submanifold of codimension $n + 1$. Let $\mathscr{F}_1$ and $\mathscr{F}_0 \subset \mathscr{F}$ be the sets of those $F \in \mathscr{F}$ for which $j^1 F$ meets transversally E and E_0, respectively. By the Transversality Theorem (see Section 10), $\mathscr{F}_0$ and $\mathscr{F}_1$ are open and dense in $\mathscr{F}$. Since $Z(F) = (j^1 F)^{-1}(E)$ and $Z_0(F) = (j^1 F)^{-1}(E_0)$, dim $Z(F) = 1$ for $F \in \mathscr{F}_1$ and dim $Z_0(F) = 0$ for $F \in \mathscr{F}_0$.

It remains to show that the condition on transversal intersections between $j^1 F$ and E or $j^1 F$ and E_0 gives precisely the statements 1 and 2 in the theorem. Transversality of the intersection of $j^1 F$ and E means that the projections of the column vectors of the matrix $(x, D_x F, D_\alpha F)$ into the linear subspace normal to the tangent space of E span that subspace, i.e. we have

$$\operatorname{rank}(-I, I)\begin{pmatrix} I & 0 \\ D_X F & D_\alpha F \end{pmatrix} = \operatorname{rank}(D_x F - I, D_\alpha F) = n.$$

From this, it follows that, if $D_X F$ has 1 as its eigenvalue, its multiplicity is 1. This proves the first part of the theorem.

If $(x_0, \alpha_0) \in Z_0(F)$, then we can choose the coordinates $x = (x_1, \ldots, x_n, \alpha)$ so that $x_0 = 0$, $\alpha = 0$ and

$$D_x F(x_0, \alpha_0) = \begin{pmatrix} \dfrac{\partial(F_1, \ldots, F_{n-1})}{\partial(x_1, \ldots, x_{n-1})} & 0 \\ & \\ 0 & \dfrac{\partial F_n}{\partial x_n} \end{pmatrix}$$

so that $(\partial F_n)/(\partial x_n) = 1$ and

$$\det \frac{\partial(F_1, \ldots, F_{n-1})}{\partial(x_1, \ldots, x_{n-1})} - I \neq 0.$$

Then $(\partial F_n)/(\partial \alpha) \neq 0$ if $F \in \mathscr{F}_1$ (because of rank $(D_X F - I, D_\alpha F) = n$).

It remains to show that for $F \in \mathscr{F}_1$ transversality of $j^1 F$ to E_0 means $(\partial^2 F)/(\partial x_n^2) \neq 0$. Since $u = D_x F(x_0, \alpha_0)$ satisfies $u_{in} = u_{ni} = 0, i = 1, \ldots, n-1$, $u_{nn} = 1$, the manifold E_0 at (x_0, α_0) is tangent to the plane $u_{nn} = 1$ in J^1. This plane is orthogonal to the vector v in J^1 with components corresponding to u_{nn} equal 1, the other components being equal to zero; the vector lies in the tangent space of E. Under our choice of coordinates, the vector $D_{x_n}(j^1 F)$ lies in F and the vectors $D_{x_i}(j^1 F)$, $D_\alpha(j^1 F)$, $i = 1, \ldots, n-1$, span a complement to the tangent space of E at (x_0, α_0). Since the vectors $D_{x_i}(j^1 F)$, $i = 1, \ldots, n$, $D_\alpha(j^1 F)$ have to span a complement to the tangent space of E_0 at (x_0, α_0), $D_{x_n}(j^1 F)$ cannot belong to the tangent space of E_0, so $\langle v, D_{x_n}(j^1 F) \rangle = (\partial^2 F_n)/(\partial x_n^2) \neq 0$. This completes our proof. $\square$

2.12. Ljusternik–Schnirelman Theory in $\mathbb{R}^n$

Let M be a smooth compact n-dimensional manifold without boundary and $f: M \to \mathbb{R}$ be C^2. Recall that a point $p \in M$ is called critical if its differential $df(p)$ is the zero map; a critical point p is called nondegenerate if its second differential $d^2 f(p)$ is a nonsingular symmetric bilinear form on the tangent space $T_p M$ of M at p. The value $f(p)$ of f at a critical point is called a critical value of f. In local coordinates, for $f: \mathbb{R}^n \to \mathbb{R}$, the origin $x = 0$ is a critical point of f if the gradient $\nabla f(0)$ vanishes; it is a nondegenerate critical point if the Hessian matrix $\partial^2 f/\partial x_i \partial x_j$ is nonsingular at $x = 0$. Clearly, nondegenerate critical points are isolated. In fact, if $p \in M$ is a nondegenerate critical point of f, then there exist local coordinates $(x_1, \ldots, x_n)$ and an integer $0 \le k \le n$ such that f has the normal form

$$f(p + x) = -x_1^2 - \cdots - x_k^2 + x_{k+1}^2 + \cdots + x_n^2 + f(p), \qquad 0 \le k \le n$$

for x small. This is the well-known *Morse lemma*. The integer k is the *Morse index* of the point p. If all critical points of f are nondegenerate and all critical values distinct, then f is called a *Morse function*. A fundamental result of Morse theory, is the following set of *Morse inequalities*.

Theorem 12.1. *Let M be a smooth n-dimensional compact manifold without boundary and $f: M \to \mathbb{R}$ be a C^2 Morse function. Let c_k, $k = 0, 1, \ldots, n$, denote the number of critical points of f of index k and β_k be the kth Betti number of M relative to some coefficient field. Then,*

$$c_0 \geq \beta_0$$
$$c_1 - c_0 \geq \beta_1 - \beta_0$$
$$c_2 - c_1 + c_0 \geq \beta_2 - \beta_1 + \beta_0$$
$$\cdots\cdots$$
$$c_n - c_{n-1} + \cdots + (-1)^n c_0 \geq \beta_n - \beta_{n-1} + \cdots + (-1)^n \beta_0.$$

We will not give a proof of Theorem 12.1.

If $M = T^2$ is the two dimensional torus, then any Morse function has at least four distinct critical points since $\beta_0 = \beta_2 = 1$ and $\beta_1 = 2$. This is important in applications since many problems can actually be reduced to finding the number of critical points of a function. For example, let $M = \{x: |x| = \varepsilon, x \in \mathbb{R}^n\}$ and $f(x) = \frac{1}{2}\langle x, Ax\rangle + 0(|x|^3)$ where A is a symmetric matrix, and $\varepsilon > 0$ is small. Then the critical points of f on M will be solutions of the nonlinear eigenvalue problem with $\lambda \in \mathbb{R}$ as the Lagrange multiplier,

$$\nabla f(x) = Ax + 0(|x|^2) = \lambda x, \qquad |x| = \varepsilon.$$

Unfortunately, Morse inequalities are often not applicable to such problems since it is often not known a-priori that the critical points will be nondegenerate. The following example on the torus T^2 shows that a function $f: T^2 \to \mathbb{R}$ may have less than four critical points.

EXAMPLE 12.2. Let T^2 be represented by a square (see Fig. 12.1) with opposite sides identified. Define $f(p) = 0$ for any p on the sides of the square or the diagonal of the square. Let $f(p) > 0$ for p in the interior of the upper triangle and $f(p) < 0$ for p in the interior of the lower triangle. We may assume that f has a unique maximum (minimum) in the upper (lower) triangle. In fact f

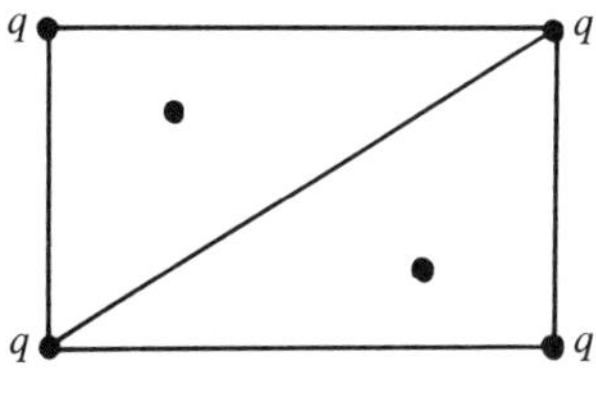

Figure 12.1

can be constructed so as to be smooth and have precisely three critical points: a maximum, a minimum, and a "monkey saddle" at the corner q. Clearly q is a degenerate critical point so the results of Morse theory are not applicable here.

The theory of Ljusternik–Schnirelman deals with critical points which may or may not be nondegenerate and in fact may not even be isolated. The basic technique is the theory of gradient flows, also known as the method of steepest descent. In fact, this is also the basic technique in Morse theory. Let $f : M \to \mathbb{R}$ be C^2. The *gradient vector field* $-\nabla f(p)$ of f on M, defined by $\langle \nabla f(p), v \rangle = df(p)v = vf(p)$ for all $v \in T_p M$, gives rise to a flow $\phi_t(p)$ on M satisfying

$$\frac{d\phi}{dt} = -\nabla f(\phi), \qquad \phi_0(p) = p,$$

where t lies on the maximal interval of existence of the solution through p. Since M is compact and without boundary, this interval is $(-\infty, \infty)$. Note that

$$(12.1) \qquad \frac{d}{dt} f(\phi_t(p)) = -|\nabla f(p)|^2.$$

Thus, any limit set for this flow consists only of critical points. As in Morse theory, we use the gradient flow to study the level sets of f as the levels pass through critical values. Let

$$f^c = \{p \in M : f(p) \le c\},$$
$$f^{-1}(c) = \{p \in M : f(p) = c\},$$
$$W = \{p \in M : df(p) = 0\}, \quad \text{and}$$
$$W_c = W \cap f^{-1}(c).$$

Lemma 12.3. *Let $p \notin W$ and $f(p) = c$. Then there exist $\varepsilon > 0$ and an open neighborhood U of p such that $\phi_1(U) \subseteq f^{c-\varepsilon}$*

Proof. Since $|\nabla f(p)| \ne 0$, integrating (12.1) we have $f(\phi_1(p)) - f(p) < 0$. By the continuity of f and ϕ_1, there exist $\varepsilon > 0$ and an open neighborhood U of p such that, for $q \in U$, $f(\phi_1(q)) - f(p) < -\varepsilon$. $\square$

Theorem 12.4. *Let c be real and U be an open neighborhood of W_c in M. Then there exists $\varepsilon > 0$ such that $\phi_1(f^{c+\varepsilon} - U) \subseteq f^{c-\varepsilon}$.*

Proof. Since M is compact, so is $S = f^{-1}(c) - U$. Let $q \in S$. By Lemma 12.3, there exist $\varepsilon(q) > 0$ and an open neighborhood V_q such that $\phi_1(V_q) \subseteq f^{c-\varepsilon(q)}$. Let $V_{q_1} \cup \cdots \cup V_{q_m}$ be a finite cover of S and $\varepsilon > 0$ be sufficiently small; in particular, $\varepsilon < \min\{\varepsilon(q_1), \ldots, \varepsilon(q_m)\}$. Then $(V_{q_1} \cup \cdots \cup V_{q_m}) \subseteq f^{c-\varepsilon}$, and

$$f^{-1}([c - \varepsilon, c + \varepsilon]) \subseteq U \cup V_{q_1} \cup \cdots \cup V_{q_m}.$$

Since $f^{c+\varepsilon} = f^{c-\varepsilon} \cup f^{-1}((c - \varepsilon, c + \varepsilon])$, we have

$$f^{c+\varepsilon} - U \subseteq f^{c-\varepsilon} \cup (V_{q_1} \cup \cdots \cup V_{q_m}).$$

Hence

$$\phi_1(f^{c+\varepsilon} - U) \subseteq \phi_1(f^{c-\varepsilon}) \cup \phi_1(V_{q_1}) \cup \cdots \cup \phi_1(V_{q_m}) \subseteq f^{c-\varepsilon}. \quad \square$$

We may now formulate the minimax principle as a characterization of critical values. For a family $\mathscr{F}$ of subsets of M we need the following property.

Property (P). If $F \in \mathscr{F}$, then $\phi_1(F) \in \mathscr{F}$.

Theorem 12.5 (Minimax Principle). *Let $\mathscr{F}$ be a nonempty family of subsets of M satisfying property (P). Then*

$$(12.2) \qquad\qquad c = \inf_{F \in \mathscr{F}} \sup\{f(p) : p \in F\}$$

is a critical value of f.

Proof. It is not difficult to see that

$$(12.3) \qquad c = \inf\{b \in R : \text{ there exists } F \in \mathscr{F} \text{ with } F \subseteq f^b\}.$$

Hence, for any $\varepsilon > 0$, there exists $F_\varepsilon \in \mathscr{F}$ such that $F_\varepsilon \subseteq f^{c+\varepsilon}$. If c is a regular value of f, then the set U in Theorem 12.4 is empty. With $\varepsilon > 0$ as in this theorem, we have $\phi_1(f^{c+\varepsilon}) \subseteq f^{c-\varepsilon}$ and so

$$\phi_1(F_\varepsilon) \subseteq f^{c-\varepsilon}.$$

By property (P), $\phi_1(F_\varepsilon) \in \mathscr{F}$, contradicting the characterization of c. $\quad \square$

Before taking up the category of Ljusternik–Schnirelman, we give some elementary applications of the minimax principle.

EXAMPLES. In the following examples, the number c is given by (12.2) or (12.3).

(a) If $\mathscr{F} = \{M\}$, then $c = \max_{p \in M} f(p)$.
(b) If $\mathscr{F} = \{\{p\} : p \in M\}$, then $c = \min_{p \in M} f(p)$.
(c) Fix a nonzero homology class $0 \neq \gamma \in H_k(M)$ of the manifold M and let $\mathscr{F} = \{F \subset M : \gamma \in i_*(H_k(F)) \subset H_k(M)\}$, where $i : F \to M$ is the

inclusion map. Then $\mathscr{F}$ satisfies property (P). This therefore associates critical points of f with each γ. For example, if $M = T^2$ is the two dimensional torus and $0 \neq \gamma \in H_1(T^2)$, then the associated critical points are neither maxima nor minima. Therefore any function on T^2 has at least three distinct critical points.

Now we consider other classes of $\mathscr{F}$'s. In particular, the basic Ljusternik–Schnirelman category theory will be introduced.

Definition 12.6. Let $F \subset M$. We say that the *category* of F in M is k, denoted by $\text{Cat}(F, M) = k$, if F can be covered by k closed sets, each contractible to a point in M, but not by $k - 1$ such sets.

Theorem 12.7. *If $F, G \subset M$, then*

(1) $\text{Cat}(F, M) = 0$ *if and only if* $F = \varnothing$.
(2) $\text{Cat}(F, M) = 1$ *if and only if its closure $\bar{F}$ is contractible to a point in M.*
(3) $\text{Cat}(F, M) = \text{Cat}(\bar{F}, M)$.
(4) $\text{Cat}(F \cup G, M) \leq \text{Cat}(F, M) + \text{Cat}(G, M)$.
(5)   *If* $F \subseteq G$, *then* $\text{Cat}(F, M) \leq \text{Cat}(G, M)$.
(6)   *If* $\eta_t : M \to M$, $0 \leq t \leq 1$, *is a deformation of M; that is, $\eta_0(p) = p$ for all $p \in M$ and $\eta_t(p)$ is continuous in t, p, then* $\text{Cat}(F, M) \leq \text{Cat}(\eta_1(F), M)$.
(7)   *If $F \subseteq M$ is compact, then there exists an open neighborhood U of F such that* $\text{Cat}(U, M) = \text{Cat}(F, M)$.

Proof. We will only prove (6) and (7). For (6), let $\text{Cat}(\eta_1(F), M) = k$. Then $\eta_1(F) \subseteq F_1 \cup \cdots \cup F_k$, where F_i is closed and contractible in M. Then each $\eta_1^{-1}(F_i)$ is also closed and contractible so (6) holds because

$$F \subseteq \eta_1^{-1}(F_1) \cup \cdots \cup \eta_1^{-1}(F_k).$$

To prove (7), let $\text{Cat}(F, M) = k$ and $F = F_1 \cup \cdots \cup F_k$ with each F_i closed and contractible in M. It suffices to show each F_i has an open neighborhood U_i with $\text{Cat}(U_i, M) = 1$, for then $U = U_1 \cup \cdots U_k$. We may assume $\text{Cat}(F, M) = 1$. Let $\eta_t : F \to M$, $0 \leq t \leq 1$, be a deformation of F to a point $p \in M$ and V be an open neighborhood of p with $\bar{V}$ contractible in M. By the homotopy extension theorem, η_t can be extended to a deformation $\bar{\eta}_t : M \to M$ of M. Hence,

$$F = \eta_1^{-1}(p) = \eta_1^{-1}(V) \subseteq \bar{\eta}_1^{-1}(V).$$

Let U be an open neighborhood of F with $U \subset \bar{\eta}_1^{-1}(V)$. Then $1 \leq \text{Cat}(U, M) \leq \text{Cat}(\bar{\eta}_1(U), M) \leq \text{Cat}(V, M) = 1$. $\square$

Theorem 12.8. *Let*

$$\mathscr{F}_k = \{F \subset M : F \text{ is compact and } \text{Cat}(F, M) \geq k\}.$$

Then $\mathscr{F}_k$ satisfies property (P). Hence,

$$c_k = \min_{F \in \mathscr{F}_k} \max_{p \in F} \{f(p)\}$$

is a critical value of f if $\mathscr{F}_k \neq \varnothing$. Moreover, if for some $m \geq 0$ and $k \geq 1$,

$$c = c_{m+1} = \cdots = c_{m+k},$$

then $\mathrm{Cat}(W_c, M) \geq k$, where $W_c = \{p \in M : f(p) = c,\ df(p) = 0\}$. In particular, if $\mathrm{Cat}(M, M) = r$, then $f : M \to \mathbb{R}$ has at least r distinct critical points.

Proof. Let $\eta_t(p)$ in (6) of Theorem 12.7 be the gradient flow $\phi_t(p)$ of the vector field $-\nabla f$. It is easy to see that $\mathscr{F}_k$ satisfies property (P) for any k. Thus, each c_k is a critical value of f if $\mathscr{F}_k$ is nonempty. Next, by (7) of Theorem 12.7, there exists an open neighborhood U of W_c such that $\mathrm{Cat}(U, M) = \mathrm{Cat}(W_c, M)$. Assume that $\mathrm{Cat}(W_c, M) \leq k - 1$. By Theorem 12.4, there exists $\varepsilon > 0$ such that $\phi_1(f^{c+\varepsilon} - U) \subseteq f^{c-\varepsilon}$. Since $c = c_{m+k}$, $\mathrm{Cat}(f^{c+\varepsilon}, M) \geq m + k$. Thus, by Theorem 12.7,

$$\begin{aligned}
\mathrm{Cat}(f^{c-\varepsilon}, M) &\geq \mathrm{Cat}(\phi_1(f^{c+\varepsilon} - U), M) \\
&\geq \mathrm{Cat}(f^{c+\varepsilon} - U, M) \\
&\geq \mathrm{Cat}(f^{c+\varepsilon}, M) - \mathrm{Cat}(U, M) \\
&\geq (m + k) - (k - 1) \\
&= m + 1.
\end{aligned}$$

By the definition of c_m,

$$c = c_{m+1} \leq \sup\{f(p) : p \in f^{c-\varepsilon}\} \leq c - \varepsilon.$$

This contradiction proves the theorem. $\square$

We now give some examples.

EXAMPLES 12.9. (a) If $M = T^2$ is the two dimensional torus, then $\mathrm{Cat}(T^2, T^2) = 3$. By Theorem 12.8 any C^2 function $f : T^2 \to \mathbb{R}$ has at least 3 distinct critical points.

(b) If $M = S^n$ is the n-dimensional sphere, then $\mathrm{Cat}(S^n, S^n) = 2$. If $f : M \to \mathbb{R}$ is C^2, then f has at least two critical points. This conclusion is certainly trivial. However, it is best possible.

(c) Let $M = S^n$ and $f : S^n \to \mathbb{R}$ be even; that is, $f(x) = f(-x)$. By identifying antipodal points, we may consider f to be a real-valued function on the projective space P^n. Since $\mathrm{Cat}(P^n, P^n) = n + 1$, $f : S^n \to \mathbb{R}$ has at least $n + 1$ distinct pairs of critical points. We will subsequently see that this is precisely the reason why some symmetries in differential equations yield more solutions.

(d) Let $M = S^{2n-1}$ be the unit sphere in $\mathbb{R}^{2n}$ and

$$J = \begin{bmatrix} 0 & I \\ -I & 0 \end{bmatrix}$$

where I is the $n \times n$ identity matrix. Then we have a rotation $e^{J\theta}, 0 \leq \theta \leq 2\pi$, on $\mathbb{R}^{2n}$ of period 2π. This defines an S^1-action on S^{2n-1}. Consider a smooth map $f: S^{2n-1} \to \mathbb{R}$ which is equivariant with respect to this action; that is, $f(e^{J\theta}x) = f(x)$ for $x \in S^{2n-1}, 0 \leq \theta \leq 2\pi$. Since the S^1-action on S^{2n-1} is free; that is, $e^{J\theta}$ has no fixed points for any $0 < \theta < 2\pi$, S^{2n-1}/S^1 has a manifold structure and is diffeomorphic to the complex projective space $\mathbb{C}P^{n-1}$. By considering f as a mapping from $\mathbb{C}P^{n-1}$ to $\mathbb{R}$, we conclude that there exist at least n distinct critical points of f on $\mathbb{C}P^{n-1}$ since $\text{Cat}(\mathbb{C}P^{n-1}, \mathbb{C}P^{n-1}) = n$. In other words, f has at least n distinct periodic orbits on S^{2n-1}.

For even functions defined on a symmetric subset of $\mathbb{R}^n$, the notion of genus is often used instead of category of a set. Even though they coincide for compact symmetric sets, it is sometimes easier to compute the genus of a set. We now define the genus and show its relation to the minimax principle.

Definition 12.10. Let Σ denote the set of all closed subsets F of $\mathbb{R}^n - \{0\}$ which are symmetric about the origin; that is, $x \in F$ implies $-x \in F$. For $F \in \Sigma$, the *genus* $\gamma(F)$ of F is defined to be the smallest integer k for which there exists an odd continuous map $\eta: F \to \mathbb{R}^k - \{0\}$. We define $\gamma(\varnothing) = 0$.

The following are elementary facts about genus.

Theorem 12.11. *If $F, G \in \Sigma$, then we have*

(1) $\gamma(F) = 0$ *if and only if* $F = \varnothing$.
(2) *If $\gamma(F) = 1$, then F is not connected.*
(3) *If $F \subset G$, then $\gamma(F) \leq \gamma(G)$.*
(4) $\gamma(F \cup G) \leq \gamma(F) + \gamma(G)$.
(5) $\gamma(\overline{F - G}) \geq \gamma(F) - \gamma(G)$.
(6) *If F is compact, then there exists an open neighborhood U of F such that $\gamma(\overline{U}) = \gamma(F)$.*
(7) *If there exists an odd continuous $\eta: F \to G$, then $\gamma(F) \leq \gamma(G)$.*
(8) $\gamma(S^{n-1}) = n$, *where $S^{n-1} \subset \mathbb{R}^n$ is the unit sphere.*

Proof. To prove (6), let $\gamma(F) = k$. By definition, there exists an odd continuous map $\eta: F \to \mathbb{R}^k - \{0\}$. By the Tietze extension theorem, there exists a continuous extension ξ of η to all of $\mathbb{R}^n$. If $\hat{\eta}(x) = [\xi(x) - \xi(-x)]/2$, then $\hat{\eta}: \mathbb{R}^n \to \mathbb{R}^k$ is an odd continuous extension of η. Since F is compact, there exists an open neighborhood U of F with $\hat{\eta}(x) \neq 0$ for all $x \in \bar{U}$. Hence, $\bar{\eta}: \bar{U} \to \mathbb{R}^k - \{0\}$ and $\gamma(\bar{U}) = k$. Part (8) is just a restatement of the Borsuk-Ulam theorem. The remaining parts of Theorem 12.11 are easily proved. $\square$

The following theorem is a standard application of the minimax principle together with the notation of genus.

Theorem 12.12. *If $f: S^{n-1} \to \mathbb{R}$ is even and C^2, then f has at least n distinct pairs of critical points.*

Proof. For each $k = 1, 2, \ldots, n$, define

$$\mathscr{F}_k = \{F \subset S^{n-1}: F \in \Sigma, \gamma(F) \geq k\},$$

$$c_k = \min_{F \in \mathscr{F}_k} \max_{x \in F} f(x)$$

Since f is even, the gradient vector field $-\nabla f$ is odd and so is the flow ϕ_t of this vector field. Since then $\mathscr{F}_k$ clearly satisfies property (P), Theorem 12.5 (the minimax principle) implies each c_k is a critical value. If, for some $m \geq 0$ and $k \geq 1$, $c = c_{m+1} = \cdots = c_{m+k}$, then it can be shown as in Theorem 12.8 that $\gamma(W_c) \geq k$, where $W_c = \{x \in S^{n-1}: f(x) = c, \, df(x) = 0\}$. This completes the proof.

Finally, we state without proof a relation between genus and category of a closed set.

Theorem 12.13. *Let $\sim$ denote the equivalence relation $x \sim -x$ on $\mathbb{R}^n - \{0\}$ and let M be the quotient space $(\mathbb{R}^n - \{0\})/\sim$. For $F \in \Sigma$ compact, let $A = F/\sim$. Then $\gamma(F) = \mathrm{Cat}(A, M)$.*

2.13. Bibliographical Notes

For a more complete discussion of calculus in Banach spaces, see Abraham and Robbin [1], Berger [1], Dieudonné [1], Lang [2], Schwartz [1]. Example 1.12 is due to Henry [1].

The Implicit Function Theorem (Theorem 2.3) is essentially due to Hildebrandt and Graves [1]. The generalization in Theorem 2.4 was first used by Hale [9] in the study of the Hopf bifurcation theorem for functional differential equations (see also Hale [12, 13], Lima [1]). In certain problems in the geometry of manifolds and in problems of the existence of invariant tori for Hamiltonian systems, the Implicit Function Theorems in Section 2 are not applicable. The inverse of the derivative may exist but be unbounded. Under certain additional hypotheses, methods of accelerated convergence (which go under the names of Kolmogorov, Arnol'd and Moser) can be used to obtain Implicit Function Theorems which are applicable to the above problems (see Arnol'd [6], Kolmogorov [1], Moser [1, 3], Schwartz [1], Sergeraert [1], Zhender [1]).

The proof of Theorem 3.5 is based on Prodi and Ambrosetti [1]. Theorem 3.9 can be found in Schwartz [1].

The method of reduction in Section 4 referred to as the method of Liapunov–Schmidt was used by Liapunov [2] for the bifurcation problem studied by Poincaré [2] concerned with the equilibrium positions of rotating liquid masses. Liapunov [1] also used the function in (4.8) for the study of the stability of the zero solution of a system of nonlinear differential equations when the linear part contained some zero eigenvalues (see Section 9.8). In this case, there is no parameter and no bifurcations, but the function is still of utmost importance and is often referred to as the bifurcation function. E. Schmidt [1] also used this method in integral equations. We do not attempt to trace the history of this method, but the underlying principle has certainly been independently discovered by a number of authors (see, for example, Antosiewicz [1], Bartle [1], Bass [1], Cesari [1–6], Cronin [2–4], Friedrichs [1, 2], Graves [1], Hale [1, 2], Lewis [1], and the extensive references in Mawhin [4], Gaines and Mawhin [1], Vainberg and Tregonin [1, 2]).

Cesari [1, 2] was the first to point out the importance of extending the method of reduction in Section 4 to alternative problems by using a projection P whose range is larger than the null space of A. In some significant situations, he showed that it was possible to reduce the discussion of the solutions of Eq. (4.1) to an alternative problem in a *finite* dimensional space X_P for an appropriate projection P. His results led to a general procedure for reducing Eq. (4.1) to an *alternative problem*. The presentation of the procedure in the text follows Bancroft, Hale and Sweet [1], but is a less general version than the one presented here.

Several methods have been used to solve Eq. (4.9a) for $z^*(y)$. The Banach contraction principle was first applied by Cesari [1, 2] to determine all solutions which lie in a ball $\{x : |x| \le a\}$. He even showed that it was possible in some cases to obtain a finite dimensional projection P for which one could reduce the determination of all solutions in this ball to Eq. (4.11) for $y \in X_P$. For the case in which X is a Hilbert space and the right inverse K of A is compact, Hale [5] showed such a reduction to a finite dimensional problem was always possible. It was later shown independently by Cesari and McKenna [1] and Cooperman [1] that the same type of reduction could be made if K is compact and X has a Schauder basis. Cooperman [1] also gave some results for k-set or α-contractions.

The basic idea behind these results is to construct operators $P_\varepsilon, Q_\varepsilon$, depending on a parameter $\varepsilon > 0$, so that each has finite dimensional range and $|KQ_\varepsilon| < \varepsilon$. In Hilbert space, finite dimensional projections are easily constructed by taking the span of a finite number of elements of a complete orthonormal system. When the alternative method can be applied, it can be considered as a way to justify the Galerkin procedure (see Cesari [2]).

Other devices have been used to solve Eq. (4.9a) for $z^*(y)$; for example, the Schauder fixed point theorem and the theory of monotone operators. The literature is extensive and we refer the reader to the survey articles of Cesari [3–5] and the book of Cesari and Kannan [1] for the discussion and references.

It is also possible to solve Eq. (4.9) in different ways. For example, one could solve (4.9b) for $y = y^*(z)$ and then determine z as a solution of

$$z - KQN(y^*(z) + z) = 0.$$

This procedure has been used by Rabinowitz [1], Hall [1, 2], deSimon and Torelli [1], Torelli [1], Sova [1].

Another approach that has been used in discussing Eq. (4.9) is to transform it to a fixed point problem to which one can apply the Leray-Schauder theory of degree. Let us rewrite Eq. (4.9) as

$$(13.1) \qquad \begin{aligned} x &= Px + KQNx \\ (I - Q)(Ax - Nx) &= 0 \end{aligned}$$

If there is a one-to-one map $S: Z_{I-Q} \to X_P$, then it is not difficult to see that Eq. (13.1) is equivalent to

$$(13.2) \qquad \begin{aligned} x &= Tx \\ Tx &\overset{\text{def}}{=} Px + KQNx + S(I - Q)(Ax - Nx). \end{aligned}$$

Finding solutions of Eq. (4.1) has thus been reduced to finding fixed points of the map $T_{P,Q}$. If one could define the degree of $I - T_{P,Q}$, then another powerful mathematical tool can be brought to bear on this problem. For a special case with $P = Q$ finite dimensional, Williams [1] obtained an interesting connection of this approach with the Brouwer degree of the map associated with the bifurcation function. Mawhin (see Gaines and Mawhin [1] for references) has considered in detail the case where $P = U$, a projection on $\mathcal{N}(A)$, $Q = E$, a projection on $\mathcal{R}(A)$ with P, Q having the same finite dimension. With some other hypotheses which assure that T is completely continuous, he defines the Leray-Schauder degree $d(I - T, \Omega, 0)$ for any bounded set Ω and shows this number depends only on A, N, if the map $S: Z_{I-E} \to X_U$ is orientation preserving. He then defines the *coincidence degree* $d[(A, N), \Omega]$ of A, N in Ω to be $d(I - T, \Omega, 0)$. Extensive applications of this concept to boundary value problems have been made (see Gaines and Mawhin [1]). Tarafdar and Teo [1] have extended this method to equation $Lx \in Nx$.

For a proof of Theorem 5.1 as well as other properties of Nemitskii operators see Krasnosel'skii [1], Vainberg [1]. For the properties of Sobolev spaces and imbedding theorems, see Adams [1], Sobolev [1], Friedman [1], Gilbarg and Trudinger [1].

The proof of the Weierstrass Preparation Theorem in the text is based on Golubitsky and Guillemin [1]. For some interesting pictures of some of the bifurcation surfaces for polynomials of degree less than 6, see Woodcock and Poston [1].

The presentation of the Malgrange Preparation Theorem in Section 7 follows Golubitsky and Schaeffer [1]. The extension lemma (Theorem 7.4) was first proved by Nirenberg [1]. Michor [1] gave the extension of the Malgrange Preparation Theorem to the case where the parameter x belongs to an arbitrary Banach space.

For a further discussion of and references to Newton's polygon, see Lefschetz [3], MacMillan [1, 2], Sather [2], Vainberg and Tregonin [2].

Lang [1], Milnor [1], Smith [1], and Sternberg [1] are good sources for material on the theory of manifolds. For transversality, see Abraham and Robbin [1], Golubitsky and Guillemin [1], Abraham and Marsden [1].

For Sard's Theorem in Section 10, see Abraham and Robbin [1], Milnor [1]. Theorem 10.5 was first proved by Chow, Mallet–Paret and Yorke [2, 3]. For further results in this direction as well as references, see the survey article of Allgower and Georg [1]. Actual numerical computations of fixed points have been performed by Watson, Li and Wang [1]. For a proof of Theorem 10.12, see Abraham and Robbin [1].

The idea of defining degree in Section 11 using Formula (11.1) originated with Nagumo [1]. Similar ideas appear in the definitions of Fuller index (see, for example, Chow and Mallet-Paret [2]). The presentation in the text differs in several respects from the above. The result in the appendix of Section 11 is due to Brunovsky [1] (see also Meyer [2], Medved [1, 2]). The proof of the Poincaré formula (Theorem 11.6) can be found in Hartman [1]. Theorem 11.8 is due to Shub and Sullivan [1]. Theorem 11.9 is a special case of a result in Chow, Mallet-Paret and Yorke [4]. The example on area preserving maps is due to Simon [1] (see also Meyer [1]) in which he used this formula to prove the Liapunov center theorem at resonance for two degrees of freedom. The proof of Theorem 11.11 can be found in Chow, Mallet–Paret and Yorke [2]. Some general references on the Brouwer and Leray–Schauder degree are Cronin [5], Lloyd [1], Nirenberg [2], Schwartz [1]. Generalizations of this concept can also be found in Nussbaum [5], and Sadovskii [1].

Theorem 12.1 may be found in Milnor [2]. The main ideas in Section 12 on Ljusternik–Schnirelman theory follows Palais [1–3] (see also Ljusternik [1], Ljusternik and Schnirelman [1]). The definition of genus was first given by Coffman [1] and was used by D. C. Clark [1–4], Rabinowitz [6]. For a proof of Theorem 12.13, see Rabinowitz [5].

Chapter 3

Applications of the Implicit Function Theorem

3.1. Existence of Solutions of Ordinary Differential Equations

Suppose Λ is a Banach space, $\Omega \subset \mathbb{R} \times \mathbb{R}^n$, $G \subset \Lambda$ are open sets, $f : \Omega \times G \to \mathbb{R}^n$ is continuous, $D_x f(t, x, \lambda)$ is continuous for $(t, x) \in \Omega$, $\lambda \in \Lambda$, and consider the initial value problem

$$\dot{x} = f(t, x, \lambda)$$
(1.1)
$$x(\sigma) = \xi$$

where $(\sigma, \xi) \in \Omega$.

Theorem 1.1. *Under the above hypotheses, there is a solution $x(\sigma, \xi, \lambda)(t)$ of (1.1) which is unique and continuous in $(\sigma, \xi, \lambda, t)$ in its domain of definition. If $f \in C^k(\Omega \times G, \mathbb{R}^n)$, then $x(\sigma, \xi, \lambda)(t)$ has continuous derivatives of order k in $(\sigma, \xi, \lambda, t)$ in its domain of definition. If f is analytic, so is $x(\sigma, \xi, \lambda)(t)$ analytic in its arguments.*

Proof. If x is a solution of (1.1) on $[\sigma - \alpha, \sigma + \alpha]$, $t - \sigma = \alpha\tau$, $x(\alpha\tau + \sigma) - \xi = z(\tau)$, then z must satisfy

$$\frac{dz(\tau)}{d\tau} - \alpha f(\alpha\tau + \sigma, \xi + z, \lambda) = 0, \qquad -1 \leq \tau \leq 1$$
(1.2)
$$z(0) = 0$$

and conversely. Let

$$X = \{\phi \in C^1([-1, 1], \mathbb{R}^n) : \phi(0) = 0\}, \ Y = C([-1, 1], \mathbb{R}^n)$$
$$F : \mathbb{R} \times \mathbb{R} \times \mathbb{R}^n \times \Lambda \times X \to Y$$

$$F(\alpha, \sigma, \xi, \lambda, \phi)(\tau) = \frac{d\phi(\tau)}{d\tau} - \alpha f(\alpha\tau + \sigma, \xi + \phi(\tau), \lambda), \ -1 \leq \tau \leq 1.$$

Then $F(0, \sigma_0, \xi_0, \lambda_0, 0) = 0$, $D_\phi F(0, \sigma_0, \xi_0, \lambda_0, 0)\psi = d\psi/d\tau$, for any $(\sigma_0, \xi_0, \lambda_0)$. If $\psi \in X$, $d\psi/d\tau = y \in Y$, then $\psi(\tau) = \int_0^\tau y(s)\,ds$ and ψ is continuous and linear in y. Therefore, the Implicit Function Theorem implies the existence of a unique solution $z^*(\alpha, \sigma, \xi, \lambda, \tau)$ of $F(\alpha, \sigma, \xi, \lambda, z) = 0$ in a neighborhood of $(0, \sigma_0, \xi_0, \lambda_0, 0)$ and this function has the same smoothness properties as f. This proves the theorem locally. The global version is obtained by the process of continuation. $\square$

3.2. Admissible Classes in Ordinary Differential Equations

Let us begin with a simple illustration. Suppose $A(t)$ is an $n \times n$ matrix, continuous in t for $t \in [0, 1]$, M, N are $n \times n$ constant matrices, $f \in C([0, 1], \mathbb{R}^n)$, $\gamma \in \mathbb{R}^n$, and consider the nonhomogeneous boundary value problem

$$\dot{x} = A(t)x + f(t)$$
(2.1)
$$Mx(0) + Nx(1) = \gamma$$

as well as the homogeneous boundary value problem

$$\dot{x} = A(t)x$$
(2.2)
$$Mx(0) + Nx(1) = 0.$$

Lemma 2.1. *If Problem (2.2) has only the solution $x = 0$, then Problem (2.1) has a unique solution $K(f, \gamma)$ where*

$$K: C([0, 1], \mathbb{R}^n) \times \mathbb{R}^n \to C^1([0, 1], \mathbb{R}^n)$$

is continuous, linear.

Proof. Let $X(t)$ be an $n \times n$ matrix solution of $\dot{x} = A(t)x$ with $X(0) = I$. Then every solution of Problem (2.2) must satisfy $x(t) = X(t)a$, $[M + NX(1)]a = 0$ for some $a \in \mathbb{R}^n$. The hypothesis of the lemma implies $[M + NX(1)]^{-1}$ exists. Every solution of Problem (2.1) must satisfy

$$x(t) = X(t)a + \int_0^t X(t)X^{-1}(s)f(s)\,ds$$

$$[M + NX(1)]a = \gamma - N \int_0^1 X(1)X^{-1}(s)f(s)\,ds.$$

Since $[M + NX(1)]^{-1}$ exists, this uniquely defines a as a continuous linear operator on (f, γ). Substituting this expression for a in the formula for x, one obtains the result stated in the lemma. $\square$

For any $g \in \mathscr{G} \stackrel{\text{def}}{=} C^1([0, 1] \times \mathbb{R}^n, \mathbb{R}^n)$, $h \in \mathscr{H} \stackrel{\text{def}}{=} C^1(C[0, 1], \mathbb{R}^n)$, one can consider the boundary value problem

$$\dot{x}(t) = A(t)x(t) + g(t, x(t)), \qquad 0 \le t \le 1,$$
(2.3)
$$Mx(0) + Nx(1) = h(x).$$

Theorem 2.2. *If Problem (2.2) has a unique solution, then there are $v > 0, \mu > 0$, such that for every g, h, $|g|_{\mathscr{G}} < v$, $|h|_{\mathscr{H}} < v$ there is a unique solution $x^*(g, h)$ of Problem (2.3) in the region $|x| < \mu$, $x^*(g, h)$ is continuously differentiable in g, h, $x^*(0, 0) = 0$.*

Proof. From Lemma 2.1, any solution of Problem (2.3) must satisfy the equation

$$T(x, g, h) \overset{\text{def}}{=} x - K(G(x), h(x)) = 0$$
$$G(x)(t) = g(t, x(t)).$$

The operator $T: C^1([0, 1], \mathbb{R}^n) \times \mathscr{G} \times \mathscr{H} \to C^1([0, 1], \mathbb{R}^n)$, $T(0, 0, 0) = 0$, $D_x T(0, 0, 0) = I$, the identity. The Implicit Function Theorem implies the conclusions stated in the theorem. $\square$

The technique used to discuss Problem (2.3) is applicable to many other types of problems. To illustrate this, let us give a brief abstract summary of the basic ideas.

Suppose I is an interval in $\mathbb{R}$, $A(t)$ is an $n \times n$ matrix function continuous on I and $\mathscr{B}, \mathscr{D}$ are given Banach spaces of continuous n-vector functions on I. For every $f \in \mathscr{D}$, suppose the equation

$$(2.4) \qquad \dot{x} = A(t)x + f(t)$$

has a unique solution $\mathscr{K}f$ in $\mathscr{B}$ and $\mathscr{K}: \mathscr{D} \to \mathscr{B}$ is a continuous linear operator. In this case, we say $(\mathscr{B}, \mathscr{D})$ is *strongly admissible* for Equation (2.4).

Let $\mathscr{C}^1(I \times \mathbb{R}^n, \mathbb{R}^n) = \{f: I \times \mathbb{R}^n \to \mathbb{R}^n, f(t, x)$ continuous and continuously differentiable in $x\}$. Let

$$|f|_1 = \sup\{|f(t, x)| + |\partial f(t, x)/\partial x|, (t, x) \text{ in } I \times \mathbb{R}^n\}.$$

If ϕ is in $\mathscr{B}$, let us suppose the function $f(\cdot, \phi(\cdot))$ is in $\mathscr{D}$ and consider the problem of the existence of solutions in $\mathscr{B}$ of the equation

$$(2.5) \qquad \dot{x} = A(t)x + f(t, x)$$

If there exists a solution x of Eq. (2.5) in $\mathscr{B}$, then x must satisfy the equation

$$(2.6) \qquad G(x, f) \overset{\text{def}}{=} x - \mathscr{K}F(x, f) = 0$$

where $\mathscr{K}: \mathscr{D} \to \mathscr{B}$ is the continuous linear operator defined above and

$$(2.7) \qquad \begin{aligned} F: \mathscr{B} \times \mathscr{C}^1(I \times \mathbb{R}^n, \mathbb{R}^n) &\to \mathscr{D} \\ F(x, f)(t) = f(t, x(t)), \qquad & t \in I. \end{aligned}$$

Let us suppose the function $F(x, f)$ is continuous together with its Fréchet derivative.

To solve Eq. (2.6) for $|f|_1$ small, one can use the Implicit Function Theorem observing that $G(0,0) = 0$, $\partial G(0,0)/\partial x = I$, where G is defined in Eq. (2.6). The results may then be summarized as

Theorem 2.3. *Suppose $(\mathscr{B}, \mathscr{D})$ is strongly admissible for system (2.4) and the function F defined in Relation (2.7) is continuous together with its Fréchet derivative. Then there is a $\delta > 0$, $\eta > 0$, and a unique function*

$$x^*: \{f \in \mathscr{C}^1(I \times \mathbb{R}^n, \mathbb{R}^n): |f|_1 < \delta\} \to \mathscr{B}$$

such that $x^*(f)$ is continuous together with its derivative in f, $x^*(0) = 0$, $x^*(f)$ satisfies Equation (2.5) and is the only solution of Equation (2.6) in $\mathscr{B}$ with norm less than η.

3.3. Global Boundary Value Problems for Ordinary Differential Equations

Consider the boundary value problem

(3.1)
$$x''(t) + \psi(x(t)) = f(t), \qquad 0 < t < \pi,$$
$$x(0) = x(\pi) = 0$$

where $f \in C([0, \pi], \mathbb{R})$, $\psi \in C^1(\mathbb{R}, \mathbb{R})$.

Under various conditions on ψ, we show there is a unique solution of Problem (3.1) for any $f \in C([0, \pi], \mathbb{R})$. The following lemmas from elementary differential equations are needed and stated without proof.

Lemma 3.1. *If ρ is measurable, bounded, positive on $[0, \pi]$, then the eigenvalues $\mu_1 < \mu_2 < \cdots$ of*

$$v'' + \mu\rho v = 0, \qquad 0 < t < \pi,$$
$$v(0) = v(\pi) = 0$$

satisfy the following properties:

    (i) $0 < \mu_1 < \mu_2 < \cdots$, $\mu_n \to \infty$ *as $n \to \infty$*
    (ii) *Each eigenvalue is simple and has $n - 1$ zeros in $(0, \pi)$*
    (iii) *μ_n is a nonincreasing function of ρ. If $\rho_1(t) < \rho_2(t)$ a.e. on $[0, \pi]$, then the corresponding eigenvalues μ_n^1, μ_n^2 satisfy $\mu_n^1 > \mu_n^2$.*
    (iv) *μ_n is continuous in ρ in the topology of $L^1[0, \pi]$.*

Lemma 3.2. *If x is a solution on (a, b) of*

$$(px')' + g_1 x = 0$$

and y is a solution on (a, b) of

$$(py')' + g_2 y = 0$$

with $p > 0$ continuous, g_1, g_2 integrable functions, $g_2 \geq g_1$ on (a, b), $g_2 > g_1$ on a set of positive measure, then there is a zero of y between any two consecutive zeros of x.

Suppose now that

$$(3.2) \qquad \psi(0) = 0, \qquad m^2 < h \leq \psi'(s) \leq k < (m + 1)^2, \qquad s \in \mathbb{R},$$

where $m \geq 0$ is an integer, h, k are constants. Recall that the eigenvalues of the linear problem

$$x'' + \lambda x = 0, \qquad 0 < t < \pi,$$
$$x(0) = x(\pi) = 0$$

are given by $\lambda = m^2$, $m \geq 1$ an integer. Thus, the hypothesis on ψ in Relation (3.2) is a nonresonance condition in the sense that zero is the only solution of the problem

$$x''(t) + \psi'(s)x(t) = 0, \qquad 0 < t < \pi,$$
$$x(0) = x(\pi) = 0$$

for any fixed $s \in \mathbb{R}$. We prove this condition is enough to ensure that Problem (3.1) has a unique solution

Theorem 3.3. *If ψ satisfies Condition (3.2), then Problem (3.1) has a unique solution.*

Proof. Let $X = \{x \in C^2([0, \pi], \mathbb{R}) : x(0) = x(\pi) = 0\}$, $Z = C([0, \pi], \mathbb{R})$. The idea of the proof is to show that the map

$$(3.3) \qquad\qquad T : X \to Z, \qquad Tx(t) = x''(t) + \psi(x(t))$$

is proper and a local homeomorphism. Corollary 2.3.6 then implies T is a global homeomorphism and, thus, the conclusion stated in the theorem.

We first show that if $Tx_n = f_n$, $\{f_n\} \subset Z$ is bounded, then $\{x_n\}$ is bounded in Z. Suppose on the contrary that $\{x_n\}$ is not bounded. Then we may assume without loss of generality that $|x_n|_Z \to \infty$ as $n \to \infty$. If

$$\omega(s) = \psi(s)/s \quad \text{for } s \neq 0; \qquad \omega(s) = 0 \quad \text{for } s = 0$$

then $h \leq \omega(s) \leq k$ for all s. If $g_n = f_n/|x_n|_Z$, $z_n = x_n/|x_n|_Z$, then z_n satisfies

$$(3.4) \qquad\qquad z_n'' + \omega(x_n)z_n = g_n.$$

Since $-\omega(x_n)z_n + g_n$ is bounded in Z, the sequence $\{z_n''\}$ is bounded in Z and there is a subsequence of $\{z_n\}$ (which we label the same) converging to z^* in $\{z \in C^1([0, \pi], \mathbb{R}) : z(0) = z(\pi) = 0\}$.

From Equation (3.4), for any $w \in C_0^\infty([0, \pi], \mathbb{R})$,

$$(3.5) \qquad -\int_0^\pi z_n'w' \, dt + \int_0^\pi \omega(x_n)z_n w \, dt = \int_0^\pi g_n w \, dt.$$

Since $\{\omega(x_n)\}$ is bounded, it is precompact in the weak*-topology of $L^1([0, \pi], \mathbb{R})$. Thus, there is an ω^* in $L^1([0, \pi], \mathbb{R})$, $h \leq \omega^* \leq k$, and a subsequence (which we label the same) so that $\omega(x_n) \to \omega^*$ in this topology. Taking the limit in Equation (3.5), one obtains

$$-\int_0^\pi z^{*'}w' \, dt + \int_0^\pi \omega^* z^* w \, dt = 0$$

for all $w \in C_0([0, \pi], \mathbb{R})$. Thus, z^* is a solution (in the generalized sense) of the equation

$$(3.6) \qquad \begin{aligned} y'' + \mu\omega^* y &= 0, \qquad 0 < t < \pi, \\ y(0) = y(\pi) &= 0 \end{aligned}$$

for $\mu = 1$. Since $m^2 < h \leq \omega^* \leq k < (m + 1)^2$ for all nonnegative integers m, the classical Sturmian theory (Lemma 3.1) implies that the eigenvalues $\{\mu_n\}$ of Problem (3.6) satisfy

$$\mu_m < m^2 < (m + 1)^2 < \mu_{m+1}.$$

Therefore, $\mu_m = 1$ for some integer m is impossible. This proves boundedness of the sequence $\{x_n\}$ in Z.

Since $\{x_n\}$ is bounded in Z, it follows that $\{x_n''\}$ is bounded in Z. Thus, one may choose a subsequence of the x_n converging in $\{x \in C^1([0, \pi], \mathbb{R}) : x(0) = x(\pi) = 0\}$. For this same subsequence, the x_n'' converge in Z. This proves that T is proper.

To show T is locally one-to-one, one needs only to show that $DT(x)$ is invertible for each $x \in X$. This is equivalent to showing that the problem

$$\begin{aligned} y''(t) + \psi'(x(t))y(t) &= 0, \qquad 0 < t < \pi, \\ y(0) = y(\pi) &= 0 \end{aligned}$$

has only the trivial solution $y = 0$. But this follows from the same type of reasoning as employed when discussing Problem (3.6). $\square$

We give a second proof of Theorem 3.3. A preliminary lemma is needed.

Let H be a real Hilbert space and $K : H \to H$ a completely continuous, symmetric, positive semidefinite operator. Let

$$0 < \lambda_1 \le \lambda_2 \le \cdots \le \lambda_n \le \cdots$$

denote those λ for which $\dim \mathcal{N}(I - \lambda K) \ge 1$.

Lemma 3.4. *Let* $\mathcal{A} \subset L(H, H)$ *be any set of symmetric, linear operators on* H. *If there exist real numbers* μ_N, μ_{N+1} *such that*

$$\lambda_N I < \mu_N I \le A \le \mu_{N+1} I < \lambda_{N+1} I$$

for each $A \in \mathcal{A}$, *then the function* $F : H \to H$ *defined by the equation* $F(x) = x - KAx$ *for each* $A \in \mathcal{A}$ *has a bounded inverse and there is a constant* k *such that*

$$|(I - KA)^{-1}| \le k \quad \text{for all } A \in \mathcal{A}.$$

Proof. Let $\mu = (\mu_{N+1} + \mu_N)/2$, $\gamma = (\mu_{N+1} - \mu_N)/2$ and define $B = A - \mu I$. Then $|B| \le \gamma$. For any $w \in H$, $A \in \mathcal{A}$, the equation $x - KAx = w$ is equivalent to the equation

$$x = (I - \mu K)^{-1} w + (I - \mu K)^{-1} K B x$$

since $\lambda_N < \mu < \lambda_{N+1}$ implies $I - \mu K$ has a bounded inverse. Let us estimate the norm of $(I - \mu K)^{-1} K$. If $\{v_m\}$ is an orthnormal sequence in H such that $v_m = \lambda_m K v_m$, then, for any $y \in H$,

$$\begin{aligned}
K(I - \mu K)^{-1} y &= K\left[y + \sum_1^\infty \frac{\mu \langle y, v_m \rangle v_m}{\lambda_m - \mu} \right] \\
&= \sum_1^\infty \frac{\langle y, v_m \rangle v_m}{\lambda_m} + \sum_1^\infty \frac{\mu \langle y, v_m \rangle v_m}{\lambda_m(\lambda_m - \mu)} \\
&= \sum_1^\infty (\lambda_m - \mu)^{-1} \langle y, v_m \rangle v_m.
\end{aligned}$$

Therefore,

$$\begin{aligned}
|K(I - \mu K)^{-1}| = |(I - \mu K)^{-1} K| &= \sup_m |\lambda_m - \mu|^{-1} \\
&= \max[(\mu - \lambda_N)^{-1}, (\lambda_{N+1} - \mu)^{-1}] \overset{\text{def}}{=} \alpha.
\end{aligned}$$

Thus, $|K(I - \mu K)^{-1}B| < \alpha\gamma$. From the definition of α, γ, it is easy to verify that $\alpha\gamma < 1$. Thus, the operator $I - K(I - \mu K)^{-1}B$ has a bounded inverse and this implies $(I - KA)^{-1}$ exists for each $A \in \mathscr{A}$,

$$(I - KA)^{-1} = [I - K(I - \mu K)^{-1}B]^{-1}(I - \mu K)^{-1}$$
$$|(I - KA)^{-1}| \le (1 - \alpha\gamma)^{-1}|(I - \mu K)^{-1}|,$$

a bound which is independent of A. This proves the lemma. $\qquad\square$

Second Proof of Theorem 3.3. Let $H = L^2[0, \pi]$ and

$$k(s, t) = \begin{cases} s(t - \pi)/\pi & 0 \le s \le t \le \pi \\ t(s - \pi)/\pi & 0 \le t \le s \le \pi. \end{cases}$$

Any solution of Problem (3.1) must satisfy

$$x = -Kf + K\Psi(x),$$

$$Kf(t) = \int_0^\pi k(s, t)f(s)\,ds$$

$$\Psi(x)(t) = \psi(x(t)), \qquad 0 \le t \le T.$$

The operator K is symmetric completely continuous, and the values λ such that $\dim \mathscr{N}(I - \lambda K) \ge 1$ are given by $\lambda_n = n^2$, $n = 1, 2, \ldots$. If ψ satisfies Condition (3.2), then take $\mathscr{A}$ to be the family of bounded linear operators on H defined by $Ay(t) = \psi'(x(t))y(t)$ for any $x \in L^2[0, \pi]$. From Lemma 3.4, the operator $I - K\Psi'(x)$ has a bounded inverse for all $x \in H$, uniformly bounded for $x \in H$. Theorem 2.3.9 implies the conclusion of Theorem 3.3. $\qquad\square$

Using this second proof, one can easily generalize Theorem 3.3 to systems as stated in the following result.

Theorem 3.5. *Suppose $x \in \mathbb{R}^n$, $H : \mathbb{R}^n \to \mathbb{R}$ is a C^2-function such that*

$$m^2 I < hI \le \frac{\partial^2 H(x)}{\partial x^2} \le kI < (m + 1)^2 I, \qquad x \in \mathbb{R}^n$$

for some integer $m \ge 0$, $h, k \in \mathbb{R}$. For any $f \in C([0, \pi], \mathbb{R}^n)$, there is a unique solution of the problem

$$x'' + DH(x) = f, \qquad 0 < t < \pi,$$
$$x(0) = x(\pi) = 0.$$

It is possible to prove an analogue of Theorem 3.3 even when some of the conditions in Relation (3.2) are not satisfied. If ψ in Equation (3.1) satisfies

(3.7) $\psi(0) = 0, \qquad \psi'(0) = 1, \qquad \psi''(0) = 0, \qquad s\psi''(s) < 0 \quad \text{for } s \neq 0$

then one can prove the following result.

Theorem 3.6. *For each $f \in C([0, \pi], \mathbb{R})$, there is a unique solution of Problem* (3.1) *provided ψ satisfies Relation* (3.7).

Proof. We are going to apply the version of the global Implicit Function Theorem given in Theorem 2.3.5. With X, Z, T defined in Equation (3.3), we first obtain a priori bounds on the solutions. For any $l < 1$ and sufficiently close to 1, there are $\xi_1 > 0$, $\xi_2 < 0$ such that $\psi'(\xi_1) = \psi'(\xi_2) = l$. Therefore,

(3.8)
$$\begin{aligned}
\psi(s) &\leq ls + k_1, & s &> 0, & k_1 &= \psi(\xi_1) - \xi_1 l \\
\psi(s) &\geq ls - k_2, & s &< 0, & k_2 &= \xi_2 l - \psi(\xi_2) \\
s\psi(s) &\leq ls^2 + k|s|, & k &= \max(k_1, k_2).
\end{aligned}$$

Furthermore $k(1 - l)^{-1} \to 0$ as $l \to 1$.

For any $f \in Z$ and solution $x \in X$ of $Tx = f$, multiplying by x and integrating from 0 to π, we obtain

(3.9)
$$\int_0^\pi \{x'(t)^2 - \psi(x(t))x(t)\}\, dt = -\int_0^\pi f(t)x(t)\, dt.$$

From Equations (3.8), (3.9),

(3.10)
$$\begin{aligned}
\int_0^\pi (x'^2 - lx^2) &\leq \int_0^\pi [x'^2 + k|x| - \psi(x)x] \\
&= \int_0^\pi (k|x| - fx) \leq \int_0^\pi (k + |f|)|x|.
\end{aligned}$$

Since $x(0) = x(\pi) = 0$, it follows from the Poincaré inequality

$$\int_0^\pi x^2 \leq \int_0^\pi x'^2$$

and Schwarz inequality that

$$(1 - l)|x|_{L^2}^2 = (1 - l)\int_0^\pi x^2 \leq \int_0^\pi (k + |f|)|x| \leq |k + |f|\,|_{L^2}|x|_{L^2}.$$

Therefore,

(3.11)
$$|x|_{L^2} \leq \frac{\pi^{1/2}}{1 - l}(k + |f|_Z).$$

Using Relation (3.10) again, one obtains

$$|x'|_{L^2}^2 \le l|x|_{L^2}^2 + |k + |f||_{L^2}|x|_{L^2}.$$

Therefore, $|x'|_{L^2}$ is bounded from Inequality (3.11) by a quantity depending only on k, l and $|f|_Z$. Using the fact that $x(t) = x(0) + \int_0^t x'$, one thus obtains $|x|_Z$ bounded by a quantity depending only on k, l and $|f|_Z$. Since $x'' = f - \psi(x)$, one obtains the same type of estimate for $|x''|_Z$ and, thus, $|Tx|_X$. This gives the necessary a priori bounds to apply the same argument as in the proof of Theorem 3.3 to show that T is proper.

Our next objective is to show that the set of singular points W of T consists only of the point $\{0\}$ and $T^{-1}(0) = 0$. Suppose $x \in X$, $x \ne 0$, $T'(x)$ does not have a bounded inverse at x. Then there is a $v \in X$, $v \ne 0$, such that

$$v'' + \psi'(x)v = 0.$$

If $s \ne 0$, then $\psi'(s) < 1$ and, thus, $\psi'(x(t))$ is less than 1 on a set of positive measure. Using the theorem of Sturm (Lemma 3.2) to compare this equation with the equation $w'' + w = 0$, $w \in X$, one observes that $v(\pi)$ cannot be zero. Therefore, v cannot be in X and $W = \{0\}$, $T(W) = 0$. To show $T^{-1}(0) = 0$, observe from Inequality (3.11) that $x \in T^{-1}(0)$ implies

$$|x|_{L^2} \le \frac{k\pi^{1/2}}{1 - l}$$

for any $l < 1$ sufficiently close to one. Since $k(1 - l)^{-1} \to 0$ as $l \to 1$, this implies $x = 0$ and $T^{-1}(0) = 0$.

To complete the proof of the theorem, observe that $X - \{0\}$ is connected, $Z - \{0\}$ is simply connected. Thus, the equation $Tx = f \ne 0$, has a unique solution from Theorem 2.3.5. We showed in the previous paragraph that $Tx = 0$ has a unique solution $x = 0$. This completes the proof. $\square$

3.4. Hopf Bifurcation Theorem

In Section 1.1.4, the Implicit Function Theorem was used to prove the Hopf Bifurcation Theorem for $n = 2$. The same ideas yield a similar theorem for arbitrary n. In fact, suppose

$$(4.1) \qquad\qquad \dot{x} = A(\alpha)x + f(\alpha, x)$$

where $x \in \mathbb{R}^n$, $\alpha \in \mathbb{R}$, $A(\alpha)$ is an $n \times n$ matrix $A(\alpha)$, $f(\alpha, x)$ have continuous derivatives up through order one, $D_{\alpha x} f(\alpha, x)$ is continuous for $|\alpha| < \alpha_0$,

$x \in \mathbb{R}^n$, $f(\alpha, 0) = 0$, $D_x f(\alpha, 0) = 0$ for $|\alpha| < \alpha_0$ and

$$
A(\alpha) = \begin{bmatrix} B(\alpha) & 0 \\ 0 & C(\alpha) \end{bmatrix}
$$

(4.2)
$$
B(\alpha) = \begin{bmatrix} \alpha & \beta(\alpha) \\ -\beta(\alpha) & \alpha \end{bmatrix}, \qquad \beta(0) = 1,
$$

$$
[e^{C(\alpha)2\pi} - I]^{-1} \quad \text{exists}
$$

for $|\alpha| < \alpha_0$.

Theorem 4.1. *Under the above hypotheses, there are constants $a_0 > 0$, $\alpha_0 > 0$, $\delta_0 > 0$, functions $\alpha(a) \in \mathbb{R}$, $\omega(a) \in \mathbb{R}$, $\alpha(0) = 0$, $\omega(0) = 2\pi$, and an $\omega(a)$-periodic function $x^*(a)$, with all functions having continuous first derivatives up through order one for $|a| < a_0$, such that $x^*(a)$ is a solution of Equation (4.1) with*

$$
x^*(a)(t) = \begin{bmatrix} a \cos \omega(a)t \\ -a \sin \omega(a)t \\ 0 \\ \vdots \\ 0 \end{bmatrix} + o(|a|)
$$

as $|a| \to 0$. Furthermore, for $|\alpha| < \alpha_0$, $|\omega - 2\pi| < \delta_0$, every ω-periodic solution x of Equation (4.1) with $|x(t)| < \delta_0$ must be given by $x^*(a)$ except for a translation in phase.

Proof. Let $x = (\rho \cos \theta, -\rho \sin \theta, \rho \tilde{x})$ in Equation (4.1) and eliminate t to obtain

$$
\frac{d\rho}{d\theta} = \frac{\alpha}{\beta(\alpha)} \rho + R(\alpha, \theta, \rho, \tilde{x})
$$

$$
\frac{d\tilde{x}}{d\theta} = \frac{1}{\beta(\alpha)} C(\alpha)\tilde{x} + \tilde{X}(\alpha, \theta, \rho, \tilde{x})
$$

where $\tilde{x} = (\tilde{x}_2, \ldots, \tilde{x}_n)$ and the functions $R, \tilde{X}$ are 2π-periodic in θ, vanish at $\rho = 0, \tilde{x}$ and $D_{\rho, \tilde{x}} R, D_{\rho, \tilde{x}} \tilde{X}$ also vanish at $\rho = 0$. The proof in Section 1.1.4 can now be repeated almost verbatum to obtain the conclusion stated in the theorem. $\square$

3.5. Liapunov Center Theorem

In this section, we use the Hopf Bifurcation Theorem to obtain an easy proof of the Liapunov Center Theorem for the special case of Hamiltonian systems.

Suppose $H \in C^2(\mathbb{R}^n \times \mathbb{R}^n, \mathbb{R})$, $H(0,0) = 0$, $DH(0,0) = 0$, and consider the Hamiltonian system

$$\dot{x} = -\frac{\partial H}{\partial y}(x, y)$$

(5.1)

$$\dot{y} = \frac{\partial H}{\partial x}(x, y).$$

If $(x(t), y(t))$ is any solution of Equation (5.1), then

(5.2)
$$H(x(t), y(t)) = H(x(0), y(0))$$

for all t. If $n = 1$ and

$$H(x, y) = \frac{\lambda}{2}(x^2 + y^2) + o((|x| + |y|)^2), \qquad \lambda > 0$$

as $(x, y) \to (0,0)$, then every solution in a neighborhood of $(x, y) = (0,0)$ is periodic. This is the Liapunov Center Theorem for $n = 1$.

If $n \geq 1$, suppose

(5.3)
$$H(x, y) = \sum_{j=1}^{n} \frac{\lambda_j}{2}(x_j^2 + y_j^2) + o((|x| + |y|)^2)$$

as $|x|, |y| \to 0$, each $\lambda_j \neq 0$, $j = 1, 2, \ldots, n$, and satisfies the nonresonance condition

(5.4)
$$\frac{\lambda_j}{\lambda_1} \neq \text{an integer}, \qquad j \geq 2.$$

Theorem 5.1. *If H satisfies Relations* (5.3), (5.4), *then there exists a one-parameter family of periodic orbits of Equation* (5.1) *containing the origin. More precisely, there are constants $a_0 > 0$, $\delta_0 > 0$, a function $\omega(a) \in \mathbb{R}$, $\omega(0) = 2\pi/|\lambda_1|^{1/2}$, and an $\omega(a)$-periodic function $(x^*(a), y^*(a))$ with all functions having continuous derivatives up through order one for $|a| < a_0$ such that $(x^*(a), y^*(a))$ is a solution of* (5.1) *with*

$$x_1^*(a)(t) = a \cos|\lambda_1|t + o(|a|^2), \qquad x_j^*(a)(t) = o(|a|), \qquad j \geq 2,$$
$$y_1^*(a)(t) = a \sin|\lambda_1|t + o(|a|^2), \qquad y_j^*(a)(t) = o(|a|^2), \qquad j \geq 2$$

as $|a| \to 0$. Furthermore, for $|\omega - 2\pi|/|\lambda_1|^{1/2} < \delta_0$, every ω-periodic solution (x, y) of Equation (5.1) *with $|(x(t), y(t))| < \delta_0$ must be given by $(x^*(a), y^*(a))$ except for a translation in phase.*

Proof. By the introduction of an artificial parameter, we convert the problem to one in which a "vertical" Hopf bifurcation exists.

For any $\alpha \in \mathbb{R}$, consider the equation

$$\dot{x} = -\frac{\partial H}{\partial y}(x, y) + \alpha \frac{\partial H}{\partial x}(x, y)$$

(5.5)

$$\dot{y} = \frac{\partial H}{\partial x}(x, y) + \alpha \frac{\partial H}{\partial y}(x, y).$$

If $(x(t), y(t))$ is a solution of Equation (5.5), then

$$\frac{d}{dt} H(x(t), y(t)) = \alpha |\text{grad } H|^2.$$

If $\alpha \neq 0$, there can be no periodic solution of (5.5) except $(x, y) = (0,0)$. In fact, if $\alpha < 0$ and a solution $(x(t), y(t))$ of Equation (5.5) remains in a bounded neighborhood V of $(0,0)$ for all t, then necessarily one must have $H(x(t), y(t)) = $ constant for all t. Thus, $(x(t), y(t))$ satisfies Equation (5.1). If $\alpha > 0$, then one replaces t by $-t$ and uses the same argument.

Thus, every periodic solution near $x = 0$ of Equation (5.5) must satisfy Equation (5.1). If we show that Equation (5.5) satisfies the conditions of the Hopf Bifurcation Theorem with respect to the parameter α, then the proof of Theorem 5.1 will be complete.

We must investigate some detailed properties of the eigenvalues of the linear variational equation of Equation (5.5) about $x = 0$. If $(x, y) = w$ and

$$J = \begin{bmatrix} 0 & I \\ -I & 0 \end{bmatrix}$$

and A is the Hessian matrix of H, then the linear variational equation has the form

(5.6)
$$\dot{w} = -JAw + \alpha Aw.$$

The function $w \cdot Aw$ is a first integral of this equation for $\alpha = 0$. If P is a real orthogonal matrix, the change of variables $w \mapsto Pw$ yields a new equation of the form

(5.7)
$$\dot{w} = -Bw + \alpha Cw$$

where $C = P^{-1}AP$ and $w \cdot Cw$ is a first integral of (5.7) for $\alpha = 0$. Thus, $B^T C + CB = 0$. One can always find a real orthogonal matrix P such that

$$B = \begin{bmatrix} 0 & \lambda_1 & 0 \\ -\lambda_1 & 0 & 0 \\ 0 & 0 & \tilde{B} \end{bmatrix}.$$

Since $B^T C + CB = 0$, it follows that

$$C = \begin{bmatrix} c & 0 & 0 \\ 0 & c & 0 \\ 0 & 0 & \tilde{C} \end{bmatrix}, \qquad c \neq 0.$$

This implies there are eigenvalues $\alpha c \pm i\lambda_1$ of (5.6) and the conditions of the Hopf Bifurcation Theorem are satisfied. This proves the theorem. $\quad\square$

3.6. Stable and Unstable Manifolds

In this section, we discuss the stable and unstable sets for a hyperbolic equilibrium point of a differential equation and prove that they are submanifolds with the same smoothness properties as the vector field.

Consider the differential equation

$$(6.1) \qquad \dot{x} = Ax + f(x),$$

where A is hyperbolic (that is, $\operatorname{Re} \sigma(A) \neq 0$) and $f : \mathbb{R}^d \to \mathbb{R}^d$ is a Lipschitz continuous function satisfying

$$(6.2) \qquad \begin{aligned} f(0) &= 0 \\ |f(x) - f(y)| &\leq \eta(\delta)|x - y| \qquad \text{if } |x|, |y| \leq \delta, \end{aligned}$$

where $\eta : [0, \infty) \to [0, \infty)$ is a continuous function with $\eta(0) = 0$.

For any $x_0 \in \mathbb{R}^d$, let $\varphi_t(x_0)$ be the solution of (6.1) through x_0. The *local unstable set* $W^u_{\text{loc}}(0)$ and the *local stable set* $W^s_{\text{loc}}(0)$ of 0 corresponding to a neighborhood U of 0 are defined by

$$\begin{aligned} W^u_{\text{loc}}(0) \equiv W^u(0, U) &= \{x_0 \in \mathbb{R}^d : \varphi_t(x_0) \text{ is defined for } t \leq 0, \\ &\qquad \varphi_t(x_0) \in U, t \leq 0, \text{ and } \varphi_t(x_0) \to 0 \text{ as } t \to -\infty\}, \\ W^s_{\text{loc}}(0) \equiv W^s(0, U) &= \{x_0 \in \mathbb{R}^d : \varphi_t(x_0) \text{ is defined for } t \geq 0, \\ &\qquad \varphi_t(x_0) \in U, t \geq 0, \text{ and } \varphi_t(x_0) \to 0 \text{ as } t \to \infty\}. \end{aligned}$$

We are going to show that these local sets are submanifolds of $\mathbb{R}^d$ in a sufficiently small neighborhood U of 0. Furthermore, if f is a C^r-function (or analytic), then so are these submanifolds.

Since the eigenvalues of A have nonzero real parts, there is a continuous projection operator P on $\mathbb{R}^d$ such that $P\mathbb{R}^d$ and $Q\mathbb{R}^d$, $Q = I - P$, are invariant under A, each element of the spectrum $\sigma(AP)$ of AP has positive real part and each element of the spectrum $\sigma(AQ)$ of AQ has negative real part. Therefore, there are positive constants k, α such that

$$
(6.3) \qquad
\begin{aligned}
|e^{At}P| &\le k e^{\alpha t}, && t \le 0, \\
|e^{At}Q| &\le k e^{-\alpha t}, && t \ge 0.
\end{aligned}
$$

A basic lemma is the following:

Lemma 6.1. *If $\varphi_t(x_0)$, $t \le 0$, is a bounded solution of (6.1), then $\varphi_t(x_0)$ must satisfy the integral equation*

$$
(6.4) \qquad y(t) = e^{At}Px_0 + \int_0^t e^{A(t-s)}Pf(y(s))\,ds + \int_{-\infty}^t e^{A(t-s)}Qf(y(s))\,ds.
$$

If $\varphi_t(x_0)$, $t \ge 0$, is a bounded solution of (6.1), then $\varphi_t(x_0)$ must satisfy the integral equation

$$
(6.5) \qquad y(t) = e^{At}Qx_0 + \int_0^t e^{A(t-s)}Qf(y(s))\,ds - \int_t^\infty e^{A(t-s)}Pf(y(s))\,ds.
$$

Conversely, if $y(t)$, $t \le 0$, (or $t \ge 0$) is a bounded solution of (6.4) or (6.5), then $y(t)$ satisfies (6.1).

Proof. Let $y(t) = \varphi_t(x_0)$, $t \le 0$, be a bounded solution of (6.1). Then, for any $\tau \in (-\infty, 0]$,

$$
Qy(t) = e^{A(t-\tau)}Qy(\tau) + \int_\tau^t e^{A(t-s)}Qf(y(s))\,ds.
$$

If we use (6.3), the fact that $y(s)$ is bounded in s for $s \le 0$, and let $\tau \to -\infty$, we obtain

$$
Qy(t) = \int_{-\infty}^t e^{A(t-s)}Qf(y(s))\,ds.
$$

Since

$$
Py(t) = e^{At}Px_0 + \int_0^t e^{A(t-s)}Pf(y(s))\,ds,
$$

we see that $y(t)$ satisfies (6.4). The proof for the case when $\varphi_t(x_0)$, $t \ge 0$, is bounded is similar and therefore omitted. The converse statement is proved by direct computation and the lemma is proved. $\square$

We say that $W^u(0, U)$ is a *Lipschitz graph over* $P\mathbb{R}^d$ if there is a neighborhood V of 0 in $P\mathbb{R}^d$ such that $W^u(0, U) = \{(u, v) \in \mathbb{R}^d : v = g(u),\ u \in V,\ \text{where}$

$g : V \to Q\mathbb{R}^d$ is Lipschitz continuous}. The set $W^u(0, U)$ is said to be *tangent to* $P\mathbb{R}^d$ at 0 if $|Qx|/|Px| \to 0$ as $x \to 0$ in $W^u(0, U)$. Similar definitions hold for $W^s(0, U)$.

A classical theorem on local unstable and stable sets which can be traced initially to the work of Poincaré and Liapunov is contained in the following result.

Theorem 6.1. *If f satisfies (6.2) and $\mathrm{Re}\,\sigma(A) \neq 0$, then there is a neighborhood U of 0 in $\mathbb{R}^d$ such that $W^u(0, U)$ (or $W^s(0, U)$) is a Lipschitz graph over $P\mathbb{R}^d$ (or $Q\mathbb{R}^d$) which is tangent to $P\mathbb{R}^d$ (or $Q\mathbb{R}^d$) at 0. Furthermore, if f is a C^k-function (or an analytic function) in a neighborhood of 0, then so are $W^u(0, U)$ and $W^s(0, U)$.*

Proof. The proof is a standard application of the contraction mapping principle (for example, see Hale [1969] or [1980]) and gives exponential decay rates of the solutions to the origin. With the function η as in (6.2) and k, α as in (6.3), choose $\delta > 0$ so that $4k\eta(\delta) < \alpha$, $8k^2\eta(\delta) < \alpha$. Let $S(\delta)$ be the set of continuous functions $x : (-\infty, 0] \to \mathbb{R}^d$ such that $|x| = \sup_{-\infty < t \leq 0} |x(t)| \leq \delta$. The set $S(\delta)$ is a complete metric space with the metric induced by the uniform topology. For any $y \in S(\delta)$ and any $x_0 \in P\mathbb{R}^d$ with $|x_0| \leq \delta/2k$, we define, for $t \leq 0$,

$$(6.6) \quad (T(y, x_0))(t) = e^{At}x_0 + \int_0^t e^{A(t-s)}Pf(y(s))\,ds + \int_{-\infty}^t e^{A(t-s)}Qf(y(s))\,ds.$$

It is easy to show that $T : S(\delta) \times \{x_0 \in P\mathbb{R}^d : |x_0| \leq \delta/2k\} \to S(\delta)$ and $T(\cdot, x_0)$ is a contraction mapping with contraction constant $1/2$. Therefore, $T(\cdot, x_0)$ has a unique fixed point $x^*(\cdot, x_0)$ in $S(\delta)$. Since the contraction constant is independent of x_0, $x^*(\cdot, x_0)$ is Lipschitz continuous in x_0. Also, since $T(y, x_0)$ is an analytic function of x_0, and is as smooth in y as f, it follows that $x^*(x_0)$ is as smooth in x_0 as f.   □

We next obtain exponential decay rates of the solutions $x^*(\cdot, x_0)$. To do this, we use the following lemma whose proof is postponed until later.

Lemma 6.2. *Suppose that $\alpha > 0$, $\gamma > 0$, K, L, M, are nonnegative constants and u is a nonnegative bounded continuous function satisfying one of the inequalities*

$$(6.7) \quad u(t) \leq Ke^{-\alpha t} + L \int_0^t e^{-\alpha(t-s)}u(s)\,ds + M \int_0^\infty e^{-\gamma s}u(t+s)\,ds, \qquad t \geq 0,$$

$$(6.8) \quad u(t) \leq Ke^{\alpha t} + L \int_0^t e^{\alpha(t-s)}u(s)\,ds + M \int_{-\infty}^0 e^{\gamma s}u(t+s)\,ds, \qquad t \leq 0.$$

If $\beta \equiv L/\alpha + M/\gamma < 1$, then, in either case,

$$u(t) \leq (1 - \beta)^{-1}Ke^{-[\alpha - (1-\beta)^{-1}L]|t|}.$$

Since $x^*(\cdot, x_0)$ is a fixed point of T in $S(x_0, \delta)$, we have the following estimate:

$$|x^*(t, x_0)| \leq k e^{\alpha t}|x_0| + k\eta(\delta)\left[\int_0^t e^{\alpha(t-s)}|x^*(s, x_0)|\,ds\right.$$

$$\left. + \int_{-\infty}^t e^{-\alpha(t-s)}|x^*(s, x_0)|\,ds\right].$$

If we now apply Lemma 6.2 to this inequality, we obtain

$$(6.9) \qquad |x^*(\cdot, x_0)| \leq 2k e^{\alpha t/2}|x_0|, \qquad t \leq 0.$$

This estimate shows that $x^*(\cdot, 0) = 0$ and $g(x_0) \equiv x^*(0, x_0) \in W^u(0)$. Similar to (6.9), we have for $t \leq 0$,

$$|x^*(s, x_0) - x^*(s, \bar{x}_0)| \leq 2k e^{\alpha s/2}|x_0 - \bar{x}_0|$$

and hence

$$|g(x_0) - g(\bar{x}_0)| \geq |x_0 - \bar{x}_0| - \int_{-\infty}^0 k\eta(\delta)e^{\alpha s}|x^*(s, x_0) - x^*(s, \bar{x}_0)|\,ds$$

$$\geq |x_0 - \bar{x}_0|\left[1 - \frac{4k^2\eta(\delta)}{3\alpha}\right] \geq \frac{1}{2}|x_0 - \bar{x}_0|.$$

Thus, the mapping $x_0 \mapsto g(x_0)$ is one-to-one with a continuous inverse. Let $W = \{x^*(t, x_0) : x_0 \in P\mathbb{R}^d, |x_0| \leq \delta/2k\}$. Then there is a neighborhood U of 0 such that $W = W^u(0, U)$.

It remains to show that $W^u(0, U)$ is tangent to $P\mathbb{R}^d$ at 0. From (6.6), the definition of $g(x)$ and (6.9), we have

$$|Qg(x_0)| = \left|\int_{-\infty}^0 e^{-As}Qf(x^*(s, x_0))\,ds\right|$$

$$\leq k\int_{-\infty}^0 e^{\alpha s}\eta(|x^*(s, x_0)|)|x^*(s, x_0|\,ds$$

$$\leq k\int_{-\infty}^0 e^{\alpha s}\eta(2k e^{\alpha s}|x_0|)2k e^{\alpha s/2}|x_0|\,ds$$

$$\leq 2k^2\eta(2k|x_0|)|x_0|\int_{-\infty}^0 e^{-(1/2)\alpha s/2}\,ds$$

$$= \frac{4k^2}{\alpha}\eta(2k|x_0|)|x_0|.$$

Therefore, $|Qg(x_0)|/|Px_0| = (4k^2/\alpha)\eta(2k|x_0|) \to 0$ as $|x_0| \to 0$.

The regularity properties in the last statement of the theorem have been proved above and the proof is complete. $\square$

Corollary 6.1. *If the hypotheses of Theorem 6.1 are satisfied and if $x \in U \setminus (W^u(0, U) \cup W^s(0, U))$, then there exist $t_1 > 0, t_2 > 0$ such that $\varphi_{-t_1}(x), \varphi_{t_2}(x) \in \partial U$.*

Proof. From the manner in which the manifolds $W^u(0, U)$, $W^s(0, U)$ were constructed, Lemma 6.1 yields the assertion of the lemma.

Proof of Lemma 6.2. We only need to prove the lemma for u satisfying the Inequality (6.7) since the transformation $t \mapsto -t, s \mapsto -s$ reduces the discussion of (6.8) to (6.7). We show first that $u(t) \to 0$ as $t \to \infty$. If $\delta = \limsup_{t \to \infty} u(t)$, then u bounded implies $\delta < \infty$. If $\beta < \theta < 1$, then $\delta > 0$ implies that there is a $t_1 \geq 0$ such that $u(t) \leq \theta^{-1}\delta$ for $t \geq t_1$. From (6.7), we have

$$u(t) \leq Ke^{-\alpha t} + Le^{-\alpha t} \int_0^{t_1} e^{\alpha s} u(s)\, ds + \left(\frac{L}{\alpha} + \frac{M}{\gamma} \right) \theta^{-1}\delta.$$

Therefore, $\limsup_{t \to \infty} u(t) \leq \beta \theta^{-1}\delta < \delta$, which is a contradiction. Thus, $\delta = 0$ and $u(t) \to 0$ as $t \to \infty$.

If $v(t) = \sup_{s \geq t} u(s)$, then $u(t) \to 0$ as $t \to \infty$ implies that, for any $t \in [0, \infty)$, there is a $t_1 \geq t$ such that $v(t) = v(s) = u(t_1)$ for $t \leq s \leq t_1$, $v(s) < v(t_1)$ for $s > t_1$. From (6.7), we deduce that

$$v(t) = u(t_1) \leq Ke^{-\alpha t_1} + L \int_0^{t_1} e^{-\alpha(t_1 - s)} u(s)\, ds + M \int_0^{\infty} e^{-\gamma s} u(t_1 + s)\, ds$$

$$\leq Ke^{-\alpha t_1} + L \left(\int_0^t + \int_t^{t_1} \right) e^{-\alpha(t_1 - s)} v(s)\, ds + M \int_0^{\infty} e^{-\gamma s} v(t + s)\, ds$$

$$\leq Ke^{-\alpha t_1} + L \int_0^t e^{-\alpha(t - s)} v(s)\, ds + \beta v(t).$$

If we let $z(t) = e^{\alpha t} v(t)$, then $t_1 \geq t$ implies that

$$z(t) \leq (1 - \beta)^{-1} K + (1 - \beta)^{-1} L \int_0^t z(s)\, ds.$$

An application of Gronwall's inequality completes the proof of the lemma. $\square$

For any $x_0 \in \mathbb{R}^d$, let $\varphi_t(x_0)$ be the solution of (6.1) through x_0. The *unstable set* $W^u(0)$ and the *stable set* $W^s(0)$ of 0 are defined as

$$W^u(0) = \{x_0 \in \mathbb{R}^d : \varphi_t(x_0) \text{ is defined for } t \leq 0$$
$$\text{and } \varphi_t(x_0) \to 0 \text{ as } t \to -\infty\},$$

$$W^s(0) = \{x_0 \in \mathbb{R}^d : \varphi_t(x_0) \text{ is defined for } t \geq 0$$
$$\text{and } \varphi_t(x_0) \to 0 \text{ as } t \to \infty\}.$$

We can now state the following result:

Theorem 6.2. *If* $\operatorname{Re} \sigma(A) \neq 0$ *and if the vector field* f *in* (6.1) *is* C^k *(or analytic), then* $W^s(0)$, $W^u(0)$ *are embedded* C^k-*manifolds (or analytic manifolds).*

Proof. This follows immediately from the relations

$$W^s(0) = \bigcup_{t \geq 0} \varphi_{-t}(W^s_{\text{loc}}(0)),$$
$$W^u(0) = \bigcup_{t \geq 0} \varphi_t(W^u_{\text{loc}}(0)).$$

Our next results are concerned with foliations over the stable and unstable manifolds.

Theorem 6.3. *There is a positive constant* β *and a neighborhood* V *of* 0 *in* $Q\mathbb{R}^d$ *such that, for each* $\xi \in W^u(0, U)$, *there is a function* $h_\xi : V \to \mathbb{R}^d$, $h_\xi(\eta)$ *is continuous in* ξ, η, *together with all derivatives in* η *up through order* k, *and a manifold* $M^s_{\text{loc}}(\xi) = \{\bar{\xi} \in \mathbb{R}^d : \bar{\xi} = h_\xi(\eta), \eta \in V \subset Q\mathbb{R}^d\}$ *which is a graph over* V, *such that the following properties are satisfied for* $\bar{\xi} \in M^s_{\text{loc}}(\xi)$:

$$\tag{6.10}
\begin{aligned}
x(t, \bar{\xi}) &= h_{x_u(t, \xi)}(Qx(t, \bar{\xi})), \qquad t \geq 0, \\
\sup_{t \geq 0} &|x(t, \bar{\xi}) - x_u(t, \xi)| e^{\beta t} < \infty,
\end{aligned}$$

as long as the solution remains in a small neighborhood U *of the origin.*

The manifolds $M^s_{\text{loc}}(\xi)$, $\xi \in W^u(0, U)$ are a foliation of a neighborhood of 0 in $\mathbb{R}^d$ with the leaves being $M^s_{\text{loc}}(\xi)$. These leaves have the property that if we begin a solution on $M^s_{\text{loc}}(\xi)$, then, after time t, the solution is on the leaf $M^s_{\text{loc}}(x_u(t, \xi))$. Also, the orbit of the solution $x(t, \bar{\xi})$ approaches the unstable manifold exponentially as $t \to \infty$. Since $x_u(t, \xi)$ represents the flow on the unstable manifold, the second relation in (6.10) implies that the solution $x(t, \bar{\xi})$ approaches a trace $x_u(t, \xi)$ on the unstable manifold exponentially as $t \to \infty$. Finally, we note that $M^s_{\text{loc}}(\xi)|_{\xi=0}$ is the local stable manifold at the origin. Using this fact, together with $h_\xi(\eta)$ being a C^k-function in η and continuous in ξ, we see that the leaves $M^s_{\text{loc}}(\xi)$ converge to $M^s_{\text{loc}}(0)$ in the C^k-sense as $\xi \to 0$.

An interesting consequence of Theorem 6.3 is a version of the λ-*Lemma*. To make this precise, fix $\delta > 0$ and define the following sets:

$$\tag{6.11}
\tilde{M}^s_{\text{loc}}(t) = M^s_{\text{loc}}(x_u(-t, \xi)) \cap \{x : |Qx| \leq \delta\}, \qquad t \geq 0.$$

Corollary 6.2 (λ-*Lemma*). *If the conditions of Theorem 6.3 are satisfied and* $\tilde{M}^s_{\text{loc}}(t)$ *is defined by* (6.11), *then*

$$\text{dist}_{C^k}(\tilde{M}^s_{\text{loc}}(t), W^s(0)) \le e^{-\beta t}, \qquad t \ge 0,$$

where dist_{C^k} denotes the C^k distance between two sets.

We now turn to the proof of Theorem 6.3. For notation, we let $B_{Q\mathbb{R}^d}(\delta) = \{x \in Q\mathbb{R}^d : |x| \le \delta\}$. Suppose $\delta_1 > 0$ is given and consider the neighborhood $U = \{x : |x| \le \delta_1\}$ of the origin in $\mathbb{R}^d$. The first step in the proof of the theorem is to replace the vector field in (6.1) by a new vector field $\tilde{f}$ with the property that it coincides with f in U and is bounded in the set $U_{\delta_1} \equiv P\mathbb{R}^d \times B_{Q\mathbb{R}^d}(\delta_1)$. Such an extension can be made in the following way. Let $\xi : P\mathbb{R}^d \to [0, 1]$ be a C^∞ function with

$$\xi(x) = \begin{cases} 1 & |x| \le 1 \\ 0 & |x| \ge 2 \end{cases}$$

and define

$$\tilde{f}(x) = f\left(Px\xi\left(\frac{Px}{\delta}\right) + Qx\right).$$

We will assume that such an extension has been made, that we are working on the region U_{δ_1} and we drop the tilde on f. The next step is to derive an integral equation for the function $h_\xi(\eta)$ in the statement of the theorem, assuming that such a function exists. The variation of constants formula implies that, for any $\tau \ge 0$,

$$x(t, \bar{\xi}) - x_u(t, \xi) = e^{A(t-\tau)}[x(\tau, \bar{\xi}) - x_u(\tau, \xi)]$$

$$+ \int_\tau^t e^{A(t-s)}[f(x(s, \bar{\xi})) - f(x_u(s, \xi))] \, ds$$

Since $x(t, \bar{\xi}) - x_u(t, \xi)$ is bounded for $t \ge 0$ and $\text{Re}\,\sigma(AP) > 0$, we have $e^{-At}P[x(t, \bar{\xi}) - x_u(t, \xi)] \to 0$ as $t \to \infty$. As a consequence, we have

$$P[x(\tau, \bar{\xi}) - x_u(\tau, \xi)] = -\int_\tau^\infty e^{A(\tau-s)}P[f(x(s, \bar{\xi})) - f(x_u(s, \xi))] \, ds.$$

Therefore, if we let $y(t, \xi) = x(t, \bar{\xi}) - x_u(t, \xi)$, then $y(t, \xi)$ must satisfy the equation

$$y(t, \xi) = e^{At}Qy(0) + \int_0^t e^{A(t-s)}Q[f(x_u(s, \xi) + y(s, \xi)) - f(x_u(s, \xi))] \, ds$$

$$- \int_t^\infty e^{A(t-s)}P[f(x_u(s, \xi) + y(s, \xi)) - f(x_u(s, \xi))] \, ds.$$

These observations lead us to formulate the problem of the existence of the manifolds $M_{\text{loc}}^s(\xi)$ as the fixed point of an operator.

Let C_b^0 be the space of bounded continuous functions from $[0, \infty)$ to $\mathbb{R}^d$. We let $|y|_0 = \sup_{t \geq 0} |y(t)|$ for $y \in C_b^0$. For a given positive constant μ, let $S(\mu) = \{ y \in C_b^0 : |y|_0 \leq \mu \}$.

For a fixed $\xi \in W^u(0, U)$, we consider the map

$$T(\xi, \cdot, \cdot) : S(\mu) \times B_{Q\mathbb{R}^d}(\delta_1) \to C([0, \infty); \mathbb{R}^d)$$

by the relation

$$(6.12) \quad (T(\xi, y, \eta))(t) = e^{At}\eta$$

$$+ \int_0^t e^{A(t-s)} Q[f(x_u(s, \xi) + y(s, \xi)) - f(x_u(s, \xi))] \, ds$$

$$- \int_t^\infty e^{A(t-s)} P[f(x_u(s, \xi) + y(s, \xi)) - f(x_u(s, \xi))] \, ds.$$

We proceed now exactly as in the proof of the existence of stable and unstable manifolds in the previous section to obtain the following estimates:

$$|T(\xi, y, \eta)(t)| \leq k\delta_1 + \frac{2k}{\alpha}|f|_1 \mu$$

(6.13)

$$|T(\xi, y, \eta)(t) - T(\xi, \bar{y}, \eta)(t)| \leq \frac{2k}{\alpha}|f|_1 |y - \bar{y}|_0$$

for every $y, \bar{y} \in S(\mu)$ and all $t \geq 0$, where $|f|_1$ is the C^1-norm of the function f on U_{δ_1}. If we choose the original neighborhood U of the origin sufficiently small and choose δ_1 sufficiently small, then we can be sure that

$$k\delta_1 + \frac{2k}{\alpha}|f|_1 \mu < \mu, \qquad \frac{2k}{\alpha}|f|_1 < \frac{1}{2}.$$

With this choice, we observe from (6.13) that $T(\xi, \cdot, \eta) : S(\mu) \to S(\mu)$ and is a contraction uniform with respect to η. Therefore, there is a unique fixed point $y^*(\xi, \eta)$ of the map $T(\xi, \cdot, \eta)$ in $S(\mu)$. Since $T(\xi, y, \eta)$ is analytic in η and C^k with respect to y, it follows that $y^*(\xi, \eta)$ is C^k in η.

If we define $h_\xi(\eta) = y^*(\xi, \eta)(0)$ and $) = \xi + y$

$$M^s(\xi) = \{ \bar{\xi} \in \mathbb{R}^d : \bar{\xi} = h_\xi(\eta), \eta \in B_{Q\mathbb{R}^d}(\delta_1) \},$$

then we have the first relationship in (6.10) from the manner in which the map $T(\xi, y, \eta)$ was constructed.

For the exponential decay rates, one proceeds exactly as in the proof for the stable and unstable manifold in the previous section.

It remains to show that the function $y^*(\cdot,\cdot)$ is continuous; that is, the following map is continuous:

$$y^*: \mathbb{R}^d \times P\mathbb{R}^d \to C([0,\infty); \mathbb{R}^d).$$

Let $\sigma > 0$ and

$$C_\sigma = C_\sigma([0,\infty); \mathbb{R}^d) = \left\{ z \in C([0,\infty); \mathbb{R}^d): |z|_\sigma = \sup_{t \geq 0} e^{\sigma t}|z(t)| < \infty \right\}.$$

It is not hard to see that if $1 \gg \sigma > 0$, then the map $T(\xi, y, \eta)$ satisfies

$$T: \mathbb{R}^d \times C_\sigma([0,\infty); \mathbb{R}^d) \times P\mathbb{R}^d \to C_\sigma([0,\infty); \mathbb{R}^d).$$

Furthermore, T is a contraction in $C_\sigma([0,\infty); \mathbb{R}^d)$ and has a unique fixed point $y_\sigma^*(\xi, \eta)$. Since $C_\sigma \subset C([0,\infty); \mathbb{R}^d)$, we have that $y_\sigma^*(\xi, \eta) = y^*(\xi, \eta)$.

Next, let $(\xi, \eta) \in \mathbb{R}^d \times P\mathbb{R}^d$ and $(\hat{\xi}, \hat{\eta}) \in \mathbb{R}^d \times P\mathbb{R}^d$ be arbitrary but fixed. Since $C_\sigma \subset C([0,\infty); \mathbb{R}^d)$, we may consider $T(\xi, y, \eta)$ as a map with range $C([0,\infty); \mathbb{R}^d)$

$$T: \mathbb{R}^d \times C_\sigma([0,\infty); \mathbb{R}^d) \times P\mathbb{R}^d \to C([0,\infty); \mathbb{R}^d).$$

Hence, if

$$\Delta T = |y^*(\xi, \eta) - y^*(\hat{\xi}, \hat{\eta})|_0,$$

then

$$\Delta T = |T(\xi, y_\sigma^*(\xi, \eta), \eta) - T(\hat{\xi}, y_\sigma^*(\hat{\xi}, \hat{\eta}), \hat{\eta})|_0.$$

We need to show that, for every $\varepsilon > 0$, there exists $\mu > 0$ such that if $|(\xi, \eta) - (\hat{\xi}, \hat{\eta})| < \mu$, then $\Delta T < \varepsilon$. Now, let $\varepsilon > 0$ be given and write

$$\Delta T = |N_1 + N_2 + N_3 - \hat{N}_1 - \hat{N}_2 - \hat{N}_3|_0$$

where

$$N_1 = e^{At}\eta,$$

$$N_2 = \int_0^t e^{A(t-s)}Q[f(x_u(s, \xi) + y_\sigma^*(\xi, \eta)(s)) - f(x_u(s, \xi))]\,ds,$$

$$N_3 = \int_t^\infty e^{A(t-s)}P[f(x_u(s, \xi) + y_\sigma^*(\xi, \eta)(s)) - f(x_u(s, \xi))]\,ds,$$

and

$$\hat{N}_1 = e^{At}\hat{\eta},$$

$$\hat{N}_2 = \int_0^t e^{A(t-s)}Q[f(x_u(s,\hat{\xi}) + y_\sigma^*(\hat{\xi},\hat{\eta})(s)) - f(x_u(s,\hat{\xi}))]\,ds,$$

$$\hat{N}_3 = \int_t^\infty e^{A(t-s)}P[f(x_u(s,\hat{\xi}) + y_\sigma^*(\hat{\xi},\hat{\eta})(s)) - f(x_u(s,\hat{\xi}))]\,ds.$$

It is easy to see that if $|\eta - \hat{\eta}| \ll 1$, then $|N_1 - \hat{N}_1|_0 < \varepsilon/3$. Next, we estimate $N_2 - \hat{N}_2$. For this purpose, it is convenient to write

$$N_2 - \hat{N}_2 = \int_0^t e^{A(t-s)}Qf^*$$

where f^* is the appropriate integrand. Let $R > 0$ be fixed. By continuous dependence on initial data, $|f^*| < \varepsilon_1$ for $0 \le t \le R$ provided that $|(\xi,\eta) - (\hat{\xi},\hat{\eta})| < \mu$, where ε_1 and μ depend only ε and R.

If $0 \le t \le R$, then $|N_2 - \hat{N}_2|_0 < \varepsilon/6$ for appropriate ε_1. If $t > R$, then

$$N_2 - \hat{N}_2 = \int_0^R e^{A(t-s)}Qf^*\,ds + \int_R^t e^{A(t-s)}Qf^*\,ds.$$

As we have just shown, $|\int_0^R e^{A(t-s)}Qf^*| < \varepsilon/6$ for an appropriate ε_1. Since f is Lipschitz, $x_u(\cdot,\xi)$, $x_u(\cdot,\hat{\xi}) \in W^u(0,U)$ and $y_\sigma^*(\xi,\eta)$, $y_\sigma^*(\hat{\xi},\hat{\eta}) \in C_\sigma([0,\infty);\mathbb{R}^d)$, there exists a constant $C > 0$ such that

$$\left| \int_R^t e^{A(t-s)}Qf^*\,ds \right| < C \int_R^t e^{-\alpha(t-s)}e^{-\sigma s}\,ds.$$

This implies that we can choose $\mu = \mu(\varepsilon) > 0$ and $R = R(\varepsilon) > 0$ so that if $|(\xi,\eta) - (\hat{\xi},\hat{\eta})| < \mu$, then $|N_2 - \hat{N}_2|_0 < \varepsilon/3$. A similar estimate gives $|N_3 - \hat{N}_3|_0 < \varepsilon/3$.

This proves the continuity of $M^s(\xi)$ in ξ.

The same type of argument as used above will yield the following analogue of Theorem 6.3 and Corollary 6.2 for an unstable foliation over $W^s(0,U)$. For $\eta \in W^s_{loc}(0)$, we let $x_s(t,\eta)$ be the solution of (6.1) through η.

Theorem 6.4. *For each $\eta \in W^s(0,U)$, there is a positive constant β, a neighborhood V of 0 in $P\mathbb{R}^d$, a function $h_\eta: V \to \mathbb{R}^d$, $h_\eta(\xi)$ is continuous in η, ξ, together with all derivatives in ξ up through order k, and a manifold $M^u_{loc}(\eta) = \{\bar{\eta} \in \mathbb{R}^d: \bar{\eta} = h_\eta(\xi), \xi \in V \subset P\mathbb{R}^d\}$ which is a graph over V, such that the following properties are satisfied for $\bar{\eta} \in M^u_{loc}(\eta)$:*

(6.14)
$$x(t, \bar{\eta}) = h_{x_s(t, \eta)}(Px(t, \bar{\eta})), \qquad t \leq 0,$$
$$\sup_{t \leq 0} |x(t, \bar{\eta}) - x_s(t, \eta)| e^{\beta t} < \infty,$$

as long as the solution remains in a small neighborhood U of the origin.

Fix $\delta > 0$ and define the following sets:

(6.15)
$$\tilde{M}^u_{\text{loc}}(t) = M^u_{\text{loc}}(x_s(t, \xi)) \cap \{x : |Px| \leq \delta\}, \qquad t \geq 0.$$

Corollary 6.3. *(λ-Lemma) If the conditions of Theorem 6.4 are satisfied and $\tilde{M}^u_{\text{loc}}(t)$ is defined by (6.15), then*

$$\text{dist}_{C^k}(\tilde{M}^u_{\text{loc}}(t), W^u(0)) \leq e^{-\beta t}, \qquad t \geq 0.$$

Similar results hold for maps $L + f$, where L is linear, hyperbolic and f satisfies (6.2). The stable and unstable sets are defined in exactly the same way using the iterates of the map.

Theorem 6.5. *If L is a linear map such that $\sigma(L) \cap \{z \in \mathbb{C} : |z| = 1\} = \varnothing$ and f is a C^k-map, $k \geq 1$, (or an analytic map) satisfying (6.2), then $W^u(0, U)$ and $W^s(0, U)$ are embedded C^k-manifolds (or analytic manifolds) which are tangent at the origin to the respective unstable and stable manifolds for the linear map L.*

The proof is essentially the same as the one for the continuous flow. The only essential difference is that the integrals in Lemma 6.1 are replaced by sums. The details are left to the reader.

The foliation theorems also hold for maps with obvious statements and obvious modifications in the proofs.

3.7. The Hartman–Grobman Theorem

As an application of the contraction mapping theorem, we will give a proof of the theorem of the Hartman–Grobman Theorem on the linearization of maps near a hyperbolic fixed point.

We say that an $d \times d$ matrix L is *hyperbolic* if it is nonsingular and $\sigma(L) \cap S^1 = \varnothing$.

Recall that $C^j = C^j(\mathbb{R}^d; \mathbb{R}^d)$ is the space of functions from $\mathbb{R}^d$ to $\mathbb{R}^d$ which are continuous together with all derivatives up through order j. Let $C^j_b \equiv C^j_b(\mathbb{R}^d; \mathbb{R}^d)$ be the set of functions in C^j for which all derivatives up through order j also are bounded. The norm $|\cdot|_j$ in C^j_b is taken to be the usual sup norm of the function and its derivatives up to order j.

Theorem 7.1. (*Hartman–Grobman*) *If L is hyperbolic, then there is a $\mu_0 > 0$ such that, for any $f \in C^1_{b\mu_0} \equiv \{ f \in C^1_b : |f|_1 < \mu_0 \}$, there is a homeomorphism $h = h(f) = I + g(f)$, $g : C^1_{b\mu_0} \to C^0_b$ is continuous and unique, $g(0) = 0$, such that $h \circ (L + f) = L \circ h$.*

Proof. Since L is hyperbolic, we may suppose, without loss of generality, that $\mathbb{R}^d = W^s \oplus W^u$, where W^s (resp. W^u) is the stable (resp. unstable) manifold of the fixed point 0; that is, W^s and W^u are subspaces of $\mathbb{R}^d$ which are invariant under L, $L_s \equiv L|W^s$ has eigenvalues with moduli less than one and $L_u \equiv L|W^u$ has eigenvalues with moduli greater than one. Furthermore, by choosing an appropriate coordinate system in $\mathbb{R}^d$, we may assume that

$$(7.1) \qquad\qquad |L_s| < 1, \qquad |L_u^{-1}| < 1.$$

Hence L_s is a contraction while L_u is an expansion.

Let $a = \max\{ |L_s|, |L_u| \}$ and choose $\mu_0 > 0$ so that $a - \mu_0 > 0$, and, for any $f \in C^1_{b\mu_0}$, the inverse $(L + f)^{-1}$ of the function $L + f$ exists and belongs to C^1. The existence of μ_0 is proved in the following way. For any $y \in \mathbb{R}^d$ and any $f \in C^1_b$, let $F(x, f, y) = Lx + f(x) - y$. Then $F(L^{-1}y, 0, y) = 0$ and $D_x F(L^{-1}y, 0, y) = L + D_x f(L^{-1}y)$. If we choose μ_0 sufficiently small, then the latter operator has a bounded inverse for any $f \in C^1_{b\mu_0}$. Therefore, there is a unique solution $x(y, f)$ of the equation $Lx + f(x) = y$ for any $y \in \mathbb{R}^d$, $f \in C^1_{b\mu_0}$. The function $x(y, f)$ is a C^1-function of y, f. We remark that $(L + f)^{-1}$ may not be in C^1_b.

Let $h = I + g$, $g \in C^0_b$. For any function $f \in C^1_{b\mu_0}$, the functional equation $h \circ (L + f) = L \circ h$ is equivalent to either of the equations

$$(7.2) \quad
\begin{array}{ll}
\text{(a)} & g = L \circ g \circ (L + f)^{-1} + L \circ (L + f)^{-1} - I \\[4pt]
\text{(b)} & g = L^{-1} \circ g \circ (L + f) + L^{-1} \circ (L + f) - I.
\end{array}$$

For any function $f \in C^j$, we let $f = f_s + f_u$, where $f_s(x) \in W^s$, $f_u(x) \in W^u$ for all $x \in \mathbb{R}^d$. We are going to use (7.2)(a) to define g_s and (7.2)(b) to define g_u.

We first observe that the functions $L \circ (L + f)^{-1} - I$ and $L^{-1} \circ (L + f)$ are in C^1_b. In fact, $L \circ (L + f)^{-1} - I = -f \circ (L + f)^{-1}$ and $L^{-1} \circ (L + f) = I + L^{-1} \circ f$. Since $(I + L^{-1} \circ f)^{-1}$ and L^{-1} are in C^1 and $f \in C^1_b$, it follows that there is a positive constant k such that $|L \circ (L + f)^{-1} - I|_1 \leq k|f|_1$, $|L^{-1} \circ (L + f) - I|_1 \leq k|f|_1$ for every $f \in C^1_{b\mu_0}$. Thus, $L \circ (L + f)^{-1} - I$ and $L^{-1} \circ (L + f) - I$ are in C^1_b.

For any $g \in C^0$, $f \in C^1_{b\mu_0}$, we define $T(g, f) = T(g, f)_s + T(g, f)_u$ by the relations

$$(7.3) \quad
\begin{array}{ll}
\text{(a)} & T(g, f)_s = L \circ g \circ (L + f)^{-1} + L \circ (L + f)^{-1} - I \\[4pt]
\text{(b)} & T(g, f)_u = L^{-1} \circ g \circ (L + f) + L^{-1} \circ (L + f) - I.
\end{array}$$

The function $T(g,f)$ is continuous and $|T(g,f)|_0 \le a|g|_0 + k|f|_1$ for every $f \in C^1_{b\mu_0}$. Therefore, $T(g,f) \in C^0_b$. Furthermore, the map $T: C^0_b \times C^1_{b\mu_0} \to C^0_b$ is continuous. Also, for any $g, \bar{g} \in C^0_b$, we have $|T(g,f) - T(\bar{g},f)|_0 \le a|g - \bar{g}|_0$. Therefore, $T(\cdot, f)$ is a contraction with contraction constant a.

Thus, there is a unique fixed point $g^* = g^*(f)$ of $T(g,f)$. Also, $g^*(0) = 0$. Since $T(\cdot, f)$ has contraction constant independent of $f \in C^1_{b\mu_0}$, it follows that $g^*(f)$ is continuous in f.

It remains to show that $h^* = I + g^*$ is a homeomorphism. To do this, we consider the equation $(L + f) \circ (I + \bar{g}) = (I + \bar{g}) \circ L$ for $\bar{g} \in C^0_b$, $f \in C^1_{b\mu_0}$. We can repeat the above argument to obtain a unique function $\bar{g}^* = \bar{g}^*(f)$, continuous in f, $\bar{g}^*(0) = 0$, such that $(L + f) \circ (I + \bar{g}^*) = (I + \bar{g}^*) \circ L$. Let $\bar{h}^* = I + \bar{g}^*$. From the definitions of h^* and $\bar{h}^*$, we deduce that

$$(7.4) \qquad L \circ h^* \circ \bar{h}^* = h^* \circ (L + f) \circ \bar{h}^* = h^* \circ \bar{h}^* \circ L.$$

The equation $p \circ L = L \circ p$ has a unique solution $p = I$ (this is our original equation with $f = 0$). It follows from (7.4) that $h^* \circ \bar{h}^* = I$; that is, $\bar{h}^*$ is the identity and $\bar{h}^*$ is the inverse of h^* and h^* is a homeomorphism. This completes the proof of the theorem. $\square$

We now give the corresponding theorem for flows. Consider the differential equations

$$(7.5) \qquad\qquad \dot{x} = Ax,$$

$$(7.6) \qquad\qquad \dot{x} = Ax + f(x),$$

where $f \in C^1$.

Theorem 7.2. *Let $\varphi_t(x)$ (resp. $\psi_t(x)$) be the flows generated by (7.5) (resp. (7.6)) and suppose that $\operatorname{Re}\sigma(A) \ne 0$. Then there is a $\mu_0 > 0$ such that, for any $f \in C^1_{b\mu_0}$, there exists a homeomorphism $h = h(f) = I + g(f)$, $g: C^1_{b\mu_0} \to C^0_b$ is continuous and unique, $g(0) = 0$, such that, for all $x \in \mathbb{R}^d$, $t \in \mathbb{R}$,*

$$(h(\psi_t(x)) = \varphi_t(h(x)).$$

Proof. Consider the time one maps $\varphi_1(x)$, $\psi_1(x)$; that is,

$$\varphi_1(x) = e^A x$$
$$\psi_1(x) = e^A x + F(x),$$

where $F(x) = \int_0^1 e^{A(1-s)} f(\psi_s(x))\,ds$ and $F \in C^1_{b\mu_1}$ for some $\mu_1 > 0$. In Theorem 7.1, replace L by e^A and $L + f$ by $e^A + F$ to obtain the homeomorphism $h = I + g$ such that $h \circ (e^A + F) = A \circ h$. We claim that h satisfies $h(\psi_t(x)) = \varphi_t(h(x))$ for all $x \in \mathbb{R}^d$, $t \in \mathbb{R}$. This is equivalent to proving that $\varphi_t(h(\psi_{-t}(x))) =$

$h(x)$. If $t \in \mathbb{R}$ is fixed and $\alpha(x) = \varphi_t(h(\psi_{-t}(x)))$, then $\alpha \in C^0$ and $\alpha \circ \psi_1 = \varphi_1 \circ \alpha$. By the uniqueness of h in Theorem 7.1, it follows that $\alpha = h$, which completes the proof of the theorem. $\square$

Remark 7.1. Theorem 7.1 (resp. Theorem 7.2) gives local results about the behavior near a fixed point (resp. equilibrium point). In fact, given any neighborhood of the fixed point (resp. equilibrium point), we always can extend the map (resp. vector field) to a smooth bounded map (resp. vector field) on the whole space and apply the above results. Of course, the resulting homeomorphism is not unique for the local results since the extension is not unique.

Remark 7.2. Theorem 7.2 is false if we assume only that f belongs to the class of functions such that 0 is an isolated equilibrium point of $A + f$, the function f is continuous, is continuously differentiable except at 0 and satisfies $f(x) = o(|x|)$ as $|x| \to 0$ (for an example, see Hale [p. 117]).

Remark 7.3. From the proof for Theorem 7.2, it is clear that the same results are valid if we assume only that f belongs to the class $C_b^{0,1}$ of functions which are continuous, bounded and uniformly Lipschitzian. The norm in $C_b^{0,1}$ is the supremum of the norm of the function and the Lipschitz constant.

3.8. An Elliptic Problem

Consider the boundary value problem

$$
\text{(8.1)} \qquad
\begin{aligned}
\Delta u + f(u) &= p(x), & x &\in G, \\
u &= 0, & x &\in \partial G,
\end{aligned}
$$

where G is a bounded open set in $\mathbb{R}^n$ with sufficiently smooth boundary ∂G. Let $C^k(\bar{G})$ denote the space of k-times continuously differentiable functions on $G = G \cup \partial G$ with the usual norm

$$
|u|_k = \sup_{0 \le r \le k} \sup_{x \in G} |D^r u(x)|.
$$

Let $C^{k,\alpha}(\bar{G})$ denote the space of functions $u \in C^k(\bar{G})$ such that $D^k u$ is Hölder-continuous with exponent $0 < \alpha < 1$ in $\bar{G}$ with the usual norm

$$
|u|_{k,\alpha} = |u|_k + \sup_{\substack{x \ne y \\ x,y \in \bar{G}}} \frac{|D^k u(x) - D^k u(y)|}{|x - y|^\alpha}
$$

and $C_0^{k,\alpha}(\bar{G})$ denote the subspace of $C^{k,\alpha}(\bar{G})$ consisting of the functions vanishing on ∂G.

Theorem 8.1. *Suppose that f is continuously differentiable and assume that there are μ_{n-1}, μ_n such that*

$$\lambda_{n-1} < \mu_{n-1} \le f'(u) \le \mu_n < \lambda_n, \qquad u \in R$$

where λ_{n-1} and λ_n are two successive distinct eigenvalues of the problem

$$\Delta u + \lambda u = 0, \qquad x \in G,$$
$$u = 0, \qquad x \in \partial G.$$

Then, for any $p \in C^\alpha(\bar{G})$, there exists a unique solution $u \in C_0^{2,\alpha}(\bar{G})$ of the Problem (8.1).

Proof. Define

$$T : C_0^{2,\alpha}(\bar{G}) \to C^{0,\alpha}(\bar{G}),$$
$$u \mapsto \Delta u + f(u).$$

It follows from the differentiability of f that T is Fréchet differentiable and

$$DT(u) : v \mapsto \Delta v + f'(u)v.$$

Since $\lambda_{n-1} < \mu_{n-1} \le f'(u) \le \mu_n < \lambda_n$, $DT(u)$ is an isomorphism for each u in $C_0^{2,\alpha}(\bar{G})$. This follows from the maximum principle. For any $p \in C^\alpha(\bar{G})$, $u \in C_0^{2,\alpha}(\bar{G})$, the equation $DT(u)v = p$ has a unique solution $v \in C_0^{2,\alpha}(\bar{G})$. Furthermore, one has the following estimates:

$$|v|_{2,\alpha} \le C |p|_{0,\alpha}$$

where $C = C(n, \sigma, \mu_{n-1}, \mu_n)$. An application of Theorem 2.3.9 completes the proof of the theorem. $\quad\square$

3.9. A Hyperbolic Problem

In this section, we discuss the problem of the existence of periodic solutions for a nonlinear nonautonomous hyperbolic equation. If $I = [0, \pi] \times [0, 2\pi]$, $S = [0, \pi] \times \mathbb{R}$, define

$$Z = L^2(I, \mathbb{R}), \qquad C_{2\pi}^k(S) = \{ u \in C^k(S, \mathbb{R}) : u(x, t + 2\pi) = u(x, t) \},$$

$$\mathscr{D}(A) = \{ u \in L^2(I, \mathbb{R}) : u_{tt} - u_{xx} \text{ is defined and is in } L^2(I, R) \},$$

$$A : \mathscr{D}(A) \subset Z \to Z, \qquad Au = u_{tt} - u_{xx}.$$

Consider the nonlinear equation,

$$(9.1) \qquad Au = u_{tt} - u_{xx} = \varepsilon f(x, t, u),$$

where ε is a small real parameter and the function f satisfies

$$(9.2)$$
$$f \in C(S \times \mathbb{R}, \mathbb{R}), \qquad f(x, t, u) = f(x, t + 2\pi, u),$$
$$0 < h \le \frac{\partial f}{\partial u}(x, t, u) \le H, \qquad \frac{\partial f}{\partial u} \text{ continuous}, \quad h, H \text{ constants}.$$

By a generalized 2π-periodic solution of the equation

$$(9.3)$$
$$Au = \varepsilon f(x, t, u),$$
$$u(0, t) = u(\pi, t) = 0,$$

we mean a function $u \in Z$ such that

$$\iint_I \varepsilon f(x, t, u(x, t)) \phi(x, t)\, dx\, dt = \iint_I u(x, t) A\phi(x, t)\, dx\, dt$$

for all $\phi \in C^2_{2\pi}(S)$, $\phi = 0$ on ∂S.

Theorem 9.1. *Under the hypotheses* (9.2), *there is an $\varepsilon_0 > 0$ such that the equation $u_{tt} - u_{xx} = \varepsilon f(x, t, u)$, $u(0, t) = u(\pi, t) = 0$, has a unique generalized 2π-periodic solution in t for $0 < |\varepsilon| < \varepsilon_0$.*

Proof. Only an outline of the proof is given. The null space $\mathcal{N}(A)$ of A is easily seen to be

$$\mathcal{N}(A) = \{u(x, t) = \phi(x + t) - \phi(-x + t)\}$$

where ϕ is 2π-periodic and square integrable on bounded intervals. Let $E: Z \to Z$ be a projection operator such that $Z_{I-E} = \mathcal{N}(A)$. Let $X_{I-U} = \mathcal{D}(A) \cap Z_E$ with the graph norm, $|u|^2_{X_{I-U}} = |Au|^2 + |u|^2$. If $X = Z_{I-E} \oplus X_{I-U}$, then $A: X \to Z$ defined on X_{I-U} as above and on Z_{I-E} as the zero operator implies $A: X_{I-U} \to Z_E$ is an isomorphism.

Let $N: Z \to Z$ be defined by $Nu(x, t) = f(x, t, u(x, t))$. The problem (9.3) is to find a generalized solution of $Au = \varepsilon Nu$. For $\varepsilon \ne 0$, this is equivalent to

$$u = v + w, \qquad v \in Z_{I-E}, \qquad w \in X_{I-U},$$
$$Aw - \varepsilon EN(v + w) = 0,$$
$$(I - E)N(v + w) = 0.$$

If we let $B_\varepsilon: Z_{I-E} \times X_{I-U} \to Z_{I-E} \times Z_E$ be the map defined by

$$B_\varepsilon(v, w) = ((I - E)N(v + w), \qquad Aw - \varepsilon EN(v + w))$$

and N' denote the derivative of N, then the following facts follow easily from the fact that $(I - E)N: Z_{I-E} \to Z_{I-E}$ and $(I - E)N': Z_{I-E} \to Z_{I-E}$ are maximal monotone and strongly monotone:

(i)   there is a unique solution v_0 of the equation

$$(I - E)N(v_0) = 0:$$

(ii) $(I - E)N': Z_{I-E} \to Z_{I-E}$ is an isomorphism.

Since $A: X_{I-U} \to Z_E$ is an isomorphism, it therefore follows that the derivative of B_ε with respect to v, w at $\varepsilon = 0$, $v = v_0$, $w = 0$ is an isomorphism of X onto Z. The Implicit Function Theorem may now be applied to complete the proof. $\square$

3.10. Bibliographical Notes

Robbin [1] was the first to prove Theorem 1.1 using the Implicit Function Theorem. For a more general discussion of admissibility in ordinary differential equations see Antosiewicz [1], Corduneanu [1], Hartman [1], Massera and Schaffer [1]. For a proof of Lemmas 3.1, 3.2, see Coddington and Levinson [1]. Theorem 3.3 is due to Lazer and Leach [1]. The generalization to systems of differential equations in Theorem 3.5 is due to Lazer and Sanchez [1]. The first proof of Theorem 3.3 is based on Prodi and Ambrosetti [1]. The second proof follows Lazer and Sanchez [1]. The proof of Lazer and Sanchez [1] also allows periodic boundary conditions.

The use of the Implicit Function Theorem on the contraction mapping principle to prove results of the above type can be found also in Chow, Hale and Mallet–Paret [1], Kannan and Locker [1], Mawhin [2]. Invernezzi and Zanolin [1] have used similar techniques for obtaining periodic solutions of nonautonomous delay differential equations.

Theorem 3.6 is due to Prodi and Ambrosetti [1]. The Hopf Bifurcation Theorem will be discussed more extensively in Chapter 9 where references are also given. Theorem 5.1 and its generalization to equations with a first integral is due to Liapunov [1]. The proof in the text is based on D. S. Schmidt [1].

Given a smooth manifold of periodic orbits emanating from an equilibrium point, it is interesting to determine how the family behaves as the amplitude increases and, in particular, what happens when the family ceases to exist. Meyer and Schmidt [1] have discussed this question for a special class of equations in celestial mechanics.

When the resonance conditions in (5.4) are not satisfied, there may be no families of periodic orbits emanating from zero or several such families. There are several perturbation schemes available for determining these families in this case (see Schmidt and Sweet [1] for results and references).

It is difficult to trace the origin of the results in Section 6 on the stable and unstable manifolds near an equilibrium point. Poincaré [1] was certainly aware of these and, for analytic systems, Liapunov [1] gave series expansions for them.

Theorem 7.1 was independently discovered by Grobman [1,2] and Hartman [1–3]. The proof in the text is based on Pugh [1].

Theorem 8.1 was first proved by Landesman and Lazer [1, 2] by a different method. These papers stimulated much research on nonlinear boundary value problems. A complete treatment of the subject requires a book. The reader may consult the books of Cesari and Kannan [1] or Gaines and Mawhin [1] for relevant literature.

A version of Theorem 9.1 was originally proved by Rabinowitz [1] using a proof based on the Galerkin procedure. The proof in the text using monotone operators is based on de Simone and Torelli [1] (see, also, Torelli [1], Hall [1]). For another proof, see Mawhin [3].

Chapter 4

Variational Method

4.1. Introduction

In this chapter, we consider problems which are derived from a functional
on a function space in variational problems. We will begin with an introduction of existence theory by using minimization techniques and some
elementary facts about monotone operators. By using a compactness condition (condition (C)), we give a minimax principle in Banach spaces. This
result will then be used to prove a theorem (mountain pass theorem) on the
existence of critical points which are not necessary minimal points. An
application to a semi-linear hyperbolic equation will be considered. In
Sections 4.8 and 4.9, we present a summary of Ljusternik–Schnirelman
theory on Banach manifolds with an application.

Bifurcation theory based on only the linear parts for variational problems
is presented in Sections 4.9–4.12. The results here depend in a very essential
way on the variational property of the equation.

4.2. Weak Lower Semicontinuity

Let X be a Banach space and X^* be its dual space. If $v^* \in X^*$ and $v \in X$,
then define $\langle v^*, v \rangle$ by $\langle v^*, v \rangle = v^*(v)$. The map $(v^*, v) \to \langle v^*, v \rangle$ is a bilinear
continuous functional on the product space $X^* \times X$.

Let $U \subset X$ be open and $f : U \to \mathbb{R}$ be C^1. The derivative of f at $p \in U$ is
a continuous linear map $\lambda : X \to \mathbb{R}$, i.e., $\lambda \in X^*$. We will denote λ by $df(p)$.
Recall that a point $p \in U$ is a critical point of f if $df(p) = 0$, i.e.,

$$\langle df(p), v \rangle = 0 \quad \text{for all } v \in X.$$

EXAMPLE 2.1. Let $X = C^1[0, 1]$ and define $f : X \to \mathbb{R}$ by

$$f(v) = \int_0^1 [\dot{v}^2 + v^2 + 2\zeta v]\, dt, \qquad v \in X$$

where $\zeta \in C[0, 1]$ is fixed. If $u(\cdot) \in X$ is a critical point of f, then

$$\langle df(u), v \rangle = 2 \int_0^1 [\dot{u}\dot{v} + uv + \zeta v]\, dt = 0 \quad \text{for all } v \in X.$$

If

$$a(t) = \int_0^t [u(s) + \zeta(s)]\, ds, \quad 0 \le t \le 1,$$

then, by integration by parts, we obtain

$$\int_0^1 [\dot{u}(t) - a(t)]\dot{v}(t)\, dt = 0$$

for all $v \in X$ satisfying $v(0) = v(1) = 0$. The above condition implies that $\dot{u}(t) - a(t) = \text{constant}$ and hence, by the definition of $a(t)$,

$$\ddot{u} = u + \zeta, \qquad 0 \le t \le 1$$

We note that the differentiability of $\dot{u}(t)$ was not assumed. Furthermore, integration by parts of the equation $\langle df(u), v \rangle = 0$ yields

$$0 = \int_0^1 [\ddot{u} - u - \zeta]v\, dt = \dot{u}(1)v(1) - \dot{u}(0)v(0).$$

Since $v(1)$ and $v(0)$ are arbitrary, the critical point u of f satisfies the Neumann boundary value problem,

$$\ddot{u} = u + \zeta, \qquad 0 \le t \le 1,$$
$$\dot{u}(1) = \dot{u}(0) = 0.$$

The above example illustrates typically the connections between solutions of boundary value problems and critical points of a functional. It also will become clear that some other boundary value problems are related in the same manner. This makes the problem of determining the existence of critical points of a functional of great importance in differential equations.

We will begin with the following simple but useful theorem.

Theorem 2.2. *Let X be a reflexive Banach space and*

$$a : X \times X \to \mathbb{R}$$

be a continuous bilinear functional. Define

$$f(v) = a(v, v) - 2L(v)$$

where $L \in X^*$. Suppose that $a(\cdot, \cdot)$ is symmetric and there exists $c > 0$ such that

$$(2.1) \qquad\qquad a(v, v) \geq c|v|^2.$$

Then there exists a unique $u \in X$ such that

$$f(u) = \inf\{f(v) : v \in X\} = m$$

and

$$df(u) = 0.$$

Proof. It is clear that, if $f(u) = m$, then $df(u) = 0$. Let $v_1, v_2 \in X$, $v_1 \neq v_2$, and $0 < \lambda < 1$. We have

$$
\begin{aligned}
a(\lambda v_1 &+ (1 - \lambda)v_2, \lambda v_1 + (1 - \lambda)v_2) \\
&= \lambda^2 a(v_1, v_1) + 2\lambda(1 - \lambda)a(v_1, v_2) + (1 - \lambda)^2 a(v_2, v_2) \\
&< \lambda^2 a(v_1, v_1) + \lambda(1 - \lambda)[a(v_1, v_1) + a(v_2, v_2)] + (1 - \lambda)^2 a(v_2, v_2) \\
&= \lambda a(v_1, v_1) + (1 - \lambda)a(v_2, v_2)
\end{aligned}
$$

This shows that $f(v)$ is strictly convex. If $u_1 \neq u_2$ and

$$f(u_1) = f(u_2) = \inf\{f(v) \mid v \in X\} = m,$$

then by the strict convexity of f

$$f\left(\frac{u_1 + u_2}{2}\right) < \frac{1}{2}f(u_1) + \frac{1}{2}f(u_2) = m.$$

This contradiction gives the uniqueness. It remains to show the existence of the critical point. Let $\{v_n\}$ be a minimizing sequence; i.e.,

$$\lim_{n \to \infty} f(v_n) = m.$$

Since

$$
\begin{aligned}
\frac{f(v)}{|v|} &= \frac{a(v, v)}{|v|} - \frac{2L(v)}{|v|} \\
&\geq c|v| - 2|L|,
\end{aligned}
$$

$f(v)/|v| \to \infty$ as $|v| \to \infty$. This implies that the minimizing sequence $\{v_n\}$ is bounded in norm. Since X is reflexive, $\{v_n\}$ contains a weakly convergent subsequence which we label again by $\{v_n\}$. Let $u \in X$ be the weak limit of the sequence $\{v_n\}$. By (2.1),

$$a(v_n, v_n) \geq 2a(u, v_n) - a(u, u).$$

Since $u \in X$ is fixed, $a(u, \cdot)$ is a continuous linear functional on X. Thus, by passing to the limit

$$\liminf_{n \to \infty} a(v_n, v_n) \geq a(u, u).$$

This implies that

$$f(u) \leq \liminf_{n \to \infty} f(v_n) = m$$

and completes the proof. $\square$

In the following, we indicate by means of examples how Theorem 1.2 is used in applications.

EXAMPLE 2.3. Let $\Omega \subset \mathbb{R}^n$ be bounded and open with smooth boundary Γ. Suppose that

$$c(x), \qquad a_{ij}(x) \in L^\infty(\Omega); \qquad i, j = 1, 2, \ldots, n, \quad a_{ij} = a_{ji}.$$

Define the bilinear functional $a(\cdot, \cdot)$ on the Sobolev space $H^1(\Omega)$,

$$a(u, v) = \int_\Omega \left(\sum_{i,j} a_{ij}(x) \frac{\partial u}{\partial x_i} \frac{\partial v}{\partial x_j} \right) dx + \int_\Omega c(x) uv \, dx$$

where $u, v \in H^1(\Omega)$. If there exists a constant $\alpha > 0$ such that

$$\sum_{i,j} a_{ij}(x) \xi_i \xi_j \geq \alpha(\xi_1^2 + \cdots + \xi_n^2), \quad \text{for all } \xi_1, \ldots, \xi_n \in \mathbb{R}$$

and $c(x) \geq \alpha$, then the bilinear functional $a(\cdot, \cdot)$ satisfies condition (2.1) in Theorem 2.2 and is symmetric and continuous. Let $\zeta \in L^2(\Omega)$ and

$$f(v) = a(v, v) - 2 \int_\Omega \zeta v \, dx.$$

By Theorem 2.2, there exists a critical point u of f and, for all $v \in H^1(\Omega)$,

$$(2.2) \qquad \int_\Omega \left[\sum_{i,j} a_{ij} \frac{\partial u}{\partial x_i} \frac{\partial v}{\partial x_j} + cuv - \zeta v \right] dx = 0.$$

If the functions a_{ij}, c, ζ and u are smooth, then we may deduce from the above equation by Green's theorem that u satisfies the Neumann boundary value problem:

$$-\sum_{i,j} \frac{\partial}{\partial x_i}\left(a_{ij}\frac{\partial u}{\partial x_j}\right) + cu = \zeta, \qquad x \in \Omega$$

$$\frac{\partial u}{\partial n_A} = 0, \qquad x \in \Gamma$$

where

$$(2.3) \qquad \frac{\partial u}{\partial n_A}(x) = \sum_{i,j} a_{ij}\frac{\partial u}{\partial x_j}\, n_i(x)$$

$$n_i = i^{\text{th}} \text{ component of the unit outward}$$
$$\text{normal to } \Gamma \text{ at } x \in \Gamma.$$

We note that, if f is considered as a functional on the subspace $H_0^1(\Omega)$ (i.e., the completion of C^∞ functions with compact supports in $H^1(\Omega)$), then Theorem 2.2 is again applicable. However, the critical point $\tilde{u} \in H_0^1(\Omega)$ satisfies (2.2) only for $v \in H_0^1(\Omega)$. In this case, $\tilde{u}$ is a weak solution of the Dirichlet boundary value problem:

$$-\sum_{i,j} \frac{\partial}{\partial x_i}\left(a_{ij}\frac{\partial \tilde{u}}{\partial x_j}\right) + c\tilde{u} = \zeta, \qquad x \in \Omega$$

$$\tilde{u} = 0, \qquad x \in \Gamma.$$

EXAMPLE 2.4. Assume that the boundary Γ of Ω consists of two connected components Γ_1 and Γ_2, and $\Gamma_1 \cap \Gamma_2 = \varnothing$. Let X be the completion of the following subspace in $H^1(\Omega)$

$$\{v \in H^1(\Omega): v:\bar{\Omega} \to \mathbb{R} \text{ is } C^\infty \text{ and } v(x) = 0 \text{ for all } x \in \Gamma_1\}$$

Define $f: X \to \mathbb{R}$ by

$$f(v) = a(v, v) - 2 \int_\Omega \zeta v\, dx$$

where $a(\cdot, \cdot)$ and ζ are as in Example 2.3. The critical point u of f is now a weak solution to the mixed boundary value problem:

$$-\sum \frac{\partial}{\partial x_i}\left(a_{ij}\frac{\partial}{\partial x_j}\right) + Cu = \zeta, \qquad x \in \Omega$$

$$u = 0, \qquad x \in \Gamma_1$$

$$\frac{\partial u}{\partial n_A} = 0, \qquad x \in \Gamma_2$$

where $\partial u/\partial n_A$ is as in (2.3).

The above examples clearly illustrate that Theorem 2.2 gives a somewhat unified approach to a class of linear boundary value problems. It is now natural to generalize Theorem 2.2 to include a class of nonlinear boundary value problems. The following theorem is a simple generalization of Theorem 2.2. It is no longer assumed that the functional f is a sum of a bilinear functional and a linear functional. However, the main features of the proof are maintained. Some applications to some nonlinear problems will be given below.

A functional $f : X \to \mathbb{R}$ is *weakly (sequentially) lower semicontinuous* at a point $v \in X$ if $f(v) \leq \lim_{n \to \infty} f(v_n)$ for any sequence $v_n \to v$ weakly.

Theorem 2.5. *Let X be a reflexive Banach space and $f : X \to \mathbb{R}$ be C^1. Suppose that*

(a) $f(v) \to \infty$ *as* $|v| \to \infty$,
(b) f *is weakly lower semicontinuous.*

Then f has a critical point $u \in X$.

Proof. By (a), there exists $R > 0$ such that

$$\inf_{|v| = R} f(v) > f(0).$$

Define

$$c = \inf_{|v| \leq R} f(v) \geq -\infty.$$

Let $\{v_n\}$ be a minimizing sequence and $|v_n| \leq R$. By the weak compactness, we may assume without loss of generality that there exists u, $|u| \leq R$, such that $v_n \to u$ (weakly) as $n \to \infty$. By (b),

$$f(u) \leq \lim_{n \to \infty} f(v_n) = c \leq f(0).$$

This implies $c > -\infty$ and $|u| < R$. Hence, u is a local minimum of $f(v)$ and $df(u) = 0$. $\square$

Remark 2.6. The condition of weak lower semicontinuity of the functional is often guaranteed by the monotonicity of df (or convexity of f). We will exploit these connections in the next section.

EXAMPLE 2.7.   Let $p > 1$ and $X = W_0^{1,p}(\Omega)$, where $\Omega \subset \mathbb{R}^n$ is open and bounded with sufficiently smooth boundary Γ. Define

$$f(v) = \frac{1}{p} \int_\Omega \sum_{i=1}^{n} \left| \frac{\partial v}{\partial x_i} \right|^p dx + \int_\Omega \zeta v \, dx,$$

where $v \in X$ and $\zeta \in L^q(\Omega)$, $q = p/(p-1)$. It is clear that f satisfies (a). It remains to show that f is weakly lower semicontinuous. We note that the functional f_1 defined by

$$v \to \frac{1}{p} \int_\Omega \sum_{i=1}^n \left| \frac{\partial v}{\partial x_i} \right|^p dx$$

is strictly convex; i.e., for all $v_1 \neq v_2$, $0 < \lambda < 1$,

$$f_1(\lambda v_1 + (1-\lambda)v_2) < \lambda f_1(v_1) + (1-\lambda)f_1(v_2).$$

Hence, the epigraph of f_1,

$$\text{epi } f_1 = \{(v,a) \in X \times \mathbb{R} : f_1(v) \leq a\},$$

is closed and convex. By Mazur's theorem, epi f_1 is also weakly closed. This implies that f_1 is weakly lower semicontinuous. By Theorem 2.5, there exists a critical point u of f which is by Green's theorem a weak solution of the nonlinear Dirichlet boundary value problem:

$$-\sum_{i=1}^n \frac{\partial}{\partial x_i}\left(\left| \frac{\partial u}{\partial x_i} \right|^{p-2} \frac{\partial u}{\partial x_i} \right) = \zeta, \qquad x \in \Omega$$

$$u = 0, \qquad x \in \Gamma.$$

We note that the above problem may not have a smooth solution $u(x)$ even though $\zeta(x)$ is smooth. Indeed, let $\zeta \in C^\infty(\bar{\Omega})$ and $\zeta(x) \geq c > 0$ for all $x \in \bar{\Omega}$. Then the solution $u \notin C^2(\Omega)$. In fact, if $u \in C^2(\Omega)$, $u = 0$ on Γ, then there exists an extremal point $\bar{x} \in \Omega$ of $u(x)$. At this point,

$$-\sum_{i=1}^n \frac{\partial}{\partial x_i}\left(\left| \frac{\partial u}{\partial x_i}(\bar{x}) \right|^{p-2} \frac{\partial u}{\partial x_i}(\bar{x}) \right) = 0 \neq \zeta(\bar{x}),$$

which is a contradiction.

EXAMPLE 2.8. Consider the second order system of ordinary differential equations

$$(2.4) \qquad \frac{d^2 x}{dt^2} = \nabla V(x,t) \qquad x \in \mathbb{R}^n, t \in \mathbb{R},$$

where $V : \mathbb{R}^n \times \mathbb{R} \to \mathbb{R}$ is a scalar-valued 2π-periodic function in t, i.e., $V(x, t + 2\pi) = V(x,t)$, for all $t \in \mathbb{R}$, and "∇" denotes the gradient of V with respect to $x \in \mathbb{R}^n$. For simplicity, we assume that V is even in $x \in \mathbb{R}^n$ i.e.,

$V(x, t) = V(-x, t)$ for all $t \in \mathbb{R}$, $x \in \mathbb{R}^n$. Suppose that V satisfies the following condition:

$$(2.5) \qquad V(x, t) \to \infty, \quad \text{as } |x| \to \infty \text{ uniformly in } t.$$

We claim that (2.4) has a 2π-periodic solution.

To prove our claim, we note that, if there exists a solution of (2.4) which satisfies the boundary conditions,

$$(2.6) \qquad x(0) = x(\pi) = 0,$$

then, by the evenness of $V(x, t)$, there exists a 2π-periodic solution of (2.4). In fact, this solution is odd in t. We now consider the nonlinear boundary value problem (2.4) and (2.6). Let $X = H_0^1(0, \pi)$ be the completion of C^∞ functions with compact supports in $(0, \pi)$ under the H^1-norm:

$$|x|_1 = \left\{ \int_0^\pi |\dot{x}(t)|^2 \, dt \right\}^{1/2}$$

Define $f: X \to \mathbb{R}$ by

$$f(x) = \int_0^\pi \left[\tfrac{1}{2} |\dot{x}(t)|^2 + V(x(t), t) \right] dt$$

By Sobolev's embedding theorem, the functional

$$\int_0^\pi V(x(t), t) \, dt$$

is weakly continuous from X to $\mathbb{R}$. Thus, $f: X \to \mathbb{R}$ is weakly lower semi-continuous. It remains to show that $f(x) \to \infty$ as $|x|_1 \to \infty$. By (2.5), $V(x, t)$ is bounded below uniformly by c_1, say. Thus,

$$f(x) \geq c_0 |x|_1^2 + c_1 \to \infty \quad \text{as } |x|_1 \to \infty$$

where $c_0 > 0$ is obtained from Friedrichs' inequality.

The above result still holds if $V(x, t)$ is not even.

EXAMPLE 2.9. Consider the nonlinear boundary value problem:

$$(2.7) \qquad \begin{aligned} -\Delta u(x) &= g(x, u(x)), & x \in \Omega \\ u(x) &= 0, & x \in \Gamma \end{aligned}$$

where $\Omega \subset \mathbb{R}^n$, $n \geq 3$, is bounded and open with smooth boundary Γ and $g(x, u)$ satisfies the following conditions:

(a) $g: \bar{\Omega} \times \mathbb{R} \to \mathbb{R}$ is continuous,

(b)   there exists $0 < \gamma < 1$ such that

$$|g(x, r)| \leq a(x) + c|r|^{\gamma}, \qquad x \in \bar{\Omega}, r \in \mathbb{R},$$

where $a \in L^{\gamma + 1/\gamma}(\Omega)$ and $c > 0$.

We claim that the nonlinear boundary value problem (2.7) has a weak solution. The proof is as follows.

Firstly, we will consider the Nemitskii operator $G(v)(x) = g(x, v(x))$. By the results in Section 2.5 and conditions (a) and (b),

$$G : L^{\gamma + 1}(\Omega) \to L^{\gamma + 1/\gamma}(\Omega)$$

is continuous and takes bounded sets into bounded sets. Since

$$\frac{1}{\gamma + 1} + \frac{1}{\dfrac{\gamma + 1}{\gamma}} = 1,$$

G is the Fréchet derivative of the functional

$$\varphi(v) = \int_0^1 d\lambda \int_\Omega g(x, \lambda v(x)) v(x)\, dx.$$

In other words, $G(v) = d\varphi(v)$, or

$$\langle d\varphi(v), w \rangle = \langle G(v), w \rangle, \qquad v, w \in L^{\gamma + 1}(\Omega),$$

where $\langle \cdot, \cdot \rangle$ denotes the L^2-inner product. We note that $\varphi(v)$ is continuous on $L^{\gamma + 1}(\Omega)$. By the Sobolev embedding theorem, the inclusion map

$$i : W_0^{1,2}(\Omega) \to L^t(\Omega), \qquad 1 \leq t < \frac{2n}{n - 2}$$

takes weakly convergent sequences into strongly convergent ones. Since $0 < \gamma < 1$, $\varphi : W_0^{1,2}(\Omega) \to \mathbb{R}$ is weakly continuous.

Next, let

$$f(v) = \frac{1}{2} \int_\Omega |\nabla v|^2\, dx - \varphi(v)$$

It is clear that $f : W_0^{1,2}(\Omega) \to \mathbb{R}$ is weakly lower semi-continuous and is (strongly) C^1. By Theorem 2.5, our claim will be proved if we show that $f(v) \to \infty$ as $|v|_{1,2} \to \infty$.

Let $|\cdot|_{m,p}$ denote the norm in the Sobolev space $W^{m,p}(\Omega)$. Note that $|\cdot|_{0,p}$ denotes the L^p-norm. By Friedrichs' inequality, there exists $c_0 > 0$ such that

$$|v|_{0,2} \leq c_0 |\nabla v|_{0,2} \quad \text{for all } v \in W_0^{1,2}(\Omega).$$

This implies that there exists $c_1 > 0$ such that

$$(2.8) \qquad \frac{f(v)}{|v|_{1,2}} \geq c_1 |v|_{1,2} - \frac{\varphi(v)}{|v|_{1,2}}$$

By Hölder's inequality and condition (b),

$$|\varphi(v)| \leq \int_0^1 d\lambda \int_\Omega |g(x, \lambda v(x))| \, |v(x)| \, dx$$

$$\leq \int_0^1 d\lambda \int_\Omega [a(x)|v(x)| + c\lambda^\gamma |v(x)|^{1+\gamma}] \, dx$$

$$= \int_\Omega a(x)|v(x)| \, dx + \frac{c}{1+\gamma} \int_\Omega |v(x)|^{1+\gamma} \, dx$$

$$\leq |a|_{0,(1+\gamma)/\gamma} |v|_{0,1+\gamma} + \frac{c}{1+\gamma} |v|_{0,1+\gamma}^{1+\gamma}.$$

By Sobolev's embedding theorem, there exists $c_2 > 0$ such that

$$|v|_{0,1+\gamma} \leq c_2 |v|_{1,2} \quad \text{for all } v \in W_0^{1,2}(\Omega),$$

$$|\varphi(v)| \leq c_2 |a|_{0,1+\gamma/\gamma} |v|_{1,2} + \frac{cc_2^2}{1+\gamma} |v|_{1,2}^{1+\gamma}.$$

Since $0 < \gamma < 1$, by (2.8), $f(v)/|v|_{1,2} \to \infty$ as $|v|_{1,2} \to \infty$. This proves the claim.

Finally, we note that, if $\gamma = 1$, then the result is still true provided c is sufficiently small. If $\gamma > 1$, then our method fails.

4.3. Monotone Operators

It is now clear that, for concrete nonlinear problems, weak continuity is generally an unrealistic condition. However, weak lower semi-continuity is realistic and is applicable. We also note that the condition of weak lower semi-continuity is closely related to the convexity of the functional. In this section, we will present some elementary facts concerning this relation and its applications and generalizations.

Definition 3.1. Let X be a Banach space and X^* be its dual space. A map $A: X \to X^*$ is said to be *monotone* if

$$\langle Au - Av, u - v \rangle \geq 0 \quad \text{for } u, v \in X.$$

Theorem 3.2. *Let $f: X \to \mathbb{R}$ be C^1 and $df = A$. Then A is monotone if and only if f is convex.*

Proof. Suppose that f is convex, then

$$f(v + \lambda(u - v)) - f(v) \le \lambda[f(u) - f(v)]$$

for $u, v \in X$ and $0 < \lambda < 1$. By letting $\lambda \to 0^+$,

$$\langle df(v), u - v \rangle \le f(u) - f(v).$$

Similarly,

$$\langle df(u), v - u \rangle \le f(v) - f(u).$$

Adding the two inequalities, $\langle Au - Av, u - v \rangle \ge 0$. Conversely, suppose A is monotone. Let $u, v \in X$ and

$$p(\lambda) = f(\lambda u + (1 - \lambda)v) - \lambda f(u) - (1 - \lambda)f(v),$$

$0 \le \lambda \le 1$. Thus, $p \in C^1$ and $p(0) = p(1) = 0$. To show that $p(\lambda) \le 0$ for $0 \le \lambda \le 1$, suppose not. Then $p(\lambda)$ has a maximum at $\lambda_0 \in (0, 1)$ and $p'(\lambda_0) = 0$. If $0 < \lambda_0 < \lambda \le 1$, then

$$\begin{aligned}
p'(\lambda) - p'(\lambda_0) &= \langle A(\lambda u + (1 - \lambda)v) - A(\lambda_0 u + (1 - \lambda_0)v), u - v \rangle \\
&= (\lambda - \lambda_0)^{-1} \langle A(\lambda u + (1 - \lambda)v) - A(\lambda_0 u + (1 - \lambda_0)v), \\
&\quad [\lambda u + (1 - \lambda)v] - [\lambda_0 u + (1 - \lambda_0)v] \rangle \\
&\ge 0.
\end{aligned}$$

Hence p is increasing for $\lambda > \lambda_0$. It follows that $p(\lambda_0) \le p(1) = 0$ and $p(\lambda) \le 0$ for all $\lambda \in [0, 1]$. This completes the proof. $\square$

Corollary 3.3. *If $f: X \to \mathbb{R}$ is C^1 and $df = A$ is monotone, then f is weakly lower semi-continuous.*

Proof. By Theorem 3.1, the epigraph of f,

$$\mathrm{epi}\, f = \{(v, a) \in X \times \mathbb{R}: f(v) \le a\},$$

is closed and convex. By Mazur's Theorem, epi f is also weakly closed. Hence, f is weakly lower semi-continuous. $\square$

Corollary 3.4. *Let $f: X \to \mathbb{R}$ be C^1 and $df = A$ be monotone. If $f(v) \to \infty$ as $|v| \to \infty$, then f has a critical point $u \in X$.*

Proof. Apply Theorem 2.5. $\square$

In many applications, the continuity condition on $df = A$ in Corollary 3.4 may be too strong. There are various generalizations of Corollary 3.4

replacing the continuity hypothesis by a weaker one. Moreover, the operator A need not be the differential of some scalar functional. We will only give the following and refer the reader to the literature for more information.

Theorem 3.5. *Let X be a separable reflexive Banach space and $A: X \to X^*$. Assume that*

(a)   *A is monotone,*
(b)   *A is coercive, i.e.,*

$$\frac{\langle Au, u \rangle}{|u|} \to \infty \quad as \ |u| \to \infty$$

(c)   *A is continuous.*

Then, for every $h \in X^*$, there exists $u \in X$ such that

$$(3.1) \qquad\qquad\qquad Au = h.$$

Proof. Since X is separable, there exists a countable dense subset $\{w_m\}$. Let X_m be the linear span of $\{w_1, \ldots, w_m\}$. The scalar product on X_m is the product $\langle \cdot, \cdot \rangle$ induced by X. For each $m \geq 1$, $B_m: X_m \to X_m^*$ is defined by

$$\langle B_m(u), v \rangle = \langle A(u) - h, v \rangle, \qquad u, v \in X_m.$$

Consider the finite dimensional problem

$$(3.2) \qquad\qquad\qquad B_m(v) = 0 \qquad v \in X_m$$

Since A is coercive and $h \in X^*$ is fixed, there exists a constant $c > 0$ independent of $m \geq 1$ such that $\langle B_m(v), v \rangle > 0$ for all $v \in X_m$, $|v| = c$. This condition (see Section 2.11) implies that there exists a $u_m \in X_m$ such that u_m satisfies (3.2), $|u_m| < c$.

Since X is reflexive, we may pass to a subsequence of $\{u_m\}$ and assume without loss of generality that there exists a $u \in X$ such that

$$(3.3) \qquad\qquad\qquad u_m \to u \quad \text{weakly as } m \to \infty$$

By the monotonicity A, it follows that for any $v \in X_m$ and $k \geq m$

$$\begin{aligned}
0 &\leq \langle Av - Au_k, v - u_k \rangle \\
&= \langle Av - h, v - u_k \rangle - \langle Au_k - h, v - u_k \rangle \\
&= \langle Av - h, v - u_k \rangle
\end{aligned}$$

By (3.3),

$$\lim_{k \to \infty} \langle Av - h, v - u_k \rangle = \langle Av - h, v - u \rangle$$

Hence, $\langle Av - h, v - u \rangle \geq 0$ for any $v \in X_m$. Since A is continuous and $\bigcup X_m$ is dense in X, $\langle Av - h, v - u \rangle \geq 0$, for all $v \in X$. Replacing v by $u + \lambda v$ in the above inequality, $\lambda \in R$, one obtains

$$\langle A(u + \lambda v) - h, \lambda v \rangle \geq 0 \quad \text{for all } v \in X.$$

Cancelling $\lambda > 0$ and letting $\lambda \to 0+$, we obtain by the continuity of A, $\langle Au - h, v \rangle \geq 0$ for all $v \in X$. This says u is a solution of (3.1) and completes the proof. $\quad \square$

EXAMPLE 3.6. Let $\Omega \subset \mathbb{R}^n$ be open and bounded with smooth boundary Γ. Consider the Dirichlet problem for the quasilinear elliptic equation:

$$(3.4) \quad \begin{cases} \sum_{|\alpha| \leq m} (-1)^{|\alpha|} D^\alpha(a_\alpha(x, u(x), Du(x), \ldots, D^\beta u(x))) \\ \qquad = \zeta(x), \qquad |\beta| \leq m, \, x \in \Omega \\ D^\alpha u(x) = 0, \qquad |\alpha| \leq m - 1, \, x \in \Gamma \end{cases}$$

where $\alpha = (\alpha_1, \ldots, \alpha_n)$, $\beta = (\beta_1, \ldots, \beta_n)$ are the usual integral multi-indices, $|\alpha| = \alpha_1 + \cdots + \alpha_n$, $D^\alpha = D_1^{\alpha_1} \cdots D_n^{\alpha_n}$ and $D_i = \partial/\partial x_i$. The functions $a_\alpha(x, \xi)$, where $\xi = (\xi_1, \ldots, \xi_d)$, are in general nonlinear and $d \geq 1$ is maximal number of all possible derivatives $D^\beta u$ with $|\beta| \leq m$.

We assume that for each α, $|\alpha| \leq m$,

(a) $a_\alpha : \bar{\Omega} \times \mathbb{R}^d \to \mathbb{R}$ is continuous,
(b) $|a_\alpha(x, \xi)| \leq c(|\xi|^{p-1} + 1)$, $c > 0$, $1 < p < \infty$ $x \in \bar{\Omega}$, $\xi \in \mathbb{R}^d$,
(c) $\sum_{|\alpha| \leq m} [a_\alpha(x, \xi) - a_\alpha(x, \eta)][\xi_\alpha - \eta_\alpha] \geq 0$, $\xi, \eta \in \mathbb{R}^d$, where ξ_α (or η_α) denotes the component of ξ which corresponds to $D^\alpha u$ in $a_\alpha(x, u, \ldots, D^\beta u)$,
(d) $\sum_{|\alpha| \leq m} a_\alpha(x, \xi)\xi_\alpha \geq c_0(|\xi|^p - 1)$, $\xi \in \mathbb{R}^d$, $c_0 > 0$.

We will apply Theorem 3.5 to show the existence of a weak solution of (3.4) if ζ belongs to an appropriate function space.

Let $W^{m,p}(\Omega)$ denote the Sobolev space of all L^p functions $v(x)$, $x \in \Omega$, whose distributional derivatives up to order m are also in $L^p(\Omega)$ with norm

$$|v|_{m,p} = \left(\sum_{|\alpha| \leq m} |D^\alpha v|_{L^p}^p \right)^{1/p}, \qquad p \neq \infty,$$

and $W_0^{m,p}(\Omega)$ be the closure in $W^{m,p}(\Omega)$ of all C^∞ functions with compact supports in Ω. Let $X = W_0^{m,p}(\Omega)$ and A be a nonlinear operator defined on X as follows:

$$\langle Au, v \rangle = \sum_{|\alpha| \leq m} \langle a_\alpha(\cdot, u, Du, \ldots, D^\beta u), D^\alpha v \rangle_0, \qquad u, v \in X,$$

where $\langle \cdot, \cdot \rangle_0$ denotes the L^2 inner product. Note that the operator A can be written as

$$Au = \sum_{|\alpha| \leq m} (-1)^{|\alpha|} D^\alpha(a_\alpha(\cdot, u, Du, \ldots, D^\beta u))$$

where D^α is taken in the distributional sense. A weak solution of (3.4) is defined to be a function $u \in X$ such that

$$\langle Au - \zeta, v \rangle = 0 \quad \text{for all } v \in X.$$

If $u \in X$, then by conditions (a) and (b) for each α, $|\alpha| \leq m$,

$$a_\alpha(\cdot, u, Du, \ldots, D^\beta u) \in L^{p'}(\Omega), \qquad p' = \frac{p}{p-1}.$$

It follows from Hölder's inequality that $\langle Au, v \rangle$ is well-defined provided $v \in X$. Moreover,

$$|\langle Au, v \rangle| \leq \varphi(|u|_{m,p})|v|_{m,p}$$

where $\varphi : [0, \infty) \to [0, \infty)$ is continuous. Thus, $Au \in X^*$. Since the a_α's are continuous, $A : X \to X^*$ is continuous. If A is monotone and coercive, then, by Theorem 3.5, there exists a weak solution $u \in X$ of (3.4) if $\zeta \in X^*$. We will now show that this is the case. Let $u, v \in X$. Then

$$\langle Au - Av, u - v \rangle$$
$$= \sum_{|\alpha| \leq m} \langle a_\alpha(\cdot, u, Du, \ldots, D^\beta u) - a_\alpha(\cdot, v, Dv, \ldots, D^\beta v), D^\alpha u - D^\alpha v \rangle_0$$
$$= \int_\Omega \sum_{|\alpha| \leq m} [a_\alpha(x, u(x), \ldots, D^\beta u(x)) - a_\alpha(x, v(x), \ldots, D^\beta v(x))]$$
$$\times [D^\alpha u(x) - D^\alpha v(x)] \, dx$$
$$\geq 0$$

since the integrand is nonnegative by condition (c). This implies that A is monotone. Finally, by condition (d)

$$\langle Av, v \rangle = \int_\Omega \sum_{|\alpha| \leq m} a_\alpha(x, v(x), \ldots, D^\beta v(x)) D^\alpha v(x) \, dx$$
$$\geq c_0 \int_\Omega \left[\left(\sum_{|\alpha| \leq m} |D^\alpha v(x)|^p \right) - 1 \right] dx$$
$$\geq c_0 |v|_{m,p}^p - c_1, \qquad c_1 > 0 \text{ constant}$$

Since $p > 1$,

$$\frac{\langle Av, v \rangle}{|v|_{m,p}} \geq c_0 |v|_{m,p}^{p-1} - \frac{c_1}{|v|_{m,p}} \to \infty, \quad \text{as } |v|_{m,p} \to \infty$$

Hence, A is coercive.

4.4. Condition (C)

Consider the nonlinear elliptic boundary value problem in $\mathbb{R}^3$.

(4.1) $-\Delta u + au^3 = 0, \quad x \in \Omega$

(4.2) $u = 0, \quad x \in \Gamma$

where $\Omega \subset \mathbb{R}^3$ is bounded and open with smooth boundary Γ and $a \in \mathbb{R}$. Let $W_0^{1,2}(\Omega)$ be the usual Sobolev space which is the completion of $C_0^\infty(\Omega)$ under the norm

$$|v|_{1,2} = \int_\Omega |\nabla v|^2 \, dx$$

A weak solution $u \in W_0^{1,2}(\Omega)$ may be defined to be a critical point of the functional $f: W_0^{1,2}(\Omega) \to \mathbb{R}$, where

$$f(v) = \tfrac{1}{2}|v|_{1,2}^2 + \tfrac{1}{4} \int_\Omega av^4 \, dx$$

By Sobolev's embedding theorem, the inclusion map $i: W_0^{1,2}(\Omega) \to L^p(\Omega)$, $1 \leq p < 6$, takes weakly convergent sequences into strongly convergent sequences. Thus, f is weakly lower semi-continuous. If $a > 0$, then $f(v) \to \infty$ as $|v|_{1,2} \to \infty$ and, by Theorem 2.5, f has a critical point $u \in W_0^{1,2}(\Omega)$. It follows from the maximal principle that the critical point u is identically zero. On the other hand, if $a < 0$, $v_0 \in W_0^{1,2}(\Omega)$ and $v_0 \not\equiv 0$; then we have

$$f(\alpha v_0) = \alpha^2 \left[\tfrac{1}{2}|v_0|_{1,2}^2 + \alpha^2 \frac{a}{4} \int_\Omega v_0^4 \, dx \right] \to -\infty \quad \text{as } \alpha \to \infty$$

However, there exists a sequence $\{v_m\} \subset W_0^{1,2}(\Omega)$ such that $|v_m|_{1,2} \to \infty$ as $m \to \infty$ and $|v_m(x)| \leq 1$ for all $x \in \Omega$ and m. (If Ω were the interval $(0, \pi)$, then let $v_m(x) = \sin(mx)$). Thus, $f(v_m) \to \infty$ as $m \to \infty$. This shows that, if $a < 0$, then f is neither bounded above nor below. We note that $u = 0$ is a critical point of f. It is natural to ask whether there are nontrivial solutions of (4.1), (4.2) (the answer will be given in Section 4.6). The method of finding critical points given in the previous sections is clearly not applicable here.

There are also many interesting variational problems in which the desired solutions may not yield the minimal value of the associated functional. In general, they are saddle points of the functionals.

In order to study the theory of existence of critical points of functionals defined on Banach spaces or Banach manifolds, we will need a compactness condition on the functionals replacing the weak lower semi-continuity condition. We will use the following type of conditions used in the study of Morse index on Hilbert manifolds.

Definition 4.1. Let X be a Banach space and $f: X \to \mathbb{R}$ be C^1. We say f satisfies *condition* (C) if, for any sequence $\{v_n\} \subset X$ such that $|f(v_n)|$ is bounded and $|df(v_n)| \to 0$ as $n \to \infty$, there exists a convergent subsequence of $\{v_n\}$. If this condition is only satisfied in the region where $f \geq \alpha > 0$ (resp. $f \leq -\alpha < 0$) for all $\alpha > 0$, we say f satisfies *condition* (C^+) *(resp.* (C^-)).

We will now give some theorems which are related to condition (C) (or C^+, C^-).

Theorem 4.2. *Let X be a Banach space and $f: X \to \mathbb{R}$ be C^1. Then f satisfies condition (C^-) if and only if the following two conditions are satisfied:*

(a) *for every $k > 0$ and $\alpha > 0$, there exist $r \geq 0$ and $\delta > 0$ such that*

$$-k < f(v) < -\alpha, \qquad |v| > r \Rightarrow |df(v)| > \delta,$$

(b) *if a bounded sequence $\{v_n\} \subset X$ is such that $f(v_n) < 0$ and $|df(v_n)| \to 0$ as $n \to \infty$, then $\{v_n\}$ contains a convergent subsequence.*

Proof. Suppose f satisfies condition (C^-). It follows that f satisfies (b). Suppose (a) is false. Then there exists $k_0 > 0$ and $\alpha_0 > 0$ such that, for any integer $n \geq 1$, there exists $v_n \in X$ which satisfies

$$-k_0 < f(v_n) < -\alpha_0, |v_n| > n, \qquad |df(v_n)| \leq \frac{1}{n}.$$

By condition (C^-), $\{v_n\}$ contains a convergent subsequence. This contradicts the fact that $|v_n| > n$. Conversely, let f satisfy (a) and (b) and choose $\alpha > 0$. Let $\{v_n\} \subset X$ be a sequence along which $f(v_n) < -\alpha$ is bounded below and $|df(v_n)| \to 0$. By (a), the sequence $\{v_n\}$ is bounded. By (b), it contains a convergent subsequence. $\square$

Corollary 4.3. *If X is finite dimensional, then condition (C^-) is equivalent to* (a) *in Theorem* 4.2.

It is interesting to note that, if X is finite dimensional and $f: X \to \mathbb{R}$ satisfies condition (C), then the gradient flow defined by

(4.3)
$$\frac{d\varphi}{dt} = -\nabla f(\varphi), \qquad \varphi(0) = p, \, p \in X$$

has the following property: for every $k > 0$, there exists $r > 0$ such that the set of all bounded solutions $\varphi(t)$ of (4.3) satisfying

$$|f(\varphi(t))| < k, \qquad |\varphi(t)| < r$$

forms an isolated invariant set in the sense of Conley [1] provided $|\nabla f(v)| \neq 0$ whenever $|f(v)| = k$. There is also an analogous property in terms of isolating invariant sets if X is infinite dimensional.

Recall that a map $f : X \to Y$, where X and Y are topological spaces, is said to be *proper* if for every compact set $K \subset Y$, $f^{-1}(K) \subset X$ is compact.

Theorem 4.4. *Let X be a Banach space and $f : X \to \mathbb{R}$ be C^1. Then, if f satisfies condition (C), then the restriction of f to the set of all critical points of f is proper.*

Proof. To prove the assertion, let $W = \{v \in X : df(v) = 0\}$, $-\infty < a < b < \infty$ and $\{v_n\} \subset W \cap f^{-1}[a, b]$. Then $a \leq f(v_n) \leq b$ and $|df(v_n)| = 0$. By condition (C), $\{v_n\}$ contains a convergent subsequence $\{v_{n_j}\}$ such that $v_{n_j} \to v$ as $n_j \to \infty$. Since f is C^1, $df(v) = 0$ and $v \in W \cap f^{-1}[a, b]$. $\square$

Corollary 4.5. *If f satisfies condition (C), then $f(W)$ is closed, where $W = \{v \in X : df(v) = 0\}$.*

We will now show the relationship between condition (C) and the existence of critical points which may or may not be local minima. For simplicity, we assume that X is a Hilbert space with inner product $\langle \cdot, \cdot \rangle$. The dual space X^* is identified with X under the inner product. Let $f : X \to \mathbb{R}$ be C^2. For each $v \in X$, by the Riesz representation theorem, there exists a unique $\nabla f(v) \in X$ such that

$$df(v)(w) = \langle \nabla f(v), w \rangle \quad \text{for all } w \in X.$$

It is also clear that $\nabla f : X \to X$ is C^1 since f is C^2. Consider the gradient flow defined by the equation

$$(4.4) \qquad \frac{d\varphi}{dt} = -\nabla f(\varphi).$$

The initial value problem for (4.4) has a unique solution $\varphi(t)$ with maximal interval of existence (ω_-, ω_+), $-\infty \leq \omega_- < \omega_+ \leq \infty$. The proof is similar to the finite dimensional case and will be omitted.

Theorem 4.6. *Let X be a Hilbert space and $f : X \to \mathbb{R}$ be C^2 and satisfy condition (C). Let $\varphi(t)$ be a solution of (4.4) with maximal interval of existence $(\omega-, \omega+)$. Then either*

$$(4.5) \qquad \lim_{t \to \omega_+} f(\varphi(t)) = -\infty \quad \left(\text{resp. } \lim_{t \to \omega_-} f(\varphi(t)) = +\infty \right)$$

or $\omega_+ = \infty$ (resp. $\omega_- = -\infty$) and there exists $q \in X$ such that, for some sequence $t_n \to \infty$ (resp. $t_n \to -\infty$),

$$(4.6) \qquad \lim_{t_n \to \infty} \varphi(t_n) = q \quad \left(resp. \lim_{t_n \to -\infty} \varphi(t_n) = q \right) \quad and \quad df(q) = 0.$$

Proof. Let $g(t) = f(\varphi(t))$. By (4.4),

$$\frac{d}{dt} g(t) = -\langle \nabla f(\varphi(t)), \nabla f(\varphi(t)) \rangle = -|\nabla f(\varphi(t))|^2.$$

So $g(t)$ is monotonically decreasing and hence

$$\lim_{t \to \omega+} g(t) = c \geq -\infty.$$

If $c = -\infty$, then (4.5) holds. Let $c > -\infty$. Since

$$g(t) = g(s) - \int_s^t |\nabla f(\varphi(r))|^2 \, dr, \qquad 0 \leq s < t < \omega_+,$$

we have

$$(4.7) \qquad \int_0^{\omega+} |\nabla f(\varphi(r))|^2 \, dr < \infty.$$

Suppose $\omega_+ < \infty$. For $0 \leq s < t < \infty$,

$$|\varphi(t) - \varphi(s)| \leq \int_s^t |\nabla f(\varphi(r))| \, dr$$

$$\leq (t-s)^{1/2} \left(\int_s^t |\nabla f(\varphi(r))|^2 \, dr \right)^{1/2}$$

This implies that $\{\varphi(t)\}$ is Cauchy for $0 \leq t < \omega_+$ and hence there exists $v \in X$ such that $\varphi(t) \to v$ as $t \to \omega_+$. Hence, $\varphi(t)$ can be extended beyond ω_+ as a solution of (4.4). This contradicts the definition of ω_+. So, $\omega_+ = +\infty$. By (4.7), there exists a sequence $t_n \to \infty$ as $n \to \infty$ such that $|\nabla f(\varphi(t_n))| \to 0$ as $t_n \to \infty$. Since $\{f(\varphi(t_n))\}$ is bounded, by condition (C), we may assume that $\varphi(t_n) \to q \in X$ as $t_n \to \infty$. Since f is C^1, $\nabla f(q) = 0$. The case where $t \to \omega^-$ is analyzed in a similar way. $\square$

Corollary 4.7. *If all the critical points of f are isolated, then (4.6) may be replaced by*

$$(4.8) \qquad \lim_{t \to \infty} \varphi(t) = q \quad \left(resp. \lim_{t \to -\infty} \varphi(t) = q \right) \quad and \quad df(q) = 0.$$

Proof. Since all the critical points of f are isolated, we can find open neighborhoods U and V of q such that $\bar{U} \subset V$ and $\bar{V} - U$ contains no critical points of f. By condition (C), there exists $\delta > 0$ such that $|\nabla f(p)| \geq \delta$ for all $p \in \bar{V} - U$. Now, suppose that $\varphi(t)$ does not converge to q as $t \to \infty$. Hence, we may assume that there exists a sequence $\{s_n\}$ such that $s_n \to \infty$ as $n \to \infty$ and $\varphi(s_n) \in \bar{V} - U$. In fact, we may assume that there are disjoint intervals $[s_n, \bar{s}_n]$ such that, for all n,

$$\varphi([s_n, \bar{s}_n]) \subset \bar{V} - U, \qquad |\varphi(s_n) - \varphi(\bar{s}_n)| \geq \varepsilon > 0,$$

where $\varepsilon > 0$ is fixed. Hence,

$$\int_0^\infty |\nabla f(\varphi(t))|^2 \, dt \geq \sum_n \int_{s_n}^{\bar{s}_n} |\nabla f(\varphi(t))|^2 \, dt$$

$$\geq \delta \sum_n \int_{s_n}^{\bar{s}_n} |\nabla f(\varphi(t))| \, dt$$

$$\geq \delta \sum_n |\varphi(s_n) - \varphi(\bar{s}_n)|$$

$$\geq \delta \sum_n \varepsilon = \infty.$$

This contradicts (4.7) and proves (4.8). $\square$

Corollary 4.8. *Let X be a Hilbert space and $f : X \to \mathbb{R}$ be C^2 and satisfy condition (C). If f is bounded below, then there exists q such that*

$$f(q) = \inf\{f(v) : v \in X\} \stackrel{\text{def}}{=} c.$$

Proof. For every integer $n \geq 1$, let $p_n \in X$ and

$$c \leq f(p_n) < c + \frac{1}{n}.$$

Let $\varphi(t)$ be the solution of (4.4) with $\varphi(0) = p_n$. By Theorem 4.6, there exists $q_n \in X$ such that $df(q_n) = 0$ and

$$c \leq f(q_n) \leq f(p_n) < c + \frac{1}{n}.$$

By condition (C), we may assume that $q_n \to q \in X$. Hence, $f(q) = c$. $\square$

Remark 4.9. It is interesting to compare Theorem 2.5 with the above Corollary.

4.5. Minimax Principle in Banach Spaces

In this section, Ljusternik–Schnirelman theory on Banach spaces will be discussed. The results will then be generalized to Banach manifolds (see Section 4.8). The basic idea in Ljusternik–Schnirelman theory is the classical idea of Morse of deformation along the gradient flow defined by the associated functional. If the Banach space X is a Hilbert space and $f : X \to \mathbb{R}$ is C^2, then we may define the gradient flow of f as in the previous section by using the duality mapping. We note that $Vf(v) \in X$ for each $v \in X$ and $\langle Vf(v), w \rangle = df(v)(w)$ for all $v, w \in X$, where $\langle \cdot, \cdot \rangle$ denotes the inner product in X. Hence,

$$|df(v)|^2 = \sup\{df(v)(w) : |w| \le |df(v)| ; w \in X\}$$

and the supremum is attained at a unique point $w = Vf(v)$. If X is a reflexive Banach space, then by the weak compactness of closed balls, the supremum will be attained. However, it may not be attained at a unique point for general Banach spaces X. This motivates the following notion of a pseudo-gradient vector and pseudo-gradient flow of f.

Definition 5.1. If X is a Banach space and $f : X \to \mathbb{R}$ is C^1, then $w \in X$ is called a *pseudo-gradient vector* of f at $v \in X$ if

    (i) $|w| \le 2|df(v)|$
    (ii) $\langle df(v), w \rangle \ge |df(v)|^2.$

A vector field $\Phi : S \to X, S \subset X$ is called a *pseudo-gradient vector field* for f on S if, for each $v \in S$, $\Phi(v)$ is a pseudo-gradient vector of f at v.

Theorem 5.2. *Let $f : X \to R$ be C^1. Let $W = \{v \in X : df(v) = 0\}$ and $\tilde{X} = X \backslash W$. Then there exists a locally Lipschitz pseudo-gradient vector field Φ of f on $\tilde{X}$.*

Proof. Given $v \in \tilde{X}$, there exists $\tilde{w} \in X$ such that $|\tilde{w}| = 1$, $df(v)(\tilde{w}) > \frac{2}{3}|df(v)|$. Then $w = \frac{3}{2}|df(v)|\tilde{w}$ is a pseudo-gradient vector of f at v satisfying

$$|w| = \tfrac{3}{2}|df(v)| < 2|df(v)|;$$
$$df(v)(w) > |df(v)|^2.$$

Since the above inequalities are strict and df is continuous, w is a pseudo-gradient vector of f for all u in an open neighborhood N_v of v. The set of all such neighborhoods forms an open cover of $\tilde{X}$. Since $\tilde{X}$ is paracompact, there exists a locally finite refinement $\{N_{v_i}\}$. Define

$$\zeta_i(v) = \inf\{|v - w| : w \in \tilde{X} \backslash N_{v_i}\}$$

If $v_1, v_2 \in \tilde{X}$ and $\varepsilon > 0$, then there exists $w \in \tilde{X} \setminus N_{v_i}$ such that $|v_2 - w| < \zeta_i(v_2) + \varepsilon$. Hence,

$$\zeta_i(v_1) \leq |v_1 - w| \leq |v_1 - v_2| + \zeta_i(v_2) + \varepsilon$$

Since $\varepsilon > 0$ is arbitrary, $\zeta_i(v_1) - \zeta_i(v_2) \leq |v_1 - v_2|$. This implies $|\zeta_i(v_1) - \zeta_i(v_2)| \leq |v_1 - v_2|$; i.e., ζ_i is globally Lipschitz. Since $\{N_{vi}\}$ is a locally finite open cover, we can define $\phi_i(v)$ and $\Phi(v)$ by

$$\phi_i(v) = \frac{\zeta_i(v)}{\sum \zeta_j(v)}$$

$$\Phi(v) = \sum v_i \phi_i(v)$$

where $v_i \in N_{v_i}$ is the pseudo-gradient vector at v_i constructed earlier. It is not difficult to see that $\Phi(v)$ is locally Lipschitz and $\Phi(v)$ is a pseudo-gradient vector field of f on $\tilde{X}$. This completes the proof. $\quad\square$

We note that the pseudo-gradient vector field Φ of f is only defined on $\tilde{X}$. In order to obtain the generalization of the deformation theorem (Theorem 2.12.4) to Banach spaces, we have to examine carefully the flow generated by Φ near the critical points of f. However, if f satisfies condition (C) (or (C^+); (C^-)), then we may redefine Φ in such a way that the flow generated by the new vector field has all the desired properties. It is also interesting to observe that even if X is a Hilbert space, the gradient vector field given by df is only continuous ($f: X \to \mathbb{R}$ is C^1) but the new vector field will be locally Lipschitz.

Let $c \in \mathbb{R}$ be fixed, $\delta > 0$ and

$$\begin{aligned}
f^c &= \{v \in X : f(v) \leq c\}, \\
f^{-1}(c) &= \{v \in X : f(v) = c\}, \\
W &= \{v \in X : df(v) = 0\}, \\
W_c &= \{v \in W : f(v) = c\}, \quad \text{and} \\
U_\delta &= \{v \in X : d(v, w_c) < \delta\}
\end{aligned}$$

where

$$d(v, W_c) = \inf\{|v - w| : w \in W_c\}.$$

Lemma 5.3. *Suppose*: $X \to \mathbb{R}$ *is* C^1. *If* f *satisfies condition* (C), *and* $c \in \mathbb{R}$, *then for every* $\delta > 0$ *there exist* $b > 0$ *and* $\varepsilon > 0$ *such that*

$$(5.1) \qquad\qquad |df(v)| \geq b, \quad \text{for all } v \in f^{c+\varepsilon} - f^{c-\varepsilon} - U_\delta.$$

Proof. If $c \notin W$, then the Lemma follows from Theorem 4.2. Let $c \in W$ and suppose the lemma is false. Then there exist sequences $b_n \to 0$, $\varepsilon_n \to 0$ and

$v_n \in f^{c+\varepsilon_n} - f^{c-\varepsilon_n} - U_{\delta/2}$ such that $|df(v_n)| < b_n$. By condition (C), a sub-sequence of $\{v_n\}$ converges to v which satisfies $f(v) = c$, $df(v) = 0$ and $v \notin U_{\delta/2}$. But $v \in W_c$ which is a contradiction. $\quad\square$

We are now ready to construct a new vector field from the pseudo-gradient vector field. There exists $\varepsilon_1 > 0$ so that (5.1) remains valid for any $0 < \varepsilon < \varepsilon_1$ and we may assume

$$(5.2) \qquad\qquad 0 < \varepsilon < \varepsilon_1 < \min\left(\frac{b\delta}{32}, \frac{b^2}{8}, \frac{1}{8}\right).$$

Define

$$A = (X \backslash f^{c+\varepsilon_1}) \cup f^{c-\varepsilon_1}$$
$$B = f^{c+\varepsilon} \backslash f^{c-\varepsilon}$$

We note that $A \cap B = \varnothing$. Let

$$g(v) = \frac{d(v, A)}{d(v, A) + d(v, B)}, \qquad d(x, A) = \inf\{|v - w| : w \in A\}$$

and

$$\zeta(t) = \begin{cases} 1, & 0 \le t \le 1 \\ \dfrac{1}{t}, & t \ge 1. \end{cases}$$

It is clear that $0 \le g(v) \le 1$, g is globally Lipschitz and

$$g(v) = \begin{cases} 0, & v \in A \\ 1, & v \in B \end{cases}$$

Similarly, there exists a Lipschitz function $h(v)$ such that

$$h(v) = \begin{cases} 0, & v \in U_{\delta/4} \\ 1, & v \in X \backslash U_{\delta/8}, \end{cases}$$
$$0 \le h(v) \le 1.$$

Finally, let

$$\tilde{\Phi}(v) = g(v) h(v) \zeta(|\Phi(v)|) \Phi(v),$$

where Φ is the pseudo-gradient vector field of f on $\tilde{X}$ obtained from Theorem 5.2. We may now summarize the above construction as a Lemma.

Lemma 5.4. $\tilde{\Phi}: X \to X$ *is a locally Lipschitz vector field on* X *satisfying*:

$$(5.3) \qquad\qquad 0 \le \tilde{\Phi}(v) \le 1 \quad \text{for all } v \in X.$$

Lemma 5.5. *Consider the differential equation*

$$(5.4) \qquad\qquad \dot{\varphi} = -\tilde{\Phi}(\varphi), \qquad \varphi(0) = v.$$

The flow φ_t *satisfies the following*:

    (i) $\varphi_t(v)$ *is defined for all* $v \in X$ *and* $t \in \mathbb{R}$,
    (ii) $\varphi_0(v) = v$ *if* $v \in X$.
    (iii) $\varphi_t(v) = v$ *if* $t \in \mathbb{R}$, $v \notin f^{-1}([c - \varepsilon, c + \varepsilon])$.
    (iv) $\varphi_t(\cdot)$ *is a homeomorphism of* X *onto* X *for all* t.
    (v) $f(\varphi_t(v)) \le f(v)$ *if* $v \in X$, $t \ge 0$.

Proof. Since $\tilde{\phi}$ is locally Lipschitz, (5.4) is uniquely solvable on its maximal interval of existence (ω^-, ω^+). If $\omega^+ < \infty$, let $t_n \to \omega^+, t_n < \omega^+$. By integrating (5.4) and using (5.3),

$$\left| \varphi_{t_{n+1}}(v) - \varphi_{t_n}(v) \right| \le \int_{t_n}^{t_{n+1}} \left| \tilde{\Phi}(\varphi_s(v)) \right| ds \le \left| t_{n+1} - t_n \right|$$

This says that $\{\varphi_{t_n}(v)\}$ is a Cauchy sequence converging to some $\tilde{v} \in X$ as $t_n \to \omega^+$. But then the solution $\varphi_t(\tilde{v})$ gives a continuation of $\varphi_t(v)$ by a time shift, which is a contradiction. Hence $\omega^+ = \infty$. Similarly, $\omega^- = -\infty$. This proves (i) and (ii). Since $g(v) = 0$ on A, (iii) is proved. Since the inverse of $\varphi_t(\cdot)$ is $\varphi_{-t}(\cdot)$, (iv) is obtained. To show (v), we note that, if $df(\varphi_t(v)) \ne 0$, then

$$\frac{d}{dt} f(\varphi_t(v)) = \left\langle df(\varphi_t(v)), \frac{d}{dt} \varphi_t(v) \right\rangle$$

$$= -\left\langle df(\varphi_t(v)), \tilde{\Phi}(\varphi_t(v)) \right\rangle$$

$$= -\left\langle df(\varphi_t(v)), g(\varphi_t(v)) h(\varphi_t(v)) \zeta(|\Phi(\varphi_t(v))|) \Phi(\varphi_t(v)) \right\rangle$$

$$= -g(\varphi_t) h(\varphi_t) \zeta(|\Phi(\varphi_t)|) \langle df(\varphi_t), \Phi(\varphi_t) \rangle$$

By the definition of pseudo-gradient vector field,

$$\frac{d}{dt} f(\varphi_t) \le -g(\varphi_t) h(\varphi_t) \zeta(|\Phi(\varphi_t)|) |df(\varphi_t)|^2$$

$$\le 0$$

This verifies (v) if $df(\varphi_t(v)) \ne 0$. The remaining case is trivial. This completes the proof. $\square$

Theorem 5.6 (Deformation Theorem). *Let $f: X \to \mathbb{R}$ be C^1 and satisfy condition (C). If $c \in R$ and U is an open neighborhood of W_c, then there exist $\varepsilon > 0$ and a flow φ_t satisfying all the conditions in Lemma 5.5 such that*

$$\varphi_1(f^{c+\varepsilon} - U) \subset f^{c-\varepsilon}$$

Proof. Since the proof for the case $W_c = \varnothing$ is similar and simpler, we assume $W_c \neq \varnothing$. By condition (C) and Theorem 4.4, W_c is compact. Hence, there exist $\delta > 0$ such that $U_{\delta'} \subset U$ where $8\delta' = \delta$. We now use the above construction (Lemma 5.3–Lemma 5.5) to define the flow φ_t and $\varepsilon_1 > \varepsilon > 0$ are exactly as in (5.2). It suffices to show

$$\varphi_1(f^{c+\varepsilon} - U_\delta) \subset f^{c-\varepsilon}$$

By (v) of Lemma 5.5, we only need to show

$$f(\varphi_1(v)) \leq c - \varepsilon, \quad \text{for all } v \in f^{c+\varepsilon} - f^{c-\varepsilon} - U_\delta$$

For fixed $v \in f^{c+\varepsilon} - f^{c-\varepsilon} - U_\delta$, let $\alpha(t) = f(\varphi_t(v))$. Hence, we need to show

(5.5) $$\alpha(1) \leq c - \varepsilon.$$

By Lemma 5.5, $\alpha'(t) \leq 0$. Since $g(w) = \tilde{\Phi}(w) = 0$ for $w \in f^{c-\varepsilon_1}$, $\varphi_t(v) \notin f^{c-\varepsilon_1}$. This implies that

(5.6) $$0 \leq \alpha(0) - \alpha(t) \leq 2\varepsilon,$$

For small $t > 0$, $\varphi_s(v) \in f^{c+\varepsilon} - f^{c-\varepsilon} - U_{\delta/2}$ for $s \in [0, t]$. By (5.1), $|df(\varphi_s(v))| \geq b > 0$ for $s \in [0, t]$. By their definitions, $g(\varphi_s(v)) = h(\varphi_s(v)) = 1$. Therefore, by dropping the dependence on v,

$$2\varepsilon_1 \geq \int_0^t \zeta(|\Phi(\varphi_s)|)\langle df(\varphi_s), \Phi(\varphi_s)\rangle \, ds$$

$$\geq \int_0^t \zeta(|\Phi(\varphi_s)|)|df(\varphi_s)|^2 \, ds, \quad \text{(by Definition 5.1)}$$

$$\geq b \int_0^t \zeta(|\Phi(\varphi_s)|)|df(\varphi_s)| \, ds, \quad \text{(by (5.1))}$$

$$\geq \frac{b}{2} \int_0^t \zeta(|\Phi(\varphi_s)|)|\Phi(\varphi_s)| \, ds \quad \text{(by Definition 5.1)}$$

$$\geq \frac{b}{2} \left| \int_0^t \zeta(|\Phi(\varphi_s)|)\Phi(\varphi_s) \, ds \right|$$

$$= \frac{b}{2} \left| \int_0^t \tilde{\Phi}(\varphi_s) \, ds \right|$$

$$= \frac{b}{2} |\varphi_t(v) - v|.$$

The above inequality and (5.2) imply

$$|\varphi_t(v) - v| < \frac{\delta}{8}$$

This says that the orbit $\varphi_t(v)$ cannot enter $U_{\delta/2}$ and therefore can only leave the set $(f^{c+\varepsilon} - f^{c-\varepsilon} - U_{\delta/2})$ by entering $f^{c-\varepsilon}$. To show this in fact occurs for some $t \in [0,1)$, i.e., (5.5) holds, we suppose otherwise. Then, $\varphi_t(v) \in f^{c+\varepsilon} - f^{c-\varepsilon} - U_{\delta/2}$ for $t \in [0,1]$. By (5.1), $|df(\varphi_t(v))| \geq b > 0$. Since $g(\varphi_s(v)) = h(\varphi_s(v)) = 0$,

$$\frac{d}{dt}\alpha(t) \leq -\zeta(|\Phi(\varphi_t)|)|df(\varphi_t)|^2$$

where $\varphi_t = \varphi_t(v)$. By the definition of ζ,

$$\zeta(|\Phi(\varphi_t)|) = \begin{cases} 1 & \text{if } |\Phi(\varphi_t)| \leq 1 \\ |\Phi(\varphi_t)|^{-1} & \text{if } |\Phi(\varphi_t)| \geq 1. \end{cases}$$

Hence, either (if $|\Phi(\varphi_t)| \leq 1$)

$$\frac{d}{dt}\alpha(t) \leq -|df(\varphi_s)|^2 \leq -b^2$$

or (if $|\Phi(\varphi_t)| \geq 1$)

$$\frac{d}{dt}\alpha(t) \leq -\frac{|df(\varphi_s)|^2}{|\Phi(\varphi_s)|} \leq -\frac{|\Phi(\varphi_s)|}{4} \leq -\frac{1}{4}.$$

Combining both inequalities,

$$\frac{d}{dt}\alpha(t) \leq -\min(b^2, \tfrac{1}{4}).$$

By (5.6),

$$\min(b^2, \tfrac{1}{4}) \leq \alpha(0) - \alpha(1) \leq 2\varepsilon_1.$$

This violates (5.2) and completes the proof. $\quad\square$

Corollary 5.7. *If $c > 0$ ($c < 0$), then Theorem 5.6 is still true when we replace condition (C) by condition (C^+) (condition (C^-)). Moreover, ε can be chosen in $(0, c)((c, 0))$.*

We may now formulate the minimax principle as a characterization of critical values. Let $\mathscr{T}$ be a family of subsets of X satisfying the following property:

Property $(P)_c$. If $F \in \mathscr{T}$, then $\varphi_1(F) \in \mathscr{T}$.

We note that the above definition depends on $c \in \mathbb{R}$ since ϕ_t does.

Theorem 5.8 (Minimax Principle). *Let $f: X \to \mathbb{R}$ be C^1 and satisfy condition (C). Let $\mathcal{F}$ be a nonempty family of subsets of E satisfying property $(P)_c$ where c is given by*

$$c = \inf_{F \in \mathcal{F}} \sup\{f(v): v \in F\}$$

and is finite. Then c is a critical value of f.

Remark 5.9. We note that $c < \infty$ if and only if there exists $F \in \mathcal{F}$ such that f is bounded above on F, and that $-\infty < c$ if f is bounded below.

Proof. It is not difficult to see that

$$(5.7) \qquad c = \inf\{b \in R: \text{there exists } F \in \mathcal{F} \text{ with } F \subseteq f^b\}$$

Hence, for any $\varepsilon > 0$, there exists $F_\varepsilon \in \mathcal{F}$ such that $F_\varepsilon \subseteq f^{c+\varepsilon}$. If c is a regular value of f, then the set U in Theorem 5.6 is empty. With $\varepsilon > 0$ exactly as in Theorem 5.8, we have $\varphi_1(f^{c+\varepsilon}) \subseteq f^{c-\varepsilon}$. Thus, $\varphi_1(F) \subseteq f^{c-\varepsilon}$. By property $(P)_c$, $\varphi_1(F) \in \mathcal{F}$, contradicting the characterization (5.7). This completes the proof. $\square$

We now give an elementary application of the minimax principle.

Theorem 5.10. *Let $f: X \to \mathbb{R}$ be C^1 and satisfy condition (C). If f is bounded above (below), then f attains its maximum (minimum).*

Proof. If f is bounded above, let $\mathcal{F} = \{X\}$. If f is bounded below, let $\mathcal{F} = \{\{v\}: v \in X\}$. $\square$

Remark 5.11. Theorem 5.10 is a generalization of Corollary 4.7.

4.6. Mountain Pass Theorem

As a first application of minimax principle, we will consider the example presented at the beginning of Section 4. First, we will present an abstract result.

Again, let X be a Banach space and $f: X \to \mathbb{R}$ be C^1. Let

(a) $f(0) = 0$ and there exist $\alpha, \beta > 0$ such that

$$f(v) > 0, \quad \text{if } 0 < |v| < \alpha$$
$$f(v) \geq \beta, \quad \text{if } |v| = \alpha$$

(b)   there exist $e \in X$ such that $e \neq 0$, $f(e) = 0$.

Theorem 6.1. *Let $f: X \to \mathbb{R}$ be C^1 and satisfy conditions (a) and (b) above. If f satisfies condition (C^+), then f has a critical value c, $\infty > c \geq \beta > 0$.*

Proof. Let

$$\Gamma = \{\gamma : [0,1] \to X : \gamma(0) = 0, \gamma(1) = e, \gamma \text{ continuous}\}$$

For each $\gamma \in \Gamma$, let

$$F_\gamma = \{\gamma(t) \in X : 0 \leq t \leq 1\} \quad \text{and} \quad \mathcal{F} = \{F_\gamma : \gamma \in \Gamma\}.$$

Define

$$c = \inf_{F_\gamma \in \mathcal{F}} \sup \{f(v) : v \in F_\gamma\}.$$

Condition (a) implies that $c \geq \beta > 0$. Let $0 < \varepsilon < c$. By (iii) of Lemma 5.5, $\mathcal{F}$ satisfies property $(P)_c$. Hence, c is a critical value of f by Theorem 5.8. $\quad\square$

As an application, recall the example at the beginning of Section 4. Consider the nonlinear elliptic boundary value problem in $\mathbb{R}^3$

$$\text{(6.1)} \qquad\qquad\qquad \Delta u + u^3 = 0, \qquad x \in \Omega$$

$$\text{(6.2)} \qquad\qquad\qquad u = 0, \qquad x \in \partial\Omega$$

where Ω, $\partial\Omega$ are as before. We will show that (6.1), (6.2) have a nontrivial solution. The method can be extended to more general nonlinearities.

By the Sobolev inequalities, the embedding $W_0^{1,2}(\Omega) \to L^p(\Omega)$, $1 \leq p \leq 6$, is continuous; i.e.,

$$\left\{ \int_\Omega |u|^p \, dx \right\}^{1/p} \leq c|u|_{1,2} \quad (c = \text{constant})$$

For $p = 4$, this implies that $u = 0$ is a local minimum for

$$f(u) = \tfrac{1}{2} \int_\Omega |\nabla u|^2 \, dx - \tfrac{1}{4} \int_\Omega |u|^4 \, dx$$

where $u \in W_0^{1,2}(\Omega)$. Let $\bar{u} \in W_0^{1,2}(\Omega)$ be fixed and satisfy:

$$\int_\Omega |\nabla \bar{u}|^2 \, dx = 1, \qquad \int_\Omega |\bar{u}|^4 \, dx = c > 0$$

For $\alpha > 0$ large, we have

$$f(\alpha \bar{u}) = \tfrac{1}{2}\alpha^2 - \tfrac{1}{4}\alpha^4 c < 0.$$

Hence, conditions (a) and (b) in Theorem 6.1 are satisfied. It remains to show that f satisfies condition (C^+). Let $\{u_m\} \subset W_0^{1,2}(\Omega)$, $m = 1, 2, \ldots$. Suppose that $0 < f(u_m) < M < \infty$ and $df(u_m) \to 0$ as $m \to \infty$. Then,

$$M > f(u_m) = \tfrac{1}{2}|u_m|_{1,2}^2 - \tfrac{1}{4}\int_\Omega |u_m|^4 \, dx$$

$$df(u_m) \cdot v = \langle u_m, v \rangle_{1,2} - \int u_m^3 v \, dx \to 0$$

where $\langle \cdot, \cdot \rangle_{1,2}$ denotes the inner product in $W_0^{1,2}(\Omega)$. Since, for m large,

$$|df(u_m) \cdot v| \le |v|_{1,2},$$

$$M \ge \tfrac{1}{4}|u_m|_{1,2}^2 + \tfrac{1}{4}\left\{|u_m|_{1,2}^2 - \int_\Omega u_m^4 \, dx\right\}$$

$$\ge \tfrac{1}{4}|u_m|_{1,2}^2 + |u_m|_{1,2},$$

it follows that $\{|u_m|_{1,2}\}$ is bounded. Note that $df(u) \cdot v = \langle u - T(u), v \rangle_{1,2}$, where $T : W_0^{1,2}(\Omega) \to W_0^{1,2}(\Omega)$ is a nonlinear compact operator. Since $\{u_m\}$ is bounded in $W_0^{1,2}(\Omega)$, T is compact and $df(u_m) = u_m - T(u_m) \to 0$ as $m \to \infty$, $\{u_m\}$ contains a convergent subsequence in $W_0^{1,2}(\Omega)$.

4.7. Periodic Solutions of a Semilinear Wave Equation

As another application of Theorem 6.1, we consider the existence of non-trivial periodic solutions of a semilinear wave equation. Let $\Omega = (0, \pi) \times (0, 2\pi)$. Consider the nonlinear scalar wave equation:

$$(7.1) \qquad u_{tt} - u_{xx} + u^3 = 0 \qquad (x, t) \in \Omega$$

with the Dirichlet boundary conditions:

$$(7.2) \qquad u(0, t) = u(\pi, t) = 0.$$

We say u is a nontrivial solution of (7.1), (7.2) if $u(x, t) \ne 0$ on a set (x, t) of positive measure.

Theorem 7.1. *Equation* (7.1) *under boundary conditions* (7.2) *has a nontrivial time periodic solution of period T for any T which is a rational multiple of π.*

The remainder of this section is devoted to the proof of Theorem 7.1.

Let the period $T = 2\pi$ be fixed. The existence of periodic solutions whose periods are rational multiples of π may be shown by the same method. Thus, we have the periodic condition:

$$(7.3) \qquad u(x, t + 2\pi) = u(x, t).$$

Consider the linear wave operator $Au = u_{tt} - u_{xx}$. The kernel N of A acting on functions in L^1 satisfying (7.2), (7.3) consists of precisely functions of the form:

$$N = \left\{ p(t + x) - p(t - x) : p \text{ is } 2\pi\text{-periodic}, p \in L^1_{\text{loc}}(\mathbb{R}) \text{ and } \int_0^{2\pi} p(s)\,ds = 0 \right\}.$$

This can be shown easily by using (7.2), (7.3) and the fact that the general solution of A has the form $h(t + x) + g(t - x)$ for some functions h and g. We note that $N \subset L^1$ is closed.

Lemma 7.2. *If $h \in L^1(\Omega)$ and*

$$(7.4) \qquad\qquad \int_\Omega h\varphi = 0 \quad \text{for all } \varphi \in N \cap L^\infty,$$

then there exists a unique continuous solution Kh of the equation

$$(7.5) \qquad\qquad Au = h$$

satisfying (7.1), (7.3) and

$$(7.6) \qquad\qquad \int_\Omega \varphi Kh = 0 \quad \text{for all } \varphi \in N.$$

Proof. Suppose $\tilde{u}$ is a smooth solution of (7.5). Integrating (7.5) over the shaded region in Fig. 7.1 and using (7.2),

$$\tilde{u}(x, t) = -\tfrac{1}{2} \int_{t+x-\pi}^{t-x+\pi} \tilde{u}_x(\pi, \tau)\,d\tau - \tfrac{1}{2} \int_x^\pi \int_{t+x-\xi}^{t-x+\xi} h(\xi, \tau)\,d\tau\,d\xi.$$

Since $\tilde{u}_x$ is 2π-periodic and (7.2) is satisfied,

$$\int_0^\pi \int_{t-\xi}^{t+\xi} h(\xi, \tau)\,d\tau\,d\xi = -\int_{t-\pi}^{t+\pi} \tilde{u}_x(\pi, \tau)\,d\tau$$

$$= \text{constant, } k.$$

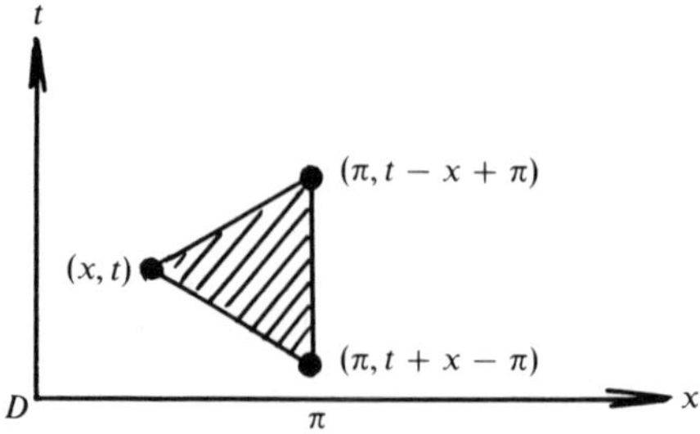

Figure 7.1

Hence, we may let

$$(7.7) \qquad \tilde{u}(x,t) = -\tfrac{1}{2} \int_x^\pi \int_{t+x-\xi}^{t-x+\xi} h(\xi,\tau)\,d\tau\,d\xi + k\,\frac{\pi - x}{\pi}$$

In order to meet the orthogonality condition (7.6), let

$$(7.8) \qquad (Kh)(x,t) = \tilde{u}(x,t) + p(t+x) - p(t-x)$$

where

$$(7.9) \qquad p(x) = \frac{1}{2\pi} \int_0^\pi \left[\tilde{u}(x, s-x) - \tilde{u}(x, s+x) \right] dx$$

We may now prove the lemma by using directly the explicit formula for Kh. $\square$

Lemma 7.3. *Let h and Kh be as in the Lemma 7.2. Then there are constants $c_1, c_2 > 0$ such that*

$$(7.10) \qquad |Kh|_{L^\infty} \le c_1 |h|_{L^1}$$

$$(7.11) \qquad |Kh|_{C^{0,\alpha}} \le c_2 |h|_{L^q} \quad \text{where } \alpha = 1 - \frac{1}{q}.$$

Proof. (7.10) is trivial. In order to show (7.11), it suffices to show $\tilde{u}$ given by (7.7) is in $C^{0,\alpha}$. In fact, we may just consider the function

$$\psi(x,t) = \int_x^\pi \int_{t+x-\xi}^{t-x+\xi} h(\xi,\tau)\,d\tau\,d\xi$$

Let h_1, h_2 be reals. By using Hölder's inequality,

$$|\psi(x+h_1, t+h_2) - \psi(x,t)| \le \left| \int_x^{x+h_1} \int_{t+x-\xi}^{t-x+\xi} h(\xi,\tau)\,d\tau\,d\xi \right|$$

$$+ \left| \int_x^\pi \int_{t+x-\xi}^{t+x-\xi+h_1+h_2} f(\xi,\tau)\,d\tau\,d\xi \right|$$

$$+ \left| \int_x^\pi \int_{t-x+\xi}^{t-x+\xi+h_2-h_1} f(\xi,\tau)\,d\tau\,d\xi \right|$$

$$\le c_2(|h_1| + |h_2|)^\alpha |f|_{L_p}.$$

This completes the proof. $\square$

Consider the space

$$X = \left\{ v \in L^{4/3}(\Omega) \colon \int_\Omega v\varphi = 0 \text{ for all } \varphi \in N \cap L^4 \right\}$$

provided with the $L^{4/3}$-norm. Thus, X is a Banach space with dual X^*. By using Lemma 7.3, we have

Lemma 7.4. $K \colon X \to L^4$ *is compact.*

Define the functional $f \colon X \to \mathbb{R}$ by

$$f(v) = \tfrac{1}{2} \int_\Omega (Kv)v + \tfrac{3}{4} \int_\Omega |v|^{4/3}$$

It is clear that f is C^1 and its derivative $df(v) \in X^*$ satisfies:

$$\langle df(v), \eta \rangle = \int_\Omega (Kv)\eta + \int_\Omega |v|^{-2/3} v\eta, \qquad \eta \in X.$$

By the Hahn-Banach theorem, we may express $df(v)$ as follows:

$$(7.12) \qquad\qquad Kv + |v|^{-2/3}v = w + \theta$$

where $w \in L^4$, $|w|_{L^4} = |df(v)|$ and $\theta \in N \cap L^4$.

Lemma 7.5. *The above functional* $f \colon X \to \mathbb{R}$ *satisfies condition* (C).

Proof. Let $c > 0$ and $\{v_j\} \subset X, j = 1, 2, \ldots,$ satisfy:

$$|f(v_j)| \le c \quad \text{for all } j = 1, 2, \ldots, \quad \text{and}$$
$$|df(v_j)| \to 0 \quad \text{as } j \to \infty.$$

By (7.12), $Kv_j + |v_j|^{-2/3}v_j = w_j + \theta_j$ where $w_j \in L^4$, $\theta_j \in N \cap L^4$ and $w_j \to 0$ as $j \to \infty$. The conditions on $\{v_j\}$ say that

$$c \ge \left| \tfrac{1}{2} \int_\Omega (Kv_j)v_j + \tfrac{3}{4} \int_\Omega |v_j|^{4/3} \right|$$

and

$$\left| \tfrac{1}{2} \int_\Omega (Kv_j)v_j + \tfrac{1}{2} \int_\Omega |v_j|^{4/3} \right| = \tfrac{1}{2} \left| \int_\Omega w_j v_j \right|$$
$$\le \tfrac{1}{2} |w_j|_{L^4} |v_j|_{L^{4/3}}$$

Hence,

$$(\tfrac{3}{4} - \tfrac{1}{2}) \int_\Omega |v_j|^{4/3} \le c + \tfrac{1}{2}|w_j|_{L^4}|v_j|_{L^{4/3}}$$

This implies that $\{|v_j|_{L^{4/3}}\}$ is bounded and we may assume v_j converges weakly in X to some $v \in X$ as $j \to \infty$. By the convexity of the mapping $t \to |t|^{4/3}$,

$$\tfrac{3}{4}|v|^{4/3} - \tfrac{3}{4}|v_j|^{4/3} \ge |v_j|^{-2/3}v_j(v - v_j)$$
$$= (w_j + \theta_j - Kv_j)(v - v_j)$$

Upon integration,

$$\tfrac{3}{4}\int_\Omega |v|^{4/3} - \tfrac{3}{4}\int_\Omega |v_j|^{4/3} \ge \int_\Omega (w_j - Kv_j)(v - v_j)$$

Since K is compact (Lemma 7.4), $|df(v_j)| \to 0$ as $j \to 0$ and $v_j \to v$ weakly in X as $j \to \infty$, the right hand side of the above inequality goes to zero as $j \to \infty$. Hence,

$$\limsup_{j \to \infty} |v_j|_{L^{4/3}} \le |v|_{L^{4/3}}$$

This implies $v_j \to v$ strongly in X as $j \to \infty$ and f satisfies condition (C). $\quad\square$

Lemma 7.6. *There exist $r > 0$, $\zeta > 0$ and $e \in X$ with $|e| > r$ such that $f(0) = f(e) = 0$ and $f(v) \ge \zeta$ for all $|v|_{L^{4/3}} = r$.*

Proof. By (7.10) in Lemma 7.3, there exist constants $c_3, c_4 > 0$ such that for all $|v|_{L^{4/3}} = r$.

$$f(v) \ge -c_3|v|_{L^1}^2 + \tfrac{3}{4}|v|_{L^{4/3}}^{4/3}$$
$$\ge -c_4|v|_{L^{4/3}}^2 + \tfrac{3}{4}|v|_{L^{4/3}}^{4/3}$$
$$= \tfrac{3}{4}r^{4/3} - c_4 r^2 > 0 \quad \text{for all } 0 < |r| \text{ sufficiently small}$$

Next, it is easy to see that K is a compact self-adjoint operator in an appropriate subspace of L^2 with eigenvalues $(j^2 - k^2)^{-1}$, $j, k = 1, 2, \ldots, j \ne k$. Thus, there exists $v_1 \in X$ such that

$$\int_\Omega (Kv_1)v_1 < 0.$$

Let $\alpha > 0$. We have, for $\alpha > 0$ large,

$$f(\alpha v_1) = \tfrac{1}{2}\alpha^2 \int_\Omega (Kv_1)v_1 + \alpha^{4/3} \cdot \tfrac{3}{4}\int_\Omega |v_1|^{4/3} < 0.$$

This proves the lemma. $\quad\square$

Proof of Theorem 7.1. By Lemmas 7.5 and 7.6, we may use Theorem 6.1 to conclude the existence of a nonzero critical point $\bar{u}$ of f. Since $df(\bar{u}) = 0$,

$$K\bar{u} + |\bar{u}|^{-2/3}\bar{u} = \theta, \qquad \theta \in N \cap L^4$$

because of (7.12). If $u = \theta - K\bar{u}$, then $Au = A(\theta - K\bar{u}) = -\bar{u}$. On the other hand, $|\bar{u}|^{-2/3}\bar{u} = \theta - K\bar{u} = u$. This implies that $u^3 = \bar{u}$. Hence, $Au + u^3 = 0$. We may also show that $u \in L^\infty$. By using regularity theorems we have $u \in C^\infty$. $\square$

The above proof will be valid even when the nonlinear term u^3 in (7.1) is replaced by a monotone function $g(u)$. Of course, there will be a restriction on the rate of growth at infinity which is compatible with the Sobolev space being used.

4.8. Ljusternik–Schnirelman Theory on Banach Manifolds

Let X be a Banach space and M be a manifold modelled on X. In applications, M will be typically the unit sphere in X or sets of the form:

$$\{v : v \in X, h(v) = c\}$$

where c is a constant and $h : X \to \mathbb{R}$ is C^1. In many interesting problems, the first derivative dh of the nonlinear function h is Lipschitz continuous but not differentiable. For example, let $g : \mathbb{R} \to \mathbb{R}$ be smooth and $\Omega \subset \mathbb{R}^n$, $n \geq 2$, be bounded and open with smooth boundary. Define the functional h on the Sobolev space $H_0^1(\Omega)$

$$h(v) = \int_\Omega g(v).$$

If $g(v) \neq v^2$, then h is in general not C^2 but its derivative is Lipschitz. This motivates the following definition.

Definition 8.1. A C^1 manifold M modelled on X is called C^{2-} if the first derivatives of the coordinate transformations are locally Lipschitz.

Let M be a C^{2-} manifold modelled on X and $f : M \to \mathbb{R}$ be C^1. Denote the tangent bundle of M by TM and the tangent space at $v \in X$ by T_vM. Recall that $v \in M$ is a critical point of f if $df(v) \cdot \theta = 0$ for all $\theta \in T_vM$, where $df(v)$ is the derivative f at v and is an element of the cotangent space T_v^*M. The following theorem relates critical points to the method of Lagrange multipliers.

Theorem 8.2. *Let* $f, h: X \to \mathbb{R}$ *be* C^1 *and*

$$M = \{v : h(v) = c\}$$

where $c \in \mathbb{R}$ *is fixed. Assume that for every* $v \in M$, $dh(v) \neq 0$. *Then*

(1) *M is a C^1 manifold and, for each $v \in M$, the tangent space $T_v M$ can be identified by the subspace*

$$\{w \in X : \langle dh(v), w \rangle = 0\}$$

(2) *if v is a critical point of the restriction $f: M \to \mathbb{R}$, then there exists $\lambda \in \mathbb{R}$ such that*

$$df(v) = \lambda dh(v).$$

Proof. (1) is a direct consequence of the implicit function theorem. By definition, $df(v) \cdot \theta = 0$ for all $\theta \in T_v M$. Hence, $\langle df(v), w \rangle = 0$ for all $w \in X$ satisfying $\langle h(v), w \rangle = 0$. Since $dh(v) \neq 0$, there exists $w_0 \in X$ such that $\langle dh(v), w_0 \rangle \neq 0$. By the implicit function theorem, each element $w \in X$ has a representation $w = \theta(w) + \alpha(w)w_0$, where $\theta(w) \in T_v M$, $\alpha(w) \in \mathbb{R}$ and $\theta(w)$ and $\alpha(w)$ are C^1. We note that, for any $w \in X$,

$$\langle df(v), w \rangle = \alpha(w)\langle df(v), w_0 \rangle$$
$$\langle dh(v), w \rangle = \alpha(w)\langle dh(v), w_0 \rangle.$$

If we let $\lambda = \langle df(v), w_0 \rangle / \langle dh(v), w_0 \rangle$, then $df(v) = \lambda dh(v)$. $\square$

Definition 8.3. Let M be a C^1 manifold modelled on X. A *Finsler structure* on M is given by a continuous map $|\cdot| : TM \to \mathbb{R}$ such that (i) for every $v \in M$, $|\cdot|_v : T_v M \to \mathbb{R}$ is a norm; (ii) there exists a neighborhood U of $v \in M$ in which TM is trivialized as $U \times X$ such that

$$\frac{1}{k} |x|_u \leq |x|_v \leq k|x|_u, \qquad u \in U, x \in X$$

for some $k > 1$.

Definition 8.4. *A* $C^k(C^{k^-})$ *Finsler manifold* is a $C^k(C^{k^-})$ Banach manifold together with a Finsler structure.

Note that X is a Banach manifold and $TX = X \times X$. The mapping $|\cdot| : TX \to \mathbb{R}$ given by $|(v, x)| = |x|$ is clearly a Finsler structure on X. Hence, if $M \subset X$ is a C^1 submanifold, then M has the induced Finsler structure on M and is a Finsler manifold.

Definition 8.5. Let M be a C^1 Finsler manifold and $\sigma:[t_1, t_2] \to M$ be a C^1 curve. We define the *length* of σ by

$$l(\sigma) = \int_{t_1}^{t_2} |\dot{\sigma}(t)| \, dt$$

If $u, v \in M$ belong to the same component of M, we define the distance between u and v by

$$(8.1) \qquad \zeta(u, v) = \inf \int_{t_1}^{t_2} |\dot{\sigma}(t)| \, dt$$

where the infimum is taken over all C^1 curves joining u and v.

Theorem 8.6. *The function ζ defined by (8.1) is a metric on M and its metric topology is consistent with the topology on M.*

Let M be a C^1 Finsler manifold and TM^* be the cotangent bundle. There exists a natural Finsler structure for TM^* defined by

$$(8.2) \qquad \|x^*\| = \sup\{\langle x^*, v \rangle : |x|_v \leq 1\}$$

for $x^* \in T_v M^*$. For simplicity, we will write $|x^*|$ for $\|x^*\|$.

Definition 8.7. Let M be a C^1 Finsler manifold and $f: M \to \mathbb{R}$ be C^1. f is said to satisfy *condition* (C) if, for any sequence $\{v_n\} \subset M$ such that $\{|f(v_n)|\}$ is bounded and $|df(v_n)| \to 0$ as $n \to \infty$, there exists a convergent subsequence. Note that $|df(v_n)|$ is defined by (8.2). Similarly, we define conditions (C^+) and (C^-) as in Definition 4.1.

We now proceed as in Section 5.4. Since most of the proofs are similar to those in 5.4, we will present only the theorems.

Definition 8.8. Let M be a C^1 Finsler manifold, $f: M \to \mathbb{R}$ be C^1 and $v \in M$. A vector $w \in T_v M$ is called a *pseudo-gradient vector* of f at v if

   (i) $|w| \leq 2|df(v)|$
   (ii) $\langle df(v), w \rangle \geq |df(v)|^2$

where $\langle \cdot, \cdot \rangle$ denotes the pairing between $T_v M$ and its dual space $T_v M^*$. If $S \subset M$ and Φ is a vector field on S, then Φ is called a *pseudo-gradient vector field* for f on S if, for each $v \in S$, $\Phi(v)$ is a pseudo-gradient vector of f at v.

Theorem 8.9. *Let M be a C^{2^-} Finsler manifold and $f: M \to \mathbb{R}$ be C^1. If*

$$W = \{v \in M : df(v) = 0\},$$

then there exists a locally Lipschitz pseudo-gradient vector field Φ for f on $M \backslash W$.

Remark. Φ may be constructed in the same way as in Theorem 4.2. The smoothness condition (C^{2-}) on M gives a locally finite partition of unity with locally Lipschitz functions on $M \backslash W$.

With minor changes of the proofs of Lemmas 5.3, 5.4 and 5.5, one proves the following theorem.

Theorem 8.10. *Let M be a C^{2-} Finsler manifold and $f : M \to \mathbb{R}$ be C^1. If f satisfies condition (C) on M and $c \in \mathbb{R}$ is fixed, then there exist a pseudo-gradient vector field $\tilde{\Phi}$ for f on M and $\varepsilon > 0$ such that the flow $\varphi_t(\cdot)$ generated by $-\tilde{\Phi}$ satisfies the following*

- (i)   *$\varphi_t(v)$ is defined for all $v \in M$ and $t \in \mathbb{R}$*
- (ii) *$\varphi_0(v) \equiv v$*
- (iii) *$\varphi_t(v) = v$ if $t \in \mathbb{R}$ and $v \notin f^{-1}([c - \varepsilon, c + \varepsilon])$*
- (iv) *$\varphi_t(\cdot)$ is a homeomorphism of M onto M for any fixed $t \in \mathbb{R}$*
- (v)   *$f(\varphi_t(v)) \leq f(v)$ if $v \in M$ and $t \geq 0$.*

Theorem 8.11 (Deformation Theorem). *Under the hypotheses of Theorem 8.10, for any open neighborhood U of $W_c = \{v \in M : df(v) = 0, f(v) = c\}$, there exists $\varepsilon \geq \varepsilon_1 > 0$ such that*

$$\varphi_1(f^{c + \varepsilon_1} - U) \subset f^{c - \varepsilon_1}$$

where $f^{\tilde{c}} = \{v \in M : f(v) \leq \tilde{c}\}$ for any $\tilde{c} \in \mathbb{R}$, and ε and $\varphi_t(\cdot)$ are given by Theorem 8.10.

Theorem 8.12 (Minimax Principle). *Let M be a C^{2-} Finsler manifold and $f : M \to \mathbb{R}$ be C^1 and satisfy condition (C). Suppose $\mathscr{F}$ is a nonempty family of subsets $\{F\}$ of M and*

$$(8.3) \qquad\qquad c = \inf_{F \in \mathscr{F}} \sup \{f(v) : v \in F\}$$

is finite. If $\mathscr{F}$ satisfies the following condition:

$$(8.4) \qquad\qquad F \in \mathscr{F} \Rightarrow \varphi_1(F) \in \mathscr{F}$$

where $\varphi_t(\cdot)$ is the flow given by Theorem 8.10 with the constant c as in (8.3), then c is a critical value of f.

By using Theorem 8.12, the minimax principle, one obtains the following theorem as in the finite dimensional case (see Section 2.12).

Theorem 8.13. *In addition to the hypotheses of Theorem 8.12, assume that f is bounded below. If*

$$\mathscr{F}_k = \{F \subset M : \mathrm{Cat}(F, M) \geq k\}, \qquad k = 1, 2, \ldots,$$

then

$$c_k = \inf_{F \in \mathscr{F}_k} \sup\{f(v) : v \in F\}$$

is a critical value of f. In particular, f has at least $\mathrm{Cat}(M, M)$ *critical points.*

It is well-known that, if $S \subset X$ is the unit sphere and $\dim X = \infty$, then $\mathrm{Cat}(P(S), P(S)) = \infty$ where $P(S)$ is the projective space obtained by identifying $x \sim -x$ on S. Hence, we have the following

Theorem 8.14. *Let $h: X \to \mathbb{R}$ be C^{2-} and $f: X \to \mathbb{R}$ be C^1 and $\dim X = \infty$. Suppose that*

   (i)   *$f(v) = f(-v)$, $h(v) = h(-v)$,*
   (ii) *$M = \{v : h(v) = 0\}$ is homeomorphic to the unit sphere in X and the homeomorphism is equivariant with respect to the reflection $v \to -v$,*
   (iii) *the restriction of f to M, $f|M$, satisfies condition (C) and is bounded below.*

Then $f|M$ has infinite many pairs of critical points.

4.9. Stationary Waves

In this section, we show how the theory in the previous section may be used in differential equations.

Consider the nonlinear elliptic equation in $\mathbb{R}^3$

$$-\Delta u + u = \tilde{g}(u), \qquad x \in \mathbb{R}^3.$$

The solutions are often referred to as stationary waves since they could be the time independent solutions of the wave equation

$$u_{tt} = \Delta u - u + \tilde{g}(u).$$

They could also be stationary solutions of a parabolic equation.

We seek radial solutions. This leads to the singular ordinary differential equation in self-adjoint form:

$$(9.1) \qquad\qquad -\frac{1}{t^2}(t^2 \dot{u})^{\cdot} + u = \tilde{g}(u) \qquad t > 0$$

where $t = |x|$ and "$\cdot$" denotes the differentiation with respect to t. We are interested in the existence of nontrivial solutions $u(t)$ of (9.1) satisfying $\dot{u}(0) = 0$ and $u(t) \to 0$ as $t \to \infty$. The existence of such solutions may be seen geometrically for some special equations. For example, let $\tilde{g}(u) = u^3$. Consider the equation

$$(9.2) \qquad -\ddot{u} - \frac{2}{t}\dot{u} + u = u^3.$$

If we drop the term $2\dot{u}/t$ from (9.1), then it is a conservative system with the first integral:

$$E(\dot{u}, u) = \tfrac{1}{4}(u^4 - 2u^2 + 2\dot{u}^2)$$

The level curves of E are shown in Fig. 9.1. If $u(t)$ is a solution of (9.1), then

$$\frac{d}{dt}E(\dot{u}(t), u(t)) = -\frac{2(\dot{u}(t))^2}{t}.$$

This says that $(u(t), \dot{u}(t))$ is decreasing along the level curves of E. Hence, $(u(t), \dot{u}(t))$ will tend to either $(1, 0)$, $(-1, 0)$ or $(0, 0)$ as $t \to \infty$. This gives the reason for expecting the existence of the desired solution.

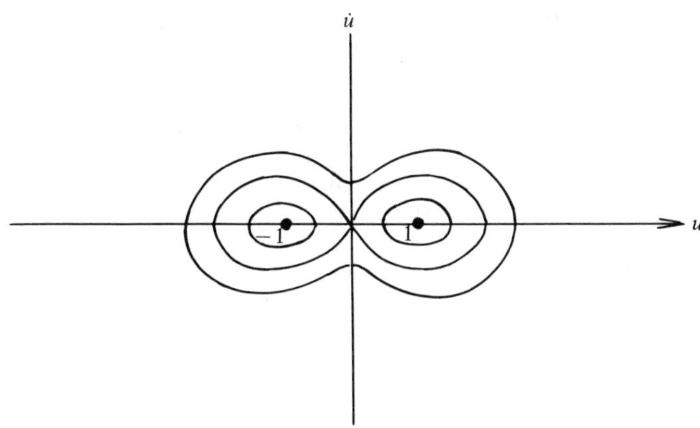

Figure 9.1

Assume that

$$(9.3) \qquad \tilde{g}(u) = g(u^2)u$$

and g satisfies

$$(9.4) \qquad g \in C^1[0, \infty), \qquad g(0) = 0, \qquad g(r) > 0 \quad \text{for } r > 0$$

$$(9.5) \qquad |g(r)| \le c|r|^\sigma, \qquad 0 < \sigma \le 1, c = \text{constant}$$

$$(9.6) \qquad r^{-\delta}g(r) \quad \text{is increasing for some} \quad \delta > 0$$

$$(9.7) \qquad rg'(r) \le cg(r).$$

Let X denote the weighted Sobolev space with norm:

$$\|u\|^2 = \int_0^\infty (u^2 + \dot{u}^2)t^2\, dt$$

Let

$$G(r) = \int_0^r g(s)\, ds$$

Define the following functionals on X

$$f(u) = \|u\|^2 - \int_0^\infty G(u^2)t^2\, dt, \quad \text{and}$$

$$h(u) = \|u\|^2 - \int_0^\infty g(u^2)u^2t^2\, dt$$

and define

$$M = \{u \in X : u \neq 0,\, h(u) = 0\}$$

Theorem 9.1. *If $\tilde{g}$ satisfies conditions (9.3)–(9.7) then there are infinitely many distinct solutions of (9.1) satisfying $\dot{u}(0) = 0$ and $u(t) \to 0$ as $t \to \infty$.*

Proof. The plan of the proof is to verify the following assertions.

- (A) f, h are well-defined even smooth functionals,
- (B) M is diffeomorphic to the unit sphere in X and is invariant under reflection $u \to -u$,
- (C) f is bounded below on M,
- (D)   the restriction $f\,|\,M$ of f to M satisfies condition (C),
- (E) $f\,|\,M$ has infinitely many distinct pairs of critical points,
- (F)   if $u \in M$ is a critical point of $f\,|\,M$, then we have $df(u) = \lambda dh(u)$ with $\lambda = 0$.

We observe first that, if $u \in M$ is a critical point of f restricted to M, then $df(u) = 0$. This says that $u \in M$ is in fact a critical point of f with no restriction and is a nontrivial solution of (9.1). The fact that $u(t)$ is a classical solution follows by the usual bootstrap argument.

To prove (A)–(F), we need some lemmas.

Lemma 9.2. *If $u \in X$, then $t^2u^2(t) \leq \|u\|^2$.*

Proof.

$$u^2(t) = -2\int_t^\infty u\dot{u}\, ds \leq \int_t^\infty (u^2 + \dot{u}^2)\, ds$$

$$\leq \frac{1}{t^2}\int_t^\infty (u^2 + \dot{u}^2)s^2\, ds \leq \frac{1}{t^2}\|u\|^2.$$

Lemma 9.3. *If $u \not\equiv 0$ and $u \in X$, then*

$$\int_0^\infty u^2 \, dt \le 4 \int_0^\infty \dot{u}^2 t^2 \, dt.$$

Proof. See Hardy, Littlewood and Polya [1] (p. 245–246).

Corollary 9.4. *For all $u \in X$,*

$$(9.8) \qquad \int_0^\infty u^2 \, dt \le 4\|u\|^2$$

Lemma 9.5. *If $c > 0$, the set*

$$\{t^2 u^2 : u \in X, \|u\| \le c\}$$

is equicontinuous on any finite interval $[0, T]$, $T > 0$.

Proof. If $0 \le t_1 < t_2 \le T$, then

$$|t_2^2 u^2(t_2) - t_1^2 u^2(t_1)| = 2 \left| \int_{t_1}^{t_2} (u\dot{u}t^2 + u^2 t) \, dt \right|$$

$$\le 2 \left(\int_{t_1}^{t_2} u^2 t^2 \, dt \right)^{1/2} \left(\int_{t_1}^{t_2} \dot{u}^2 t^2 \, dt \right)^{1/2} + 2 \int_{t_1}^{t_2} u^2 t \, dt$$

$$\le 2 \left[\|u\|^2 (t_2 - t_1)^{1/2} + \int_{t_1}^{t_2} u^2 t \, dt \right],$$

from Lemma 9.2. By Corollary 9.4 and Lemma 9.2,

$$\int_{t_1}^{t_2} u^2 t \, dt \le \left(\int_{t_1}^{t_2} u^2 \, dt \right)^{1/2} \left(\int_{t_1}^{t_2} u^2 t^2 \, dt \right)^{1/2}$$

$$\le 2\|u\|^2 (t_2 - t_1)^{1/2}.$$

Hence,

$$(9.9) \qquad |t_2^2 u^2(t_2) - t_1^2 u^2(t_1)| \le 2c^2 (t_2 - t_1)^{1/2}.$$

Corollary 9.6. $t^2 u^2(t) \to 0$ *as* $t \to 0+$ *uniformly for* $\|u\| \le c$.

Proof. It follows from (9.9).

Proof of (A). By (9.5), $|G(u^2)| \le c|u|^{2+2\sigma}$. If $u \in X$, this inequality and Lemma 9.2 imply

$$\left| \int_0^\infty G(u^2) t^2 \, dt \right| \le \int_0^\infty u^{2+2\sigma} t^2 \, dt \le \|u\|^{2\sigma} \int_0^\infty u^2 t^{2-2\sigma} \, dt < \infty.$$

This implies f is well-defined on X. Similarly, h is well-defined. It is easy to see that df and dh are smooth.

Proof of (B). If $u \in M$, we have

$$
\begin{aligned}
\|u\|^2 &= \int_0^\infty g(u^2)u^2t^2\, dt \\
&\leq c \int_0^\infty u^{2+2\sigma}t^2\, dt \\
&\leq c\|u\|^{2\sigma} \int_0^\infty u^2 t^{2-2\sigma}\, dt \\
&\leq c\|u\|^{2\sigma}\left(\int_0^\infty u^2\, dt + \int_0^\infty u^2 t^2\, dt \right) \\
&\leq 5c\|u\|^{2\sigma}\|u\|^2.
\end{aligned}
$$

from Lemma 9.2 and Corollary 9.4. Hence,

$$
(9.10) \qquad \|u\| \geq \left(\frac{1}{5c}\right)^{1/2\sigma} > 0 \quad \text{if } u \in M.
$$

This implies $0 \notin \bar{M}$. Consider the mapping $\Phi: M \to S$, where S is the unit sphere in X, defined by $\Phi(u) = u/|u|$. If $v \in S$ and $\alpha > 0$, we have as in the proof of (9.10),

$$
\begin{aligned}
h(\alpha v) &= \alpha^2\|v\|^2 - \alpha^2 \int_0^\infty g(\alpha^2 v^2)v^2 t^2\, dt \\
&= \alpha^2\left[1 - \int_0^\infty g(\alpha^2 v^2)v^2 t^2\, dt \right] \\
&\geq \alpha^2[1 - \alpha^{2\sigma}\|v\|^{2+2\sigma}] \\
&= \alpha^2[1 - \alpha^{2\sigma}] > 0
\end{aligned}
$$

provided $\alpha > 0$ is sufficiently small. On the other hand, by (9.6), if $\alpha > 1$,

$$
\begin{aligned}
h(\alpha v) &= \alpha^2 - \alpha^2 \int_0^\infty g(\alpha^2 v^2)v^2 t^2\, dt \\
&\leq \alpha^2 - \alpha^2 \int_0^\infty \alpha^{2\delta}g(v^2)v^2 t^2\, dt \\
&= \alpha^2[1 - \alpha^{2\delta}c_1]
\end{aligned}
$$

where $c_1 \neq 0$. If $\alpha > 0$ is large, $h(\alpha v) < 0$. This implies there exists $\alpha_0 > 0$ such that $h(\alpha_0 v) = 0$. Thus, Φ is onto. To show that Φ is one-to-one, let $u_1, u_2 \in M$ and $u_1/|u_1| = u_2/|u_2|$. Hence, $u_2 = tu_1$ for some $t \neq 0$. We may assume $t > 0$. By the mean value theorem,

$$
(9.11) \qquad 0 = h(tu_1) - h(u_1) = \int_1^t dh(su_1) \cdot u_1\, ds
$$

Since $u_1 \in M$,

$$dh(u_1) \cdot u_1 = 2\|u_1\|^2 - 2 \int_0^\infty [g'(u_1^2)u_1^4 + g(u_1^2)u_1^2]t^2 \, dt$$

$$= -2 \int_0^\infty g'(u_1^2)u_1^4 t^2 \, dt$$

$$< 0$$

and, for $s > 1$,

$$dh(su_1) \cdot su_1 \le 2s^2 \left[\|u_1\|^2 - \int_0^\infty s^{2\delta} g(u_1^2)u_1^2 t^2 \, dt \right.$$

$$\left. - \int_0^\infty s^2 g'(u_1^2)u_1^4 t^2 \, dt \right] \quad \text{(by (9.6))}$$

$$= -2s^2 \left[(s^{2\delta} - 1) \int_0^\infty g(u_1^2)u_1^2 t^2 \, dt \right.$$

$$\left. + \int_1^\infty s^2 g'(u_1^2)u_1^4 t^2 \, dt \right]$$

$$< 0.$$

This contradicts (9.11) and proves that Φ is a diffeomorphism.

Proof of (C). Let $u \in M$. By (9.6),

$$\int_0^\infty G(u^2)t^2 \, dt = \int_0^\infty \left(\int_0^{u^2} s^\delta [s^{-\delta} g(s)] \, ds \right) t^2 \, dt$$

$$\le \int_0^\infty u^{-2\delta} g(u^2)t^2 \, dt \int_0^{u^2} s^\delta \, ds \, dt$$

$$= (1 + \delta)^{-1} \|u\|^2.$$

It follows that $f(u) \ge \delta(1 + \delta)^{-1}\|u\|^2$. By (9.10), f is bounded below on M.

Proof of (D). Let $\{u_m\} \subset M$, $\{f(u_m)\}$ be bounded and $\|df|_M(u_m)\| \to 0$ as $m \to \infty$. We will show that $\{u_m\}$ contains a convergent subsequence. We note that $\{u_m\}$ is bounded in X and we may assume $\{u_m\}$ converges weakly to $u \in X$. By the definition of M, $df(v) \cdot v = 0$ for any $v \in M$. Thus, $\|df(u_m)\| \to 0$ as $m \to \infty$. In other words, $\|2u_m - d\varphi(u_m)\| \to 0$ as $m \to \infty$, where $\varphi(v) = \int_0^\infty G(v^2)t^2 \, dt$. It suffices to show $\{d\varphi(u_m)\}$ contains a convergent subsequence. We have

$$\|d\varphi(u_m) - d\varphi(u)\| \le 2 \sup_{\|v\| \le 1} \left| \int_0^\infty \{g(u_m^2)u_m - g(u^2)u\} v t^2 \, dt \right|$$

By Lemma 9.2 and Corollaries 9.4 and 9.6,

$$\int_0^\varepsilon \{g(u_m^2)u_m - g(u^2)u\}v\,dt = 0(1) \quad \text{as } \varepsilon \to 0+.$$

By Lemmas 9.2 and 9.5, $\{u_m\}$ contains a subsequence which converges uniformly on $[\varepsilon, T]$, $T > 0$, to u. By (9.5),

$$\left| \int_T^\infty \{g(u_m^2)u_m - g(u^2)u\}vt^2\,dt \right| \le \frac{c}{T^{2\sigma}}$$

where $c > 0$. This completes the proof.

Proof of (E). Use Theorem 8.14.

Proof of (F). Let $u \in M$ be a critical point of $f\,|\,M$. By Theorem 8.2, $df(u) \cdot v = \lambda dh(u) \cdot v$ for all $v \in X$. Since $df(u) \cdot u = 2h(u) = 0$,

$$0 = \lambda dh(u) \cdot u$$
$$= 2\lambda \left\{ \|u\|^2 - \int_0^\infty g(u^2)u^2 t^2\,dt - \int_0^\infty g'(u^2)u^4 t^2\,dt \right\}$$
$$= -2\lambda \int_0^\infty g'(u^2)u^4 t^2\,dt.$$

This implies $\lambda = 0$. $\quad\square$

4.10. The Krasnoselski Theorems

Variational methods for bifurcation problems will be discussed in the next several sections. We will begin with the Krasnoselski theorems on the existence of bifurcating solutions. By using alternative methods, the existence problem is essentially finite dimensional. Hence, we will consider only mappings in finite dimensional spaces.

Let $G : \mathbb{R}^{n+1} \to \mathbb{R}^n$ be C^2. Suppose that

$$G(\lambda, a) = \lambda a + H(\lambda, a), \qquad a \in \mathbb{R}^n, \lambda \in \mathbb{R},$$

where $H(a, \lambda) = 0(|a|^2)$ as $|a| \to 0$ uniformly in a neighborhood of $\lambda = 0$. In applications, $G(\lambda, a)$ is the bifurcation equation for a nonlinear problem whose linearized equation has $\lambda = 0$ as an isolated simple eigenvalue of multiplicity $n \ge 1$. The following result does not assume G is the gradient of some function. It will be proved in Section 5.5 as a consequence of a much more general proposition.

Theorem 10.1. *If n is odd, then $(\lambda, a) = (0,0)$ is a bifurcation point of $G(\lambda, a)$:* i.e., *there are solutions (λ, a) with $a \neq 0$ of the equation*

$$(10.1) \qquad\qquad\qquad G(\lambda, a) = 0$$

in every neighborhood of $(\lambda, a) = (0,0)$.

Theorem 10.1 is false in general if n is even. We give an example illustrating the fact. Consider the equation in $\mathbb{R}^2$

$$(10.2) \qquad\qquad \begin{aligned} \lambda a_1 + (a_1^2 + a_2^2)a_2 &= 0 \\ \lambda a_2 - (a_1^2 + a_2^2)a_1 &= 0 \end{aligned}$$

where λ, a_1 and $a_2 \in \mathbb{R}$. By multiplying the first equation in (10.2) by a_1, the second one by a_2 and adding the resulting equations, we obtain

$$\lambda(a_1^2 + a_2^2) = 0.$$

Hence, any solution $(\lambda, a) \neq (0,0)$ with $a \neq 0$ must have $\lambda = 0$. However, if $\lambda = 0$, then $a = 0$ is the only solution of (10.2).

Proof of Theorem 10.1. Consider the scalar-valued function

$$\varphi(\lambda, a) = \frac{1}{|a|^2} \langle a, \lambda a + H(\lambda, a) \rangle.$$

We have $\varphi(0,0) = 0$ and $\partial\varphi(0,0)/\partial\lambda = 1$. By the implicit function theorem, there exists a unique function $\lambda(a)$ for $|a|$ small such that $\varphi(\lambda(a), a) = 0$. Let $\tilde{G}(a) = G(\lambda(a), a)$. The condition $\varphi(\lambda(a), a) = 0$ says that $\tilde{G}(a)$ is a vector field on the sphere $\{a : |a| = \varepsilon\}$ for any $|\varepsilon|$ small. Since n is odd, the sphere is even dimensional. Since every vector field on a even dimensional sphere has a zero vector (see Section 2.11), this gives the nontrivial solutions of $G(\lambda, a)$ for (λ, a) near $(0,0)$. $\qquad\square$

If the bifurcation equation is variational, then we have the following.

Theorem 10.2. *Suppose there exists a C^2 scalar-valued function $h : \mathbb{R}^n \to \mathbb{R}$ such that $h(a) = o(|a|^2)$ as $|a| \to 0$ and*

$$G(\lambda, a) = \nabla\{\tfrac{1}{2}\lambda\langle a, a \rangle + h(a)\}.$$

Then $(\lambda, a) = (0,0)$ is a bifurcation point for equation (10.1).

Proof. Consider the restriction of h to the sphere of radius ε, $\varepsilon > 0$ small. Thus, h has at least two critical points on the sphere. Let a_0 be a critical point. By the Langrange multiplier rule, there exists a $\mu \in \mathbb{R}$ such that $\nabla h(a_0) = \mu a_0$, $|a_0| = \varepsilon$. By using the Implicit Function Theorem as in the proof of Theorem 10.1, we have that $\mu \to 0$ as $\varepsilon \to 0$. This completes the proof. $\square$

In the above theorem, there is no hypothesis on the dimension of the space. However, the perturbation h is not allowed to depend on λ. We chose to present the theorem in this way in order to have a simple proof. By using different methods, we can prove the following more general result.

Theorem 10.3. *Theorem 10.2 holds if h is allowed to depend on λ, if there is a $\lambda_0 > 0$ such that $h(\lambda, a) = o(|a|^2)$ as $|a| \to 0$ uniformly for $|\lambda| < \lambda_0$.*

Outline of Proof. Consider the gradient flow

$$(10.3)_\lambda \qquad \dot{a} = -V\{\tfrac{1}{2}\lambda\langle a, a\rangle + h(\lambda, a)\}$$

Let $\lambda = 0$. If $a = 0$ is not an isolated critical point of $h(0, a)$, then the theorem is proved. Assume that $a = 0$ is an isolated critical point of $h(0, a)$. There are three cases: (i) $a = 0$ is a strict maximum of $h(0, a)$; (ii) $a = 0$ is a strict minimum of $h(0, a)$; (iii) $a = 0$ is a saddle point of $h(0, a)$. Consider case (i). Since $a = 0$ is a strict maximum for $h(0, a)$, one can show that there exists a neighborhood N of $a = 0$ whose boundary ∂N is diffeomorphic to S^{n-1} $((n-1) -$ dimensional sphere) such that the flow $\varphi_t(a, 0)$ generated by $(10.3)_0$ is transversal to ∂N. In particular, the flow $\varphi_t(a; 0)$ is leaving ∂N in positive time. (See Fig. 10.1). This transversality property is preserved under small perturbations. Thus, for $|\lambda|$ small, the flow $\varphi_t(a; \lambda)$ generated by $(10.3)_\lambda$ is also leaving ∂N in positive time. However, if λ is small, negative and fixed, there exists $\varepsilon > 0$ such that the flow $\varphi_t(a, \lambda)$ is tending towards 0 as $t \to \infty$ provided $|a| = \varepsilon$ (see Fig. 10.2). Hence, the region $\{a : a \in N, |a| > \varepsilon\}$ is negatively invariant. The existence of α-limit sets will give the desired solutions since the

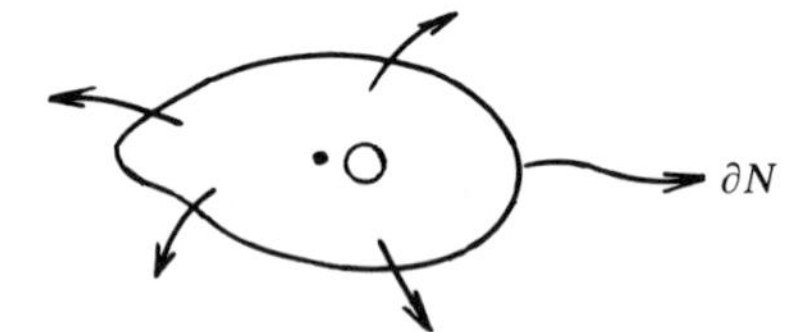

Figure 10.1

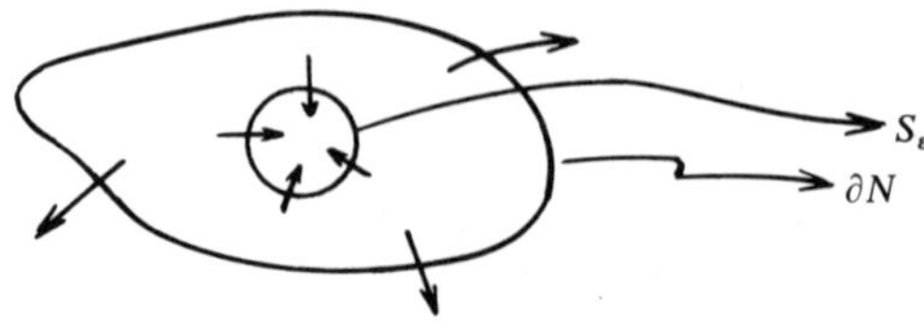

Figure 10.2. $S_\varepsilon = \{a : |a| = \varepsilon\}$

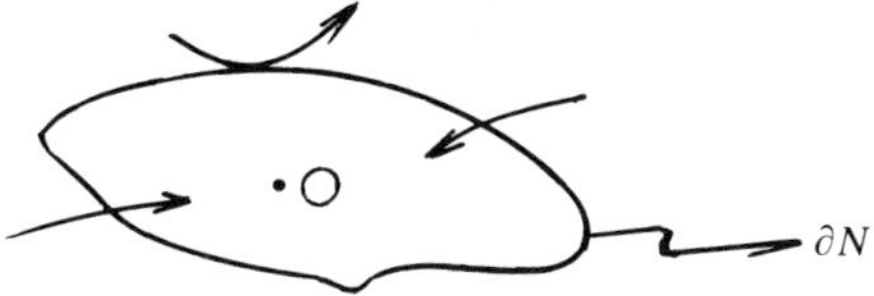

Figure 10.3

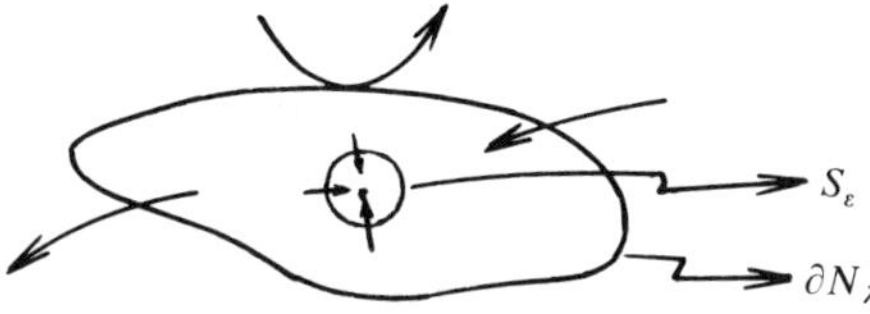

Figure 10.4

flow $(10.3)_\lambda$ is a gradient flow. Case (ii) is similar to (i). Now, we consider case (iii). It can be shown (but not trivial) that there exists a neighborhood N as before such that, if $a \in \partial N$, then either $\varphi_t(a, 0)$ is transversal to ∂N at $t = 0$ or $\varphi_t(a, 0) \notin \bar{N}$ for $|t| > 0$ small (see Fig. 10.3). In other words, N is an isolating block for the flow. Again, for $|\lambda|$ small, there are such neighborhoods N_λ whose boundaries ∂N_λ are near N. If λ is small and negative, then the flow $\varphi_t(a; \lambda)$ is as in Fig. 10.4. It is not difficult to show the existence of critical points in the region $\{a : a \in N, |a| > \varepsilon\}$. $\square$

4.11. Variational Property of Bifurcation Equation

In this section, we will show that the bifurcation equation for an infinite dimensional variational problem is in general also variational. This makes Theorem 10.3 applicable to some infinite dimensional problems.

Let X be a Hilbert space with inner product $\langle \cdot, \cdot \rangle$. Consider the nonlinear variational eigenvalue problem:

$$(11.1) \qquad df(u) = Au + H(u) = \lambda u$$

where $u \in X$, $\lambda \in \mathbb{R}$, $f : X \to \mathbb{R}$ is C^2, A is a bounded selfadjoint linear operator and $H = 0(|u|^2)$ as $|u| \to 0$. Suppose that $\lambda = 0$ is an eigenvalue of A and N_0 is the null space A. Let N_1 be the orthogonal complement of N_0 and P_0 and P_1 denote the orthogonal projections onto N_0 and N_1 respectively. For $u \in X$,

$$u = u_0 + u_1, \qquad P_0 u = u_0, \qquad P_1 u = u_1$$

By using the projections, (11.1) is equivalent to:

$$(11.2) \qquad P_0 H(u_0 + u_1) = \lambda u_0$$

$$(11.3) \qquad Au_1 + P_1 H(u_0 + u_1) = \lambda u_1.$$

The bifurcation equation is given by

$$(11.4) \qquad G(\lambda, u_0) = P_0 H(u_0 + u_1^*(\lambda, u_0)) - \lambda u_0 = 0$$

where $u_1^*(\lambda, u_0)$ is the unique solution of (11.3) for given $|\lambda|$, $|u_0|$ sufficiently small. Note that

$$(11.5) \qquad u_1^*(\lambda, u_0) = 0(|u_0|^2) \quad \text{as } |u_0| \to 0$$

uniformly in small λ. By the Implicit Function Theorem, there exists a unique C^1 function $\lambda(u_0)$, defined for $|u_0|$ sufficiently small such that $\lambda(0) = 0$ and

$$(11.6) \qquad \langle u_0, G(\lambda(u_0), u_0) \rangle = 0$$

By (11.5), $u_1^*(\lambda(u_0), u_0) = 0(|u_0|^2)$ as $|u_0| \to 0$. Consider the functionals φ: $N_0 \to \mathbb{R}$ and $\psi : N_0 \times \mathbb{R} \to \mathbb{R}$ defined by

$$(11.7) \qquad \varphi(u_0) = f(u_0 + u_1^*(\lambda(u_0), u_0))$$

$$(11.8) \qquad \psi(\lambda, u_0) = f(u_0 + u_1^*(\lambda, u_0)).$$

Theorem 11.1. *Let $|\lambda|$ be fixed and small. Then u_0 is a solution of the bifurcation equation (11.4) if and only if u_0 is a critical point of the functional $\psi(\lambda, u_0) - \frac{1}{2}\lambda|u_0 + u_1^*|^2$.*

Proof. Let u_0 be a critical point of the functional. For every $v_0 \in N_0$, we have

$$(11.9) \qquad \langle d\psi(\lambda, u_0), v_0 \rangle = \lambda \langle u_0 + u_1^*(\lambda, u_0), v_0 + du_1^*(\lambda, u_0) \cdot v_0 \rangle,$$

where "d" denotes differentiation with respect to u_0. By the chain rule, we obtain

$$\begin{aligned}
\langle d\psi(\lambda, u_0), v_0 \rangle &= \langle df(u_0 + u_1^*(\lambda, u_0)), v_0 + du_1^*(\lambda, u_0) \cdot v_0 \rangle \\
&= \langle Au_1^* + H(u_0 + u_1^*), v_0 + du_1^* \cdot v_0 \rangle,
\end{aligned}$$

where $u_1^* = u_1^*(\lambda, u_0)$. By (11.3), (11.9) and $du_1^*(\lambda, u_0) \cdot v_0 \in N_1$, we obtain

$$\begin{aligned}
\langle d\psi(\lambda, u_0), v_0 \rangle &= \langle P_0 H(u_0 + u_1^*), v_0 + du_1^* \cdot v_0 \rangle \\
&\quad + \langle Au_1^* + P_1 H(u_0 + u_1^*), v_0 + du_1^* \cdot v_0 \rangle \\
&= \langle P_0 H(u_0 + u_1^*), v_0 \rangle + \langle Au_1^* + P_1 H(u_0 + u_1^*), du_1^* \cdot v_0 \rangle \\
&= \lambda \langle u_0, v_0 \rangle + \lambda \langle u_1^*, du_1^* \cdot v_0 \rangle.
\end{aligned}$$

Hence, (11.4) holds. The converse may be shown by reversing the above steps. $\square$

Corollary 11.2. $d\psi(\lambda, u) = P_0 H(u_0 + u_1^*) - \lambda_0.$

Remarks. (1) Theorem 11.1 holds even if H depends on λ. (2) Theorem 11.1 shows exactly how one can apply Theorem 10.3.

Theorem 11.3. *Let* $0 < |\varepsilon|$ *be sufficiently small and*

$$S_\varepsilon = \{u_0 : |u_0 + u_1^*(\lambda(u_0), u_0)|^2 = \varepsilon\}.$$

If $u_0 \in S_\varepsilon$ *is a critical point of the restriction of* $\varphi(u_0)$ *defined by* (11.7) *to* S_ε, *then* u_0 *is a solution of* (11.4) *with* $\lambda = \lambda(u_0)$.

Proof. By using (11.5) and (11.6), we may proceed exactly as in the proof of Theorem 11.1. $\square$

Theorem 11.4. *Suppose* f *is even in* u; *i.e.,* df *is odd, and* $\dim N_0 = n < \infty$. *Then, for every* $\varepsilon > 0$ *sufficiently small, there are at least* n *distinct pairs of solutions of* (11.1) *having norm* ε.

Proof. If f is even, then $\varphi(u_0)$ is odd. Hence, Ljusternik–Schnirelman theory in R^n (Chapter 2) is applicable here. $\square$

4.12. Liapunov Center Theorem at Resonance

To illustrate the advantages of knowing that the bifurcation equation is variational, we consider the Liapunov center theorem. Let $z \in \mathbb{R}^{2n}$ and $H : \mathbb{R}^{2n} \to \mathbb{R}$ be sufficiently smooth. Suppose that

$$(12.1) \qquad H(z) = \frac{1}{2} \sum_{k=1}^{n} \lambda_k(z_k^2 + z_{k+n}^2) + 0(|z|^3) \quad \text{as } z \to 0$$

where $\lambda_k \in \mathbb{R}$ for $k = 1, 2, \ldots, n$. Consider the Hamiltonian system:

$$(12.2) \qquad \dot{z} = JVH(z)$$

where "$\cdot$" denotes d/dt and

$$J = \begin{pmatrix} 0 & I \\ -I & 0 \end{pmatrix}, \qquad I = n \times n \text{ identity matrix.}$$

The following result was proved in Section 4.4.

Theorem 12.1 (Liapunov Center Theorem). *If*

$$(12.3) \qquad \frac{\lambda_k}{\lambda_l} \neq integer \ for \ all \ k \neq l,$$

then there exist n one-parameter families of periodic solutions of (12.2) emanating from $z = 0$.

The condition (12.3) in this classical theorem is sometimes referred to as the resonance conditions. We will now use the varational character of Hamiltonian systems to discuss the Liapunov Center Theorem when the resonance conditions are not met. As we will see from the proof, the main idea is to apply Ljusternik–Schnirelman theory to the bifurcation equation. For simplicity, we present only the following

Theorem 12.2. *If*

$$H(z) = \frac{1}{2} \sum_{k=1}^{n} (z_k^2 + z_{k+n}^2) + 0(|z|^3) \quad \text{as } z \to 0$$

then there are n distinct periodic solutions of

$$(12.4) \qquad \dot{z} = J\nabla H(z)$$

on each energy surface $H(z) = \varepsilon$, $\varepsilon > 0$ sufficiently small.

Proof. We first derive the bifurcation equation. Since the equation is autonomous, the change of time scale $t \to \mu t$ with μ near 1 reduces our discussion to 2π-periodic solutions for the parameterized system

$$(12.8) \qquad \dot{z} = \mu J\nabla H(z).$$

If $z = e^{Jt}u$, then Equation (12.8) becomes

$$(12.9) \qquad \dot{u} = (\mu - 1)Ju + 0(|u|^2),$$

where the right hand side of (12.9) is non-autonomous and of period 2π. The linearized equation at $\mu = 1$ is the equation

$$(12.10) \qquad \dot{u} = 0$$

which has a $2n$ dimensional null space in the space of 2π-periodic functions; namely, the set of all constant functions. A projection onto this subspace is the mean value operator

$$M[f] = \frac{1}{2\pi} \int_0^{2\pi} f(t)\, dt$$

where f is a 2π-periodic function. Let I be the identity map on the space of 2π-periodic functions. Then the existence of 2π-periodic solutions of (12.9) is equivalent to the following alternative problem:

$$(12.11) \qquad \dot{u} = (I - M)[(\mu - 1)Ju + 0(|u|^2)]$$

$$(12.12) \qquad M[(\mu - 1)Ju + 0(|u|^2)] = 0$$

It follows from the Implicit Function Theorem that (12.11) has a unique 2π-periodic solution $u^*(t; a, \mu)$ for any $|\mu - 1|$ and $|a|$ sufficiently small such that $M[u^*(t; a, \mu)] = a$. Upon substitution of u^* into (12.12), one obtains a a bifurcation equation

$$(12.13) \qquad v(a, \mu) = M[(\mu - 1)Ju^* + 0(|u^*|^2)] = 0$$

The zeros of $v(a, \mu)$ will now yield the desired 2π-periodic solutions. It is not difficult to see that

$$(12.14) \qquad v(a, \mu) = (\mu - 1)Ja + 0(|a|^3)$$

and, for any $0 \le \theta < 2\pi$, $u^*(t; e^{J\theta}a, \mu) = e^{J\theta}u^*(t + \theta; a, \mu)$. Since the higher order terms $0(|u^*|^2)$ in (12.13) are of the form $e^{-Jt}h(e^{Jt}u^*)$, $h(z) = 0(|z|^2)$, we have, for $0 \le \theta < 2\pi$,

$$v(e^{J\theta}a, \mu) = M[(\mu - 1)Je^{J\theta}u^*(\cdot + \theta; a, \mu) + e^{-J\cdot}h(e^{J(\cdot + \theta)}u^*(\cdot + \theta; a, \mu))]$$
$$= e^{J\theta}M[(\mu - 1)Ju^*(\cdot + \theta; a, \mu) + e^{-J(\cdot + \theta)}h(e^{J(\cdot + \theta)}u^*(\cdot + \theta; a, \mu))].$$

Hence,

$$(12.15) \qquad v(e^{J\theta}a, \mu) = e^{J\theta}v(a, \mu), \quad \text{for } 0 \le \theta < 2\pi,$$

From (12.13), we can choose $\mu = \mu(a)$ uniquely and continuously so that $v(a, \mu(a))$ is orthogonal to Ja; that is,

$$(12.16) \qquad \langle Ja, v(a, \mu(a)) \rangle = 0.$$

The function $\mu(a)$ is obtained from the Implicit Function Theorem by writing $a = \rho\tilde{\theta}$, $|a| = \rho$, in polar coordinates and using the formula (12.13) to solve (12.16) as in the previous section. Clearly,

$$(12.17) \qquad \mu(e^{J\theta}a) = \mu(a), \qquad 0 \le \theta < 2\pi.$$

Now the existence of periodic solutions of (12.7) near those of the linearized equation is reduced to the finite dimensional problem $v(a, \mu(a)) = 0$. For simplicity, write $v(a) = v(a, \mu(a))$ and $w(t) = u^*(t; a, \mu(a))$. Define

$$f(a) = \frac{1}{2\pi} \int_0^{2\pi} \frac{1}{2} \left\langle Je^{Jt}w(t), \frac{d}{dt}(e^{Jt}w(t)) \right\rangle dt$$

$$h(a) = \frac{1}{2\pi} \int_0^{2\pi} H(e^{Jt}w(t)) \, dt$$

$$S(a) = f(a) - \mu(a)[h(a) - c], \quad c = \text{constant}.$$

It is not difficult to see that

$$f(a) - h(a) = 0(|a|^4)$$

(12.18)
$$h(a) = \tfrac{1}{2}\langle a, a\rangle + 0(|a|^4).$$

If a is a point on the surface $h(a) = c$ for $c > 0$ small, we have

$$\nabla S(a) = \nabla f(a) - \mu'(a)[h(a) - c] - \mu(a)\nabla h(a)$$

$$= \frac{1}{2\pi} \int_0^{2\pi} \langle Je^{Jt}\nabla w(t), \mu(a)J\nabla H(e^{Jt}w(t)) - e^{Jt}v(a)\rangle \, dt$$

$$- \frac{1}{2\pi} \mu(a) \int_0^{2\pi} \langle e^{Jt}\nabla w(t), \nabla H(e^{Jt}w(t))\rangle \, dt$$

$$= \frac{1}{2\pi} \int_0^{2\pi} \langle Je^{Jt}\nabla w(t), e^{Jt}\rangle \, dt \, v(a).$$

Since the mean value of $w(t)$ is precisely equal to a,

$$\frac{1}{2\pi} \int_0^{2\pi} \nabla w(t) \, dt = \text{identity matrix}.$$

Hence,

(12.19)
$$Jv(a) = \nabla S(a) \quad \text{provided } h(a) = c.$$

This also shows the variational characterization of the bifurcation equation. In fact, $S(a)$ is the well-known action integral in classical mechanics. It follows from (12.15) and (12.17) that $S(a)$ is equivariant under the circle action $e^{J\theta}$; i.e.,

(12.20)
$$S(e^{J\theta}a) = S(a)$$

By using Ljusternik–Schnirelman theory as in Example 2.12.9(c), we obtain n distinct critical circles of S restricted to $h(a) = c$ provided $c > 0$ is small. If $\bar{a}$ is one of these critical points, we have $VS(\bar{a}) = \lambda V h(\bar{a})$, where $\lambda \in \mathbb{R}$ is the Langrange multiplier. It remains to show $\lambda = 0$. By (12.16), (12.18) and (12.19),

$$
\begin{aligned}
0 &= \langle J\bar{a}, v(\bar{a}) \rangle \\
&= \langle J\bar{a}, J^{-1} V S(\bar{a}) \rangle \\
&= \lambda \langle J\bar{a}, J^{-1} V h(\bar{a}) \rangle \\
&= -\tfrac{1}{2}\lambda \langle \bar{a}, \bar{a} \rangle + 0(|\bar{a}|^4)
\end{aligned}
$$

Since $\bar{a} \neq 0$, $\lambda = 0$. This completes the proof. $\square$

4.13. Bibliographical Notes

For general methods in calculus of variations, see, for example, the books Courant and Hilbert [1, 2], Weinberger [3], Duvaut and Lions [1], Lions and Magenes [1], Ekeland and Temam [1], Rockafeller [1], Vainberg [1]. Basic regularity theory may be found in Agmon [1], Agmon, Douglis and Nirenberg [1], and Gilbarg and Trudinger [1]. Theorem 2.2 is from Lions [2] where one can find many other interesting applications. Results and applications related to Theorem 2.5 may be found in Berger [1], Dubinskii [1] and Skrypnik [1]. For further results related to Examples 2.7 and 2.8, see Berger [1], Berger and Berger [1], Clark [2, 4], Gordon [1], Jacobowitz [1], Krasnoselskii [1], Lazer [1], Temme [1].

For the theory of monotone operators, see Brezis [1], Opial [1], Haraux [1], Barbu [1], Browder [3]. Theorem 3.5 is from Lions [1]. Problems related to Example 3.6 may be found in Dubinskii [1], Browder [2].

Condition (C) was first introduced in Palais and Smale [1] and is sometimes called the *Palais–Smale condition*. Our presentation of the minimax principle in Section 4.5 follows the basic ideas of Palais [1–3], but the proof is very close to that of Rabinowitz [6]. The notion of a pseudo-gradient vector field was first introduced by Palais [1]. In our presentation, we follow the approach given by Clark [1, 3], Rabinowitz [6, 9]. The theory of isolating blocks can be found in Conley [1].

Theorem 6.1 is due to Ambrosetti and Rabinowitz [1]. For related results, see Amann [1], Benci [1], Berger and Schechter [1], Castro and Lazer [1, 2], Ahmad, Lazer and Paul [1], Berger [1], Coffman [1], Hempel [1], Zeidler [1], Fucik, Necas, Soucek and Soucek [1].

Theorem 7.1 is originally due to Rabinowitz [11]. The proof we used is given by Brezis, Caron and Nirenberg [1]. In Chow [1], the problem was considered as a bifurcation problem by using the methods in Sections 4.11, 4.12. There are also results for the nonautonomous case; see, for example,

Brezis and Nirenberg [1] and the references therein. Lemma 7.2 is due to Lovicarova [1].

Complete proofs of the results in Section 4.8 can be found in Browder [1]. Theorem 9.1 is taken from Chow and Kurtz [1] where related references can be found.

Theorem 10.1 and 10.2 are essentially due to Krasnoselskii [1]. In Böhme [1], Marino [1], Marino and Prodi [1], Rabinowitz [5, 7], various generalizations and extensions are given together with applications to differential equations. The complete proof of Theorem 10.3 may be found in Rabinowitz [7].

Theorems 11.1–3 are known to many authors; see, for example, Rothe [1], Sather [1], Rabinowitz [6]. It is also used in Section 5.12.

Theorem 12.2 was first obtained by Weinstein [1, 2]. The proof given in Section 4.12 is essentially due to Moser [2]. However, we used the proof given in Chow and Mallet–Paret [3] where the Hamiltonian is not necessarily positive definite. A more general result may be found in Fadell and Rabinowitz [1] (see also Bottkol [1]). For existence of periodic solutions on a fixed energy surface, see, for example, Rabinowitz [10–12, 13], Seifert [1], Clark and Ekeland [1].

Chapter 5

The Linear Approximation and Bifurcation

5.1. Introduction

This chapter is devoted to the most elementary parts of the analytic theory
of bifurcation of the zeros of a function. In particular, suppose $\Lambda \subseteq \mathbb{R}$, X is
a Banach space, $M : \Lambda \times X \to X$ and

$$M(\lambda, x) = Bx - \lambda x + N(x)$$

where B is a bounded linear operator on X and $N(0) = 0$, $D_x N(0) = 0$. If
λ_0 is a simple eigenvalue of B, then it is shown that the point $(\lambda_0, 0)$ is always
a bifurcation point regardless of the nonlinear function N. This is generalized
to the case where λ is an eigenvalue of B of odd multiplicity. These results
on bifurcation have the special character of implying that the existence of
the bifurcation is determined only by the linear approximation of the function M. Of course, the specific nature of the bifurcation will depend on the
nonlinear terms.

In the applications, it often happens that $M : \Lambda \times X \to Z$ where Z is a
Banach space, $X \subseteq Z$, and

$$M(\lambda, x) = Bx - \lambda A x + N(x)$$

where $B, A : X \to Z$ are bounded linear operators, $N(0) = 0$, $D_x N(0) = 0$. A
point λ_0 is a simple eigenvalue of the pair (B, A) if $\mathscr{N}(B - \lambda_0 A)$ has dimension
one, the codimension of $\mathscr{R}(B - \lambda_0 A)$ has dimension one and $A \mathscr{N}(B - \lambda_0 A) \cap
\mathscr{R}(B - \lambda_0 A) = \varnothing$. It is then shown that there is always bifurcation at a
simple eigenvalue of (B, A) (Section 5). This generalizes the previous result
when $A = I$.

An appropriate generalization of the result for $A = I$ on odd multiplicity
requires a deeper understanding of the meaning of a simple eigenvalue of
(B, A). For λ_0 a normal eigenvalue of (B, A), it is shown in Section 3 that one
can associate a certain determinant $D(\lambda)$ in a neighborhood of $\lambda = \lambda_0$. The
point λ_0 is a simple eigenvalue of (B, A) if and only if $D(\lambda_0) = 0$ and $dD(\lambda_0)/
d\lambda \neq 0$. The appropriate generalization to nonsimple eigenvalues is to re-
quire that this determinant change sign at $\lambda = \lambda_0$. The proof of this fact is
given in Section 7.

The proofs in Section 7 are given in a constructive way and would yield a general result on the existence of global branches of solutions provided that one could always use the alternative method in Section 2.4 to obtain a finite dimensional equation whose solutions would yield all solutions of $M(\lambda, x) = 0$ which lie in a compact set. This is known to be true if the original spaces have Schauder bases, but not known in general. Thus, the proof of global bifurcation in Section 8 relies on degree theory.

In Section 4, we consider the case where $\lambda \in \mathbb{R}^N$, $\lambda = (\lambda_1, \ldots, \lambda_N)$, and the linear part of $M(\lambda, x)$ is

$$Bx - \sum_{j=1}^{N} \lambda_j A_j x.$$

A simple eigenvalue is defined in such a way that bifurcation always occurs at simple eigenvalues.

Sections 6 and 9 are devoted to applications of the results.

5.2. Eigenvalues of B

Let X, Z be Banach spaces and assume that X is continuously embedded in Z. In other words, there is a continuous linear injection $I: X \to Z$ which we call the identity. The inverse of the injection map I will be denoted by j.

Our objective is to discuss some elementary properties of bounded linear maps $B: X \to Z$. Of course, one could also consider B as a mapping from $Z \to Z$ with the domain of B as $X \subset Z$. In an earlier chapter, we have seen some advantages of considering the domain of B as a Banach space in its own right. It is beneficial to continue the discussion in this more general setting.

If $B: X \to Z$ is a bounded linear map, then the *resolvent set $\rho(B)$ of B* is the set of $\lambda \in \mathbb{C}$ such that $B - \lambda I \overset{\text{def}}{=} B - \lambda$ has a bounded inverse. The *spectrum $\sigma(B)$ of B* is $\mathbb{C} \backslash \rho(B)$. A point $\lambda \in \mathbb{C}$ is called an *eigenvalue of B* if there is a nonzero $x \in X$ such that $(B - \lambda)x = 0$; that is, the null space $\mathcal{N}(B - \lambda)$ of $B - \lambda$ is at least one dimensional. Any $x \in \mathcal{N}(B - \lambda)$ is called an eigenvector. A point $\lambda \in \mathbb{C}$ is said to be an *isolated eigenvalue* of B if λ is an eigenvalue of B and λ is an isolated point of the spectrum of B.

EXAMPLE 2.1. If $X = Z = \mathbb{R}^n$, then any bounded linear operator $B: X \to Z$ can be identified with an $n \times n$ real matrix. A point $\lambda \in \mathbb{C}$ is an eigenvalue of B if and only if

$$(2.1) \qquad\qquad\qquad \det(B - \lambda) = 0.$$

Furthermore, if condition (2.1) is not satisfied, then $\lambda \in \rho(B)$. Therefore, $\sigma(B)$ consists only of eigenvalues.

EXAMPLE 2.2. Suppose $Z = C([0, \pi], \mathbb{R})$ and $X = \{x \in C^2([0, \pi], \mathbb{R}): x(0) = x(\pi) = 0\}$. If $B: X \to Z$ is defined by the differential operator $(Bx)(t) = -\ddot{x}(t)$, then the eigenvalues and eigenvectors of B are respectively those values of λ and $x \neq 0$ which satisfy the two point boundary value problem

$$
\ddot{x}(t) + \lambda x(t) = 0, \qquad 0 \le t \le \pi,
$$
$$
x(0) = x(\pi) = 0.
$$
(2.2)

The only values of λ for which there exist nontrivial solutions of this problem are $\lambda = \lambda_n = n^2$, $n = 1, 2, \ldots$, and the corresponding solutions u are a constant multiple of $x_n = \sin nt$.

If λ is an eigenvalue of B, then the null space $\mathcal{N}(B - \lambda)^k$ and range $\mathcal{R}(B - \lambda)^k$ of powers of $B - \lambda$ are defined as those of the operator $(B - \lambda)^k$ considered on its natural domain of definition in X. To be more precise, let $j: I(X) \to X$ be the inverse of the injection I and inductively set

$$
\mathcal{D}(B - \lambda) = X
$$
$$
\mathcal{D}(B - \lambda)^k = \{x \in X \mid jBx \in \mathcal{D}(B - \lambda)^{k-1}\}.
$$

Then

$$
(B - \lambda)^k : \mathcal{D}(B - \lambda)^k \subseteq X \to Z
$$

is well-defined if by this we mean $(B - \lambda)[j(B - \lambda)]^{k-1}$. We then let $\mathcal{N}(B - \lambda)^k$ and $\mathcal{R}(B - \lambda)^k$ be the null space and range of this operator.

If λ is an eigenvalue of B, then the *generalized eigenspace* $\mathcal{M}_\lambda(B)$ of B is defined to be $\mathcal{M}_\lambda(B) = \mathrm{cl}(\bigcup_{k \ge 1} \mathcal{N}(B - \lambda)^k)$. It often happens there is an integer $\delta(B) < \infty$, called the *ascent of* B, such that

$$
\mathcal{N}(B - \lambda)^{\delta(B)-1} \subsetneqq \mathcal{N}(B - \lambda)^{\delta(B)} = \mathcal{N}(B - \lambda)^{\delta(B)+1}.
$$

In this case, $\mathcal{M}_\lambda(B) = \mathcal{N}(B - \lambda)^{\delta(B)}$.

A point $\lambda \in \mathbb{C}$ is called a *normal eigenvalue* of B if

$$
\begin{array}{ll}
\text{(i)} & \delta(B) < \infty \\
\text{(ii)} & \dim \mathcal{N}(B - \lambda)^{\delta(B)} < \infty \\
\text{(iii)} & Z = I[\mathcal{N}(B - \lambda)^{\delta(B)}] \oplus \mathcal{R}(B - \lambda)^{\delta(B)}.
\end{array}
$$
(2.3)

If λ is a normal eigenvalue of B, then it follows that

$$
X = \mathcal{N}(B - \lambda)^{\delta(B)} \oplus j[\mathcal{R}(B - \lambda)^{\delta(B)}].
$$
(2.4)

We refer to the decompositions of X, Z in (2.4), (2.3) as *normal decompositions*. The subspace $\mathcal{R}(B - \lambda)^{\delta(B)}$ of Z is called the *generalized range* of $B - \lambda$. The *geometric and algebraic multiplicities* of the eigenvalue λ are respectively the dimensions of $\mathcal{N}(B - \lambda)$ and $\mathcal{N}(B - \lambda)^{\delta(B)}$. An eigenvalue λ of B is said to be *simple* if the algebraic and geometric multiplicities are equal to one. This is equivalent to saying that $\dim \mathcal{N}(B - \lambda) = 1 = \operatorname{codim} \mathcal{R}(B - \lambda)$ and $I[\mathcal{N}(B - \lambda)] \oplus \mathcal{R}(B - \lambda) = Z$.

If the operator B is represented as a matrix in block form relative to the normal decomposition, then

$$B = \begin{pmatrix} B_0 & 0 \\ 0 & B_1 \end{pmatrix};$$

$$B_0 : R^d \to R^d, \qquad d = \dim \mathcal{N}(B - \lambda)^{\delta(B)};$$

$$\sigma(B_0) = \{\lambda\}, \qquad (B_0 - \lambda)^{\delta(B)} = 0;$$

$$B_1 : j[\mathcal{R}(B - \lambda)^{\delta(B)}] \to \mathcal{R}(B - \lambda)^{\delta(B)}; \qquad \lambda \in \rho(B_1).$$

If $X = Z$, then the injection mapping I is unnecessary and the operator $(B - \lambda)^k$ can be defined in the usual way. Furthermore, the definition of a normal eigenvalue is easier to state. In fact, if $X = Z$, then λ is a normal eigenvalue of B if there is an integer $\delta(B)$ such that

(2.5)
$$\mathcal{N}(B - \lambda)^{\delta(B) - 1} \subsetneqq \mathcal{N}(B - \lambda)^{\delta(B)} = \mathcal{N}(B - \lambda)^{\delta(B) + 1}$$
$$\dim \mathcal{N}(B - \lambda)^{\delta(B)} < \infty.$$

It is a consequence of relations (2.5) that

(2.6)
$$X = \mathcal{N}(B - \lambda)^{\delta(B)} \oplus \mathcal{R}(B - \lambda)^{\delta(B)}.$$

EXAMPLE 2.3. If $X = Z$ and $B : X \to X$ is a compact operator, then every $\lambda \in \sigma(B)$, $\lambda \neq 0$, is a normal eigenvalue of B. This is the classical result in the spectral theory of compact operators.

EXAMPLE 2.4. If $I : X \to Z$ is compact, $B : X \to Z$ is a bounded linear operator with $\rho(B) \neq \phi$, then every nonzero element in $\sigma(B)$ is a normal eigenvalue of B. The proof is supplied following the same reasoning as in Example 2.3.

Of course, noncompact operators may also have eigenvalues. For example, if $B = \mu I + C$ where $C : X \to X$ is compact, then every eigenvalue $\lambda + \mu \neq 0$ of C is normal and λ is therefore a normal eigenvalue of B.

One can actually give an important characterization of normal eigenvalues in another way. For any bounded operator $B : X \to X$, define the *essential spectrum* $\sigma_e(B)$ of B as $\sigma_e(B) = \bigcap_C \sigma(B + C)$, where $C : X \to X$ is compact. If $\lambda \in \sigma(B) \backslash \sigma_e(B)$, then λ is a normal eigenvalue of B. Let rad $\sigma_e(B)$, rad $\sigma(B)$ be respectively the radii of the smallest closed disk with center zero which contains $\sigma_e(B)$, $\sigma(B)$.

Generally, boundary value problems for differential equations on compact sets correspond to compact operators (thus, rad $\sigma_e(B) = 0$) while those on noncompact sets will have rad $\sigma_e(B) > 0$. Example 2.2 is typical for compact sets, whereas the following example is typical for noncompact sets.

EXAMPLE 2.5. Suppose $q:[0, \infty) \to \mathbb{R}$ is continuous

$$(2.7) \qquad 0 < Q_1 \le q(t), \qquad t \in [0, \infty),$$

and let

$$(2.8) \qquad \begin{aligned} \mathscr{D}(B) &= \{x \in L^2[0, \infty): x' \text{ is absolutely continuous on all} \\ &\quad \text{compact sets of } [0, \infty), \, x'' \in L^2[0, \infty), \, x(0) = 0\} \\ B&: \mathscr{D}(B) \to L^2[0, \infty), \\ (Bx)(t) &= -x''(t) + q(t)x(t). \end{aligned}$$

If $\langle , \rangle$ is the inner product in $L^2[0, \infty)$, then the operator B satisfies the following property: $\langle Bx, y \rangle = \langle x, By \rangle$ for all $x \in \mathscr{D}(B)$, $y \in \mathscr{D}(B)$, y vanishing outside any compact set $[0, G] \subset [0, \infty)$. Thus, B is formally self-adjoint. Furthermore, B has a closed self-adjoint extension which we again call B, and

$$(2.9) \qquad \sigma_e(B) \subset [Q_1, \infty)$$

If $\lambda \in \sigma(B) \backslash \sigma_e(B)$, then dim $\mathscr{N}(B - \lambda I) = 1$. The ascent of any self-adjoint operator is one. If we define $X = \mathscr{D}(B)$ with the graph norm and let $Z = L^2[0, \infty)$, then λ is a simple eigenvalue of B.

In general, it is difficult to determine which elements of $\sigma(B)$ are normal eigenvalues of B. This is not surprising since it is difficult to determine any information about $\sigma(B)$. On the other hand, there is a very convenient relation expressing rad $\sigma(B)$ as

$$(2.10) \qquad \operatorname{rad} \sigma(B) = \lim_{n \to \infty} \|B^n\|^{1/n},$$

provided $B: X \to X$ is a bounded linear operator.

There is also a very interesting way to characterize rad $\sigma_e(B)$. For any bounded set $U \subset X$, the *Kuratowski measure of noncompactness* $\alpha(U)$ of U is

$$(2.11) \qquad \alpha(U) = \inf\{d: U \text{ has a finite cover of open sets with} \\ \text{diameters less than } d\}.$$

For any bounded operator $B: X \to X$, one can define

$$(2.12) \qquad \alpha(B) = \inf\{k: \alpha(BU) \le k\alpha(U) \text{ for all bounded } U \subset X\}.$$

It can then be shown that

$$(2.13) \qquad\qquad \text{rad}(\sigma_e(B)) = \lim_{n \to \infty} \left[\alpha(B^n)\right]^{1/n}.$$

Relation (2.13) can be used to assert at least that some elements of $\sigma(B)$ are normal eigenvalues since $\alpha(B^n) \leq (\alpha(B))^n$ and, thus,

$$(2.14) \qquad\qquad \text{rad}(\sigma_e(B)) \leq \alpha(B).$$

This remark can be useful in the applications.

Of course, to use the above formulas for $\text{rad}\,\sigma(B)$ and $\text{rad}\,\sigma_e(B)$, one must be discussing an operator $B: X \to X$ and not an operator from one Banach space X to another Banach space Z. This eliminates the use of the graph norm for a closed operator $B: \mathscr{D}(B) \subset Z \to Z$. The results can be applied, however, if there is a λ_0 such that $(B - \lambda_0 I)^{-1}$ exists and is bounded. In fact, the equation

$$(B - \lambda I)x = 0, \qquad x \in \mathscr{D}(B),$$

is equivalent to the equation

$$x - \mu(B - \lambda_0 I)^{-1}x = 0 \qquad x \in Z$$
$$\mu = \lambda - \lambda_0.$$

For differential operators, this corresponds to using Green's function. The following remark will be useful in a later section.

Lemma 2.6. *If H is a complex Hilbert space and $L: \mathscr{D}(L) \subset H \to H$ is a self-adjoint operator, then*

(a)   *if L is bounded, then $\alpha(L) = \text{rad}\,\sigma_e(L)$.*
(b)   *if $\sigma_e(L) \subset \{\lambda \in \mathbb{R} : |\lambda| \geq Q > 0\}$ and $0 \notin \sigma(L)$, then $\alpha(L^{-1}) \leq Q^{-1}$.*

Proof. Let $E(\lambda)$ be the spectral function for L. Suppose L is bounded. If $k = \text{rad}\,\sigma_e(L)$, $\varepsilon > 0$, then

$$L = \int_{|\lambda| < k + \varepsilon} \lambda dE(\lambda) + \int_{|\lambda| \geq k + \varepsilon} \lambda dE(\lambda) \overset{\text{def}}{=} U + C$$

where $|U| \leq k + \varepsilon$ and C is compact since it has finite rank. Therefore, $\alpha(L) \leq k + \varepsilon$ for every $\varepsilon > 0$. Thus, $\alpha(L) \leq k$. Since we know $\alpha(L) \geq k$, it follows that $\alpha(L) = k$ and part (a) is proved.

If L satisfies the hypothesis in (b), then L^{-1} exists and is bounded

$$L^{-1} = \int_{|\lambda| > Q - \varepsilon} \lambda^{-1} dE(\lambda) + \int_{|\lambda| \leq Q - \varepsilon} \lambda^{-1} dE(\lambda) \overset{\text{def}}{=} U + C$$

for any $0 < \varepsilon < Q$. Clearly, $|U| \le (Q - \varepsilon)^{-1}$ and C is compact since it has finite rank. Hence, $\alpha(L^{-1}) \le (Q - \varepsilon)^{-1}$ and, thus, $\alpha(L^{-1}) \le Q^{-1}$. This proves the lemma. $\square$

If $B: X \to Z$ is a bounded linear operator, the (Fredholm) *index $i(B)$ of B* is defined as

$$i(B) = \dim \mathcal{N}(B) - \operatorname{codim} \mathcal{R}(B).$$

If λ is a normal eigenvalue of B, then necessarily $i(B - \lambda) = 0$.

The next example is an illustration of an operator which has eigenvalues which are not normal because the index is $\ne 0$.

EXAMPLE 2.7. If $B: l^2 \to l^2$ is defined by

$$Ba = (a_2, a_3, \ldots)$$

then $\mathcal{R}(B) = l^2$ and $\dim \mathcal{N}(B) = 1$. Therefore, $i(B) = 1$ and zero is not a normal eigenvalue of B. In this case $\delta(B)$ does not exist since

$$\mathcal{N}(B) \subsetneqq \mathcal{N}(B^2) \subsetneqq \mathcal{N}(B^3) \cdots$$

5.3. Eigenvalues of (B, A)

Suppose $B, A: X \to Z$ are bounded linear operators. We may define the *resolvent set* $\rho(B, A)$ of the pair (B, A) as the set of $\lambda \in \mathbb{C}$ such that $B - \lambda A$ has a bounded inverse. The spectrum $\sigma(B, A)$ of the pair (B, A) is $\sigma(B, A) = \mathbb{C} - \rho(B, A)$. A point $\lambda \in \sigma(B, A)$ is an *eigenvalue of (B, A)* if zero is an eigenvalue of $B - \lambda A$; that is, $\dim \mathcal{N}(B - \lambda A) \ge 1$.

If $A = I$, we have seen in the previous section that λ is a simple eigenvalue of (B, I) if and only if zero is a simple eigenvalue of $B - \lambda I$; that is,

$$\dim \mathcal{N}(B - \lambda) = 1 = \operatorname{codim} \mathcal{R}(B - \lambda)$$
$$[I\mathcal{N}(B - \lambda)] \oplus \mathcal{R}(B - \lambda) = Z.$$

This concept may be generalized in the following way. A point $\lambda \in \mathbb{C}$ is a *simple eigenvalue of the pair (B, A)* if

$$\dim \mathcal{N}(B - \lambda A) = 1 = \operatorname{codim} \mathcal{R}(B - \lambda A)$$
$$[A\mathcal{N}(B - \lambda A)] \oplus \mathcal{R}(B - \lambda A) = Z.$$

The following examples show that this concept is independent of the nature of the eigenvalue zero of the operator $B - \lambda A$.

EXAMPLE 3.1. If $X = Z = \mathbb{R}^2$,

$$B = \begin{bmatrix} 0 & 1 \\ 0 & 0 \end{bmatrix}, \qquad A = \begin{bmatrix} 0 & 0 \\ 1 & 0 \end{bmatrix}, \qquad \lambda = 0,$$

then $\mathcal{N}(B - \lambda A) = \mathcal{R}(B - \lambda A) = [e_1]$, where $[\ ]$ denotes span and $e_1 = \mathrm{col}(1, 0)$. Also, $A\mathcal{N}(B - \lambda A) = [e_2]$, $e_2 = \mathrm{col}(0, 1)$ and zero is a simple eigenvalue of (B, A). On the other hand, zero is a double eigenvalue of $B - \lambda A$.

EXAMPLE 3.2. If $X = Z = \mathbb{R}^2$,

$$B = \begin{bmatrix} 1 & 0 \\ 0 & 0 \end{bmatrix}, \qquad A = \begin{bmatrix} 0 & 1 \\ 0 & 0 \end{bmatrix}, \qquad \lambda = 0,$$

then zero is a simple eigenvalue of $B - \lambda A$. Also, the fact that $\mathcal{N}(B - \lambda A) = [e_2]$, $Ae_2 = e_1 \in \mathcal{R}(B - \lambda A)$ implies that zero is not a simple eigenvalue of (B, A).

EXAMPLE 3.3. Suppose $X = Z$ is the space of doubly infinite sequences $a = (\ldots, a_{-1}, a_0, a_1, \ldots)$ such that $\sum_{k=-\infty}^{\infty} |a_k|^2 < \infty$. If e_j is the element of X with components zero except for a one in the jth place define the linear operators B, A on X which satisfy $Be_j = e_{j-1}, j \neq 1$, $Be_1 = 0$, $Aa = a_1 e_0$ for all $a \in X$. Then, for $\lambda = 0$,

$$\mathcal{N}(B - \lambda A) = [e_1]$$
$$[e_0] \oplus \mathcal{R}(B - \lambda A) = X$$
$$Ae_1 = e_0.$$

Therefore, zero is a simple eigenvalue of (B, A). On the other hand, the generalized eigenspace of $B - \lambda A$ is infinite dimensional and given by the span $[e_1, e_2, \ldots]$. Therefore, zero is not a normal eigenvalue of $B - \lambda A$ even though zero is a simple eigenvalue of (B, A). This cannot happen for the case when $A = I$ since a simple eigenvalue λ of (B, I) is always a normal eigenvalue of $(B - \lambda I)$.

The previous example showed that one can have a simple eigenvalue λ_0 of (B, A) and yet zero is not a normal eigenvalue of $B - \lambda_0 A$. On the other hand, if zero is a normal eigenvalue of $B - \lambda_0 A$, then one can obtain an interesting characterization of the fact that λ_0 is a simple eigenvalue of (B, A). We proceed to derive this characterization. We say λ_0 is a *normal eigenvalue of (B, A)* if zero is a normal eigenvalue of $B - \lambda_0 A$. If λ_0 is a normal eigenvalue of (B, A), then there is an integer $\delta(B - \lambda_0 A) \overset{\text{def}}{=} k$ such that

$$\dim \mathcal{N}(B - \lambda_0 A)^k < \infty$$

(3.1)
$$X = \mathcal{N}(B - \lambda_0 A)^k \oplus j[\mathcal{R}(B - \lambda_0 A)^k]$$
$$Z = I\mathcal{N}(B - \lambda_0 A)^k \oplus \mathcal{R}(B - \lambda_0 A)^k.$$

Relative to the normal decomposition (3.1), the operator $B - \lambda_0 A$ can be represented in the block form of a matrix as

$$
(3.2)\quad
\begin{cases}
B - \lambda_0 A = \begin{pmatrix} C_0 & 0 \\ 0 & C_1 \end{pmatrix}, \qquad A = \begin{pmatrix} A_{00} & A_{01} \\ A_{10} & A_{11} \end{pmatrix}; \\[2mm]
C_0 : R^d \to R^d, \qquad d = \dim \mathcal{N}(B - \lambda_0 A)^k; \\[2mm]
\sigma(C_0) = \{0\}, \qquad C_0^k = 0 \\[2mm]
C_1 : j[\mathcal{R}(B - \lambda_0 A)^k] \to \mathcal{R}(B - \lambda_0 A)^k \text{ an isomorphism.}
\end{cases}
$$

For λ near λ_0 we may describe the "determinant" of $B - \lambda A$ in the following way, even if X and Z are infinite dimensional. Decompose $x = (y, z)$ for $x \in X$ according to the normal decomposition (3.1) and set $\mu = \lambda - \lambda_0$. Because C_1 is an isomorphism we see that, for small μ,

$$
(3.3)\quad B - \lambda A = \begin{pmatrix} B_0(\lambda) & -\mu A_{01} \\ 0 & C_1 - \mu A_{11} \end{pmatrix} \begin{pmatrix} I & 0 \\ -\mu(C_1 - \mu A_{11})^{-1} A_{10} & I \end{pmatrix},
$$

where

$$
(3.4)\quad B_0(\lambda_0 + \mu) = C_0 - \mu A_{00} - \mu^2 A_{01}(C_1 - \mu A_{11})^{-1} A_{10}.
$$

In the finite dimensional case, therefore,

$$
\det(B - \lambda A) = \det B_0(\lambda) \det(C_1 - \mu A_{11}),
$$

where $\det(C_1 - \mu A_{11})$ is non-zero. In general, we may regard the quantity

$$
(3.5)\quad \Delta(\lambda) = \det B_0(\lambda)
$$

as representing the determinant of $B - \lambda A$ on the invariant subspace corresponding to the eigenvalues of $B - \lambda A$ near zero for λ near λ_0. In particular, for $\lambda - \lambda_0$ sufficiently small, $\Delta(\lambda)$ is positive (negative) precisely when $B - \lambda A$ has an even (odd) number of eigenvalues near zero in the left-half plane. Note that we are discussing determinants only, and it is generally not true that the eigenvalues of $B_0(\lambda)$ are eigenvalues of $B - \lambda A$.

If zero is a normal eigenvalue of $B - \lambda_0 A$ and λ_0 is a simple eigenvalue of (B, A), then we may assume that the matrix C_0 in (3.2) is given by

$$
C_0 = \begin{bmatrix}
0 & 1 & \cdots & 0 & 0 \\
0 & 0 & \cdots & 0 & 0 \\
\multicolumn{5}{c}{\cdots\cdots\cdots\cdots\cdots\cdots} \\
0 & 0 & \cdots & 0 & 1 \\
0 & 0 & \cdots & 0 & 0
\end{bmatrix}
$$

The condition that λ_0 is simple is equivalent to saying that the lower left-hand corner of the matrix A_{00} is not zero. If $B_0(\lambda)$ is defined as in Relations (3.3), (3.4), then this latter remark is equivalent to

$$(3.6) \qquad \frac{d}{d\lambda} \det B_0(\lambda)\Big|_{\lambda=\lambda_0} \neq 0.$$

This result is stated as the following lemma.

Lemma 3.4. *If zero is a normal eigenvalue of $B - \lambda_0 A$, then λ_0 is a simple eigenvalue of (B, A) if and only if Relation (3.6) is satisfied.*

Specific applications where it is necessary to consider eigenvalues λ of (B, A) with $A \neq I$ are postponed until later sections when more theory is available.

5.4. Eigenvalues of $(B, A_1, \ldots, A_N)$

Suppose $B, A_1, \ldots, A_N : X \to Z$ are bounded linear operators. If $\lambda = (\lambda_1, \ldots, \lambda_N)$ and

$$(4.1) \qquad L(\lambda) = B - \sum_{j=1}^{N} \lambda_j A_j$$

then we define the *resolvent set* $\rho(B, A_1, \ldots, A_N)$ of $(B, A_1, \ldots, A_N)$ as the set of $\lambda \in \mathbb{C}^N$ such that $L(\lambda)$ has a bounded inverse. The *spectrum* $\sigma(B, A_1, \ldots, A_N)$ of $(B, A_1, \ldots, A_N)$ is

$$(4.2) \qquad \sigma(B, A_1, \ldots, A_N) = \mathbb{C}^N \backslash \rho(B, A_1, \ldots, A_N).$$

A point λ is an *eigenvalue of* $(B, A_1, \ldots, A_N)$ if zero is an eigenvalue of $L(\lambda)$; that is, $\dim \mathcal{N}(L(\lambda)) \geq 1$.

We say $\lambda \in \mathbb{C}^N$ is a *simple eigenvalue of* $(B, A_1, \ldots, A_N)$ if

(i) $L(\lambda)$ has Fredholm index $1 - N$
(ii) $\dim \mathcal{N}(L(\lambda)) = 1$
(iii) $[A_1 \mathcal{N}(L(\lambda)), \ldots, A_N \mathcal{N}(L(\lambda))] \oplus \mathcal{R}(L(\lambda)) = Z$

where $[\ ]$ denotes the span of the subspaces specified.

If $X = Z = \mathbb{R}^n$, then every linear operator $L(\lambda) : \mathbb{R}^n \to \mathbb{R}^n$ has index zero. Thus, the concept of a simple eigenvalue requires that $N = 1$. On the other hand if $X = \mathbb{R}^n$, $Z = \mathbb{R}^m$, $m > n$, the concept is meaningful for $N = m - n + 1$. The next example illustrates this fact.

EXAMPLE 4.1. If $X = \mathbb{R}^2$, $Z = \mathbb{R}^3$

$$B = \begin{bmatrix} 0 & 1 \\ 0 & 0 \\ 0 & 0 \end{bmatrix}, \qquad A_1 = \begin{bmatrix} 0 & 0 \\ 1 & 0 \\ 0 & 0 \end{bmatrix}, \qquad A_2 = \begin{bmatrix} 0 & 0 \\ 0 & 0 \\ 1 & 0 \end{bmatrix}$$

then $y_0 = \text{col}(1, 0)$ is a basis for $\mathcal{N}(B)$, $\{A_1 y_0, A_2 y_0\}$ is a basis for the complement of $\mathcal{R}(B)$. Therefore, the point $\lambda = (0, 0) \in \mathbb{R}^2$ is a simple eigenvalue of (B, A_1, A_2).

More interesting examples occur in infinite dimensional spaces.

EXAMPLE 4.2. Consider the following three point boundary value problem

$$(4.3) \qquad \ddot{x} + \lambda_1 x + \lambda_2 b(t) x = 0 \qquad 0 \le t \le \pi$$

$$(4.4) \qquad x(0) = 0$$

$$(4.5) \qquad x\left(\frac{\pi}{2}\right) = 0$$

$$(4.6) \qquad x(\pi) = 0$$

where

$$(4.7) \qquad \begin{aligned} b(t) &> 0 \quad \text{on } (0, \pi/2) \\ &< 0 \quad \text{on } (\pi/2, \pi) \end{aligned}$$

and $\lambda = (\lambda_1, \lambda_2) \in \mathbb{R}^2$. If

$$(4.8) \qquad \begin{aligned} Z &= C([0, \pi], \mathbb{R}), \\ X &= \{x \in C^2([0, \pi], \mathbb{R}), x(0) = x(\pi/2) = x(\pi) = 0\} \\ B &: X \to Z, \qquad (Bx)(t) = \ddot{x}(t) \\ A_1 &: X \to Z, \qquad (A_1 x)(t) = -x(t) \\ A_2 &: X \to Z, \qquad (A_2 x)(t) = -b(t) x(t) \end{aligned}$$

then problem (4.3)–(4.6) is equivalent to the abstract equation

$$(4.9) \qquad L(\lambda) x \overset{\text{def}}{=} (B - \lambda_1 A_1 - \lambda_2 A_2) x = 0$$

If λ_0 is an eigenvalue of (B, A_1, A_2), then $\dim \mathcal{N}(L(\lambda_0)) = 1$. In fact, if x_0, x_1 are eigenvectors corresponding to λ_0, then $x_1 = k_0 x_0$ on $[0, \pi/2]$ for some constant k_0 since x_0, x_1 must satisfy the boundary value problem (4.3), (4.4), (4.5) on $[0, \pi/2]$. Likewise, $x_1 = k_1 x_0$ on $(\pi/2, \pi)$ for some constant k_1. Also, the zeros of x_0, x_1 are simple for the same reason. Since $\dot{x}_1(\pi/2) = k_0 \dot{x}_0(\pi/2) = k_1 \dot{x}_0(\pi/2)$ and $\dot{x}_0(\pi/2) \ne 0$, it follows that $k_1 = k_0$. This proves $\dim \mathcal{N}(L(\lambda_0)) = 1$.

Let x_0 be a unit vector in $\mathcal{N}(L(\lambda_0))$. We prove that $\mathcal{R}(L(\lambda_0))$ coincides with the set of functions f in Z satisfying

$$(4.10) \qquad \int_0^{\pi/2} fx_0 = 0, \quad \int_{\pi/2}^{\pi} fx_0 = 0$$

and, therefore, $L(\lambda_0)$ has Fredholm index $1 - 2 = -1$. It is clear that f must satisfy (4.10) to be in $\mathcal{R}(L(\lambda_0))$. To prove sufficiency, suppose f satisfies (4.10) and let

$$x_1 = \delta x_0 + K_1 f \quad \text{on } [0, \pi/2]$$
$$x_2 = K_2 f \qquad\quad \text{on } [\pi/2, \pi]$$

where δ is an arbitrary constant and $K_1 f$, $K_2 f$ are solutions of problem (4.3), (4.4), (4.5) and problem (4.3), (4.5), (4.6) respectively. We wish to choose δ so that the function $x = x_1$ on $[0, \pi/2]$, $x = x_2$ on $[\pi/2, \pi]$ is a solution of (4.3)–(4.6). Since $x_1(\pi/2) = x_2(\pi/2) = 0$, we have x is continuous on $[0, \pi]$. Also, $\dot{x}_1(\pi/2) = \dot{x}_2(\pi/2)$ if and only if

$$\delta \dot{x}_0\left(\frac{\pi}{2}\right) = (K_2^{\cdot} f)\left(\frac{\pi}{2}\right) - (K_1^{\cdot} f)\left(\frac{\pi}{2}\right).$$

Since $\dot{x}_0(\pi/2) \neq 0$, this defines δ. Thus, for this δ, $\dot{x}$ is also continuous on $[0, \pi]$. The differential equation implies x is continuous on $[0, \pi]$ and we have proved that the range is characterized by (4.10).

It remains to show that $A_1 x_0, A_2 x_0$ are linearly independent and do not belong to $\mathcal{R}(L(\lambda_0))$. From the characterization (4.10) of $\mathcal{R}(L(\lambda_0))$ and the fact that b has fixed sign on both of the intervals $(0, \pi/2)$ and $(\pi/2, \pi)$, it is clear $A_1 x_0, A_2 x_0 \notin \mathcal{R}(L(\lambda_0))$. If there are constants c_1, c_2 such that

$$c_1 A_1 x_0 + c_2 A_2 x_0 = 0$$

then

$$(4.11) \qquad \begin{aligned} c_1 \int_0^{\pi/2} A_1 x_0^2 + c_2 \int_0^{\pi/2} A_2 x_0^2 &= 0 \\ c_1 \int_{\pi/2}^{\pi} A_1 x_0^2 + c_2 \int_{\pi/2}^{\pi} A_2 x_0^2 &= 0. \end{aligned}$$

Since condition (4.7) on b implies the coefficient matrix in (4.11) has nonzero determinant, it follows that $c_1 = c_2 = 0$ and $A_1 x_0, A_2 x_0$ are linearly independent.

This result is summarized in

Lemma 4.3. *For the boundary value problem* (4.3)–(4.7), *any eigenvalue of* (B, A_1, A_2) *in* (4.8) *is simple.*

Remark 4.4. In the previous example, the characteristics of the eigenvalue λ of (B, A_1, A_2) depend significantly on the choice of the space X, Z. If one considers problem (4.3)–(4.6) and considers a solution to be any function x which is C^2 on the open intervals $(0, \pi/2)$, $(\pi/2, 0)$ and satisfies the boundary conditions, then one obtains a larger number of solutions. In fact, for the problem (4.3), (4.4), (4.5) there is a curve $\gamma(\lambda_1, \lambda_2) = 0$ in $\mathbb{R}^2$ along which there exists a nontrivial solution $x_1(\lambda)$. If one defines $x = x_1$ on $[0, \pi/2]$, $x = 0$ on $[\pi/2, \pi]$, then x is a nontrivial solution of problem (4.3)–(4.6).

EXAMPLE 4.5. Consider the linear equation

$$(4.12) \qquad \dot{x} = \lambda_1 A_1 x + \lambda_2 A_2 x$$

where $x \in \mathbb{R}^n$, $\lambda = (\lambda_1, \lambda_2) \in \mathbb{R}^2$ and A_1, A_2 are constant $n \times n$ matrices. Assume λ_1^0 is such that

$$(4.13) \qquad \lambda_1^0 A_1 \equiv \begin{bmatrix} 0 & 1 & 0 \\ -1 & 0 & 0 \\ 0 & 0 & \bar{A}_1 \end{bmatrix}$$

and $[\exp \bar{A}_1 t] c \neq [\exp \bar{A}_1(t + 2\pi)] c$ for every $(n-2)$-vector $c \neq 0$; that is, the equation

$$(4.14) \qquad \dot{y} = \lambda_1^0 A_1 y$$

has exactly a two dimensional subspace of 2π-periodic solutions and the eigenvalues $\pm i$ of $\lambda_1^0 A_1$ have simple elementary divisors. The problem is to determine values of λ near $(\lambda_1^0, 0)$ for which equation (4.12) has 2π-periodic solutions; that is, solutions which satisfy the boundary condition

$$(4.15) \qquad x(0) - x(2\pi) = 0.$$

Since equation (4.12) is autonomous, if $x(t)$ is a solution satisfying (4.15), then so is $x(t + \alpha)$ for any constant α. In order to take out this indeterminacy in the solution, one can require that $x(0)$ be orthogonal to some vector e which is not orthogonal to the two dimension subspace of $\mathbb{R}^n$ satisfying

$$\{c \in \mathbb{R}^n : (e^{2\pi \lambda_1^0 A_1} - I) c = 0\}.$$

These vectors c are the initial values of the 2π-periodic solutions of (4.14). Thus, we require every solution of (4.12) also satisfy

$$(4.16) \qquad e \cdot x(0) = 0.$$

If

$$Z = \{x \in C([0, 2\pi], \mathbb{R}), \, x(0) = x(2\pi)\}$$

(4.17) $\qquad X = \{x \in C^1([0, 2\pi], \mathbb{R}), \, x(0) = x(2\pi), \, e \cdot x(0) = 0\}$

$$B: X \to Z, \qquad (Bx)(t) = \dot{x}(t)$$

and $A_1, A_2 : X \to Z$ are identified in the obvious way, then problem (4.12), (4.15), (4.16) is equivalent to the equation

$$L(\lambda)x \overset{\text{def}}{=} (B - \lambda_1 A_1 - \lambda_2 A_2)x = 0$$

for λ near $\lambda_0 = (\lambda_1^0, 0)$.

Let $x_0(t)$, $e \cdot x_0(0) = 0$, be a nonzero 2π-periodic solution of (4.12) for $\lambda_0 = (\lambda_{10}, 0)$. Then $x_0 \in X$ and all 2π-periodic solutions of (4.12) for $\lambda = (\lambda_{10}, 0)$ are given by $x_0(t + \alpha)$, $\alpha \in \mathbb{R}$. Furthermore, $\mathcal{N}(L(\lambda_0)) = [x_0]$. The Fredholm alternative implies

$$\mathcal{R}(L(\lambda_0)) = \left\{ x : \int_0^{2\pi} x(t)x_0(t + \alpha)\, dt = 0 \text{ for all } \alpha \in \mathbb{R} \right\}.$$

Therefore, codim $\mathcal{R}(L(\lambda_0)) = 2$ and $L(\lambda_0)$ has Fredholm index -1. Since $[x_0] = \mathcal{N}(L(\lambda_0))$, then obviously $A_1 x_0 \notin \mathcal{R}(L(\lambda_0))$. To proceed further, we suppose

(4.18) $$A_2 = \begin{bmatrix} \overline{A}_2 & 0 \\ 0 & \overline{A}_3 \end{bmatrix}$$

where $\overline{A}_2$ is 2×2 and nonsingular. Then $A_2 x_0 \subseteq [x_0(t + \alpha)]$ for some $\alpha \in \mathbb{R}$ and $A_2 x_0 \notin \mathcal{R}(L(\lambda_0)$. If we make the additional assumption that rank$[A_1 x_0(0), A_2 x_0(0)] = 2$, then $A_1 x_0, A_2 x_0$ are linearly independent. In particular, this is true if

(4.19) $$\overline{A}_2 = \begin{bmatrix} 1 & 0 \\ 0 & 1 \end{bmatrix}.$$

These results are summarized in the following lemma.

Lemma 4.6. *If A_1, A_2 satisfy (4.13), (4.18), (4.19) and the boundary value problem (4.12), (4.15), (4.16) is formulated as a solution of $L(\lambda)x \overset{\text{def}}{=} (B - \lambda_1 A_1 - \lambda_2 A_2)x = 0$ with $B: X \to Z$ defined in (4.17), then $\lambda_0 = (\lambda_1^0, 0)$ is a simple eigenvalue of (B, A_1, A_2).*

This lemma will be used to give another proof of the Hopf bifurcation theorem in a later section.

If $L(\lambda)$ has Fredholm index $1 - N$, then $L^k(\lambda)$ has index $k(1 - N)$ for any integer k. Therefore, either the ascent or descent of $L(\lambda)$ is infinity. Thus, for any λ_0, zero will never be a normal eigenvalue of $L(\lambda_0)$. On the other hand, there are situations where $L^k(\lambda_0)$ gives a decomposition of the domain space X in a manner similar to the decomposition for a normal eigenvalue.

A point $\lambda_0 \in \mathbb{C}$ is a *normal eigenvalue of* $(B, A_1, \ldots, A_N)$ if

(i) $L(\lambda)$ has Fredholm index $1 - N$

(ii)   there is an integer $\delta(L(\lambda_0)) \overset{\text{def}}{=} k$ such that

$$\mathcal{N}(L^{k-1}(\lambda_0)) \subsetneqq \mathcal{N}(L^k(\lambda_0)) = \mathcal{N}(L^{k+1}(\lambda_0))$$

(iii) $X = X_0 \oplus X_1$, $X_0 = \mathcal{N}(L^k(\lambda_0))$, $X_1 = j[\mathcal{R}(L^k(\lambda_0))]$

(iv) $Z = Z_0 \oplus Z_1$, $Z_1 = \mathcal{R}(L^k(\lambda_0))$, $Z_0 \subseteqq A_1 \mathcal{N}(L^k(\lambda_0)) \oplus \cdots \oplus \cdots \oplus A_N \mathcal{N}(L^k(\lambda_0))$

If we write the operators $L(\lambda)$, A_j, in matrix form according to the normal decomposition (iii), (iv), then

$$L(\lambda_0) = \begin{bmatrix} C_0 & 0 \\ 0 & C_1 \end{bmatrix}, \qquad A_j = \begin{bmatrix} A_{00}^j & A_{01}^j \\ A_{10}^j & A_{11}^j \end{bmatrix}$$

$$C_0 : \mathbb{R}^k \to \mathbb{R}^{Nk}$$

$$C_1 : X_1 \to Z_1 \quad \text{is an isomorphism}$$

$$C_0 = \begin{bmatrix} C_0^* \\ \vdots \\ C_0^* \end{bmatrix}, \qquad C_0^* = \begin{bmatrix} 0 & 1 & \cdots & 0 \\ 0 & 0 & \cdots & 0 \\ \cdots\cdots\cdots\cdots \\ 0 & 0 & \cdots & 1 \\ 0 & 0 & \cdots & 0 \end{bmatrix} \quad \text{is a } k \times k \text{ matrix}$$

For $\lambda = \lambda_0 + \mu$, $\mu = (\mu_1, \ldots, \mu_N)$ small, we may also factor the operator $L(\lambda) = L(\lambda_0) + \sum_{j=1}^N \mu_j A_j$ as

$$(4.20) \quad L(\lambda) = \begin{bmatrix} B_0(\lambda) & -\sum \mu_j A_{01}^j \\ 0 & C_1 - \sum \mu_j A_{11}^j \end{bmatrix} \begin{bmatrix} I & 0 \\ -(C_1 - \mu_j A_{11}^j)^{-1} \sum \mu_j A_{10}^j & I \end{bmatrix}$$

where

$$B_0(\lambda) = C_0 - \sum_{j=1}^N \mu_j A_{00}^j - (\sum \mu_j A_{01}^j)(C_1 - \sum \mu_j A_{11}^j)(\sum \mu_j A_{10}^j)$$

$$B_0(\lambda) : \mathbb{R}^k \to \mathbb{R}^{kN}.$$

Let us now suppose that λ_0 is a simple and normal eigenvalue of $(B, A_1, \ldots, A_N)$. Then $\dim \mathcal{N}(L^k(\lambda)) = k$ and the subspaces $A_1 \mathcal{N}(L^k(\lambda_0)), \ldots, A_N \mathcal{N}(L^k(\lambda_0))$ are linearly independent of dimension k. By an appropriate choice of bases, one can further assume that

$$(4.21) \qquad B_0(\lambda) = \begin{bmatrix} C_0^* + \mu_1 C_{11} \\ C_0^* + \mu_2 C_{12} \\ \vdots \\ C_0^* + \mu_N C_{1N} \end{bmatrix} + 0(|\mu|^2)$$

where $C_{1j} = (c_{rs}^{1j}, r, s = 1, 2, \ldots, k)$ and $c_{ki}^{1j} \neq 0$ for $j = 1, 2, \ldots, N$, since λ_0 is a simple eigenvalue of $(B, A_1, \ldots, A_N)$. These results are summarized in the following lemma.

Lemma 4.7. *If $\lambda_0 \in \mathbb{C}$ is a simple and normal eigenvalue of $(B, A_1, \ldots, A_N)$ and the matrix*

$$B_0(\lambda): \mathbb{R}^k \to \mathbb{R}^{kN}$$

is defined as in (4.21),

$$B_0(\lambda) = \begin{bmatrix} B_{01}(\lambda) \\ \vdots \\ B_{0N}(\lambda) \end{bmatrix}, \qquad B_{0j}(\lambda) \qquad a\ k \times k\ \text{matrix}$$

then

$$\frac{d}{d\lambda_j} \det B_{0j}(\lambda)\Big|_{\lambda = \lambda_0} \neq 0, \qquad j = 1, 2, \ldots, N.$$

For certain classes of operators $L(\lambda) = B - \sum \lambda_j A_j$, it is possible to give another interesting characterization of a simple eigenvalue of $(B, A_1, \ldots, A_N)$. To motivate the notation, let us reexamine Example 4.2. Let

$$\begin{aligned} Z &= C([0, \pi], \mathbb{R}) \\ X &= \{x \in C^2([0, \pi], \mathbb{R}): x(\pi/2) = 0\} \\ X_1 &= \{x \in X: x(0) = 0\} \\ X_2 &= \{x \in X: x(\pi) = 0\} \\ X_0 &= X_1 \cap X_2 \\ L(\lambda) &= B - \lambda_1 A_1 - \lambda_2 A_2 \\ L_j(\lambda) &= L(\lambda)|X_j, \qquad j = 0, 1, 2, \end{aligned}$$

where B, A_1, A_2 are defined in Equation (4.8). The equation $L_1(\lambda)x = 0$ corresponds to problem (4.3), (4.4), (4.5), the equation $L_2(\lambda)x = 0$ to problem (4.3), (4.5), (4.6) and the equation $L_0(\lambda)x = 0$ to problem (4.3–(4.6). Also, $L_1(\lambda), L_2(\lambda)$ have index 0, $L_0(\lambda)$ has index -1. Furthermore, if $x_0^j \in \mathcal{N}(L_j(\lambda))$, $x_0^j \neq 0$, then $A_1 x_0^j, A_2 x_0^j \notin \mathcal{R}(L_j(\lambda)), j = 1, 2$.

Our next objective is to generalize this example to obtain a characterization of a simple eigenvalue of $(B, A_1, \ldots, A_N)$.

Suppose $B, A_1, \ldots, A_N: X \to Z$ are bounded linear operators, $X_j \subset X$, $j = 1, 2, \ldots, N$ are closed subspaces, $X_0 = \bigcap_{j=1}^N X_j$. Let

$$L(\lambda) = B - \sum_{j=1}^N \lambda_j A_j$$

$$(4.22) \qquad L_j(\lambda) = L(\lambda) | X_j, \qquad j = 0, 1, 2, \ldots, N.$$

We will make the following hypothesis:

(4.23) (i) $L_j(\lambda)$ has index zero

         (ii) $[A_1 \mathcal{N}(L_j(\lambda)), \ldots, A_N \mathcal{N}(L_j(\lambda))] \oplus \mathcal{R}(L_j(\lambda)) = Z$
             $j = 1, 2, \ldots, N.$

Lemma 4.8. *If Hypothesis* (4.23) *is satisfied and* $\dim \mathcal{N}(L_j(\lambda_0)) = 1$, *then, for each* $j = 1, 2, \ldots, N$, *there exists a unique curve* $C_j \subset \mathbb{R}^N$ *through* λ_0, *analytic in a neighborhood of* λ_0, *which is given by* $C_j = \{\lambda : \lambda = \lambda_0 + \mu + 0(|\mu|^2)$ *where* $\mu = (\mu_1, \ldots, \mu_N)$ *satisfies*

$$u \cdot \alpha^j = 0, \alpha^j = \alpha_1^j, \ldots, \alpha_N^j$$
$$\alpha_k^j w_0^j = Q_0^j A_k x_0^j,$$
$$[w_0^j] = \operatorname{coker} L_j(\lambda_0), [x_0^j] = \mathcal{N}(L_j(\lambda_0))$$

and $Q_0^j : Z \to [w_0^j]$ *is the projection associated with the decomposition* $[w_0^j] \oplus \mathcal{R}(L_j(\lambda_0)) = Z\}$.

Proof. Fix $j \in \{1, 2, \ldots, N\}$ and define w_0^j, x_0^j, Q_0^j as in the statement of the lemma. If we apply the method of Liapunov–Schmidt to obtain the solutions x of the equation $L_j(\lambda)x = 0$ in the form $x = u x_0^j + y^j$ with y^j in the complement of $[x_0^j]$, then the bifurcation equation has the form $f_j(\lambda, \lambda_0)u = 0$,

$$f_j(\lambda_0 + \mu, \lambda_0) = \mu \cdot \alpha^j + 0(|\mu|^2)$$

as $|\mu| \to 0$. By Hypothesis (4.23)(ii), at least one of the components of α^j is $\neq 0$, say $\alpha_1^j \neq 0$. One can then solve for $\mu_1 = \mu_1^*(\mu_2, \ldots, \mu_N)$ where μ_1^* is analytic in $\mu_2, \ldots, \mu_N$ in a neighborhood of $\mu_2 = \cdots = \mu_N = 0$, $\mu_1^*(0, \ldots, 0) = 0$. This proves the lemma. $\square$

Lemma 4.9. *If the curves* $C_j, j = 1, 2, \ldots, N$ *are defined as in Lemma* 4.8 *and* $\lambda_0 \in \bigcap_{j=1}^N C_j$, *then the curves* $C_j, j = 1, 2, \ldots, N$, *intersect transversally at* λ_0 *if and only if*

$$\det(\alpha_k^j, j, k = 1, 2, \ldots, N) \neq 0.$$

Proof. This is obvious from the formulas for the C_j. $\quad\square$

Lemma 4.10. *If the curves* $C_1, \ldots, C_N$ *in Lemma* 4.8 *intersect transversally,* $[x_0] = \mathcal{N}(L_j(\lambda_0)), j = 1, 2, \ldots, N,$ *then*

$$\dim[A_1 x_0, \ldots, A_N x_0] = N$$
$$A_j x_0 \notin \mathcal{R}(L_0(\lambda_0)), \qquad j = 1, 2, \ldots, N.$$

Proof. If $A_1 x_0 \in \mathcal{R}(L_0(\lambda_0))$, then $A_1 x_0 \in \mathcal{R}(L_j(\lambda_0))$, $j = 1, 2, \ldots, N$, and $\alpha_k^1 = 0$, $k = 1, 2, \ldots, N$. This contradicts Lemma 4.9.

If there are numbers $v_1, \ldots, v_N$ such that

$$v_1 A_1 x_0 + \cdots + v_N A_N x_0 = 0,$$

then

$$v_1 Q_0^j A_1 x_0 + \cdots + v_N Q_0^j A_N y_0 = 0, \qquad j = 1, 2, . \qquad , N$$

and

$$v_1 \alpha_1^j + \cdots + v_N \alpha_N^j = 0, \qquad j = 1, 2, \ldots, N.$$

Lemma 4.9 implies $v_1 = \cdots = v_N = 0$ and the lemma is proved. $\quad\square$

As an immediate consequence, we have

Theorem 4.11. *If Hypothesis* (4.23) *is satisfied,* $\dim \mathcal{N}(L_j(\lambda_0)) = 1$, $j = 1, 2, \ldots, N$, *and* $L_0(\lambda_0)$ *has index* $1 - N$, *then* λ_0 *is a simple eigenvalue of* $(B, A_1, \ldots, A_N)$ *if and only if the curves* $C_1, \ldots, C_N$ *of Lemma* 4.8 *intersect transversally.*

EXAMPLE 4.12. This example is a generalization of Example 4.1. Suppose $\alpha < \beta < \gamma$ are real numbers, $p > 0$ is continuously differentiable on $[\alpha, \gamma]$, q, a, b are continuous on $[\alpha, \gamma]$, $\lambda = (\lambda_1, \lambda_2) \in \mathbb{R}^2$ and define

$$
\begin{aligned}
Bx &= -(px')' + qx, \qquad \left('= \frac{d}{dt}\right) \\
L(\lambda)x &= Bx + \lambda_1 ax + \lambda_2 bx.
\end{aligned}
\tag{4.24}
$$

For given $\alpha_1, \beta_1, \gamma_1 \in [0, \pi)$, consider the boundary value problem

$$L(\lambda)x = 0 \quad \text{on } (\alpha, \gamma) \tag{4.25}$$
$$x(\alpha)\cos \alpha_1 - x'(\alpha)\sin \alpha_1 = 0 \tag{4.26}$$
$$x(\beta)\cos \beta_1 - x'(\beta)\sin \beta_1 = 0 \tag{4.27}$$
$$x(\gamma)\cos \gamma_1 - x'(\gamma)\sin \gamma_1 = 0 \tag{4.28}$$

Let

$$Z = C([\alpha, \gamma], \mathbb{R})$$
$$X = \{x \in C^2([\alpha, \gamma], \mathbb{R}) : x \text{ satisfies } (4.26)\}$$
$$X_1 = \{x \in X : x \text{ satisfies } (4.25)\}$$
$$X_2 = \{x \in X : x \text{ satisfies } (4.27)\}$$
$$A_1 x = -ax, \qquad A_2 x = -bx.$$

Define the operators $L_j(\lambda) = L(\lambda)|X_j, j = 0, 1, 2, X_0 = X_1 \cap X_2$.
Suppose

$$(4.29) \qquad \mathcal{N}(L_0(\lambda_0)) = [x_0], \quad x_0 \text{ has simple zeros}$$

$$(4.30) \qquad \det \begin{bmatrix} \int_\alpha^\beta ax_0^2 & \int_\alpha^\beta bx_0^2 \\ \int_\beta^\gamma ax_0^2 & \int_\beta^\gamma bx_0^2 \end{bmatrix} \neq 0.$$

Except for a few technical details one can repeat the proof in Example 4.1 to show that λ_0 is a simple eigenvalue of (B, A_1, A_2). The condition (4.29) is used to show $L_0(\lambda_0)$ has index -1. The condition (4.30) shows that $\dim[A_1 x_0, A_2 x_0] = 2$ and $A_1 x_0, A_2 x_0 \notin \mathcal{R}(L(\lambda_0))$. Condition (4.30) is also equivalent to the transversal intersection of the curves C_1, C_2 in Theorem 4.11.

5.5. Bifurcation from a Simple Eigenvalue

Consider the equation

$$(5.1) \qquad M(\lambda, x) = Bx - \lambda x + N(\lambda, x)$$

where $B : X \to Z$ is a bounded linear operator, $N(\lambda, x) \in C^2(\Lambda \times X, Z)$, $N(\lambda, 0) = 0$, $D_x N(\lambda, 0) = 0$, $\Lambda \subset \mathbb{R}$ (or $\mathbb{C}$) is an open set. If λ_0 is a simple eigenvalue of B, then the spaces X, Z can be decomposed as

$$X = X_0 \oplus X_1$$
$$Z = Z_0 \oplus Z_1.$$

Here we may consider $X_0 = Z_0 = \mathcal{N}(B - \lambda_0)$ as a one dimensional vector space. If $y_0 \in X_0$ is a nonzero vector, then any $y \in X_0$ can be written as $y = uy_0$ for some $u \in \mathbb{R}$.

We apply the method of Liapunov–Schmidt to show λ_0 is always a bifurcation point for $M(\lambda, x)$. In block form relative to our decomposition,

$$B = \begin{pmatrix} \lambda_0 & 0 \\ 0 & B_1 \end{pmatrix}$$

where $B_1 - \lambda_0$ is an isomorphism from X_1 to Z_1. Upon decomposing $x = y + z$, $y \in X_0$, $z \in X$, and using the fact that $(B - \lambda_0)y = 0$, the auxiliary equation

$$(B_1 - \lambda)z + Q_1 N(\lambda, y + z) = 0$$

is solved for a unique $z = z^*(\lambda, y) \in C^2(\Omega)$ where Q_1 is the projection onto Z_1 and Ω is a neighborhood of $(\lambda_0, 0) \in \mathbb{R} \times X_0$. Furthermore, $z^*(\lambda, 0) = 0$, $D_y z^*(\lambda, 0) = 0$. This can be seen from the fact that $z^*(\lambda, y)$ is the unique solution of the equation

$$z^*(\lambda, y) = (B_1 - \lambda)^{-1} Q_1 N(\lambda, y + z^*(\lambda, y))$$

for (λ, y) in a sufficiently small neighborhood of $(\lambda_0, 0)$. Uniqueness clearly implies $z^*(\lambda, 0) = 0$. Implicit differentiation shows $D_y z^*(\lambda, 0) = 0$. It follows, therefore, that $z^*(\lambda, 0) = 0(|y|^2)$.

The bifurcation equation $Q_0 M(\lambda, y + z^*(\lambda, y)) = 0$ is one dimensional where Q_0 is the projection onto Z_0. Using the basis vector y_0 and letting $y = u y_0$, we obtain the bifurcation equation

$$F(\lambda, u) = 0$$

for λ, u and F real, where F is defined as

$$F(\lambda, u)y_0 = (\lambda_0 - \lambda)u y_0 + Q_0 N(\lambda, u y_0 + z^*(\lambda, u y_0)).$$

The nonlinear term here is of order $0(u^2)$ and so F has the form

$$(5.2) \qquad F(\lambda, u) = -(\lambda - \lambda_0)u + G(\lambda, u)$$

where $G(\lambda, u) = O(u^2)$. If $G(\lambda, u) = u\bar{G}(\lambda, u)$, then $\bar{G}(\lambda, 0) = 0$ and the zeros of F in (5.2) are obtained by solving

$$-(\lambda - \lambda_0) + \bar{G}(\lambda, u) = 0.$$

By the Implicit Function Theorem, these solutions form a C^1 curve

$$\lambda = \lambda^*(u) = \lambda_0 + 0(|u|)$$

passing through $(\lambda, u) = (\lambda_0, 0)$. The solution $x^*(u)$ represented here has the form

$$x^*(u) = u y_0 + z^*(\lambda^*(u), u y_0)$$
$$= u y_0 + O(u^2).$$

In particular, the curve of solutions is tangent to the eigenspace X_0.

A careful analysis shows that the only smoothness necessary to obtain the same result is $D_\lambda N, D_x N, D_{\lambda x}$ continuous on $\Lambda \times X$ with $N(\lambda, 0) = 0$, $D_\lambda N(\Lambda, 0) = 0$.

Also, if $N \in C^m(\Lambda, X)$, $m \geq 2$, then the bifurcation curve $\lambda = \lambda^*(u)$ and solution $x^*(u)$ are C^{m-1} in a neighborhood of $0 \in \mathbb{R}$. If N is analytic, then λ^*, x^* are analytic.

These remarks are summarized in the following statement.

Theorem 5.1. *Let X, Z be real (or complex) Banach spaces, Λ be an open set in $\mathbb{R}$ (or $\mathbb{C}$) and $M \in C^m(\Lambda \times X, Z)$, $m \geq 2$. Suppose that*

$$(5.3) \qquad \begin{aligned} M(\lambda, x) &= Bx - \lambda x + N(\lambda, x) \\ N(\lambda, 0) &= 0, \quad D_x N(\lambda, 0) = 0. \end{aligned}$$

If λ_0 is a simple eigenvalue of B with eigenvector $y_0 \neq 0$, then $(\lambda, x) = (\lambda_0, 0)$ is a bifurcation point of $M(\lambda, x) = 0$. Moreover, there exist C^{m-1} functions

$$(5.4) \qquad \begin{aligned} \lambda^*(u) &= \lambda_0 + O(|u|) \\ x^*(u) &= u y_0 + O(u^2) \end{aligned}$$

for real u near zero such that

$$M(\lambda^*(u), x^*(u)) \equiv 0.$$

All zeros of M near $(\lambda_0, 0)$ are either the trivial solution $x = 0$ or given by (5.4). Finally, if M is an analytic function of λ, x near $(\lambda_0, 0)$, then λ^*, x^* are analytic near zero.

In the applications, the function N in (5.3) depends on some other parameters w and $N = 0((|w| + |x|)|x|)$ as $w, x \to 0$. The proof of Theorem 1 is easily modified to apply to this case. For later reference, this is stated in the following result.

Theorem 5.2. *Let X, Z, W be real (or complex) Banach spaces, Λ be an open set in $\mathbb{R}$ (or $\mathbb{C}$) and $M \in C^m(\Lambda \times W \times X, Z)$, $m \geq 2$. Suppose*

$$(5.5) \qquad \begin{aligned} M(\lambda, w, x) &= Bx - \lambda x + N(\lambda, w, x) \\ N(\lambda, w, 0) &= 0, \qquad D_x N(\lambda, 0, 0) = 0. \end{aligned}$$

If λ_0 is a simple eigenvalue of B with eigenvector $y_0 \neq 0$, then there is a $\delta > 0$ and C^{m-1} functions

$$(5.6) \qquad \begin{aligned} \lambda^*(w, u) &= \lambda_0 + O(|w| + |u|) \\ x^*(w, u) &= u y_0 + O((|w| + |u|)|u|) \end{aligned}$$

for $|u| < \delta$, $|w| < \delta$ *such that* $M(\lambda^*(w, u), w, x^*(w, u)) = 0$. *All zeros of* M *near* $(\lambda_0, 0, 0)$ *are either* $(\lambda, w, 0)$ *or given by* $(\lambda^*(w, u), w, x^*(w, u))$.

The proof of the above result is easily generalized to the more general situation where λ_0 is a simple eigenvalue of (B, A) and

$$M(\lambda, x) = Bx - \lambda Ax + N(\lambda, x)$$
$$N(\lambda, 0) = 0, \qquad D_x N(\lambda, 0) = 0.$$

If λ_0 is a simple eigenvalue of (B, A) and $y_0 \in \mathcal{N}(B - \lambda_0 A)$, $y_0 \neq 0$, then the spaces X, Z can be decomposed as

$$X = X_0 \oplus X_1 \qquad X_0 = \mathcal{N}(B - \lambda_0 A) = [y_0]$$
$$Z = Z_0 \oplus Z_1 \qquad Z_0 = [Ay_0], \quad Z_1 = \mathcal{R}(B - \lambda_0 A).$$

Applying the method of Liapunov–Schmidt with $x = y + z$, $y \in X_0$, $z \in X_1$ and using the fact that $(B - \lambda_0 A)y = 0$, $Q_1 Ay = 0$, $x = y + z$, $y \in X_0$, $z \in X_1$, one observes that the equation

$$Q_1(B - \lambda A)(y + z) + Q_1 N(\lambda, y + z) = Q_1(B - \lambda A)z + Q_1 N(\lambda, y + z) = 0$$

can be solved for a unique $z = z^*(\lambda, y)$ in a neighborhood Ω of $(\lambda_0, 0)$. As in the proof of Theorem 5.1, if $N \in C^m(\Lambda \times X, Z)$, $m \geq 2$, then $z^* \in C^m(\Omega, X)$, $z^*(\lambda, 0) = 0$, $D_y z^*(\lambda, 0) = 0$ and $z^*(\lambda, 0) = O(|y|^2)$ as $y \to 0$.

The bifurcation equation $Q_0 M(\lambda, y + z^*(\lambda, y)) = 0$ is one-dimensional. Using the basis vector Ay_0 of Z_0 and letting $y = uy_0$, the bifurcation function $F(\lambda, u)$ satisfies

$$F(\lambda, u)Ay_0 = Q_0(B - \lambda A)(uy_0 + z^*(\lambda, uy_0)) + Q_0 N(\lambda, uy_0 + z^*(\lambda, uy_0))$$
$$= -(\lambda - \lambda_0)uAy_0 + Q_0(B - \lambda A)z^*(\lambda, uy_0)$$
$$+ Q_0 N(\lambda, uy_0 + z^*(\lambda, uy_0)).$$

Therefore,

$$F(\lambda, u) = -(\lambda - \lambda_0)u + G(\lambda, u)$$

where $G(\lambda, 0) = 0$, $D_u G(\lambda, 0) = 0$. If $G(\lambda, u) = u\bar{G}(\lambda, u)$, then the analysis proceeds exactly as in the proof of Theorem 5.1 to yield the following result.

Theorem 5.3. *Assume* $M \in C^m(\Lambda \times X, Z)$, $m \geq 2$, $\Lambda \subset \mathbb{R}$ *an interval,* X, Z *real Banach spaces,*

(5.7)
$$M(\lambda, x) = Bx - \lambda Ax + N(\lambda, x)$$
$$N(\lambda, 0) = 0, \qquad D_\lambda N(\lambda, 0) = 0.$$

If λ_0 is a simple eigenvalue of (B, A), then the conclusions of Theorem 5.1 hold. If $N = N(\lambda, w, x)$ as in (5.5) and λ_0 is a simple eigenvalue of (B, A), then the conclusion of Theorem 5.2 holds.

The preceding ideas also can be used to prove that bifurcation occurs at a simple eigenvalue $\lambda_0 \in \mathbb{R}^N$ of $(B, A_1, \ldots, A_n)$. Suppose

$$\lambda = (\lambda_1, \ldots, \lambda_N) \in \mathbb{R}^N$$
$$M(\lambda, x) = L(\lambda)x + N(\lambda, x)$$
$$L(\lambda) = B - \sum_{j=1}^{N} \lambda_j A_j$$
$$N(\lambda, 0) = 0, \qquad D_x N(\lambda, 0) = 0.$$

If $\lambda_0 \in \mathbb{R}^N$ is a simple eigenvalue of $(B, A_1, \ldots, A_N)$, $y_0 \in \mathcal{N}(L(\lambda_0))$, $y_0 \neq 0$, then the spaces X, Z can be decomposed as

$$X = X_0 \oplus X_1, \qquad X_0 = [y_0]$$
$$Z = Z_0 \oplus Z_1, \qquad Z_0 = [A_1 y_0, \ldots, A_N y_0], \quad Z_1 = \mathcal{R}(L(\lambda_0)).$$

Applying the method of Liapunov–Schmidt with $x = y + z$, $y \in X_0$, $z \in X_1$ and using the fact that $L(\lambda_0)y_0 = 0$, $Q_1 A_j y_0 = 0$, $j = 1, 2, \ldots, N$, the auxiliary equation

$$Q_1 L(\lambda)(y + z) + Q_1 N(\lambda, y + z) = Q_1 L(\lambda)z + Q_1 N(\lambda, y + z) = 0$$

can be solved for a unique $z^*(\lambda, y)$ in a neighborhood Ω of $(\lambda_0, 0)$. As in the proof of Theorem 5.1, if $N \in C^m(\Lambda \times X, Z)$, $m \geq 2$, then $z^* \in C^m(\Omega, X)$, $z^*(\lambda, 0) = 0$, $D_y z^*(\lambda, 0) = 0$ and $z^*(\lambda, 0) = O(|u|^2)$ as $u \to 0$.

The bifurcation equation $Q_0 M(\lambda, y + z^*(\lambda, y))$ is N-dimensional. Using the basis vectors $A_1 y_0, \ldots, A_N y_0$ of Z_0 and letting $y = u y_0$, the bifurcation functions $F(\lambda, u) \in \mathbb{R}^N$ satisfy $F = (F_1, \ldots, F_N)$, $\lambda = \lambda_0 + \mu$

$$\sum_{j=1}^{N} F_j A_j y_0 = Q_0 L(\lambda_0 + \mu)(u y_0 + z^*(\mu, u y_0))$$
$$+ Q_0 N(\lambda_0 + \mu, u y_0 + z^*(\mu, u y_0))$$
$$= u \sum_{j=1}^{N} \mu_j A_j y_0 + Q_0 L(\lambda_0 + \mu)z^*(\mu, u y_0)$$
$$+ Q_0 N(\lambda_0 + \mu, u y_0 + z^*(\mu, u y_0)).$$
$$F(\lambda, u) = -(\lambda - \lambda_0)u + G(\lambda, u)$$

where $G(\lambda, 0) = 0$, $D_u G(\lambda, 0) = 0$. If $G(\lambda, u) = u\bar{G}(\lambda, u)$, then $\bar{G}(\lambda, 0) = 0$, $D_\lambda \bar{G}(\lambda_0, 0) = 0$. The Implicit Function Theorem implies the equations

$$-(\lambda - \lambda_0) + \bar{G}(\lambda, u) = 0$$

can be solved for a unique $\lambda^*(u)$ in a neighborhood of $u = 0$, the functions λ^* in C^{m-1} and

$$\lambda^*(u) = \lambda_0 + O(|u|) \quad \text{as } u \to 0.$$

These results are summarized in the following statement.

Theorem 5.4. *Let X, Z be Banach spaces, $\Lambda \subset \mathbb{R}^N$ an open interval, $M \in C^m(\Lambda \times X, Z)$, $m \geq 2$. Suppose that*

$$M(\lambda, x) = Bx - \sum_{j=1}^{N} \lambda_j A_j x + N(\lambda, x)$$

(5.8)
$$N(\lambda, 0) = D_x N(\lambda, 0) = 0$$
$$B, A_1, \ldots, A_N : X \to Z \text{ bounded, linear}$$
$$\lambda = (\lambda_1, \ldots, \lambda_N).$$

If $\lambda_0 \in \mathbb{R}^N$ is a simple eigenvalue of $(B, A_1, \ldots, A_N)$ with eigenvector $y_0 \neq 0$, then $(\lambda, x) = (\lambda_0, 0)$ is a bifurcation point of $M(\lambda, x) = 0$. Moreover, there exist C^{m-1} functions

(5.9)
$$\lambda^*(u) = \lambda_0 + O(|u|)$$
$$x^*(u) = uy_0 + o(|u|) \quad \text{as } u \to 0$$

for real u near zero such that

(5.10)
$$M(\lambda^*(u), x(u)) = 0.$$

All zeros of M near $(\lambda, 0)$ are either the trivial solution $x = 0$ or given by (5.9). If M is analytic in a neighborhood of $(\lambda_0, 0)$, then so are $\lambda^*(u), x^*(u)$ near $u = 0$. If $N = N(\lambda, w, x)$ satisfies (5.5), then the same conclusions hold for $\lambda^*(w, u) = O(|w| + |u|)$, $x^*(w, u) = uy_0 + O((|w| + |u|)|u|)$.

5.6. Applications of Simple Eigenvalues

In this section, we give some applications of the results on bifurcation from a simple eigenvalue given in Theorems 5.1–5.4.

EXAMPLE 6.1. Suppose $X = Z = \mathbb{R}^3$,

$$B = \begin{bmatrix} 0 & 1 & 0 \\ 0 & 0 & 0 \\ 0 & 0 & 1 \end{bmatrix}$$

$N: \mathbb{R} \times \mathbb{R}^3 \to \mathbb{R}^3$ is a C^2-function such that $N(\lambda, 0) = 0$, $D_x N(\lambda, 0) = 0$. Since $\lambda_0 = 1$ is a simple eigenvalue of B, Theorem 5.1 implies the point $(1, 0) \in \mathbb{R} \times \mathbb{R}^3$ is a bifurcation point of the equation $Bx - \lambda x - N(\lambda, x) = 0$.

EXAMPLE 6.2. Suppose B, N as in Example 6.1. The point $\lambda_0 = 0$ is a double eigenvalue of B and, thus, Theorem 5.1 does not apply. On the other hand, if

$$A = \begin{bmatrix} 0 & 0 & 0 \\ 1 & 0 & 0 \\ 0 & 0 & 1 \end{bmatrix}$$

it was shown in Example 3.1 that $\lambda_0 = 0$ is a simple eigenvalue of (B, A). Thus, Theorem 5.3 implies the point $(1, 0) \in \mathbb{R} \times \mathbb{R}^3$ is a bifurcation point for the equation $Bx - \lambda Ax + N(\lambda, x) = 0$.

EXAMPLE 6.3. Suppose B, A_1, A_2 are defined as in Example 4.1. Then $N = 2$, $\lambda_0 = (0, 0)$ is a simple eigenvalue of (B, A_1, A_2). From Theorem 5.4, for any $N \in C^m(\mathbb{R}^2 \times \mathbb{R}^3, \mathbb{R}^3)$, the point $(\lambda_0, 0) \in \mathbb{R}^2 \times \mathbb{R}^3$ is a bifurcation point for the equation $Bx - \lambda_1 A_1 x - \lambda_2 A_2 x + N(\lambda, x) = 0$.

The remaining examples deal with infinite dimensional problems.

EXAMPLE 6.4. Let $Z = C([0, \pi], \mathbb{R})$, $X = \{x \in C^2([0, \pi], \mathbb{R}): x(0) = x(\pi) = 0\}$, $M: \mathbb{R} \times X \to Z$ be defined by

$$M(\lambda, x)(t) = -\ddot{x}(t) - \lambda x(t) + f(\lambda, x(t), \dot{x}(t)), \qquad 0 \le t \le 2\pi$$

where $f: \mathbb{R}^3 \to \mathbb{R}$ is a C^2 function, $f(\lambda, 0, 0) = 0$, $D_x f(\lambda, 0, 0) = 0$, $D_{\dot{x}} f(\lambda, 0, 0) = 0$. The equation $M(\lambda, 0) = 0$ is equivalent to the boundary value problem

$$\begin{aligned} \ddot{x} + \lambda x - f(\lambda, x, \dot{x}) = 0, \qquad 0 \le t \le \pi \\ x(0) = x(\pi) = 0. \end{aligned}$$

(6.1)

If $B: X \to Z$ is defined by $(Bx)(t) = -\ddot{x}(t)$, then it was shown in Example 2.2 that the only eigenvalues of B are $\lambda_n = n^2, n = 1, 2, \ldots$ with $\dim \mathcal{N}(B - \lambda_n I) = 1$ and $x_n(t) = \sin nt$ a basis for $\mathcal{N}(B - \lambda_n I)$. It is easy to observe that $x_n \in \mathcal{R}(B - \lambda_n I)$. Therefore, $-n^2$ is a simple eigenvalue of B. Theorem 5.1 implies the point $(n^2, 0) \in R \times X$ is a bifurcation point for $M(\lambda, x) = 0$; that is, there is a neighborhood U of $(n^2, 0) \in \mathbb{R} \times X$ such that there is a pair

$(\lambda^*, x^*) \in U$ such that $x^* \neq 0$ and x^* satisfies the boundary value problem (6.1) for $\lambda = \lambda^*$. In addition, all nonzero solutions can be described parametrically by a real parameter u by the formulas

$$\lambda^* = n^2 + O(|u|)$$
$$x^* = u \sin nt + O(|u|^2) \quad \text{as } |u| \to 0.$$

EXAMPLE 6.5. Suppose B is the differential operator defined in Example 2.5, Formula 2.7 and consider the boundary value problem

$$(Bx)(t) \overset{\text{def}}{=} -x'' + q(t)x = \lambda x + f(t, x), \qquad t \in [0, \infty)$$
$$x(0) = 0$$

for $x \in L^2[0, \infty)$, where the function f satisfies enough conditions to ensure that

$$F: L^2[0, \infty) \to L^2[0, \infty), \quad (Fx)(t) = f(t, x(t))$$

is C^2 in x, $F(0) = 0$, $DF(0) = 0$. It was noted in Example 2.5 that any eigenvalue of B is simple. Thus, there is always a bifurcation at each eigenvalue of B. If $0 < Q_1 \leq q(t)$ for all $t \in [0, \infty)$, then for any $0 < \lambda < Q_1$, it follows from Relation (2.8) that λ is either in the resolvent set of B or an eigenvalue of B.

EXAMPLE 6.6. The following boundary value problem arises in connection with chemical reaction and diffusion:

$$\ddot{x}_1 + (\lambda - 1)x_1 + \alpha^2 x_2 + f_1(\lambda, x_1, x_2) = 0$$
(6.2)
$$\ddot{x}_2 - \lambda x_1 - \alpha^2 x_2 + f_2(\lambda, x_1, x_2) = 0, \qquad 0 \leq t \leq 1$$
$$x_1(0) = x_1(1) = \dot{x}_2(0) = \dot{x}_2(1) = 0$$

where f_1, f_2 are C^2 functions satisfying $f_j(\lambda, 0, 0) = 0$, $D_{x_k} f_j(\lambda, 0, 0) = 0$, $j, k = 1, 2$.

Let

$$Z = C([0, 1], \mathbb{R}^2)$$
$$X = \{x \in C^2([0, 1], \mathbb{R}^2): x_1(0) = x_1(1) = \dot{x}_2(0) = \dot{x}_2(1) = 0\}.$$

If $x = (x_1, x_2)$, then Problem (6.2) can be written in the form

$$(B - \lambda A)x + N(x, \lambda) = 0$$

where

$$B = \frac{d^2}{dt^2} + \begin{bmatrix} -1 & \alpha^2 \\ 0 & -\alpha^2 \end{bmatrix}, \qquad A = \begin{bmatrix} -1 & 0 \\ 1 & 0 \end{bmatrix}, \qquad N = \begin{bmatrix} f_1 \\ f_2 \end{bmatrix}.$$

The eigenvalues λ_n of (B, A) are given by

$$\lambda_n = 1 + \alpha^2 + n^2\pi^2 + \frac{\alpha^2}{n^2\pi^2}, \qquad n = 1, 2, \ldots$$

with the corresponding eigenvectors

$$x^{(n)}(t) = \begin{bmatrix} \sin n\pi t \\ -\left(1 + \dfrac{1}{n^2\pi^2}\right)\cos n\pi t \end{bmatrix}.$$

If $Ax^{(n)} \in \mathscr{R}(B - \lambda_n A)$, then there is a solution of the boundary value problem

$$\ddot{x}_1 + (\lambda_n - 1)x_1 + \alpha^2 x_2 = -\sin n\pi t$$
$$\ddot{x}_2 - \lambda_n x_1 - \alpha^2 x_2 = \sin n\pi t, \qquad 0 \le t \le 1$$
$$x_1(0) = x_1(1) = \dot{x}_2(0) = \dot{x}_2(1) = 0.$$

Inner product this equation with the vector

$$\begin{bmatrix} \sin n\pi t \\ \dfrac{\alpha^2}{\alpha^2 + n^2\pi^2}\cos n\pi t \end{bmatrix}$$

integrate from a to 1 and perform some integrations by parts to obtain

$$0 = -\int_0^\pi \sin^2 n\pi t \, dt \neq 0.$$

This contradiction implies λ_n is a simple eigenvalue of (B, A), $n = 1, 2, \ldots$. Theorem 5.3 implies the point $(\lambda_n, 0) \in \mathbb{R} \times X$ is a bifurcation point for Equation (6.2).

Our next application concerns a generalization of the *Liapunov Center Theorem* to equations which depend on a parameter λ. The form of the equations arise in studying the stationary solutions in reaction-diffusion problems.

Suppose E is a real constant nonsingular $n \times n$ matrix, F, G are constant $n \times n$ matrices, λ is a real parameter and $h: \mathbb{R}^n \to \mathbb{R}^n$ is a C^2 function with $h(0) = 0$, $D_x h(0) = 0$. The problem is to determine the nature of the non-constant periodic solutions of the system of second order differential equations

$$(6.3) \qquad\qquad E\ddot{x} + (F + \lambda G)x + h(x) = 0, \qquad x \in \mathbb{R}^n,$$

which are close to zero when λ is close to some critical value to be specified below.

If (6.3) has a periodic solution of least period $p > 0$, then the change of variables $t \to pt$ reduces the discussion to the determination of 2π-periodic solutions of the equation

$$(6.4) \qquad p^2 E\ddot{x} + (F + \lambda G)x + h(x) = 0.$$

We make the following hypotheses:

(H_1)   *There exist $p_0 > 0$, $\lambda_0 \in \mathbb{R}$ such that*

$$(6.5) \qquad \begin{aligned} \det[-p_0^2 E + F + \lambda_0 G] &= 0 \\ \dim \mathcal{N}(-p_0^2 E + F + \lambda_0 G) &= 1. \end{aligned}$$

(H_2)   *If $\phi \in \mathbb{R}^n$, $\phi \neq 0$, is such that*

$$(6.6) \qquad (-p_0^2 E + F + \lambda_0 G)\phi = 0$$

then the only 2π-periodic solutions of the differential equation

$$(6.7) \qquad p_0^2 E\ddot{x} + (F + \lambda_0 G)x = 0$$

are linear combinations of $\phi \cos t$ and $\phi \sin t$.

If ψ is an n-dimensional row vector such that

$$(6.8) \qquad \psi[-p_0^2 E + F + \lambda_0 G] = 0$$

then we can prove the following results.

Theorem 6.7. *If* (H_1), (H_2) *are satisfied and*

$$(H_3) \quad \psi(F + \lambda_0 G)\phi \neq 0 \quad (equivalently, \ \psi E\phi \neq 0)$$

then there is a $\delta > 0$ such that, for each $(\lambda, a) \in \mathbb{R}^2$, $|\lambda - \lambda_0| < \delta$, $|a| < \delta$, there is a periodic solution $x^*(\lambda, a)$ of Equation (6.3) of period $p^*(\lambda, a)$, where $x^*(\lambda, a)$, $p^*(\lambda, a)$ are C^1 in (λ, a) with

$$p^*(\lambda, 0) = p_0$$
$$x^*(\lambda, 0) = 0, \qquad D_a x^*(\lambda, 0) = \phi \cos t.$$

Furthermore, each periodic solution of Equation (6.3) with minimal period close to p_0 and norm less than δ is given (except for a translation in phase) by $x^*(\lambda, a)$ for some a.

Theorem 6.8. *If* (H_1), (H_2) *are satisfied and*

$$(H_4) \quad \psi G\phi \neq 0$$

then there is a $\delta > 0$ *such that for each* $(p, a) \in \mathbb{R}^2$, $|a| < \delta$, *there is a* C^1 *function* $\lambda^*(p, a)$, $\lambda^*(p_0, 0) = \lambda_0$ *and a* C^1 *function* $x^*(p, a)$, *periodic of period* p, *such that* $x^*(p_0, 0) = 0$, $D_a x^*(p_0, 0) = \phi \cos t$ *and* $x^*(p, a)$ *satisfies* (6.3) *for* $\lambda = \lambda^*(p, a)$. *Furthermore,* $x^*(p, a)$ *is unique in the same sense as described in the statement of Theorem* 6.7.

Proof. As remarked earlier, it is only necessary to consider 2π-periodic solutions of (6.4) which we write in a more convenient form

$$(6.9) \qquad E\ddot{x} + (\beta_1 F + \beta_2 G)x + \beta_1 h(x) = 0$$

where $\beta_1 = 1/p^2$, $\beta_2 = \lambda/p^2$. Hypothesis (H_2) states that every 2π-periodic solution of

$$E\ddot{x} + (\beta_1 F + \beta_2 G)x = 0$$

has the form $a\phi \cos(t + \alpha)$ for some constants $(a, \alpha) \in \mathbb{R}^2$. Since the equation is autonomous, we may choose $\alpha = 0$. Furthermore, an application of Liapunov–Schmidt will obviously lead to even functions of t if we choose $\alpha = 0$. Let

$$X = \{x \in C^2(\mathbb{R}, \mathbb{R}^n) : 2\pi\text{-periodic, even}\}$$
$$Z = \{x \in C(\mathbb{R}, \mathbb{R}^n) : 2\pi\text{-periodic, even}\}$$

and define

$$N_1(\beta_1, x)(t) = \beta_1 h(x(t)), \qquad (Bx)(t) = \ddot{x}(t),$$
$$(Fx)(t) = Fx(t), (Gx)(t) = Gx(t), \qquad t \in \mathbb{R}$$

with all operators taking X into Z.

We make the following simple observations: for $\beta_1 = \beta_{10} = 1/p_0^2$, $\beta_2 = \beta_{20} = \lambda_0/p_0^2$,

Hypothesis $(H_3) \Leftrightarrow \beta_{10}$ is a simple eigenvalue $(E + \beta_{20} G, F)$

Hypothesis $(H_4) \Leftrightarrow \beta_{20}$ is a simple eigenvalue of $(E + \beta_{10} F, G)$.

Therefore, the point $(\beta_{10}, 0) \in \mathbb{R} \times X$ is a bifurcation point for the equation

$$(6.10) \qquad (E - \beta_1 F - \beta_2 G)x + N_1(\beta_1, x) = 0$$

if Hypothesis (H_3) is satisfied. The point $(\beta_{20}, 0) \in \mathbb{R} \times X$ is a bifurcation point for Equation (6.10) if Hypothesis (H_4) is satisfied. Theorem 5.3 applies directly since

$$N(x, \beta_1, \beta_2) = N_1(\beta_1, x) - (\beta_2 - \beta_{20})Gx$$

under (H_3) and

$$N(x, \beta_1, \beta_2) = N_1(\beta_1, x) - (\beta_1 - \beta_{10})Fx$$

under (H_4). Since β_1, β_2 are equivalent to p, λ for $p_0 > 0$ and p near p_0, this proves both Theorems 6.1 and 6.2. $\square$

EXAMPLE 6.9. As a specific application of Theorems 6.7, 6.8, suppose

$$(6.11) \qquad E = \begin{bmatrix} 1 & 0 \\ 0 & 2 \end{bmatrix}, \qquad F = \begin{bmatrix} 0 & -2 \\ 2 & -1 \end{bmatrix}, \qquad G = \begin{bmatrix} 1 & 0 \\ 0 & 0 \end{bmatrix}.$$

Equation (6.5) is

$$0 = \det[-p^2 E + F + \lambda G] = (p^2 - \lambda)(2p^2 + 1) + 4 \overset{\text{def}}{=} Q(p^2, \lambda).$$

We discuss only those values of (p^2, λ) near $(p_0^2, \lambda_0) = (\frac{7}{2}, 4)$. A simple computation shows that

$$-p_0^2 E + F + \lambda_0 G = \begin{bmatrix} \frac{1}{2} & -2 \\ 2 & -8 \end{bmatrix}, \qquad \phi = \begin{pmatrix} 4 \\ 1 \end{pmatrix}, \qquad \psi = (4, -1)$$

$$\psi E \phi = -31 \neq 0, \qquad \psi G \phi = -1 \neq 0.$$

Thus, Hypotheses (H_1), (H_3), (H_4) are satisfied. Hypothesis (H_1) implies the only periodic solution of

$$(6.12) \qquad\qquad\qquad p_0^2 E\ddot{x} + (F + \lambda_0 G)x = 0$$

of least period 2π is a linear combination of $\phi \cos t$, $\phi \sin t$, any other 2π-periodic solution of this equation must have period $2\pi/m$ for some integer $m > 1$. If $t \mapsto mt$, this implies there is an integer $m > 1$ such that

$$\det[-m^2 p_0^2 E + F + \lambda_0 G] = Q(m^2 p_0^2, \lambda_0) = 0.$$

But, for any $m > 1$, $Q(m^2 p_0^2, \lambda_0) > 0$. Thus, there does not exist any such $(2\pi/m)$ periodic solutions of Equation (6.12) for any $m > 1$. This proves (H_2) is satisfied and the conclusions of Theorems 6.7, 6.8 hold for Equation (6.3) with E, F, G given in (6.11).

EXAMPLE 6.10. As another application of Theorem 6.8, let us consider the same equation as in Example 6.6 which is a special case of (6.3) with

$$E = I \quad \text{the identity}$$

(6.13)
$$F = \begin{bmatrix} -1 & \alpha^2 \\ 0 & -\alpha^2 \end{bmatrix}, \qquad G = \begin{bmatrix} 1 & 0 \\ -1 & 0 \end{bmatrix}$$

One easily obtains

$$Q(p^2, \lambda) \stackrel{\text{def}}{=} \det[-p^2 I + F + \lambda G] = p^4 + p^2(\alpha^2 + 1 - \lambda) + \alpha^2$$

$$\min_{p^2} Q(p^2, \lambda) = -\frac{(\lambda - 1 - \alpha^2)^2}{4} + \alpha^2 \quad \text{at } p^2 = \frac{\lambda - 1 - \alpha^2}{2}.$$

Thus, there exist solutions (p^2, λ) of (6.12) if and only if

$$\lambda \geq 1 + \alpha^2 + 2\alpha$$

and the smallest value λ_0 for which there is a solution is $\lambda_0 = 1 + \alpha^2 + 2\alpha$. The corresponding value of p is $p_0^2 = \alpha$. Furthermore,

$$-p_0^2 I + F + \lambda_0 G = \begin{bmatrix} \alpha^2 + \alpha & \alpha^2 \\ -1 - \alpha - 2\alpha & -\alpha - \alpha^2 \end{bmatrix}$$

$$\phi = \begin{pmatrix} 1 \\ -1 - \dfrac{1}{\alpha} \end{pmatrix} \qquad \psi = \left(1 + \frac{1}{\alpha}, 1\right)$$

$$\psi\phi = 0, \qquad \psi G\phi = \frac{1}{\alpha} \neq 0.$$

Then, Hypotheses (H_1), (H_4) are satisfied. In the same way as in Example 6.9, one shows that (H_2) is satisfied. Thus, the conclusions of Theorem 6.8 hold for the equation (6.3) with E, F, G given in (6.13).

EXAMPLE 6.11. Suppose $f_1 : \mathbb{R} \to \mathbb{R}$, $f_2 : \mathbb{R} \to \mathbb{R}$, $h : \mathbb{R}^2 \to \mathbb{R}$ are C^2 functions $f_1(x), f_2(x), h(\lambda, x)$ vanishing together with the first derivative in x at $x = 0$. Consider the boundary value problem

(6.14)
$$\begin{aligned} &\text{(a)} & -\ddot{x} - \lambda x + h(\lambda, x) = 0, && 0 \leq t \leq 1 \\ &\text{(b)} & \dot{x}(0) + f_1(x(0)) = 0, && \dot{x}(1) + f_2(x(1)) = 0. \end{aligned}$$

Any solution of (6.14a) must satisfy the equation

$$\text{(6.15)} \quad -x(t) + at + b - \lambda \int_0^t (t-s)x(s)\,ds + \int_0^t (t-s)h(\lambda, x(s))\,ds = 0$$

$$0 \leq t \leq 1$$

for some constants a, b and conversely. Let

$$X = Z = C^1([0,1], \mathbb{R}) \times \mathbb{R}^2$$

$$B(x, a, b) = (-x + a\cdot + b, \dot{x}(0), \dot{x}(1))$$

$$A(x, a, b) = \left(\int_0^{\cdot} (\cdot - s)x(s)\,ds, 0, 0 \right)$$

$$N(x, a, b) = \left(\int_0^{\cdot} (\cdot - s)h(\lambda, x(s))\,ds, f_1(x(0)), f_2(x(1)) \right).$$

Problem (6.14) is equivalent to

$$(B - \lambda A)x + N(\lambda, x) = 0$$

for $x \in X$. It can be shown that $\lambda_n = (2n + 1)^2 \pi^2/4$, $n = 0, 1, 2, \ldots$ is a simple eigenvalue of (B, A). Therefore, Theorem 5.3 implies the point $(\lambda_n, 0) \in \mathbb{R} \times X$ is a bifurcation point for (6.14).

EXAMPLE 6.12. Suppose the operator $L(\lambda)$ is defined in (4.24), let $f : [\alpha, \gamma] \times \mathbb{R}^2 \to \mathbb{R}$ be a C^2 function with $f(t, 0, 0) = 0$, $\partial f(t, 0, 0)/\partial(x, y) = 0$ and consider the equation

$$\text{(6.16)} \qquad L(\lambda)x + f(t, x, \dot{x}) = 0, \qquad \alpha \leq t \leq \gamma$$

together the boundary conditions at the three points $\alpha < \beta < \gamma$ given by (4.26)–(4.28). If there is a $\lambda_0 = (\lambda_{10}, \lambda_{20})$ such that Relations (4.29), (4.30) are satisfied, then $(\lambda_0, 0)$ is a bifurcation point for equation (6.16) with boundary conditions (4.26–(4.28).

EXAMPLE 6.13. In this example, we give another proof of the Hopf bifurcation theorem using the concept of a simple eigenvalue for a two parameter family together with the remarks in Example 4.5.

Consider the equation

$$\text{(6.17)} \qquad \dot{x} = A(\alpha)x + f(\alpha, x)$$

where $x \in \mathbb{R}^n$, $\alpha \in \mathbb{R}$, $f : \mathbb{R} \times \mathbb{R}^n \to \mathbb{R}^n$ is C^2 with $f(\alpha, 0) = 0$, $D_x f(\alpha, 0) = 0$, $A(\alpha)$ is a C^2 $n \times n$ matrix function of α which has two simple eigenvalues

$$(6.18) \qquad \mu(\alpha) \pm iv(\alpha)$$
$$v(0) = v_0 > 0, \qquad \mu'(0) \neq 0$$

for α in a neighborhood of zero and each eigenvalue $\lambda_j \neq iv_0$ satisfies $\lambda_j \neq imv_0$ for all integers $m = 0, \pm 1, \pm 2, \dots$.

To obtain the periodic solutions of (6.17) near $x = 0$ which have period β close to $2\pi/v_0$, let $t \mapsto \beta t$ and investigate the equivalent problem of the existence of 2π-periodic solutions of the equation

$$\dot{x} = \beta A(\alpha) x + \beta f(\alpha, x)$$

for (β, α) near $(2\pi/v_0, 0)$. The assumptions on $A(\alpha)$ imply that $A(\alpha)$ has the form

$$\beta A(\alpha) = \beta A(0) + \beta \alpha A'(0) + 0(|\alpha|^2)$$

as $|\alpha| \to 0$ and in an appropriate coordinate system

$$\frac{v_0}{2\pi} A(0) = \begin{pmatrix} \begin{matrix} 0 & 1 \\ -1 & 0 \end{matrix} & O \\ O & \bar{A}_1 \end{pmatrix}, \qquad A'(0) = \begin{pmatrix} \begin{matrix} 1 & 0 \\ 0 & 1 \end{matrix} & O \\ O & \bar{A}_3 \end{pmatrix}$$

and $(\exp \bar{A}_1 t)c \neq (\exp \bar{A}_1(t + 2\pi))c$ for all $(n-2)$-vectors c. Therefore, the conditions of Example 4.3 are satisfied and $(v_0(2\pi, 0)$ is a simple eigenvalue of $(\partial/\partial t, A(0), A'(0))$ relative to the spaces X, Z in (4.17). Theorem 5.3 implies $(v_0/2\pi, 0, 0) \in \mathbb{R}^2 \times X$ is a bifurcation point for Equation (6.17). This is the Hopf bifurcation theorem which is stated precisely below.

Theorem 6.14. *Under Hypothesis* (6.18), *there are* $\delta_1 > 0$, $\delta > 0$, *and* C^1 *functions* $\alpha^*, \omega^* : (-\delta, \delta) \to \mathbb{R}$ *such that* $\alpha^*(0) = 0$, $\omega^*(0) = 1$, *and, for every* $a \in \mathbb{R}$, $|a| < \delta$, *Equation* (6.17) *for* $\alpha = \alpha^*(a)$ *has a* $2\pi/\omega^*(a)$ *periodic solution* $x^*(t, a)$ *with*

$$x^*(t, a) = a(\cos t, -\sin t, 0, \dots, 0) + 0(|a|^2)$$

as $a \to 0$. *Furthermore, this is the only periodic solution, except for a translation in phase, which lies in the* δ_1-*neighborhood of zero.*

5.7. Bifurcation Based on the Linear Equation

In this section, we present a general condition on the linear part $B - \lambda A$ at $x = 0$ of equation (5.7) which ensures bifurcation must occur. Our conditions will always imply that the bifurcation equations obtained from an application of the method of Liapunov–Schmidt are finite dimensional. We, therefore, discuss a general finite dimensional problem and deduce results for equation (5.7) as immediate consequences.

Suppose $\Omega \subseteq \mathbb{R} \times \mathbb{R}^d$ is an open neighborhood of $(\lambda_0, 0)$,

$$F : \Omega \to \mathbb{R}^d$$
$$(7.1) \qquad F(\lambda, y) = B_0(\lambda)y + F_1(\lambda, y)$$

where $y \in \mathbb{R}^d$, $B_0(\lambda)$ is a $d \times d$ C^m, $m \geq 2$, matrix function of λ, F_1 is a C^m vector function of λ, y,

$$(7.2) \qquad F_1(\lambda, 0) = 0, \qquad D_y F_1(\lambda, 0) = 0.$$

Theorem 7.1. *If $\lambda_0 \in \mathbb{R}$ is such that*

$$(7.3) \qquad \sigma(B_0(\lambda_0)) = \{0\}$$

$$(7.4) \qquad \Delta(\lambda) \overset{\text{def}}{=} \det B_0(\lambda) \quad \text{changes sign at } \lambda = \lambda_0$$

then $(\lambda_0, 0)$ is a bifurcation point for the equation

$$(7.5) \qquad F(\lambda, y) = 0.$$

Also, there is a connected set $C \subseteq \mathbb{R} \times (\mathbb{R}^d \backslash \{0\})$ of zeros of F with $(\lambda_0, 0) \in \bar{C}$, the closure of C.

Theorem 7.1 has several proofs. The one immediately following is perhaps the most geometric and most clearly shows how the hypotheses arise. It also provides a natural motivation for problems of global bifurcation.

Proof of Theorem 7.1. Let $\mathcal{O}$ be the set

$$\mathcal{O} = \{(\lambda, y) \in \mathbb{R} \times \mathbb{R}^d : |\lambda - \lambda_0| < \alpha \text{ and } |y| < \beta\}$$

for some fixed α and β small enough that

$$\bar{\mathcal{O}} \subseteq \Omega;$$

$$\Delta(\lambda) \neq 0 \quad \text{on } 0 < |\lambda - \lambda_0| \leq \alpha$$
$$F(\lambda, y) \neq 0 \quad \text{on } \{(\lambda, y) \in \bar{\mathcal{O}} : |\lambda - \lambda_0| = \alpha \text{ and } y \neq 0\}.$$

Also, for any $c \in \mathbb{R}^d$ set

$$S(c) = \{(\lambda, y) \in \bar{\mathcal{O}} : F(\lambda, y) = c\}.$$

By the Implicit Function Theorem, for small $|c|$, there are unique points $(\lambda_0 \pm \alpha, y_\pm^*(c)) \in S(c)$ on the end surfaces $\lambda = \lambda_0 \pm \alpha$, $|y| \leq \beta$ of $\bar{\mathcal{O}}$. The set $S(c)$ is transverse to the end surfaces of $\bar{\mathcal{O}}$.

A point $c \in \mathbb{R}^d$ is a regular value of F if, for any $(\lambda_1, y_1) \in S(c)$, the map

$$DF(\lambda_1, y_1): \mathbb{R} \times \mathbb{R}^d \to \mathbb{R}^d$$

has rank d. If c is a regular value of F, then the solution set of the equation

$$F(\lambda, y) = c$$

near c is a smooth arc given by the Implicit Function Theorem.

In the following, we suppose $c_m \to 0$ is a sequence of regular values of F. Sard's theorem implies such a sequence exists.

The following lemma is needed to complete the proof.

Lemma 7.2. *Let $S_0(c_m)$ be the maximal connected component in $\bar{\mathcal{O}}$ of $S(c_m)$ containing $(\lambda_0 - \alpha, y_-^*(c_m))$. Then $S_0(c_m)$ contains a point $(\lambda, y) \in \partial\mathcal{O}$ where $|y| = \beta$.*

Proof. At $\lambda = \lambda_0 \pm \alpha$, the values of $\Delta(\lambda)$ have opposite signs, say

$$(7.6) \qquad\qquad \pm \Delta(\lambda_0 \pm \alpha) > 0$$

We cannot have $S_0(c_m) - \{(\lambda_0 - \alpha, y_-^*(c_m))\}$ disjoint from $\partial\mathcal{O}$. If this were so, then $S_0(c_m)$ would be a compact connected smooth arc with only one endpoint $(\lambda_0 - \alpha, y_-^*(c_m))$, which is certainly impossible. If the lemma fails, then necessarily $(\lambda_0 + \alpha, y_+^*(c_m)) \in S(c_m)$ is the other endpoint of this arc. To show this is impossible, we orient the arcs comprising $S(c_m)$ and use the fact that $\Delta(\lambda)$ changes sign at λ_0 to show this orientation is inconsistent at the endpoints of $S_0(c_m)$.

An orientation of $S(c_m)$ is specified by choosing in a continuous manner one of the two unit vectors in the kernel of $DF(\lambda, y)$. Such vectors represent tangents to $S(c_m)$. Do this as follows: if $(\lambda, y) \in S(c_m)$, let $v(\lambda, y) \in \mathbb{R} \times \mathbb{R}^d$ satisfy

$$|v(\lambda, y)| = 1;$$
$$DF(\lambda, y)v(\lambda, y) = 0;$$
$$\det \begin{pmatrix} v(\lambda, y)^T \\ DF(\lambda, y) \end{pmatrix} > 0$$

for the $d \times (d + 1)$ matrix $DF = (D_\lambda F, D_y F)$. At the points $(\lambda_0 \pm \alpha, y_\pm^*(c_m))$ we have, in particular,

$$(D_\lambda F, D_y F) = (0, B_0(\lambda_0 \pm \alpha)) + 0(c_m)$$

and so from (7.6)

$$v(\lambda_0 \pm \alpha, y_\pm^*(c_m)) = (\pm 1, 0) + 0(c_m).$$

Thus, the vectors $v(\lambda^0 \pm \alpha, y_\pm^*(c_m))$ point approximately in opposite directions. In particular, both point exterior to $\overline{\mathcal{O}}$, which implies $(\lambda_0 \pm \alpha, y_\pm^*(c_m))$ cannot be endpoints of the same oriented arc in $S(c_m)$. This proves Lemma 7.2.

$\square$

We note that the above proof is similar to the constructive proof of the Brouwer fixed point theorem in Chapter 2. In fact, $S_0(c_m)$ is a smooth 1-manifold (curve) in $\mathcal{O}$. Suppose that $S_0(c_m)$ has end points $(\lambda_0 \pm \alpha, y_\pm^*(c_m))$ and is the unique solution of the initial value problem:

$$D_\lambda F \lambda' + D_y F y' = 0$$
$$\lambda(0) = \lambda_0 + \alpha$$
$$y(0) = y_+^*(c_m)$$

where "$'$" denotes the differentiation with respect to arc length s. Since $(\lambda(s), y(s))$ describes the curve $S_0(c_m)$ with s being the arc length, $|(\lambda'(s), y'(s))| = 1$. Hence, we may take $v(\lambda, y) = (\lambda'(s), y'(s))$. Since $DF(\lambda, y)$ has maximal rank and $(\lambda'(s), y'(s))$ belongs to its null space, the matrix

$$\begin{bmatrix} v(\lambda, y)^T \\ DF(\lambda, y) \end{bmatrix}$$

is nonsingular along $S_0(c_m)$ and hence

$$\det \begin{bmatrix} v(\lambda, y)^T \\ DF(\lambda, y) \end{bmatrix}$$

is of constant sign along $S_0(c_m)$. This contradicts the fact that $\pm \Delta(\lambda_0 \pm \alpha) > 0$.

It is very easy now to complete the proof of Theorem 7.1. Simply let Γ be the set of limit points of $S_0(c_m)$ as $m \to \infty$:

$$\Gamma = \{(\lambda, y) \in \mathcal{O} \colon \text{there exists } (\lambda_{m'}, y_{m'}) \in S_0(c_{m'})$$
$$\text{with } (\lambda_{m'}, y_{m'}) \to (\lambda, y) \text{ for some subsequence } c_{m'} \text{ of } c_m\}$$

Then Γ is a connected set of solutions of $F(\lambda, y) = 0$ and contains both the point $(\lambda_0 - \alpha, 0) = (\lambda_0 - \alpha, y^*(0))$ and some other point $(\lambda_1, y_1) \in \partial\mathcal{O}$ with

$|y_1| = \beta$. It is quite easy to see that the set

$$\Gamma \backslash ([\lambda_0 - \alpha, \lambda_0 + \alpha] \times \{0\})$$

yields the continuum C of solutions. The theorem is proved. $\square$

Suppose $B, A: X \to Z$ are bounded linear operators. We say λ_0 is a *generic eigenvalue* of (B, A) if

(i) $B - \lambda_0 A$ is Fredholm of index zero
(ii) $A\mathcal{N}(B - \lambda_0 A) \oplus \mathcal{R}(B - \lambda_0 A) = Z$.

If $A = I$, then (B, I) is generic if and only if λ_0 is a normal eigenvalue of B with $\delta(B - \lambda_0) = 1$. A pair (B, A) can be generic at λ_0 and yet zero is not a normal eigenvalue of $B - \lambda_0 A$.

Theorem 7.3. *If λ_0 is a generic eigenvalue of (B, A) and dim $\mathcal{N}(B - \lambda_0 A)$ is odd, then the point $(\lambda_0, 0)$ is a bifurcation point for the equation*

$$(7.7) \qquad M(\lambda, x) \overset{\text{def}}{=} Bx - \lambda Ax + N(\lambda, x) = 0$$

where $N \in C^m(\mathbb{R} \times X, Z)$, $m \geq 2$, $N(\lambda, 0) = 0$, $D_x N(\lambda, 0) = 0$.

Proof. Decompose X, Z as

$$X = X_0 \oplus X_1, \qquad X_0 = \mathcal{N}(B - \lambda_0 A), \qquad X_1 = j\mathcal{R}(B - \lambda_0 A)$$
$$Z = Z_0 \oplus Z_1, \qquad Z_0 = A\mathcal{N}(B - \lambda_0 A), \qquad Z_1 = \mathcal{R}(B - \lambda_0 A).$$

Apply the method of Liapunov–Schmidt to obtain the bifurcation equations (7.5) with F given in (7.1). For this case,

$$B_0(\lambda) = (\lambda - \lambda_0)B_{01} + 0(|\lambda - \lambda_0|^2)$$

where B_{01} is a nonsingular matrix obtained from the isomorphism defined by the mapping A taking $\mathcal{N}(B - \lambda_0 A)$ onto the coker$(B - \lambda_0 A)$. Theorem 7.1 may now be directly applied to complete the proof. $\square$

Suppose now that λ_0 is an isolated normal eigenvalue of (B, A); that is, zero is a normal eigenvalue of $B - \lambda_0 A$ and $B - \lambda A$ is an isomorphism for $0 < |\lambda - \lambda_0| < \delta$ for some δ sufficiently small.

Theorem 7.4. *Consider equation (7.7) and suppose λ_0 is an isolated normal eigenvalue of (B, A). Let $r_+ \in \{0, 1\}$ denote the number (counting algebraic multiplicity) of eigenvalues of $B - \lambda A$ near zero with positive real part for $0 < \lambda - \lambda_0 \ll 1$, counted mod 2. Similarly, let r_- denote the corresponding* *number for* $0 < -(\lambda - \lambda_0)$ *small. If* $r_+ - r_- \not\equiv 0$ (mod 2), *then* $(\lambda_0, 0)$ *is a bifurcation point for Equation* (7.7). *There is, in fact, a connected set* $C \subseteq \mathbb{R} \times (X - \{0\})$ *of zeros of* M *with* $(\lambda_0, 0) \in \bar{C}$, *the closure of* C.

Proof. Decompose X, Z according to the normal decomposition. The bifurcation equations take the form (7.5) with F given in (7.1). Furthermore, $\Delta(\lambda) = \det B_0(\lambda)$ has an isolated zero. The hypothesis $r_+ - r_- \not\equiv 0$ (mod 2) is equivalent to $\Delta(\lambda)$ changes sign at λ_0. Theorem 7.1 now may be applied to complete the proof of the theorem. $\quad\square$

Corollary 7.5. *If* λ_0 *is a normal eigenvalue of* (B, A) *of odd multiplicity and* $A = I$, *then the conclusion of Theorem 7.4 holds.*

The assumption in this corollary that the multiplicity of λ_0 be odd is necessary. The assumption $A = I$ is also needed as is seen in the following example where no bifurcation occurs:

$$\lambda x_1 + x_2 = 0$$
$$\lambda x_1 + \lambda x_2 + x_3 = 0$$
$$\lambda x_3 = x_1^3.$$

Here $r \equiv 0$ (mod 2).

The bifurcation at a simple eigenvalue λ_0 of (B, A) can now be described in a different way. For such an eigenvalue, zero is a simple eigenvalue of $B - \lambda_0 A$. If $\mu(\lambda), \mu(\lambda_0) = 0$, is an eigenvalue of $B - \lambda A$ for λ near λ_0, then λ_0 being simple for (B, A) is equivalent to $\mu(\lambda)$ crossing zero transversally (thereby changing sign) as λ increases past λ_0. In this sense Theorem 7.4 can be regarded as a generalization of this previous result. However, it is very important to note that when Theorem 7.4 does apply, the solutions C branching from $(\lambda_0, 0)$ can have a very complicated structure. They may not admit a clear description as for bifurcation from a simple eigenvalue. There may be many branches and they may not be smooth. For example,

$$\lambda x_1 + x_2 = 0$$
$$\lambda x_2 = x_1^4$$
$$\lambda x_3 = x_3^2$$

possesses non-zero solution arcs given by

$$x_1 = 0 \quad \text{or} \quad -\lambda^{2/3}, \qquad x_2 = -\lambda x_1,$$
$$x_3 = 0 \quad \text{or} \quad \lambda.$$

The arcs along which $x_1 \neq 0$ have a non-smooth cusp at the origin.

5.8. Global Bifurcation

One may observe that the arguments of the previous section are not limited by the size of the neighborhood $\mathcal{O}$, but only by the disposition of solutions on $\partial\mathcal{O}$. If $\mathcal{O}$ is a very large (but still bounded) subset of $R \times R^d$ which as before intersects the λ axis in only a small interval $[\lambda_0 - \alpha, \lambda_0 + \alpha]$, then it would still follow that the arc $S_0(c_m)$ would intersect $\partial\mathcal{O}$ at a point other than $(\lambda_0 \pm \alpha, y_\pm^*(c_m))$. The continuum C of non-zero solutions would, therefore, still extend from $(\lambda_0, 0)$ to a point $(\lambda_1, y_1) \in \partial\mathcal{O}$ with $y_1 \neq 0$. If F were defined on all of $\mathbb{R} \times \mathbb{R}^d$, then the maximal connected component of non-zero solutions emanating from $(\lambda_0, 0)$ would either be an unbounded subset of $\mathbb{R} \times \mathbb{R}^d$, or would contain some other bifurcation point $(\lambda_1, 0) \neq (\lambda_0, 0)$ in its closure.

The above observations give the following global bifurcation theorem.

Theorem 8.1. *Consider the equation*

$$(8.1) \qquad M(\lambda, x) = Bx - \lambda Ax + N(\lambda, x) = 0$$

where $\lambda \in \mathbb{R}$, $x \in \mathbb{R}^n$, $B, A = \mathbb{R}^n \to \mathbb{R}^n$ are linear, N is C^2, $N(\lambda, 0) = 0$ and $D_x N(\lambda, 0) = 0$. Suppose zero is an eigenvalue of $B - \lambda_0 A$, λ_0 fixed, of odd (algebraic) multiplicity, then $(\lambda_0, 0)$ is a bifurcation point of (8.1). Moreover, if

$$K = closure \; \{(\lambda, x) : M(\lambda, x) = 0, \; x \neq 0\}$$
$$K_0 = the \; maximal \; connected \; component \; of \; K \; containing \; (\lambda_0, 0)$$

then at least one of the following holds:

(i)   *K_0 is an unbounded subset of $\mathbb{R} \times \mathbb{R}^n$*
(ii) *K_0 contains a point $(\lambda_1, 0) \neq (\lambda_0, 0)$.*

We consider now the infinite dimensional analog of Theorem 8.1. Let X and Z be real Banach spaces and $M : \mathbb{R} \times X \to Z$ be C^r, $r \geq 2$. Assume that

$$(8.2) \qquad M(\lambda, x) = Bx - \lambda Ax + N(\lambda, x)$$

where $B, A : X \to Z$ are bounded linear operators, $N(\lambda, 0) = 0$ and $D_x N(\lambda, 0) = 0$. Consider the equation

$$(8.3) \qquad M(\lambda, x) = 0.$$

Suppose λ_0 is an isolated normal eigenvalue of (B, A) and the hypotheses of Theorem 7.4 are satisfied at $\lambda = \lambda_0$. In the following, we consider the global bifurcation problem for Equation (8.3) at $\lambda = \lambda_0$.

Theorem 8.1 would give a global bifurcation theorem in infinite dimensions if it were known that the solutions (λ, x) of Equation (8.3) which belong to an arbitrary compact set could be obtained by solving an equivalent finite dimensional problem; that is, a global reduction to an equivalent finite dimensional problem through an application of the Alternative Method in Section 2.4 could be made. It is not known if this is possible for general Banach spaces (see Section 2.13 for some references). Therefore, we work directly with the infinite dimensional system. Because of the infinite dimensional nature of the problem it is necessary to impose certain compactness conditions on the function $M(\lambda, x)$. Results can be obtained from an adaptation of the previous proof for finite dimensions. However, there are several technical difficulties (which can be overcome) in orienting the curves in $S(c_m)$. For simplicity, we will give a proof using degree theory.

We shall assume that $B: X \to Z$ is an isomorphism and let

$$G(\lambda, x) = B^{-1}M(\lambda, x) = (I - \lambda L)x + H(\lambda, x)$$

where $L: X \to X$ is bounded and linear, $H(\lambda, 0) = 0$ and $D_x H(\lambda, 0) = 0$. Consider the equation:

$$(8.4) \qquad\qquad\qquad G(\lambda, x) = 0.$$

Theorem 8.2. *Assume that*

(i)   *L and H are compact maps*
(ii) *λ_0 is a normal eigenvalue of (I, L) with odd multiplicity. If*

$$K = closure\ \{\lambda, x): G(\lambda, x) = 0,\ x \neq 0\}$$
$$K_0 = the\ maximal\ connected\ component\ of\ K\ containing\ (\lambda_0, 0)$$

then one of the following must hold:

(1)   *K_0 is an unbounded subset of $R \times X$,*
(2)   *K_0 contains a point $(\lambda_1, 0) \neq (\lambda_0, 0)$.*

The following lemma on continua is needed in the proof of Theorem 8.2.

Lemma 8.3. *Let C be a compact metric space, and $C_1, C_2 \subset C$ are disjoint and closed. Then, either there exists a connected set $\tilde{C}$ such that $\tilde{C} \cap C_1$ and $\tilde{C} \cap C_2$ are nonempty or there exists disjoint closed sets C_a and C_b such that $C_1 \subset C_a$, $C_2 \subset C_b$ and $C_a \cup C_b = C$.*

Proof of Theorem 8.2. Suppose (1) and (2) are false. Let $N(K_0)$ be an open neighborhood of K_0. Since (1) and (2) are false, we may assume that $N(K_0)$ is bounded and $N(K_0)$ contains no solutions of (8.4) for $|\lambda - \lambda_0| > \delta$, where $\delta > 0$ is small. By compactness of L and H, $C = \overline{N(K_0)} \cap K$ is compact and $K_0 \cap \partial N(K_0) = \varnothing$. Let $C_1 = K_0$ and $C_2 = K \cap \partial N(K_0)$. By Lemma 8.3 and the definition of K_0, there exist disjoint compact subsets C_a and C_b such that $K_0 = C_1 \subset C_a$ and $K \cap \partial N(K_0) = C_2 \subset C_b$. Let $\mathcal{O}$ be a small neighborhood of C_a such that $\mathcal{O} \cap C_b = \varnothing$. Hence, $K \cap \partial \mathcal{O} = \varnothing$. For each

λ, define $\mathcal{O}_\lambda = \{x:(\lambda, x) \in \mathcal{O}\}$. Let deg denote the Leray–Schauder degree. We have $\deg(G(\lambda, \cdot), \mathcal{O}_\lambda, 0) = \text{constant}$ for all λ. If $B_\rho = \{x:|x| < \rho\}$, then there exists a continuous function $\rho(\lambda)$ such that $\rho(\lambda) > 0$ for $\lambda \neq \lambda_0$, $\rho(\lambda_0) = 0$ and $x = 0$ is the only solution of $G(\lambda, x) = 0$ with $|x| \leq \rho(\lambda)$ because (2) is false. Since, for large $|\lambda|$, $\mathcal{O}_\lambda = \varnothing$, this implies $\deg(G(\lambda, 0), \mathcal{O}_\lambda \backslash \bar{B}_{p\lambda}, 0) = 0$, for $\lambda \neq \lambda_0$. If $\lambda_- < \lambda_0 < \lambda_+$ are sufficiently close to λ_0,

$$\deg(G(\lambda_\pm, 0), \mathcal{O}_{\lambda_\pm}, 0) = \deg(G(\lambda_\pm, 0), \mathcal{O}_{\lambda_\pm} \backslash \bar{B}_{\rho(\lambda_\pm)}, 0)$$
$$+ \deg(G(\lambda_\pm, 0), B_{\rho(\lambda_\pm)}, 0)$$
$$= \deg(G(\lambda_\pm, 0), B_{\rho(\lambda_\pm)}, 0).$$

But

$$\deg(G(\lambda_+, 0), B_{\rho(\lambda_+)}, 0) = -\deg(G(\lambda_-, 0), B_{\rho(\lambda_-)}, 0) \neq 0$$

because λ_0 is of odd multiplicity. This is a contradiction. $\quad\square$

As an application, consider the scalar second order equation:

$$(8.5) \qquad \ddot{x} + \lambda x = \phi(t, x, \dot{x})$$

$$(8.6) \qquad x(0) = x(\pi) = 0$$

where $\phi = O(|x|^2 + |\dot{x}|^2)$ uniformly in $t \in [0, \pi]$. The above boundary value problem may be written as Equation (8.3) with $Bx = \ddot{x}$. By inverting B, we obtain an integral equation

$$G(\lambda, x) = x - \lambda L x + N(\lambda, x) = 0$$

where L and N satisfy the conditions in Theorem 8.2. The eigenvalues of (8.5), (8.6), namely, $\lambda_n = n^2$, $n = 1, 2, \ldots$, are all simple. Any eigenfunction $x_n(t)$ corresponding to λ_n has exactly $(n - 1)$ simple zeros in the open interval $(0, \pi)$. Let K_0^n be the set K_0 in Theorem 8.2 corresponding to $(\lambda_n, 0)$.

Theorem 8.4. *For each $n = 1, 2, \ldots$, there exists a continuum of solutions of (8.5), (8.6), K_0^n which contains $(\lambda_n, 0)$ and is unbounded in $R \times X$, where $X = \{x \in C^2[0, \pi]: x(0) = x(\pi) = 0\}$.*

Proof. Since λ_n is a simple eigenvalue, $K_0^n \neq \varnothing$ and one of the alternatives in Theorem 8.2 must hold. Let S^n denote the set of all functions $x \in X$ having exactly $(n - 1)$ simple zeros in $(0, \pi)$ and $x'(0) \neq 0$, $x'(\pi) \neq 0$. It is clear that S^n is open in X and $S^n \cap S^m = \varnothing$ for $n \neq m$. By using the method of Liapunov–Schmidt, it follows that, if $(\lambda, x) \in K_0^n$ and $0 < |\lambda - \lambda_n|$ small, then

$$x = \alpha x_n + O(|\alpha|^2)$$

as $\alpha \to 0$. Since the eigenfunction x_n has exactly $(n-1)$ simple zeros in $(0, \pi)$, we have

$$\{(\lambda, x) \in K_0^n : (\lambda, x) \neq (\lambda_n, 0), \; |\lambda_n - \lambda| + |x| \text{ small}\} \subset \mathbb{R} \times S^n.$$

If $(\lambda, x) \in K_0^n \cap (\mathbb{R} \times \partial S^n)$, then x must have a double zero in $(0, \pi)$. By the uniqueness of the initial value problem, $x \equiv 0$. Therefore, the only way for K_0^n to leave $\mathbb{R} \times S^n$ from the bifurcation point $(\lambda_n, 0)$ is to go to $(\lambda_n, 0)$ again. Hence, it is impossible for K_0^n to contain $(\lambda, 0) \neq (\lambda_n, 0)$. This completes the proof. $\square$

5.9. An Application to a Delay Differential Equation

The idea of global bifurcation may be extended to operators which may not be differentiable at the bifurcation point. In this section, we will give an abstract theorem with an application to delay differential equations.

First, we need some definitions. Let $C \subset X$ and $x \in C$ be fixed. Given a map $g: C\backslash\{x\} \to C$, a point x is called an *ejective point* of g if there is an open neighborhood U of x such that for every $y \in U \cap C$, $y \neq x$, there is an integer $m = m(y)$ such that $g^m(y) \notin C \cap U$. A point x is called an *attractive point* of g if there exists an open neighborhood U of x such that, for every open neighborhood V of x, there exists an integer $m = m(V)$ such that $g^j(U \cap C) \subset V \cap C$ for all $j \geq m(V)$.

Theorem 9.1. *Let $C \subset X$ be closed and convex, $0 \in C$ be an extreme point of C ($C \neq \{0\}$) and $a \geq -\infty$. Let $g: (a, \infty) \times C \to C$ be compact, $g(\lambda, 0) = 0$ for all $\lambda \in (a, \infty)$. Suppose that $\lambda_0 \in (a, \infty)$ and $g_\lambda(\cdot) = g(\lambda, \cdot)$. Assume*

$\quad$ (i)$\quad$ *0 is an attractive point of g_λ for all $a < \lambda < \lambda_0$,*
$\quad$ (ii)$\quad$ *0 is an ejective point of g_λ for all $\lambda > \lambda_0$,*
$\quad$ (iii)$\quad$ *if $g(\lambda_k, x_k) = x_k$ and $\lambda_k \to a$, then $|x_k| \to \infty$.*

If

$$K = closure \; \{(\lambda, x) : g(\lambda, x) = x, \; x \neq 0, \; x \in C\}$$
$$K_0 = the \; maximal \; connected \; component \; of \; K_0 \; which \; contains \; (\lambda_0, 0),$$

then $K_0 \neq \varnothing$ is unbounded in $(a, \infty) \times C$.

The proof of Theorem 9.1 will not be given here. For references, see Section 10.

As an application, consider the scalar equation

$$(9.1) \qquad\qquad \dot{x}(t) = -\alpha x(t-1)[1 + x(t)]$$

where $\alpha > 0$. Let $X = C[-1,0]$. For every $\phi \in X$, there exists a unique solution $x(\phi, \alpha)(t)$ of (9.1). Note that if $\phi(0) > -1$, then

$$x(\phi, \alpha)(t) > -1, \quad \text{for all } t \geq 0.$$

Also, there is no $t_0 > 0$ such that $x(\phi, \alpha)(t) = 0$ for all $t \geq t_0$ unless $\phi(0) = 0$.

By linearizing (9.1) at $x = 0$, it can be shown that Equation (9.1) has a Hopf bifurcation at $\alpha = \pi/2$. We are interested in the existence of periodic solutions of (9.1) which can be continued from the Hopf bifurcation point. The following lemmas will be useful.

Lemma 9.2.

    (i)   *If $\phi(0) > -1$ and the zeros of $x(\phi, \alpha)$ are bounded, then $x(\phi, \alpha)(t) \to 0$ as $t \to \infty$.*

    (ii) *If $\phi(0) > -1$, then $x(\phi, \alpha)(t)$ is bounded. Furthermore, if the zeros of $x(\phi, \alpha)$ are unbounded, then any maximum of $x(\phi, \alpha)(t)$, $t > 0$, is less than $e^{\alpha} - 1$.*

    (iii) *If $\phi(0) > -1$ and $\alpha > 1$, then the zeros of $x(\phi, \alpha)$ are unbounded.*

    (iv) *If $\phi(\theta) > 0$, $-1 < \theta < 0$ [or if $\phi(0) > -1$, $\phi(\theta) < 0$, $-1 < \theta < 0$], then the zeros (if any) of $x(\phi, \alpha)(t)$ are simple and the distance from a zero of $x(\phi, \alpha)(t)$ to the next maximum or minimum is ≥ 1.*

Proof.

(i) Suppose there is a $t_1 > 0$ such that

$$x(t) \stackrel{\text{def}}{=} x(\phi, \alpha)(t)$$

is of constant sign for $t \geq t_1 - 1$. Since $x(t) > -1$ for all $t \geq 0$, $\dot{x}(t)x(t-1) < 0$, $t \geq t_1$. Therefore, $x(t)$ is bounded and approaches a limit monotonically. This implies $\dot{x}(t)$ is bounded and therefore $\dot{x}(t) \to 0$ as $t \to \infty$. This implies $x(t) \to 0$ or -1, but -1 is obviously excluded.

(ii) x satisfies

$$(9.2) \qquad 1 + x(t) = [1 + x(t_0)]\exp\left[-\alpha \int_{t_0-1}^{t-1} x(\xi)\,d\xi\right]$$

for any $t \geq t_0 \geq 0$. If the zeros of $x(t)$ are bounded, then Part (i) implies x is bounded. If there is a sequence of nonoverlapping intervals I_k of $[0, \infty)$ such that x is zero at the endpoints of each I_k and has constant sign on I_k, then there is a t_k such that $\dot{x}(t_k) = 0$. Thus, $x(t_k - 1) = 0$. Consequently, Equation (9.2) implies for $t_0 = t_k - 1$, $t = t_k$,

$$\ln(1 + x(t_k)) = -\alpha \int_{t_k-2}^{t_k-1} x(\xi)\,d\xi < \alpha$$

since $x(t) > -1$, $t \geq 0$. Finally, $x(t_k) \leq e^{\alpha} - 1$ for all t_k. This proves Part (ii).

(iii) If the zeros of x are bounded, then Part (i) implies $x(t) \to 0$ as $t \to \infty$ and, thus, the existence of a $t_0 > 0$ such that $\alpha(1 + x(t)) > 1$ for $t \geq t_0$ and $x(t)$ has constant sign for $t \geq t_0$. Thus,

$$\dot{x}(t)x(t-1) = -\alpha x^2(t-1)(1 + x(t)) < -x^2(t-1) < 0, \qquad t \geq t_0 + 1,$$

and $x(t) \to 0$ monotonically as $t \to \infty$. If x is positive on $[t_0, \infty)$, then

$$x(t_0 + 3) - x(t_0 + 2) = \int_{t_0+2}^{t_0+3} \dot{x}(t)\,dt < \int_{t_0+1}^{t_0+2} x(t)\,dt < -x(t_0 + 2)$$

and $x(t_0 + 3) < 0$. This is a contradiction. If $x(t)$ is negative on $[t_0, \infty)$, then a similar contradiction is obtained.

(iv) Suppose $x(t_0) = 0$ and $x(t) > 0$, $t_0 - 1 < t < t_0$. For $t_0 < t < t_0 + 1$, $\dot{x}(t) < 0$. Similarly, if $x(t) < 0$ for $t_0 - 1 < t < t_0$ and $x(t_0) = 0$, then $\dot{x}(t) > 0$, $t_0 < t < t_0 + 1$. Thus, the assertions of (iv) are obvious and the lemma is proved. $\square$

Let C be the class of all functions $\phi \in X$ such that $\phi(\theta) \geq 0$, $-1 < \theta \leq 0$, $\phi(-1) = 0$, ϕ nondecreasing. Then C is a *cone*; that is, C is a closed convex X set in C with the following properties: if $\phi \in C$, then $\lambda\phi \in C$ for all $\lambda > 0$ and if $\phi \in C$, $\phi \neq 0$, then $-\phi \notin C$. If $\alpha > 1$, $\phi \in C$, $\phi \neq 0$, let

$$z(\phi, \alpha) = \min\{t : x(\phi, \alpha)(t) = 0, \ \dot{x}(\phi, \alpha)(t) > 0\}.$$

This minimum exists from Lemma 9.2, Parts (iii) and (iv). Also $z(\phi, \alpha) > 2$. Furthermore, Lemma 9.2, Part (iv) implies $x(\phi, \alpha)(t)$ is positive and non-decreasing on $(z(\phi, \alpha), z(\phi, \alpha) + 1)$. Consequently, if $\tau(\phi, \alpha) = z(\phi, \alpha) + 1$, then the mapping

$$g(\alpha, 0) = 0$$
$$g(\alpha, \phi) = x_{\tau(\phi,\alpha)}(\phi, \alpha)$$

is a mapping of C into itself, where $x_\tau(\theta) = x(\phi, \alpha)(\tau + \theta)$ for $-1 \leq \theta \leq 0$. Since $\dot{x}(\phi, \alpha)(\tau(\phi, \alpha) - 1) > 0$, τ is continuous in $(1, \infty) \times (C \backslash \{0\})$.

Lemma 9.3.   *The map* $\tau : (C \backslash \{0\}) \times (1, \infty) \to (0, \infty)$ *is completely continuous.*

Proof. First of all, we claim a solution $x = x(\phi, \alpha)$, $\phi \in C$, cannot take a time longer than 2 to become negative because, if $x(1) = \eta > 0$, we have

$$1 + x(2) = (1 + \eta)\exp\left[-\alpha \int_0^1 x(s)\,ds\right] \leq (1 + \eta)e^{-\alpha\eta}$$

and so $x(2) \leq (1 + \eta)e^{-\alpha\eta} - 1$, and this quantity is negative because the function $h(\eta) = (1 + \eta)e^{-\alpha\eta} - 1$ satisfies $h(0) = 0$ and $h'(\eta) = (-\alpha - \alpha\eta + 1)e^{-\alpha\eta} < 0$ for $\eta > 0$.

For any bounded set $B \subseteq C$ and any $\psi \in B$, $\alpha \in (1, \infty)$, let $t_0(\phi, \alpha) \le 3$ denote the point where the solution $x = x(\phi, \alpha)$ has a minimum. Since $t_0(\phi, \alpha) \ge 1$, the set $H(\alpha) = \mathrm{Cl} \bigcup_{\phi \in B} x_{t_0(\phi,\alpha)}(\phi, \alpha)$ is compact and

$$H(\alpha) \subseteq C_1 \overset{\text{def}}{=} \{\psi \in C : -1 < \psi(\theta) \le 0, \ -1 \le \theta \le 0, \ \psi \text{ nonincreasing}\}.$$

For any $\psi \in C_1$, define the continuous function $\tau_1 : [C_1 \backslash \{0\}] \times (1, \infty) \to (0, \infty)$ by the relation $\tau_1(\psi, \alpha) = \min\{t > 0 : x(t, \psi, \alpha) = 0\}$. If we prove $\tau_1(H(\alpha)\backslash\{0\}, \alpha)$ is bounded for each $\alpha \in (1, \infty)$, then $\tau(B\backslash\{0\}, \alpha\}$ is bounded for each α. Since $H(\alpha)$ is compact, it is therefore only necessary to prove that τ_1 is bounded on a neighborhood of zero in C_1. This is easy to verify in the following manner. If $\psi \in C_1\backslash\{0\}$ and $x(1, \psi, \alpha) = \beta < 0$, that is, $\tau_1(\psi, \alpha) > 1$, then

$$1 + x(2, \psi, \alpha) = (1 + \beta)\exp\left[-\alpha \int_0^1 x(s, \psi, \alpha)\, ds \right] \ge (1 + \beta)e^{-\alpha\beta} \quad ,$$

and

$$x(2, \psi, \alpha) \ge h(\beta) \overset{\text{def}}{=} (1 + \beta)e^{-\alpha\beta} - 1.$$

If β is small and negative, then this function is positive since $h(0) = 0$ and $h'(0) = 1 - \alpha < 0$.

Since $\tau(\phi, \alpha)$ is continuous for $(\phi, \alpha) \in [K\backslash\{0\}] \times (1, \infty)$, and $\mathbb{R}$ is locally compact, one obtains the conclusion stated in the lemma. $\quad\square$

Lemma 9.4. $g : (1, \infty) \times C \to C$ *is compact.*

Proof. From parts (ii) and (iv) of Lemma 9.2, it follows that $|g(\alpha, \phi)| \le e^\alpha - 1$ for each $\phi \in C$. For any bounded set $B \subset C\backslash\{0\}$, $g(\alpha, B)$ is bounded. By Lemma 9.3, $g(\alpha, B)$ is precompact. Also, if $\phi_k \in C\backslash\{0\}$, $\phi_k \to 0$, then we may assume $\tau(\phi_k, \alpha) \to \tau_0(\alpha)$ as $k \to \infty$. By continuity of solutions with respect to initial data,

$$x_{\tau(\phi_k,\alpha)}(\phi_k, \alpha) \to x_{\tau_0(\alpha)}(0, \alpha) = 0.$$

Therefore $g(\alpha, \phi)$ is continuous at 0 and g is compact. $\quad\square$

By using the theory of linear functional differential equations, we obtain the following.

Lemma 9.5. 0 *is an ejective point of $g(\alpha, \phi)$ for $\alpha > \pi/2$ and 0 is an attractive point of $g(\alpha, \phi)$ for $1 < \alpha < \pi/2$.*

Lemma 9.6. *There is an $\alpha_1 > 1$ such that, for any $\alpha \in (1, \alpha_1)$, the only solution of $g(\alpha, \phi) = \phi$ in C is $\phi = 0$.*

Proof. Suppose $\phi \in C$, $g(\alpha, \phi) = \phi$ and let $z_1(\phi, \alpha) = z_1$ and $z_2(\phi, \alpha) = z_2$ be the first and second zeros of the periodic solution $x(\phi, \alpha)$ of Equation (9.1). From Equation (9.2) and the fact that $|\phi| = |\phi(0)| = |x(z_2 + 1)|$, $x(\xi) \geq x(z_1 + 1)$ for $\xi \in [z_2 - 1, z_2]$, we have

$$-x(z_1 + 1) = 1 - \exp\left[-\alpha \int_{z_1 - 1}^{z_1} x(\xi)\, d\xi\right] \leq 1 - e^{-\alpha|\phi|} \overset{\text{def}}{=} \gamma(|\phi|)$$

$$|\phi| = x(z_2 + 1) = \exp\left[-\alpha \int_{z_2 - 1}^{z_2} x(\xi)\, d\xi\right] - 1 \leq e^{-\alpha x(z_1 + 1)} - 1$$

$$\leq e^{\alpha \gamma(|\phi|)} - 1.$$

If $f(\beta) = \exp(+\alpha\gamma(\beta)) - 1 - \beta$, then there is a unique solution $\beta(\alpha)$ of $f(\beta) = 0$ for each $\alpha > 1$ and $\beta(\alpha) > 0$ as $\alpha \to 1$. Furthermore, $f(\beta) < 0$ if $\beta > \beta(\alpha)$. Therefore, if there is a ϕ such that $g(\alpha, \phi) = \phi$, $\phi \in C$, then $|\phi| \leq \beta(\phi)$. Since the zero solution of Equation (9.1) is asymptotically stable for $0 < \alpha < \pi/2$, there are $\delta > 0$ and $\varepsilon > 0$, such that the only periodic solution of Equation (9.1) with $\alpha \in (1 - \varepsilon, 1 + \varepsilon)$, $|\phi| < \delta$, is $\phi = 0$. Therefore, if α_1 is chosen such that $\beta(\alpha_1) < \delta$, the lemma is proved. $\square$

By applying Theorem 9.1 to the mapping $g(\alpha, \phi)$, we obtain the following:

Theorem 9.7. *There exists a connected set $K_0 \subset (1, \infty) \times C$ such that the α-component of K_0 is unbounded and if $(\alpha, \phi) \in K_0$, then $x(\phi, \alpha)$ is a periodic solution of (9.1) with least period strictly greater than 4.*

5.10. Bibliographical Notes

The proof that the operator in Example 2.5 has a self-adjoint extension can be found in Dunford and Schwartz [2, Theorem 10, p. 1294, Corollary 3.1, p. 1308]. The fact that the essential spectrum satisfies (1.9) follows from Corollary 19, p. 1451 and that any normal eigenvalue is simple from Theorems 10, p. 1400 and 14, p. 1405 of the same reference.

For properties of the Kuratowskii measure of noncompactness, see Martin [1], Sadovskii [1]. The characterization in Formula (2.13) of rad $\sigma_e(B)$ is due to Nussbaum [4]. Lemma 2.1 is due to C. A. Stuart [2].

The definition of simple eigenvalue of (B, A) in Section 3 is due to Crandall and Rabinowitz [1,2]. Lemma 3.1 is essentially due to Hale [15] (see also Howard and Koppel [2]). The definition of simple eigenvalue for several parameters in Section 4 is due to Hale [11]. Example 4.5 was motivated by Langford [2]. The proof that L^k has Fredholm index kr if L has Fredholm index k may be found in Lang [2]. Example 4.12 has been considered also by Thomas and Zachmann [1].

Theorem 5.3 is due to Crandall and Rabinowitz [1, 2] (see also Hale [5]). For other presentations, see Nirenberg [2], Stuart [8]. Theorem 5.4 is due to Hale [11].

For the role of Equations 6.2 in chemical reactions, see Boa and Cohen [1]. Fife [2] has proved a result similar to Theorem 6.8, but with much more restrictive hypotheses. Example 6.12 is due to Thomas and Zachman [1], Hale [11]. The proof in Example 6.13 was motivated by Langford [2].

The method of proof of Theorem 7.1 is similar to ideas used by Alexander [1], Alexander and Yorke [1–3], Chow, Mallet–Paret and Yorke [1, 2], Peitgen and Prüfer [1]. Corollary 7.5 is due to Krasnoselskii [1] (see also Weistrich [1], Nirenberg [2]). Hetzler [1] has proved results similar to Theorem 7.4.

Global bifurcation theorems of the type given in Section 8 were first proved by Rabinowitz [2, 3]. Several generalizations have been given and the reader may consult the following papers for these generalizations as well as applications: Antman [1, 2], Bazley and Pimbley [1], Bazley and Zwalen [2], Dancer [4], Ize [1, 2], Kolodner [1], Pimbley [1], Reeken [1], C. A. Stuart [1, 3–6, 8], Weistrich [2–4]. The proof of Lemma 8.3 may be found in Whyburn [1].

Theorem 9.1 is due to Nussbaum [1] (see also Turner [1]). More details on some of the computations in the example in Section 9 can be found in Hale [9]. For other applications of Theorem 9.1, see Nussbaum [2].

Chapter 6

Bifurcation with One Dimensional Null Space

6.1. Introduction

In Chapter 4, we have discussed bifurcation of solutions of an equation

$$(1.1) \qquad\qquad M(\lambda, x) = 0$$

under the assumption that $M(\lambda, 0) = 0$ for all $\lambda \in \Lambda$. The parameter could be a scalar or an n-dimensional vector, but the criterion for bifurcation was based only upon the linear approximation to $M(\lambda, x)$ near $x = 0$, namely, the operator $D_x M(\lambda, 0)$. If $D_x M(\lambda, 0) = B - \sum_{j=1}^{N} \lambda_j A_j$ and λ_0 is an eigenvalue of $(B, A_1, \ldots, A_N)$ possessing certain properties, then it was shown that λ_0 is always a bifurcation point for $M(\lambda, x)$ regardless of the nature of the nonlinear terms $N(\lambda, x)$ defined by

$$(1.2) \qquad\qquad M(\lambda, x) = D_x M(\lambda, 0)x + N(\lambda, x).$$

If λ_0 is a simple eigenvalue of $(B, A_1, \ldots, A_N)$, then a constructive method was also given for obtaining the unique curve of bifurcating nonzero solutions. If λ_0 is not simple, the existence of bifurcation was given by qualitative rather than constructive methods.

Examples were also given to show that bifurcation may or may not occur if the conditions on the linear terms are violated. Whether or not bifurcation occurs in these situations depends very strongly on the nature of the nonlinear terms. Techniques must be devised which are more quantitative in nature and make use of the nonlinear nature of the problem.

Another extremely important problem to which none of the theory in Chapter 4 is applicable concerns the case when

$$(1.3) \qquad\qquad M(\lambda_0, 0) = 0$$

but $M(\lambda, 0) \neq 0$ for some values of λ near λ_0. There is no known solution for all λ near λ_0! In this case, it is essential to know some properties of the nonlinearities in x in $M(\lambda, x)$.

Our objective in this and the following chapter is to discuss problems where the nonlinearities in x in $M(\lambda, x)$ are needed and to emphasize especially the case where λ is a vector parameter of dimension > 1. We also emphasize constructive methods which are elementary in nature—depending only on a knowledge of calculus and the Implicit Function Theorem. Most of the methods also do not require that $M(\lambda, x)$ be the gradient of some function.

To state the problem more precisely, suppose $\lambda_0 \in \Lambda$ is such that Equation (1.3) is satisfied. For some open neighborhood U of $0 \in X$ and some open neighborhood V of $\lambda_0 \in \Lambda$, find all solutions (λ, x) of Equation (1.1) for which $(\lambda, x) \in V \times U$. As remarked before, this will require specific knowledge of the nonlinearities in x in $M(\lambda, x)$ as well as the dependence of $M(\lambda, x)$ upon the parameters λ.

We emphasize that the problem as stated requires that λ be allowed to vary in a full neighborhood of the particular value λ_0. This is to be contrasted with the approach often taken in specific applications in the literature especially methods based on asymptotic expansions. To obtain, for example, all solutions $x(\lambda)$ of Equation (1.1) for a vector $\lambda = (\lambda_1, \lambda_2) \in \mathbb{R}^2$ by series, one generally assumes λ_1, λ_2 are functions of some small scalar parameter μ, say

$$\lambda_1 = \lambda_1(\mu) = \lambda_{10} + a\mu^\alpha + \cdots,$$
$$\lambda_2 = \lambda_2(\mu) = \lambda_{20} + b\mu^\beta + \cdots$$

and then determines a, b and $x(\lambda_1(\mu), \lambda_2(\mu))$ so that the equation $M(\lambda(\mu), x(\lambda(\mu))) = 0$ is satisfied for $|\mu| < \mu_0$, where μ_0 will be in general a function $\mu_0(a, b)$. By varying a, b, one obtains all solutions in a full neighborhood of λ_0 provided $\mu_0(a, b)$ remains bounded away from zero. This latter property is very difficult to verify by this method. The reason for this fact is that estimation of $\mu_0(a, b)$ almost always involves an application of the Implicit Function Theorem and the Implicit Function Theorem fails to apply when one reaches the value of a, b where the number of solutions changes; namely, a bifurcation surface in the parameter space. If one cannot verify $\mu_0(a, b)$ is bounded away from zero, there is always the possibility that all solutions of Equation (1.1) have not been obtained.

The methods described below will also use certain parametrizations of neighborhoods of λ_0, but series solutions will not be attempted until we have discussed the qualitative properties of all solutions as well as the bifurcation surfaces. In this manner, we are sure that every solution near $(\lambda_0, 0)$ will be obtained. The qualitative discussion of the solutions of Equation (1.1) is specific enough in order to lead to explicit methods for constructing the solutions, either numerical or as series.

Throughout this chapter, we assume

$$\text{(1.4)} \qquad \dim \mathcal{N}(D_x M(\lambda_0, 0)) = 1 = \text{codim } \mathcal{R}(D_x M(\lambda_0, 0))$$

and $x = 0$ is an isolated solution of

$$\text{(1.5)} \qquad M(\lambda_0, x) = 0.$$

The case where relation (1.4) is not satisfied will be considered in Chapter 7. The case in which the Equation (1.5) has a family of solutions will be considered in Chapter 11.

6.2. Quadratic Nonlinearities

Suppose $M: \Lambda \times X \to Z$ is C^k, $k \geq 2$,

$$\begin{aligned} M(0,0) &= 0, \qquad D_x M(0,0) = B \\ \dim \mathcal{N}(B) &= 1 = \operatorname{codim} \mathcal{R}(B) \end{aligned} \tag{2.1}$$

In this section, we impose certain conditions on $D_x^2 M(0,0)$ and give a complete description of the number of solutions of

$$M(\lambda, x) = 0 \tag{2.2}$$

which belong to a neighborhood $V \times U \subset \Lambda \times X$ of $(0,0)$. Applications will also be given.

If Hypothesis (2.1) is satisfied, then there exist continuous projections

$$\begin{aligned} P: X \to X, \qquad \mathcal{R}(P) &= \mathcal{N}(B), \\ Q: Z \to Z, \qquad \mathcal{R}(Q) &= \mathcal{R}(B), \end{aligned}$$

and $y_0 \in X$, $w_0 \in Z$ such that

$$[y_0] = \mathcal{N}(B), \qquad [w_0] = (I - Q)Z.$$

where $[y_0]$, $[w_0]$ denote the subspaces generated by y_0, w_0. To solve the Equation (2.2), one can apply the method of Liapunov–Schmidt to obtain a C^k function $z^*(a, \lambda)$, $a \in \mathbb{R}$, $\lambda \in \Lambda$, $|a| < \delta$, $|\lambda| < \delta$, $z^*(0,0) = 0$, $D_a z^*(0,0) = 0$ and $0 < \delta$ sufficiently small, such that

$$QM(\lambda, ay_0 + z^*(a, \lambda)) = 0. \tag{2.3}$$

Furthermore, in a sufficiently small neighborhood of $(\lambda, x) = (0,0)$, the pair (λ, x), $x = ay_0 + z$, $z \in (I - P)X$, is a solution of Equation (2.2) if and only if $z = z^*(a, \lambda)$ and (a, λ) satisfies the *bifurcation equation*

$$f(a, \lambda) = 0 \tag{2.4}$$

where

$$f(a, \lambda)w_0 = (I - Q)M(\lambda, ay_0 + z^*(a, \lambda)). \tag{2.5}$$

If

$$(2.6) \qquad (I - Q)D_a^2 M(0, ay_0)\big|_{a=0} \neq 0$$

then

$$(2.7) \qquad f(0,0) = 0, \qquad D_a f(0,0) = 0$$
$$D_a^2 f(0,0) \neq 0$$

The Implicit Function Theorem implies there is a unique C^{k-1} function $a^*(\lambda)$ defined for λ in a sufficiently small neighborhood V of zero such that

$$(2.8) \qquad a^*(0) = 0, \qquad D_a f(a^*(\lambda), \lambda) = 0, \qquad \lambda \in V$$

Let

$$(2.9) \qquad \gamma(\lambda) = f(a^*(\lambda), \lambda)\operatorname{sgn} D_a^2 f(0,0), \qquad \lambda \in V.$$

We can now prove the following result.

Theorem 2.1. *If relation (2.6) is satisfied, then there is a neighborhood $U \subseteq X$ of $x = 0$, a neighborhood $V \subseteq \Lambda$ of $\lambda = 0$, and C^{k-1} functions*

$$a^* : V \to \mathbb{R}, \qquad \gamma : V \to \mathbb{R}$$

defined by relations (2.8), (2.9), respectively, such that the following relations hold for each $\lambda \in V$:

(i)   *$\gamma(\lambda) > 0$ implies no solutions (λ, x) of equation (2.2) with $x \in U$,*

(ii)   *$\gamma(\lambda) = 0$ implies exactly one solution (λ, x) of equation (2.2) with $x \in U$,*

(iii) *$\gamma(\lambda) < 0$ implies two solutions (λ, x) of equation (2.2) with $x \in U$.*

If $M_\lambda(x) = M(\lambda, x)$, $f_\lambda(a) = f(\lambda, a)$ for $\lambda \in V$, then the sets

$$(2.10) \qquad \begin{aligned} &\text{(a)} \quad \{a^*(\lambda)y_0 + z^*(a^*(\lambda), \lambda) : \lambda \in V,\ \gamma(\lambda) = 0\} \\ &\text{(b)} \quad \{a^*(\lambda) : \lambda \in V,\ \gamma(\lambda) = 0\} \end{aligned}$$

correspond, respectively, to the set of critical points of the family of mappings $M_\lambda, f_\lambda, \lambda \in V$. If, in addition, there is a $\lambda \in \Lambda$ such that

$$(2.11) \qquad (I - Q)D_\lambda M(0,0)\lambda \neq 0$$

then $\gamma^{-1}(0)$ is a C^{k-1} submanifold of codimension one.

Proof. For definiteness in the proof, suppose $D_a^2 f(0,0) = \alpha > 0$. With $a^*(\lambda)$, $\gamma(\lambda)$ as defined in relations (2.8), (2.9), we have

$$f(a, \lambda) = \gamma(\lambda) + \tfrac{1}{2}D_a^2 f(a^*(\lambda), \lambda)(a - a^*(\lambda))^2 + o(|a - a^*(\lambda)|^2)$$

for $\lambda \in V$ and a in a sufficiently small neighborhood of zero. Since $\alpha > 0$ it follows that $D_a^2(a^*(\lambda), \lambda) > 0$ for $\lambda \in V$ if V is a sufficiently small neighborhood of zero. Therefore, γ satisfies the properties (i)–(iii) stated in the theorem, $\gamma(0) = 0$ and

$$(2.12) \qquad \frac{d}{dt}\gamma(t\lambda)\Big|_{t=0} w_0 = (I - Q)D_\lambda M(0,0)\lambda$$

If condition (2.11) is satisfied for some λ, then it follows from (2.12) that $\gamma^{-1}(0)$ is a C^{k-1} submanifold of codimension one.

It is clear that the set (2.10b) is the set of critical points of f_λ for $\lambda \in V$. It remains only to show that the set (2.10a) is the set of critical points for M_λ. From (2.5), we have

$$(I - Q)D_a M(\lambda, ay_0 + z^*(a, \lambda)) = D_a f(a, \lambda)w_0.$$

Therefore, a is a singular point of f_λ if and only if $ay_0 + z^*(a, \lambda)$ is a singular point of M_λ. This completes the proof of the theorem. $\square$

Remark 2.2. The method above also gives a way of determining the approximate value of the solutions.

Remark 2.3. In Theorem 2.1, the parameter λ can be of any dimension ≥ 1. It can even be so large as to permit a complete classification of all maps in a neighborhood a given map. In fact, suppose Ω is an open neighborhood of zero in X, $T \in C^k(\Omega, Z)$, $k \geq 2$ is a given mapping satisfying

$$(2.13) \qquad T(0) = 0, \qquad B \overset{\text{def}}{=} DT(0) \quad \text{satisfies relation (2.1)}$$
$$(I - Q)D_a^2 T(ay_0)\big|_{a=0} \neq 0$$

where the projection operator Q and element $y_0 \in \mathcal{N}(B)$ have the usual meaning. For any $S \in C^k(\Omega, Z)$, define

$$\lambda = S - T, \qquad M(\lambda, x) = Sx.$$

Then $M(\lambda, x) = 0$ is equivalent to $Sx = 0$ and $M(0, x) = 0$ is equivalent to $Tx = 0$. Theorem 2.1 may therefore be applied to discuss the zeros of $Sx = 0$ near $x = 0$ for each S near T. In this case, condition (2.11) may be verified. In fact, if $S_0 \in C^k(\Omega, Z)$ is chosen so that $(I - Q)S_0(0) \neq 0$, then condition (2.11) is satisfied. Thus, we obtain the following consequence of Theorem 2.1.

Corollary 2.4. *If* $T \in C^k(\Omega, Z)$, $k \geq 2$ *satisfies relation (2.13), then there is an* $\varepsilon > 0$ *and a function*

$$\gamma : B(T, \varepsilon) \to \mathbb{R},$$

where

$$B(T, \varepsilon) = \{S : |S - T|_{C^2(\Omega, Z)} < \varepsilon\}$$

such that the following relations hold for $|x| < \varepsilon$, $S \in B(T, \varepsilon)$:

- (i) *the set $M_\gamma \overset{\text{def}}{=} \{S : |T - S|_{C^2(\Omega, Z)} < \varepsilon, \gamma(S) = 0\}$ is a submanifold of codimension one,*
- (ii) $\gamma(S) > 0$ *implies no zeros of Sx,*
- (iii) $\gamma(S) = 0$ *implies one double zero of Sx,*
- (iv) $\gamma(S) < 0$ *implies two simple zeros of Sx.*

Remark 2.5. From the previous remark, the zeros of all maps in a full neighborhood of a C^2 map can be obtained. In many applications, one has only a family of mappings which depend on a few parameters. This family defines a surface in $C^2(\Omega, Z)$ and one is interested in how this surface intersects the set M_γ in Corollary 2.4. The statement in Theorem 2.1 says that it is unnecessary to go through the corollary to obtain the information needed for the special family.

As an immediate corollary Theorem 2.1 gives the following local result. Suppose $T : X \to Z$ is C^k, $k \geq 2$ and there is an $x_0 \in X$ such that

$$\dim \mathcal{N}(DT(x_0)) = 1 = \operatorname{codim} \mathcal{R}(DT(x_0))$$
$$[y_0] = \mathcal{N}(DT(x_0)),$$

(2.14)
$$\mathcal{R}(DT(x_0)) = QZ, \qquad [w_0] = (I - Q)Z$$

$$(I - Q) \left. \frac{\partial^2 T(a y_0)}{\partial a^2} \right|_{a = 0} \neq 0$$

If T satisfies the above condition (2.14) at $x = x_0$, we say x_0 is a *quadratic singular point* of T.

Theorem 2.6. *If x_0 is a quadratic singular point of T and W is the set of critical points of T, then there is a neighborhood $U \subseteq X$, of x_0 such that $T(W \cap U)$ is a C^{k-1} manifold of codimension 1 in Z. Furthermore, there is a neighborhood $V \subseteq Z$ of $T(x_0)$ such that $V \setminus T(W \cap U) = A_1 \cup A_2$ where A_1, A_2 are open connected sets such that the equation*

(2.15)
$$T(x) - \lambda = 0$$

has

- (i) *no solutions for $\lambda \in A_1$,*
- (ii) *exactly one solution for $\lambda \in T(W \cap U)$,*
- (iii) *exactly two solutions for $\lambda \in A_2$.*

Corollary 2.7. *If the conditions of Theorem 2.6 are satisfied, then for any $z_0 \in Z$ such that the manifold $\{T(x_0) + tz_0 : t \in \mathbb{R}\}$ is transversal to $T(W \cap U)$ at $T(x_0)$, there is an $\varepsilon > 0$ such that*

(i)   *for any $\lambda = T(x_0) + tz_0$, $0 < t \leq \varepsilon$, there are no solutions of (2.15),*
(ii) *for any $\lambda = T(x_0) - tz_0$, $0 < t \leq \varepsilon$, there are exactly two solutions of (2.15).*

Proofs. Define $\lambda_0 = T(x_0)$,

$$M(\lambda, x) = T(x) - \lambda - \lambda_0, \qquad x \in X, \lambda \in Z.$$

All of the conditions of Theorem 2.1 are satisfied and we can determine the function $\gamma(\lambda)$ with the set $\gamma^{-1}(0)$ being a C^{k-1} submanifold of codimension 1. The critical points of T in a neighborhood of x_0 are determined by the solutions of the equation

$$T(x) - \lambda - \lambda_0 = 0, \qquad \gamma(\lambda) = 0.$$

Thus $T(W \cap U)$ is a C^{k-1} manifold of codimension one given by $\gamma^{-1}(0)$. The remaining statements in the theorem are consequences of Theorem 2.1. The corollary is a consequence of (2.11) and the proofs are complete. $\square$

Under some additional conditions, it is possible to prove an interesting global version of Theorem 2.6. Compare this with the Global Implicit Function Theorem 2.3.5.

Theorem 2.8. *Suppose $T: X \to Z$ is a C^k mapping, $k \geq 2$ satisfying*

(i)   *T is proper; that is, the inverse image of any compact set is compact,*
(ii) *the set of critical points W of T is nonempty, closed, connected and contains only quadratic singular points,*
(iii) *$T^{-1}(z)$ is a singleton for each $z \in T(W)$.*

Then $T(W)$ is a closed connected C^{k-1} manifold of codimension one and $Z \backslash T(W)$ contains exactly two connected components A_1, A_2 with the property that

(a)   *if $z \in A_1$, then $T^{-1}(z)$ is empty,*
(b)   *if $z \in A_2$, then $T^{-1}(z)$ consists of two points.*

Proof. Since T is proper and W is closed and connected, it follows that $T(W)$ is closed and connected. For otherwise, if $T(W) = N_1 \cup N_2, N_1, N_2$ closed, $N_1 \cap N_2 = \varnothing$, then $T^{-1}(N_1) \cap T^{-1}(N_2) = \varnothing$, $W = T^{-1}(N_1) \cup T^{-1}(N_2)$. If $x^n \to x$, $x^n \in T^{-1}(N_1)$, then $T(x^n) \to N(x)$ implies $T(x) \in N_1$ and $U \overset{\text{def}}{=} \{T(x^n) : n = 1, 2, \ldots\} \cup \{T(x)\}$, is compact. Therefore, $T^{-1}(U)$ is compact which implies $x \in T^{-1}(N_1)$. Thus, $T^{-1}(N_1)$ is closed. Similarly, $T^{-1}(N_2)$

is closed which implies W is not connected, which is a contradiction. It follows from (ii) and Theorem 2.6 that $T(W)$ is a C^{k-1} manifold of codimension 1. From (iii) and (i), T is a homeomorphism of W onto $T(W)$.

For any $z \in Z$, let $n(z)$ be the cardinal number of $T^{-1}(z)$. Since T is proper, $n(z)$ is finite provided $z \in Z\backslash T(W)$. It is not difficult to see that $n(z)$ is upper and lower semicontinuous on $Z\backslash T(W)$. This implies $n(z)$ is constant on each connected component of $Z\backslash T(W)$.

By Theorem 2.6, for every $z_0 \in T(W)$, there exists a neighborhood V of z_0 such that $V \cap (Z\backslash T(W))$ has exactly two connected components and $n(z) = 0$ or 2 for $z \in V \cap (Z\backslash T(W))$. This implies that the theorem is proved if $Z\backslash T(W)$ has at most two connected components. Suppose not, then there are nonempty open sets A_1, A_2, A_3 mutually disjoint such that

$$Z\backslash T(W) = A_1 \cup A_2 \cup A_3.$$

Since $Z\backslash T(W)$ is open in Z, the A_i's are also open in Z. Let ∂A_i be the boundary of A_i. Since A_i is not open and closed, $\partial A_i \neq \varnothing$. In fact, $\partial A_i \subset T(W)$ and ∂A_i is closed in $T(W)$. By Theorem 2.6, $V \cap (Z\backslash T(W))$ has exactly two components. This implies that without loss of generality,

$$(V \cap (Z\backslash T(W))) \cap \partial A_1 = \varnothing$$

Let $z_i \in \partial A_1 \subset T(W)$. There exists a neighborhood V_1 of z_1 such that $V_1 \cap (Z\backslash T(W))$ has exactly two components. This implies that one of the components is a subset of A_1 and $\partial A_1 \subset V_1 \cap T(W)$. Hence, ∂A_1 is open and closed in $T(W)$. We have $\partial A_1 = T(W)$. This is a contradiction. $\square$

6.3. Applications

As a first application, consider the boundary value problem,

$$
\begin{aligned}
x'' + \psi(x) &= f(t), \qquad 0 \le t \le \pi \\
x(0) &= x(\pi) = 0
\end{aligned}
$$

(3.1)

where $f \in C([0, \pi], \mathbb{R})$ and $\psi \in C^2(\mathbb{R}, \mathbb{R})$. As an application of Theorem 2.6, we prove

Theorem 3.1. *If* $\psi(0) = 0$, $\psi'(0) = 1$, $\psi''(0) > 0$, *then there is a neighborhood* U *of zero in* $X \overset{\text{def}}{=} \{x \in C^2([0, \pi], \mathbb{R}) : x(0) = x(\pi) = 0\}$ *and a neighborhood* V *of zero in* $Z \overset{\text{def}}{=} C([0, \pi], \mathbb{R})$ *and a closed, connected submanifold* M *of codimension 1 in* Z *such that* $V\backslash M$ *consists of exactly two connected components* A_1, A_2 *such that Problem* (3.1) *has*

    (i)   *no solutions in* U *for* $f \in A_1$,
    (ii) *two solutions in* U *for* $f \in A_2$,
    (iii) *one solution in* U *for* $f \in M \cap V$.

Proof. If $T: X \to Z$ is defined by $Tx(t) = x''(t) + \psi(x(t))$, then $T(0) = 0$, $DT(0)y = y'' + y = 0$ if and only if $y = a \sin t$ for some constant a. Therefore, $\dim \mathcal{N}(DT(0)) = 1$. Also,

$$\mathcal{R}(T'(0)) = \left\{ f \in Z : \int_0^\pi f(t) \sin t\, dt = 0 \right\}$$

has codimension 1. Furthermore,

$$\int_0^\pi (\sin t)\, \frac{\partial^2 T(a \sin \cdot)}{\partial a^2}(t)\bigg|_{a=0}\, dt = \psi''(0) \int_0^\pi \sin^3 t\, dt \neq 0.$$

Therefore, zero is a quadratic singular point of T. Theorem 2.6 implies the conclusion stated in the theorem. $\square$

The local result in Theorem 3.1 can be generalized if ψ satisfies the following properties:

$$(3.2) \qquad \psi(0) = 0, \qquad \psi'(0) = 1, \qquad \psi''(s) > 0$$

$$\lim_{s \to -\infty} \psi'(s) = l', \qquad 0 < l' < 1$$

$$(3.3) \qquad \lim_{s \to +\infty} \psi'(s) = l'', \qquad 1 < l'' < 4$$

In fact, we prove the following result.

Theorem 3.2. *If ψ satisfies relations* (3.2), (3.3), *then there exists a closed connected manifold M of codimension 1 in $C([0, \pi], \mathbb{R})$, actually homeomorphic to a linear subspace of codimension 1, such that $C([0, \pi], \mathbb{R}) \setminus M$ consists of exactly two connected components A_1, A_2 such that Problem* (3.1) *has*

    (a)   *no solutions for $f \in A_1$,*
    (b)   *two solutions for $f \in A_2$,*
    (c)   *one solution for $f \in M$.*

Proof. We apply Theorem 2.8. Let $X = \{x \in C^2([0, \pi], \mathbb{R}) : x(0) = x(\pi) = 0\}$, $Z = C([0, \pi], \mathbb{R})$, $T: X \to Z$, $Tx(t) = x''(t) + \psi(x(t))$. Following the proof of Theorem 3.3.3 to obtain a priori bounds, let $\omega(s) = \psi(s)/s$, $s \neq 0$, $\omega(0) = \psi'(0)$. If $\{f_n\} \subset Z$ is bounded, we first show that $Tx_n = f_n$ implies $\{x_n\} \subset Z$ is bounded. If not, then we may assume without loss of generality that $|x_n|_Z \to \infty$ as $n \to \infty$. If $z_n = x_n/|x_n|_Z$, $g_n = f_n/|x_n|_Z$, then z_n satisfies the equation

$$(3.4) \qquad z_n'' + \omega(x_n)z_n = g_n.$$

If $w \in C^\infty([0, \pi], \mathbb{R})$, $w(0) = w(\pi) = 0$, then

$$(3.5) \qquad \int_0^\pi z_n' w' + \int_0^\pi \omega(x_n)z_n w = \int_0^\pi g_n w$$

As in the proof of Theorem 3.3.3, we may assume $z_n \to \bar{z}$ in the $C^1([0,\pi], \mathbb{R})$, $|\bar{z}|_Z = 1$. If

$$\alpha(t) = \begin{cases} l' & \text{if } \bar{z}(t) < 0 \\ l'' & \text{if } \bar{z}(t) > 0 \\ \psi'(0) & \text{if } \bar{z}(t) = 0 \end{cases}$$

then

$$\lim_{n \to \infty} \omega(x_n(t))z_n(t) = \alpha(t)\bar{z}(t), \qquad t \in [0,\pi]$$

since $x_n(t) \to -\infty$, $\omega(x_n(t)) \to l'$ if $z(t) < 0$; $x_n(t) \to \infty$, $\omega(x_n(t)) \to l''$ if $\bar{z}(t) > 0$. From Eq. (3.5), this implies

$$-\int_0^\pi \bar{z}'w' + \int_0^\pi \alpha\bar{z}w = 0$$

for all w as above. Thus, $\bar{z}$ is a generalized solution of the equation

$$(3.6) \qquad \bar{z}'' + \alpha\bar{z} = 0, \qquad \bar{z}(0) = \bar{z}(\pi) = 0.$$

Since $1 < l'' < 4$ by hypothesis, then Lemma 3.3.1 implies that $\lambda = 1$ is the first eigenvalue of the problem

$$v'' + \lambda\alpha v = 0, \qquad v(0) = v(\pi) = 0.$$

Therefore, $\bar{z}(t)$ has a fixed sign on $(0,\pi)$ and $\alpha(t)$ is either l' or l'' on $(0,\pi)$. But relation (3.6) would then imply $\bar{z}(t) = 0$ for all $t \in [0,\pi]$, contradicting the fact that $|\bar{z}|_Z = 1$. Therefore, the sequence $\{x_n\}$ defined by $Tx_n = f_n$ is bounded in Z if the sequence $\{f_n\}$ is bounded in Z.

As in the proof of Theorem 3.3.3, one shows that T is proper.

The next observation is that for any $x_0 \in W$, the set of critical points of T, is a quadratic singular point. A point $x_0 \in W$ if and only if the problem

$$v'' + \psi'(x_0)v = 0, \qquad v(0) = v(\pi) = 0$$

has a nonzero solution $v_0 \in X$; that is, if and only if $\mu = 1$ is an eigenvalue of

$$(3.7) \qquad v'' + \mu\psi'(x_0)v = 0, \qquad v(0) = v(\pi) = 0$$

Since $0 < l' < \psi'(x_0) < l''$, $l' < 1 < l'' < 4$, the point $\mu = 1$ is the first eigenvalue of Problem (3.7) and v_0 has constant sign on $(0,\pi)$ from Lemma 3.3.1. Also, Lemma 3.3.1 implies $\mathcal{N}(T'(x_0)) = [v_0]$ has dimension 1 and the Fredholm alternative implies $\mathcal{R}(T'(x_0)) = \{z : \langle z, v_0 \rangle = 0\}$, where

$$\langle z, v_0 \rangle = \int_0^\pi z(t)v_0(t)\, dt.$$

Therefore, codim $\mathscr{R}(T'(x_0)) = 1$. Finally, the expression $(I - Q)\partial^2 T(av_0)/\partial a^2|_{a=0} \neq 0$ in expression (2.13) if and only if

$$\int_0^\pi \psi''(x_0(t))v_0^3(t)\,dt \neq 0.$$

This latter relation is obviously true since $\psi'' > 0$ and v_0 has fixed sign. This proves that each point in W is a quadratic critical point.

Next, we show that W is nonempty and connected; even more, W can be represented as the homeomorphic image of a linear subspace of Z of codimension 1. In fact, choose $u \in X$, $u(t) > 0$ for $t \in (0, \pi)$ and let L be any linear subspace of X of codimension 1 such that $u \notin L$. If $x \in X$, then x has a unique representation $x = z + vu$, $z \in L$, $v \in \mathbb{R}$. For a fixed z the problem

$$v'' + \lambda\psi'(z + vu)v = 0, \qquad v(0) = v(\pi) = 0.$$

has as its first eigenvalue a function $\lambda(v)$ which depends continuously upon v and is monotone decreasing from Lemma 3.3.1. Since $u > 0$ on $(0, \pi)$, $\lim_{v \to -\infty} \psi'(z(t) + vu(t)) = l'$, $\lim_{v \to \infty} \psi'(z(t) + vu(t)) = l''$. Since $l' < \psi'(t) < l''$, the preceeding limits also exist in the L^1 norm by the dominated convergence theorem. Thus, Lemma 3.3.1 implies

$$\lim_{v \to -\infty} \lambda(v) = 1/l' > 1; \qquad \lim_{v \to +\infty} \lambda(v) = 1/l'' < 1.$$

Therefore, there is a $\bar{v}$ such that $\lambda(\bar{v}) = 1$ and it is unique since $\lambda(v)$ is monotone decreasing. Thus, the ray defined by $\{z + vu, v \in \mathbb{R}\}$ meets W in a unique point. The observation that this point varies continuously with x completes the proof asserted about W.

If $f \in T(W)$, we now show that $T^{-1}(f)$ is a single point. If there are $x \neq \bar{x}$ in X such that $Tx = T\bar{x} = f$, define

$$\varphi(t) = \begin{cases} \dfrac{\psi(\bar{x}(t)) - \psi(x(t))}{\bar{x}(t) - x(t)} & \text{when } \bar{x}(t) \neq x(t) \\[2mm] \psi'(x(t)) & \text{when } \bar{x}(t) = x(t) \end{cases}$$

If $z = \bar{x} - x$, then

$$z'' + \varphi z = 0, \qquad z(0) = z(\pi) = 0.$$

Since $0 < l' < \varphi(t) < l'' < 4$, it follows that z is an eigenfunction for the first eigenvalue $\mu = 1$ for

$$v'' + \mu\varphi v = 0, \qquad v(0) = v(\pi) = 0.$$

Thus, $z = \bar{x} - x$ has constant sign on $(0, \pi)$ from Lemma 3.3.1. Since $\psi'' > 0$, it follows that $\varphi(t) > \psi'(x(t))$ for $t \in (0, \pi)$.

In the proof above that W contained only quadratic critical points, we showed that $x \in W$ implies $\mu = 1$ is the first eigenvalue of

$$v'' + \mu \psi'(x)v = 0, \qquad v(0) = v(\pi) = 0.$$

Lemma 3.3.1 and $\varphi(t) > \psi'(x(t))$ leads to a contradiction.

We have shown that all hypotheses of Theorem 2.8 are satisfied. The proof is completed by applying this theorem. $\qquad \square$

6.4. Cubic Nonlinearities

Suppose M satisfies relations (2.1), $M \in C^k(\Lambda \times X, Z)$, $k \geq 3$. With the notation of Section 2, and $f(a, \lambda)$ defined in relation (2.3), we suppose in this section that M satisfies the condition

$$
\begin{aligned}
M(0,0) &= 0, \\
(I - Q)D_a M(0, ay_0)|_{a=0} &= 0, \\
(I - Q)D_a^2 M(0, ay_0)|_{a=0} &= 0 \\
(I - Q)D_a^3 M(0, ay_0)|_{a=0} &\neq 0
\end{aligned}
$$

(4.1)

or, equivalently,

(4.2)
$$f(0,0) = 0, \qquad D_a f(0,0) = 0, \qquad D_a^2 f(0,0) = 0$$
$$D_a^3 f(0,0) \neq 0$$

Our objective is to obtain the analogue of Theorem 2.1 under Hypothesis (4.2). More specifically, we characterize the number of solutions of $M(\lambda, x) = 0$ in a neighborhood of $\lambda = 0$, $x = 0$.

Hypothesis (4.2) and the Implicit Function Theorem imply there is a unique C^{k-2} function $a^*(\lambda)$ defined for λ in a sufficiently small neighborhood V of zero such that

(4.3)
$$a^*(0) = 0, \qquad D_a^2 f(a^*(\lambda), \lambda) = 0 \quad \text{for all } \lambda \in V$$

The function $f(a, \lambda)$ has a Taylor series

(4.4)
$$f(a, \lambda) = f(a^*(\lambda), \lambda) + D_a f(a^*(\lambda), \lambda)(a - a^*(\lambda))$$
$$+ \frac{1}{3!} D_a^3 f(a^*(\lambda), \lambda)(a - a^*(\lambda))^3 + h_1(\lambda, a)$$

where $h_1(\lambda, a) = o(|a - a^*(\lambda)|^3)$ as $a - a^*(\lambda) \to 0$.

If

$$\gamma_0(\lambda) = \frac{1}{3!} D_a^3 f(a^*(\lambda), \lambda), \quad \gamma_1(\lambda) = D_a f(a^*(\lambda), \lambda)/\gamma_0(\lambda)$$

(4.5)

$$\gamma_2(\lambda) = f(a^*(\lambda), \lambda)/\gamma_0(\lambda), \quad g(u, \lambda) = f(a^*(\lambda) + u, \lambda)/\gamma_0(\lambda)$$

then $f(a, \lambda) = 0$ in a neighborhood of $\lambda = 0$, $a = 0$ is equivalent to the equation

$$
\begin{aligned}
g(u, \lambda) &= 0 \\
g(u, \lambda) &= \gamma_2(\lambda) + \gamma_1(\lambda)u + u^3 + h(u, \lambda) \\
h(u, \lambda) &= o(|u|^3) \quad \text{as } u \to 0.
\end{aligned}
$$

(4.6)

We can now prove the following result.

Theorem 4.1. *If relation (4.1) is satisfied, then there is a neighborhood $U \subseteq X$ of $x = 0$, a neighborhood $V \subseteq \Lambda$ of $\lambda = 0$ and functions*

$$a^*: V \to \mathbb{R}, \qquad \gamma: V \to \mathbb{R},$$

which are C^{k-1} except where the function γ_1 defined in relation (4.5) vanishes, such that the following properties hold for $\lambda \in V$:

   (i)   *$\gamma(\lambda) > 0$ implies one solution (λ, x) of Eq. (2.2) with $x \in U$,*
   (ii) *$\gamma(\lambda) = 0$ implies two solutions (λ, x) of Eq. (2.2) with $x \in U$,*
   (iii) *$\gamma(\lambda) < 0$ implies three solutions (λ, x) of Eq. (2.2) with $x \in U$.*

If $M_\lambda(x) = M(\lambda, x)$, $f_\lambda(a) = f(\lambda, a)$ for $\lambda \in V$, then the sets

(4.7)

$$\{a^*(\lambda) + z^*(a^*(\lambda), \lambda): \lambda \in V, \gamma(\lambda) = 0\}$$
$$\{a^*(\lambda): \lambda \in V, \gamma(\lambda) = 0\}$$

correspond, respectively, to the set of critical points of the family of mappings $M_\lambda, f_\lambda, \lambda \in V$. If, in addition,

(4.8)

$$\text{rank}[D_\lambda \gamma_1(0), D_\lambda \gamma_2(0)] = 2$$

then $\gamma^{-1}(0)$ is a C^{k-1} manifold of codimension 1 except at $\gamma_1(\lambda) = 0$, $\gamma_2(\lambda) = 0$, where there is a cusp.

Remark 4.2. Condition (4.8) can be checked directly from the function $M(\lambda, x)$ from the relations

(4.9)

$$
\begin{aligned}
D_\lambda \gamma_1(0) w_0 &= (I - Q) D_{\lambda u}^2 M(\lambda, u y_0)_{\lambda=0, u=0} \\
D_\lambda \gamma_2(0) w_0 &= (I - Q) D_\lambda M(0, 0)
\end{aligned}
$$

Proof of Theorem 4.1. As remarked earlier, we need only consider equation (4.6). We actually discuss a more general equation

$$(4.10) \qquad G(u, \lambda, \lambda_1, \lambda_2) \overset{\text{def}}{=} \lambda_2 + \lambda_1 u + u^3 + h(u, \lambda) = 0$$

where λ_1, λ_2 are real numbers varying arbitrarily in a neighborhood of zero. After complete information is obtained for equation (4.10), the results for equation (4.6) will follow by setting $\lambda_1 = \gamma_1(\lambda)$, $\lambda_2 = \gamma_2(\lambda)$.

The multiple solutions of equation (4.10) are the simultaneous solutions of equation (4.10) and the equation

$$\lambda_1 + 3u^2 + D_u h(u, \lambda) = 0.$$

From the Implicit Function Theorem, these two equations have a unique solution $\lambda_1(u, \lambda)$, $\lambda_2(u, \lambda)$ in a neighborhood of zero with

$$\lambda_1(u, \lambda) = -3u^2 + o(|u|^2)$$
$$\lambda_2(u, \lambda) = 2u^3 + o(|u|^3).$$

For each fixed λ, this is the parametric representation of a cusp in the (λ_1, λ_2)-plane. Eliminating the parameter u, one obtains a function $\gamma^*(\lambda_1, \lambda_2, \lambda)$ for which equation (4.10) has a multiple solution u if and only if $\gamma^*(\lambda_1, \lambda_2, \lambda) = 0$. The function γ^* is given as

$$(4.11) \qquad \gamma^*(\lambda_1, \lambda_2, \lambda) = \lambda_1 + 3\left(\frac{\lambda_2}{2}\right)^{2/3} v(\lambda, \lambda_2)$$

$$v(\lambda, \lambda_2) = 1 + O(|\lambda_2|^{1/3})$$

as $\lambda_2 \to 0$.

It is easy to verify that there are no triple solutions of Eq. (4.10) except at $\lambda_1 = \lambda_2 = 0$. Therefore, the surface $\gamma^*(\lambda_1, \lambda_2, \lambda) = 0$ is a bifurcation surface for which the number of solutions change by exactly two as this surface is crossed. Also, it is obvious that this change is from one solution to three solutions or vice versa.

If we define

$$\gamma(\lambda) = \gamma^*(\gamma_1(\lambda), \gamma_2(\lambda), \lambda) = \gamma_1(\lambda) - 3\left(\frac{\gamma_2(\lambda)}{2}\right)^{2/3} v(\lambda, \gamma_2(\lambda))$$

then $\gamma(\lambda)$ satisfies the assertions in the first part of the theorem. From the construction of γ, the assertions concerning the sets in relation (4.7) are obvious. Since the map

$$[D_\lambda \gamma_1(0), D_\lambda \gamma_2(0)] : \Lambda \to \mathbb{R}^2$$

is onto, one easily verifies the last assertion in the theorem. $\square$

Remark 4.3. Approximate computations of $\gamma(\lambda)$. It is very easy to obtain the approximate values of all functions in Theorem 4.1 by using the Taylor series for $f(a, \lambda)$. In fact,

$$f(a, \lambda) = \alpha_0(\lambda) + \alpha_1(\lambda)a + \alpha_2(\lambda)\frac{a^2}{2} + \alpha_3(\lambda)\frac{a^3}{6} + o(|a|^3)$$

$$\alpha_j(\lambda) = \beta_j(\lambda) + O(|\lambda|^2), \qquad j = 0, 1, 2$$

where $\beta_j(\lambda)$ is a linear functional of λ. Also $\alpha_3(\lambda) = \beta_3 + O(|\lambda|)$ where $\beta_3 \neq 0$ by hypothesis.

If we let

$$a = u - \frac{\alpha_2(\lambda)}{\alpha_3(\lambda)},$$

then

$$f\left(u - \frac{\alpha_2(\lambda)}{\alpha_3(\lambda)}, \lambda\right) = \delta_0(\lambda) + \delta_1(\lambda)u + \alpha_3(\lambda)\frac{u^3}{6} + o(|u|^3)$$

$$\delta_0(\lambda) = \alpha_0(\lambda) - \frac{\alpha_1(\lambda)\alpha_2(\lambda)}{\alpha_3(\lambda)} + \frac{1}{3}\frac{\alpha_2^3(\lambda)}{6\alpha_3^2(\lambda)}$$

$$\delta_1(\lambda) = \alpha_1(\lambda) - \frac{1}{2}\frac{\alpha_2^2(\lambda)}{\alpha_3(\lambda)}.$$

In relation (4.6),

$$\gamma_2(\lambda) = \frac{6\delta_0(\lambda)}{\alpha_3(\lambda)}, \qquad \gamma_1(\lambda) = \frac{6\delta_1(\lambda)}{\alpha_3(\lambda)}$$

The function $a^*(\lambda)$ in Theorem 4.1 is given by

$$a^*(\lambda) = -\frac{\alpha_2(\lambda)}{\alpha_3(\lambda)}$$

and the function $\gamma(\lambda)$ is given by

$$\gamma(\lambda) = \gamma_1(\lambda) - 3\left(\frac{\gamma_2(\lambda)}{2}\right)^{2/3}[1 + O(|\gamma_2(\lambda)|^{1/2})].$$

As in Remark 2.3, the parameter λ can be of any dimension ≥ 1. Suppose Ω is an open neighborhood of zero in X, $T \in C^k(\Omega, Z)$, $k \geq 3$, is a given mapping satisfying

(4.12)
$$T(0) = 0, \qquad B \stackrel{\text{def}}{=} T^{-1}(0) \quad \text{satisfies relation (2.1)}$$
$$(I - Q)D_a^2 T(ay_0) = 0, \qquad (I - Q)D_a^3 T(ay_0) \neq 0$$

where the projection operator Q and element $y_0 \in \mathcal{N}(B)$ have the usual meaning. When relation (4.12) is satisfied, we say 0 is a *cubic singular point of T*. For any $S \in C^k(\Omega, Z)$ define

$$\lambda = S - T, \qquad M(\lambda, x) = Sx.$$

Then $M(\lambda, x) = 0$ is equivalent to $Tx = 0$. Theorem 4.1 may thus be applied to discuss the zeros of $Sx = 0$ near $x = 0$ for each S near T. Condition (4.8) may always be verified since we are considering a full neighborhood of T. Thus, we obtain the following consequence of Theorem 4.1.

Corollary 4.4. *Suppose $T \in C^k(\Omega, Z)$, $k \geq 3$, satisfies relations (2.1), (4.12) and, for $\lambda = S - T$, let $\gamma_1(\lambda) = \gamma_1(S)$, $\gamma_2(\lambda) = \gamma_2(S)$ be defined by relation (4.5). Then there is an $\varepsilon > 0$ and functions*

$$\gamma: B(T, \varepsilon) \to \mathbb{R}, \qquad a^*: B(T, \varepsilon) \to \mathbb{R}$$
$$B(T, \varepsilon) = \{S: |S - T|_{C^3(\Omega, Z)} < \varepsilon\}$$

which are C^{k-1} except where the function $\gamma_1(S)$ vanishes, $\gamma^{-1}(0)$ is a C^{k-1} manifold of codimension 1 except at $\gamma_1(S) = \gamma_2(S) = 0$, where there is a cusp, and the following relations hold for $|x| < \varepsilon$, $S \subset B(T, \varepsilon)$:

- (i)   *$\gamma(S) > 0$ implies one zero of Sx,*
- (ii) *$\gamma(S) = 0$ implies two zeros of Sx,*
- (iii) *$\gamma(S) < 0$ implies three zeros of Sx.*

The sets

$$\{a^*(S) + z^*(a^*(S), S): S \in B(T, \varepsilon), \gamma(S) = 0\}$$
$$\{a^*(S): S \in B(T, \varepsilon), \gamma(S) = 0\}$$

correspond, respectively, to the set of critical points of the mappings S, f_S, $S \in B(T, \varepsilon)$, where $z^*(a, S) \in (I - P)X$ is the function obtained by the application of the Liapunov–Schmidt procedure and $f_S(a) \overset{\text{def}}{=} f(a, S)$ is the bifurcation function.

6.5. Applications

Consider the boundary value problem

$$\text{(5.1)} \qquad \begin{aligned} x'' + \psi(x) &= q(t)x + p(t), \qquad 0 < t < \pi \\ x(0) &= x(\pi) = 0 \end{aligned}$$

where $p, q \in C([0, \pi], \mathbb{R})$ and $\psi \in C^3(R, \mathbb{R})$.

Theorem 5.1. *If* $\psi(0) = 0$, $\psi'(0) = 1$, $\psi''(0) = 0$, $\psi'''(0) \neq 0$, *then there is a neighborhood* U *of zero in* $X = \{x \in C^2([0, \pi], \mathbb{R}) : x(0) = x(\pi) = 0\}$ *and a neighborhood* V *of zero in* $\mathscr{C} = C([0, \pi], \mathbb{R}) \times C([0, \pi], \mathbb{R})$ *and a closed connected cusp manifold* M *of codimension 1 in* $\mathscr{C}$ *such that* $V \backslash M$ *consists of exactly two connected components* A_1, A_2 *such that Problem* (5.1) *has*

- (i)   *one solution in* U *for* $(p, q) \in A_1$,
- (ii)   *three solutions in* U *for* $(p, q) \in A_2$,
- (iii)   *two solutions in* U *for* $(p, q) \in M \cap V$.

Proof. If $Z = C([0, \pi], \mathbb{R})$, $T : X \to Z$ is defined by $Tx(t) = x''(t) + \psi(x(t))$, then $T(0) = 0$, $T'(0)y = y'' + y = 0$ if and only if $y = a \sin t$ for some constant a. Therefore, dim $\mathscr{N}(T'(0)) = 1$. Also,

$$\mathscr{R}(T'(0)) = \left\{ p \in Z : \int_0^\pi p(t) \sin t \, dt = 0 \right\}$$

has codimension one. Furthermore,

$$\int_0^\pi (\sin t) \frac{\partial^2 T(a \sin \cdot)}{\partial a^2}(t) \bigg|_{a=0} dt = 0$$

$$\int_0^\pi (\sin t) \frac{\partial^3 T(a \sin \cdot)}{\partial a^3}(t) \bigg|_{a=0} dt = \psi'''(0) \int_0^\pi \sin^4 t \, dt \neq 0$$

Therefore, zero is a cubic singularity of T. Theorem 4.1 may now be applied to $M(\lambda, x) = Tx - qx - p$, $(p, q) = \lambda$. The verification of the hypothesis (4.8) is left for the reader. $\square$

The above result for Problem 5.1 with $\psi(x)$ similar to $x + \gamma x^3$ at $x = 0$ is completely different from Theorem 3.1 where $\psi(x)$ was assumed to be like $x + \beta x^2$ at $x = 0$. We now discuss a result which includes both Theorems 3.1 and 5.1. This example will also bring out the importance of additional parameters in the problem.

Suppose $\psi \in C^3(\mathbb{R}, \mathbb{R})$

$$\psi(0) = 0, \qquad \psi'(0) = 1, \qquad \psi''(0) = 0, \qquad \psi'''(0) \neq 0$$

and consider the problem

$$(5.2) \qquad \begin{aligned} x'' + \psi(x) + \beta x^2 &= q(t)x + p(t), \qquad 0 < t < \pi \\ x(0) &= x(\pi) = 0 \end{aligned}$$

where $(p, q) \in \mathscr{C} = C([0, \pi], \mathbb{R}) \times C([0, \pi], \mathbb{R})$ and $\beta \in \mathbb{R}$.

Let $X = \{x \in C^2([0, \pi], \mathbb{R}) : x(0) = x(\pi) = 0\}$, $Z = C([0, \pi], \mathbb{R})$, $\lambda = (\beta, p, q) \in \mathbb{R} \times \mathscr{C}$,

$$M : \mathbb{R} \times X \to Z$$

(5.3)

$$M(\lambda, x)(t) = x''(t) + \psi(x(t)) + \beta x^2(t) - q(t)x(t) - p(t)$$

Then

$$[D_x M(0, 0)y](t) = y''(t) + \psi'(0)y$$

has $[y_0] = \mathscr{N}(D_x M(0, 0))$, $[y_0] \oplus \mathscr{R}(D_x M(0, 0)) = Z$ where $y_0(t) = \sin t$. The computations in the proof of Theorem 5.1 show that condition (4.2) is satisfied. The bifurcation function $f(a, \lambda)$ has the form

$$f(a, \lambda) = \alpha_0(\lambda) + \alpha_1(\lambda)a + \alpha_2(\lambda)a^2 + \alpha_3(\lambda)a^3 + o(|a|^3)$$

as $a \to 0$. Furthermore,

$$\alpha_0(\lambda) = \left(\frac{2}{\pi}\right)^{1/2} \int_0^\pi p(t)\sin t \, dt + o(|\lambda|)$$

$$\alpha_1(\lambda) = \left(\frac{2}{\pi}\right)^{1/2} \int_0^\pi q(t)\sin^2 t \, dt + o(|\lambda|)$$

$$\alpha_2(\lambda) = \left(\frac{2}{\pi}\right)^{1/2} \left(\frac{4}{3}\right)\beta + o(|\lambda|),$$

$$\alpha_3(\lambda) = \frac{3\pi}{8} \psi'''(0) + O(|\lambda|).$$

Since $\psi'''(0) \neq 0$, one can obtain the complete bifurcation diagram in $\lambda = (\beta, q, p)$ space by a manner similar to the one used for the case in which $\beta = 0$. More specifically, one considers the more general problem where $\alpha_0(\lambda)$, $\alpha_1(\lambda)$, $\alpha_2(\lambda)$ are independent parameters μ_0, μ_1, μ_2 and then replaces μ_j by $\alpha_j(\lambda)$. In any neighborhood of zero, there will still be one or three solutions. If $\beta \neq 0$, is fixed, and y is fixed, then a functional of the solutions plotted as a functional of p has the form shown in Figure 5.1. In the discussion in

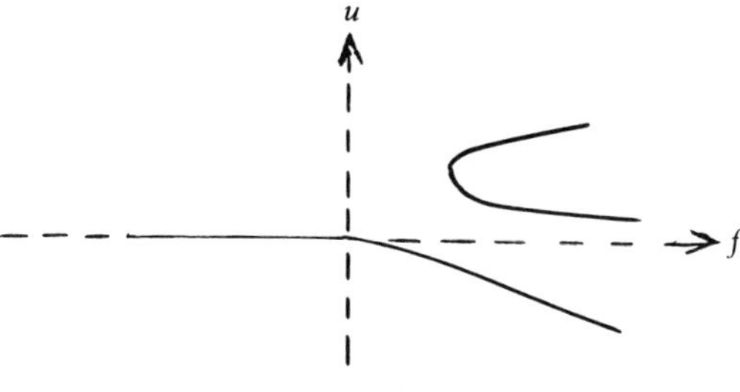

Figure 5.1

Section 3, there was only 0 or two solutions. That analysis was based on the lowest order terms being quadratic and is applicable only to the parabola in Figure 5.1. The analysis in this section based on the cubic terms is more global and gives the more complete behavior of the solutions.

6.6. Bifurcation from Known Solutions

Suppose that $M: \Lambda \times X \to Z$ is C^k, $k \geq 2$, and

$$
\text{(6.1)} \qquad
\begin{aligned}
M(0,0) &= 0, \qquad D_x M(0,0) = B \\
\dim \mathcal{N}(B) &= 1 = \operatorname{codim} \mathcal{R}(B).
\end{aligned}
$$

In previous sections, we imposed certain generic conditions on the nonlinear terms in $M(\lambda, x)$ and obtained complete descriptions of the number of solutions of the equation

$$
\text{(6.2)} \qquad M(\lambda, x) = 0.
$$

These conditions were not always satisfied if there is a function $\phi(\lambda)$ such that $M(\lambda, \phi(\lambda)) = 0$ for all λ in a neighborhood of zero. In applications, this condition says that there is a known branch of solutions of (6.2) through $(\lambda, x) = (0, 0)$. By a change of coordinates, we may assume that $x = 0$ on this branch; that is,

$$
\text{(6.3)} \qquad M(\lambda, 0) = 0 \quad \text{for all } |\lambda| \text{ sufficiently small.}
$$

In this section, we illustrate how the methods used in previous sections may be adapted to this situation.

Let $f(a, \lambda)$ be the bifurcation function for equation (6.2). It follows from (2.5) that

$$
\text{(6.4)} \qquad f(a, \lambda) w_0 = (I - Q) M(\lambda, a y_0 + z^*(a, \lambda))
$$

where the notations are exactly as those in Section 6.2. If (6.3) is satisfied, then $f(a, \lambda) = a g(a, \lambda)$ for some C^{k-1} function g. Thus, nonzero solutions of the bifurcation equation

$$
\text{(6.5)} \qquad f(a, \lambda) = 0
$$

correspond to precisely those of $g = 0$.

Theorem 6.1. *If*

$$
\text{(6.6)} \qquad \left. \frac{\partial g}{\partial \lambda} \right|_{\lambda = 0} w_0 = D^2_{a\lambda}(I - Q) M(\lambda, a y_0) \big|_{(\lambda, a) = (0,0)} \neq 0,
$$

then there are solutions (λ, x) of (6.2) with $x \neq 0$ in every sufficiently small neighborhood of $(\lambda, x) = (0, 0)$.

Proof. By (6.6), there exists $\lambda_0 \in \Lambda$ such that

$$\left.\frac{\partial g}{\partial \lambda}\right|_{\lambda=0} \cdot \lambda_0 \neq 0.$$

Since every element $\lambda \in \Lambda$ has a unique representation $\lambda_0 + \tilde{\lambda}$, where $\tilde{\lambda}$ belongs to the complimentary subspace of the space $[\lambda_0]$, we obtain from the implicit function theorem that

$$g(a, \tilde{\lambda} + \alpha(a, \tilde{\lambda})\lambda_0) = 0$$

where $\alpha(a, \tilde{\lambda})$ is C^{k-1} and $\alpha(0, 0) = 0$. This completes the proof. $\square$

Corollary 6.2. *If, in addition to the hypotheses in Theorem 6.1, we have*

$$\left.\frac{\partial g}{\partial a}\right|_{(\lambda,a)=(0,0)} w_0 = D_a^2(I - Q)M(0, ay_0)\big|_{a=0} \neq 0$$

then there exists a C^{k-1} function $\gamma(\lambda)$ defined for $|\lambda|$ small, $\gamma(0) = 0$, such that $W = \gamma^{-1}(0)$ is a C^{k-1} manifold of codimension 1 and the following holds for $|x|$ sufficiently small:

(i) *If $\lambda \in W$, then $(\lambda, 0)$ is the only solution of (6.2);*
(ii) *If $\lambda \notin W$, then there exists a unique solution (λ, x) of (6.2) with $x \neq 0$.*

Proof. The function $\gamma(\lambda)$ is in fact given by $g(0, \lambda)$. $\square$

Remark 6.3. If dim $\Lambda = 1$, then dim $W = 0$. The above corollary gives the familiar bifurcation diagram in Fig. 6.1.

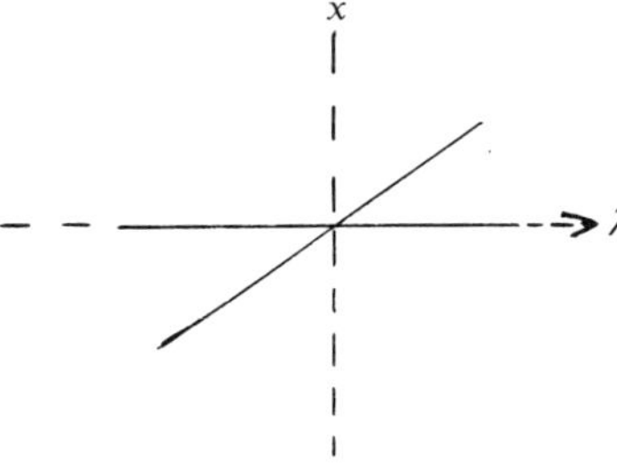

Figure 6.1

By using the methods in Section 6.4, we have the following:

Corollary 6.4. *If, in addition to the hypotheses of Theorem 6.1, we have*

$$\left.\frac{\partial g}{\partial a}\right|_{(\lambda,a)=0} = 0, \qquad \left.\frac{\partial^2 g}{\partial a^2}\right|_{(\lambda,a)=(0,0)} \neq 0$$

then there exists a C^{k-1} function $\gamma(\lambda)$ defined for $|\lambda|$ small, $\gamma(0) = 0$, such that $W = \gamma^{-1}(0)$ is a C^{k-1} manifold of codimension 1 and the following holds for $|x|$ sufficiently small:

(i) *if $\gamma(\lambda) \leq 0$, then $(\lambda, 0)$ is the only solution of (6.2);*
(ii) *if $\gamma(\lambda) > 0$, then there are exactly two distinct solutions (λ, x_1), (λ, x_2) of (6.2), $x_1, x_2 \neq 0$.*

Remark 6.5. If $\dim \Lambda = 1$, then the above Corollary gives the bifurcation diagram in Fig. 6.2.

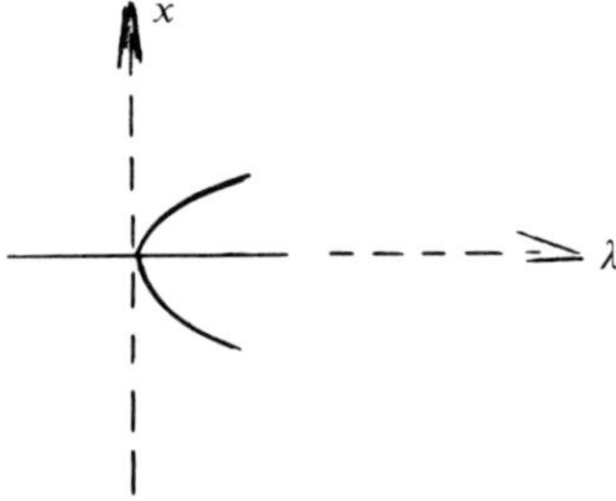

Figure 6.2

As an application, consider the following boundary value problem:

$$(6.7) \qquad \ddot{x} + (1 + \lambda)x - \psi(x) = 0, \qquad 0 \leq t \leq \pi,$$
$$x(0) = x(\pi) = 0$$

where $\psi \in C^3(R)$. Assume that

$$(6.8) \qquad \psi(0) = \psi'(0) = \psi''(0) = 0, \qquad \psi'''(0) > 0.$$

If we proceed as in Section (6.5), we have

$$f(a, \lambda) = a(\lambda - \alpha a^2 + \cdots)$$

where "$\cdots$" denotes higher order terms and

$$\alpha = \tfrac{1}{6}\psi'''(0) \int_0^\pi \sin^4 t \, dt > 0.$$

This shows that Corollary 6.4 is applicable here and the bifurcation diagram is exactly as in Fig. 6.2. In particular, there are exactly two nonzero solutions of (6.7) for $\lambda > 0$ small.

Remark 6.6. If we assume that

$$\lim_{t \to \pm \infty} \psi'(t) = \pm \infty$$

then it can be shown that there are precisely two nonzero solutions of (6.7) for $1 < \lambda < 3$ and there are no nonzero solutions of (6.7) for $0 < \lambda \le 1$.

Remark 6.7. If a forcing term $\varepsilon p(t)$, $\varepsilon \in R$, is added to the right hand side of the equation in (6.7) and if

$$\int_0^\pi p(t)\sin t \, dt \ne 0$$

then, for $\varepsilon \ne 0$, Theorem 4.1 is applicable. This gives the bifurcation diagrams in Fig. 6.3.

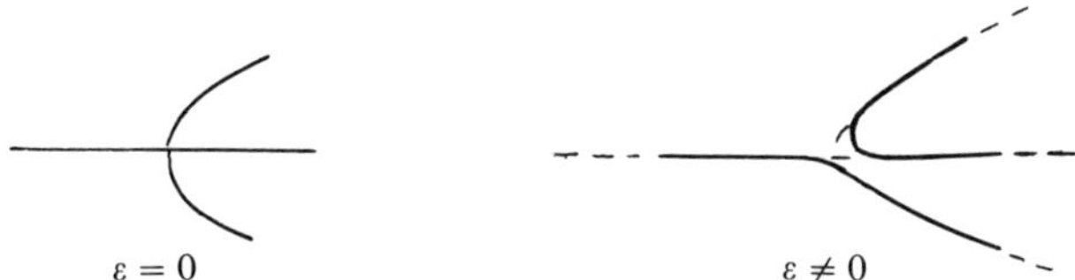

Figure 6.3

6.7. Effects of Symmetry

In this section, we discuss the effects of symmetry properties on the bifurcation equation.

Let $G = \{s_1, \ldots, s_N\}$ be a finite group with the discrete topology. Thus, G is a compact topological group. A representation of G over X is a map

$$\Gamma : G \to L(X)$$

where $L(X)$ is the Banach space of bounded linear operators from X into X, satisfying:

$$\Gamma(s_i s_j) = \Gamma(s_i)\Gamma(s_j), \qquad s_i, s_j \in G.$$

Consider the equation in Section 6.2:

(7.1) $$M(\lambda, x) = 0$$

and its auxiliary and bifurcation equations

(7.2) $$QM(\lambda, x) = 0$$

(7.3) $$(I - Q)M(\lambda, x) = 0.$$

Suppose that there are linear representations Γ_x and Γ_z of G over X and Z respectively such that

(7.4) $$\Gamma_z(s_i)M(\lambda, x) = M(\lambda, \Gamma_x(s_i)x), \qquad i = 1, \ldots, N.$$

Define

(7.5) $$\bar{P} = \frac{1}{N} \sum_{i=1}^{N} \Gamma_x(s_i^{-1})P\Gamma_x(s_i)$$

and

(7.6) $$\bar{Q} = \frac{1}{N} \sum_{i=1}^{N} \Gamma_z(s_i^{-1})Q\Gamma_z(s_i)$$

where P is as in Section 6.2.

Theorem 7.1. *By using the above notations, we have*

(7.7) $$\Gamma_x(s_i)\bar{P} = \bar{P}\Gamma_x(s_i), \qquad i = 1, \ldots, N;$$

and

(7.8) $$\Gamma_z(s_i)\bar{Q} = \bar{Q}\Gamma_z(s_i), \qquad i = 1, \ldots, N.$$

Proof. By substituting (7.5) and (7.6) into the Equations (7.7) and (7.8), we obtain the desired result. $\square$

We note that $\bar{P}$ and $\bar{Q}$ are again projection operators. Since $PX = \mathscr{N}(D_x M(0,0))$, $\Gamma_x(s_i)\bar{P}X = \bar{P}X$. This gives the following.

Theorem 7.2. *If (7.4) is satisfied, we may choose projections P and Q such that the bifurcation equation (7.3) satisfies:*

(7.9) $$\Gamma_z(s_i)(I - Q)M(\lambda, x) = (I - Q)M(\lambda, \Gamma_x(s_i)x).$$

Remark 7.3. Theorem 7.2 says that the bifurcation equation inherits all the symmetry properties from the original equation by choosing appropriate projections. We note that Theorem 7.2 is true even if dim $\mathscr{N}(B) > 1$, where B is defined in (2.1).

As we assumed that dim $\mathscr{R}(P) = 1 = \text{codim } \mathscr{R}(Q)$ and G is a finite group, this implies

$$\Gamma_x(s_i)y_0 = \pm y_0, \qquad \Gamma_z(s_i)w_0 = \pm w_0, \qquad i = 1, \ldots, N.$$

Let

$$G_x = \{s_i : \Gamma_x(s_i)y_0 = y_0\} \quad \text{and} \quad G_z = \{s_i : \Gamma_z(s_i)w_0 = w_0\}.$$

Theorem 7.4. *Let*

$$f(a, \lambda)w_0 \equiv (I - Q)M(\lambda, ay_0 + z^*(a, \lambda))$$

where $z^*(a, \lambda)$ is the unique solution of $QM(\lambda, ay_0 + z) = 0$. If $f(a, \lambda) \not\equiv 0$ near $(a, \lambda) = (0, 0)$, then $G_x = G_z$.

Proof. If $s_i \in G_x$ and $s_i \notin G_z$, then

$$\begin{aligned}
f(a, \lambda)w_0 &= -\Gamma_z(s_i)f(a, \lambda)w_0 \\
&= -\Gamma_z(s_i)(I - Q)M(\lambda, ay_0 + z^*(a, \lambda)) \\
&= -(I - Q)M(\lambda, \Gamma_x(s_i)(ay_0 + z^*(a, \lambda))).
\end{aligned}$$

By the uniqueness of z^*, $\Gamma_x(s_i)z^*(a, \lambda) = z^*(a, \lambda)$. This implies $f(a, \lambda) = -f(a, \lambda)$ for all (a, λ) near $(0, 0)$. This gives a contraction. Hence, $G_x \subset G_z$. Similarly, $G_z \subset G_x$. $\square$

Theorem 7.5. *If $G_x \neq G$, then $f(a, \lambda) = -f(-a, \lambda)$.*

Proof. It follows easily from the proof of Theorem 7.4. $\square$

We now give an example to illustrate the effects of symmetry. Consider the buckling of a simply supported rectangular plate under a compressive thrust at its short edges, and under a normal load. The mathematical description is as follows.

Let $D = \{(x, y): -l < x < l, -1 < y < 1\}$ be the region in $\mathbb{R}^2$ occupied by the unstressed plate. Let $X = Z$, with

$$X = H^2(D) \cap H_0^1(D) = \{u \in H^2(D) : u \in C(\bar{D}) \text{ and } u = 0 \text{ on } \partial D\}.$$

The space X is given a Hilbert space structure using the inner product and associated norm,

$$\langle u, v \rangle_X = \int_D \Delta u \cdot \Delta v, \qquad \|u\|_X^2 = \int_D (\Delta u)^2.$$

We assume the equilibrium positions of the plate are described by the von Kármán equations (see Section 8.2, Equation (8.2.1)) with the stress function and deflection satisfying the boundary conditions (8.2.2), (8.2.4), respectively. These equations can then be reduced to the following equation in the space X:

$$(7.10) \qquad\qquad (I - \lambda L)w + C(w) = vp,$$

where $v \in \mathbb{R}$ is small,

$w \in X$ describes the deflection of the plate;
$\lambda \in \mathbb{R}$ is the magnitude of thrust at $x = \pm l$;
$L: X \to X$, $Lw = -\Delta^{-2}w_{xx}$, is a compact self-adjoint linear operator;
Δ^{-1} is the inverse of the Laplacian, with zero Dirichlet data;
$C: X \to X$, $C(w) = \frac{1}{2}B(w, B(w, w))$;
$B: X \times X \to X$, $B(u, v) = \Delta^{-2}[u, v]$ is a bounded bilinear operator on X;
$[u, v] = u_{xx}v_{yy} + u_{yy}v_{xx} - 2u_{xy}v_{xy}$;
$p = \Delta^{-2}p'$ where p' is the given normal load.

Let $T(w) = (I - \lambda_0 L)w + C(w)$ for some fixed λ_0. Then

$$DT(0) = I - \lambda_0 L.$$

This has a nontrivial kernel when λ_0 is a characteristic value of the operator L. These characteristic values (see Section 8.3) are given by

$$\lambda_{mn} = \frac{\pi^2(m^2 + n^2 l^2)^2}{4l^2 m^2}, \qquad m, n = 1, 2, \ldots,$$

with the corresponding complete orthonormal set of eigenfunctions,

$$\phi_{mn} = C_{mn} \sin \frac{m\pi}{2l}(x + l)\sin \frac{n\pi}{2}(y + 1)$$

$$C_{mn} = \frac{4l^{3/2}}{\pi^2(m^2 + n^2 l^2)}.$$

In what follows, we will study the bifurcation problem for (7.10) in the neighborhood of simple characteristic values of L; that is, we will take $\lambda_0 = \lambda_{mn}$ for some (m, n), which is such that $\lambda_{mn} \neq \lambda_{m'n'}$ for all $(m', n') \neq (m, n)$. We write then also $\phi_0 = \phi_{mn}$ for the corresponding eigenfunction. For $p = 0$, (7.10) remains invariant under the following group of operators on X:

$$G = \{I, \Gamma_x, \Gamma_y, \Gamma_{xy}, -I, -\Gamma_x, -\Gamma_y, -\Gamma_{xy}\}$$

where

$$(\Gamma_x u)(x, y) = u(-x, y),$$
$$(\Gamma_y u)(x, y) = u(x, -y),$$
$$\Gamma_{xy} = \Gamma_x \Gamma_y = \Gamma_{yx}.$$

We have, for each $\Gamma \in G$:

$$L\Gamma = \Gamma L$$
$$C(\Gamma w) = \Gamma C(w)$$
$$B(\Gamma u, \Gamma v) = \Gamma B(u, v)$$
$$\langle \Gamma u, \Gamma v \rangle_X = \langle u, v \rangle_X.$$

Define the projections P and Q by $P = Q = \langle w, \phi_0 \rangle_X \phi_0$. Since $\Gamma \phi_0 = \pm \phi_0$ for every $\Gamma \in G$, every Γ commutes with P. Let $G_0 = \{\Gamma \in G : \Gamma \phi_0 = \phi_0\}$. It is easily shown that $G_x = \{\Gamma_1, \Gamma_2, \Gamma_3, \Gamma_4\}$, where

$$\Gamma_1 = I, \qquad \Gamma_2 = (-1)^{m+1}\Gamma_x, \qquad \Gamma_3 = (-1)^{n+1}\Gamma_y, \qquad \Gamma_4 = (-1)^{m+n}\Gamma_{xy}.$$

Note that $G_0 \cup (-G_0) = G$. By Theorem 7.5, the bifurcation equation is odd in a when $p = 0$. In fact, the bifurcation equation has the following form:

$$f(a, \eta, p) = -\eta a + \langle C(a\phi_0 + v^*(a, \eta, v)), \phi_0 \rangle_X - v\langle p, \phi_0 \rangle_X$$

where $\eta = \lambda/\lambda_0 - 1$. Moreover,

$$\frac{\partial f}{\partial a}(0,0,0) = \frac{\partial^2 f}{\partial a^2}(0,0,0) = 0$$

$$\frac{\partial^3 f}{\partial a^3}(0,0,0) = 3\langle C(\phi_0), \phi_0 \rangle_X$$

$$= 3\|B(\phi_0, \phi_0)\|^2 > 0.$$

Let $X_0 = \{u \in X : \Gamma u = u \text{ for all } \Gamma \in G_0\}$ and X_1 be its orthogonal complement. If $p = \varepsilon p_0 + p_1$, $p_0 \in X_0$, $p_1 \in X_1$, $\langle p_0, \phi_0 \rangle_X > 0$ and $\varepsilon \in \mathbb{R}$ is small, then

$$f(a, \eta, p) = f(a, \eta, \varepsilon, v)$$
$$\cong \tfrac{1}{2}\|B(\phi_0, \phi_0)\|^2 a^3 - \eta a + v\varepsilon\langle p_0, \phi_0 \rangle$$

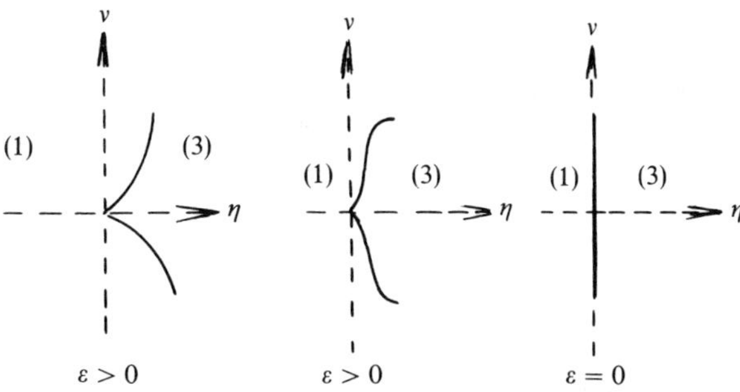

Figure 7.1

This gives a bifurcation diagram as in Fig. 7.1 in which "(m)" denotes that there are m solutions in the region indicated. It is interesting to see how the cusp on the left in Fig. 7.1 deforms to a straight line as ε decreases to 0.

6.8. Universal Unfoldings

For the case dim $\mathcal{N}(B) = 1 = \operatorname{codim} \mathcal{R}(B)$, we have seen from the results of previous sections that the solution of the bifurcation problem reduces to the discussion of the manner in which the solutions $a \in \mathbb{R}$ of a scalar equation

$$(8.1) \qquad\qquad f(a, \lambda) = 0$$

depend on $\lambda \in \Lambda$. If f is C^∞,

$$f(0,0) = \frac{\partial}{\partial a} f(0,0) = \cdots = \frac{\partial^{k-1}}{\partial a^{k-1}} f(0,0) = 0, \qquad \frac{\partial^k}{\partial^k a} f(0,0) \neq 0,$$

where $k \geq 2$, then, by the Malgrange preparation theorem, the bifurcation problem (8.1) is equivalent to the following polynomial equation in a neighborhood of $(a, \lambda) = (0, 0)$:

$$(8.2) \qquad\qquad \bar{f}(\bar{a}, \bar{\lambda}) = \bar{a}^k + \sum_{i=0}^{k-2} \bar{\lambda}_{i+1} \bar{a}^i$$

where $\bar{a} \in \mathbb{R}$, $\bar{\lambda} \in \mathbb{R}^{k-1}$ and $(\bar{\lambda}_1, \ldots, \bar{\lambda}_{k-1}) = \bar{\lambda}$. In fact, $\bar{f}$ is obtained from f by a C^∞ change of variables

$$(8.3) \qquad\qquad \bar{\lambda} = \eta(\lambda), \qquad \bar{a} = \xi(a, \lambda)$$

in which $\bar{\lambda}$ depends on λ. The function $\bar{f}$ in (8.2) is called the *universal unfolding* of f.

The function $\bar{f}(\bar{a}, \bar{\lambda})$ carries all of the information necessary to solve the bifurcation problem. It contains only a finite number of parameters $\bar{\lambda}$ and the bifurcation surfaces in the $\bar{\lambda}$-space determine those in the Banach space Λ via the transformation (8.3). For example, consider Theorem 2.1. The universal unfolding of the bifurcation function in Theorem 2.1 is simply the quadratic function

$$\bar{f}(\bar{a}, \bar{\lambda}) = \bar{a}^2 + \bar{\lambda}, \qquad \bar{\lambda} \in \mathbb{R}.$$

The bifurcation surface is therefore $\bar{\lambda} = 0$ with no solutions for $\bar{\lambda} > 0$ and two solutions with $\bar{\lambda} < 0$. Also, $\bar{\lambda} = \eta(\lambda) = C(\lambda)\gamma(\lambda)$, where $\gamma(\lambda)$ is the function in Theorem 2.1 and $C(0) > 0$ is C^∞. The condition (2.11) in Theorem 2.1 implies that the transformation (8.3), $\bar{\lambda} = \eta(\lambda)$, is nondegenerate at $\lambda = 0$

and hence defines a bifurcating surface which is a local C^∞ manifold of codimension 1 in the Banach space Λ. If the universal unfolding $\bar{f}$ has k large, then the bifurcation surfaces are very complicated.

For the case where dim $\mathcal{N}(B) > 1$, the theory of universal unfoldings has been developed for those functions $M(\lambda, x)$ which are the gradient of some function $V(\lambda, x)$. In applications, it often occurs that $M(\lambda, x)$ is not a gradient and, thus, other techniques are necessary. In the next chapters, we show how the ideas presented in the previous sections may be adapted to discuss such problems.

6.9. Bibliographical Notes

Theorem 2.1 was proved by Chow, Hale and Mallet–Parret [1]. Theorem 2.6 has also been discussed by Ambrosetti and Prodi [1], Gromoll and Meyer [1], Chillingworth [2]. Theorem 2.8 was first proved by Ambrosetti and Prodi [1]. Theorem 3.2 is due to Prodi and Ambrosetti [1] where one can also find an extension to elliptic equations. McKenna and Shaw [1] have also discussed when the solutions of an elliptic boundary value problem form a manifold. Theorem 6.1 is sometimes referred to as the theorem of bifurcation from a simple eigenvalue and was first stated in this way by Crandall and Rabinowitz [1] (see also Hale [5]). For another proof of this theorem using Morse theory, see Nirenberg [2]. Remark 6.6 is due to Ambrosetti and Mancini [1] and shows there is a possibility of a global version of Theorem 4.1.

It often happens that the Fredholm operator $DT(0)$ has dim $\mathcal{N}DT(0) = 1$ and codim $\mathcal{R}D(T(0)) = k > 1$. The index is therefore negative. In such a case, there are k bifurcation equations and parameters other than the element in $\mathcal{N}D(T(0))$ are needed to obtain solutions. For a few results in this direction, see Boucherif [1], Hale [11], Thomas and Zachman [1]. The results of Boucherif [1] also apply to a three point boundary value problem for a second order equation of the type discussed in Section 6.3. In this case, there are two bifurcation equations and the additional parameter was chosen to be one of the boundary conditions. Thus, a free boundary arose in a natural way. Free boundary value problems of this type arise in a similar way when the nonlinearities in the differential equations are not continuous. For example $\ddot{x} = \lambda$ on $[0, A)$, $\ddot{x} = 0$ on $(A, 1]$ with boundary conditions $x(0) = x(A) = 0$, $x(1) = -1$. For bifurcation problems of this type, see Alexander and Fleischman [1], Berger and Fraenkel [1].

When the mapping $T(x, \lambda)$ is continuous, but not differentiable, the bifurcation curves can be discontinuous. A theory for this case is in the process of being developed (see McLeod and Turner [1], K. Schmidt [1], Stuart [9, 10] for results and references).

For results and references on bifurcation from infinity, see Bazley and McLeod [1], Nussbaum and Stuart [1], Rabinowitz [15].

For results and references on the role of symmetries in bifurcation theory, see Dancer [5], Golubitsky and Schaeffer [2], Lewis [2], Loginov and Trenogin [1, 2], Poenaru [1], Ruelle [1], Sattinger [5–10], Vanderbanwhede [1–6], Shearer [3]. The example in Section 7 is due to Vanderbanwhede [5, 6].

For a further discussion of the singularity theory in Section 8 (sometimes referred to as castastrophe theory), proofs of the results, and applications to bifurcation theory, see Arnol'd [1–3], Brocker and Lander [1], Chillingworth [1, 2], Golubitsky [1], Golubitsky and Guillemin [1], Lu [1], Magnus [1–4], Poston and Stuart [1], Thom and Zeeman [1], Wasserman [1]. For the corresponding results with some of the parameters restricted, see Golubitsky and Keyfitz [1], Golubitsky, Keyfitz and Schaeffer [1], Golubitsky and Schaeffer [1–2]. In Chafee [3], Takens [2], Hale [7], Bernfeld, Negrini and Salvadori [1] and Golubitsky and Langford [1], the relationship between singularity theory and Hopf bifurcation are discussed.

The ideas behind the development of the theory of unfolding of singularities has also been used in the theory of asymptotic expansions of integrals and solutions of partial differential equations. These results have been especially informative in the theory of caustics in optics (see Arnol'd [4], Duistermaat [1], Guillemin and Sternberg [1]).

For an application of the singularity theory to Hill's equation, see Lazutkin and Paukratova [1] and to integral equations, see Likhtarinkov [1].

Chapter 7

Bifurcation with Higher Dimensional Null Spaces

7.1. Introduction

To motivate the problems to be discussed in this section, consider M: $\Lambda \times X \to Z$, $M(0,0) = 0$, and suppose $D_x M(0,0)$ is Fredholm of index zero. The method of Liapunov–Schmidt reduces the study of local bifurcation to the study of a system of nonlinear equations $f(\lambda, u) = 0$ for $u \in \mathbb{R}^k$, $f \in \mathbb{R}^k$, where $k = \dim \mathcal{N}(D_x M(0,0))$. If $k = 1$ and $\partial^2 f(0,0)/\partial u^2 \neq 0$, we have seen in Section 6.2 that the local bifurcations are determined by a scalar function of λ. If $\partial^2 f(0,0)/\partial u^2 = 0$, $\partial^3 f(0,0)/\partial u^3 \neq 0$, we have also seen in Section 6.4 that the local bifurcations are determined by two scalar functions of λ. In Section 6.8, we have also discussed the situation for arbitrary k—the basic result being that the local bifurcations are determined by a polynomial of degree k whose coefficients are functions of λ.

One primary objective in this chapter is to consider problems for which $\dim \mathcal{N}(D_x M(0,0)) > 1$. In this case, if M is the gradient with respect to x of some functional $V(\lambda, x)$, then $f(\lambda, u)$ will also be the gradient with respect to u of some scalar function $U(\lambda, u)$. Also, for gradients, we have indicated that some results are available to permit one to discuss completely the local bifurcations. For $\dim \mathcal{N}(D_x M(0,0)) > 1$, the number of parameters necessary for a complete description of the bifurcations is very large. Also, if M is not the gradient of some functional, no general results are available.

In this chapter, we use scaling techniques to solve problems with $\dim \mathcal{N}(D_x M(0,0)) > 1$ when the specific dependence on the vector parameter λ is known precisely and it is not required that M be a gradient of some functional. The discussion centers around the case where λ is a two vector. We choose different families of mappings M to indicate the different types of bifurcations that can occur in several parameter problems. The form of the families chosen is dictated by the frequency with which they occur in applications. It is our hope that a systematic repetition of the scaling procedures for specific examples will improve the intuition of the reader and assist him in solving more complicated problems that may arise in particular applications. We emphasize again that the analysis is not restricted to equations which are the gradient of some function. Also, the methods are constructive in the sense that only the Implicit Function Theorem is used. Thus, all bifurcation curves and solutions can be obtained by successive approximations.

The importance of these results will become clear in Chapter 8 where applications are given to several specific problems.

7.2. The Quadratic Revisited

In this section, we consider scaling techniques for the scalar equation

$$(2.1) \qquad f(\alpha, u) = \alpha_0 + \alpha_1 u + u^2 + \text{h.o.t.} = 0$$

where $\alpha = (\alpha_0, \alpha_1) \in \mathbb{R}^2$, $u \in \mathbb{R}$, $f \in \mathbb{R}$, $f(0, \alpha_1, 0) = 0$ and

$$(2.2) \qquad \text{h.o.t.} = O(|u|^3 + |\alpha_0|^2 + |\alpha_0 u| + |\alpha_0 \alpha_1| + |\alpha_1 u|^2)$$

as $\alpha, u \to 0$.

It is possible to discuss the bifurcation in Equation (2.1) from the theory previously discussed. The first step is to make the transformation $u = v - \alpha_1/2$ to obtain an equation without the linear term in v, $v^2 + \bar{\alpha}_0 + \text{h.o.t.} = 0$. The left-hand side of this equation has a minimum. If the minimum is positive, there are no solutions, if it is negative, there are two solutions.

If $f \in \mathbb{R}^2$, $u \in \mathbb{R}^2$ in Equation (2.1); such a simple procedure is not possible. In order to understand the procedure for this more complicated case, let us consider Equation (2.1) by a different method. The discussion will be made in such a way as to be directly applicable to the higher dimensional case.

Lemma 2.1. *There is a neighborhood V of $(\alpha, u) = (0,0)$ and a constant $\beta > 0$ such that any solution of Equation (2.1) in V must satisfy*

$$|u| \leq \beta(|\alpha_1| + |\alpha_0|^{1/2}).$$

Proof. If this is not the case, then there exists a sequence of solutions $(\alpha_{0n}, \alpha_{1n}, u_n) \to 0$ as $n \to \infty$ such that $|\alpha_{0n}|^{1/2}/|u_n|$, $|\alpha_{1n}|/|u_n| \to 0$ as $n \to \infty$. Divide Equation (2.1) by $|u_n|^2$ to obtain

$$0 = \frac{f(\alpha_{0n}, \alpha_{1n}, u_n)}{|u_n|^2} = 1 + O\left(|u_n| + \left|\frac{\alpha_{0n}}{u_n^2}\right| + \left|\frac{\alpha_{1n}}{u_n}\right|\right).$$

Since the right-hand side approaches 1 as $n \to \infty$, this gives a contradiction and proves the lemma. $\square$

Lemma 2.1 justifies scaling procedures in Equation (2.1) which will have the effect of reducing the number of parameters in the equation by one. To illustrate, suppose first that $\alpha_1 = 0$, $\alpha_0 \neq 0$; that is, consider the equation

$$
\begin{aligned}
f(\alpha_0, u) &= \alpha_0 + u^2 + \text{h.o.t.} = 0 \\
\text{h.o.t.} &= O(|u|^3 + |\alpha_0|^2 + |\alpha_0 u|).
\end{aligned}
\tag{2.3}
$$

From Lemma 2.1, determining small solutions (α_0, u) of Equation (2.3) is equivalent to letting $u = |\alpha_0|^{1/2} v$ and determining all possible solutions in $\mathbb{R}$ of the equations

$$
1 + v^2 + O(|\alpha_0|^{1/2}) = 0 \quad \text{for } \alpha_0 > 0 \text{ small}
\tag{2.4}
$$

$$
-1 + v^2 + O(|\alpha_0|^{1/2}) = 0 \quad \text{for } \alpha_0 < 0 \text{ small.}
\tag{2.5}
$$

Since Equation (2.4) has no solutions in $\mathbb{R}$ for $\alpha_0 = 0$ and Equation (2.5) has two distinct solutions ± 1, for $\alpha_0 = 0$, the Implicit Function Theorem implies $\alpha_0 = 0$ is a bifurcation point for Equation (2.3). Also, there are no real solutions of Equation (2.3) for $\alpha_0 < 0$ and two distinct solutions for $\alpha_0 > 0$. Finally, the solutions $u_j(\alpha_0) = \alpha_0^{1/2} v_j(\alpha_0)$, $v_j(0) = (-1)^j, j = 1, 2$, where the functions $v_j(\alpha_0)$ are as smooth in $|\alpha_0|^{1/2}$ as the function f. These functions may be obtained by successive approximations since they are determined by an application of the Implicit Function Theorem.

If $\alpha_0 = 0$, $\alpha_1 \neq 0$, then Equation (2.1) becomes

$$
\begin{aligned}
\alpha_1 u + u^2 + \text{h.o.t.} &= 0 \\
\text{h.o.t.} &= O(|u|^3 + |\alpha_1^2| + |\alpha_1 u|^2).
\end{aligned}
\tag{2.6}
$$

From Lemma 2.1, determining small solutions (α_1, u) of Equation (2.6) is equivalent to letting $u = \alpha_1 w$ and determining all solutions in $\mathbb{R}$ of the equation

$$
w + w^2 + O(|\alpha_1|) = 0
\tag{2.7}
$$

For $\alpha_1 = 0$ this equation has two distinct solutions $w = 0$, $w = -1$. The Implicit Function Theorem therefore implies there are two solutions for α_1 small—positive or negative. The point $\alpha_1 = 0$ is again a bifurcation point, but there are the same number of solutions for $\alpha_1 > 0$ as for $\alpha_1 < 0$—in contrast to what happens for $\alpha_1 = 0$, $\alpha_0 \neq 0$.

In Figure 2.1, we have indicated the solutions of Equation (2.1) for several values of α_0, α_1. As we have seen above, one does not obtain the complete Figure 2.1 by studying only a one parameter problem. Let us now obtain Figure 2.1 by scaling techniques. From Lemma 2.1, determining small solutions (α_0, α_1, u) of Equation (2.1) is equivalent to letting $u = |\alpha_0|^{1/2} v$, $\alpha_1 = |\alpha_0|^{1/2} \beta_1$, and determining all solutions $(\beta_1, v) \in \mathbb{R}^2$ of

$$
1 + \beta_1 v + v^2 + O(|\alpha_0|^{1/2}) = 0 \quad \text{for } \alpha_0 > 0 \text{ small}
\tag{2.8}
$$

$$
-1 + \beta_1 v + v^2 + O(|\alpha_0|^{1/2}) = 0 \quad \text{for } \alpha_0 < 0 \text{ small}
\tag{2.9}
$$

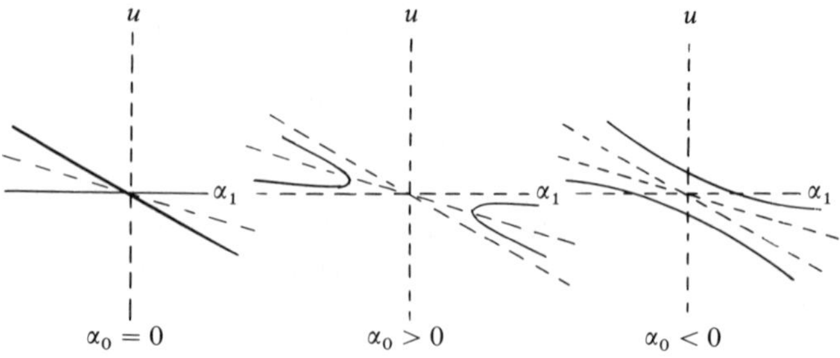

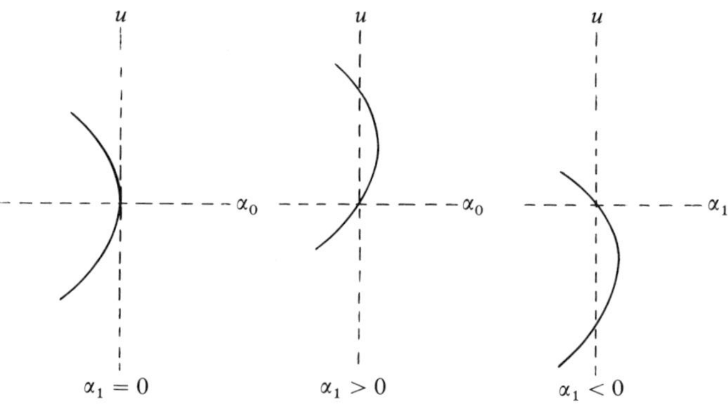

Figure 2.1

The basic idea that is being exploited is the following. We are parameterizing the neighborhood of $(\alpha_0, \alpha_1) = (0,0) \in \mathbb{R}^2$ in the parameter space by the parabolas $\alpha_1 = |\alpha_0|^{1/2}\beta_1$, $\beta_1 \in \mathbb{R}$. The bifurcation curves in parameter space are obtained at those β_1 for which either Equation (2.8) or (2.9) has a multiple solution. The Implicit Function Theorem implies neither of these equations has a multiple solution for α_0, β_1 small. Thus, no bifurcation can occur near the α_0-axis in parameter space; that is, $|\beta_1| \geq \delta > 0$ is necessary for bifurcation. This means that Equations (2.8), (2.9) need only be considered for $v \in \mathbb{R}$, $|\beta_1| \geq \delta > 0$.

To avoid the noncompact region $|\beta_1| \geq \delta$, let us reparametrize the neighborhood of $(\alpha_0, \alpha_1) = (0,0)$ of interest by $\alpha_0 = \beta_0 \alpha_1^2$, $|\beta_0| \leq 1/\delta^2$. This gives a compact region for β_0. The parametrization of the solution in terms of α_1 should be $u = \alpha_1 w$ by Lemma 2.1. The Equation (2.1) is then equivalent to

$$(2.10) \qquad\qquad \beta_0 + w + w^2 + O(|\alpha_1|) = 0$$

and must be considered for $w \in \mathbb{R}$, $|\beta_0| \leq 1/\delta^2$ and α_1 small. The bifurcation curves will be $\alpha_0 = \beta_0 \alpha_1^2$ at those $\beta_0 = \beta_0(\alpha_1)$ for which Equation 2.10 has multiple solutions; that is, we must satisfy Equation (2.10) as well as

$$(2.11) \qquad\qquad 1 + 2w + O(|\alpha_1|) = 0.$$

For $\alpha_1 = 0$, these equations have the unique solution $w = -\frac{1}{2}$, $\beta_0 = \frac{1}{4}$. Thus, for α_1 small there is a unique solution $\beta_0^*(\alpha_1)$, $w^*(\alpha_1)$ of Equation (2.10), (2.11), $\beta_0^*(0) = \frac{1}{4}$, $\alpha^*(0) = -\frac{1}{2}$, and the bifurcation curve is $\alpha_0 = \beta_0^*(\alpha_1)\alpha_1^2$ with the solution along the bifurcation curve being $u = \alpha_1 w^*(\alpha_1)$. In Figure 2.2, we have drawn the curve $\Gamma : \alpha_0 = \beta_0^*(\alpha_1)\alpha_1^2$. On the left of this curve, there are two solutions of Equation (2.1) and on the right there are no solutions. In Figure 2.1, the dotted curve represents $u = \alpha_1 w^*(\alpha_1)$. Notice that all functions may be computed by successive approximations since they were obtained as an application of the Implicit Function Theorem.

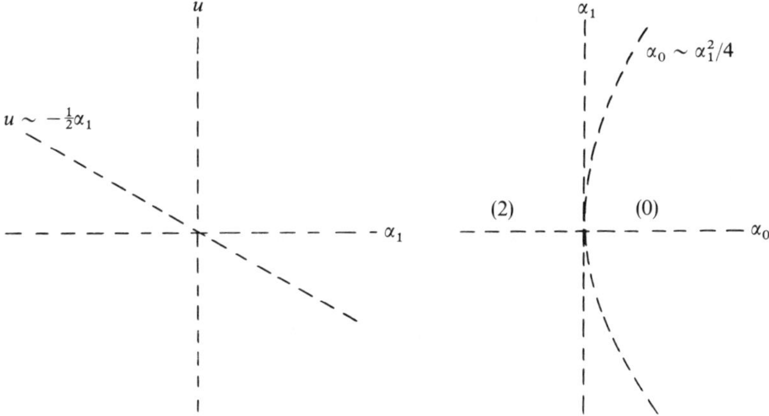

Figure 2.2

The solutions of Equation (2.1) away from the bifurcation curve may also be computed by successive approximations. In fact, if β_0 is not equal to $\beta_0^*(\alpha_0)$, then Equation (2.10) has two simple roots and these may be computed by successive approximations.

In this rather complicated analysis, we have obtained no more qualitative information than could have been obtained more easily by the previous methods. However, the same method used here is easily adapted to higher order systems as we indicate in the next section.

7.3. Quadratic Nonlinearities I

In this section, we consider the system of equations

$$(3.1) \qquad \begin{aligned} &f(\alpha, u) = \alpha_0 k + \alpha_1 Lu + Q(u) + \text{h.o.t.} = 0, \qquad f(0, \alpha_1, 0) = 0, \\ &\text{h.o.t.} = O(|u|^3 + |\alpha_1 u|^2 + |\alpha_0|^2 + |\alpha_0 \alpha_1| + |\alpha_0 u|) \end{aligned}$$

as $\alpha_0, \alpha_1, u \to 0$, where $\alpha = (\alpha_0, \alpha_1) \in \mathbb{R}^2$, $u = (u_1, u_2) \in \mathbb{R}^2$, $f = (f_1, f_2) \in \mathbb{R}^2$,

$$
\begin{aligned}
&Q: \mathbb{R}^2 \to \mathbb{R}^2 \text{ is homogeneous quadratic} \\
&L \text{ is a } 2 \times 2 \text{ real matrix} \\
&k \in \mathbb{R}^2 \text{ is given.}
\end{aligned}
$$
(3.2)

Our objective is to determine all solutions of Equation (3.1) near $(\alpha_1, \alpha_2, u) = 0$ by using the scaling methods of the previous section. To obtain the analogue of Lemma 2.1, we need

$$(\text{H}_1) \quad Q(u) = 0 \quad \text{implies} \quad u = 0.$$

Lemma 3.1. *There is a neighborhood* $V \subset \mathbb{R}^4$ *of* $(\alpha, u) = (0,0)$ *and a constant* $\beta > 0$ *such that any solution of Equation* (3.1) *in* V *must satisfy*

$$|u| \le \beta(|\alpha_1| + |\alpha_0|^{1/2}).$$

Proof. As in the proof of Lemma 2.1, if the lemma is not true, there exist sequences $(\alpha_{0n}, \alpha_{1n}, u^n) \to 0$ as $n \to \infty$, such that $|\alpha_{0n}|^{1/2}/|u^n|$, $|\alpha_{1n}|/|u^n| \to 0$ as $n \to \infty$. Divide Equation (3.1) by $|u^n|^2$ to obtain $Q(u^n/|u^n|) \to 0$ as $n \to \infty$. Since $u^n/|u^n|$ has norm one, we may assume $u^n/|u^n| \to v$ as $n \to \infty$, $|v| = 1$. Thus, $Q(v) = 0$. Hypothesis (H_1) implies $v = 0$. This is a contradiction. $\square$

As for the scalar case, we consider special cases. Suppose $\alpha_1 = 0$; that is, consider the equation

$$
\begin{aligned}
&\alpha_0 k + Q(u) + \text{h.o.t.} = 0 \\
&\text{h.o.t.} = O(|u|^3 + |\alpha_0 u| + |\alpha_0|^2)
\end{aligned}
$$
(3.3)

If $u = |\alpha_0|^{1/2} v$, then Lemma 3.1 implies Equation (3.3) is equivalent to finding all solutions $v \in \mathbb{R}^2$ of

$$(3.4) \qquad k + Q(v) + O(|\alpha_0|^{1/2}) = 0 \quad \text{for } \alpha_0 > 0 \text{ small}$$
$$(3.5) \qquad -k + Q(v) + O(|\alpha_0|^{1/2}) = 0 \quad \text{for } \alpha_0 < 0 \text{ small.}$$

If we hope to find all solutions of Equations (3.4), (3.5) for α_0 small without imposing some further conditions on the cubic and higher order terms in u, we must be sure that all solutions of Equations (3.4), (3.5) for $\alpha_0 = 0$ are simple in order to be able to apply the Implicit Function Theorem. Therefore, we impose

$$(\text{H}_2) \quad \text{If } Q(v) \pm k = 0, \text{ then } \det \partial Q(v)/\partial v \ne 0.$$

We have now proved the following result.

Lemma 3.2. *If Hypotheses* (H_1), (H_2) *are satisfied, then there is a neighborhood* $V \subset \mathbb{R}^3$ *of* $(\alpha_0, u) = (0, 0)$ *such that any solution of Equation* (3.3) *in* V *is given by* $|\alpha_0|^{1/2} v_j^*(\alpha_0)$ *where* $v_j^*(\alpha_0)$ *are the solutions of Equations* (3.4), (3.5).

It is no longer possible to say that α_0 is a bifurcation point because $u \in \mathbb{R}^2$. The number of solutions for $\alpha_0 > 0$ depends on the number of solutions of

$$(3.6) \qquad\qquad Q(v) + k = 0$$

and the number of solutions for $\alpha_0 < 0$ depends on the number of solutions of

$$(3.7) \qquad\qquad Q(v) - k = 0.$$

For example, if $k = (1, 1)$ and $Q(v) = (v_1^2 + \beta v_2^2, -\beta v_1^2 - v_2^2), \beta > 0$, then there are going to be no solutions of either Equation (3.6) or Equation (3.7). If $k = (1, 1)$ and $Q(v) = (v_1^2 + \beta v_2^2, \beta v_1^2 + v_2^2)$, then there are no solutions for $\alpha_0 > 0$ and four solutions for $\alpha_0 < 0$. If $k = (1, 1)$ and $Q(v) = (v_1 v_2, (v_1 + v_2)(v_1 - v_2))$, then there are four solutions for $\alpha_0 > 0$ and $\alpha_0 < 0$. Some typical solution versus α_0 curves are shown in Figure 3.1.

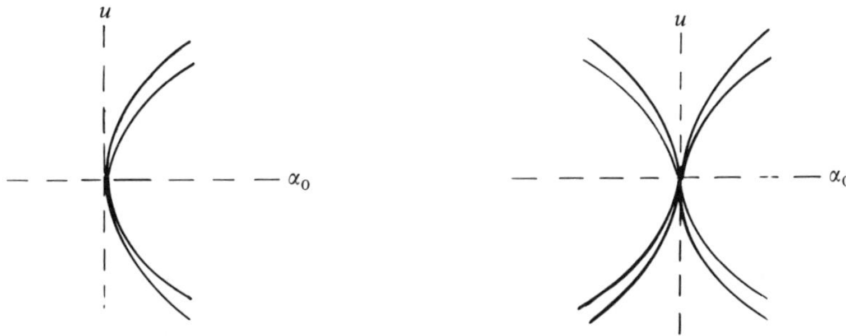

Figure 3.1

If $\alpha_0 = 0$, then Equation (3.1) is equivalent to the equation

$$(3.8) \qquad \begin{aligned} \alpha_1 L u + Q(u) + \text{h.o.t.} &= 0 \\ \text{h.o.t.} &= O(|u|^3 + |\alpha_1 u|^2 + |\alpha_1|^2). \end{aligned}$$

If $u = \alpha_1 w$, then Lemma 3.1 implies Equation (3.8) is equivalent to finding all solutions $w \in \mathbb{R}^2$ of the equation

$$(3.9) \qquad Lw + Q(w) + O(|\alpha_1|) = 0, \quad \alpha_1 \text{ small.}$$

As before, if we hope to find all solutions of Equation (3.9) for α_1 small without imposing conditions on the cubic and higher order terms in w, we must require

(H_2')   If $Lw + Q(w) = 0$, then $\det[L + \partial Q(w)/\partial w] \neq 0$.

We can then prove

Lemma 3.3. *If Hypotheses* (H_1), (H_2') *are satisfied, then there is a neighborhood* $V \subset \mathbb{R}^3$ *of* $(\alpha_1, u) = (0, 0)$ *such that any solution of Equation* (3.8) *in* V *is given by* $\alpha_1 w_j^*(\alpha_1)$ *where* $w_j^*(\alpha_1)$ *are the solutions of Equation* (3.9).

As for Equation (3.3), the point $\alpha_1 = 0$ may or may not be a bifurcation point. However, in contrast to Equation (3.3), the number of solutions will not depend on the sign of α_1.

A typical solution versus α_1 curve for Equation (3.8) is shown in Figure 3.2.

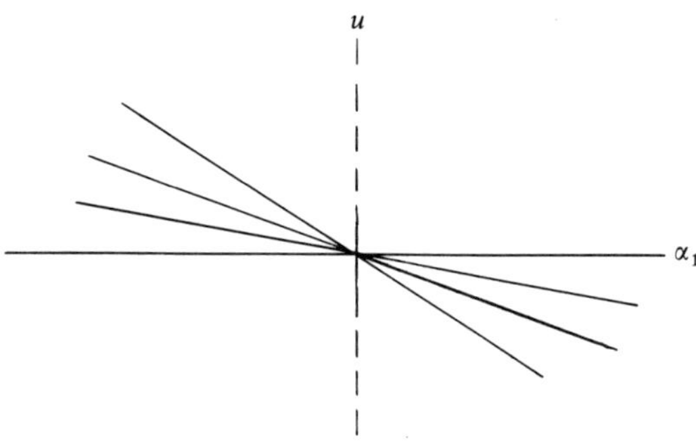

Figure 3.2

Let us now treat the general case allowing both α_0 and α_1 to vary. As in the previous section, we first parametrize a neighborhood of (α_0, α_1) by the parabolas $\alpha_1 = |\alpha_0|^{1/2}\beta_1, \beta_1 \in \mathbb{R}$ and let $u = |\alpha_0|^{1/2}v$. From Lemma 3.1, determining small solutions $(\alpha_0, \alpha_1, u) \in \mathbb{R}^4$ of Equation (3.1) is equivalent to determining all solutions $(\beta_1, v) \in \mathbb{R}^3$ of

$$(3.10a) \qquad k + \beta_1 Lv + Q(v) + O(|\alpha_0|^{1/2}) = 0 \quad \text{for } \alpha_0 > 0 \text{ small,}$$

$$(3.10b) \qquad -k + \beta_1 Lv + Q(v) + O(|\alpha_0|^{1/2}) = 0 \quad \text{for } \alpha_0 < 0 \text{ small.}$$

The bifurcation curves in parameter space are determined at those values of β_1 for which either Equation (3.10a) or Equation (3.10b) has a multiple solution. To avoid having infinitely many bifurcations, these equations should not have a sequence of solutions $\beta_{1j} \to 0$ as $j \to \infty$. Therefore, we assume (H_2) is satisfied. In this case, the Implicit Function Theorem implies every solution of Equation (3.10a), (3.10b) are simple for β_1, α_0 sufficiently small. Thus, no bifurcations occur in parameter space in the region $|\beta_1| \leq \delta, \delta > 0$ small.

As before, to avoid the noncompact region $|\beta_1| \geq \delta > 0$, we rescale by $\alpha_0 = \beta_0 \alpha_1^2$, $u = \alpha_1 w$ to see that Equation (3.1) is equivalent to

$$(3.11) \qquad h(\beta_0, w, \alpha_1) \stackrel{\text{def}}{=} \beta_0 k + Lw + Q(w) + O(|\alpha_1|)) = 0.$$

For Equation (3.11), we must determine all solutions for $w \in \mathbb{R}^2$, $|\beta_0| \leq 1/\delta^2$ and α_1 small. The bifurcation curves are determined by the multiple solutions of Equation (3.11); that is, the simultaneous solutions of Equation (3.11) and the equation

$$(3.12) \qquad \det[\partial h(\beta_0, w, \alpha_1)/\partial w] = 0.$$

Our next hypothesis is

(H$_3$)   If $h(\beta_0, w, 0) = 0$, $\Delta(w) \stackrel{\text{def}}{=} \det(L + \partial Q(w)/\partial w) = 0$, then
$\det \partial(h, \Delta)/\partial(\beta_0, w)|_{(\beta_0, w, 0)} \neq 0$.

If (H$_3$) is satisfied and (β_0^0, w^0) is a solution of Equations (3.11), (3.12) for $\alpha_1 = 0$, then the Implicit Function Theorem implies there is an $\varepsilon > 0$ such that Equations (3.11), (3.12) have a unique C^1-solution $\beta_0^*(\alpha_1)$, $w^*(\alpha_1)$ for $|\alpha_1| < \varepsilon$, $\beta_0^*(0) = \beta_0$, $w_0^*(0) = w^0$. Furthermore, there are only a finite number of such solutions for $|\alpha_1| < \varepsilon$.

Notice that Hypothesis (H$_3$) implies Hypothesis (H$_2'$). We also need the following implication of Hypothesis (H$_3$). By (H$_3$), some element of the matrix

$$\frac{\partial h(\beta_0, w, 0)}{\partial w} = L + \frac{\partial Q(w)}{\partial w}$$

is nonzero at each solution β_0^0, w^0 of Equations (3.11), (3.12). Suppose without loss of generality that

$$(3.13) \qquad \frac{\partial h_1(\beta_0^0, w^0, 0)}{\partial w_1} \neq 0$$

where $h = (h_1, h_2)$, $w = (w_1, w_2)$. By the Implicit Function Theorem, there is a unique C_1-function $\phi(\beta_0, w_2, \alpha_1)$ defined in a neighborhood of $(\beta_0, w, \alpha_1) = (\beta_0^0, w^0, 0)$, $\phi(\beta_0^0, w_2^0, 0) = w_1^0$, such that

$$(3.14) \qquad h_1(\beta_0, \phi(\beta_0, w_2, \alpha_1), w_2, \alpha_1) = 0.$$

Thus, the Equation (3.11) is equivalent to the equation

$$(3.15) \qquad H(\beta_0, w_2, \alpha_1) \stackrel{\text{def}}{=} h_2(\beta_0, \phi(\beta_0, w_2, \alpha_1), w_2, \alpha_1) = 0$$

in a neighborhood of $(\beta_0^0, w^0, 0)$.

In the remainder of this section, Figure 3.3 will be very useful to understand the very simple ideas involved in the technical discussion. In the figure for

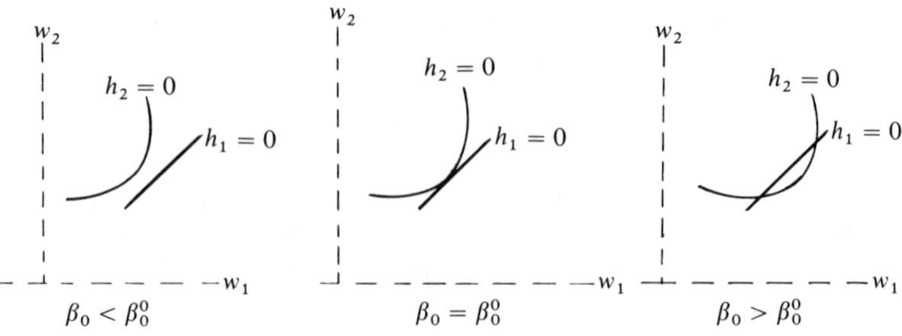

Figure 3.3

$\beta = \beta_0^0$, we have drawn the simultaneous intersection of the curves $h_1 = 0$, $h_2 = 0$. By the introduction of the function ϕ, we are able to use $h_1 = 0$ as a new coordinate and study the variation in h_2 as we move along $h_1 = 0$. The next lemma and the proof of Theorem 3.7 below show that these curves vary qualitatively in β_0 as indicated in the figure. The sign of some constant determines whether the picture for $\beta_0 < \beta_0^0$ and $\beta_0 > \beta_0^0$ should be reversed.

The following lemma uses the above notation, but does not require (H$_3$) to be satisfied.

Lemma 3.4. *Suppose β_0^0, w^0 is a solution of Equations (3.11), (3.12) and condition (3.13) is satisfied. If $H(\beta_0, w_2, \alpha_1)$ is defined in (3.15), then $H(\beta_0^0, w_2^0, 0) = 0$ and*

$$(3.16) \qquad \det \frac{\partial(h, \Delta)}{\partial(\beta_0, w)} = -\left(\frac{\partial h_1}{\partial w_1}\right)^2 \frac{\partial H}{\partial \beta_0} \frac{\partial^2 H}{\partial w_2^2}$$

at the point $(\beta_0^0, w_2^0, 0)$.

Proof. Let $\psi(\beta_0, w_2, \alpha_1) = (\beta_0, \phi(\beta_0, w_2, \alpha_1), w_2, \alpha_1)$. Since $\det(\partial h/\partial w) = 0$ and $\partial \phi/\partial w_2 = -(\partial h_1/\partial w_2)(\partial h_1/\partial w_1)^{-1}$ at $\psi(\beta_0^0, w_2^0, 0)$, it follows that

$$\det \frac{\partial(h, \Delta)}{\partial(\beta_0, w)} = \det \begin{pmatrix} \dfrac{\partial h_1}{\partial w_1} & \dfrac{\partial h_1}{\partial w_2} & k_1 \\[2mm] \dfrac{\partial h_2}{\partial w_1} & \dfrac{\partial h_2}{\partial w_2} & k_2 \\[2mm] \dfrac{\partial \Delta}{\partial w_1} & \dfrac{\partial \Delta}{\partial w_2} & 0 \end{pmatrix}$$

$$= k_1 \det \begin{pmatrix} \dfrac{\partial h_2}{\partial w_1} & \dfrac{\partial h_2}{\partial w_2} \\[2mm] \dfrac{\partial \Delta}{\partial w_1} & \dfrac{\partial \Delta}{\partial w_2} \end{pmatrix} - k_2 \det \frac{\partial(h_1, \Delta)}{\partial w}$$

at $\psi(\beta_0^0, w_2^0, 0)$. Since $\det \partial h/\partial w = 0$, this implies

$$\det \frac{\partial(h, \Delta)}{\partial(\beta_0, w)} = \left[k_1 \left(\frac{\partial h_2}{\partial w_1} \right) \left(\frac{\partial h_1}{\partial w_1} \right)^{-1} - k_2 \right] \det \frac{\partial(h_1, \Delta)}{\partial w}$$

at $\psi(\beta_0^0, w_2^0, 0)$. By direct calculation, one shows that

$$
\begin{aligned}
\partial^2 H/\partial w_2^2 &= (\partial h_1/\partial w_1)^{-2} \det[\partial(h_1, \Delta)/\partial w] \\
\partial H/\partial \beta_0 &= (\partial h_1/\partial w_1)^{-1} \det[\partial h/\partial(w_1, \beta_0)]
\end{aligned}
\tag{3.17}
$$

at $(\beta_0^0, w_2^0, 0)$. These relations imply relation (3.16) is satisfied at $\psi(\beta_0^0, w_2^0, 0)$. This proves the lemma. $\square$

Corollary 3.5. *Suppose β_0^0, w_0^0 is a solution of Equations* (3.11), (3.12), *and condition* (3.13) *is satisfied. If $H(\beta_0, w_2, \alpha_1)$ is defined in* (3.15), *then condition* (H_3) *is equivalent to $\partial H/\partial \beta_0 \neq 0$, $\partial^2 H/\partial w_2^2 \neq 0$ at the point $(\beta_0^0, w_2^0, 0)$.*

Proof. This is a consequence of formula (3.16). $\square$

Remark 3.6. Corollary 3.5 has the following interesting interpretation. If $h_1 = 0$ is used as one of the coordinate axes in the plane in a neighborhood of $(\beta_0^0, w_2^0, 0)$, then Condition (H_3) is equivalent to the fact that the function $H(\beta_0, w_2, \alpha_1)$ has an extreme value at $(\beta_0^0, w_2^0, 0)$ which is generic in the sense that $\partial^2 H/\partial w_2^2 \neq 0$ at this point and furthermore the derivative of this extreme value with respect to β_0 is different from zero at $(\beta_0^0, w_2^0, 0)$. This geometric interpretation of Condition (H_3) is very useful in the verification of the condition.

We are now in a position to prove the following theorem.

Theorem 3.7. *For Equation* (3.1), *suppose*

(H_1) $Q(u) = 0$ *implies* $u = 0$
(H_2) *If $Q(v) \pm k = 0$ then* $\det \partial Q(v)/\partial v \neq 0$
(H_3) *If*

$$
\begin{aligned}
h^0(\beta_0, w) &\overset{\text{def}}{=} \beta_0 k + Lw + Q(w) = 0 \\
\Delta(w) &\overset{\text{def}}{=} \det(L + \partial Q/\partial w) = 0
\end{aligned}
\tag{3.18}
$$

then $\det \partial(h^0, \Delta)/\partial(\beta_0, w) \neq 0$.

Under these hypotheses, there are a finite number of curves of the approximate form

$$\alpha_0 \sim \beta_0^0 \alpha_1^2$$

on which a bifurcation occurs for Equation (3.1). Along each bifurcation curve, the solution is given approximately by

$$u \sim w^0 \alpha_1$$

where (β_0^0, w^0) are solutions of Equation (3.18). Furthermore, if each of the bifurcation curves is distinct in the (α_0, α_1)-plane, the number of solutions of Equation (3.1) changes by exactly two as a curve is crossed.

Proof. Apply the above scaling technique to obtain Equation (3.11). Let (β_0^0, w^0) be a solution of Equation (3.18). Hypothesis (H_3) implies some element of the matrix $\partial h(\beta_0^0, w^0, 0)/\partial w$ is nonzero. Without loss in generality, we may assume Relation (3.13) is satisfied, and reduce the discussion of the solutions of Equation (3.11) to Equation (3.15). For definiteness, let us also suppose $\det \partial(h_1, \Delta)/\partial w > 0$. Then relation (3.17) implies the function H in Equation (3.15) has a unique minimum near $(\beta_0^0, w_2^0, 0)$. In fact, from Corollary 3.5, the Implicit Function Theorem implies there is a C^1-function $\gamma(\beta_0, \alpha_1)$, $\gamma(\beta_0^0, 0) = w_2^0$, such that

$$\frac{\partial H}{\partial w_2}(\beta_0, \gamma(\beta_0, \alpha_1), \alpha_1) = 0$$

for (β_0, α_1) near $(\beta_0^0, 0)$. If $G(\beta_0, \alpha_1) = H(\beta_0, \gamma(\beta_0, \alpha_1), \alpha_1)$, then G is the unique minimum of H near $(\beta_0^0, w^0, 0)$. If $G(\beta_0, \alpha_1) > 0$, there are no solutions of $H(\beta_0, w_2, \alpha_1)$ for w_2 near w_2^0, and, thus, no solutions of Equation (3.11) for w near w^0. If $G(\beta_0, \alpha_1) < 0$, there are two solutions of $H(\beta_0, w_2, \alpha_1)$ for w_2 near w_2^0 and, thus, two solutions of Equation (3.11) for w near w^0. Hence, the values of the parameter for which a bifurcation could possibly occur near w^0 are those values of (β_0, α_1) for which $G(\beta_0, \alpha_1) = 0$. From Corollary 3.5,

$$(3.19) \qquad \frac{\partial G}{\partial \beta_0}(\beta_0^0, 0) = \frac{\partial H}{\partial \beta_0}(\beta_0^0, w_2^0, 0) \neq 0.$$

The Implicit Function Theorem implies there is a C^1-function $\delta(\alpha_1)$ such that $\delta(0) = \beta_0^0$, $G(\delta(\alpha_1), \alpha_1) = 0$ for α_1 small. The solution of Equation (3.11) along the curve $\beta_0 = \delta(\alpha_1)$ is given by

$$(\phi(\beta_0, \gamma(\delta(\alpha_1), \alpha_1), \alpha_1), \gamma(\delta(\alpha_1), \alpha_1)) \overset{\text{def}}{=} \psi(\alpha_1)$$

where ϕ is given in Equation (3.14). That each point of the curve $\beta_0 = \delta(\alpha_1)$ is a bifurcation point follows from Relation (3.19). This completes the proof of the theorem. $\square$

Remark 3.8. The direction of bifurcation (i.e. for β_0 above or below the curve $\delta(\alpha_1)$) may also be determined from the proof of Theorem 3.7. If the extreme value of H is a minimum, then there are two bifurcating solutions when $\partial G/\partial\beta_0$ and $\beta_0 - \delta(\alpha_1)$ have opposite signs. If the extreme value of H is a maximum, then two solutions exist when $\partial G/\partial\beta_0$ and $\beta_0 - \delta(\alpha_1)$ have the same sign. Also, Relations 3.17 imply the extreme value is a minimum when $\det \partial(h_1, \Delta)/\partial w > 0$ and a maximum when $\det \partial(h_1, \Delta)/\partial w < 0$.

Remark 3.9. Theorem 3.7 was proved by a systematic application of the implicit function theorem. Therefore, all functions can be computed by successive approximations.

Remark 3.10. The application of Theorem 3.7 is very simple. The Hypotheses (H_1), (H_2), (H_3) involve only the quadratic terms $Q(u)$, the linear terms Lu and the vector k in Equation (3.1). The first step in the application is to verify these hypotheses. After this is done, one draws the bifurcation diagram for the simpler equations

$$(3.20) \qquad Q(u) + \alpha_1 Lu + \alpha_0 k = 0.$$

Theorem 3.7 asserts that the bifurcation diagram for Equation (3.1) is qualitatively and approximately the same as the one for the truncated Equation (3.20). Moreover, the bifurcation curves in the parameter space (α_0, α_1) for Equation (3.1) are tangent at the origin to the ones for Equation (3.20).

To illustrate the application of the method, consider the situation where Equation (3.1) without the higher order terms is given by

$$(3.21) \qquad \begin{aligned} u_1^2 + vu_2^2 - \alpha_1 u_1 + \alpha_0 &= 0 \\ \mu u_1 u_2 - \alpha_1 u_2 + \alpha_0 &= 0 \end{aligned}$$

with $v > 0$, $\mu > 0$ given constants. Hypotheses (H_1) is always satisfied. To verify (H_2), one first observes that only the minus sign in (H_2) needs to be considered. An easy computation shows that all solutions of the equations

$$\begin{aligned} u_1^2 + vu_2^2 - 1 &= 0 \\ \mu u_1 u_2 - 1 &= 0 \end{aligned}$$

are simple if and only if $v \neq \mu^2/4$. Thus, (H_2) is equivalent to $v \neq \mu^2/4$. Therefore, assume (v, μ) satisfy this inequality. Hypothesis (H_3) is always satisfied since one of the equations defines an ellipse, the other one defines a hyperbola, and β_0 changes the axes. The bifurcation curves are approximately $\alpha_0 = \beta_0 \alpha_1^2$ for those β_0 for which the curves defined by the

$$\begin{aligned} (u_1 - \tfrac{1}{2})^2 + vu_2^2 - \tfrac{1}{4} + \beta_0 &= 0 \\ u_2(\mu u_1 - 1) + \beta_0 &= 0 \end{aligned}$$

are tangent. If $v > \mu^2/4$, $\mu > 1$, it is not difficult to see that the bifurcation diagrams are those given in Figure 3.4. The discussions of the other case is left as an exercise.

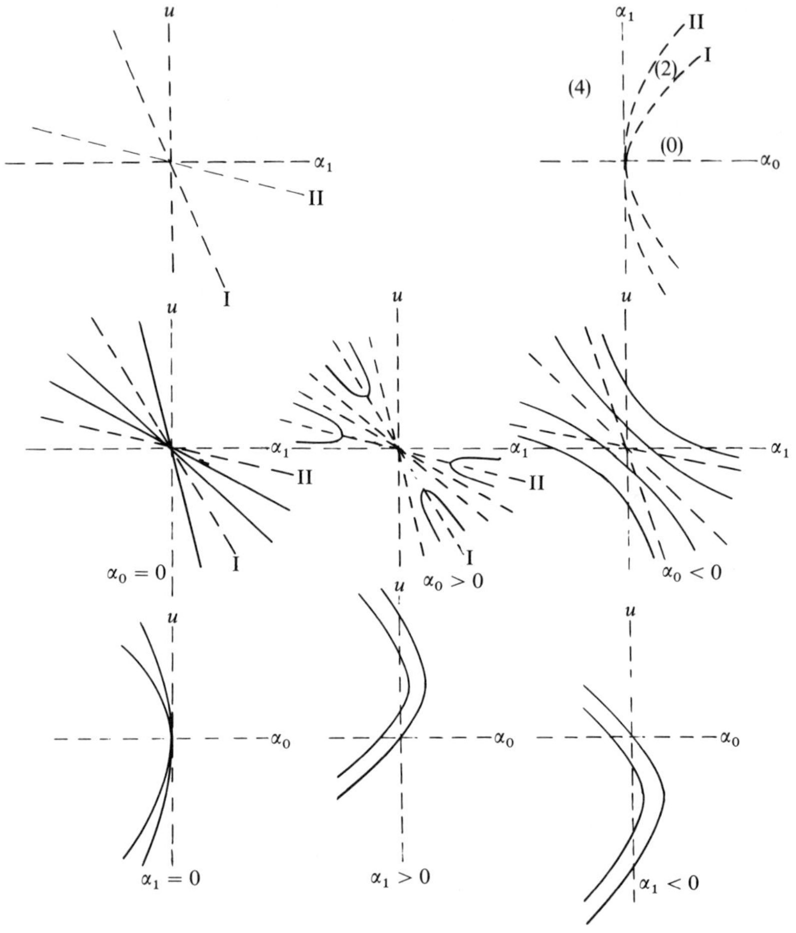

Figure 3.4

We conclude this section with a few remarks about other hypotheses which are equivalent to (H_1)–(H_3). Hypothesis (H_2) was motivated by scaling the parameters by $\alpha_1 = |\alpha_0|^{1/2}\beta_1$, $u = |\alpha_0|^{1/2}v$, and requiring that no bifurcation occur near $\beta_1 = 0$. After this, one rescales with $\alpha_0 = \beta_0\alpha_1^2$, $u = \alpha_1 w$. Hypothesis (H_3) requires a generic type of tangency of the resulting curves as described in Remark 3.6.

Suppose the process is reversed. First scale as $\alpha_0 = \beta_0\alpha_1^2$, $u = \alpha_1 w$ and require that no bifurcations occur near $\beta_0 = 0$. This leads to Hypotheses (H_2') stated before Lemma 3.2. Now scale $\alpha_1 = |\alpha_0|^{1/2}\beta_1$, $u = |\alpha_0|^{1/2}v$ and require that the resulting curves have the same generic tangency as before. This leads to the following Hypothesis (H_3'):

(H_3') If

$$h^0(\beta_1, v) \overset{\text{def}}{=} Q(v) + \beta_1 Lv \pm k = 0$$

$$\Delta(\beta_1, v) \overset{\text{def}}{=} \det(\partial Q(v)/\partial v + \beta_1 L) = 0$$

then $\det \partial(h^0, \Delta)/\partial(\beta_1, v) \neq 0$.

It is not difficult to show that (H_1), (H_2), (H_3) are equivalent to (H_1'), (H_2'), (H_3'). This remark can be important in the applications because the computations can be much easier for verifying one set of conditions as opposed to the other. This same remark applies to the remaining illustrations in this chapter and will not be stated explicitly.

7.4. Quadratic Nonlinearities II

In this section, we consider the system of equations

(4.1)
$$f(\alpha, u) = \alpha_1 L_1 u + \alpha_2 L_2 u + Q(u) + \text{h.o.t.} = 0$$
$$\text{h.o.t.} = O(|u|[|\alpha_1|^2 + |\alpha_2|^2 + |\alpha_1 \alpha_2| + |\alpha_1 u| + |\alpha_2 u| + |u|^2])$$

as $\alpha_1, \alpha_2, u \to 0$ where $\alpha = (\alpha_1, \alpha_2) \in \mathbb{R}^2$, $u = (u_1, u_2) \in \mathbb{R}^2$, $f = (f_1, f_2) \in \mathbb{R}^2$

(4.2)
$$Q: \mathbb{R}^2 \to \mathbb{R}^2 \text{ is homogeneous quadratic}$$
$$L_1, L_2 \text{ are } 2 \times 2 \text{ real matrices.}$$

Notice that $f(\alpha, 0) = 0$ for all α. Our objective is to apply scaling techniques to determine all solutions of Equation (4.1) near $(\alpha, u) = (0, 0)$. Since the proofs of the results are very similar to the ones of the previous section, many details will be omitted.

Lemma 4.1. *If*

(H_1) $Q(u) = 0$ *implies* $u = 0,$

then there is a neighborhood $V \subset \mathbb{R}^4$ of $(\alpha, u) = (0, 0)$ and a constant $\beta > 0$ such that all solutions of Equation (4.1) in V must satisfy

$$|u| \leq \beta(|\alpha_1| + |\alpha_2|).$$

Proof. The proof is almost the same as the proof of Lemma 3.1. $\square$

If $u = \alpha_1 v, \alpha_2 = \alpha_1 \beta_2$, then finding solutions of Equation (4.1) near $(u, \alpha) = (0,0)$ is equivalent to finding solutions of the equation

$$(4.3) \qquad L_1 v + \beta_2 L_2 v + Q(v) + O(|\alpha_1|) = 0$$

for $\beta_2 \in \mathbb{R}$, $v \in \mathbb{R}^2$, α_1 small.

The bifurcation curves in parameter space (α_1, α_2) are determined by $(\alpha_1, \alpha_1 \beta_2)$ at those values of β_2 for which Equation (4.3) has a multiple solution. To avoid having infinitely many bifurcations, these equations should not have a sequence of solutions $\beta_{2j} \to 0$ as $j \to \infty$. Therefore, we assume

$$(H_2) \quad \text{If } L_1 v + Q(v) = 0, \text{ then } det[L_1 + \partial Q(v)/\partial v] \neq 0.$$

In this case, the Implicit Function Theorem implies every solution of Equation (4.3) is simple for (β_2, α_1) sufficiently small. Thus, no bifurcations occur in parameter space (α_1, α_2) in the region $|\beta_2| \leq \delta$, $\delta > 0$ small.

As before, to avoid the noncompact region $|\beta_2| \geq \delta > 0$, we rescale by $u = \alpha_2 w$, $\alpha_1 = \alpha_2 \beta_1$ to obtain the equivalent equation

$$(4.4) \qquad h(\beta_1, w, \alpha_2) \overset{\text{def}}{=} \beta_1 L_1 w + L_2 w + Q(w) + O(|\alpha_2|) = 0.$$

For Equation (4.4), we must determine all solutions for $w \in \mathbb{R}^2$, $|\beta_1| \leq 1/\delta$ and α_2 small. The bifurcation curves in parameter space are determined at those points $(\alpha_2 \beta_1, \alpha_2)$ for which Equation (4.4) has a multiple solution. Thus, we must determine the simultaneous solutions of Equation (4.4) and the equation

$$(4.5) \qquad \det \partial h(\beta_1, w, \alpha_2)/\partial w = 0.$$

Following exactly the same reasoning as in the proof of Theorem 3.7, one obtains

Theorem 4.2. *For Equation* (4.1), *suppose*

$(H_1) \quad Q(u) = 0$ *implies* $u = 0$.

$(H_2) \quad$ *If* $L_1 v + Q(v) = 0$, *then* $\det(L_1 + \partial Q(v)/\partial v] \neq 0$.

$(H_3) \quad$ *If*

$$h^0(\beta_1, w) \overset{\text{def}}{=} \beta_1 L_1 w + L_2 w + Q(w) = 0$$

$$(4.6) \qquad \Delta(\beta_1, w) \overset{\text{def}}{=} \det[\beta_1 L_1 + L_2 + \partial Q(w)/\partial w] = 0$$

then $\det[\partial(h^0, \Delta)/\partial(\beta_1, w)] \neq 0$.

$(H_4) \quad$ *The eigenvalues of* (L_2, L_1) *are simple. If* β_1 *is an eigenvalue of* (L_2, L_1) *with eigenvector* z^0, *then* $\langle L_1 z^0, Q(z^0) \rangle \neq 0$ *where* $\langle , \rangle$ *is the inner product in* $\mathbb{R}^2$.

If (β_1^0, w^0) is a solution of Equation (4.6), then there is a bifurcation curve for Equation (4.1) of the approximate form

$$\alpha_1 \sim \beta_1^0 \alpha_2$$

If the bifurcation curves are distinct, the number of solutions changes by two as the curve is crossed. Along the bifurcation curve, the solution is given approximately by

$$u \sim w^0 \alpha_2$$

If (H_4) is satisfied, there are two curves of solutions of Equation (4.4) through $(\beta_1, w, \alpha_2) = (\beta_1, 0, 0)$.

Proof. Only a sketch of the proof is given since it is so similar to the proof of Theorem 3.7. Suppose (β_1^0, w^0) satisfy Equation (4.6). From (H_3), without loss in generality, we may assume $\partial h_1(\beta_1^0, w^0, 0)/\partial w_1 \neq 0$. Determine $\phi(\beta_1, w_2, \alpha_2)$, $\phi(\beta_1^0, w_2^0, 0) = w_1^0$, such that

$$h_1(\beta_1, \phi(\beta_1, w_2, \alpha_2), w_2, \alpha_2) = 0$$

and define

$$H(\beta_1, w_2, \alpha_2) \stackrel{\text{def}}{=} h_2(\beta_1, \phi(\beta_1, w_2, \alpha_2), w_2, \alpha_2).$$

At $(\beta_1, w_2, \alpha_2) = (\beta_1^0, w_2^0, 0)$, one shows that

$$
\begin{aligned}
H &= \partial H/\partial w_2 = 0 \\
\det[\partial(h, \Delta)/\partial(\beta_1, w)] &= -(\partial h_1/\partial w_1)^2(\partial H/\partial \beta_1)(\partial^2 H/\partial w_2^2) \\
\partial^2 H/\partial w_2^2 &= (\partial h_1/\partial w_1)^{-2}\det[\partial(h_1, \Delta)/\partial w] \neq 0 \\
\partial H/\partial \beta_1 &= (\partial h_1/\partial w_1)^{-1}\det[\partial h/\partial(w_1, \beta_1)] \neq 0.
\end{aligned}
$$

(4.6)′

The proof now follows exactly as in the proof of Theorem 3.7. There is a function $\gamma(\beta_1, \alpha_2)$, $\gamma(\beta_1^0, 0) = w_2^0$ such that

$$\frac{\partial H}{\partial w_2}(\beta_1, \gamma(\beta_1, \alpha_2), \alpha_2) = 0$$

for (β_1, α_2) near $(\beta_1^0, 0)$. If $G(\beta_1, \alpha_2) = H(\beta_1, \gamma(\beta_1, \alpha_2), \alpha_2)$, then G is the unique extreme value of H near $(\beta_1^0, w_2^0, 0)$. Suppose it is a minimum. This minimum being positive or negative determines whether there are no solutions or two solutions of Equation (4.4) near $(\beta_1^0, w^0, 0)$. Since

$$\frac{\partial G}{\partial \beta_1}(\beta_1^0, 0) = \frac{\partial H}{\partial \beta_1}(\beta_1^0, w_2^0, 0) \neq 0,$$

there is a function $\delta(\alpha_2)$, $\delta(0) = \beta_1^0$ such that $G(\delta(\alpha_2), \alpha_2) = 0$ near $\alpha_2 = 0$. Each point of the curve $\beta_1 = \delta(\alpha_2)$ is thus a bifurcation point. If $-\beta_1^0$ is an eigenvalue of (L_2, L_1), then (H_3) is not satisfied at $w = 0$, $\beta_1 = \beta_1^0$. From (H_4), one can apply the results of Section 5.5 (Theorem 5.3) in a neighborhood of $(\beta_1, w, \alpha_2) = (\beta_1^0, 0, 0)$. There will always be a bifurcation at β_1^0 with two curves of solutions through $(\beta_1^0, 0, 0)$. This completes the proof of the theorem. $\square$

Remark 4.3. Relation (4.6) shows that Hypothesis (H_3) has the same geometric interpretation as the one mentioned in Remark 3.6. As noted there, this can be useful in the verification of (H_3).

Remark 4.4. As in Remark 3.8, the direction of bifurcation is determined from the sign of $\partial G/\partial \beta_1$ and $\beta_1 - \delta(\alpha_2)$ where G, δ were given in the proof of Theorem 4.2.

Remark 4.5. Since the Implicit Function Theorem was used to prove Theorem 4.2, all functions can be computed by successive approximations. The approximate bifurcation diagrams are obtained from the truncated equations

$$(4.7) \qquad\qquad Q(u) + \alpha_1 L_1 u + \alpha_2 L_2 u = 0$$

as noted in Remark 3.10.

Remark 4.6. The bifurcation diagrams for Equation (4.1) are completely different from the ones for Equation (3.1). In fact, the bifurcation curves in (α_1, α_2)-space are approximately straight lines and the solution of Equation (4.1) along a bifurcation curve is almost a linear function.

To illustrate the above result, let us consider the example

$$(4.8) \qquad \begin{aligned} u_1^2 + v u_2^2 - \alpha_1 u_1 - \alpha_2 v \sigma u_2 &= 0 \\ \mu u_1 u_2 - \alpha_1 u_2 - \alpha_2 u_2 &= 0 \end{aligned}$$

for $\mu > 0$, $v > 0$, $\sigma > 0$. Hypothesis (H_1) is always satisfied and (H_2) is equivalent to $\mu \neq 1$.

To verify (H_3), we must investigate the multiple solutions of the equations

$$\left(u_1 - \frac{\beta_0}{2}\right)^2 + v\left(u_2 - \frac{\sigma}{2}\right)^2 - \frac{\beta_0^2}{4} - \frac{v^2 \sigma^2}{4} = 0$$

$$(4.9) \qquad\qquad u_2(\mu u_1 - \beta_0 - 1) = 0.$$

If $\mu > 1$, the complete bifurcation diagram for the truncated equations (4.8) is shown in Figure 4.1.

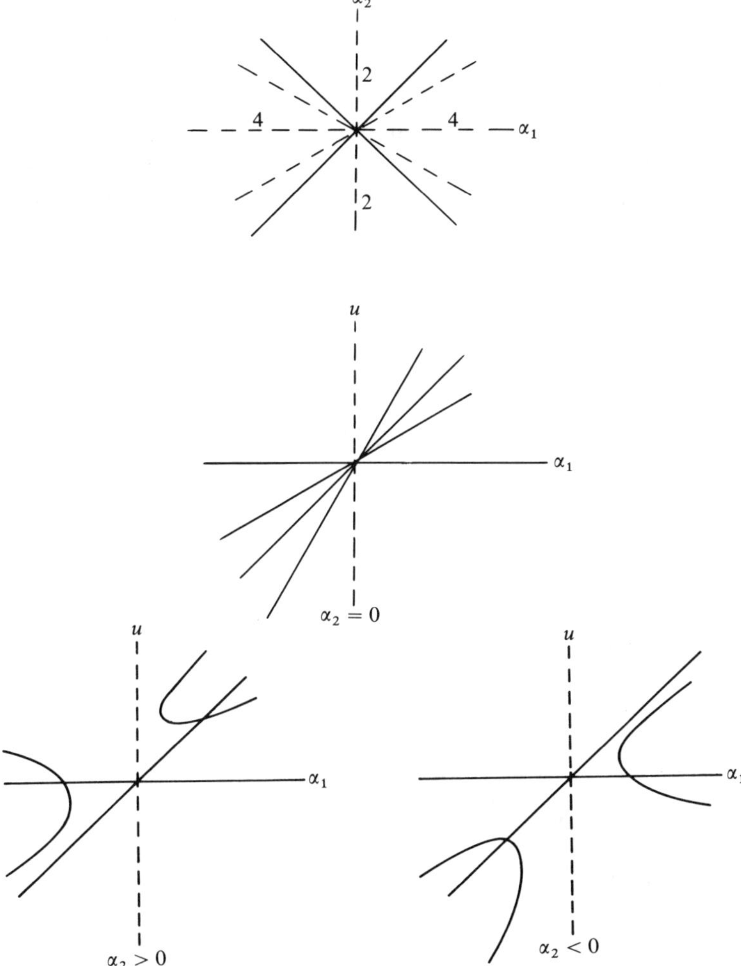

Figure 4.1

It is not difficult to verify that (H_3) for $\alpha_2 \neq 0$ is satisfied everywhere except at those β_0 corresponding to the point where the straight line and parabolic arcs meet and where the two straight lines intersect. From the hypothesis on the perturbation in f, the Equation (4.1) has the solution $u = 0$. Thus, the point at which the straight lines intersect remains after the higher order terms are included. The other point of intersection will not be preserved under perturbations required only to vanish at $u = 0$ and be of higher order than the ones in (4.8). To preserve this part of the diagram, one must require that the second function in (4.1) vanish when $u_2 = 0$. When $u_2 = 0$ is factored from the second equation, then the problem is generic. The dotted line represents the points (α_1, α_2) corresponding to the transversal intersections of the lines $u_1 = u_2 = 0$ and $\mu u_1 = \alpha_1 + \alpha_2$ and the parabolic arcs.

This example illustrates the peculiarities that occur in bifurcation theory. One cannot ignore the special structure of the equations. The Hypotheses (H_1)–(H_3) may fail at isolated points and auxiliary arguments are needed to verify that the truncated equations also give the correct results at these special points. Similar remarks apply when there is symmetry in the equations as we shall see later.

This example is easily modified to eliminate the intersection of the parabola with the line in the bifurcation diagram. It is left to the reader to show that this occurs for the equation

$$u_1^2 + vu_2^2 - \alpha_1 u_2 = 0$$
$$\mu u_1 u_2 - \alpha_1 u_2 - \alpha_2 u_1 = 0.$$

Of course, one must assume $u = 0$ satisfies the equations with higher order terms.

7.5. Cubic Nonlinearities I

In this section, we consider the system of equations

$$(5.1) \quad \begin{aligned} f(\alpha, u) &= \alpha_1 L_1 u + \alpha_2 L_2 u + C(u) + \text{h.o.t.} = 0 \\ \text{h.o.t.} &= O(|u|[|\alpha_1|^2 + |\alpha_2|^2 + |\alpha_1\alpha_2| + |\alpha_1 u| + |\alpha_2 u| + |u|^3]) \end{aligned}$$

as $\alpha_1, \alpha_2, u \to 0$ where $\alpha = (\alpha_1, \alpha_2) \in \mathbb{R}^2$, $u = (u_1, u_2) \in \mathbb{R}^2$, $f = (f_1, f_2) \in \mathbb{R}^2$,

$$(5.2) \quad \begin{aligned} C &: \mathbb{R}^2 \to \mathbb{R}^2 \text{ is a homogeneous cubic} \\ L_1, L_2 &\text{ are } 2 \times 2 \text{ real matrices.} \end{aligned}$$

Notice that $f(\alpha, 0) = 0$. Following the same ideas as before, one can prove the following result.

Theorem 5.1. *For Equation* (5.1), *suppose*

(H_1) $C(u) = 0$ *implies* $u = 0$.
(H_2) *If* $\pm L_1 v + C(v) = 0$, *then* $\det[\pm L_1 + \partial C(v)/\partial v] \neq 0$.
(H_3) *If*

$$(5.3) \quad \begin{aligned} h_\pm^0(\beta_1, w) &= \pm \beta_1 L_1 w \pm L_2 w + C(w) = 0 \\ \Delta_\pm(\beta_1, w) &\overset{\text{def}}{=} \det[\pm \beta_1 L_1 \pm L_2 + \partial C(w)/\partial w] = 0 \end{aligned}$$

then $\det[\partial(h_\pm^0, \Delta_\pm)/\partial(\beta_1, w)] \neq 0$.

Under these hypotheses, there are a finite number of curves of the approximate form

$$\alpha_1 \sim \beta_1^0 \alpha_2$$

on which a bifurcation occurs for Equation (5.1). Along each bifurcation curve, the solution is given approximately by

$$u \sim w^0 |\alpha_2|^{1/2}$$

where (β_1^0, w_0) are solutions of Equations (5.3) using (h_+, Δ_+) if $\alpha_1 > 0$ and (h_-, Δ_-) if $\alpha_1 < 0$. Furthermore, if each of these bifurcation curves are distinct in the (α_1, α_2)-plane, the number of solutions of Equation (5.1) changes by exactly two as a curve is crossed.

Proof. It is the same type as before except using the a priori bounds $|u| \le \beta(|\alpha_1|^{1/2} + |\alpha_2|^{1/2})$ and the scalings $u = |\alpha_1|^{1/2} v$, $\alpha_2 = \beta_2 \alpha_1$ and $u = |\alpha_2|^{1/2} w$, $\alpha_1 = \beta_1 \alpha_2$. $\square$

If β_1^0 is an eigenvalue of (L_2, L_1), then (H_3) is not satisfied at $\beta_1 = \beta_1^0$, $w = 0$. It is natural to assume that β_1^0 is a simple eigenvalue of (L_2, L_1), but this is not enough information to describe the nature of the bifurcations for (β_1, w, α_2) near $(\beta_1^0, 0, 0)$ since there are no quadratic terms in w. The higher order terms may play an important role unless additional symmetry conditions prevail.

Let us analyze a specific example in detail. Consider the equations (5.1) with

$$(5.4) \quad C(u) = \begin{bmatrix} u_1^3 + \mu u_1 u_2^2 \\ v u_1^2 u_2 + u_2^3 \end{bmatrix}, \qquad L_1 = -\begin{bmatrix} 1 & 0 \\ 0 & 1 \end{bmatrix}, \qquad L_2 = -\begin{bmatrix} 0 & 0 \\ 0 & 1 \end{bmatrix}$$

where $\mu > 0$, $v > 0$. Theorem 5.1 asserts that the complete bifurcation diagram is approximately the same as the one obtained by considering the bifurcation diagram for the equation

$$(5.5) \qquad C(u) + \alpha_1 L_1 u + \alpha_2 L_2 u = 0$$

provided the hypotheses (H_1), (H_2), (H_3) are satisfied. For the special case (5.4), Equations (5.5) are the same as

$$(5.6) \qquad \begin{aligned} u_1(u_1^2 + \mu u_2^2 - \alpha_1) &= 0 \\ u_2(v u_1^2 + u_2^2 - \alpha_1 - \alpha_2) &= 0. \end{aligned}$$

Since $\mu > 0$, $v > 0$, Hypothesis (H_1) is satisfied. To verify (H_2), consider the equations

$$
\begin{aligned}
u_1(u_1^2 + \mu u_2^2 \pm 1) &= 0 \\
u_2(v u_1^2 + u_2^2 \pm 1) &= 0.
\end{aligned}
\tag{5.7}
$$

It is now easy to verify that (H_2) is satisfied if $v \neq 1$ or $\mu \neq 1$. Thus, we assume $\mu \neq 1$, $v \neq 1$. To investigate the bifurcation curves, let us suppose first that $\alpha_2 \geq 0$. To verify (H_3), we must consider multiple solutions of the equations

$$
\begin{aligned}
w_1(w_1^2 + \mu w_2^2 \pm \beta_1) &= 0 \\
w_2(v w_1^2 + w_2^2 - \beta_1 \pm 1) &= 0.
\end{aligned}
\tag{5.8}
$$

To see that (H_3) imposes some essential restrictions on μ, v, let us consider the simultaneous intersection of the ellipses

$$
\begin{aligned}
w_1^2 + \mu w_2^2 \pm \beta_1 &= 0 \\
v w_1^2 + w_2^2 - \beta_1 \pm 1 &= 0.
\end{aligned}
\tag{5.9}
$$

To have isolated multiple solutions of these equations, it is necessary and sufficient that $v\mu \neq 1$. Thus, for (H_3) to be valid, we must necessarily have $v\mu \neq 1$. If $v\mu \neq 1$, then the multiple solutions of (5.8) that arise from (5.9) will satisfy the conditions imposed in (H_3).

Summarizing the above remarks, we see that we must have μ, v satisfy the condition

$$
\mu \neq 1, \qquad v \neq 1, \qquad \mu v \neq 1.
\tag{5.10}
$$

The inequalities (5.10) divide the positive quadrant of the parameter space (μ, v) into five regions as shown in Figure 5.1.

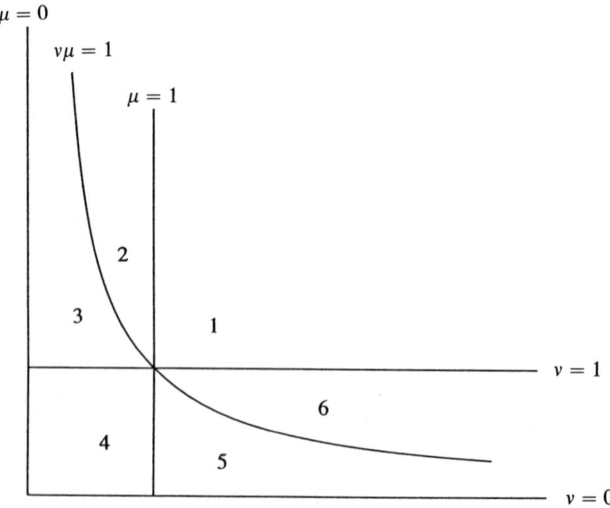

Figure 5.1

Region 1: $\mu > 1$, $v > 1$.

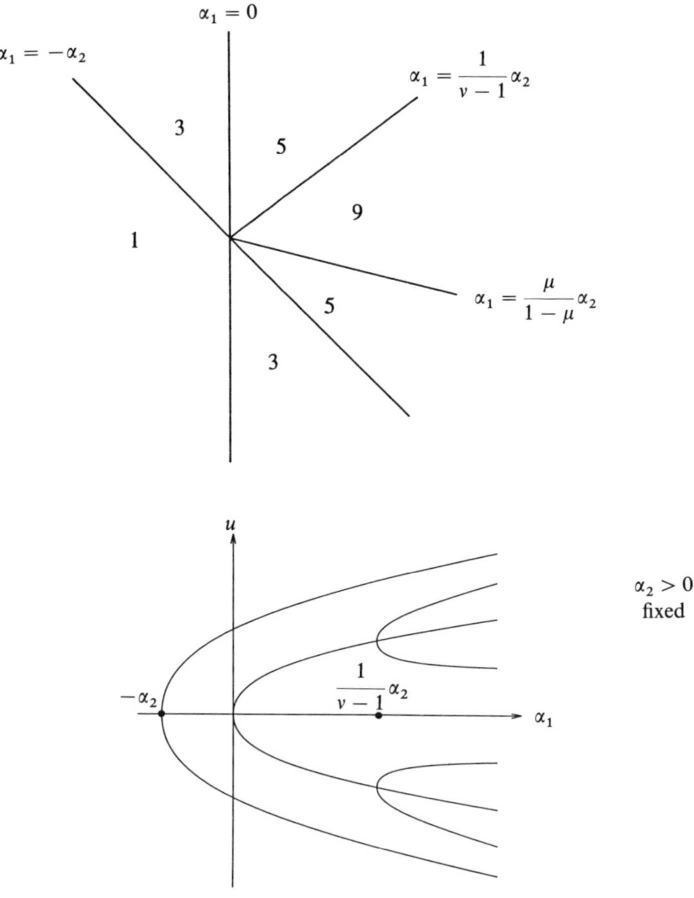

Figure 5.2

For the truncated equations, the complete bifurcation diagrams for each of the regions 1 through 5 are shown in Figures 5.3–5.7. As in the example in Section 4, none of the points in these bifurcation diagrams corresponding to the pitchforks can satisfy (H$_3$). However, since we assume that $u = 0$ satisfies the complete equations (6.1), the ones on the α_1 axis will be preserved for the complete equations (5.1). To preserve the others, we must impose the condition that the first equation in (5.1) vanish for $u_1 = 0$ and the second one vanish for $u_2 = 0$. This will be satisfied if the function $f = (f_1, f_2)$ satisfies the symmetry condition

$$f_1(-u_1, u_2) = -f_1(u_1, u_2)$$
$$f_2(u_1, -u_2) = -f_2(u_1, u_2).$$

Region 2: $v > 1, 0 < \mu < 1, \mu v > 1$.

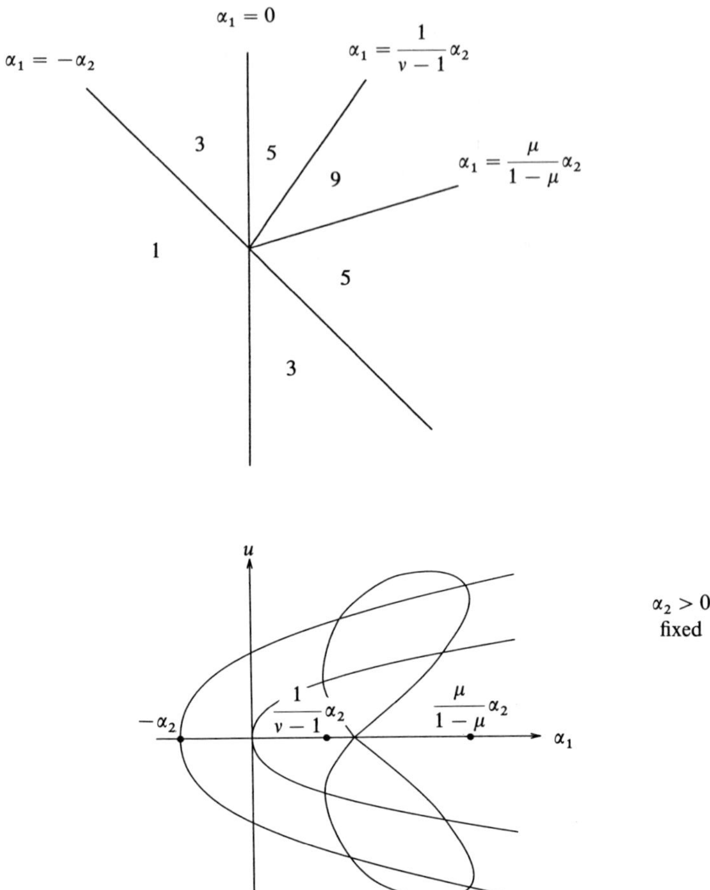

Figure 5.3

Under the above symmetry conditions, the complete bifurcation diagrams for Equation (5.1) are approximately the same ones as for Equation (5.8).

If $\alpha_2 \leq 0$, then the verification of (H_3) involves the discussion of the multiple solutions of the equations

$$w_1(w_1^2 + \mu w_2^2 + \beta_1) = 0$$
$$w_2(v w_1^2 + w_2^2 + \beta_1 + 1) = 0.$$

The conditions on v, μ become

Region 3: $v > 1, 0 < \mu < 1, \mu v < 1$.

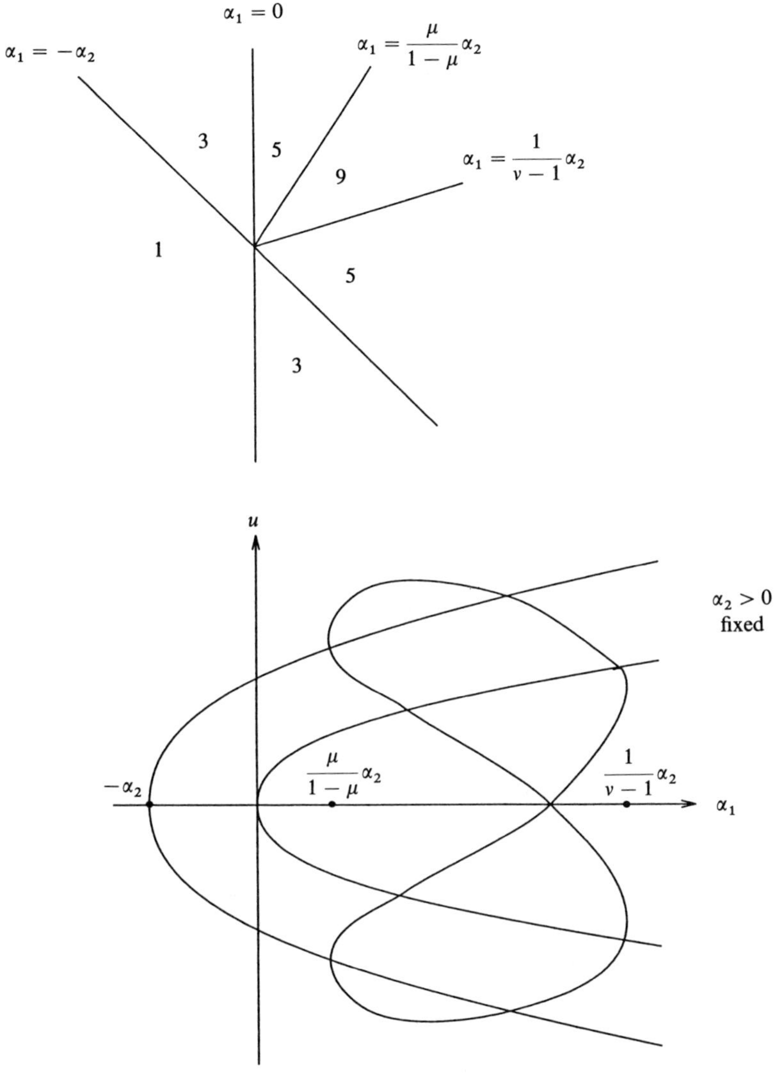

Figure 5.4

(5.11) $\mu \neq 1, \qquad v \neq 1$ and if $\alpha_2 \leq 0$,   then $v\mu \neq 1$ when $v < 1$

which divide the first quadrant in (μ, v)-space into the five regions shown in Figure 5.2. For the truncated equations (5.6), the complete bifurcation diagram for (μ, v) in region j' is the same as the one in region j. With the above symmetry conditions on f, the same conclusion holds for 5.1.

Region 4: $\mu < 1$, $\nu < 1$.

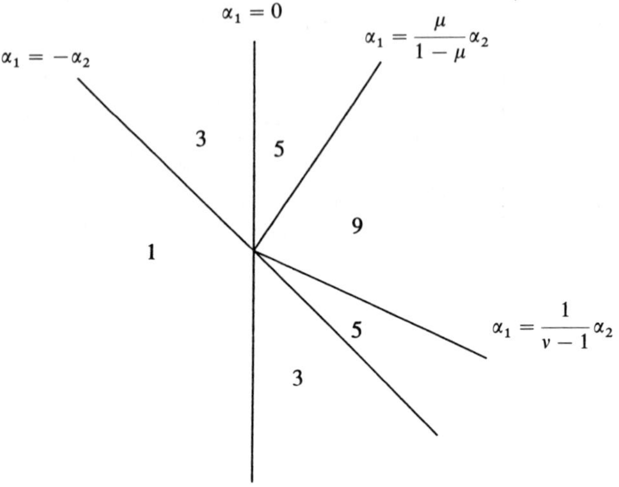

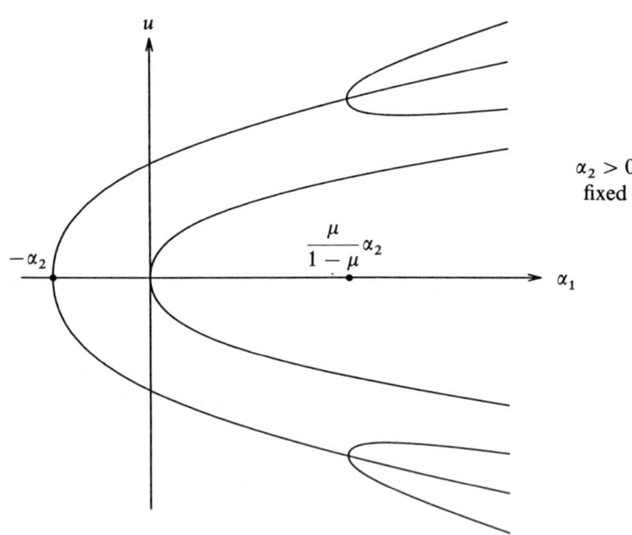

Figure 5.5

Region 5: $\mu > 1, 0 < v < 1, \mu v < 1$.

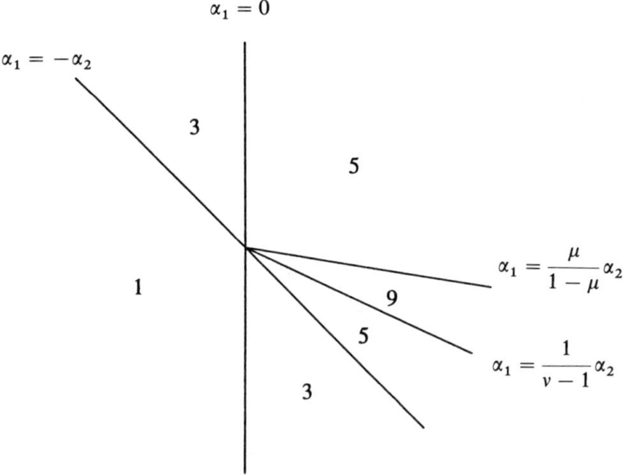

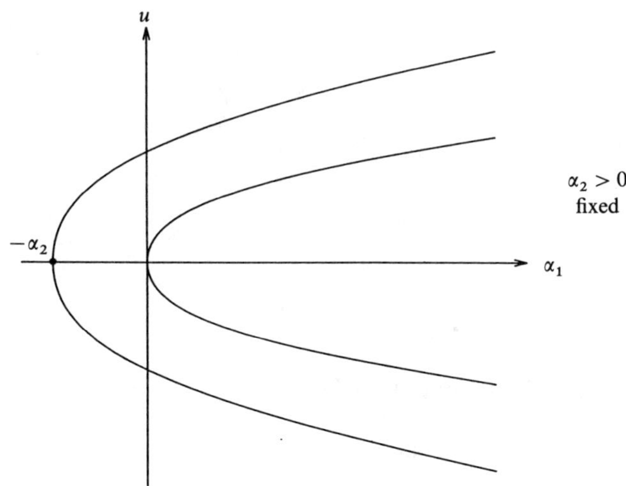

Figure 5.6

Region 6: $\mu > 1, 0 < v < 1, \mu v > 1.$

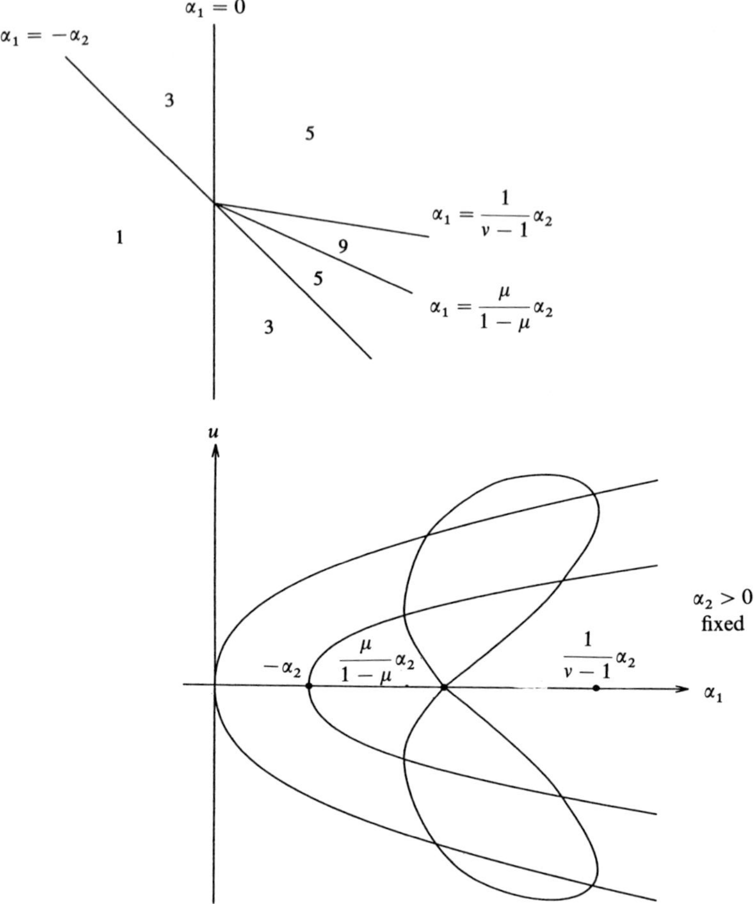

Figure 5.7

7.6. Cubic Nonlinearities II

In this section, we consider the system of equations

(6.1)
$$f(\alpha, u) = \alpha_0 k + \alpha_1 Lu + C(u) + \text{h.o.t.} = 0, \qquad f(0, \alpha_1, 0) = 0,$$
$$\text{h.o.t.} = O(|u|^4 + |\alpha_1 u^2| + |\alpha_1^2 u| + |\alpha_0 \alpha_1| + |\alpha_0|^2 + |\alpha_0 u|)$$

as $\alpha_0, \alpha_1, u \to 0$ where $\alpha = (\alpha_0, \alpha_1) \in \mathbb{R}^2$, $u = (u_1, u_2) \in \mathbb{R}^2$, $f = (f_1, f_2) \in \mathbb{R}^2$,

(6.2)
$$C: \mathbb{R}^2 \to \mathbb{R}^2 \text{ is a homogeneous cubic}$$
$$L \text{ is a } 2 \times 2 \text{ real matrix}$$
$$k \in \mathbb{R}^2 \text{ is given.}$$

Our objective is to determine all solutions of Equation (6.1) near $(\alpha, u) = (0, 0)$ by using the scaling techniques of the previous sections. The methods of proof of the results as well as the motivation for the hypotheses are similar to the ones given before. The first step is to obtain a priori bounds on the solutions.

Lemma 6.1. *If*

$$(\mathrm{H}_1) \qquad\qquad C(u) = 0 \quad \textit{implies} \quad u = 0$$

then there is a neighborhood V in $\mathbb{R}^4$ of $(\alpha, u) = (0, 0)$ and a constant $\beta > 0$ such that any solution of Equation (6.1) in V must satisfy

$$|u| \le \beta(|\alpha_0|^{1/3} + |\alpha_1|^{1/2}).$$

Proof. The proof is essentially the same as the proof of Lemma 3.1. $\square$

Theorem 6.2. *Suppose the following hypotheses are satisfied:*

(H_1) $C(u) = 0$ *implies* $u = 0$.
(H_2) *If* $k + C(v) = 0$, *then* $\det \partial C(v)/\partial v \ne 0$.
(H_3) *If*

$$(6.3) \qquad \begin{aligned} h^0_\pm(\beta_0, w) &\overset{\text{def}}{=} \beta_0 k \pm Lw + C(w) = 0 \\ \Delta_\pm(w) &\overset{\text{def}}{=} \det[\pm L + \partial C(w)/\partial w] = 0 \end{aligned}$$

then $\det[\partial(h_\pm, \Delta_\pm)/\partial(\beta_0, w)] \ne 0$.

Under these hypotheses, there are a finite number of curves of the approximate form

$$\alpha_0 \sim |\alpha_1|^{3/2} \beta_0^0$$

on which a bifurcation occurs for Equation (6.1). Along each bifurcation curve, the solution is given approximately by

$$u \sim |\alpha_1|^{1/2} w^0$$

where (β_0^0, w^0) are solutions of Equation (6.3) using (h_+, Δ_+) if $\alpha_1 > 0$ and (h_-, Δ_-) if $\alpha_1 > 0$. Furthermore, if each of these bifurcation curves are distinct in the (α_0, α_1)-plane, the number of solutions of Equation (6.1) changes by exactly two as a curve is crossed.

Proof. Only a sketch of the proof is given since it is so similar to the previous ones. Applying the scaling

$$u = \alpha_0^{1/3} v, \qquad \alpha_1 = \alpha_0^{2/3} \beta_1,$$

Equation (6.1) becomes equivalent to the equation

$$k + \beta_1 Lv + C(v) + O(|\alpha_0|^{1/3}) = 0.$$

The Hypotheses (H_2) and the Implicit Function Theorem imply all solutions of this equation are simple for $|\beta_1| \leq \delta$, $|\alpha_0| \leq \delta$, $\delta > 0$. Thus, no bifurcation occurs in this region. The scaling

$$u = |\alpha_1|^{1/2} w, \qquad \alpha_0 = |\alpha_1|^{3/2}\beta_0$$

implies Equation (6.1) is equivalent to the equation

$$(6.4) \qquad h_\pm(\beta_0, w, \alpha_1) \overset{\text{def}}{=} \beta_0 k \pm Lw + C(w) + O(|\alpha_1|^{1/2}) = 0$$

where the $+$ corresponds to $\alpha_1 > 0$, $-$ to $\alpha_1 < 0$. The bifurcation curves are determined by the simultaneous solutions of Equation (6.4) and the equation

$$(6.5) \qquad \det[\partial h_\pm(\beta_0, w, \alpha_1)/\partial w] = 0.$$

Suppose (b_0^0, w^0) satisfy Equation (6.3) and, to simplify notation, let us consider only $\alpha_1 > 0$ and let $h_+ = h$, $\Delta_+ = \Delta$. From (H_3), without loss in generality, we may assume $\partial h_1(\beta_0^0, w^0, 0)/\partial w_1 \neq 0$. Determine $\phi(\beta_0, w_2, \alpha_1)$, $\phi(\beta_0^0, w_2^0, 0) = w_1^0$, such that

$$h_1(\beta_0, \phi(\beta_0, w_2, \alpha_1), \alpha_1) = 0$$

and define

$$H(\beta_0, w_2, \alpha_1) = h_2(\beta_0, \phi(\beta_0, w_2, \alpha_1), w_2, \alpha_1).$$

At $(\beta_0, w_2, \alpha_1) = (\beta_0^0, w_2^0, 0)$, one shows that

$$H = \partial H/\partial w_2 = 0$$
$$\partial^2 H/\partial w_2^2 = (\partial h_1/\partial w_1)^{-1} \det[\partial(h_1, \Delta)/\partial w] \neq 0$$
$$\partial H/\partial \beta_0 = (\partial h_1/\partial w_1)^{-2} \det[\partial h/\partial(w_1, \beta_0)] \neq 0$$
$$\det[\partial(h, \Delta)/\partial(\beta_0, w)] = -(\partial h_1/\partial w_1)^2(\partial H/\partial \beta_0)(\partial^2 H/\partial w_2^2).$$

The proof now follows exactly as in the proof of Theorem 3.7. There is a function $\gamma(\beta_0, \alpha_1)$, $\gamma(\beta_0^0, 0) = w_2^0$, such that

$$\frac{\partial H}{\partial w_2}(\beta_0, \gamma(\beta_0, \alpha_1), \alpha_1) = 0$$

for (β_0, α_1) near $(\beta_0^0, 0)$. If $G(\beta_0, \alpha_1) = H(\beta_0, \gamma(\beta_0, \alpha_1), \alpha_1)$, then G is the unique extreme value of H near $(\beta_0^0, w_2^0, 0)$. Suppose it is a minimum. The minimum being positive or negative determines whether there are no solutions or two solutions of the equation $h(\beta_0, w, \alpha_1) = 0$ near $(\beta_0^0, w^0, 0)$. Since

$$\partial G(\beta_0^0, 0)/\partial \beta_0 = \partial H(\beta_0^0, w_2^0, 0)/\partial \beta_0 \neq 0,$$

there is a function $\delta(\alpha_1)$, $\delta(0) = \beta_0^0$ such that $G(\delta(\alpha_1), \alpha_1) = 0$ near $\alpha_1 = 0$. Each point of the curve $\beta_0 = \delta(\alpha_1)$ is thus a bifurcation point. $\quad\square$

Remark 6.3. The direction of bifurcation across the bifurcation curve is determined from the sign of $\partial G/\partial \beta_0$ and $\beta_0 - \delta(\alpha_1)$ as in Remark 3.8.

Remark 6.4. The bifurcation curves for Equation (6.1) are cusps in contrast to the parabolas for Equation (3.1) and straight lines for Equations (4.1), (5.1).

As an example, consider Equation (6.1) for which the lower order terms are

$$\text{(6.6)} \qquad \begin{aligned} u_1^3 + \mu u_1 u_2^2 - \alpha_1 u_1 + \alpha_0 &= 0 \\ v u_1^2 u_2 + u_2^3 - \alpha_1 u_2 + \alpha_0 &= 0 \end{aligned}$$

with $v > 0$, $\mu > 0$. Hypothesis (H_1) is always satisfied. To verify (H_2) and (H_3), it is convenient to use the fact that (H_1), (H_2), (H_3) are equivalent to (H_1) and

(H_2') If $C(v) \pm Lv = 0$, then $\det[\partial C(v)/\partial v \pm L] \neq 0$

(H_3') If

$$h^0(\beta_1, w) \stackrel{\text{def}}{=} C(w) + \beta_1 Lw + k = 0$$

$$\Delta(\beta_1, w) \stackrel{\text{def}}{=} \det[\partial C(w)/\partial w + \beta_1 L] = 0$$

then $\det \partial(h^0, \Delta)/\partial(\beta_1, w) \neq 0$.

This remark only asserts that one can interchange the order in which the variables are scaled.

As in Section 5, (H_2') is equivalent to $\mu \neq 1$, $v \neq 1$. To verify (H_3'), we must discuss the multiple solutions of the equations

$$\text{(6.7)} \qquad \begin{aligned} w_1^3 + \mu w_1 w_2^2 - \beta_1 w_1 + 1 &= 0 \\ v w_1^2 w_2 + w_2^3 - \beta_1 w_2 + 1 &= 0. \end{aligned}$$

These equations are very difficult to analyze. First assume that $f = (f_1, f_2)$ in (6.1) satisfies, for $\alpha_0 = 0$, $f_1 = u_1 g_1(u, \alpha_1)$, $f_2 = u_2 g_2(u, \alpha_1)$, where $g_j(-u, \alpha_1) = g_j(u, \alpha_1)$. Then, for $\mu > 1$, $v > 1$ (or $\mu < 1$, $v < 1$), the bifurcation diagram has the same qualitative behavior as the ones shown in Figure 6.1. The other regions $\mu > 1$, $v > 1$, $\mu < 1$), have bifurcation diagrams similar to the ones shown in Figure 6.2. If one does not make the symmetry hypothesis, the diagram for $\alpha_0 > 0$ will be a single line together with parabolic arcs, but the ordering may not be the one shown in Figures 6.1, 6.2. Also, for $\alpha_0 = 0$, one could have the same diagram as the one for $\alpha_0 > 0$ because of the higher order terms in Equation (6.1).

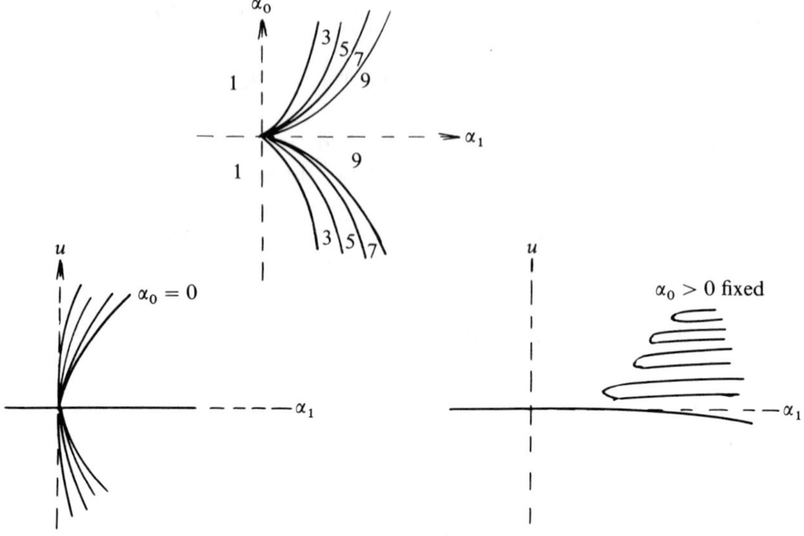

Figure 6.1

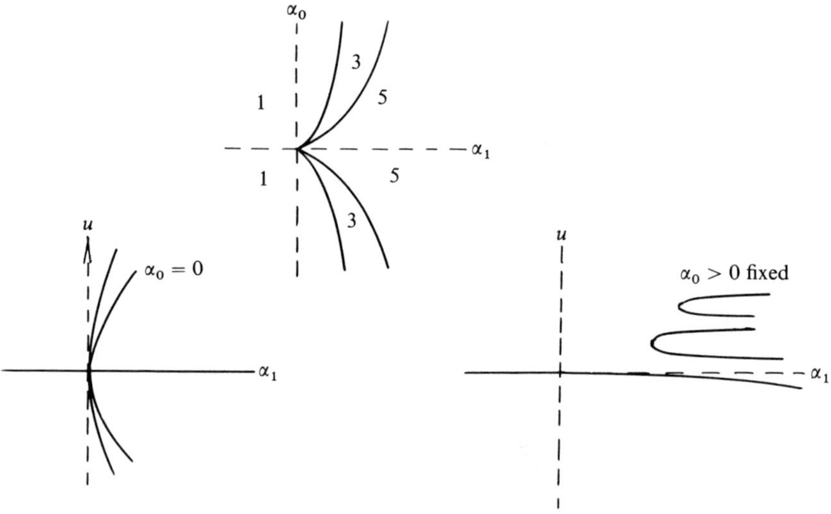

Figure 6.2

7.7. Cubic Nonlinearities III

In this section, we consider the system of equations

$$(7.1) \quad \begin{aligned} f(\alpha, u) &= \alpha_1 L u + \alpha_3^2 M u + \alpha_3 Q(u) + C(u) + \text{h.o.t.} = 0 \\ \text{h.o.t.} &= O(|u|[|\alpha_1 u| + |\alpha_1|^2 + |\alpha_3^2 u| + |\alpha_3 u^2| + |u|^3) \end{aligned}$$

as $\alpha_1, \alpha_3, u \to 0$ where $\alpha = (\alpha_1, \alpha_3) \in \mathbb{R}^2$, $u = (u_1, u_2) \in \mathbb{R}^2$, $f = (f_1, f_2) \in \mathbb{R}^2$,

$$(7.2) \qquad \begin{aligned} &C : \mathbb{R}^2 \to \mathbb{R}^2 \text{ is a homogeneous cubic} \\ &Q : \mathbb{R}^2 \to \mathbb{R}^2 \text{ is a homogeneous quadratic} \\ &L \text{ is a } 2 \times 2 \text{ real matrix.} \end{aligned}$$

Our objective is to apply the scaling techniques of the preceding sections to determine the solutions of Equation (7.1) near $(\alpha, u) = (0, 0)$. It is slightly more complicated because of the presence of the quadratic terms. On the other hand, the scaling techniques will suggest the appropriate hypotheses.

Lemma 7.1. *If*

$$(\text{H}_1) \qquad\qquad C(u) = 0 \quad \text{implies} \quad u = 0,$$

then there is a neighborhood V in $\mathbb{R}^4$ of $(\alpha, u) = (0, 0)$ and a constant $\beta > 0$ such that any solution of Equation (7.1) in V must satisfy

$$|u| \leq \beta(|\alpha_1|^{1/2} + |\alpha_3|).$$

Proof. The proof is essentially the same as the proof of Lemma 3.1. $\quad\square$

If $u = |\alpha_1|^{1/2} v$, $\alpha_3 = |\alpha_1|^{1/2} \beta_2$, then Equation (7.1) is equivalent to the system of equations

$$\pm Lv + \beta_2^2 Mv + \beta_2 Q(v) + C(v) + O(|\alpha_1|^{1/2}) = 0.$$

To avoid infinitely many bifurcations near the α_1-axis, we suppose

$$(\text{H}_2) \quad \text{If } \pm Lv + C(v) = 0, \text{ then } \det[\pm L + \partial C/\partial v] \neq 0.$$

The hypothesis implies no bifurcation for $|\beta_2| \leq \delta$, $|\alpha_1| \leq \delta$, $\delta > 0$.

If we rescale by $u = \alpha_3 w$, $\alpha_1 = \alpha_3^2 \beta_1$, then Equation (7.1) is equivalent to the system of equations

$$(7.3) \qquad h(\beta_1, w, \alpha_3) \overset{\text{def}}{=} \beta_1 Lw + Mw + Q(w) + C(w) + O(|\alpha_3|) = 0.$$

In Equation (7.3), we need only consider β_1, w bounded and α_3 small. The bifurcation curves in parameter space are determined by those β_1 for which there is a multiple solution of Equation (7.3); that is, a simultaneous solution of Equation (7.3) and the equation

$$(7.4) \qquad\qquad \det[\partial h(\beta_1, w, \alpha_3)/\partial w] \neq 0.$$

The procedure of the previous sections suggests that our next hypothesis should be obtained in the following manner. Put $\alpha_3 = 0$ in Equation (7.3),

(7.4). These represent three equations in three unknowns (β_1, w). If there is a solution of these equations, require that it be simple. Such an hypothesis is very reasonable, but it will not permit a complete discussion of the solutions of Equations (7.3), (7.4) since, for $w = 0$, it eliminates the consideration of β_1 as an eigenvalue of the pair of matrices (M, L). The point $w = 0$, β_1 an eigenvalue of (M, L) must be treated separately and the quadratic terms will be more important than the cubic terms. Consequently, we assume

(H_3) If

$$h^0(\beta_1, w) \stackrel{\text{def}}{=} \beta_1 L w + M w + Q(w) + C(w) = 0$$

(7.5)

$$\Delta(\beta_1, w) \stackrel{\text{def}}{=} \det[\beta_1 L + M + \partial Q(w)/\partial w + \partial C(w)/\partial w] = 0$$

and $w \neq 0$, then $\det[\partial(h^0, \Delta)/\partial(\beta_1, w)] \neq 0$.

With Hypotheses (H_1)–(H_3), we can determine all the bifurcation curves and solutions on these curves corresponding to solutions (β_1^0, w^0) of Equation (7.5) for $w^0 \neq 0$.

For the solutions near $w = 0$, let us first suppose

$(H_4)_0$ $M = 0$, $\det L \neq 0$.

In this case, Equation (7.5) has a solution $(\beta_1, 0)$ if and only if $\beta_1 = 0$. Thus, we need only discuss Equations (7.3), (7.4) near $(\beta_1, w, \alpha_3) = (0, 0, 0)$. As remarked earlier, near $w = 0$, the quadratic terms $Q(w)$ should dominate the cubic terms $C(w)$. This certainly will be the case if we assume

$(H_5)_0$ $Q(w) = 0$ implies $w = 0$.

With Hypothesis $(H_5)_0$ and the particular form of the higher order terms h.o.t. in Equation (7.1), one obtains an a priori bound on all solutions of Equation (7.3) for $M = 0$ in a neighborhood of zero in the form

(7.6) $$|w| \leq k|\beta_1|.$$

If $w = \beta_1 z$, then the small solutions of Equation (7.3) for $M = 0$ are obtained by solving the equivalent equation

(7.7) $$Lz + Q(z) + \beta_1 C(z) + O(|\alpha_3|) = 0$$

for $z \in \mathbb{R}^2$, $\beta_1 \in \mathbb{R}$, α_3 small. The natural hypotheses for Equation (7.7) is

$(H_6)_0$ If $Lz + Q(z) = 0$, then $\det(L + \partial Q(z)/\partial z) \neq 0$.

The Implicit Function Theorem then implies there are no bifurcations near $(\beta_1, \alpha_3) = (0, 0)$; that is, there are no bifurcation curves in the parameter space (α_1, α_3) near the α_3-axis.

We have thus proved the following result.

Theorem 7.2. *Consider the equation*

$$
\text{(7.8)} \qquad
\begin{aligned}
\alpha_1 Lu + \alpha_3 Q(u) + C(u) + \text{h.o.t.} &= 0 \\
\text{h.o.t.} = O\big(|u|[|\alpha_1 u| + |\alpha_1|^2 &+ |\alpha_3^2 u| + |\alpha_3 u^2| + |u^3|\big)
\end{aligned}
$$

as $\alpha_1, \alpha_3, u = (u_1, u_2) \to 0$ and suppose the following conditions hold:

(H_1) $C(u) = 0$ *implies* $u = 0$

(H_2)   *If* $\pm Lv + C(v) = 0$*, then* $\det[\pm L + \partial C(v)/\partial v] \neq 0$

(H_3)$_0$ *If*

$$
\text{(7.9)} \qquad
\begin{aligned}
h^0(\beta_1, w) &\overset{\text{def}}{=} \beta_1 Lw + Q(w) + C(w) = 0 \\
\varDelta(\beta_1, w) &\overset{\text{def}}{=} \det[\partial h^0(\beta_1, w)/\partial w] \neq 0
\end{aligned}
$$

    then $\det[\partial(h^0, \varDelta)/\partial(\beta_1, w)] \neq 0$.

(H_4)$_0$ $\det L \neq 0$

(H_5)$_0$ $Q(w) = 0$ *implies* $w = 0$

(H_6)$_0$ *If* $Lz + Q(z) = 0$*, then* $\det[L + \partial Q(z)/\partial z] \neq 0$.

Under these hypotheses, there are a finite number of curves of the approximate form

$$
\alpha_1 \sim \alpha_3^2 \beta_1^0
$$

on which a bifurcation occurs for Equation (7.8). Along each bifurcation curve, the solution is given approximately by

$$
u \sim \alpha_3 w_1^0
$$

where (β_1^0, w_1^0) are solutions of Equation (7.9). Furthermore, if each of these bifurcation curves are distinct in the (α_1, α_3)-plane, then the number of solutions of Equation (7.8) changes by exactly two as a curve is crossed.

A typical bifurcation diagram is shown in Figure 7.1. The solution versus α_1 curve is shown in Figure 6.1 for $\alpha_3 = 0$. Most of the points of intersection in this bifurcation diagram are not generic since a little asymmetry will break the intersection of the solution curves.

Let us now consider the case where $M \neq 0$. As remarked earlier the difficulty occurs in Equation (7.3) and (7.4) when $w = 0$ and β_1 is an eigenvalue of (M, L). We have seen in Chapter 4 that a bifurcation always occurs at a simple eigenvalue of (M, L) and the bifurcation is generic if the second order

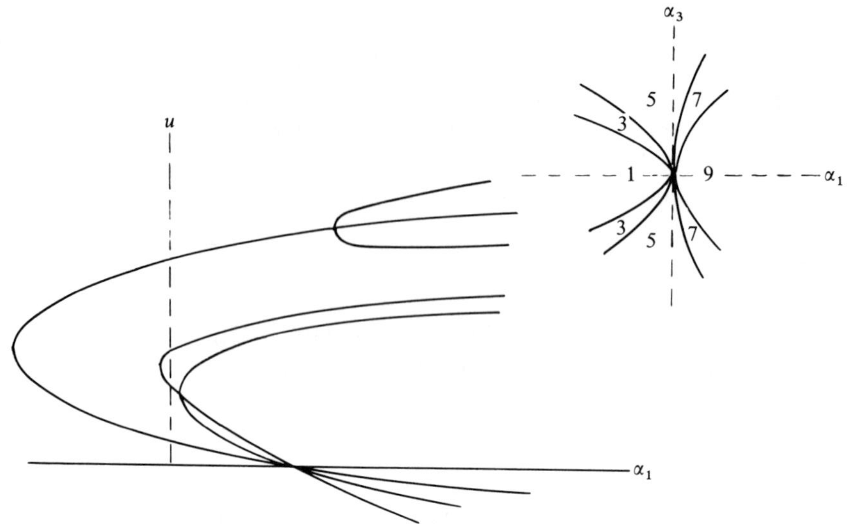

Figure 7.1. ($\alpha_3 \neq 0$)

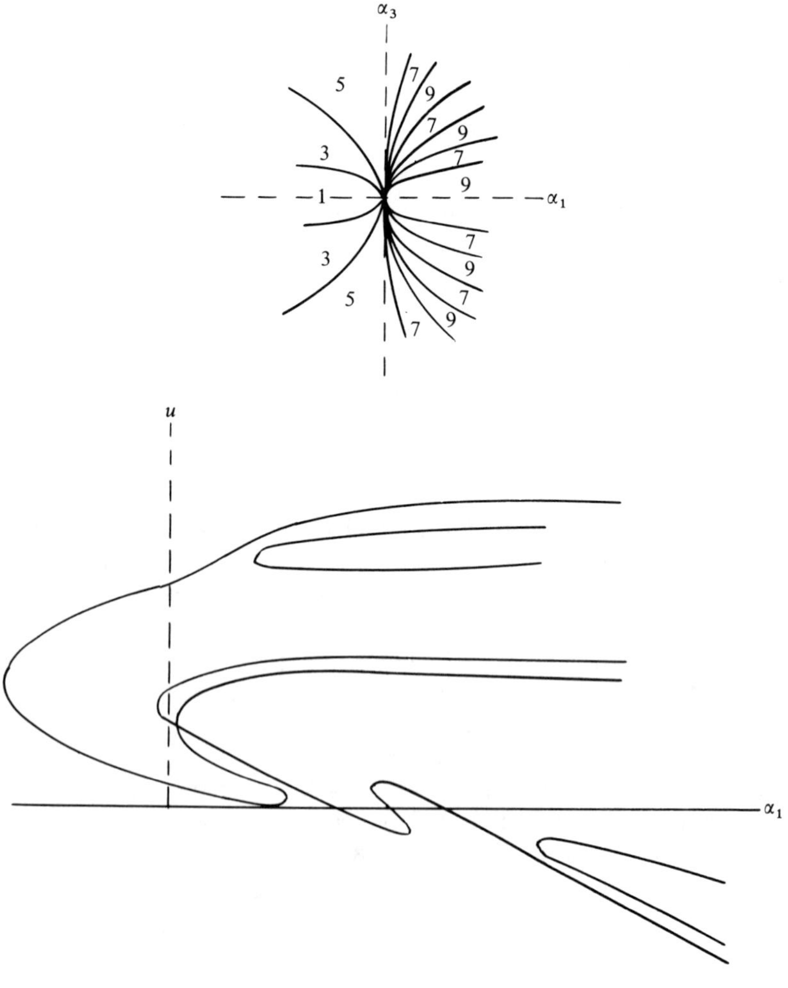

Figure 7.2

terms in the corresponding bifurcation equations do not vanish. The next hypothesis is imposed to ensure that these conditions are satisfied.

(H$_4$)   *The eigenvalues of (M, L) are simple. If β_1^0 is an eigenvalue of (M, L) with eigenvector z^0, then $\langle Lz^0, Q(z^0)\rangle \neq 0$ where $\langle, \rangle$ is the inner product in $\mathbb{R}^2$.*

Suppose Hypothesis (H$_4$) is satisfied. To determine the solution of Equation (7.3), (7.4) near $(\beta_1, w, \alpha_3) = (\beta_1^0, 0, 0)$, we can apply the results of Section 5.5. There will always be a bifurcation at β_1^0 with two curves of solutions through $(\beta_1^0, 0, 0)$. We do not state a detailed theorem for this case since it is so complicated. However, it should be clear that if L is nonsingular and M is small but not zero and Hypotheses (H$_1$)–(H$_4$) are satisfied, then the bifurcation diagram will be similar to the one in Figure 7.1 but there will, in general, be no intersections of the curves representing the solution versus α_1. In fact, the typical picture will be as indicated in Figure 7.2.

7.8. Bibliographical Notes

The use of scaling techniques in the manner described in this chapter is based on Chow, Hale and Mallet–Paret [1, 2]. When several small parameters are involved in a problem, it is very natural to express all of the parameters in terms of powers or fractional powers of a single small parameter. The particular powers to use are usually determined in a very natural way from the nonlinearities in the equation and the manner in which the parameters enter into the problem. The lemmas in the text which give a priori bounds on the solutions in a neighborhood of zero are a precise way to show that the correct fractional powers have been chosen.

Once an a priori bound has been obtained on the solution, the scaling techniques can be applied. It is worthwhile to summarize the ideas another time. Suppose we are trying to solve a vector equation

$$(8.1) \qquad\qquad F(u, \lambda) = 0$$

where u is an n-vector, λ is a p-vector, $F(0, 0) = 0$, F is analytic in u, λ,

$$F(u, 0) = F^{(k)}(u) + F^{(k+1)}(u) + \cdots$$

where each component of $F^{(j)}(u)$ is a homogeneous polynomial in the components of u of degree j. If

$$(8.2) \qquad\qquad F^{(k)}(u) = 0 \quad \text{implies} \quad u = 0$$

then one can obtain a priori bounds on the solutions in a neighborhood of $u = 0$, $\lambda = 0$ of the form

$$(8.3) \qquad |u| \leq (\text{constant})(|\lambda_1|^{\alpha_1}, \ldots, |\lambda_p|^{\alpha_p})$$

where $\lambda = (\lambda_1, \ldots, \lambda_p)$ and each α_j is a rational number. If we let $u = \varepsilon v$, $\lambda_j = \mu_j |\varepsilon|^{1/\alpha_j}$, $j = 1, 2, \ldots, p - 1$, $\lambda_p = |\varepsilon|^{1/\alpha_p}$, then (8.1) is equivalent to the equation

$$(8.4) \qquad G(v, \mu, \varepsilon) \stackrel{\text{def}}{=} \varepsilon^{-k} F(\varepsilon v, \mu_1 |\varepsilon|^{1/\alpha_1}, \ldots, |\varepsilon|^{1/\alpha_p}) = 0$$

The Eq. (8.4) must now be discussed for all $v \in \mathbb{R}^n$, $\mu \in \mathbb{R}^{p-1}$, and ε in a neighborhood of zero. The function $G(v, \mu, \varepsilon)$ now has the form

$$G(v, \mu, 0) = F^{(k)}(v) + H^{(k)}(v, \mu)$$

where $H^{(k)}(v, 0) = 0$ and $H^{(k)}(v, \mu)$ is a polynomial in the components of μ. The function $G(v, \mu, \varepsilon)$ is analytic in all variables.

If there is a (v_0, μ_0) such that $G(v_0, \mu_0, 0) = 0$ and $\det \partial G(v_0, \mu_0, 0)/\partial v \neq 0$, then the Implicit Function Theorem implies there is a unique solution of (8.4) in a neighborhood of $(v_0, \mu_0, 0)$ and it is analytic in μ, ε. One can thus obtain the solution by expanding the solution v as a power series in ε with coefficients depending on μ, substituting this power series in the function $G(v, \mu, \varepsilon)$ and equating the coefficients of powers of ε equal to zero. These series are precisely the same as the ones obtained by the usual perturbation techniques.

If there is a (v_0, μ_0) such that $G(v_0, \mu_0, 0) = 0$ and $\det \partial G(v_0, \mu_0, 0)/\partial v = 0$, then the Implicit Function Theorem cannot be applied. One now wants to determine the bifurcation surface as well as the solution on the bifurcation surface. To do this, one must impose an auxiliary condition analogous to the Hypothesis (H_3) in this chapter which for the vector parameter μ is expressible in terms of the condition

$$\text{rank}[\partial[G, \partial G/\partial v]/\partial(v, \mu)] = n + 1 \quad \text{at } (v, \mu, \varepsilon) = (v_0, \mu_0, 0)$$

With this condition, one can select one coordinate, say μ_1, of μ as a distinguished coordinate so that $\det[\partial[G, \partial G/\partial v]/\partial(v, \mu_1)] \neq 0$ at $(v_0, \mu_0, 0)$. It is then possible to obtain the bifurcation surface and the solution on the bifurcation surface by the Implicit Function Theorem as analytic functions in ε, μ_2, $\ldots$, μ_{p-1}. The specific form of the solution can then be obtained by equation coefficients of power series in ε as described before except here one uses the equation

$$G(v, \mu, \varepsilon) = 0,$$

$$\det \frac{\partial G}{\partial v}(v, \mu, \varepsilon) = 0$$

Again, these series coincide with the usual ones obtained by classical perturbation techniques.

For a discussion of the classical perturbation techniques, see the book of Nayfeh [1], some of the articles in Keller and Antman [1], the papers of Keener [1], Keller [3], Matkowsky and Reiss [1], Knightly and Sather [1–6] and the references therein.

For other papers dealing with one parameter problems which represent perturbations of a bifurcation from a function which begins with a homogeneous polynomial of degree $k \geq 2$, see Aizenlander [1], Dancer [1–3], Greenelee [1], Kirchgassner [4], Krasnoselskii [2–4], Sather [1], Stakgold [1], Sattinger [10], Shearer [2].

Even when one has two independent parameters in a problem, it is sometimes advantageous from the point of view of applications to treat one of them is a fixed parameter while the other is varied. The fixed parameter could represent a given experiment. For a discussion of these types of situations, see Golubitsky and Schaeffer [1].

The situation in Section 4 of two eigenvalue parameters led in a natural way to secondary bifurcations; that is, bifurcations from solutions other than the zero solution. The importance of this type of bifurcation in the explanation of the buckling of plates was first pointed out by Bauer, Keller and Reiss [2], Bauer and Reiss [2].

Golubitsky and Schaeffer [3] were the first to point out clearly the effect of the modal parameters μ, v on the bifurcation diagrams in Section 7.5. In the buckling theory of plates, different modal parameters correspond to different boundary conditions.

The case of the eigenvalue and imperfection parameter in Section 7.6 was partially discussed by Keener [1–3], Keener and Keller [1, 2]. The case discussed in Section 7.7 is based on Mallet–Paret [1]. List [1] has given a complete discussion of the situation of cubic nonlinearities with three bifurcation parameters—two eigenvalue parameters and an imperfection parameter. Applications were given to rectangular plates. The paper of Papanicolaou [1] contains a discussion of an interesting two parameter bifurcation problem with stochastic perturbations.

It is possible to use the methods of degree theory to obtain some information about bifurcation equations (see, for example, Cronin [1, 5, 6], Sattinger [4] and the references therein). These methods give less specific information than the analytic methods and are restricted essentially to one parameter problems. On the other hand, less hypotheses are needed to use degree theory.

When the original function for which zeros are desired is the gradient of some function, it is possible to give in some cases a more complete discussion of all possible bifurcations that can occur in a neighborhood of zero. When the null space of $D_x M(0,0)$ has dimension one and the Taylor series of the bifurcation function for $\lambda = 0$ begins with terms of order k, then the Malgrange preparation theorem implies only k parameters are needed to describe all possible bifurcations. When the null space has dimension greater than one, many more parameters are needed. In fact, if the dimension of the null space is two and $k = 3$, one needs nine parameters. This is one of the reasons we chose to discuss the effects of two parameters in the text. The other reason was to have a procedure which is valid for nongradient systems. For interesting pictures of the bifurcation surfaces of gradient systems when the number of parameters needed to describe all bifurcations is small, see Woodcock and Poston [1].

The underlying ideas of this chapter remain valid when the number of parameters is more than two or the dimension of the null space is more than two. Even after the a priori bounds on the solutions are obtained to justify scaling, the difficulties increase very rapidly as can be seen from the papers of List [1], Mallet–Paret [1] on buckling of plates.

If the null space of $D_x M(0, 0)$ has a very large dimension, it often happens that the function itself is invariant under some groups of transformations. This invariance will be shared by the bifurcation equations if the projections used in the method of Liapunov–Schmidt are chosen appropriately. (See Section 6.7). This implies the Taylor series of the bifurcation functions can only involve certain types of terms. Also, if the elements from the null space are chosen to be invariant under these groups, then some of the bifurcation functions will be linearly independent. Thus, the number of essential bifurcation functions have been reduced and one has a more reasonable problem in this restricted class of solutions. We have used this idea systematically in this chapter as well as various other places even though it was not formalized. It is an important concept which has received much attention in the literature (see, for example, Dancer [5], Golubitsky and Schaeffer [2], Poenaru [1], Rabinowitz [4], Ruelle [1], Sattinger [5–10], Shearer [1], Vanderbauwhede [1–6] and the references therein). A more systematic treatment of this subject was not given because it would be a monograph in itself (see, for example, Sattinger [10], Vanderbauwhede [1]).

In the above discussion, we gave a procedure for determining the bifurcation surfaces and solutions if all functions involved are analytic. If the functions are only C^k, then the procedure is still constructive since everything is obtained by an application of the Implicit Function Theorem and this latter theorem was proved in Section 2.2 by using contraction mappings. Thus, we can say that the bifurcation surfaces are known (approximately) in a neighborhood of zero. It is very important in the applications to know the global structure of these bifurcation surfaces. For the case in which the bifurcation parameter is a scalar, several procedures have been devised for the construction of the bifurcation diagrams. Keller [2] and Georg [1] give methods to extend the bifurcation surfaces by using approximation methods based on the tangent plane of the bifurcation surface at a point. References for other iterative methods are contained in Demoulin and Chen [1]. It is also possible to give methods based on Sperner's lemma which are more topological in nature and are being widely used for the approximation of fixed points of maps (see Allgower and Georg [1], Peitgen and Prüfer [2]).

In this chapter, we have determined the complete bifurcations of certain families of functions by imposing hypotheses which were motivated by some simple geometric ideas. The hypotheses were determined from the approximate bifurcation equations. Magnus [1–3] has been investigating bifurcation theory from a different point of view. If $T: \Lambda \times X \to Z$, he attempts to characterize the topological structure of $T^{-1}(0)$ in the space $\Lambda \times X$ by imposing some generic hypotheses on the map. For a discussion of the relation between these hypotheses and the ones imposed in this chapter, see Hale [15].

Chapter 8

Some Applications

8.1. Introduction

In this chapter, we give several applications of the results of Chapter 6 and
the scaling techniques in Chapter 7. The first four sections are devoted to the
buckling of thin rectangular plates and cylindrical shells with small curvature.
The bifurcation parameters are chosen to be external loading, imperfections,
curvature and the ratio of the sides of the plate. The ratio of the sides is
allowed to be in a neighborhood of a value for which the first eigenvalue is
simple or double. Several types of boundary conditions are imposed and
the effects of these conditions on the bifurcation behavior are discussed.
It is always assumed that the solutions of the equations of von Kármán are
a reasonable approximation to the equilibrium positions of the plate.

The second example is a system of reaction diffusion equations which
occurs in modeling the chemical reactions of two substances. The bifurcation
parameters are the relative strength of diffusion and the strength of interaction of the two chemicals. The parameters are considered in the neighborhood of a point on the curve separating the region of stability of equilibrium
from the region of instability. The case of a simple eigenvalue zero as well
as a double eigenvalue are considered.

In the third example, we discuss harmonic periodic solutions of the forced
Duffing equation. The bifurcation parameters are damping forcing frequency and amplitude. The results in Chapter 7 do not apply directly to
this example because there is too much symmetry. However, it is shown that
the principle behind the ideas in Chapter 7 can be used to give a complete
solution to the problem.

8.2. The von Kármán Equations

In the next few sections, we indicate the applications of the previous results
to the buckling of thin rectangular plates and cylindrical shells with small
curvature. We consider the buckling behavior as a function of external
loading, imperfections, curvature, the ratio of the lengths of the sides of the
plate and variations in the boundary conditions. It is always assumed that

the solutions of the equations of von Kármán are a reasonable approximation to the equilibrium positions of the plate.

Suppose the plate is described by the rectangular domain

$$\Omega = (0, \ell\pi) \times (0, \pi) \subseteq \mathbb{R}^2$$

in the (x, y)-plane. In the absence of external forces, suppose the plate is not perfectly flat so that it has a small imperfection which represents a displacement $\alpha w_0(x, y)$, where α is a small parameter and $w_0 : \Omega \to R$ is a known function. Let λ be a lateral force applied to the edges of the plate at $x = 0$, $x = \ell\pi$, F be the Airy stress function produced in the plate when the normal loading is zero and when the plate is artificially prevented from buckling, $F = -y^2/2$, and let $\varepsilon h(x, y)$ be a vertical load on the plate, where ε is a small parameter and h is a known function. If w and f are the additional deflection and stress caused by the normal loading, the vertical loading and the imperfection, the von Kármán equations are

$$(2.1) \qquad \begin{aligned} \Delta^2 f &= -\tfrac{1}{2}[w, w] - \alpha[w, w_0], \\ \Delta^2 w &= [w + \alpha w_0, f + \lambda F] + \varepsilon h \quad \text{in } \Omega \end{aligned}$$

where $[u, v] = u_{xx}v_{yy} - 2u_{xy}v_{xy} + u_{yy}v_{xx}$ and Δ is the Laplacian. The final shape of the plate is given by $w + \alpha w_0 + \lambda F$.

To solve these equations, one must impose boundary conditions on the stress function f and the deflection w. The usual boundary conditions for f are

$$(2.2) \qquad f = 0, \qquad \Delta f = 0 \quad \text{on } \partial\Omega.$$

However, there are very legitimate reasons for considering also the conditions

$$(2.3) \qquad \frac{\partial f}{\partial n} = 0, \qquad \frac{\partial(\Delta f)}{\partial n} = 0 \quad \text{on } \partial\Omega$$

where $\partial/\partial n$ is the normal derivative.

The simply supported plate should satisfy

$$(2.4) \qquad w = 0, \qquad \Delta w = 0 \quad \text{on } \partial\Omega.$$

If the plate is simply supported at $y = 0$, $y = \pi$ and clamped at $x = 0$, $x = \ell\pi$, then w satisfies

$$(2.5) \qquad \begin{aligned} w &= 0, & \Delta w &= 0 \quad \text{at } y = 0, y = \pi \\ w &= 0, & w_x &= 0 \quad \text{at } x = 0, x = \ell\pi \end{aligned}$$

If the plate is clamped on all sides, then w satisfies

$$(2.6) \qquad w = 0, \qquad \partial w/\partial n = 0 \quad \text{on } \partial\Omega.$$

Suppose a particular set of boundary conditions are chosen for both the stress and deflection. Let

$$Y = \{f \in H^2(\Omega): \text{boundary conditions for the stress function are satisfied}\}$$

$$X = \{w \in H^2(\Omega): \text{boundary conditions for the deflection are satisfied}\}$$

and let Δ_Y^{-1}, Δ_X^{-1} be the inverse of the Laplacian in the corresponding spaces Y, X. With this notation, one can solve for f in (2.1) as

$$(2.7) \qquad f = -\tfrac{1}{2}\Delta_Y^{-2}[w,w] - \alpha\Delta_Y^{-2}[w, w_0].$$

If this relation is substituted into the equation for w in (2.1) and the operator Δ_X^{-2} is applied, one obtains

$$(2.8) \qquad C(w) + \alpha Q(w) + (I - \lambda L + \alpha^2\Lambda)w - \alpha\lambda p + \varepsilon q = 0$$

where

$$
(2.9) \qquad
\begin{aligned}
Lw &= \Delta_X^{-2}[w, F] = -\Delta_X^{-2}w_{xx} \\
\Lambda w &= \Delta_X^{-2}[w_0, \Delta_Y^{-2}[w, w_0]] \\
Q(w) &= \Delta_X^{-2}\{[w, \Delta_Y^{-2}[w, w_0]] + [w_0, \Delta_Y^{-2}[w, w]]\} \\
C(w) &= \tfrac{1}{2}\Delta_X^{-2}[w, \Delta_Y^{-2}[w, w]] \\
p &= \Delta_X^{-2}[w_0, F] = -\Delta_X^{-2}w_{0xx} \\
q &= -\Delta_X^{-2}h
\end{aligned}
$$

One is interested in studying the behavior of the solutions of Equation (2.8) for $(\lambda, \alpha, \varepsilon)$ near $(\lambda_0, 0, 0)$, where λ_0^{-1} is the *first* eigenvalue of the linear operator L. This eigenvalue λ_0^{-1} depends on the parameter l. If λ_0^{-1} is a simple eigenvalue, the bifurcation diagrams are insensitive to small variations in l. However, if there is an l_0 such that λ_0^{-1} is a multiple eigenvalue, this is no longer the case and the parameter l also must be considered as one of the bifurcation parameters which varies in a neighborhood of l_0.

We consider only the first eigenvalue λ_0^{-1} because we are interested in the buckled states that can be stable. We will always understand stable to mean that the eigenvalues of the linear approximation have negative real parts.

8.3. The Linearized Problem

In this section, we discuss the eigenvalues of the linear operator L corresponding to the boundary conditions (2.4) and (2.5) on the deflection. This is equivalent to discussing the nontrivial solutions of the equation

$$(3.1) \qquad \Delta^2 w + \lambda w_{xx} = 0 \quad \text{in } \Omega$$

subject to boundary conditions on w.

If $w(x, y) = u(x)v(y)$, then

$$(3.2) \qquad \frac{u^{(4)} + \lambda u''}{u} + 2\frac{u''}{u}\frac{v''}{v} + \frac{v^{(4)}}{v} = 0$$

If the plate is simply supported; that is, w satisfies (2.4), then u, v must satisfy

$$(3.3) \qquad \begin{aligned} u(0) &= u(\ell\pi) = u''(0) = u''(\ell\pi) = 0 \\ v(0) &= v(\pi) = v''(0) = v''(\pi) = 0 \end{aligned}$$

The only nontrivial functions u, v consistent with these boundary conditions are

$$(3.4) \qquad u(x) = \sin\frac{k}{l}x, \qquad v(y) = \sin my$$

where k, m are positive integers. Furthermore, the smallest eigenvalue will correspond to $m = 1$, so we consider only this case. If $u(x) = \sin kx/l$, $v(x) = \sin y$, then (3.2) implies

$$(3.5) \qquad \lambda_k = \left(\frac{k}{l} + \frac{l}{k}\right)^2, \qquad k = 1, 2, \ldots$$

If k is considered as a continuous parameter, this function has a unique minimum at $k = l$. Thus, the first eigenvalue is simple unless l is chosen such that $\lambda_k = \lambda_{k+1}, k < l < k + 1$. This implies

$$(3.6) \qquad l^2 = k(k + 1)$$

We refer to l as a *noncritical length* if the first eigenvalue is simple and a *critical length* if the first eigenvalue is double.

If the plate is clamped at $x = 0$, $x = \ell\pi$ and simply supported otherwise, then w satisfies (2.5) and u, v satisfy

$$(3.7) \quad \begin{aligned} \text{(a)} \qquad & u(0) = u(\ell\pi) = u'(0) = u'(\ell\pi) = 0 \\ \text{(b)} \qquad & v(0) = v(\pi) = v''(0) = v''(\pi) = 0 \end{aligned}$$

For the first eigenvalue, we again have $v(y) = \sin y$ and Equation (3.2) implies

$$(3.8) \qquad u^{(4)} + (\lambda - 2)u'' + u = 0.$$

For $\lambda \leq 4$, the solutions to Equation (3.8) are nonoscillatory and it is impossible to satisfy the boundary conditions (3.7). For $\lambda > 4$, the solutions of (3.8) are linear combinations of $\cos ax$, $\sin ax$, $\cos bx$, $\sin bx$, where

$$a = (\alpha - \beta)^{1/2}, \qquad b = (\alpha + \beta)^{1/2}, \qquad \alpha = \frac{\lambda}{2} - 1, \qquad \beta = \left(\frac{\lambda^2}{4} - \lambda\right)^{1/2}$$

The boundary condition $u(0) = u'(0) = 0$ implies that u must be a linear combination of

$$\varphi = b \sin ax - a \sin bx, \qquad \psi = \cos ax - \cos bx$$

In order for Equation (3.8) to have a solution $u = A\varphi + B\psi$ with $A^2 + B^2 > 0$ and satisfying $u(\ell\pi) = u'(\ell\pi) = 0$, it is necessary and sufficient that

$$(3.9) \qquad \det\begin{bmatrix} \varphi(\ell\pi), & \psi(\ell\pi) \\ \varphi'(\ell\pi) & \psi'(\ell\pi) \end{bmatrix} = 0$$

Equation (3.9) gives the eigenvalue λ as a function of l.

If this eigenvalue is double then both φ and ψ are eigenfunctions. Using the boundary conditions in (3.7b), one easily observes that

$$a = \frac{k}{l}, \qquad b = \frac{k + 2n}{l}$$

where k, n are positive integers. Since $a^2 + b^2 = \lambda - 2, b^2 - a^2 = (\lambda^2 - 4\lambda)^{1/2}$, elimination of λ from these two expressions yields $ab = 1$ and, thus,

$$(3.10) \qquad l^2 = k(k + 2n)$$

Finally, the length is noncritical (the first eigenvalue is simple) if $l^2 \neq k(k + 2n)$ and is critical (the first eigenvalue is double if $l^2 = k(k + 2n)$.

8.4. Noncritical Length

In this section, we consider the buckling of a rectangular plate under the assumption that the first eigenvalue λ_0^{-1} of the linear operator L in (2.9) is simple. If φ is a corresponding unit eigenvector, then one can apply the method of Liapunov–Schmidt for the solutions of Equation (2.8) in a neighborhood of $(w, \alpha, \lambda - \lambda_0, \varepsilon) = (0,0,0,0)$ which have the form $w = u\varphi + v$ where v is orthogonal to φ and $u \in \mathbb{R}$. The bifurcation function $f = f(u, \alpha, \lambda - \lambda_0, \varepsilon)$ has the form

$$(4.1) \quad f = c_0 u^3 + \alpha c_1 u^2 + [(\lambda - \lambda_0)c_2 + \alpha^2 c_3]u + \alpha c_4 + \varepsilon c_5 + \text{h.o.t.}$$

where each c_j is a constant and

$$(4.2) \quad \begin{aligned} \text{h.o.t.} = O(|u|^4 &+ (|\alpha|^2 + |\lambda - \lambda_0|)u^2 + (|\lambda - \lambda_0|^2 + |\alpha^3| + |\varepsilon|)|u| \\ &+ (|\alpha| + |\lambda - \lambda_0| + |\varepsilon|)^2) \end{aligned}$$

If $\langle , \rangle$ designates the inner product in X, then the constants c_j are given explicitly as

$$(4.3) \quad \begin{aligned} c_0 &= \langle \varphi, C(\varphi) \rangle, & c_1 &= \langle \varphi, Q(\varphi) \rangle, & c_2 &= -\langle \varphi, L\varphi \rangle, \\ c_3 &= \langle \varphi, \Lambda\varphi \rangle, & c_4 &= -\lambda_0 \langle \varphi, p \rangle, & c_5 &= \langle \varphi, g \rangle. \end{aligned}$$

It is possible to show that the constants c_0, c_1, c_2, c_3 are not zero. Furthermore, the constants c_4, c_5 will be nonzero for an open dense set of functions p, g. Thus, we assume they are also nonzero. The bifurcation diagrams are now easy to obtain. We do not go into detail on the derivations, but simply illustrate the results in Figures 4.1, 4.2, 4.3. In drawing these figures, we have

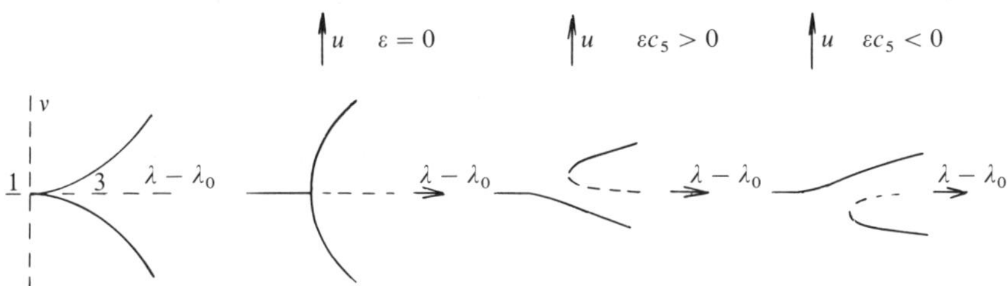

Figure 4.1. $\alpha = 0$

Figure 4.2. $\varepsilon = 0$

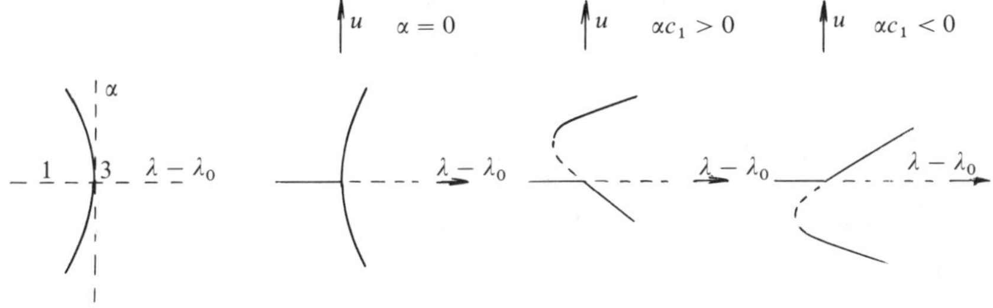

Figure 4.3

used the fact that the bifurcation function $f(u, \alpha, \lambda - \lambda_0, \varepsilon)$ satisfies the symmetry conditions

$$f(0, \alpha, \lambda - \lambda_0, 0) = 0$$
$$f(-u, 0, \lambda - \lambda_0, 0) = -f(u, 0, \lambda - \lambda_0, 0)$$

In the diagrams, we have also labeled the stable branches as solid lines and the unstable ones by dashed lines.

8.5. Critical Length

In this section, we assume that the first eigenvalue λ_0^{-1} of the operator L in (2.9) is double; that is, the side of length l assumes a critical value, say l_0. Let φ_1, φ_2 be orthogonal unit eigenvectors corresponding to λ_0^{-1}. We can now apply the method of Liapunov–Schmidt for the solutions of Equation (2.8) in a neighborhood of $(w, \alpha, \lambda - \lambda_0, \varepsilon, l - l_0) = (0, 0, 0, 0, 0)$ which have the form

$$w = u_1\varphi_1 + u_2\varphi_2 + v$$

with v orthogonal to φ_1, φ_2 and $u = (u_1, u_2) \in \mathbb{R}^2$. Let

$$\lambda - \lambda_0 = \delta\lambda_0 \qquad l - l_0 = \tau$$

The bifurcation function $f(u, \alpha, \delta, \varepsilon, \tau) \in \mathbb{R}^2$ has the form

$$(5.1) \quad f(u, \alpha, \delta, \varepsilon, \tau) = c(u) + \alpha q(u) - \delta u + \alpha^2 Mu - \tau Nu - \alpha m + \varepsilon k + \text{h.o.t.}$$

where

$$\text{h.o.t.} = O(|u|^4 + (|\alpha|^2 + |\delta| + |\tau|)|u|^2 + (|\alpha| + |\delta| + |\tau| + |\varepsilon|)^2)$$

If $\langle, \rangle$ is the inner product in X, then the coefficients in this function are given explicitly by the relations

$$(5.2) \quad
\begin{aligned}
c &= (c_1, c_2), & c_j(u) &= \langle \varphi_j, C(u_1\varphi_1 + u_2\varphi_2) \rangle \\
q &= (q_1, q_2), & q_j(u) &= \langle \varphi_j, Q(u_1\varphi_1 + u_2\varphi_2) \rangle \\
M &= (M_{ij}), & M_{ij} &= \langle \varphi_i, \Lambda\varphi_j \rangle \\
m &= (m_1, m_2), & m_j &= \lambda_0\langle \varphi_j, p \rangle \\
k &= (k_1, k_2), & k_j &= \langle \varphi_j, g \rangle
\end{aligned}$$

$$N = \begin{bmatrix} \sigma_1 & 0 \\ 0 & \sigma_2 \end{bmatrix}, \qquad \sigma_1 \neq 0, \sigma_2 \neq 0.$$

The computation of the constants σ_1, σ_2 is more difficult than the others since it requires knowledge about the manner in which the double eigenvalue λ_0^{-1} splits into two simple eigenvalues as one changes the parameter l.

It is not difficult to verify that the function $f = (f_1, f_2)$ has the following symmetry properties

$$
\begin{aligned}
f_1(u, 0, \delta, 0, \tau) &= u_1 g_1(u, \delta, \tau) \\
f_2(u, 0, \delta, 0, \tau) &= u_2 g_2(u, \delta, \tau) \\
g_j(-u, \delta, \tau) &= g_j(u, \delta, \tau), \qquad j = 1, 2.
\end{aligned}
$$
(5.3)

Using (5.2), these symmetry relations and replacing f_j by $\eta_j f_j$ where η_j is a nonzero constant, one can show that

$$
\begin{aligned}
c(u) &= (u_1^3 + \mu u_1 u_2^2, v u_1^2 u_2 + u_2^3) \\
&\mu > 0, \ v > 0 \text{ constants.}
\end{aligned}
$$
(5.4)

The constants μ, v depend upon the boundary conditions since they depend upon the eigenfunctions. For the simply supported plate, one can verify that $\mu > 1$, $v > 1$. For the plate clamped at $x = 0$, $x = l$ and simply supported otherwise, one can show that $v < 1$, $\mu v > 1$. From Section 7.5, this suggests that the boundary conditions can have an important effect on the bifurcation phenomena.

To amplify the latter remark, let us consider in more detail the case where there is no curvature and no imperfection; that is, $\alpha = 0$, $v = 0$. The bifurcation functions are then given by (5.3) and the truncated form for $f(u, 0, \delta, 0, \tau) = 0$ is

$$
c(u) - \delta u - \tau N u = 0;
$$

or, if $\delta_1 = \delta + \tau \sigma_1, \sigma = \sigma_2 - \sigma_1$,

$$
\begin{aligned}
u_1^3 + \mu u_1 u_2^2 - \delta_1 u_1 &= 0 \\
v u_1^2 u_2 + u_2^3 - \delta_1 u_2 - \tau \sigma u_2 &= 0
\end{aligned}
$$

These equations are the same as the ones considered in Section 7.5. Since the symmetry conditions (5.3) are valid for f, the complete bifurcation diagrams are shown in Figures 7.5.3 through 7.5.7.

For $\tau \sigma < 0$ and the simply supported plate, the appropriate diagram is Figure 7.5.3 since $\mu > 1$, $v > 1$ and is Figure 7.5.6 for $\tau \sigma < 0$. For $\tau \sigma > 0$ and the plate clamped at $x = 0$, $x = l$, and simply supported otherwise the appropriate diagram is Figure 7.5.7 since $v < 1$, $\mu v > 1$ and is Figure 7.5.5 for $\tau \sigma < 0$. These curves are reproduced in Figures 5.1, 5.2, 5.3, where we have also designated the branches which are stable by solid lines and the ones which are unstable by dashed lines. The manner in which the plate

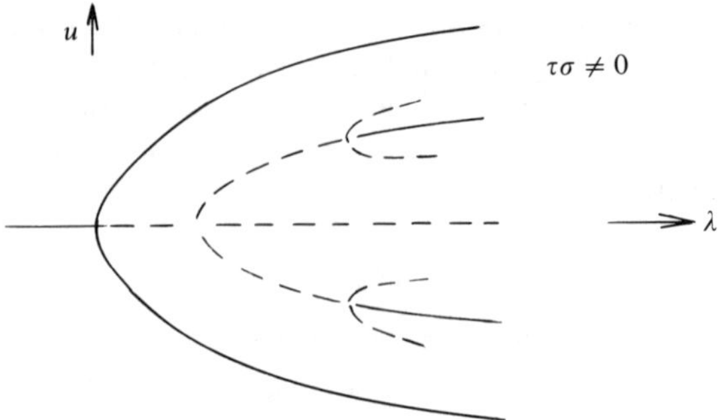

Figure 5.1. Simply supported.

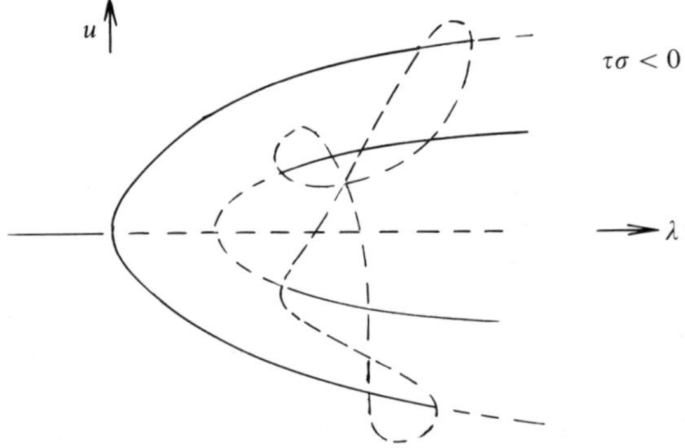

Figure 5.2. Clamped at $x = 0$, $x = l$.

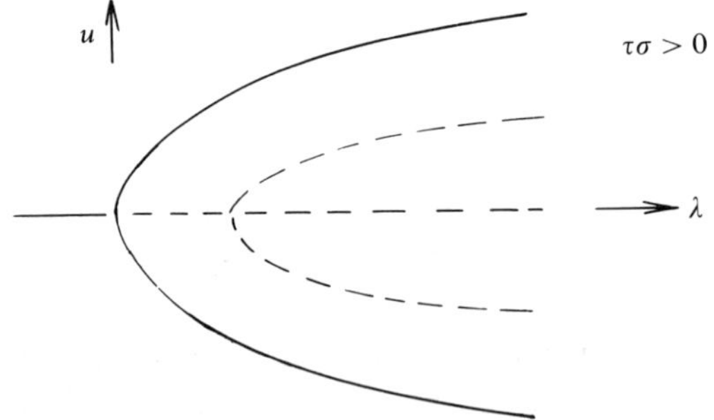

Figure 5.3. Clamped at $x = 0$, $x = l$

buckles from one state to another as the loading parameter λ changes clearly is completely different depending upon the boundary conditions.

As another illustration let us keep the length l fixed at a critical value, assume no curvature and consider the effect of varying the loading parameter and the imperfection parameter; that is, consider the case where $l = l_0$,

$\alpha = 0$. The truncated bifurcation equation is given by

$$c(u) - \delta u + \varepsilon k = 0$$

or

$$u_1^3 + \mu u_1 u_2^2 - \delta u_1 + \varepsilon k_1 = 0$$
$$v u_1^2 u_2 + u_2^3 - \delta u_2 + \varepsilon k_2 = 0$$

This is precisely the equation considered in Section 7.6. There were two possible bifurcation diagrams depending on whether (μ, v) belonged to the region $U_1 = \{v > 1, \mu > 1\} \cup \{v < 1, \mu < 1\}$ or the region $U_2 = \{v > 1, \mu < 1\} \cup \{v < 1, \mu > 1\}$. The simply supported plate has the bifurcation diagram corresponding to U_1 and the plate clamped at $x = 0$, $x = l$ and simply supported otherwise has the bifurcation diagram corresponding to the region U_2. These diagrams are reproduced in Figures 5.4, 5.5, with the stable branches labeled with solid lines and the unstable ones with dashed lines and for $k_1 > 0$, $k_2 > 0$.

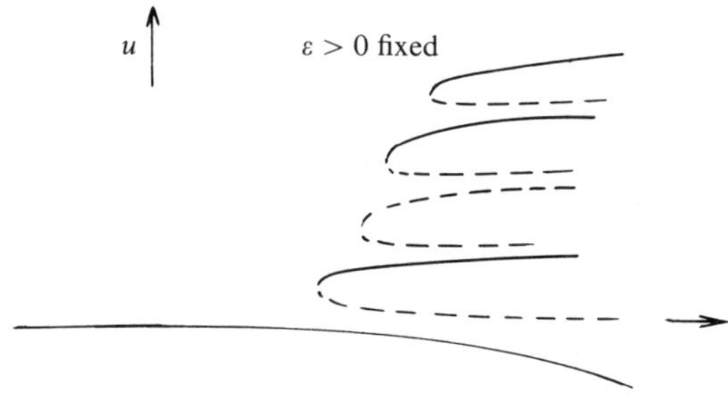

Figure 5.4

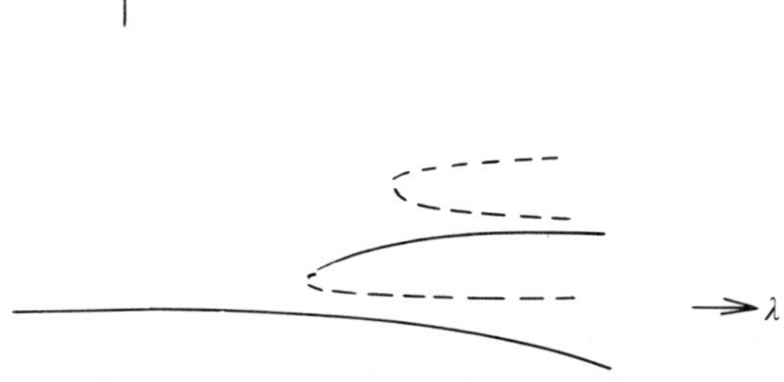

Figure 5.5. Simply supported at $x = 0$, $x = l$

8.6. An Example in Chemical Reactions

In modeling certain chemical reactions, one obtains the equation

$$\frac{\partial x}{\partial t} = A - (B+1)x + x^2 y + D\,\frac{\partial^2 x}{\partial r^2}, \qquad 0 < r < 1, t > 0$$

$$\frac{\partial y}{\partial t} = Bx - x^2 y + vD\,\frac{\partial^2 y}{\partial r^2}, \qquad 0 < r < 1, t > 0$$

with boundary conditions

$$x(0,t) = x(1,t) = A,$$
$$y(0,t) = y(1,t) = B/A,$$

for the concentrations x, y of the chemicals, where A, B, D, v are constants.

This set of equations with the above boundary conditions has the solution $x_0 = A$, $y_0 = B/A$. The problem is to determine how the steady state solutions of this problem in a neighborhood of (x_0, y_0) depend upon the relative diffusion coefficient v and the strength B of the interaction.

If $x = x_0 + u$, $y = y_0 + v$, then u, v satisfy the equations

$$\frac{\partial u}{\partial t} = [B - 1]u + A^2 v + D\,\frac{\partial^2 u}{\partial r^2} + \frac{B}{A}u^2 + 2Auv + u^2 v$$

$$\frac{\partial v}{\partial t} = -Bu - A^2 v + vD\,\frac{\partial^2 v}{\partial r^2} - \frac{B}{A}u^2 - 2Auv - u^2 v$$

with homogeneous boundary conditions at $r = 0$, $r = 1$. The steady state solutions must satisfy the equations

$$0 = [B - 1]u + A^2 v + D\,\frac{d^2 u}{dr^2} + \frac{B}{A}u^2 + 2Auv + u^2 v$$

(6.1)

$$0 = -Bu - A^2 v + vD\,\frac{d^2 v}{\partial r^2} - \frac{B}{A}u^2 - 2Auv - u^2 v$$

with boundary conditions

(6.2)
$$u(0) = v(0) = u(1) = v(1) = 0$$

Let Y be the Banach space of twice continuously differentiable functions from $[0, 1]$ to $\mathbb{R}^2$ which satisfy (6.2) and $u''(0) = v''(0) = u''(1) = v''(1) = 0$. The norm is chosen to be the usual C^2-norm. Any solution of (6.1), (6.2) must be in Y. Let X be the Banach space of continuous functions from $[0, 1]$ to $\mathbb{R}^2$ which satisfy (6.2) with the norm the usual C^0-norm.

If

$$
A(v, B)(u, v) =
\begin{bmatrix}
(B - 1) + D\dfrac{d^2}{dr^2} & A^2 \\[2ex]
-B & -A^2 + vD\dfrac{d^2}{dr^2}
\end{bmatrix}
\begin{bmatrix} u \\ v \end{bmatrix}
$$

$$
N(B, u, v) =
\begin{bmatrix}
BA^{-1}u^2 + 2Auv + u^2v \\
-BA^{-1}u^2 - 2Auv - u^2v
\end{bmatrix}
$$

Then $A(v, B): Y \to X$ is a continuous linear operator for each (v, B) in $\mathbb{R}^2$, and $N: \mathbb{R} \times Y \to X$ is continuous.

In this notation, problem (6.1), (6.2) is equivalent to the equation

$$
(6.3) \qquad\qquad 0 = A(v, B)(u, v) + N(B, u, v)
$$

If we replace $\mathbb{R}^2$ by $\mathbb{C}^2$ in the above, we see that $A(v, B)$ depends on v and B analytically and N depends on B, u and v analytically.

The eigenvalues of the linear operator $A(v, B)$ are simple and given by

$$
\sigma_n = \tfrac{1}{2}\left[L_n + (L_n^2 - 4A^2B + 4(B - 1 - n^2\pi^2D)(A^2 + n^2\pi^2vD))^{1/2}\right]
$$
$$
(6.4) \quad \mu_n = \tfrac{1}{2}\left[L_n - (L_n^2 - 4A^2B + 4(B - 1 - n^2\pi^2D)(A^2 + n^2\pi^2vD))^{1/2}\right]
$$
$$
L_n = B - 1 - A^2 - n^2\pi^2D(1 + v), \qquad n = 1, 2, \ldots
$$

The corresponding eigenfunctions are vector multiples of $\sin n\pi r$ and will be computed explicitly later.

We now determine a curve in the real (v, B)-plane such that all the eigenvalues of $A(v, B)$ have negative real part when (v, B) lies below this curve, that is, the *curve of neutral stability*. We can see that $\operatorname{Re} \sigma_n \geq \operatorname{Re} \mu_n$ for all n. If σ_n is complex, then the curve $\operatorname{Re} \sigma_n = 0$ is given by the straight line

$$
(6.5) \qquad\qquad B = 1 + A^2 + n^2\pi^2D(1 + v).
$$

If σ_n is real, the curve $\sigma_n = 0$ is given by the hyperbola H_n,

$$
(6.6) \qquad\qquad B = 1 + n^2\pi^2D + \frac{A^2}{v}\left\{1 + \frac{1}{n^2\pi^2D}\right\}.
$$

A typical curve $\operatorname{Re} \sigma_n = 0$ is given in Figure 6.1. The curve we seek is obtained by joining portions of the curves $\operatorname{Re} \sigma_n = 0$ which lie lowest for each (v, B). For sufficiently small v, it is always the case that the neutral stability curve corresponds to the case where σ_1 is complex, $\operatorname{Re} \sigma_1 = 0$; that is, it is a portion of the line (6.5) for $n = 1$. For v sufficiently large, the neutral stability curve always corresponds to the case where $\sigma_1 = 0$; that is, it is a portion of the

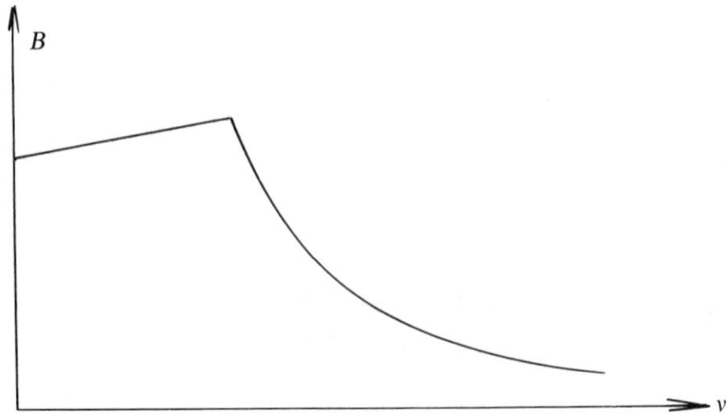

Figure 6.1. The curve Re $\sigma_n = 0$.

hyperbola (6.6) for $n = 1$. The remaining portions of the curve are portions of hyperbolas H_j corresponding to $\sigma_j = 0, j = 2, 3, \ldots, n_0(A, D)$. The number $n_0(A, D)$ of hyperbolas H_j depends on A, D and they always have the ordering shown in Figure 6.2 for $n_0(A, D) = 4$. A typical curve is always scalloped. Each point on the line segment except the point of intersection with a hyperbola corresponds to a pair of purely imaginary eigenvalues of the operator $A(v, B)$. Generically, this will correspond to a Hopf bifurcation in the original dynamic equations and will be discussed in Section 9.7. The other points of nonintersection correspond to a simple eigenvalue zero of the operator $A(v, B)$. An application of the method of Liapunov–Schmidt leads to a bifurcation function $h(\alpha, v, B)$, $\alpha \in \mathbb{R}$, which satisfies $h(\alpha, 0, 0) = \beta(A, D)\alpha^3 + O(|\alpha|^4)$ as $|\alpha| \to 0$ and generically $\beta(A, D) \neq 0$. Thus, there is a bifurcation into three equilibrium solutions at this point. This situation is very simple and will not be discussed further.

The point of intersection of the straight line with a hyperbola corresponds either to a double eigenvalue zero of the operator $A(v, B)$ with nonsimple elementary divisors or to two purely imaginary and one zero eigenvalue of the operator $A(v, B)$. A complete discussion of the nature of the bifurcation

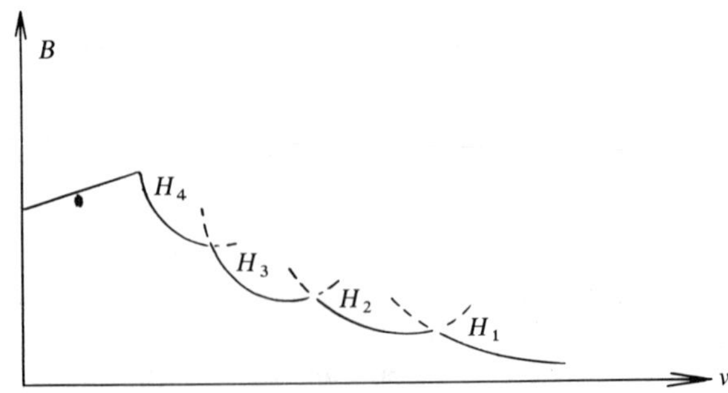

Figure 6.2. A typical curve of neutral stability.

at this point involves the original dynamical equations. It is rather complicated and can be given following the ideas in Sections 13.2–13.4.

The points of intersection (v_c, B_c) of two hyperbolas H_n and H_{n+1} correspond to a double zero eigenvalue with simple elementary divisors. It is
our objective to discuss the bifurcations near these points. In the remainder
of the discussion, we suppose that (v, B) are at the intersection (v_c, B_c) of H_n
and H_{n+1}. This point is given by

$$v_c = \frac{A^2}{n^2(n+1)^2(\pi^2 D)^2}$$

$$(6.7) \qquad B_c = 1 + n^2\pi^2 D + n^2(n+1)^2(\pi^2 D)^2 \left\{ 1 + \frac{1}{n^2\pi^2 D} \right\}$$

$$= 1 + n^2\pi^2 D + (n+1)^2\pi^2 D + n^2(n+1)^2(\pi^2 D)^2$$

$$= (1 + n^2\pi^2 D)(1 + (n+1)^2\pi^2 D)$$

We systematically apply the method of Liapunov–Schmidt. To do this,
we need explicit expressions for an eigenvector Φ_n of the operator $A(v, B)$
corresponding to σ_n as well as an eigenvector Ψ_n of the adjoint operator
corresponding to σ_n:

$$\begin{bmatrix} (B-1) + D\dfrac{\partial}{\partial r^2} & -B \\[2ex] A^2 & -A^2 + vD\dfrac{\partial^2}{\partial r^2} \end{bmatrix} \Psi_n = \sigma_n \Psi_n$$

with $\Psi_n(0) = \Psi_n(1) = 0$. A few calculations using the above expression for
B_c in (6.7) show that

$$(6.8) \qquad \Phi_n(r) = (1, M_n)\sin n\pi r, \qquad \Psi_n = (1, N_n)\sin n\pi r$$

where

$$(6.9) \qquad \begin{aligned} M_n &= -A^{-2}[B_c + 1 - n^2\pi^2 D] \\ M_{n+1} &= -A^{-2}[B_c + 1 - (n+1)^2\pi^2 D] \\ N_n &= B_c^{-1}[B_c - 1 - n^2\pi^2 D] \\ N_{n+1} &= B_c^{-1}[B_c - 1 - (n-1)^2\pi^2 D]. \end{aligned}$$

Let P_n, Q_n be respectively a projection onto the span of Φ_n, Ψ_n defined by

$$(6.10) \qquad \begin{aligned} P_n(u, v) &= \frac{2}{1 + M_n N_n}\left(\int_0^1 [u(r) + N_n v(r)]\sin n\pi r \, dr \right)\Phi_n \\[2ex] Q_n(u, v) &= \frac{2}{1 + N_n^2}\left(\int_0^1 [u(r) + N_n v(r)]\sin n\pi r \, dr \right)\Psi_n \end{aligned}$$

With these projections, define $P_X = (I - P_n - P_{n+1})X$, $Q_Y = (I - Q_n - Q_{n+1})Y$. For any $(u, v) \in X$, there are unique scalars x, y and $w \in P_X$ such that

$$(6.11) \qquad (u, v) = x\Phi_n + y\Phi_{n+1} + w, \qquad w \in P_X$$

Using the above projections and the decomposition (6.11), we may apply the method of Liapunov–Schmidt. Since the range of $A(v, B)$ is Q_Y, the auxiliary equation in the application of this method is

$$(6.12) \qquad 0 = (I - Q_n - Q_{n+1})C(v, B, x\Phi_n + y\Phi_{n+1} + w)$$
$$C(v, B, (u, v)) = A(v, B)(u, v) + N(B, u, v)$$

Equation (6.12) has a unique solution $w = w(x, y, v, B)$ analytic in a neighborhood of $(0, 0, v_c, B_c)$ and satisfying $w(0, 0, v, B) = 0$. If we define functions F, G by the relation

$$(6.13) \qquad F(x, y, v, B)\Psi_n = Q_n C(v, B, x\Phi_n + y\Phi_{n+1} + w(x, y, v, B))$$
$$G(x, y, v, B)\Psi_{n+1} = Q_{n+1}C(v, B, x\Phi_n + y\Phi_{n+1} + w(x, y, v, B))$$

then the function (u, v) in (6.11) with $w = w(x, y, v, B)$ is a solution of (6.3) if and only if (x, y, v, B) satisfy the bifurcation equations

$$(6.14) \qquad F(x, y, v, B) = 0$$
$$G(x, y, v, B) = 0$$

The functions F, G satisfy

$$(6.15) \qquad F(0, 0, v, B) = G(0, 0, v, B) = 0$$
$$\partial F(0, 0, v_c, B_c)/\partial(x, y) = 0, \qquad \partial G(0, 0, v_c, B_c)/\partial(x, y) = 0.$$

The bifurcation functions in (6.14) satisfying certain symmetry properties which will be needed to obtain information about the bifurcation curves.

Proposition 6.1. *If n is odd, then*

$$(6.16) \qquad F(x, -y, v, B) = F(x, y, v, B)$$
$$G(x, -y, v, B) = -G(x, y, v, B)$$

If n is even, then

$$(6.17) \qquad F(-x, y, v, B) = -F(x, y, v, B)$$
$$G(-x, y, v, B) = G(x, y, v, B)$$

Proof. We only prove this for the case n odd since the other case is similar. For any $(u, v) \in X$, let $\theta(u, v)(r) = (u(1 - r), v(1 - r))$, $0 \le r \le 1$. The operator θ commutes with $A(v, r)$ and $N(B, \cdot)$ and

$$\theta(x\Phi_n + y\Phi_{n+1} + w) = x\Phi_n - y\Phi_{n+1} + \theta w$$

Since the solution $w(x, y, v, B)$ of (6.12) is unique, this implies $w(x, -y, v, B) = w(x, y, v, B)$. From the definition of F, G in (6.13), this implies (6.16). $\quad\square$

The remainder of the discussion deals only with the case where n is odd. The even case is treated in a similar way. From Relations (6.15) and (6.16), the bifurcations equations (6.14) have the form

$$(6.18) \quad \begin{aligned} 0 &= F(x, y, v, B) = \tau x + ax^2 + by^2 + F_1(x, y^2, \tau, \eta) \\ 0 &= G(x, y, v, B) = \eta y + cxy + yF_2(x, y^2, \tau, \eta) \end{aligned}$$

where $\eta = \eta(v, B)$, $\tau = \tau(v, B)$ vanish for $(v, B) = (v_c, B_c)$ and $F_1, F_2 = O((|\tau| + |\eta|)^2(|x| + |y|) + (|\tau| + |\eta|)(|x| + |y|)^2 + (|x| + |y|)^3)$ as $x, y, \tau, \eta \to 0$.
One can show that the functions η, τ satisfy

$$\tau(v, B) = \frac{\partial \sigma_n}{\partial B}(v_c, B_c)(B - B_c) + \frac{\partial \sigma_n}{\partial v}(v_c, B_c)(v - v_c)$$

$$+ O(|v - v_c| + |B - B_c|)$$

$$\eta(v, B) = \frac{\partial \sigma_{n+1}}{\partial B}(v_c, B_c)(B - B_c) + \frac{\partial \sigma_{n+1}}{\partial v}(v_c, B_c)(v - v_c)$$

$$+ O(|v - v_c| + |B - B_c|)$$

as $v - v_c, B - B_c \to 0$. Thus, we conclude that $\tau(v, B)$ has the same sign as σ_n, $\eta(v, B)$ has the same sign as σ_{n+1} in a neighborhood of (v_c, B_c).
It is not difficult to show that

$$a = \frac{8}{2n\pi A}(n^3\pi^3 D - 1)\frac{1}{1 + N_n}$$

$$b = \frac{2}{\pi A}\left(\frac{1}{n} - \frac{1}{6n + 4} + \frac{2}{n + 4}\right)((n + 1)^2\pi^2 D - 1)\frac{1}{1 + N_n}$$

$$c = \frac{2}{\pi A}\left(\frac{1}{n} - \frac{1}{6n + 4} + \frac{2}{n + 4}\right)(\pi^4 D^2 n^2(n + 1)^2 - 4)\frac{1}{1 + N_{n+1}}$$

Also, $1 + N_n > 0$ for all n.

The lower order terms in (6.18) are precisely the same type as the example considered in Section 7.4 and the analysis of the bifurcation diagrams are carried out in the same way. We do not consider all possibilities, but assume $ab > 0, c \neq 0$. Then we obtain the equivalent equations

$$
\begin{aligned}
(6.19) \quad 0 &= -\lambda_1 x + x^2 + vy^2 + \tilde{F}_1(x, y^2, \lambda_1, \lambda_2) \\
0 &= -(\lambda_1 + \lambda_2)x + \mu xy + y\tilde{F}_2(x, y^2, \lambda_1, \lambda_2)
\end{aligned}
$$

where $v = b/a$, $\mu = |c|$, $\lambda_1 = -\tau/a$, $\lambda_1 + \lambda_2 = -\eta \operatorname{sgn} c$. The analysis in Section 7.4 is applicable to this equation even though it does not contain the asymmetry term σy in the first equation. This term was included in Section 7.4 because we could not analyze the solutions in a neighborhood of the λ_2-axis. However, the symmetry in y implies that there is no bifurcation taking place in a sector containing the λ_2-axis. Thus, the results in Section 7.4 carry over to this case. The bifurcation diagrams are shown in Figure 6.3 in terms of the original variables τ, η, a, b, c. In case $ab < 0$, the analysis in section 7.4 is easily extended.

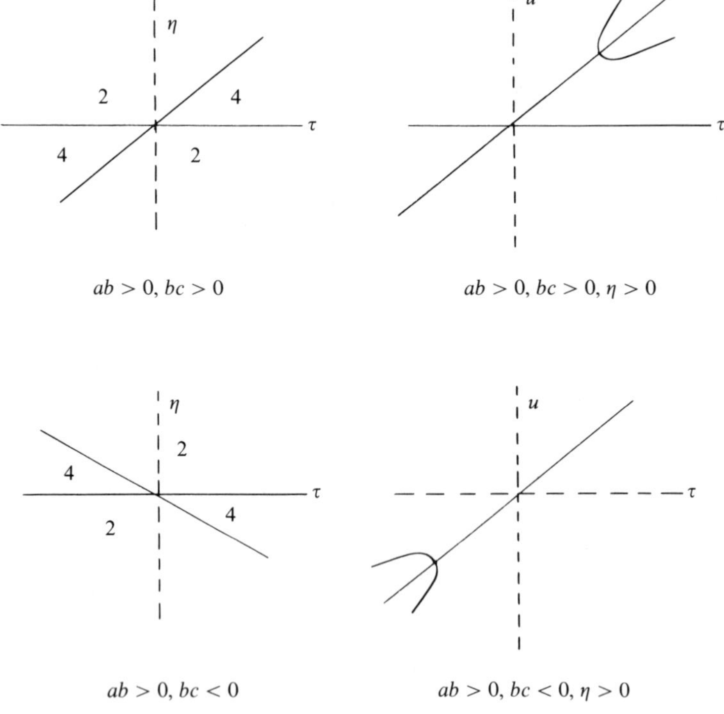

$$ab > 0, bc > 0 \qquad\qquad ab > 0, bc > 0, \eta > 0$$

$$ab > 0, bc < 0 \qquad\qquad ab > 0, bc < 0, \eta > 0$$

Figure 6.3

8.7. The Duffing Equation with Harmonic Forcing

Suppose $\lambda = (\lambda_1, \lambda_2, \lambda_3)$ is in $\mathbb{R}^3$ and x is a scalar. Our objective in this section is to discuss the number of 2π-periodic solutions of the Duffing equation

$$\ddot{x} + x = \lambda_1 x + \lambda_2 \dot{x} - x^3 + \lambda_3 \cos t \tag{7.1}$$

in a neighborhood of zero for λ in a neighborhood of zero.

We are going to apply the method of Liapunov–Schmidt, but include the details since a few modifications are required. Any 2π-periodic solution of Equation (7.1) for $\lambda = 0$ must be equal to $r \cos(t - \phi) + O(r^2)$ as $r \to 0$ for some constants $r, \phi, -\pi/2 \le \phi \le \pi/2$. Therefore, by letting $t \to t + \phi$, we will obtain a solution of our problem by considering the 2π-periodic solutions of the equation

$$\ddot{x} + x = \lambda_1 x + \lambda_2 \dot{x} - x^3 + \lambda_3 \cos(t + \phi) \tag{7.2}$$

which, for $\lambda = 0$, are equal to $r \cos t + O(r^2)$ as $r \to 0$.

Let $\mathcal{H} = \{h : R \to R : h \text{ is continuous } h(t + 2\pi) = h(t)\}$ and for any $h \in \mathcal{H}$, let $|h| = \sup_t |h(t)|$. Let $P : \mathcal{H} \to \mathcal{H}$ be the projection defined by

$$(Ph)(t) = \frac{1}{\pi} \cos t \int_0^{2\pi} h(s)\cos s \, ds + \frac{1}{\pi} \sin t \int_0^{2\pi} h(s)\sin s \, ds. \tag{7.3}$$

For any $h \in \mathcal{H}$, the equation

$$\ddot{x} + x = h \tag{7.4}$$

has a solution in $\mathcal{H}$ if and only if $Ph = 0$. Furthermore, there is a continuous linear operator $K : (I - P)\mathcal{H} \to (I - P)\mathcal{H}$ such that $K(I - P)h$ is the unique solution of

$$\ddot{x} + x = (I - P)h \tag{7.5}$$

which satisfies $PK(I - P)h = 0$; that is, $K(I - P)h$ is simply the 2π-periodic solution of Equation (7.4) which does not contain $\cos t$, $\sin t$ in its Fourier series. To this solution $K(I - P)h$, one can add an arbitrary linear combination of $\sin t$ and $\cos t$ to obtain the general solution of Equation (7.5). As remarked earlier, it is only necessary for us to add a term $r \cos t$ since the phase shift of the solution is included in the forcing function.

If r is fixed and we define

$$P_r = \{h \in P : (Ph)t = r \cos t\},$$
$$f(x, \lambda_1, \lambda_2) = \lambda_1 x + \lambda_2 \dot{x} - x^3$$

then x is a solution of Equation (7.2) in P_r if and only if

$$(7.6a) \qquad x = r \cos t + w, \qquad w \in (I - P)\mathscr{H},$$

$$(7.6b) \qquad \ddot{w} + w = (I - P)f(r \cos(\cdot) + w, \lambda_1, \lambda_2),$$

$$(7.6c) \qquad P[f(r \cos(\cdot) + w, \lambda_1, \lambda_2) + \lambda_3 \cos(\cdot + \phi)] = 0,$$

since $(I - P)\cos(\cdot + \phi) = 0$.

By an application of the implicit function theorem, there are $\delta > 0, \varepsilon > 0$, such that for $|r| < \delta, |\lambda_1| + |\lambda_2| < \varepsilon$, there is a unique solution $w^*(r, \lambda_1, \lambda_2)$ of Equation (7.6b) in $(I - P)\mathscr{H}$, the function $w^*(r, \lambda_1, \lambda_2)$ is analytic in $(r, \lambda_1, \lambda_2)$ and $w^*(0, \lambda_1, \lambda_2) = 0$. Furthermore, it is very easy to see that $w^*(r, \lambda_1, 0)$ is an even function of t since $x = r \cos t + w$ and only odd powers of x occur on the right hand side of Equation (7.6b) for $\lambda_2 = 0$.

Since $w^*(r, \lambda_1, \lambda_2)$ is uniquely determined, it follows that there is a solution $x = r \cos t + w \in \mathscr{H}$ of Equation (7.2) which lies in a sufficiently small neighborhood of zero for λ in a sufficiently small neighborhood of zero if and only if $x = r \cos t + w^*(r, \lambda_1, \lambda_2)$ and the vector (r, ϕ, λ) satisfies the bifurcation equations

$$P[f(r \cos(\cdot) + w^*(r, \lambda_1, \lambda_2), \lambda_1, \lambda_2) + \lambda_3 \cos(\cdot + \phi)] = 0.$$

From the definition of P in Equation (7.3), these latter equations are equivalent to the system of equations

$$G_1(r, \phi, \lambda) \overset{\text{def}}{=} \lambda_3 \cos \phi + \frac{1}{\pi} \int_0^{2\pi} f(r \cos t + w^*(r, \lambda_1, \lambda_2)(t), \lambda_1, \lambda_2)\cos t \, dt = 0$$

$$G_2(r, \phi, \lambda) \overset{\text{def}}{=} \lambda_3 \sin \phi - \frac{1}{\pi} \int_0^{2\pi} f(r \cos t + w^*(r, \lambda_1, \lambda_2)(t), \lambda_1, \lambda_2)\sin t \, dt = 0$$

Since $w^*(r, \lambda_1, 0)$ is an even function, it is clear that $G_2(r, \phi, \lambda_1, 0, \lambda_3) = \lambda_3 \sin \phi$. For $w^*(r, \lambda_1, \lambda_2) = 0$, it is easy to evaluate the above integrals. If this computation is made and one uses the Taylor series to obtain the order estimate $O(|\lambda_1 r| + |\lambda_2 r| + |r|^3)$ on $w^*(r, \lambda_1, \lambda_2)$, the bifurcation equations become

$$(7.7a) \qquad G_1(r, \phi, \lambda) = \lambda_3 \cos \phi + \lambda_1 r - \tfrac{3}{4}r^3 + rg_1(r, \lambda_1, \lambda_2) = 0,$$

$$(7.7b) \qquad G_2(r, \phi, \lambda) = \lambda_3 \sin \phi + \lambda_2 r + \lambda_2 rg_2(r, \lambda_1, \lambda_2) = 0,$$

where

$$(7.8) \qquad \begin{aligned} g_1(r, \lambda_1, \lambda_2) &= O(|\lambda_1|^2 + |\lambda_2|^2 + |\lambda_1\lambda_2| + r^2|\lambda_1| + r^2|\lambda_2| + r^4), \\ g_2(r, \lambda_1, \lambda_2) &= O(r^2 + |\lambda_1| + |\lambda_2|). \end{aligned}$$

These results are summarized in the following lemma.

Lemma 7.1. *There is a neighborhood $U \subset R^4$ of $(r, \lambda) = 0$ and a neighborhood $V \subset \mathcal{H}$ of $x = 0$ such that Equation (7.2) has a solution $x \in \mathcal{H} \cap V$ for $(r, \lambda) \in U$ and a given ϕ if and only if $x = r \cos t + w^*(r, \lambda_1, \lambda_2)$ and (r, λ, ϕ) satisfy the bifurcation Equations (7.7) where g_1, g_2 satisfy (7.8).*

An immediate corollary is the following result on the undamped Duffing equation.

Corollary 7.2. *There is a neighborhood $U \subset R^2$ of $(\lambda_1, \lambda_3) = 0$ and a neighborhood $V \subset \mathcal{H}$ of $x = 0$ such that the only 2π-periodic solutions in V of the undamped Duffing equation*

$$(7.9) \qquad\qquad \ddot{x} + x = \lambda_1 x - x^3 + \lambda_3 \cos t$$

are even functions of t if $\lambda_3 \neq 0$.

Proof. Suppose the neighborhoods U, V are as in Lemma 7.1. Let $U_1 = \{(\lambda_1, \lambda_3): (\lambda_1, 0, \lambda_3) \in U\}$, let $(\lambda_1, \lambda_3) \in U_1$ and let x be a 2π-periodic solution of (7.9) in V. Then $x(t - \phi)$ is a solution of (7.2) for $\lambda_2 = 0$ for some ϕ in $[-\pi/2, \pi/2]$. From Lemma 7.1, $x(t - \phi) = r \cos t + w^*(r, \lambda, 0)(t)$ and $(r, \lambda_1, \lambda_3, \phi)$ satisfy the bifurcation equations (7.7) with $\lambda_2 = 0$. Thus, $\lambda_3 \sin \phi = 0$. If $\lambda_3 \neq 0$, then $\phi = 0$. This means $x(t) = r \cos t + w^*(r, \lambda_1, 0)(t)$. Since this latter function is even, the result is proved. $\square$

Using this remark, we can state the following result for the undamped equation.

Theorem 7.3. *There is a neighborhood $U \subset \mathbb{R}^2$ of $(\lambda_1, \lambda_3) = 0$ and a neighborhood $V \subset \mathcal{H}$ of $x = 0$ such that the only 2π-periodic solutions in V of Equation (7.9) for $(\lambda_1, \lambda_3) \in U$ are even in t. Furthermore, there is a curve $\Gamma \subset U$ given approximately by the cusp $\lambda_1^3 = 81\lambda_3^2/16$ such that $U \backslash \Gamma = U_1 \cup U_2$ with U_1, U_2 disjoint and Equation (7.9) has one solution in V for $(\lambda_1, \lambda_3) \in U_1$ and 3 solutions in V for $(\lambda_1, \lambda_3) \in U_2$ (see Figure 7.1).*

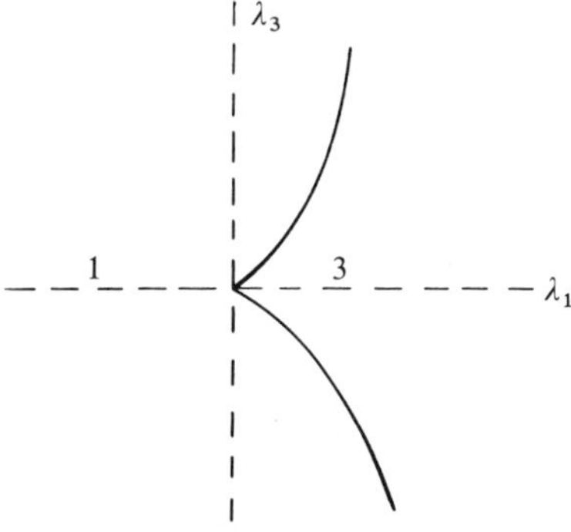

Figure 7.1

Proof. From Corollary 7.2, for $\lambda_2 = 0$, the solutions are even and, for $\phi = 0$, the bifurcation equations reduce to the single equation

$$H(r, \lambda_1, \lambda_3) = \lambda_3 + \lambda_1 r - \tfrac{3}{4} r^3 + r g_1(r, \lambda_1, 0) = 0$$

The equations $H(r, \lambda_1, \lambda_3) = 0$, $\partial H(r, \lambda_1, \lambda_3)/\partial r = 0$ define a curve Γ with the properties stated in the theorem. $\square$

The analysis of the bifurcation diagram for $\lambda_2 \neq 0$ is much more difficult. It follows the spirit of the scaling techniques used in Chapter 7, but there are a number of technical difficulties that arise because of symmetries in the bifurcation functions. Therefore, we give some of the details.

The following lemma is proved in the usual way.

Lemma 7.4. *There is a neighborhood U of zero in $\mathbb{R}^4$ and a constant $c > 0$ such that all solutions (r, ϕ) of Equations (7.7) with $(r, \lambda) \in U$ satisfy*

$$|r| \leq c(|\lambda_1|^{1/2} + |\lambda_2|^{1/2} + |\lambda_3|^{1/3})$$

Lemma 7.4 justifies the scaling $r = \lambda_3^{1/3}\rho$, $\lambda_1 = \mu_1 \lambda_3^{2/3}$, $\lambda_2 = \mu_2 \lambda_3^{2/3}$ in Equations (7.7). Performing this scaling and dividing by λ_3, it is very easy to see that the only solutions of the resulting equations for $(\mu_1, \mu_2, \lambda_3) = (0, 0, 0)$ is $(\rho, \phi) = ((\tfrac{4}{3})^{1/3}, 0)$. Furthermore, this is a simple solution of the equations. Consequently, the Implicit Function Theorem implies there can be no bifurcation in the region in λ-space,

$$S_{\alpha, \varepsilon_0} = \{\lambda : |\lambda_3| \leq \varepsilon_0, \ |\lambda_1|^3 + |\lambda_2|^3 \leq \lambda_3^2/\alpha\}$$

for some positive constants α, ε_0. This implies that the bifurcation surfaces in an ε_0-neighborhood of $\lambda = 0$ must be in the region

$$R_{\alpha, \varepsilon_0} = \{\lambda : \lambda_1^2 + \lambda_2^2 \leq \varepsilon_0^2, \ \lambda_3^2 \leq \alpha(|\lambda_1|^3 + |\lambda_2|^3)\}$$

Let us now introduce the scaling

$$\text{(7.10)} \qquad r = \mu\rho, \qquad \lambda_1 = m_1\mu^2, \qquad \lambda_2 = m_2\mu^2, \qquad \lambda_3 = m_3\mu^3$$
$$|m_3| \leq \alpha(|m_1| + |m_2|)$$

with μ a small real parameter. The resulting equations are

$$\text{(7.11)} \qquad \begin{aligned} \bar{G}_1 &= m_1\rho - \tfrac{3}{4}\rho^3 + m_3 \cos\varphi + \rho O(\mu^2) = 0 \\ \bar{G}_2 &= m_2\rho + m_3 \sin\varphi + \rho O(\mu^2) = 0 \end{aligned}$$

To obtain the bifurcation curves, we must investigate the multiple solutions of (7.11). For $\mu = 0$, one obtains

$$\Delta \overset{\text{def}}{=} \det \frac{\partial(\bar{G}_1, \bar{G}_2)}{\partial(\rho, \varphi)} = -m_3[(m_1 - \tfrac{9}{4}\rho^2)\cos\phi + m_2 \sin\phi]$$

$$= \rho[(-m_1 + \tfrac{3}{4}\rho^2)(-m_1 + \tfrac{9}{4}\rho^2) + m_2^2]$$

(7.12)

for any solution of (7.11) for $\mu = 0$. If $m_1 < 0$, the only solution of $\Delta = 0$ is $\rho = 0$. Also, one finds that the only zero is $\rho = 0$ for $m_1 > 0$, $m_1^2 < 3m_2^2$. If $\rho = 0$ is a solution of Equation (7.11), then $m_3 = 0$ and our original Equation (7.2) is autonomous. But, we know this equation has only the solution $v = 0$ if $\lambda_2 \neq 0$; that is, $m_2 \neq 0$. Therefore, no bifurcation can occur in the region

$$R_1 = R_{\alpha, \varepsilon_0} \cap \{\lambda : \lambda_2 \neq 0, |\lambda_1| < (3)^{1/2}|\lambda_2|\} \cup \{\lambda : \lambda_1 < 0\}$$

If we analyze the region

$$R_2 = R_{\alpha, \varepsilon_0} \cap \{\lambda : \lambda_1 > 0, |\lambda_2| \leq \lambda_1\}.$$

then the analysis will be complete. For this region, we can use the scaling (7.10) with $m_1 = 1$. To simplify notation, let us write the scaling as

$$r = \mu\rho, \qquad \lambda_1 = \mu^2, \qquad \lambda_2 = m\mu^2, \qquad \lambda_3 = v\mu^3,$$

$$|m| \leq 1, \qquad |v| \leq \alpha(1 + |m|)$$

(7.13)

Then using (7.11), (7.12), $v \neq 0$ (that is, $\lambda_3 \neq 0$), the multiple solutions of our scaled bifurcation equations are the solutions of the three equations

$$h_1 \overset{\text{def}}{=} \rho - \tfrac{3}{4}\rho^3 + v\cos\phi = \rho O(\mu^2)$$

$$h_2 \overset{\text{def}}{=} m\rho + v\sin\phi = \rho O(\mu^2)$$

(7.14)

$$h_3 \overset{\text{def}}{=} (1 - \tfrac{9}{4}\rho^2)\cos\phi + m\sin\phi = O(\mu^2)$$

We must investigate the solution of these equations for $|m| \leq 1$, $|v| \leq \alpha(1 + |m|)$, $|\rho| \leq 2c(1 + \alpha^{1/3})$.

The outline of the program is now as follows.

Treating ρ as a parameter, we determine ϕ, v, m as functions of ρ such that Equations (7.14) for $\mu = 0$ are satisfied. If $\det \partial(h_1, h_2, h_3)/\partial(\phi, v, m) \neq 0$, then we can determine solutions of Equations (7.14) for μ sufficiently small. Elimination of the parameter ρ from the resulting solutions $v^*(\rho, \mu)$, $m^*(\rho, \mu)$ will give the possible bifurcation surfaces v as a function of m, or perhaps m as a function of v, μ. We then verify that the number of solutions (ρ, ϕ) changes by two as this surface is crossed. This proves that it is a bifurcation surface.

If $h_1 = h_2 = h_3 = 0$, $\mu = 0$ then

$$(7.15) \qquad \cos \phi = \frac{1 - \frac{3}{4}\rho^2}{m} = \frac{-m}{1 - \frac{9}{4}\rho^2}$$

and so

$$(7.16) \qquad m^2 = [\tfrac{9}{4}\rho^2 - 1][1 - \tfrac{3}{4}\rho^2]$$

Also, $h_1 = h_2 = 0$ further implies

$$(7.17) \qquad v^2 = \rho^2[\tfrac{3}{4}\rho^2 - 1]^2 + m^2\rho^2$$

Equations (7.15), (7.16), (7.17) are the parametric forms of the solutions ϕ, m, v as functions of ρ if we substitute Equation (7.16) into Equations (7.15) and (7.17). Eliminating ρ from Equations (7.16), (7.17), we have

$$(7.18) \qquad v^2 = \tfrac{8}{81}[1 + 9m^2 \pm (1 - 3m^2)^{3/2}].$$

The locus of the points in the (v, m) plane described by Equation (7.18) are plotted in Figure 7.2.

If $\varDelta_1 = \det \partial(h_1, h_2, h_3)/\partial(\phi, m, v)$, then

$$\varDelta_1 = v \sin \phi - \rho[\tfrac{9}{4}\rho^2 - 1]\cos \phi \sin \phi - m\rho \cos^2 \phi.$$

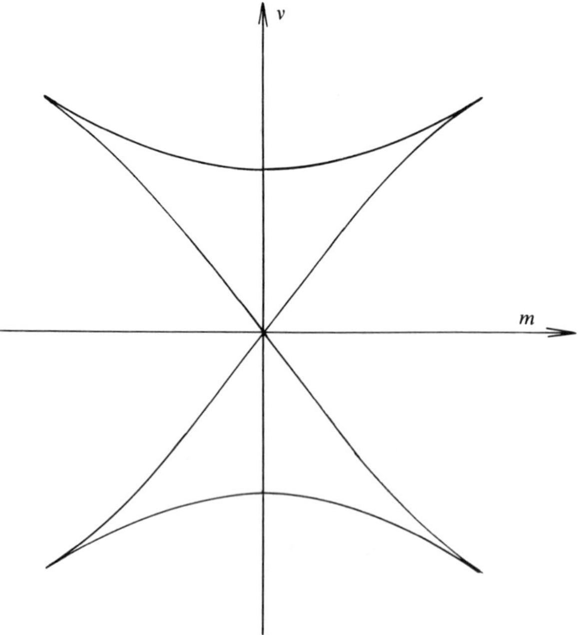

Figure 7.2

As remarked earlier, if $\Delta_1 \neq 0$ at a solution $(\phi_0, m_0, v_0, \rho_0)$ of $h_1 = h_2 = h_3 = 0$, then we can obtain a solution (ϕ, m, v) of Equations (7.14) for $|\rho - \rho_0|$ and $|\mu|$ sufficiently small by employing the Implicit Function Theorem.

If $h_1 = h_2 = h_3 = 0$, then

$$\Delta_1 = -2m[\tfrac{9}{4}(2 \pm (1 - 3m^2)^{1/2})]^{1/2}$$

and $\Delta_1 = 0$ implies $m = 0$ since $m^2 \leq 1$. If $m = 0$, then $(v^2, \rho^2, \phi) = (\tfrac{16}{81}, \tfrac{4}{9}, 0)$ or $(0, \tfrac{4}{3}, \pi/2)$.

Therefore, at each solution $(\phi_0, m_0, v_0, \rho_0)$ of $h_1 = h_2 = h_3 = 0$ for which $(v_0, m_0) \neq (\pm\tfrac{4}{9}, 0)$ or $(0, 0)$, there is an $\varepsilon > 0$ and unique solutions $\phi(\rho, \mu)$, $m(\rho, \mu)$, $v(\rho, \mu)$ of Equations (7.14) for $|\rho - \rho_0|$, $|\mu| < \varepsilon$, $\phi(\rho_0, 0) = \rho_0$, $m(\rho_0, 0) = m_0$, $v(\rho_0, 0) = v_0$. Thus, if we exclude a neighborhood V of the points $(\pm\tfrac{4}{9}, 0)$ and $(0, 0)$ in the (v, m) plane, then we can find a $\mu_0 > 0$ such that the Equations (7.14) can be uniquely solved for (ϕ, m, v) as functions of (ρ, μ) for $|\mu| \leq \mu_0$ and $|\rho| \leq 2c(1 + \alpha^{1/3})$, the a priori bound on ρ. Elimination of ρ from the corresponding functions m, v gives part of the surface indicated in Figure 7.3 in terms of the unscaled variables $(\lambda_1, \lambda_2, \lambda_3)$. The first approximation to the explicit formula for this surface is obtained from the scaling and Equation (7.18) and is given by

$$(7.19) \qquad \lambda_3^2 = \tfrac{8}{81}[\lambda_1^3 + 9\lambda_1\lambda_2^2 \pm (\lambda_1^2 - 3\lambda_2^2)^{3/2}]$$

At points $(v, m) = (\pm\tfrac{4}{9}, 0)$ in the (v, m) plane where $\Delta_1 = 0$, we compute another Jacobian and apply the implicit function theorem. At these points

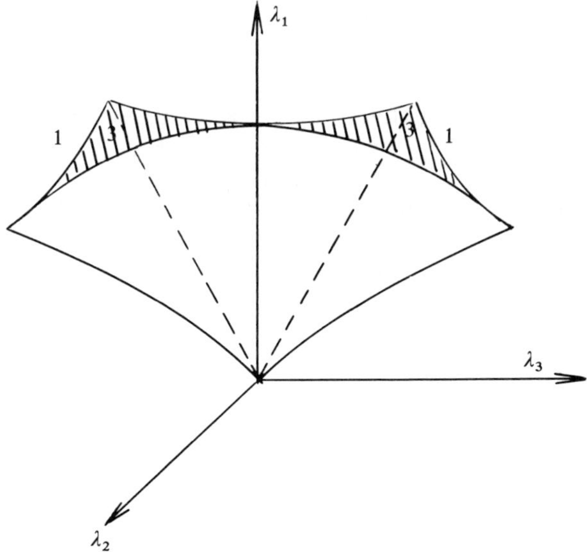

Figure 7.3

and for $(\rho, \phi) = (\pm\frac{2}{3}, 0)$, we see that

$$\Delta_2 = \det \frac{\partial(h_1, h_2, h_3)}{\partial(\rho, m, v)} = -2 \neq 0.$$

Therefore, we can solve the Equations (7.18) for ρ, m, v as functions of ϕ, μ for ϕ, μ sufficiently small. Eliminating the parameter ϕ gives the bifurcation surface near the points $(\pm\frac{4}{9}, 0)$.

At the point $(v, m) = (0, 0)$, the corresponding solutions (ρ, ϕ) of $h_1 = h_2 = h_3 = 0$ are $(\pm 2/3^{1/2}, \pi/2)$. If we let $v = \alpha m$, then the bifurcation equations are

$$h_1 = \rho - \tfrac{3}{4}\rho^3 + \alpha m \cos \phi = 0(\mu^2)$$

(7.20)
$$\frac{h_2}{m} = \rho + \alpha \sin \phi = 0(\mu^2).$$

For (ρ, ϕ) either of the above points, we have $h_2/m = 0$ implies $\alpha = \pm 2/3^{1/2}$. We wish to determine the multiple solutions of Equation (7.20). The analysis proceeds as before. If $\tilde{\Delta} = \det \partial(h_1, h_2/m)/\partial(\rho, \phi)$, then

$$\tilde{\Delta} = \alpha(1 - 9\rho^2/4)\cos \phi + \alpha m \sin \phi.$$

We now consider the equations $h_1 = 0$, $h_2/m = 0$, $\tilde{\Delta} = 0$ as defining functions ρ, m, α as functions of ϕ. Along the solution of these equations, we have

$$\det \frac{\partial(h_1, h_2/m, \tilde{\Delta})}{\partial(\rho, m, \alpha)} = \pm\frac{4}{3^{3/2}} \neq 0.$$

Therefore, we can find functions $m(\phi, \mu)$, $\alpha(\phi, \mu)$ and $\rho(\phi, \mu)$ that will satisfy the Equations (7.20) for μ close to zero and ϕ close to $\pi/2$. Eliminating ϕ from the functions $m(\phi, \mu)$ and $\alpha(\phi, \mu)$ completes the discussion of the possible bifurcation surface shown in Figure 7.3.

It remains to show that this surface is a bifurcation surface; that is, we must show that the number of solutions (ρ, ϕ) of the equations changes as this surface is crossed. However, we need not check all points on the surface. It is sufficient, for example, to show that the number of solutions changes as we cross the surface in the plane $\lambda_2 = 0$ since the only possible way for the number of solutions to change is to pass through a multiple solution. For $\lambda_2 = 0$, Theorem 7.3 implies three solutions for $\lambda_3 = 0$, $\lambda_1 > 0$ and one solution for $\lambda_1 = 0$, $\lambda_3 \neq 0$. Therefore, the surface in Figure 7.3 is a bifurcation surface with the number of solutions as indicated.

We summarize these results in the following Theorem.

Theorem 7.5. *There is a neighborhood U in $\mathbb{R}^3$ of zero such that the bifurcation surface Γ for Equation (7.1) with $\lambda \in U$ is depicted in Figure 7.3 and the surface is approximately given by Equation (7.19). The number of 2π-periodic solutions of Equation (7.1) at a point $\lambda \in U$ is shown in Figure 7.3.*

8.8. Bibliographical Notes

For a derivation of the von Kármán equations, the limitations of their validity and results on existence of solutions, see Berger [3, 4], Berger and Fife [1, 2], Fife [3], Freidrichs and Stoker [1], Golubitsky and Schaeffer [3], Knightly [1], Knightly and Sather [1, 2], von Kármán [1], Ciarlet [1].

We give only a few references on the buckling of plates and shells since the literature is so enormous. For perturbation techniques, see Bauer, Keller and Reiss [1, 2], Bauer and Reiss [1, 2], Kenner [1–3], Keener and Keller [1, 2], Keller [1], Keller, Keller and Reiss [1], Knightly [2], Knightly and Sather [1–6], Lange and Newell [1, 2], Matkowsky and Putnik [1], Matkowsky and Reiss [1], Reiss [1], Sather [3, 4], Thompson [1], Wolkowisky [1].

For techniques which make explicit use of the fact that the equilibrium positions of a plate must be an extreme point of a functional, see Ambrosetti [2, 3], Antman [1, 2], Chillingworth [3], Golubitsky and Schaeffer [3, 4], Koiter [1–3], Naumann [1–3], Plant [1], Potier–Ferry [2], Thompson and Hunt [1].

Golubitsky and Schaeffer [3] were the first to point out so clearly the effect of boundary conditions on the structure of the bifurcation diagrams. In this paper, they also made another important observation. For the von Kármán equations with the symmetries that go with a rectangular plate with fixed boundary conditions and loading parameter λ, one only needs one more parameter to describe all possible bifurcations that can occur for the plate. Furthermore, this parameter can be taken to be the ratio of the lengths of the edges of the plate. The computations in Section 8.3 were based on Golubitsky and Schaeffer [1].

For the evaluation of the constants σ_1, σ_2 in Relation (5.2), see Golubitsky and Schaeffer [3], List [1]. For the simply supported plate, it was asserted in Section 5 that the constants μ, v in Equation (5.4) satisfy $\mu > 1, v > 1$. For a verification of this fact, see Matkowsky and Putnik [1], Chow, Hale and Mallet–Paret [2]. For the clamped plate, the proof that $v < 1, \mu v > 1$ may be found in Golubitsky and Schaeffer [3].

It is possible to consider other choices for the parameters in Section 5. For example, for the simply supported plate, the interaction of small curvature α and the external loading λ with no imperfection ($\varepsilon = 0$) and the length l fixed at the critical value l_0 leads to bifurcation diagrams similar to the ones shown in Section 7.7. The computations in the specific example are very complicated and the details are in Mallet–Paret [1]. The situation for the plate clamped at $x = 0$, $x = l$, has not been discussed.

List [1] has considered the bifurcation diagrams for the simply supported plate with no curvature ($\alpha = 0$), but allowing variations in the length l about the critical value l_0, the loading parameter λ and the imperfection parameter ε. The method is a systematic application of the scaling techniques of Section 7 generalized appropriately for the three parameter problems.

Boa and Cohen [1], Cohen [1, 2] obtained part of the results in Section 6. The presentation in the text is based on the Ph.D. thesis of Waller [1]. The details of the analysis in Section 6 for $ab < 0$ may be found in Waller [1]. A complete discussion may also be found in Golubitsky and Schaeffer [5].

The presentation in Section 7 is based on Hale and Rodrigues [1]. Some partial results on Duffing's equation had been previously obtained by Holmes and Rand [1].

The conclusion of Corollary 7.2 is also valid for the equation

$$\ddot{x} + x = \lambda_1 x - x^3 + \lambda_3 f(t)$$

where $f(t)$ is even in t and $\int_0^{2\pi} f(t)\cos t\, dt = \pi$. To prove this, one considers the same equation with t replaced by $t + \phi$ and observes that the bifurcation equation obtained by projecting onto $\sin t$ has the form $\sigma(\sin \phi)h(r, \phi, \lambda_1) = 0$ with $h(0, \phi, 0) = 1$. One can now repeat the same argument as in the proof of Corollary 7.2 to complete the proof.

The ideas used in the proof of the evenness of the solutions in Corollary 7.2 has been abstracted to obtain interesting information about the solutions of equations in Banach spaces which remain invariant under certain groups of transformations (see Rodrigues and Vanderbauwhede [1]). This latter paper also contains interesting applications to oridinary and partial differential equations.

There are many other applications where interesting bifurcation problems occur which are of the type discussed in Chapters 6, 7, 8. For some problems in fluids, see Fige [1], Fife and Joseph [1], Rabinowitz [4], Segal and Stuart [1], J. T. Stuart [1], DiPrima and Stuart [1], Zachmann [1]. For many other problems and references on waves, see Zeidler [1]. For crystals, see Raveché and Stuart [1, 2]. Many results and problems can be found in the collection of papers given in the volumes edited by Gurel and Rössler [1], Holmes [5], Keller and Antman [1], Robinowitz [14]. Some interesting elementary examples are also in Sewell [1].

It is natural to inquire if it is of interest to discuss problems for which the null space has dimension greater than one. In problems which have more than one independent parameter, it is always true that there will be values of the parameters where the dimension of the null space is more than one. Also, it occurs often in the optimal design of systems (see, for example, Haug [1], Olhoff and Rasmussen [1]).

Chapter 9

Bifurcation near Equilibrium

9.1. Introduction

In this chapter, we discuss various types of dynamic behavior when a bifurcation arises from the existence of a simple eigenvalue. More specificially, we consider an equation

$$(1.1) \qquad Cx + N(x,\mu) = 0$$

in a Banach space X for μ in a Banach space E, $N(0,0) = 0$, $\partial N(0,0)/\partial x = 0$ under the assumption that the linear operator C has zero as a simple eigenvalue. The method of Liapunov–Schmidt gives a scalar bifurcation function $G(a,\mu)$ defined for (a,μ) in neighborhood of $(0,0) \in \mathbb{R} \times E$. Suppose $Cx + N(x,\mu)$ is the vector field for an evolutionary equation

$$(1.2) \qquad \frac{dx}{dt} = Cx + N(x,\mu).$$

If $x \in \mathbb{R}^n$ and all other eigenvalues of C have negative real parts, then the main result in Section 3 is that the dynamic behavior of (1.2) in a neighborhood of $(x,\mu) = (0,0)$ is completely determined by the dynamics of the scalar equation

$$(1.3) \qquad \frac{da}{dt} = G(a,\mu)$$

in a neighborhood of $(a,\mu) = (0,0)$. The result is also valid if x is an infinite dimensional space if C,N satisfy certain properties. This is mentioned in Section 6.

The next problem is concerned with the case where the operator equation (1.1) arises from a dynamic problem in a different way. More specifically, consider the evolutionary equation

$$(1.4) \qquad \frac{dx}{dt} = Ax + f(t,x,\mu)$$

where $x \in Z$, a Banach space, $\mu \in E$, a Banach space, $f(t, x, \mu) = f(t + 2\pi, x, \mu)$, $f(t, 0, 0) = 0$, $\partial f(t, 0, 0)/\partial x = 0$. The problem is to determine the existence of 2π-periodic solutions of (1.4) in a neighborhood of $(x, \mu) = (0, 0)$. If we let $Z_{2\pi} = \{x : \mathbb{R} \to Z : x \text{ continuous}, x(t + 2\pi) = x(t)\}$ and define Cx, $N(x, \mu)$ on an appropriate domain $\mathscr{D}$ in $Z_{2\pi}$ by

$$(Cx)(t) = -\frac{dx(t)}{dt} + Ax(t)$$

$$(Nx)(t) = f(t, x(t), \mu)$$

then the existence of 2π-periodic solutions of (1.4) is equivalent to finding an $x \in \mathscr{D}$ such that Eq. (1.1) is satisfied. With a few other assumptions, one can use the graph norm of C to define a Banach space X such that $C, N : X \to Z$. The situation now becomes the bifurcation problem discussed in Chapter 6.

Assuming that C has zero as a simple eigenvalue, there is a scalar bifurcation function $G(a, \mu)$ defined for (a, μ) in a neighborhood of $(0, 0) \in \mathbb{R} \times E$ which has the property that there is a one-to-one correspondence between the zeros of G and the 2π-periodic solutions of the original equation (1.4) in a neighborhood of zero. With the function $G(a, \mu)$, one can define the scalar ordinary differential equation (1.3) for which the stability properties of the zeros of G are easy to analyze.

In Section 4, we prove that these stability properties coincide with the stability properties of the 2π-periodic solution of (1.4) provided $x \in \mathbb{R}^n$ and A has one zero eigenvalue and the remaining ones have negative real parts. Some remarks on extensions to infinite dimensional systems are given in Section 6.

Section 5 is devoted to an application to the problem of bifurcation from a focus for m-dimensional systems. Using a priori bounds on the center manifold to justify scaling of variables, polar coordinates are introduced to reduce the discussion to the vector periodic case in Section 4.

All of the results are stated independently of the existence of center manifolds. However, the proofs of the results use this fact. The relevant material on center manifolds is given in Section 2.

9.2. Center Manifolds

In this section, we consider the situation where the linearization near an equilibrium point has m eigenvalues on the imaginary axis. We show that the qualitative properties of the flow near the equilibrium point can be reduced to the discussion of a differential equation on a manifold of dimension m, called a center manifold.

We begin with the simplest situation. We assume that we have chosen the coordinate system in such a way that the equations can be written as

$$(2.1) \qquad \begin{aligned} \dot{x} &= Ax + u(x, y) \\ \dot{y} &= By + v(x, y) \end{aligned}$$

where $x \in \mathbb{R}^m$, $y \in \mathbb{R}^n$, $\operatorname{Re} \sigma(A) = 0$, $\operatorname{Re} \sigma(B) \neq 0$ and u, v are C^k-functions, $k \geq 1$ which vanish together with their first derivatives at the origin.

Definition 2.1. A C^k-manifold $W^c_{\text{loc}}(0) \equiv W^c(0, U)$ in a neighborhood U of 0 is said to be a *center manifold* of (2.1) if $W^c(0, U)$ is invariant under the flow as long as the solution remains in U and $W^c(0, U)$ is the graph of a C^k function $h(x) = y$ which is tangent at $(0, 0) \in \mathbb{R}^m \times \mathbb{R}^n$ to the x-axis.

If $y = h(x)$ is a C^k center manifold given by $y = h(x)$, then flow invariance implies that h must satisfy the partial differential equation

$$(2.2) \qquad \mathcal{M}(h)(x) \equiv \frac{\partial h}{\partial x}(Ax + u(x, h)) - (Bh + v(x, h)) = 0.$$

The equation (2.2) can be used to uniquely determine the Taylor series of a function which describes a center manifold. In fact, we have the following result whose proof is left to the reader.

Theorem 2.1. *Suppose that* $y = h(x)$ *is a* C^k *center manifold of (2.1). If g satisfies* $\mathcal{M}(g)(x) = O(|x|^k)$ *as* $|x| \to 0$, *then*

$$|h(x) - g(x)| = O(|x|^k) \qquad \text{as } |x| \to 0.$$

EXAMPLE 2.1. *Center Manifolds are not unique.* Consider the planar equation

$$(2.3) \qquad \dot{x} = -x^3, \qquad \dot{y} = -y.$$

For any constants c_1, c_2, the curve $y = h(x; c_1, c_2)$, where

$$h(x; c_1, c_2) = \begin{cases} -c_1 e^{-(1/2x^2)}, & x < 0 \\ 0, & x = 0 \\ c_2 e^{-(1/2x^2)}, & x > 0 \end{cases}$$

consists of orbits of (2.3) and is therefore invariant. Also, this curve is tangent to the x-axis at the origin, which implies that it is a center manifold. The Taylor series of each center manifold is uniquely determined from Theorem 2.1 and, therefore, must be the Taylor series of the zero function. Of course, there is no contradiction with the assertion of non-uniqueness.

EXAMPLE 2.2. *Analytic vector fields may not have analytic center manifolds.* Consider the equation

$$\dot{x} = -x^3, \qquad \dot{y} = -y + x^2.$$

If $y = h(x)$ is an analytic local center manifold given by $h(x) = \sum_{n=0}^{\infty} a_n x^n$, then it is not difficult to verify that $a_{2n+1} = 0$ for all n, $a_2 = 1$ and $a_{2n+2} = 2na_{2n}$ for all n, which is a contradiction.

EXAMPLE 2.3. *Analytic vector fields may not have C^∞ center manifolds.* Consider the equation

$$\dot{\varepsilon} = 0, \qquad \dot{z} = -\varepsilon z - z^3, \qquad \dot{y} = -y + z^2.$$

This is special case of (2.1) with $x = \text{col}(\varepsilon, x)$, $A = 0$, $B = (-1)$. Suppose that there is a C^∞ local center manifold $y = h(z, \varepsilon)$ for $|z| < \delta$, $|\varepsilon| < \delta$. Choose $n > \delta^{-1}$. Since $h(z, (2n)^{-1})$ is C^∞ in z, there are constants $a_1, a_2, \ldots, a_n$ such that

$$h(z, (2n)^{-1}) = \sum_{j=1}^{2n} a_j z^j + O(z^{2n+1})$$

as $|z| \to 0$. One can show that $a_j = 0$ for j odd and, if $n > 1$,

$$[1 - (2j)(2n)^{-1}]a_{2j} = (2j - 2)a_{2j-2},$$

for all j and $a_2 \neq 0$. It is not difficult to show that this contradicts the hypothesis that h is C^∞.

We now state one of the main results of this section.

Theorem 2.2. *There exists a neighborhood U of $0 \in \mathbb{R}^m \times \mathbb{R}^n$ such that there exists a local center manifold $W^c(0, U)$ of (2.1) which is the graph of a C^k function $h(x) = y$.*

We remark that the neighborhood U in which the manifold $W^c(0, U)$ is C^k depends upon k as well as $\sigma(A)$, $\sigma(B)$.

If $y = h(x)$ is the graph of a local C^k center manifold of (2.1), then the flow on $W^c(0, U)$ is given by the differential equation

$$(2.4) \qquad\qquad \dot{x} = Ax + u(x, h(x)).$$

EXAMPLE 2.4. Consider the equation

$$\dot{u} = \varepsilon u - u^3 + uy$$
$$\dot{\varepsilon} = 0$$
$$\dot{y} = -y + y^2 - u^2.$$

If $x = \text{col}(u, \varepsilon)$, then this is a special case of (2.1). If we use (2.2), then we deduce that a local center manifold has the form $y = h(x) = -u^2 + O((|u| + |\varepsilon|)^3)$ as $x \to 0$. The flow on the center manifold is given by

$$\dot{u} = \varepsilon u - 2u^3 + O(|u|(|u| + |\varepsilon|)^3)$$

as $x \to 0$. The structure of the flow is determined by the equation $\dot{u} = \varepsilon u - 2u^3$ if $|x|$ is small enough. For $\varepsilon > 0$, there are three equilibrium points with two stable and one unstable, and, for $\varepsilon < 0$, there is one equilibrium point which is stable. The bifurcation at $\varepsilon = 0$ is observed on the center manifold. Later, we will show that the stability properties of these equilibrium points as a solution of the full equations are the same as the stability properties of the equilibrium point on the center manifold.

We now turn to the proof of Theorem 2.2. Suppose $\delta > 0$ is given and consider the neighborhood $U = \max\{|x|, |y|\} \le \delta$ of the origin in $\mathbb{R}^m \times \mathbb{R}^n$. The first step in the proof of the theorem is to replace the vector field in (2.1) by a new vector field $(\tilde{u}, \tilde{v})$ with the property that it coincides with (u, v) in U and is bounded in the set $U_\delta \equiv \mathbb{R}^m \times \{y : |y| \le \delta\}$. Such an extension can be made in the following way. Let $\xi : \mathbb{R}^m \to [0, 1]$ be a C^∞ function with

$$\xi(x) = \begin{cases} 1, & |x| \le 1 \\ 0, & |x| \ge 2 \end{cases}$$

and define

$$\tilde{u}(x, y) = u\left(x\xi\left(\frac{x}{\delta}\right), y\right), \qquad \tilde{v}(x, y) = v\left(x\xi\left(\frac{x}{\delta}\right), y\right).$$

We will assume that such an extension has been made, that we are working on the region U_δ and that we drop the tildes. For the extended system (2.1), we define a *global center manifold* $W^c(0)$ as follows:

$$W^c(0) = \left\{(x, y) \in \mathbb{R}^m \times \mathbb{R}^n : \sup_{t \in \mathbb{R}} |\tilde{y}(t, x, y)| < \infty\right\}$$

where $(\tilde{x}(t, x, y), \tilde{y}(t, x, y))$ is the solution of (2.1) with initial value (x, y) at $t = 0$. Note that, in the definition, it is assumed that the solution on $W^c(0)$ be defined for all $t \in \mathbb{R}$.

The plan is to obtain a global center manifold for the extended system (2.1) and then show that this manifold is a local center manifold for the original system in a sufficiently small neighborhood U of the origin. The restrictions on the size of the neighborhood U depend upon k and $\sigma(A)$, $\sigma(B)$.

The next step is to derive the integral equation for the global center manifold of the extended system (2.1). If $y = h(x)$ is the graph of the global center manifold of (2.1), then

$$\dot{x} = Ax + u(x, h(x)), \qquad x(0) = x_0,$$

$$h(x)(t) = e^{B(t-\tau)}h(x_0) + \int_\tau^t e^{B(t-s)}v(x(s), h(x(s)))\,ds.$$

There exist a continuous projection $P: \mathbb{R}^n \to \mathbb{R}^n$ such that $\mathscr{R}(P)$, $\mathscr{R}(Q)$, $Q = I - P$ are invariant under B, $\sigma(BP) = \{z \in \mathbb{C}: \operatorname{Re} z > 0\}$, $\sigma(BQ) = \{z \in \mathbb{C}: \operatorname{Re} z < 0\}$. If we use the fact that $h(x)$ is bounded, then it follows that we must have

(2.5)
$$\dot{x} = Ax + u(x, h(x)), \qquad x(0) = x_0,$$

$$h(x)(t) = \int_{-\infty}^\infty K(t - s)v(x(s), h(x(s)))\,ds,$$

where

(2.6)
$$K(t) = \begin{cases} -e^{Bt}P, & t \leq 0 \\ e^{Bt}Q, & t > 0. \end{cases}$$

Also, there are positive constants k, α such that

(2.7)
$$|K(t)| \leq ke^{-\alpha|t|}, \qquad \text{for all } t \in \mathbb{R}.$$

For given positive constants μ, v, let

$$S(\mu, v) = \{h \in C(\mathbb{R}^m; \mathbb{R}^n): |h|_0 \leq \mu, \operatorname{Lip} h \leq v\}$$

For any $h \in S(\mu, v)$, let $x(t, x_0, h)$ be the solution of the equation

(2.8)
$$\dot{x} = Ax + u(x, h(x)), \qquad x(0) = x_0,$$

and define the operator $\mathscr{T}: S(\mu, v) \to C(\mathbb{R}^m; \mathbb{R}^n)$ by the relation

(2.9)
$$(\mathscr{T}h)(x_0) = \int_{-\infty}^\infty K(-s)v(x(s, x_0, h), h(x(s, x_0, h)))\,ds.$$

We want to choose the constants δ, μ, v so that $\mathscr{T}: S(\mu, v) \to S(\mu, v)$ and is a contraction. Let us first estimate $\mathscr{T}h$. From (2.7), (2.9), we have

$$|(\mathscr{T}h)(x_0)| \leq k \int_{-\infty}^\infty e^{-\alpha|t|}|v|_0\,dt \leq \frac{k|v|_0}{\alpha} \leq \mu,$$

if $|v|_0 \le \alpha\mu/k$. To estimate the Lipschitz constant of $\mathscr{T}h$, we need to estimate the Lipschitz constant of $x(\cdot, x_0, h)$ with respect to x_0, h. Since $\text{Re}\,\sigma(A) = 0$, for any $\varepsilon > 0$, there is a constant $K = K(\varepsilon) > 0$ such that

$$(2.10) \qquad\qquad |e^{At}| \le Ke^{\varepsilon|t|}, \qquad \text{for } t \in \mathbb{R}.$$

Using the variation of constants formula, we have

$$|x(t, x_0, h) - x(t, x_0, \bar{h})|$$

$$\le K\left|\int_0^t e^{\varepsilon|t-s|}\big|u(x(s, x_0, h), h(x(s, x_0, h))) - u(x(s, x_0, h), h(x(s, x_0, \bar{h})))\big||ds|\right.$$

$$+ K\left|\int_0^t e^{\varepsilon|t-s|}\big|u(x(s, x_0, h), h(x(s, x_0, \bar{h}))) - u(x(s, x_0, \bar{h}), h(x(s, x_0, \bar{h})))\big||ds|\right.$$

$$+ K\left|\int_0^t e^{\varepsilon|t-s|}\big|u(x(s, x_0, \bar{h}), h(x(s, x_0, \bar{h}))) - u(x(s, x_0, \bar{h}), \bar{h}(x(s, x_0, \bar{h})))\big||ds|\right.$$

$$\le K\left|\int_0^t e^{\varepsilon|t-s|}(\text{Lip}_x u)\big|x(s, x_0, h) - x(s, x_0, \bar{h})\big||ds|\right.$$

$$+ K\left|\int_0^t e^{\varepsilon|t-s|}(\text{Lip}_y u)\big|h(x(s, x_0, \bar{h})) - \bar{h}(x(s, x_0, \bar{h}))\big||ds|\right.$$

If we apply the Gronwall inequality, we obtain

$$(2.11)$$
$$|x(t, x_0, h) - x(t, x_0, \bar{h})| \le K_1(\text{Lip}_y u)e^{(3\varepsilon/2 + K(\text{Lip}_x u + \text{Lip}_y u)v)|t|}|h - \bar{h}|_0, \qquad t \in \mathbb{R}.$$

$$(2.12)$$
$$|x(t, x_0, h) - x(t, \bar{x}_0, h)| \le Ke^{(\varepsilon + K(\text{Lip}_x u + \text{Lip}_y u\,\text{Lip}\,h))|t|}|x_0 - \bar{x}_0|, \qquad t \in \mathbb{R}.$$

If we now use (2.9), (2.12) and (2.7), we deduce that

$$|(\mathscr{T}h)(x_0) - (\mathscr{T}h)(\bar{x}_0)| \le \frac{k|v|_1(v + 1)}{\alpha - (\varepsilon + K(\text{Lip}_x u + v\,\text{Lip}_y u))} \le v,$$

if $|u|_1, |v|_1$ are sufficiently small. This shows that $\mathscr{T} : S(\mu, v) \to S(\mu, v)$.

If we use (2.11) and similar estimates, we conclude that the map $\mathscr{T}$ is a contraction map on $S(\mu, v)$. As a consequence, $\mathscr{T}$ has a unique fixed point $h^* \in S(\mu, v)$. We remark that the restrictions imposed upon $|u|_1, |v|_1$ can be achieved by choosing the above neighborhood U of the origin sufficiently small.

We now show that the function $h^* \in S(\mu, v)$ is a C^1 function. This can be proved by first guessing the formula that the derivative (if it exists) of the fixed point h^* should satisfy and then showing that it actually is the derivative. This also can be achieved by the following general result of Henry [3] of independent interest which is based on a remark in Hirsch, Pugh and Shub [2, p. 35].

Lemma 2.1. *Let X, Y be Banach spaces, $U \subset X$ be an open set and $g: U \to Y$ be locally Lipschitzian. Then g is continuously differentiable if and only if, for each $x_0 \in U$,*

$$(2.13) \qquad |g(x + h) - g(x) - g(x_0 + h) + g(x_0)|_Y = o(|h|_X)$$

as $(x, h) \to (x_0, 0)$.

Proof. It is easy to see that (2.13) holds if g is a C^1-function. Without loss of generality, we take U to be a ball and g to be Lipschitzian in U. If the derivative g' of g exists at each point of U and (2.13) is satisfied, then g' is continuous. Thus, it is enough to prove that g' exists at each point of U.

Case 1. Let us first suppose that $X = Y = \mathbb{R}$. Since g is absolutely continuous, it is differentiable almost everywhere. For any $x_0 \in U$, $\varepsilon > 0$, there is a $\delta > 0$ such that

$$|g(x + h) - g(x) - g(x_0 + h) + g(x_0)| \le \varepsilon|h|, \qquad \text{if } |x - x_0| + |h| < \delta.$$

There is an x^* in $(x_0 - \delta, x_0 + \delta)$ such that $g'(x^*)$ exists. Thus, for $h \ne 0$ sufficiently small, we have

$$\left| \frac{g(x_0 + h) - g(x_0)}{h} - g'(x^*) \right| \le 2\varepsilon,$$

$$0 \le \left\{ \limsup_{h \to 0} - \liminf_{h \to 0} \right\} \frac{g(x_0 + h) - g(x_0)}{h} \le 4\varepsilon.$$

Since ε is arbitrary, this implies that $g'(x_0)$ exists.

Case 2. Now suppose that $X = R$ and Y is a Banach space with Y^* being the dual space. If $\eta \in Y^*$, then Case 1 implies that ηg is a C^1-function and $|(\eta g)'(x)| \le |\eta|_{Y^*} \cdot \operatorname{Lip} g$, where $\operatorname{Lip} g$ is the Lipschitz constant for g. If $D(x): \eta \mapsto (\eta g)'(x)$, then $D(x) \in Y^{**}$ is continuous in x from (2.13).

For $x_0 \in U$, $\eta \in Y^*$, $|\eta|_{Y^*} \le 1$, we have

$$\eta\left(\frac{g(x_0 + h) - g(x_0)}{h} \right) - D(x_0)\eta = \frac{1}{h} \int_{x_0}^{x_0 + h} (D(x) - D(x_0))\eta\, dx \to 0$$

as $h \to 0$ uniformly for $|\eta|_{Y^*} \le 1$. Let $\tau: Y \to Y^{**}$ be the canonical inclusion. Then $\tau[g(x_0 + h) - g(x_0)]/h \to D(x_0)$ as $h \to 0$. Since τ is an isometry, this implies that $[g(x_0 + h) - g(x_0)]/h \to$ a limit in Y as $h \to 0$; that is, $g'(x_0)$ exists.

Case 3. Finally, let X, Y be arbitrary Banach spaces. From Case 2, for any $x \in U$, $h \in X$, the map $t \mapsto g(x + th)$ taking $\mathbb{R}$ to Y is C^1 if t is small. Thus, the Gateaux derivative $dg(x, h) = dg(x + th)/dt|_{t=0}$ exists. Condition (2.13) implies that $dg(x, h) \to dg(x_0, h)$ in Y as $x \to x_0 \in U$, uniformly for $|h|_X \le 1$.

This implies that the map $h \mapsto dg(x, h)$ is linear and continuous. Thus, $dg(x, h)$ is the Fréchet derivative at h of g and the proof is complete.

The verification of the hypotheses of Lemma 2.1 for our situation uses estimates very similar to the ones that we have performed above and are left for the reader.

To prove that the center manifold that we have obtained is C^k, one can use induction, assuming that the manifold is C^{k-1} with Lipschitz continuous $(k-1)^{th}$ derivative, and then use Lemma 2.1 (see, for example, Chow and Lu [1]). This completes the proof of the existence of the global center manifold for the extended system (2.1).

If we appropriately restrict the neighborhood U of the origin, the global center manifold restricted to the neighborhood U will be a local center manifold for the original equation (2.1). The size of U depends upon k and the spectral sets $\sigma(A)$, $\sigma(B)$. These quantities become involved when the precise estimates are made in the induction hypothesis mentioned above. This completes the proof of Theorem 2.2. $\square$

Our next objective is to obtain more precise information about the behaviour of the solutions in a neighborhood of a local center manifold. To do this, we will make use of center-unstable and center-stable manifolds. The center-unstable manifold will be defined for the system

$$\dot{x} = Ax + u(x, y)$$
$$(2.14) \qquad \dot{y} = By + v(x, y)$$
$$\mathrm{Re}\,\sigma(A) \geq 0, \qquad \mathrm{Re}\,\sigma(B) < 0$$

and the center-stable manifold for the system

$$\dot{x} = Ax + u(x, y)$$
$$(2.15) \qquad \dot{y} = By + v(x, y)$$
$$\mathrm{Re}\,\sigma(A) \leq 0, \qquad \mathrm{Re}\,\sigma(B) > 0.$$

Definition 2.2. A C^k-manifold $W_{\mathrm{loc}}^{cu}(0) = W^{cu}(0, U)$ (resp. $W_{\mathrm{loc}}^{cs}(0) = W^{cs}(0, U)$) in a neighborhood U of 0 is said to be a *local center-unstable* (resp. *center-stable*) manifold of (2.14) (resp. (2.15)) if $W^{cu}(0, U)$ (resp. $W^{cs}(0, U)$) is invariant under the flow defined by (2.14) (resp. (2.15)) as long as the solution remains in U and is the graph of a C^k function $h(x) = y$ which is tangent at $(0, 0) \in \mathbb{R}^m \times \mathbb{R}^n$ to the x-axis (resp. y-axis).

We have the following result.

Theorem 2.3. *There exists a C^k local center-unstable (resp. center-stable) manifold of (2.14) (resp. (2.15)).*

Proof. For (2.14), assume that we have extended the vector field as in the proof of the center manifold theorem. We now apply the same proof as in the proof of Theorem 2.2, replacing the map $\mathcal{T}$ in (2.9) by the map

$$(\mathcal{T}h)(x_0) = \int_{-\infty}^{0} e^{-Bs}v(x(s, x_0, h), h(s, x_0, h))\, ds.$$

We leave the details to the reader.

In an analogous way, we consider (2.15) and assume that we have extended the vector field as in the proof of the center manifold theorem. We now proceed as before replacing the map $\mathcal{T}$ in (2.9) by the map

$$(\mathcal{T}h)(x_0) = -\int_{0}^{\infty} e^{-Bs}v(x(s, x_0, h), h(s, x_0, h))\, ds.$$

We leave the details to the reader. $\square$

Assume that we have extended the vector field as in the proof of the center manifold theorem. It is possible to show that the global center-unstable manifold $W^{cu}(0)$ is unique. Furthermore, there exists a constant $\beta > 0$ such that

$$W^{cu}(0) = \left\{ \xi \in \mathbb{R}^m \times \mathbb{R}^n : \sup_{t \leq 0} |y(t, \xi)| < \infty \right\}$$

$$= \left\{ \xi \in \mathbb{R}^m \times \mathbb{R}^n : \sup_{t \leq 0} e^{\beta t}|(x(t, \xi), y(t, \xi))| < \infty \right\}.$$

We now consider the foliations over the center-unstable and center-stable manifolds. We first consider equation (2.14) and assume that we have extended the vector field in such a way that it is bounded in $\mathbb{R}^m \times \mathbb{R}^n$. We could prove a similar result as below by only making the extension to $U_\delta = \mathbb{R}^m \times \{y : |y| < \delta\}$, but the statements would require more notation. Let $W^{cu}(0)$ be the global center-unstable manifold of this extended system. Let $z = (x, y) \in \mathbb{R}^m \times \mathbb{R}^n$ and let $z(t, \xi) = (x(t, \xi), y(t, \xi))$ be the solution of (2.14) with initial data $\xi \in \mathbb{R}^m \times \mathbb{R}^n$. Let

$$\beta_- = \min\{|\operatorname{Re}\lambda| : \lambda \in \sigma(B)\},$$

choose β so that $0 < \beta < \beta_-$ and define the Banach space

$$C_\beta^+ = \left\{ z : [0, \infty) \times \mathbb{R}^m \times \mathbb{R}^n \to \mathbb{R}^m \times \mathbb{R}^n : z \in C^0, |z|_\beta \equiv \sup_{t \geq 0} e^{\beta t}|z(t)| < \infty \right\}.$$

A *stable leaf* $M^s(\xi)$ *for* $\xi \in \mathbb{R}^m \times \mathbb{R}^n$ is defined as

$$(2.16) \qquad M^s(\xi) = \{\bar{\xi} \in \mathbb{R}^m \times \mathbb{R}^n : z(t, \bar{\xi}) - z(t, \xi) \in C_\beta^+\}.$$

Theorem 2.4. *For the extended system (2.14), there exists a continuous function* $h:(\mathbb{R}^m \times \mathbb{R}^n) \times \mathbb{R}^n \to \mathbb{R}^m$, $(\xi, y) \mapsto h(\xi, y)$, *with continuous partial derivatives of order* k *with respect to* y, *such that*

$$M^s(\xi) = \{y + h(\xi, y) : y \in \mathbb{R}^m\}.$$

Proof. For any $\tilde{z} \in \mathbb{R}^m \times \mathbb{R}^n$, $\xi \in \mathbb{R}^m \times \mathbb{R}^n$, let

$$\tilde{u}(t, \xi; \tilde{z}) = u(z_{cu}(t, \xi) + \tilde{z}) - u(z_{cu}(t, \xi)),$$
$$\tilde{v}(t, \xi; \tilde{z}) = v(z_{cu}(t, \xi) + \tilde{z}) - v(z_{cu}(t, \xi)).$$

For any $\xi \in \mathbb{R}^m \times \mathbb{R}^n$, $y \in \mathbb{R}^n$, $z \in C_\beta^+$, define the operator

$$\mathscr{T} : (\mathbb{R}^m \times \mathbb{R}^n) \times \mathbb{R}^n \times C_\beta^+ \to C_\beta^+$$

by the relation

$$\tilde{\mathscr{T}}(\xi, y, z(\cdot)) = \begin{bmatrix} \int_\infty^t e^{A(t-s)} \tilde{u}(s, \xi; z(s))\, ds \\ e^{Bt} y + \int_0^t e^{B(t-s)} \tilde{v}(s, \xi; z(s))\, ds \end{bmatrix}.$$

If the Lipschitz constants for u, v are small enough; that is, if we choose the original neighborhood U of the origin small enough, then the uniform contraction mapping principle can be used to show that there is a unique fixed point $z^*(\xi, y) \in C_\beta^+$ of $\tilde{\mathscr{T}}$. Furthermore, this function has the smoothness properties stated in y.

To show the continuity in ξ, let $\sigma > 0$ be sufficiently small so that $C_{\sigma+\beta}^+ \subset C_\beta^+$. If the neighborhood U is further restricted, then

$$\mathscr{T} : (\mathbb{R}^m \times \mathbb{R}^n) \times \mathbb{R}^n \times C_{\sigma+\beta}^+ \to C_{\sigma+\beta}^+$$

and has a unique fixed point $z_\sigma^*(\xi, y) \in C_{\sigma+\beta}^+$. The fact that $|z|_\beta \le |z|_{\sigma+\beta}$ for all $z \in C_{\sigma+\beta}^+$ implies that $z_\sigma^*(\xi, y) = z^*(\xi, y)$.

Now, consider the map

$$\mathscr{T} : (\mathbb{R}^m \times \mathbb{R}^n) \times \mathbb{R}^n \times C_{\sigma+\beta}^+ \to C_\beta^+$$

and the difference

$$\Delta\mathscr{T} \equiv |\mathscr{T}(\xi, y, z_\sigma^*(\xi, y)(\cdot)) - \mathscr{T}(\hat{\xi}, \hat{y}, z_\sigma^*(\hat{\xi}, \hat{y})(\cdot))|_\beta.$$

We remark that

$$\Delta\mathscr{T} = |z_\sigma^*(\xi, y) - z_\sigma^*(\hat{\xi}, \hat{y})|_\beta.$$

As a consequence, the proof of the theorem is complete if we show that, for any $\varepsilon > 0$, there exists a $\mu > 0$ such that $|\xi - \hat{\xi}| + |y - \hat{y}| < \mu$ implies that $\Delta\mathcal{F} < \varepsilon$. To show this, we use the fact that $z_\sigma^*(\xi, y) = z^*(\xi, y) \in C_{\sigma+\beta}^+$ and estimate the above integrals separately on the interval $[0, R]$ and $[R, \infty)$. The fact the functions $\tilde{u}(s, \xi, z(s)), \tilde{v}(s, \xi, z(s))$ are continuous in ξ uniformly over finite s-intervals and that the kernels in the integrals are exponential decaying as $t \to \infty$ implies that we can choose R so that we obtain the above estimate on $\Delta\mathcal{F}$. The details are supplied as in the proof of Theorem 3.6.3. $\square$

We remark that Theorem 2.4 gives a foliation of the local center-unstable manifold in an appropriate neighborhood of the origin.

For (2.15), we proceed in a similar manner. We extended the vector field in such a way that it is bounded in $\mathbb{R}^m \times \mathbb{R}^n$, let $W^{cs}(0)$ be the global center-stable manifold of this extended system, let $z = (x, y) \in \mathbb{R}^m \times \mathbb{R}^n$ and let $z(t, \xi) = (x(t, \xi), y(t, \xi))$ be the solution of (2.15) with initial data $\xi \in \mathbb{R}^m \times \mathbb{R}^n$. Let

$$\beta_- = \min\{|\operatorname{Re}\lambda| : \lambda \in \sigma(B)\},$$

choose β so that $0 < \beta < \beta_-$ and define the Banach space

$$C_\beta^+ = \left\{z : [0, \infty) \times \mathbb{R}^m \times \mathbb{R}^n \to \mathbb{R}^m \times \mathbb{R}^n : z \in C^0, |z|_\beta \equiv \sup_{t \le 0} e^{\beta t}|z(t)| < \infty\right\}.$$

An *unstable leaf* $M^u(\xi)$ *for* $\xi \in \mathbb{R}^m \times \mathbb{R}^n$ is defined as

$$(2.17) \qquad M^u(\xi) = \{\bar{\xi} \in \mathbb{R}^m \times \mathbb{R}^n : z(t, \bar{\xi}) - z(t, \xi) \in C_\beta^+\}.$$

Theorem 2.5. *For the extended system (2.15), there exists a continuous function $h : (\mathbb{R}^m \times \mathbb{R}^n) \times \mathbb{R}^n \to \mathbb{R}^m, (\xi, y) \mapsto h(\xi, y)$, with continuous partial derivatives of order k with respect to y, such that*

$$M^u(\xi) = \{y + h(\xi, y) : y \in \mathbb{R}^m\}.$$

The proof is supplied in the same way as the proof of Theorem 2.4, except replacing the '$\int_\infty^t$' by '$\int_{-\infty}^t$.'

An important consequence of Theorem 2.2–2.5 and the estimates (2.16), (2.17) is the following result whose proof is left to the reader.

Theorem 2.6. *Suppose that $y = h(x)$ is the graph of a local C^k center manifold $W^c(0, U)$ of (2.1). Then any invariant set of (2.1) which belongs to U must lie on $W^c(0, U)$. In particular, any equilibrium point in U must be given by $(x_0, h(x_0))$, where x_0 is an equilibrium point of (2.5). Furthermore, if $\operatorname{Re}\sigma(B) < 0$, then the stability properties of $(x_0, h(x_0))$ as a solution of (2.1) are the same as the stability properties of the equilibrium point x_0 of (2.5).*

Other important concepts in the local theory near an equilibrium point are strongly stable and strongly unstable manifolds.

Definition 2.3. For a differential equation $\dot{x} = f(x)$ for which $f(0) = 0$ and a given neighborhood U of 0, we define the *local strongly stable set* $W^{ss}_{\text{loc}}(0) \equiv W^{ss}(0, U)$ (resp. *local strongly unstable set* $W^{su}_{\text{loc}}(0) \equiv W^{su}(0, U)$) as the collection of points whose orbits lie in U for $t \geq 0$ and approach zero exponentially as $t \to \infty$ (resp. $t \to -\infty$).

Consider the system (2.1) under the hypothesis that $\operatorname{Re}\sigma(B) \neq 0$ and $\operatorname{Re}\sigma(A) = 0$. For the linear equation, the strongly stable set $W^{ss}_{\text{linear}}(0)$ is a linear subspace of $\mathbb{R}^m \times \mathbb{R}^n$ given by the solution $x = 0$ and those solutions y which approach zero exponentially. Similar remarks hold for the strongly unstable set $W^{su}_{\text{linear}}(0)$. Let $P: \mathbb{R}^m \times \mathbb{R}^n \to \mathbb{R}^m \times \mathbb{R}^n$, $Q = I - P$ be projections onto these subspaces.

Theorem 2.7. *There is a neighborhood U of 0 in $\mathbb{R}^m \times \mathbb{R}^n$ such that $W^{ss}(0, U)$ (resp. $W^{su}(0, U)$) is a C^k-graph over $W^{ss}_{\text{linear}}(0)$ (resp. $W^{su}_{\text{linear}}(0)$) if u, v are C^k functions. If u, v are analytic, then so are the graphs.*

We do not give the details of the proof, but only indicate the integral equation that can be used to reduce the discussion to a fixed point theorem. Let $Q: \mathbb{R}^n \to \mathbb{R}^n$, $P = I - Q$, respectively, be projections onto the subspaces $W^{ss}_{\text{linear}}(0) \cap \mathbb{R}^n$, $W^{su}_{\text{linear}}(0)\mathbb{R}^n$. If $(x(t), y(t))$ is a solution of (2.1) in $W^{ss}(0, U)$, then they must satisfy the integral equation

$$x(t) = -\int_t^\infty e^{A(t-s)} u(x(s), y(s))\, ds$$

$$y(t) = e^{At} Qy(0) + \int_0^t e^{B(t-s)} Qv(x(s), y(s))\, ds - \int_t^\infty e^{B(t-s)} Pv(x(s), y(s))\, ds.$$

One can now use the right hand side of these equations to define a mapping on an appropriate space of continuous functions from $[0, \infty)$ to $\mathbb{R}^m \times \mathbb{R}^n$ with a weighted norm $|z(\cdot)|_\beta = \sup_{[0, \infty)} |z(t)| e^{\beta t}$, where β is a specially chosen positive constant.

The proof for the local strongly unstable set is similar.

9.3. Autonomous Case

Consider the $(n + 1)$-*dimensional system*

(3.1) $$\dot{z} = Cz + w(z) + f(z)$$

where $w, f : \mathbb{R}^{n+1} \to \mathbb{R}^{n+1}$ are continuous together with derivatives up through order $k \geq 1$, $w(0) = 0$, $\partial w(0)/\partial z = 0$,

$$C = \begin{bmatrix} 0 & 0 \\ 0 & B \end{bmatrix}$$

where B is an $n \times n$ matrix with eigenvalues with negative real parts. The function f is to be considered as a parameter varying in a neighborhood of zero in the space C^k with the C^k topology.

If $z = (x, y)$, $w = (u, v)$, $f = (g, h)$, $x, u, g \in \mathbb{R}$, $y, v, h \in \mathbb{R}^n$, then (3.1) is the same as

$$
\begin{aligned}
\dot{x} &= u(x, y) + g(x, y) \overset{\text{def}}{=} U(x, y, g) \\
\dot{y} &= By + v(x, y) + h(x, y) = By + V(x, y, h)
\end{aligned}
$$

(3.2)

The equilibrium points of (3.2) are the zeros of the right hand side. The method of Liapunov–Schmidt can be used to obtain these zeros. An application of this method shows that the bifurcation function is

$$(3.3) \qquad\qquad G(x, g, h) \overset{\text{def}}{=} U(x, \varphi(x, h), g)$$

where $\varphi(x, h)$ is the unique solution in a neighborhood of zero of the equation

$$(3.4) \qquad\qquad B\varphi + V(x, \varphi, h) = 0.$$

Our objective is to show that the bifurcation function $G(x, g, h)$ not only locates the equilibrium points of (3.1) in a neighborhood of zero, but it also determines the stability properties of the equilibrium points. To state the result precisely, we need the following information on center manifolds which is a special case of the results in the previous section.

Theorem 3.1. *If w, f in (3.1) are C^k-functions, $k \geq 2$, then there is a $\delta > 0$ and a C^k-function $\psi(x, g, h)$ defined for $|x| < \delta$, $|g|_k < \delta$, $|h|_k < \delta$, such that $y = \psi(x, g, h)$ is a center manifold for (3.2). This center manifold is exponentially asymptotically stable, the flow on the center manifold is given by the scalar differential equation*

$$
\begin{aligned}
\dot{x} &= \tilde{G}(x, g, h) \\
\tilde{G}(x, g, h) &= U(x, \psi(x, g, h), g)
\end{aligned}
$$

(3.5)

and the stability properties of an equilibrium point (x_0, y_0) of (3.2) (which must lie on the center manifold) coincide with the stability properties of the solution x_0 of (3.5). In particular, (x_0, y_0) is stable (asymptotically stable) (unstable) if and only if x_0 is stable (asymptotically stable) (unstable).

Using Theorem 3.1 and other basic information about differential equations, one can give a precise description of the flow defined by (3.2) near zero. In fact, from Theorem 2.14 and the corresponding result for the center stable manifold, for any equilibrium point (x_0, y_0) of (3.2), there is a stable manifold of dimension $\geq n$ and there is an n-dimensional manifold such that the solutions starting on this manifold approach (x_0, y_0) at an exponential rate determined by the eigenvalues of B. The center manifold theorem implies there is only a one dimensional direction remaining to determine the stability properties of (x_0, y_0) and this is along the center manifold. On the center manifold, the stability properties of x_0 are determined by elementary properties of the function in (3.5). For example, x_0 is asymptotically stable from the left (right) if and only if there is an $\varepsilon > 0$ such that this function is positive (negative) for $-\varepsilon < x - x_0 < 0$ $(0 < x - x_0 < \varepsilon)$. The point x_0 is unstable from the left (right) if and only if this function is negative (positive) for $-\varepsilon < x - x_0 < 0$ $(0 < x - x_0 < \varepsilon)$. If the equilibrium point x_0 on the center manifold is asymptotically stable from the left and unstable from the right, then the above remarks imply that the flow defined by (3.2) in a neighborhood of the equilibrium point $(x_0, \psi(x_0, g, h))$ is the one shown in Fig. 3.1. With more work, one can actually show that there is a change of variables such that the flow for (3.2) near zero is topologically equivalent to $\dot{y} = By$ and x satisfying (3.5).

We now have two ways of characterizing the equilibrium points of (3.2) near zero; namely, as

$$\{(x_0, \psi(x_0, g, h)) : \tilde{G}(x_0, g, h) = 0\},$$

where $\tilde{G}(x, g, h) = U(x, \psi(x, g, h), g)$ with $\psi(x, g, h)$ a representation of a center manifold and as

$$\{(x_0, \phi(x_0, h)) : G(x_0, g, h) = 0\},$$

where $G(x, g, h) = U(x, \phi(x, h), g)$ is the bifurcation function and ϕ satisfies (3.4).

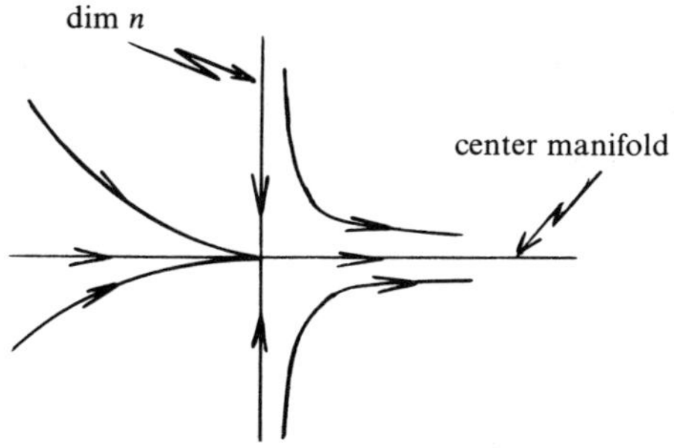

Figure 3.1

In a sufficiently small neighborhood of zero, it is clear that the zeros of $\tilde{G}(x, g, h)$ and $G(x, g, h)$ are the same. However, the functions themselves are quite different since $\phi(x, h)$ depends only on h and $\psi(x, g, h)$ depends on both g, h (see Formula (2.10) for obtaining the center manifold).

The main result of this section asserts that the stability properties of the solution x_0 of an equilibrium point of Equation (3.5) on the center manifold is the same as the stability properties of x_0 as a solution of the vector field defined by the bifurcation function

$$(3.6) \qquad\qquad\qquad \dot{x} = G(x, g, h).$$

Thus, the stability properties of the solution $(x_0, \phi(x_0, h))$ of (3.2) are determined by Equation (3.6). This implies that the center manifold is not needed for the determination of the qualitative properties of the solutions of (3.2) near zero. The advantage, from the theoretical point of view, is that $G(x, g, h)$ has the same smoothness properties as the original vector field, whereas, the center manifold in general does not. The advantage from the practical point of view is that $\phi(x, h)$ is easier to calculate than $\psi(x, g, h)$.

The precise statement of the main result is

Theorem 3.2. *There is a neighborhood V of $h = 0$, a neighborhood $\mathcal{O}_1$ of $x = 0$, $\mathcal{O}_n$ of $y = 0$ and a C^1-function $\phi : \mathcal{O}_1 \times V \to \mathcal{O}_n$ such that the only equilibrium solutions of (3.2) in $\mathcal{O}_1 \times \mathcal{O}_n$ are given by*

$$(3.7) \qquad y_0 = \phi(x_0, h), \qquad G(x_0, g, h) \stackrel{\text{def}}{=} U(x_0, \phi(x_0, h), g) = 0.$$

Furthermore, the stability properties of the solution of Equation (3.6) are the same as the stability properties of the same solution x_0 of (3.5). In particular, x_0 is stable (asymptotically stable) (unstable) for (3.6) if and only if $(x_0, y_0) = (x_0, \phi(x_0, h))$ is stable (asymptotically stable) (unstable) for (3.2).

Proof. In a neighborhood of zero, we have noted that $G(x, g, h)$, $\tilde{G}(x, g, h)$ have the same set of zeros. If x_0 is an isolated zero of these functions, then the stability properties of x_0 as a solution of (3.6), (3.5) are determined respectively by the sign of $G, \tilde{G}$ between zeros. If we prove $G, \tilde{G}$ have the same sign between zeros, then Theorem 3.2 will be a consequence of Theorem 3.1. Choose two consecutive common zeros of $G, \tilde{G}$ and suppose $G, \tilde{G}$ have opposite signs between them. Without loss in generality, we may assume these zeros are simple since we can make a small perturbation in g, h to accomplish this. By a translation in x, we may therefore suppose $G(0, g, h) = 0$, $\tilde{G}(0, g, h) = 0$, $\partial G(0, g, h)/\partial x > 0$, $\partial \tilde{G}(0, g, h) \partial x < 0$. Now replace (3.2) by the equation

$$(3.8) \qquad \begin{aligned} \dot{x} &= U(x, g, h) + \varepsilon \\ \dot{y} &= By + V(x, g, h) \end{aligned}$$

where ε is a small scalar parameter. For Eq. (3.8), we may compute the bifurcation function

$$G(x, g + \varepsilon, h) = G(x, g, h) + \varepsilon$$

by the method of Liapunov–Schmidt. We may also obtain a center manifold for each ε sufficiently small. By checking the way a center manifold is constructed in Section 2, the new function $\tilde{G}(x, g + \varepsilon, h)$ representing the vector field on the center manifold must satisfy

$$\tilde{G}(x, g + \varepsilon, h) = \tilde{G}(x, g, h) + \varepsilon\delta(x, g, h) + O(|\varepsilon|^2)$$

as $|\varepsilon| \to 0$, where $\delta(x, g, h) > 0$ in a neighborhood of $x = 0$. Since $\partial G(0, g, h)/\partial x$ and $\partial \tilde{G}(0, g, h)/\partial x$ have opposite signs, this shows that, for ε sufficiently small, there are real numbers $x_1(\varepsilon)$, $x_2(\varepsilon)$, $x_1(0) = x_2(0) = 0$, $x_1(\varepsilon) \neq x_2(\varepsilon)$ if $\varepsilon \neq 0$, such that $G(x_1(\varepsilon), g + \varepsilon, h) = 0$, $\tilde{G}(x_2(\varepsilon), g + \varepsilon, h) = 0$ and these are the only solutions of these equations in a neighborhood of $x = 0$, $\varepsilon = 0$. This is a contradiction since the zeros of G and $\tilde{G}$ must coincide. The case where x_0 is not isolated is much easier and the details are left to the reader. This proves Theorem 3.2. $\square$

As a first example, suppose there is a $\beta \neq 0$ such that the bifurcation function $G(x, g, h)$ for (3.2) satisfies

$$(3.9) \qquad G(x, 0, 0) = \beta x^2 + O(|x|^3) \quad \text{as } |x| \to 0.$$

Since u, v are second order in x, y it follows that $\phi(x, h)$ is second order in x and, thus, β can be found explicitly as

$$(3.10) \qquad 2\beta = \frac{\partial^2 u}{\partial x^2}(0, 0).$$

If $\beta > 0$, the function $G(x, g, h)$ has a minimum $\gamma(g, h)$ near $x = 0$, $g = 0$, $h = 0$, $\gamma(0, 0) = 0$. As was observed in Section 6.2, the set $\Gamma = \{(g, h) : \gamma(g, h) = 0\}$ is a submanifold of codimension one. On one side of Γ, there are no zeros of $G(x, g, h)$ and, on the other side, there are two zeros. With the information in Theorem 3.2, we can now describe the stability properties of the solutions in the following way:

(i) $\gamma(g, h) > 0$ is equivalent to no equilibrium of (3.2).
(ii) $\gamma(g, h) = 0$ is equivalent to one equilibrium of (3.2) which is asymptotically stable from the left and unstable from the right as in Fig. (3.1).
(iii) $\gamma(g, h) < 0$ is equivalent to two equilibriums of (3.2).

The case $\beta < 0$ is discussed in a similar way. When equation (3.2) satisfies relation (3.9), we say the equilibrium point of (3.2) for $(g, h) = (0, 0)$ is a *saddle-node*. The term saddle-node is used because as the vector field (g, h) moves across the manifold Γ, a saddle and node coalesce and disappear.

As another illustration, suppose there is a $\beta \neq 0$ such that

$$(3.11) \qquad\qquad G(x, 0, 0) = \beta x^3 + O(|x|^4) \quad \text{as } |x| \to 0.$$

The constant β is explicitly given from the expansion

$$(3.12) \qquad\qquad u(x, -B^{-1}\alpha x^2) = \beta x^3 + O(|x|^4) \quad \text{as } |x| \to 0$$
$$\alpha = \tfrac{1}{2}\partial^2 v(0, 0)/\partial x^2.$$

In Section 6.4, we discussed the nature of the zeros of the function $G(x, g, h)$ for (x, g, h) in a neighborhood of $(0, 0, 0)$. In the function space containing (g, h), there was a cusp manifold Γ of codimension one such that $G(x, g, h)$ has one simple zero on one side of Γ and three simple zeros on the other side of Γ. Theorem 3.2 gives the stability properties of these solutions in terms of the bifurcation function $G(x, g, h)$ as well as the stability properties of the solutions on Γ.

The reader will find it instructive to consider in detail the implications of the above remarks for one parameter families of perturbed vector fields when G satisfies (3.9) and two parameter families when G satisfies (3.11).

Theorem 3.2 also has interesting implications for the classical problem of stability in the critical case of one zero root; namely, the problem of stability of the solution $(x, y) = (0, 0)$ of Equation (3.2) for $g = 0$, $h = 0$,

$$(3.13) \qquad\qquad \begin{aligned} \dot{x} &= u(x, y) \\ \dot{y} &= By + v(x, y). \end{aligned}$$

If

$$(3.14) \qquad\qquad \begin{aligned} G(x) &= u(x, \phi(x)), \\ B\phi + v(x, \phi) &= 0, \end{aligned}$$

then the following result is an immediate consequence of Theorem 3.2.

Corollary 3.3. *The zero solution of* (3.13) *is asymptotically stable if and only if there is an $\varepsilon > 0$ such that $aG(a) < 0$ for $0 < |a| < \varepsilon$. It is unstable if and only if there is an $\varepsilon > 0$ such that $aG(a) > 0$ for either $0 < a < \varepsilon$ or $-\varepsilon < a < 0$. If there is an $\varepsilon > 0$ such that $G(a) = 0$ for $0 \le |a| < \varepsilon$, then the zero solution is stable and there is a first integral in a neighborhood of zero.*

Proof. Everything is obvious except the existence of the first integral. Suppose $G(a) = 0$ for $|a| < \varepsilon$ and let $y = h(x)$ be the parametric representation of a center manifold M at zero: $M = \{(x, y): y = h(x)\}$. We know that

$u(x, h(x)) = 0$ for $|x| < \varepsilon$. Also, each equilibrium point $(x_0, h(x_0))$ of (3.13) has an n-dimensional smooth stable manifold $S(x_0)$. The curve M and these manifolds $S(x_0)$ can be used for a coordinate system in a neighborhood of zero to transform $(x, y) \mapsto (\bar{x}, \bar{y})$ so that $\dot{\bar{x}} = 0$, $\dot{\bar{y}} = A\bar{y} + \bar{g}(\bar{x}, \bar{y})$. These latter equations have a first integral $V(\bar{x}, \bar{y}) = \bar{x}$ so that the original equation has a first integral. This proves the result. $\square$

Corollary 3.4. *If there is a scalar function $H(x, y, w)$, continuous for $(x, y, w) \in \mathbb{R} \times \mathbb{R}^n \times \mathbb{R}^n$ such that $H(x, y, 0) = 0$ and*

$$u(x, y) = H(x, y, By + v(x, y)),$$

then the zero solution of (3.13) is stable and there is a first integral near zero.

Proof. This is a direct consequence of Corollary 3.3. $\square$

Corollary 3.5. *If there is an integer $q \geq 2$ and $\beta \neq 0$ such that*

$$(3.15) \qquad G(x) = \beta x^q + O(|x|^{q+1}) \quad as \ |x| \to 0,$$

then the zero solution of (3.13) is asymptotically stable if and only if $\beta < 0$, q odd. Otherwise, it is unstable.

If u, v in (3.13) are analytic, then the function $G(x)$ in (3.14) is analytic. Thus, either there is an integer $q \geq 2$ and a nonzero constant β such that $G(x)$ satisfies (3.15) or $G(x) \equiv 0$ in a neighborhood of $x = 0$. From Corollary 3.3 and the proof of that result, we thus have

Corollary 3.6. *If u, v in (3.13) are analytic, then one of the following alternatives hold:*

   (i)   *There is an integer $q \geq 2$ and a nonzero constant β such that $G(x)$ satisfies (3.15) and the solution $x = 0$, $y = 0$ of Eq. (3.13) is asymptotically stable if q is odd, $\beta < 0$, and unstable otherwise.*

   (ii) *There is an analytic first integral of Eq. (3.13) and the solution $x = 0$, $y = 0$ is stable.*

Case (i) is referred to as the *algebraic case* and case (ii) as the *transcendental case*.

Let us illustrate with an example. Consider the 3-dimensional system

$$
\begin{aligned}
\dot{x} = X(x, y, z) &\stackrel{\text{def}}{=} (3m - 1)x^2 - (m - 1)y^2 - (n - 1)z^2 + (3n - 1)yz \\
&\quad - 2mzx - 2nxy \\
\dot{y} = Y(x, y, z) &\stackrel{\text{def}}{=} -y + x + (x - y + 2z)(y + z - x) \\
\dot{z} = Z(x, y, z) &\stackrel{\text{def}}{=} -z - x + (x + 2y - z)(y + z - x).
\end{aligned}
$$

(3.16)

Solving the equations $Y = Z = 0$ for y, z as a function of x near zero and putting the solution in X, one obtains that the bifurcation function $G(x)$ satisfies

$$G(x) = 4(5m - 7n)x^4 + 24(m - n)x^5 + \cdots$$

Thus, $5m \neq 7n$ implies instability. If $5m = 7n > 0$, then $m > 0$, $n > 0$ and $m > n$ which implies instability. If $5m = 7n < 0$, then $m < 0$, $n < 0$, $m < n$ and the origin is asymptotically stable. If $m = n = 0$, then $2X = (z - 2y - x)Y + (y - 2z - x)Z$ and Corollary 3.4 implies stability.

9.4. Periodic Case

Consider the $(n + 1)$-dimensional system

(4.1) $$\dot{z} = Cz + w(t, z) + f(t, z)$$

where $w, f: \mathbb{R}^{n+1} \to \mathbb{R}^{n+1}$ are continuous together with derivatives in z up through order $k \geq 2$, 2π-periodic in t, $w(t, 0) = 0$, $\partial w(t, 0)/\partial z = 0$ and

$$C = \begin{bmatrix} 0 & 0 \\ 0 & B \end{bmatrix}$$

where the eigenvalues of B have negative real parts. The function w is assumed to be fixed and the function f is to be considered as a parameter varying in a neighborhood of zero in the space of functions $\mathscr{F}_k \overset{\text{def}}{=} \{f: \mathbb{R} \times \mathbb{R}^{n+1} \to \mathbb{R}^{n+1} : f$ is continuous together with derivatives in z up through order $k \geq 2$, 2π-periodic in $t\}$ with the C^k-topology. We let $z = (x, y)$, $x \in \mathbb{R}$, $y \in \mathbb{R}^n$, $f = (g, h)$, $g \in \mathbb{R}$, $h \in \mathbb{R}^n$, $w = (u, v)$, $u \in \mathbb{R}$, $v \in \mathbb{R}^n$.

The method of Liapunov–Schmidt implies there is an $\varepsilon > 0$ and a neighborhood W of 0 in $\mathscr{P}_{2\pi} = \{z: \mathbb{R} \to \mathbb{R}^{n+1} : z$ is 2π-periodic and continuous$\}$ such that, for any $f \in V(\varepsilon) = \{f: |f|_k < \varepsilon\}$, every 2π-periodic solution of (4.1) in W must be of the form $z(t, a, f)$, $a \in U(\varepsilon) = (-\varepsilon, \varepsilon)$, where $z(\cdot, a, f) = (x(\cdot, a, f), y(\cdot, a, f))$ satisfies the equations

$$\frac{1}{2\pi} \int_0^{2\pi} x(t, a, f)\, dt = a$$

(4.2)
$$\dot{x} = u(t, z) + g(t, z) - G(a, f)$$
$$\dot{y} = By + v(t, z) + h(t, z)$$

where

(4.3) $$G(a, f) = \frac{1}{2\pi} \int_0^{2\pi} [u(t, z(t, a, f)) + g(t, z(t, a, f))]\, dt$$

and a is a zero of the *bifurcation function* $G(a, f)$; that is,

$$(4.4) \qquad\qquad G(a, f) = 0.$$

Conversely, any solution of Eq. (4.4) gives a 2π-periodic solution of (4.1) in W. The function $G(a, f)$ is C^k in a, f.

We remark that the function $z(\cdot, a, f)$ can be obtained from equations involving only the Fourier coefficients of z.

Our objective is to show that the bifurcation function $G(a, f)$ not only permits one to determine the periodic solutions of (4.1), but gives precise information about the stability of the periodic solutions. As in Section 3, we need to recall results on center manifolds. The results in Section 2 on center manifolds were given for autonomous systems, but are easily extended to the periodic case to yield the following.

Theorem 4.1. *If w, f in (4.1) are in $\mathscr{F}_k$, $k \geq 2$, then there is a $\delta > 0$ and a C^k-function $\psi(t, x, f) \in \mathbb{R}^n$ defined for $t \in \mathbb{R}$, $|x| < \delta$, $|f|_k < \delta$, 2π-periodic in t, such that $\psi(t, 0, 0) = 0$, $\partial\psi(t, 0, 0)/\partial x = 0$ and the set*

$$\{(t, x, y) : y = \psi(t, x, f), t \in \mathbb{R}, |x| < \delta\}$$

is a center manifold of (4.1). This center manifold is exponentially asymptotically stable, the flow on this manifold is given by the scalar differential equation

$$(4.5) \qquad\qquad \dot{x} = u(t, x, \psi(t, x, f)) + g(t, x, \psi(t, x, f))$$

and the stability properties of a 2π-periodic solution of (4.1) (which must lie on the center manifold) coincide with the stability properties of this solution as a solution of (4.5).

Since any 2π-periodic solution of (4.1) is obtained by the above method of Liapunov–Schmidt and is related in a unique way to a zero of the bifurcation function $G(a, f)$, it follows that the 2π-periodic solutions on the center manifold are in one-to-one correspondence with the zeros of $G(a, f)$. The main result of this section is to show that the stability properties of the periodic solutions of (4.5) are also determined directly from the bifurcation function. This, in turn, means that the stability properties of the periodic solutions of Equation (4.1) are determined from the bifurcation function.

Theorem 4.2. *There is a neighborhood W of zero in $\mathscr{P}_{2\pi}$, an $\varepsilon > 0$, and functions*

$$z : \mathbb{R} \times U(\varepsilon) \times V(\varepsilon) \to \mathbb{R}^{n+1}$$
$$G : U(\varepsilon) \times V(\varepsilon) \to \mathbb{R}$$

$U(\varepsilon) = (-\varepsilon, \varepsilon)$, $V(\varepsilon) = \{f \in \mathscr{F}_k : |f|_k < \varepsilon\}$, $z(t, a, f) = z(t + 2\pi, a, f)$, $G(a, f)$ *are* C^k*-functions, such that Eq.* (4.1) *has a 2π-periodic solution $z(t)$ in W if and only if $z(t) = z(t, a, f)$ and a satisfies the bifurcation equation* (4.4). *Furthermore, the stability properties of the 2π-periodic solution $z(t, a_0, f)$, $G(a_0, f) = 0$, of* (4.1) *as a solution on the center manifold coincide with the stability properties of a_0 as a solution of the autonomous scalar equation*

$$(4.6) \qquad\qquad \dot{a} = G(a, f).$$

In particular, a_0 is stable (asymptotically stable) (unstable) for (4.6) *if and only if $z(t, a, f)$ is stable (asymptotically stable) (unstable) for* (4.1).

Proof. The first step of the proof is to discuss the case when $z \in \mathbb{R}$; that is, the variable y is absent. Then the equation on the center manifold is Eq. (4.1). Choose $\varepsilon > 0$ so small that $2 > \partial x(t, a, f)/\partial a > \frac{1}{2}$ for all $a \in U(\varepsilon)$, $f \in V(\varepsilon)$. Then the 2π-periodic transformation of variables from x to a given by

$$x = x(t, a, f) = a + \xi(t, a, f), \qquad \frac{1}{2\pi} \int_0^{2\pi} \xi(t, a, f)\,dt = 0,$$

is well defined for each t. Furthermore, if x is a solution of (4.1) and $x(t) = x(t, a(t), f)$, then a is a solution of the differential equation

$$\dot{a} = \left(1 + \frac{\partial \xi(t, a, f)}{\partial a}\right)^{-1} G(a, f).$$

The stability properties of any 2π-periodic solution of (4.1) are the same as the stability properties of any 2π-periodic solution of this equation since these solutions are the zeros of $G(a, f)$ and $2 > 1 + \partial \xi(t, a, f)/\partial a > \frac{1}{2}$, the stability properties of these 2π-periodic solutions are determined from (4.6). This proves the theorem for the case $z \in \mathbb{R}$.

Suppose now that $z \in \mathbb{R}^{n+1}$ and the vector field on the center manifold is given by Equation (4.5). Applying the method of Liapunov–Schmidt to Eq. (4.5), we obtain a bifurcation function $\tilde{G}(a, f)$ for Eq. (4.5) and a 2π-periodic function $\tilde{x}(t, a, f)$ which have the property that Eq. (4.5) has a 2π-periodic solution $x(t)$ if and only if $x(t) = \tilde{x}(t, a, f)$ and $\tilde{G}(a, f) = 0$. Thus, Eq. (4.1) has a 2π-periodic solution $z(t)$ if and only if $\tilde{G}(a, f) = 0$ and $z(t) = (\tilde{x}(t, a, f), \psi(t, \tilde{x}(t, a, f), f))$. From the first part of the proof, the stability properties of the 2π-periodic solutions $x(t, a, f)$ of (4.5) coincide with the stability properties of a_0 as a solution of

$$(4.7) \qquad\qquad \dot{a} = \tilde{G}(a, f).$$

We now show that a_0 is also a zero of $G(a, f)$ and its stability properties as a solution of (4.6) coincide with those of (4.7).

Equation (4.1) has a 2π-periodic solution $z(t)$ if and only if $G(a, f) = 0$ and $z(t) = z(t, a, f)$, where $z(t, a, f)$ satisfies (4.2). In both of these processes of constructing 2π-periodic solutions, the function $\tilde{z}(t, a, f), z(t, a, f)$ are uniquely determined by a. Therefore, it follows that the zeros of $\tilde{G}(a, f), G(a, f)$ are the same. If a_0 is an isolated zero of $\tilde{G}(a, f), G(a, f)$ and these functions have the same sign between zeros, then the stability properties of any zero as a solution of either $\dot{a} = G(a, f)$ or $\dot{a} = \tilde{G}(a, f)$ are the same and Theorem 4.2 is proved. To prove they have the same sign between zeros, we may assume by making a small perturbation in f in (4.1) that the zeros of both functions are simple. Suppose this is done and there is an a_0 such that $G(a_0, f) = 0$, $\tilde{G}(a_0, f) = 0$ and the derivatives in a at a_0 have opposite signs. If $f = (g, h)$, $g \in \mathbb{R}$, replace g by $g + \varepsilon$ where $\varepsilon \in \mathbb{R}$ and obtain new bifurcation functions $G(a, f, \varepsilon), \tilde{G}(a, f, \varepsilon)$. It is not difficult to see that there is a $\delta(a, f) > 0$ such that

$$G(a, f, \varepsilon) = G(a, f) + \varepsilon$$
$$\tilde{G}(a, f, \varepsilon) = \tilde{G}(a, f) + \delta(a, f)\varepsilon + O(|\varepsilon|^2)$$

as $|\varepsilon| \to 0$ where $\delta(a, f) > 0$. Each of these new functions has a unique zero in a neighborhood of a_0 for ε small. They are distinct for $\varepsilon \neq 0$ since the signs of the derivatives in a at a_0 of $G, \tilde{G}$ are distinct. This contradicts the fact that $G(a, f, \varepsilon), \tilde{G}(a, f, \varepsilon)$ must have the same set of zeros. The discussion of the case where a_0 is not isolated is left for the reader. This completes the proof of Theorem 4.2. $\square$

Theorem 4.2 has very interesting implications. For example, suppose the bifurcation function $G(a, f)$ for Equation (4.1) is C^2 and satisfies

$$(4.8) \qquad G(a, 0) = \beta a^k + O(|a|^{k+1}) \quad \text{as } |a| \to 0$$

with $\beta \neq 0$. As a special case, suppose $k = 2$, $\beta < 0$. The function $G(a, f)$ has a maximum $\gamma(f)$ near $x = 0, f = 0, \gamma(0) = 0$. As was observed in Section 6.2, the set $\Gamma = \{f : \gamma(f) = 0\}$ is a submanifold of codimension one. On one side of γ, there are no solutions and, on the other side, there are two. This means Equation (4.1) has no 2π-periodic solutions on one side of Γ and two 2π-periodic solutions on the other side of Γ. With Theorem 4.2, one can now state the following result:

$\gamma(f) > 0$ implies two 2π-periodic solutions of (4.1), one asymptotically stable and one unstable.

$\gamma(f) = 0$ implies one semistable 2π-periodic solution of (4.1).

$\gamma(f) < 0$ implies no 2π-periodic solutions of (4.1).

Suppose now that $k = 3$. In Section 6.4, we have shown that there is a cusp manifold in the space of perturbations f described by the zeros of a function $\gamma(f)$ such that, in a neighborhood of $a = 0$, $f = 0$, there is exactly one simple zero of G one side of this manifold and exactly three simple zeros on the other side. If $\beta < 0$, one can then apply Theorem 4.2 to obtain the following conclusion regarding the stability of the corresponding periodic orbits.

$\gamma(f) > 0$ implies three 2π-periodic solutions of (4.1), two asymptotically stable and one unstable,

$\gamma(f) = 0$, f not on the cusp, implies two 2π-periodic solutions of (4.1), one semistable and the other hyperbolic.

$\gamma(f) < 0$ implies one asymptotically stable 2π-periodic solution of (4.1).

The integer k in (4.8) characterizes the maximal number of periodic solutions that are possible when the differential equation (4.1) is perturbed by any small function f. This integer and the sign of β also determine the stability properties of the solution $x = 0$ of the unperturbed equation

$$(4.9) \qquad \dot{z} = Cz + w(t, z).$$

In fact, $x = 0$ is asymptotically stable if and only if k is odd and $\beta < 0$ and unstable otherwise.

The 2π-periodic function $x(t, a, f)$ and the bifurcation function $G(a, f)$ can be calculated to any accuracy required. If $w(t, x)$, $f(t, x)$ in (4.1) are analytic functions of x in a neighborhood of $x = 0$, then $x(t, a, f)$, $G(a, f)$ are analytic in a in a neighborhood of $a = 0$. Thus,

$$(4.10) \qquad \begin{aligned} x(t, a, f) &= \sum_{j=0}^{\infty} x_j(t, f) a^j \\ G(a, f) &= \sum_{j=0}^{\infty} \beta_j(f) a^j. \end{aligned}$$

If

$$(4.11) \qquad \begin{aligned} w(t, x) &= \sum_{j=2}^{\infty} w_j(t) x^j \\ f(t, x) &= \sum_{j=0}^{\infty} f_j(t) x^j \end{aligned}$$

then the coefficients $x_j(t, f)$, $\beta_j(f)$, $j \geq 0$, can be successively determined by substituting all series in Eq. (4.2) and equating coefficients of a^j, $j \geq 0$. For any fixed integer $q \geq 0$, it is easy to verify that $x_q(t, f)$, $\beta_q(f)$ depend only upon $w_j(t)$, $f_j(t)$, for $j \leq q$.

If w, f are only C^k, then the Taylor series of $x(t, a, f)$, $G(a, f)$ are obtained in the same way.

9.5. Bifurcation from a Focus

In this section, we show how the results of Section 4 have immediate applications to the problem of bifurcation from a focus. Consider the $(n + 2)$-dimensional system

$$(5.1) \qquad \dot{z} = Cz + Z(z, \mu)$$

where $\mu \in E$, a finite or infinite dimensional Banach space, Z is a C^k-function, $k \geq 1$, satisfying

$$(5.2) \qquad Z(0, \mu) = 0, \qquad \partial Z(0, 0)/\partial z = 0$$

for all μ and the matrix C has a block diagonal representation

$$(5.3) \qquad C = \begin{bmatrix} A & 0 \\ 0 & B \end{bmatrix}, \qquad A = \begin{bmatrix} 0 & 1 \\ -1 & 0 \end{bmatrix}$$

with the eigenvalues of the $n \times n$ matrix B having negative real parts.

The problem is to discuss the behavior of all solutions of (5.1) in a neighborhood of $z = 0$, $\mu = 0$. Let $z = (x, y)$, $Z = (X, Y)$ with x, $X \in \mathbb{R}^2$, y, $Y \in \mathbb{R}^n$. Periodic orbits of autonomous systems are closed curves. It is therefore natural to try to reduce the discussion to the one considered in Section 4 by introducing appropriate polar coordinates and replace t by the angle. To show that such a coordinate system can be found, we need an a priori bound on the center manifolds of Eq. (5.1). Since (5.2) is satisfied, for any $\varepsilon > 0$, $\delta > 0$, there is a constant $K = K(\varepsilon, \delta)$ such that, if $y = \psi(x, \mu)$ is a center manifold of (5.1) for $|x| < \varepsilon$, $|\mu| < \delta$, then

$$(5.4) \qquad |\psi(x, \mu)| \leq K|x|.$$

This estimate follows from the way the center manifold is constructed in Section 2. Since every periodic orbit must lie on the center manifold, it follows that every periodic orbit of (5.1) in a neighborhood of $z = 0$, $\mu = 0$ must have the y coordinate satisfying (5.4). This justifies the transformation of variables

$$(5.5) \qquad x_1 = p \cos \theta, \qquad x_2 = -p \sin \theta, \qquad y = pw$$

in Equation (5.1). The new equations for (θ, p, w) are

$$(5.6) \quad \begin{aligned} \dot{\theta} &= 1 - [X_1 \sin \theta + X_2 \cos \theta]/p \overset{\text{def}}{=} 1 + \Theta(\theta, p, w, \mu) \\ \dot{p} &= X_1 \cos \theta - X_2 \sin \theta \\ \dot{w} &= Bw + Y/p - (X_1 \cos \theta - X_2 \sin \theta)w/p \end{aligned}$$

where all functions are evaluated at $(p \cos \theta, -p \sin \theta, pw, \mu)$.

Since the function Θ satisfies $\Theta(\theta,0,w,0) = 0$, it follows that $\dot\theta > 1/2$ for w in an arbitrary fixed compact set and (p,μ) in a sufficiently small neighborhood of zero. Thus, we may replace t by θ to obtain

$$\frac{dp}{d\theta} = R(\theta,p,w,\mu)$$

(5.7)

$$\frac{dw}{d\theta} = Bw + W(\theta,p,w,\mu)$$

The functions R, W are 2π-periodic in θ and

(5.8)
$$R(\theta,0,w,\mu) = 0, \qquad \partial R(\theta,0,w,0)/\partial p = 0$$
$$W(\theta,0,w,0) = 0.$$

For equation (5.7) satisfying (5.8), we can now apply the method of Laipunov–Schmidt as was done in Section 4 to determine the 2π-periodic functions of the form

$$p = p^*(\theta,a,\mu), \qquad w = w^*(\theta,a,\mu)$$

(5.9) $\quad p^*(\theta,0,\mu) = 0, \qquad \partial p^*(\theta,0,0)/\partial a = 1, \qquad \dfrac{1}{2\pi}\displaystyle\int_0^{2\pi} p^*(\theta,a,\mu)\,d\theta = a,$

$$w^*(\theta,0,0) = 0$$

and the bifurcation function

(5.10)
$$G(a,\mu) = \int_0^{2\pi} R(\theta,\, a + p^*(\theta,a,\mu),\, w^*(\theta,a,\mu),\, \mu)\,d\theta$$

for (a,μ) in a neighborhood of zero.

From the general theory in Section 4, the existence and stability properties of the 2π-periodic solutions of (5.7) (and thus the orbital stability properties of the periodic orbits of (5.1)) are determined by the equilibrium solutions of

(5.11)
$$\dot a = G(a,\mu).$$

We summarize these remarks in

Theorem 5.1. *Suppose the transformation (5.5) is applied to Equation (5.1) to obtain Eq. (5.7). Let $(p^*(\theta,a,\mu),\, w^*(\theta,a,\mu))$, $G(a,\mu)$ be respectively the 2π-periodic function and bifurcation function obtained by applying the method of Liapunov–Schmidt to Equation (5.7). Then there is a neighborhood U of $(p,\mu) = (0,0)$ such that Equation (5.7) has a 2π-periodic solution $(p(\theta),w(\theta))$ with $(p(\theta),\mu)) \in U$ if and only if $p(\theta) = p^*(\theta,a_0,\mu)$, $w(\theta) = w^*(\theta,a_0,\mu)$ where (a_0,μ)*

satisfy $G(a_0, \mu) = 0$. Furthermore, the stability properties of $(p(\theta), w(\theta))$ coincide with the stability properties of the solution a_0 of Eq. (5.11).

As an immediate application of Theorem 5.1, we obtain the following version of the Hopf Bifurcation Theorem for a scalar parameter.

Theorem 5.2. *If $\mu \in \mathbb{R}$,*

$$C(\mu) = C + \mu D, \qquad D = \frac{\partial^2 Z(0, 0)}{\partial z \, \partial \mu}$$

and $\lambda_1(\mu)$, $\lambda_2(\mu) = \bar{\lambda}_1(\mu)$ are the eigenvalues of $C(\mu)$ near $\mu = 0$, $\lambda_1(0) = i$, then $d\,\mathrm{Re}\,\lambda_1(0)/d\mu \neq 0$ implies there is a $\delta > 0$ and a C^1-function $\mu^* : (-\delta, \delta) \to \mathbb{R}$ such that $\mu^*(0) = 0$ and, for every a, $|a| < \delta$, Eq. (5.7) for $\mu = \mu^*(a)$ has a 2π-periodic solution $p^*(\theta, a, \mu^*(a))$, $w^*(\theta, a, \mu^*(a))$, where p^*, w^* are given in Theorem 5.1 and $(a, \mu^*(a))$ satisfy $G(a, \mu^*(a)) = 0$. If $d\,\mathrm{Re}\,\lambda_1(0)/d\mu > 0$ and there is an $\varepsilon > 0$ such that $\mu^* : (-\varepsilon, \varepsilon) \to \mathbb{R}$ is one-to-one, then the periodic solution corresponding to $(a, \mu^*(a))$ is unstable if $\mu^*(a) < 0$ (subcritical case) and asymptotically stable if $\mu^*(a) > 0$ (supercritical case). Using the polar coordinate change of variables (5.5), one obtains the corresponding results for (5.1).

Proof. Using the change of variables (5.5), one sees that the bifurcation function $G(a, \mu)$ is given by

$$G(a, \mu) = aF(a, \mu)$$

where $F(a, \mu) = \alpha(\mu) + O(|a|)$ as $|a| \to 0$, $\alpha(0) = 0$, $\alpha'(0) = d\,\mathrm{Re}\,\lambda_1(0)/d\mu \neq 0$. The Implicit Function Theorem implies the first part of the Theorem. To prove stability, we note that $\partial G(0, 0)/\partial a$ has the same sign as $d\,\mathrm{Re}\,\lambda_1(0)/d\mu$ and the solution $x = 0$ is asymptotically stable for $\mu < 0$ and unstable for $\mu > 0$. Since we are assuming there is at most one other zero in a sufficiently small neighborhood of zero, the result follows from Theorem 5.1. $\square$

In the discussion below, we need the following observation whose proof is left to the reader.

Lemma 5.3. *$G(a, \mu)$ is an odd function of a.*

Theorem 5.1 together with Lemma 5.3 give sufficient information to take care of most applications. For example, suppose we have computed the bifurcation function $G(a, \mu)$ approximately and obtained

(5.12) $$G(a, \mu) = \sum_{j=0}^{k} \beta_{k-j}(\mu) a^{2j+1} + o(|a|^{2k+1})$$

as $|a| \to 0$ and suppose that

$$(5.13) \qquad \begin{aligned} \beta_0(0) &= \beta, & \beta \neq 0, \\ \beta_j(0) &= 0, & j = 0, 1, 2, \ldots, k - 1. \end{aligned}$$

If

$$(5.14) \qquad G(a, \mu) = aF(a^2, \mu),$$

then

$$(5.15) \qquad F(r, \mu) = \sum_{j=0}^{k} \beta_{k-j}(\mu) r^j + o(|r|^k)$$

as $|r| \to 0$. The problem of the existence of 2π-periodic solutions of (5.7) is then equivalent to the existence of nonnegative solutions of

$$(5.16) \qquad F(r, \mu) = 0.$$

The stability of the periodic solutions is also easily related to the behavior of F near its zeros.

Corollary 5.4. *If $G(a, \mu)$ satisfies (5.12), (5.13), then there is a neighborhood U of $z = 0$, V of $\mu = 0$ such that Equation (5.1) has at most k periodic orbits in U for $\mu \in V$. Furthermore, for any integer j, $0 \leq j \leq k$, there is a vector field $Y \in C^{2k+1}(\mathbb{R}^2, \mathbb{R}^{n+2})$ such that the equation*

$$(5.17) \qquad \dot{z} = Az + Z(z, 0) + Y(x), \qquad z = (x, y), \qquad x \in \mathbb{R}^2, \, y \in \mathbb{R}^n,$$

has exactly j periodic orbits in U.

Proof. The first part is obvious. To prove the second part, it is only necessary to choose Y so that the bifurcation function $G(a, Y)$ has the form

$$G(a, Y) = \sum_{j=0}^{k} \beta_{k-j}(Y) a^{2j+1} + o(|a|^{2k+1})$$

as $a \to 0$ and the polynomial consisting of the first $2k + 1$ terms has j positive zeros. Thus, it is sufficient to show that the mapping $(\beta_1(Y), \ldots, \beta_k(Y))$ takes a neighborhood of $Y = 0$ into a neighborhood of zero in $\mathbb{R}^k$. To see that this is true, we make a special perturbation in the following way:

$$\dot{x} = Ax + X(x, y, 0) + (\text{grad}|x|^2) \sum_{j=0}^{k} \lambda_j |x|^{2j}$$

$$\dot{y} = By + Y(x, y, 0),$$

where $|x|^2 = x_1^2 + x_1^2$. The bifurcation function $G(a, \lambda)$, $\lambda = (\lambda_0, \lambda_1, \ldots, \lambda_k)$, has the form

$$G(a, \lambda) = \sum_{j=0}^{k} \beta_{k-j}(\lambda) a^{2j+1} + o(|a|^{2k+1})$$

as $|a| \to 0$, where $\beta_0(0) = \beta \neq 0$ and, for all j, $\beta_{k-j}(\lambda) = \lambda_j + o(|\lambda_0, \lambda_1, \ldots, \lambda_{j-1}|)$ as $\lambda \to 0$. This last formula is clear from the manner in which the bifurcation function is constructed. Therefore, the transformation $\lambda_j \mapsto \mu_j = \beta_j(\lambda)$, $j = 0, 1, \ldots, k$, is a diffeormorphism in a neighborhood of 0. This completes the proof. $\square$

We now discuss some particular cases in more detail.

Theorem 5.5 (Generic Hopf Bifurcation). *Suppose X in (5.1) is C^r, $r \geq 3$, and*

(5.18)
$$G(a, \mu) = \beta_0(\mu)a^3 + \beta_1(\mu)a + o(|a|^3) \quad as \ |a| \to 0$$
$$\beta_0(0) = \beta_0 \neq 0, \qquad \beta_1(0) = 0.$$

Then there is a neighborhood U of $\mu = 0$, V of $z = 0$ such that, for any $\mu \in U$, Eq. (5.1) has a periodic orbit in V if and only if $\beta_1(\mu)\beta_0 < 0$. When this condition is satisfied, the orbit is unique and is asymptotically stable (unstable) if and only if $\beta_0 < 0$ ($\beta_0 > 0$). The bifurcation surface

$$\Gamma_0 = \{\mu \in U : \beta_1(\mu) = 0\}$$

is a submanifold of codimension one if the derivative $D_\mu(\beta_1) \neq 0$ at $\mu = 0$.

Proof. In a neighborhood of zero, $G(a, \mu) = 0$ either has one positive solution or no positive solution. The condition for the existence of one is $\beta_1(\mu)\beta_0 < 0$. The stability follows from Theorem 5.1. The last statement of the theorem follows from the definition of a submanifold of codimension one. $\square$

The condition $D_\mu \beta_1 \neq 0$ at $\mu = 0$ says there is some direction in μ space with the property that the eigenvalues of the linear approximation will cross the imaginary axis transversally as one passes through zero. If μ is a scalar, this is just saying $\beta_1'(0) \neq 0$.

Theorem 5.5 is the simplest possible situation that can occur. If we let

$$\Gamma_+ = \{\mu \in U : \beta_1(\mu)\beta_0 > 0\}$$
$$\Gamma_- = \{\mu \in U : \beta_1(\mu)\beta_0 < 0\}$$

and $D_\mu \beta_1(0) \neq 0$, then $U = \Gamma_+ \cup \Gamma_0 \cup \Gamma_-$ and there is a unique periodic orbit in Γ_+ which is stable (unstable) if $\beta_0 < 0$ ($\beta_0 > 0$) and no periodic

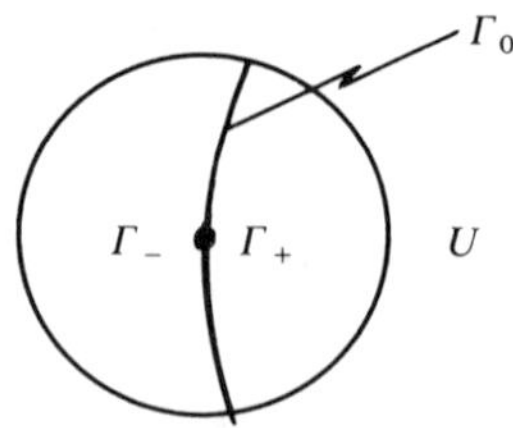

Figure 5.1

orbit in Γ_- (see Figure 5.1). When μ is a scalar, $G(a,\mu) = aF(a^2,\mu)$, then

$$F(r,\mu) = \beta_0(\mu)r + \beta_1(\mu) + o(|r|)$$

as $r \to 0$. The equation $F(r,\mu) = 0$ has a unique solution $r = r^*(\mu)$ for $|\mu| < \delta$, $r^*(0) = 0$,

$$r^*(\mu) = -\beta_0^{-1}\beta_1'(0)\mu + o(|\mu|)$$

as $|\mu| \to 0$. We must have $r^*(\mu) \geq 0$ and, thus, there exists a nontrivial periodic solution in a neighborhood of $\mu = 0$, $a = 0$ if and only if $\beta_0^{-1}\beta_1'(0)\mu < 0$. Using Theorem 5.5, for $\beta_0 < 0$, we have $\beta_0^{-1}\beta_1'(0) < 0$ implies the solution exists and is asymptotically stable for $\mu > 0$, and $\beta_0^{-1}\beta_1'(0) > 0$ implies the solution exists and is unstable for $\mu < 0$.

Let us now consider the case $k = 2$; that is,

$$(5.19) \qquad G(a,\mu) = \beta_0(\mu)a^5 + \beta_1(\mu)a^3 + \beta_2(\mu)a + o(|a|^5)$$

as $|a| \to 0$, $\beta_0(0) = \beta \neq 0$, $\beta_1(0) = \beta_2(0) = 0$. If $G(a,\mu) = aF(a^2,\mu)$, then

$$(5.20) \qquad F(r,\mu) = \beta_0(\mu)r^2 + \beta_1(\mu)r + \beta_2(\mu) + o(|r|^2)$$

as $|r| \to 0$. We must find conditions on μ to ensure that $F(r,\mu) = 0$ has nonnegative solutions.

For definiteness, let us suppose $\beta > 0$. Then the function $F(r,\mu)$ has a unique minimum $\gamma(\mu)$ in a neighborhood of $r = 0$, $\mu = 0$, with $\gamma(0) = 0$. If $\gamma(\mu) > 0$, there are no real solutions of $F(r,\mu) = 0$. If $\gamma(\mu) = 0$, there is one real solution and, if $\gamma(\mu) < 0$, there are two real solutions. The requirement that there is a nonnegative solution restricts μ. It is notationally difficult to discuss this in general so let us be content with discussing the generic case of two parameters:

$$(5.21) \quad \mu = (\mu_1, \mu_2), \qquad \beta_0(0) = -1, \qquad \beta_j(\mu) = \mu_j + O(|\mu|^2), \qquad j = 1, 2,$$

as $|\mu| \to 0$.

The bifurcation curves are going to be approximately the same as the bifurcation curves for the equation

$$-r^2 + \mu_1 r + \mu_2 = 0$$

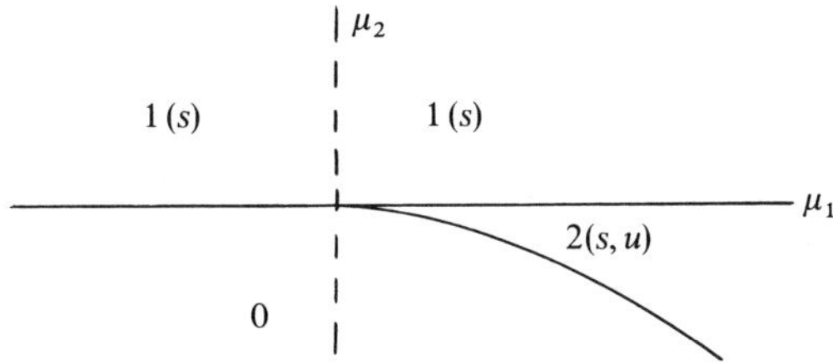

Figure 5.2. (s) = stable, (u) = unstable

The solutions of this equation are

$$2r = \mu_1 \pm (\mu_1^2 + 4\mu_2)^{1/2}$$

and it is easy to see that the number of nonnegative solutions for a given μ_1, μ_2 are as shown in Fig. 5.2. The stability of these solutions is also labeled with s denoting asymptotically stable, u denoting unstable.

This example also brings out an important property of the difference between one parameter problems and two parameter problems.

In the original differential equation, suppose the vector field depends only upon a scalar parameter α. In the generic two parameter case given by our example for $k = 2$ with $\mu = (\mu_1, \mu_2)$ and (5.20) satisfied, this means $\mu_1 = \mu_1(\alpha)$, $\mu_2 = \mu_2(\alpha)$ and we are moving along a curve in the parameter space. Suppose now the conditions of Theorem 5.2 are satisfied; that is, the matrix $C + \partial X(0, \mu(\alpha))/\partial x$ has two eigenvalues $\lambda_1(\alpha)$, $\bar{\lambda}_1(\alpha)$, $\lambda_1(0) = i \neq 0$, $d \operatorname{Re} \lambda_1(0)/d\alpha \neq 0$. This implies that the bifurcation function

$$G(a, \alpha) = aF(a^2, \alpha)$$

satisfies

$$F(r, \alpha) = -r^2 + \delta_0(\alpha)r + \delta_1(\alpha) + o(|r|^2)$$

as $r \to 0$ with $\delta_1(0) = 0$, $d\delta_1(0)/d\alpha \neq 0$, $\delta_0(0) = 0$. The properties of the zeros of this function near zero are determined solely by the leading terms $-r^2 + \delta_1(\alpha)$ and this function either has one nonnegative zero or none. Thus, it is impossible to ever have two nontrivial periodic solutions when two eigenvalues cross the imaginary axis transversally even though the bifurcation function $G(a, \alpha)$ satisfies $G(a, \alpha) = -a^5 + o(|a|^5)$ as $|a| \to 0$.

The above situation is a general phenomena. To see this, suppose that $G(a, \mu)$ satisfies (5.12), (5.13) and let

$$\Gamma_0 = \{\mu : \beta_k(\mu) = 0\}$$
$$\Gamma_+ = \{\mu : \beta_k(\mu) > 0\}$$
$$\Gamma_- = \{\mu : \beta_k(\mu) < 0\}.$$

If $\lambda_1(\mu), \lambda_2(\mu) = \bar{\lambda}_1(\mu)$ are the eigenvalues of $C + \partial Z(0,\mu)/\partial z$ with the property that $\lambda_1(0) = i$, then $\beta_k(\mu) = c\,\mathrm{Re}\,\lambda_1(\mu)$ for some positive constant c. Thus, $\mu \in \Gamma_+(\Gamma_-)$ if and only if $\mathrm{Re}\,\lambda_1(\mu) > 0\ (<0)$. If $G(a,\mu) = aF(a^2,\mu)$, then the zeros of $F(r,\mu)$ in a neighborhood of zero are determined by the terms $\beta r^k + \beta_k(\mu)$ in the Taylor series of $F(r,\mu)$. If $\beta\beta_k(\mu) > 0$, there is no zero and if $\beta\beta_k(\mu) < 0$, there is a unique positive zero. In particular, if $\mu \in \mathbb{R}$ and $d\,\mathrm{Re}\,\lambda_1(0)/d\mu \neq 0$ (that is, two eigenvalues cross the imaginary axis transversally at $\mu = 0$) (that is, Γ_0 above is crossed transversally at $\mu = 0$) then, in a neighborhood of zero, $\beta\beta_k(\mu) \neq 0$. This implies there is never more than one periodic orbit in a neighborhood of zero. The stability properties of the orbit are determined by the sign of β. This result is summarized in

Theorem 5.6. *Suppose* $\mu \in \mathbb{R}$, $G(0,\mu)$ *satisfies* (5.12), (5.13), $\lambda_1(\mu) = \bar{\lambda}_2(\mu)$, *are the eigenvalues of* $C + \partial Z(0,\mu)/\partial z$ *with* $\lambda_1(0) = i$. *If* $d\,\mathrm{Re}\,\lambda_1(0)/d\mu \neq 0$, *then there is a neighborhood* V *of* $z = 0$ *and a* $\mu_0 > 0$ *such that, if* $0 < |\mu| < \mu_0$, *then Equation* (5.1) *has a periodic orbit in* V *if and only if* $\beta d\,\mathrm{Re}\,\lambda_1(0)/d\mu < 0$. *When this condition is satisfied, the periodic orbit is unique, asymptotically stable if* $\beta < 0$ *and unstable if* $\beta > 0$.

Theorem 5.6 shows that at most one periodic orbit of (5.1) can exist in a neighborhood of zero if the eigenvalues cross the imaginary axis transversally. This is independent of the integer k in (5.15). If the crossing is not transversal, then more than one periodic can exist if $k > 1$. For example, suppose $F(r,\mu)$, $\mu \in \mathbb{R}$, has the form

$$F(r,\mu) = -r^2 + p_1\mu r + p_2\mu^2 + o((|\mu| + |r|)^2)$$

as $|\mu|, |r| \to 0$. In this case the term $p_1\mu r$ can play an important role and allow the equation to have two nonnegative solutions. In fact, if $\mu > 0$, $p_1 > 0$, $p_2 < 0$, the equation has two solutions.

For the case $k = 1, 2$, we have given above a complete description of the periodic orbits of (5.1) near zero. If the original vector field in (5.1) is C^∞, then the case of general k can at least be treated theoretically. In fact, in this case, the function $F(r,\mu)$ in (5.12) is C^∞ and one can assert from the Malgrange Preparation Theorem (see Section 2.7) that

$$F(r,\mu) = [\beta r^k + \alpha_1(\mu)r^{k-1} + \cdots + \alpha_k(\mu)]H(r,\mu)$$

with $H(0,0) > 0$ and $\alpha_j(\mu) = \mu_j + O(|\mu|^2)$ as $|\mu| \to 0$.

Thus, the complete classification of the bifurcation surfaces in μ space are possible theoretically. Obtaining these surfaces on a practical scale is more difficult since it requires knowing the $\alpha_j(\mu)$. However, approximations to these surfaces can be obtained since approximate values of the $\alpha_j(\mu)$ are obtained from approximate values of the $\beta_j(\mu)$ in (5.12).

The integer k and constant β in (5.12), (5.13) are related to the stability properties of the solution $z = 0$ of Equation (5.1) for $\mu = 0$. In fact, if

$$(5.22) \qquad G(a, 0) = \beta a^{2k+1} + O(|a|^{2k+3}), \qquad \beta \neq 0,$$

as $a \to 0$, then the solution $a = 0$ of $\dot{a} = G(a, 0)$ is asymptotically stable if $\beta < 0$ and unstable if $\beta > 0$. Thus, Theorem 5.1 implies the solution $x = 0$, $y = 0$ of

$$(5.23) \qquad \begin{aligned} \dot{x} &= Ax + X(x, y, 0) \\ \dot{y} &= By + Y(x, y, 0) \end{aligned}$$

is asymptotically stable if $\beta < 0$ and unstable if $\beta > 0$.

It is interesting to relate β and k to the stability properties of (5.23) under perturbations in a certain class. More specifically, consider the system

$$(5.24) \qquad \begin{aligned} \dot{x} &= Ax + X(x, y, 0) + f(x, y) \\ \dot{y} &= By + Y(x, y, 0) + g(x, y) \end{aligned}$$

where f, g vanish at $x = 0$, $y = 0$ and belong to a certain class to be described. Let $H(a, f, g)$ be the bifurcation function associated with (5.24) obtained through Theorem 5.1.

If $G(a, 0)$ satisfies (5.22), then, for any f, g in the class of functions for which

$$(5.25) \qquad H(a, f, g) = \beta a^{2k+1} + O(|a|^{2k+3})$$

the stability properties of the solution $x = 0$, $y = 0$ of Eq. (5.24) are the same as the ones for Eq. (5.23).

Conversely, suppose $q > 1$ is a fixed integer and consider the class of perturbations f, g which have the property that

$$(5.26) \qquad H(a, f, g) = \sum_{j=0}^{q} \beta_j a^{2j+1} + O(|a|^{2q+3})$$

as $a \to 0$, where each β_j is independent of f, g. That is, the perturbations affect only the terms in the bifurcation function of order $\geq 2q + 3$. If the stability properties of the solution $x = 0$, $y = 0$ of Equation (5.23) are the same as the ones for the solution $x = 0$, $y = 0$ of Equation (5.24) for every f, g in the above class, then there is a $\beta \neq 0$ and integer $k \leq q$ such that $H(a, 0, 0) = G(a, 0)$ satisfies (5.22). In fact, Theorem 5.1 implies these stability properties are determined by the stability properties of the solution $a = 0$ of

$$(5.27) \qquad \dot{a} = H(a, f, g).$$

Thus, the stability properties of this equation are independent of perturbations of order $2q + 3$. The assertion is now immediate.

The above remarks imply that it is possible to relate the stability properties of the solution $x = 0$, $y = 0$ of (5.23) to perturbations from a certain class to the number β and the integer k in (5.22), which in turn determines the maximal number of periodic solutions of Eq. (5.1) in a neighborhood of $z = 0$, $\mu = 0$. It is difficult to determine precisely the class of perturbations f, g to ensure that (5.22) is satisfied. However, if the y variable is absent; that is, $z \in \mathbb{R}^2$, then it is not difficult to see that the class of perturbations are precisely those of order $2k + 2$.

Due to the importance of studying the stability of the solution $z = 0$ of Eq. (5.1) for $\mu = 0$; that is, Eq. (5.23), let us summarize the information that is obtained from the bifurcation function as we did for the case of one zero root in Section 3. The proof is essentially the same.

Corollary 5.7. *Let $F(a) \overset{\text{def}}{=} G(a, 0)$ be the function given in Theorem 5.1. The solution $z = 0$ of the equation*

$$(5.28) \qquad\qquad \dot{z} = Az + Z(z, 0)$$

is asymptotically stable if and only if there is an $\varepsilon > 0$ such that $aF(a) < 0$ for $0 < |a| < \varepsilon$. It is unstable if and only if there is an $\varepsilon > 0$ such that $aF(a) > 0$ for either $0 < a < \varepsilon$ or $-\varepsilon < a < 0$. If there is an $\varepsilon > 0$ such that $F(a) = 0$ for $|a| < \varepsilon$, then the zero solution is stable and there is a first integral of (5.28) in a neighborhood of $z = 0$.

If $Z(z, 0)$ is analytic, then $F(a)$ in Corollary 5.7 is analytic. Thus, either there is a $\beta \neq 0$ and an integer $k \geq 1$ such that

$$(5.29) \qquad\qquad F(a) = \beta a^{2k+1} + O(|a|^{2k+3})$$

as $a \to 0$, or $F(a) \equiv 0$ in a neighborhood of $a = 0$. Thus, we have the following

Corollary 5.8. *If $Z(z, 0)$ is analytic, then one of the following alternatives prevail:*

(i) *$F(a)$ satisfies (5.29) with $\beta \neq 0$ and the solution $z = 0$ of (5.28) is asymptotically stable (unstable) if $\beta < 0$ ($\beta > 0$).*
(ii) *There is an analytic first integral of (5.28) and $z = 0$ is stable.*

9.6. Bibliographical Notes

For relevant material on center manifolds, see Kelley [1], Marsden and McCracken [1], Palmer [1, 2], Chafee [1, 2], Carr [1], Pugh and Shub [1], Fenichel [1, 2, 3, 4], Shoshitaishvili [1], Hirsch, Pugh and Shub [2], Pliss [1], Sijbrand [1]. The proof of Remark 2.16 may be found in Palmer [2], Shoshitaishvili [1]. Examples 2.4, 2.5 and 2.6 are taken from Carr [1].

Theorems 3.2 and 4.2 are essentially due to DeOliveira and Hale [1]. These results have other interesting implications. For example, with C as in (3.1), consider the equation $z = Cz + w(z, \lambda)$ where $\lambda \in \mathbb{R}$, $w(0, \lambda) = 0$. Suppose the solution $z = 0$ is stable for $\lambda < 0$ and unstable for $\lambda > 0$. Then Theorem 3.2 implies nontrivial solutions bifurcate from $(z, \lambda) = (0, 0)$. This result is due to Itoh [1]. He used the same type of proof and even considered evolutionary equations which include certain parabolic equations. Theorem 4.2 shows that the same result is true for periodic systems and Theorem 5.1 implies that a change in stability implies a Hopf bifurcation for Eq. (5.1).

The theory in Sections 3, 4 has meaning for equation (3.2) with the sole requirement that the eigenvalues of B have nonzero real parts. In this way we obtain a bifurcation function G. Using the results from Section 9.2, the qualitative properties of the flow near $(x, y) = (0, 0)$ is reduced to the discussion of the flow on the center manifold.

In the case where the bifurcation function G is a scalar, we have observed that the flow on the center manifold is determined by the bifurcation function. It is natural to enquire if similar results are possible in the case where the bifurcation function has dimension greater than one. Vegas [1] has shown that the bifurcation function can be used to determine the stability of equilibrium points for gradient systems. With some restrictions on the eigenvalues, the same fact is true for general systems. However, there are counterexamples in Vegas [1], showing that the flow on the center manifold and the flow defined by the bifurcation function are different.

One may also consider equations of the form

$$\dot{x} = Cx + \varepsilon f(t, x)$$

where C has one zero eigenvalue and all others with nonzero real parts, ε is a small parameter and $f(t + 2\pi, x) = f(t, x)$. In this case, for any $r > 0$, there is an $\varepsilon_0(r)$ such one can apply the same procedure to obtain the bifurcation function for $G(a, \varepsilon)$ for $|a| < r$, $|\varepsilon| < \varepsilon_0(r)$. The stability results are the same as before.

It is possible to treat the problem of bifurcation from a focus independent of the discussion of stability. The derivation of the bifurcation function by the method of Liapunov–Schmidt does not require that the eigenvalues of B in (5.3) have nonzero real parts. It is only necessary that the eigenvalues of B are not in resonance with the eigenvalues of A. More specifically, it is only necessary to have $\det[I - \exp 2\pi B] \neq 0$.

The results in all of the previous sections have implications for certain types of evolutionary equations. One important restriction is that the operator corresponding to the linear approximation generates a strongly continuous semigroup $U(t)$, and the spectrum of $U(1)$ can be decomposed into two spectral sets σ_1, σ_2 with σ_1 disjoint from the unit circle with center zero and either σ_2 has an eigenvalue one or a pair of complex eigenvalues with modulus one. Of course, one also must be able to obtain good estimates in the variation of constant formula in order to be able to discuss detailed properties of the nonhomogeneous linear system.

Modulo some nontrivial technical difficulties, the theory can be extended verbatim to retarded functional differential equations with finite delay using the theory in Hale [9, 12], Chow and Mallet–Paret [3], to retarded equations with infinite delay using Hale and Kato [1], Naito [1, 2], to functional equations including difference-integral equations using Hale and de Oliveira [1], to neutral functional differential equations with a stable D-operator using Hale [9], de Oliveira [1], to parabolic partial differential equations following Henry [1], Kielhofer [2], Marsden and McCracken [1]. Certain types of hyperbolic systems are also amenable to the theory if the spectrum satisfies the above mentioned property.

Liapunov [1] was the first to prove results in the spirit of Section 3. More specifically, he proved Corollary 3.6 (see, also Bibikov [1]). Corollary 3.3 can be considered as a generalization of the results of Liapunov in two ways. The vector field is not required to be analytic and the stability properties of an equilibrium point depend only on the bifurcation function even if this function satisfies no condition of the form (3.15). Some aspects of the proof of Liapunov are similar to the one above. He used general transformation theory to put the equation in a form where it was easy to discover the center manifold and the flow on the center manifold. We used the abstract center manifold theory and properties of the stable manifold. In addition, a small amount of perturbation theory is used in an abstract way to prove the bifurcation function determines the stability properties of the solutions.

It is instructive to review the ideas of Liapunov because it is a good motivation for further results. Suppose $g = 0$, $h = 0$ in Equation (3.2),

$$(6.1) \qquad \dot{x} = u(x, y), \qquad \dot{y} = Ay + v(x, y)$$

and $A\phi + v(x, \phi) = 0$, $\phi(0) = 0$. The transformation $x \mapsto x$, $y \mapsto \phi(x) + y$ yields the system

$$\dot{x} = u(x, \phi(x) + y)$$

$$(6.2)$$

$$\dot{y} = Ay + v(x, \phi(x) + y) - v(x, \phi(x)) - \frac{\partial \phi(x)}{\partial x} u(x, \phi(x) + y).$$

The function $G(x) = u(x, \phi(x))$ is the bifurcation function $G(x, 0, 0)$ defined in Relation (3.3).

Theorem 3.1 and, in particular, Corollary 3.3 asserts that the stability properties cf the zero solution of (6.2) or (6.3) are the same as the stability properties of the scalar equation $\dot{a} = u(a, \phi(a))$. This latter equation is just the right hand side of the equation in x in (6.2) with $y = 0$. Liapunov [1] proved part of this result based on the following observation. The function $\phi(x)$ is $O(|x|^2)$ as $|x| \to 0$. If $u(x, \phi(x)) = O(|x|^k)$ as $x \to 0$, then this implies the right hand side of the equation for y in (6.2) for $y = 0$ is $O(|x|^{k+1})$ as $|x| \to 0$. If $u(x, \phi(x)) = \beta x^k + O(|x|^{k+1})$, $\beta \neq 0$, as $|x| \to 0$, then Liapunov [1] proved the zero solution of (6.2) was stable by using a special Liapunov function of (x, y).

Even if x is not a scalar, the above argument is valid. In fact, if $x \in \mathbb{R}^k$ and $u(x, \phi(x)) = U^{(k)}(x) + O(|x|^{k+1})$ as $|x| \to 0$, $U^{(k)}(x)$ is homogeneous of degree k and if the stability properties of the equation $\dot{a} = U^{(k)}(a)$ are insensitive to perturbations of order $(k + 1)$, then the technique of Liapunov functions can be employed to show that the stability of the zero solution of (6.2) is the same as for $\dot{a} = U^{(k)}(a)$ (see, for example, Liapunov [1], Malkin [1], Zubov [9], Bibikov [1], Lefschetz [1, 2]). The same result can be obtained by using the center manifold theorem since the terms of order $\leq k$ of the vector field on the center manifold are precisely the terms in $U^{(k)}(x)$. The last method of proof is more general since it permits $U^{(k)}$ to contain arbitrary terms of order $\leq k$ provided terms of order $(k + 1)$ do not disturb the stability properties of the equation $\dot{a} = U^{(k)}(a)$.

If x is not a scalar, it is still possible to speak of the flow defined by the differential equation (3.6), $\dot{x} = G(x, g, h)$, involving the perturbation functions g, h. As (g, h) vary in a neighborhood of $(0, 0)$, many types of bifurcations can occur in (3.6). For the equation (3.5) on the center manifold, several types of bifurcations are also occuring. It is reasonable to try to correlate the bifurcation for the two equations. In Theorem 3.2, we have shown they are exactly the same when x is a scalar. Vegas [1] has shown that the bifurcation function can be used for the stability of gradient systems. This also is true in the general case under some restrictions on the eigenvalues of the linearization. To make further progress in higher dimensions, more hypotheses are needed. Let us suppose that λ is a scalar and consider a one parameter family of vector fields

$$\dot{x} = X(x, y, \lambda)$$
$$\dot{y} = Ay + Y(x, y, \lambda)$$

where $X(x, y, 0)$, $Y(x, y, 0)$ are $O(|x| + |y|)^2)$ as $|x|, |y| \to 0$. Suppose there is a neighborhood V of $\lambda = 0$ such that each bifurcation point of the equation $\dot{a} = G(a, \lambda)$, $\lambda \in V$, is a bifurcation point of degree 1 (see Section 10.5 for the definition of degree 1). Is it true that the only bifurcation for the flow on the center manifold are degree one? In the case where $x \in \mathbb{R}^2$, the bifurcations of degree 1 are known (see Section 10.5). The results to be presented in the next chapters should lead to an affirmative answer to this question when $x \in \mathbb{R}^2$.

Example (3.16) is due to Liapunov [1]. The discussion there is the same as given by him since he only used the bifurcation function.

Theorem 5.1 is due to de Oliveira and Hale [1]. Theorem 5.2 in a less general form was first stated precisely by Hopf [1]. As Hopf remarked in his paper, it must have been essentially known to a number of persons, especially

Poincaré [1]. In fact, one can find a discussion of the effect of a change in stability of a focus on the creation of a limit cycle. In his notes on nonlinear mechanics of the early 1940's, Minorsky [1] (see also Minorsky [2]) makes similar remarks including even the bifurcation diagram. Of course, it certainly was part of the vocabulary of the Russian school in differential equations as can be seen by reading Andronov and Leontovich [1] (see also Andronov, Leontovich, Gordon and Maier [1]) or the book on nonlinear oscillations of Andronov and Chaiken [1] (see also Andronov, Khaiken and Vitt [1]).

It would be a very difficult task and probably not too interesting to trace the origin of the Hopf bifurcation theorem. However, it is an important result because it is the simplest type of bifurcation that occurs in a differential equation that uses the dynamics of the problem. The theory of bifurcation of equilibrium points can be discussed without any consideration of the differential equation. It is a question of the manner in which the zeros of a vector field change with parameters. This was the way this problem was discussed in Chapters 4–7 and is usually referred to as *static bifurcation theory*. Of course, the stability of the equilibrium solutions must involve the differential equation. On the other hand, the periodic orbits obtained from the Hopf bifurcation theorem cannot exist without the dynamics. This topic belongs to the general subject of *dynamic bifurcation theory* which will be the thrust of Chapters 10–13.

When the dynamics is taken into account, each bifurcation is intimately connected with some type of exchange of stability. For the case in which the bifurcation is from a focus and there is a neighborhood of the origin which contains at most one periodic orbit, this is stated precisely in Theorem 5.2. For example, in the supercritical case, the stable origin gives up its stability to the periodic orbit after the bifurcation. The same type of exchange of stability occurs at the bifurcation of equilibrium points. For a further discussion as well as extensions and applications to finite and infinite dimensional problems, see Andronov, Leontovich, Gordon and Maier [1], Andronov and Chaiken [1], Andronov, Vitt and Khaiken [1], Auchmuty and Nicolis [1], Ashkenazi and Chow [1], Bruslinskaya [1], Carr [1], Chafee [1–4], Chafee and Infante [1], Cesari [7], Chow and Mallet–Paret [3], Cooke and Yorke [1], Crandall and Rabinowitz [2, 3, 4], Cronin [1], Cushing [1, 2], de Oliveira [1], Fife [1], Flockerzi [1], Friedrichs [1, 2], Gavalas [1], Granero, Poreti and Zanacca [1], Hale [2, 4, 6, 9, 10, 12, 13, 14], Hale and de Oliveira [1], Hausrath [1], Hassard, Kazarinoff and Wan [1], Henry [1], Hopf [2], Howard [1, 2], Iooss [1–4], Iooss and Joseph [1, 2], Iudovich [1–6], Joseph [1, 2], Joseph and Nield [1], Joseph and Sattinger [1], Jost and Zhender [1], Kielhofer [1–3], Kirchgässner [1–3], Kirchgässner and Kielhöfer [1], Kirchgässner and Sorger [1], Lima [1], Marsden [1–3], Marsden and McCracken [1], Matano [1], May [1], McLeod and Sattinger [1], Palmer [1], Poore [1], M. Potier–Ferry [1], Pyartli [1], Ruiz–Claeyssen [1], Sattinger [1–5, 10], Saut and Scheurer [1], Schmidt [1],

Schuster, Sigmund and Wolff [1], Smoller and Wasserman [1], Stakgold, Joseph and Sattinger [1], DiPrima and Stuart [1], Takens [2, 4], Temme [1], Thompson and Hunt [1], Troy [1], Uppal, Ray and Poore [1], Wan [1], Weinberger [1, 2].

Corollary 5.4 is due to Chafee [3] (see, also, Bernfeld, Negrini and Salvadori [1]). Theorem 5.5 is essentially due to Sotomayor [1, 2]. For the case where the parameter μ in (5.1) is finite dimensional, the classification of the surfaces of bifurcation of periodic orbits from a focus was first given by Takens [2]. The use of the Malgrange theorem and the bifurcation function as in the text was given by Hale [7]. Chafee [3] also used the bifurcation function to give a partial discussion where the parameter is infinite dimensional. Chafee [3], Negrini and Salvadori [1] were the first to point out the important fact in Theorem 5.6 that there can be at most one periodic orbit bifurcating from a focus when the eigenvalues cross the imaginary axis transversally (see also Hale [13]). Flockerzi [1] has studied, in detail, the bifurcation theory from a focus for one parameter problems for which the eigenvalues do not cross the imaginary axis transversally. The results of Flockerizi were extended to parabolic equations by Kielhöfer [2].

Corollary 5.8 is due to Liapunov [1] (see also Bibikov [1]). For an alternative discussion of the implications of $\beta \neq 0$ in (5.23) to stability under perturbations, see Bernfeld, Negrini and Salvadori [1].

For the bifurcation of a periodic orbit from a focus, the theory of the previous sections is rather complete. Once the local bifurcation to a periodic orbit has taken place, it is interesting to investigate the global existence of periodic orbits. This is more difficult than the global bifurcation from equilibrium because of an additional parameter in the problem; namely, the period. The development requires a new concept of index which generalizes the index of Fuller [1] for periodic orbits. See Alexander and Yorke [1], [2], Chow and Mallet–Paret [1], Chow, Mallet–Paret and Yorke [1], Mallet–Paret and Yorke [1, 2], Alexander and Auchmuty [1]. When several eigenvalues lie on the imaginary axis, this index also gives new results in local bifurcation theory similar in spirit to the ones in Section 5.7 (see also Jost and Zhender [1]). For other types of global Hopf theorems, see Nussbaum [1–3], Turner [1] for results and references.

For a discussion of the bifurcation of limit cycles of the second kind for second order equations, see Silova [1].

Ashkenazi and Chow [1] have discussed the existence of periodic orbits for a multiple eigenvalue on the imaginary axis with nonsimple elementary divisors using methods similar to the ones in the text. The dynamic behavior is not considered so that the result is more closely related to the discussion in Chapter 5 of bifurcation from a simple eigenvalue. To state the result, suppose

$$J = \begin{bmatrix} 0 & 1 \\ -1 & 0 \end{bmatrix}, \qquad I = \begin{bmatrix} 1 & 0 \\ 0 & 1 \end{bmatrix}$$

are 2×2 matrices, x^j are two-vectors, $j = 1, 2, \ldots, k$, $x = (x^1, \ldots, x^k)$ and consider the $2k$-dimensional system

$$\dot{x}^j = Jx^j + x^{j+1} + \alpha \sum_{i=1}^{k} a_{ij}x^i + g_j(x, \alpha)$$

(6.3)

$$j = 1, 2, \ldots, k, \qquad x^{k+1} = 0,$$

where $\alpha \in \mathbb{R}$, each a_{ij} is a 2×2 matrix, and $g_j(0, \alpha) = 0$, $\partial g_j(0,0)/\partial x = 0$ for all j. Ashkenazi and Chow [1] prove there is a unique branch of periodic orbits bifurcating from $(x, \alpha) = (0, 0)$ if tr $J^{k-1}a_{k1} \neq 0$. This is the analogue of the condition on the determinant in Section 5.3. (See also Vanderbauwhede [9] and Caprino, Maffei and Negrini [1]).

Chapter 10

Bifurcation of Autonomous Planar Equations

10.1. Introduction

In this chapter, we consider three different types of perturbation problems
for the two dimensional system

$$\dot{x} = g(x). \tag{1.1}$$

In Section 2, we assume (1.1) has a periodic orbit Γ which is not hyperbolic. By using a simple change of variables, the behavior of solutions of a
perturbed equation

$$\dot{x} = f(x, \mu), \qquad f(x, 0) = g(x) \tag{1.2}$$

for (x, μ) near $\Gamma \times \{0\}$ is reduced to a special case of Section 9.4. The number
of periodic orbits for (1.2) as well as their stability is discussed.

In Section 3, we suppose the invariant set Γ of (1.1) is given by the origin
0 which is assumed to be a saddle and an orbit whose α- and ω-limit set is
also 0; that is, a homoclinic orbit. Either inside Γ or outside Γ, one can
define a Poincaré map for (1.1) which can be used to study the flow of (1.2)
near $\Gamma \times \{0\}$. Periodic orbits again can appear as bifurcations from Γ.

In Section 4, we suppose the invariant set Γ of (1.1) is a smooth closed
curve containing only one equilibrium point which is a saddle-node. Integral
manifold theory is used to show that a periodic orbit of (1.2) bifurcates from
Γ if and only if the equilibrium point disappears.

10.2. Periodic Orbit

In this section, we consider a two dimensional system

$$\dot{x} = f(x, \mu) \tag{2.1}$$

where $x = (x_1, x_2) \in \mathbb{R}^2$, $\mu \in E$, a Banach space, $f: \mathbb{R}^2 \times E \to \mathbb{R}^2$ is C^k, $k \geq 1$, and, for $\mu = 0$, the system

$$(2.2) \qquad\qquad \dot{x} = f(x, 0)$$

has a periodic orbit Γ. The objective is to discuss the behavior of the orbits of (2.1) for (x, μ) in a neighborhood of $\Gamma \times \{0\}$.

For this discussion, it is convenient to use a different coordinate system near Γ. Suppose

$$(2.3) \qquad\qquad \Gamma = \{u(t) : 0 \leq t < \omega\}$$

where ω is a positive constant, $u(t)$ is ω-periodic with least period ω and $u(t)$ is a solution of (2.2), $\dot{u}(t) \neq 0$ for all t. If $u = (u_1, u_2)$, let $v(t) = (\dot{u}_2(t), -\dot{u}_1(t))$ and consider the transformation of variables from x to (θ, p) defined by the relation

$$(2.4) \qquad\qquad x = u(\theta) + pv(\theta).$$

An application of the Implicit Function Theorem and the compactness of Γ implies this is a diffeomorphism of a neighborhood of Γ onto $[0, \omega) \times \{$a neighborhood of $p = 0\}$. Identifying the point $\theta = 0$ with $\theta = \omega$, one can define the transformation (2.4) for all $\theta \in \mathbb{R}$. It is ω-periodic in θ.

Let us apply the transformation (2.4) to Eq. (2.1). If x^T is the transpose of x and $|x|^2 = x^T x$, then the equations for θ, p are

$$(2.5) \quad \begin{aligned} \dot{\theta} &= \frac{\partial u^T}{\partial \theta} f(u + pv, \mu) \left[\left| \frac{\partial u}{\partial \theta} \right|^2 + p\, \frac{\partial u^T}{\partial \theta} \frac{\partial v}{\partial \theta} \right]^{-1} \stackrel{\text{def}}{=} 1 + \Theta(\theta, p, \mu) \\ \dot{p} &= \left[-pv^T \frac{\partial v}{\partial \theta} \dot{\theta} + v^T f(u + pv, \mu) \right] \bigg/ |v|^2 \end{aligned}$$

for p in a sufficiently small neighborhood of zero. The function $\Theta(\theta, p, \mu)$ vanishes for $(p, \mu) = (0, 0)$ and is 2π-periodic in θ. Therefore, for (p, μ) is a sufficiently small neighborhood of zero, we may eliminate t in (2.5) to obtain

$$(2.6) \qquad\qquad \frac{dp}{d\theta} = R(\theta, p, \mu)$$

where $R(\theta, p, \mu) = R(\theta + \omega, p, \mu)$ and $R(\theta, 0, 0) = 0$, $\partial R(\theta, 0, 0)/\partial p = 0$. The periodic orbits of (2.1) in a neighborhood of $\Gamma \times \{0\}$ are in one-to-one correspondence with the ω-periodic solutions of (2.6) in a neighborhood of $(p, \mu) = (0, 0)$.

The problem of finding ω-periodic solutions of (2.6) in a neighborhood of $(p, \mu) = (0, 0)$ was discussed in detail in Section 9.4. From those results, one obtains

Theorem 2.1. *Let $G(a, \mu)$ be the bifurcation function of Section 9.4 obtained by the method of Liapunov–Schmidt. There are neighborhoods $U \subset \mathbb{R}^2 \times E$ of $\Gamma \times \{0\}$, $V \subset \mathbb{R} \times E$ of $(a, \mu) = (0, 0)$ such that the periodic solutions of (2.1) in U are in one-to-one correspondence with the zeros of $G(a, \mu)$ in V. Furthermore, the stability properties of these periodic solutions are the same as the stability properties of the zeros of G as solutions of the scalar equation*

$$\dot{a} = G(a, \mu). \tag{2.7}$$

The complete bifurcation diagram in a neighborhood of $\mu = 0$ is obtained by discussing the manner in which the zeros of $G(a, \mu)$ depend on μ. We discussed this problem in some detail in Section 9.4 for the case in which

$$G(a, 0) = \beta a^k + O(|a|^{k+1}), \qquad \beta \neq 0. \tag{2.8}$$

as $a \to 0$. In particular, we can state the following result.

Theorem 2.2. *If $G(a, \mu)$ satisfies (2.8), then there is a neighborhood U of $\Gamma \times \{0\}$ in which Eq. (2.1) has at most k periodic orbits. If*

$$f(x, \mu) = v(x) + \mu(x) \tag{2.9}$$

with $\mu \in E = C^{k+1}(\mathbb{R}^2, \mathbb{R}^2)$, then either, for k odd and any $j = 1, 2, \ldots, k$ or, for k even and any $j = 0, 1, \ldots, k$, there is a $\mu \in E$ and exactly j periodic solutions $\{\varphi_i\}$ of (2.1) with $(\varphi_i, \mu) \in U$ for each i.

Proof. This first part of the theorem is a trivial consequence of Rolle's theorem. The second part is conceptually very simple, but technically messy. The idea is to choose a k parameter family of vector fields μ in (2.9),

$$\mu(x) = \mu_1 g_1(x) + \cdots + \mu_k g_k(x), \qquad \mu_j \in \mathbb{R},$$

where each $g_j(x)$ is chosen in such a way as to make

$$G(a, \mu) = \beta_0(\mu)a^k + \beta_1(\mu)a^{k-1} + \cdots + \beta_k(\mu) + O(|a|^{k+1})$$

as $a \to 0$ with $\beta_0(0) = \beta$, $\beta_j(\mu) = \mu_j + O(|(\mu_1, \ldots, \mu_{j-1})|)$, $j = 1, 2, \ldots, k$. The perturbations $g_k(x)$ must be given explicitly. One can take $g_j(x) = (\nabla F)F^{j-1}$, where $F(x)$ is the distance of a point x from Γ. In our notation, $F(x) = p$, where $x = u(\theta) + pv(\theta)$. With this choice, the bifurcation function becomes

$$G(a, \mu) = \beta_k(\mu)a^k + \beta_{k-1}(\mu)a^{k-1} + \cdots + \beta_0(\mu),$$

where $\beta_0(\mu) = \mu_1$, $\beta_j(\mu) = \mu_j + O(|(\mu_1, \ldots, \mu_{j-1})|)$ as $\mu \to 0$ for $j \geq 1$. The argument is now completed as in the proof of Corollary 5.4, Chapter 9. $\square$

Remark 2.3. If $p(\theta, \theta_0, p_0)$ is the solution of (2.6) with $p(\theta_0, \theta_0, p_0) = p_0$, then $p(\theta_0 + \omega, \theta_0, p_0)$ is called the *Poincaré map* of Eq. (2.1) relative to the transversal L_0 defined by

$$L_0 = \{u(\theta_0) + pv(\theta_0) : p \in \mathbb{R}\}.$$

We say the *periodic orbit Γ of* (2.1) *has order k* if the Poincaré map satisfies

$$d(p_0) \stackrel{\text{def}}{=} p(\omega, 0, p_0) = p_0 + \beta_0 p_0^k + O(|p_0|^{k+1}), \qquad \beta_0 \neq 0.$$

Theorem 2.1 implies that saying that Γ has order k is equivalent to saying that $G(a, 0)$ satisfies (2.8). In fact, if this were not so, we could make a small perturbation of (2.2) to an Eq. (2.1) which would destroy the one-to-one correspondence between the zeros of $G(a, \mu)$ and the periodic orbits of (2.1) near $\Gamma \times \{0\}$.

A simple example of a system with a periodic orbit of order k is easy to construct. For example, consider the system

$$
\begin{aligned}
\dot{x} &= -y + x(x^2 + y^2 - 1)^k \\
\dot{y} &= x + y(x^2 + y^2 - 1)^k.
\end{aligned}
\tag{2.10}
$$

The orbit $\Gamma = \{x^2 + y^2 = 1\}$ is periodic and defined by

$$\Gamma = \{(\cos\theta, \sin\theta) : 0 \leq \theta < 2\pi\}.$$

The transformation (2.4) becomes

$$
\begin{aligned}
x &= (1 + p)\cos\theta \\
y &= (1 + p)\sin\theta.
\end{aligned}
\tag{2.11}
$$

The equations for θ, p are

$$\dot{\theta} = 1, \qquad \dot{p} = (1 + p)((1 + p)^2 - 1)^k$$

or

$$\frac{dp}{d\theta} = 2^k p^k (1 + p)\left(1 + \frac{p}{2}\right)^k. \tag{2.12}$$

For any perturbation of (2.10), the bifurcation function $G(a, \mu)$ obviously satisfies

$$G(a, 0) = 2^k a^k (1 + a)\left(1 + \frac{a}{2}\right)^k = 2^k a^k + O(|a|^{k+1})$$

as $a \to 0$. Thus, G satisfies (2.8) and Γ is of order k.

For later reference, we state the following result on periodic orbits of order 2.

Theorem 2.4. *If Γ is a periodic orbit of Eq. (2.2) of order two, then there is a neighborhood $U \subset E$ of $\mu = 0$ and a neighborhood $V \subset \mathbb{R}^2$ of Γ and a function $\gamma: U \to \mathbb{R}$ such that Eq. (2.2) has two periodic orbits in V if $\gamma > 0$ and no periodic orbits in V if $\gamma < 0$. If $E = C^2(\mathbb{R}^2, \mathbb{R}^2)$, then the set $\Gamma_0 = \{\mu : \gamma(\mu) = 0\}$ is a submanifold of codimension one. If $\mu \in \Gamma_0$, there is a unique periodic orbit in V which is asymptotically stable from one side and unstable from the other.*

Proof. This is an immediate consequence of the results in Section 9.3. $\square$

The classification of orbits of order k can be accomplished by applying the Malgrange preparation theorem to the bifurcation function $G(a, \mu)$.

10.3. Homoclinic Orbit

In this section, we consider the two dimensional system (2.1) under the assumption that (2.2) has a hyperbolic saddle point at $x = 0$ and there is an orbit other than $\{0\}$ of (2.2) whose α- and ω-limit sets are $x = 0$; that is, the stable and unstable manifolds of $x = 0$ intersect at a point other than zero. The problem will be to discuss the behavior of the solutions of (2.1) near this orbit for μ near zero.

To be specific, suppose $f(0, 0) = 0$, the eigenvalues of the matrix

$$(3.1) \qquad\qquad\qquad A = \frac{\partial f}{\partial x}(0, 0)$$

are real and opposite in sign; that is, the solution $x = 0$ of (2.2) is a hyperbolic saddle point. Let S_0, U_0 be respectively the stable and unstable manifolds of $x = 0$ and suppose

$$(3.2) \qquad\qquad\qquad (S_0 \cap U_0) \backslash \{0\} \neq \phi$$

If $p_0 \in S_0 \cap U_0$, $p_0 \neq 0$, then any solution on the orbit $0(p_0)$ through p_0 of (2.2) approaches zero exponentially as $t \to \pm \infty$. The orbit $0(p_0)$ is called a *homoclinic orbit* for the equilibrium point zero. Let

$$(3.3) \qquad \Gamma = 0(p_0) \cup \{0\}, \qquad p_0 \in S_0 \cap U_0, \qquad p_0 \neq 0.$$

Then the invariant set Γ can have either of the configurations in Fig. 3.1 with respect to the position of the stable and unstable manifolds of zero. In order

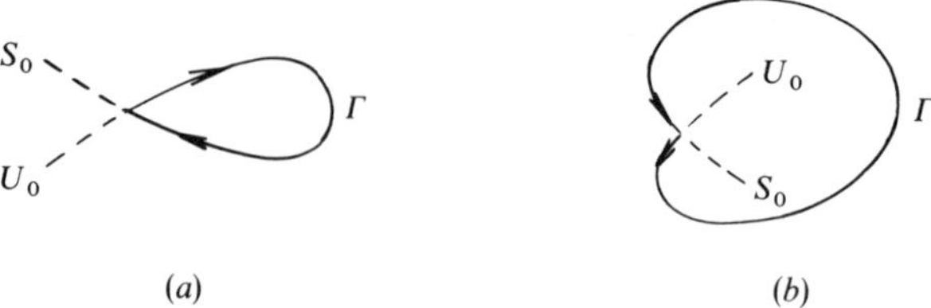

(a)   (b)

Figure 3.1

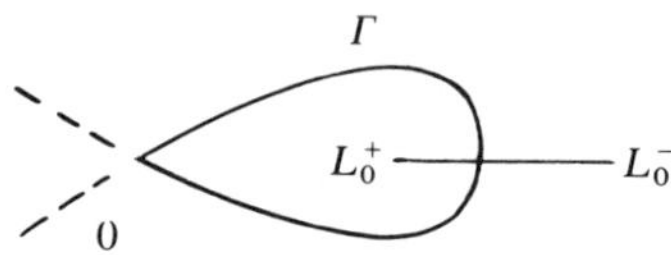

Figure 3.2

to be specific in the discussion to follow, we assume that the situation in Fig. 3.1(a) prevails. The other case can be treated in a similar manner.

It is clear that the differential equation (2.2) can be perturbed a small amount to (2.1) in such a way that the stable and unstable manifolds at zero no longer intersect. Our objective is to determine what new phenomena can arise when this occurs. We also want to impose some conditions on (2.1) which will permit us to describe everything that happens near Γ for μ near zero.

For $p_0 \in S_0 \cap U_0$, $p_0 \neq 0$, let L_0 be a transversal to Γ at p_0 relative to Eq. (2.2). Suppose the situation in Fig. 3.1(a) prevails and let L_0^+ be that part of L_0 belonging to the interior of Γ (see Fig. 3.2), L_0^- that part exterior to Γ. Using continuity with respect to initial conditions and the fact that $\Gamma\backslash\{0\}$ belongs to the stable and unstable manifold of 0, one can show that there is a neighborhood U of p_0 such that, for any $p \in U \cap L_0^+$, there is $\tau = \tau(p) > 0$ with $x(\tau(p), p) \in L_0^+, x(t, p) \notin L_0^+, 0 < t < \tau(p)$, where $x(t, p)$ is the solution of (2.2) through p. One can therefore define the Poincaré map π_0 by $\pi_0(p) = x(\tau(p), p)$ for $p \in L_0^+ \cap U$. For any point q in the range of π_0, there is a unique point p and time $\tau(p)$ such that $q = x(\tau(p), p)$, $x(t, p) \notin L_0^+$ for $0 < t < \tau(p)$. Define $\pi_0^{-1}(q) = x(-\tau(p), q)$. Similarly, one can define $\pi_0^k(p)$ for any integer k.

If $p \in L_0^-$, then the saddle point property again implies that the solution $x(t, p)$ leaves any neighborhood of Γ in both the forward and backward direction. Thus, the map π_0 determines the properties of the orbits of (2.2) near Γ. In particular, periodic orbits near Γ correspond to fixed points of π_0. The ω-limit set (α-limit set) of an orbit is Γ if and only if this orbit intersects L_0 at some point p in the domain of π_0 and $\pi_0^k(p) \to p_0$ as $k \to \infty$ ($k \to -\infty$). This last property is important later so we make

Definition 3.1. The orbit Γ is *asymptotically stable* (*unstable*) if $\pi^k(p) \to p_0$ as $k \to \infty$ ($k \to -\infty$).

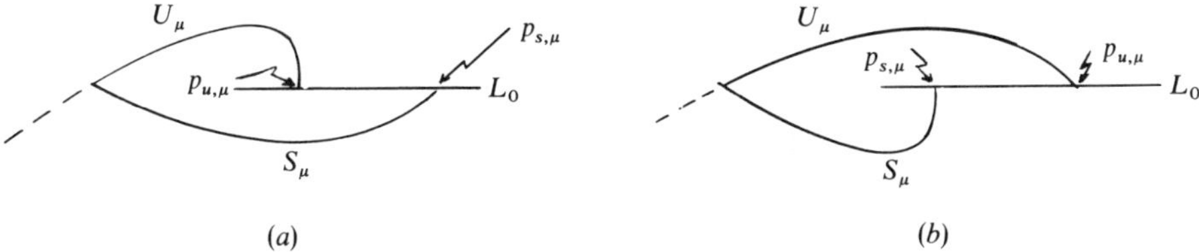

Figure 3.3

Now let us consider Eq. (2.1). Since det $A \neq 0$ for A in (3.1), the Implicit Function Theorem implies there is a unique equilibrium point in a neighborhood of $x = 0$ for μ in a neighborhood of zero and this equilibrium is as smooth in μ as the vector field in (2.1). By a translation of variables, one may thus suppose without loss in generality that (2.1) satisfies

$$(3.4) \qquad f(0, \mu) = 0 \quad \text{for all } \mu.$$

Let S_μ, U_μ be the stable and unstable manifolds of the solution zero of (2.1). These sets depend smoothly upon μ since they are obtained from an application of the Implicit Function Theorem.

Thus, there is a $\delta > 0$ and points $p_{s\mu} \in S_\mu \cap L_0$, $p_{u\mu} \in U_\mu \cap L_0$, $|\mu| < \delta$, with the property that $x(t, p_{s\mu}) \notin L_0$, $x(-t, p_{u\mu}) \notin L_0$ for $0 < t < \infty$ (see Fig. 3.3). These points $p_{s\mu}, p_{u\mu}$ are smooth functions of μ, approach p_0 as $\mu \to 0$ and must be situated on L_0 as in one of the diagrams in Fig. 3.3 if $p_{s\mu} \neq p_{u\mu}$.

Let I_μ be the open segment on L_0 between $p_{u\mu}$ and $p_{s\mu}$ and suppose Fig. 3.3(a) prevails. Then, there is a neighborhood V of Γ, $\delta > 0$, such that, for every neighborhood W of Γ, $W \subset V$, $\mu \in E$, $|\mu| < \delta$ and $p \in I_\mu$, there is a $\tau > 0$ such that the solution $x(t, p, \mu)$ of (2.1) through p satisfies $x(-\tau, p, \mu) \notin W$. Thus, any new behavior of (2.1) must arise from the behavior of the positive orbit $0^+(p_{u\mu}) = \{x(t, p_{u\mu}, \mu) : t \geq 0\}$ through the point $p_{u\mu} \in U_\mu \cap L_0$. If for every neighborhood W of Γ, $W \subset V$, $|\mu| < \delta$, the orbit $0^+(p_{u\mu})$ has points not in W, then every trajectory leaves W in both the forward and backward directions. If there is a neighborhood W of Γ, $W \subset V$, $\mu \in E$, $|\mu| < \delta$, for which $0^+(p_{u\mu}) \subset W$, then the Poincaré–Bendixson theorem implies there is a periodic orbit of (2.1) in W. If we assume Γ is asymptotically stable, continuity of the solutions of (2.1) with respect to μ implies this latter situation always prevails and there must be a periodic orbit. There may be more than one periodic orbit and the number can only be determined by knowing more about the function $f(x, \mu)$ in (2.1).

If the situation in Fig. 3.3(b) prevails, the behavior is determined by the negative orbit $0^-(p_{s\mu})$ of (2.1) through $p_{s\mu}$. If, for every neighborhood W of Γ, $W \subset V$, $|\mu| < \delta$, the orbit $0^-(p_{s\mu})$ has points not in W, then every trajectory leaves W in both the forward and backward directions. If there is a neighborhood W of Γ, $W \subset V$, $\mu \in E$, $|\mu| < \delta$, for which $0^-(p_{s\mu}) \subset W$, then there is a periodic orbit of (2.1) in W. The latter situation always prevails if Γ is unstable.

These results are summarized in

Theorem 3.2. *If $p_{s\mu} \neq p_{u\mu}$, then there is a neighborhood V of Γ, $\delta > 0$, such that, for any neighborhood W of Γ, $W \subset V$, $\mu \in E$, $|\mu| < \delta$, either*

    (i)   *every orbit of* (2.1) *leaves W in both the negative and positive directions*
or
    (ii) *there is a periodic orbit of* (2.1) *in W.*

If either Fig. 3.3(a) prevails and the orbit Γ is asymptotically stable or if Fig. 3.3(b) prevails and the orbit Γ is unstable, then case (ii) *occurs.*

Our next objective is to give specific criteria for the determination of the stability properties of Γ. At the same time, this criteria will imply that there can be at most one periodic orbit which can bifurcate from Γ by a small change in the vector field in (2.2). The bifurcation near Γ is as simple as it could possibly be.

For motivation, let us suppose that we have a vector field $f(x, \mu)$ in R^2 depending on a scalar parameter μ with the property that there is a hyperbolic saddle point at 0 for (2.1) with the eigenvalues bounded away from zero for all μ. Also, suppose there is a hyperbolic periodic orbit P_μ of period $\omega(\mu)$ of (2.1) which has the property that $\text{dist}(0, P_\mu) \to 0$ as $\mu \to 0$. This creates a homoclinic orbit Γ at $\mu = 0$ (see Fig. 3.4). For such a situation to occur, the period $\omega(\mu)$ must approach ∞ as $\mu \to 0$. If this one parameter family of vector fields is generic, then one cannot expect to have other bad things happen as we change the vector field since we use the parameter to make $\omega(\mu) \to \infty$ as $\mu \to 0$. In particular, the rate of attraction or repulsion of the periodic orbit P_μ should be exponential and uniform in μ. If we keep this uniformity as $\mu \to 0$, then the orbit Γ should have the property that the Poincaré map on the part of a transversal inside Γ should have a nonzero derivative with a nonzero limit as $\mu \to 0$; that is, the orbit Γ is either asymptotically stable or unstable.

To find a quantitative expression for this uniformity, let us recall the formula for the characteristic exponents of the linear variational equation for a periodic orbit in the plane. If $P_\mu = \{p_\mu(t) : 0 \leq t < \omega(\mu)\}$ and $x = p_\mu + y$ in (2.1), then the linear variational equation for p_μ is

$$(3.5) \qquad\qquad \dot{y} = \frac{\partial f}{\partial x}(p_\mu(t), \mu)y$$

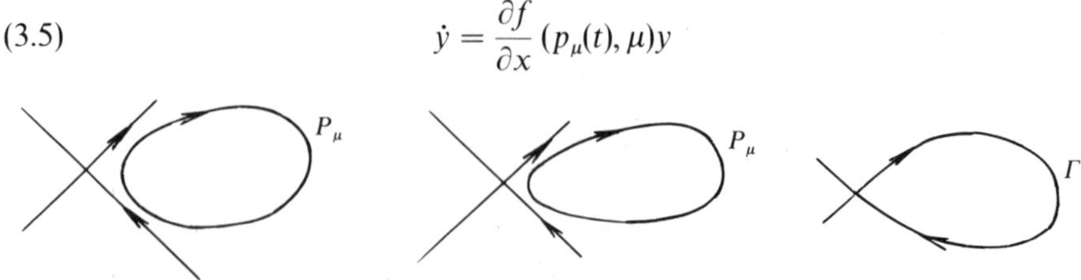

Figure 3.4

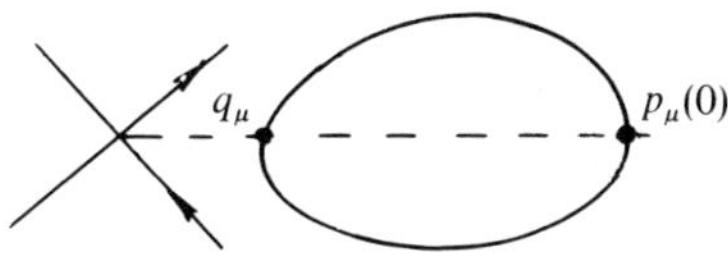

Figure 3.5

The nontrivial periodic function $\dot{p}_\mu$ satisfies this equation and thus one characteristic exponent may be taken to be zero. Let $\lambda(\mu)$ be the other characteristic exponent. The sum $\lambda(\mu) + 0 = \lambda(\mu)$ of the characteristic exponents must be

$$(3.6) \qquad \lambda(\mu) = \frac{1}{\omega(\mu)} \int_0^{\omega(\mu)} \left[\operatorname{tr} \partial f(p_\mu(t), \mu)/dx \right] dt$$

Let us choose the value $p_\mu(0)$ (by a shift in phase) so that the point q_μ in Fig. 3.5 corresponds to $p_\mu(-\omega(\mu)/2)$ as well as to $p_\mu(\omega(\mu)/2)$. Then it is not difficult to see that

$$\lambda(\mu) = \frac{1}{\omega(\mu)} \int_{-\omega(\mu)/2}^{\omega(\mu)/2} \operatorname{tr} \partial f(p_\mu(t), \mu)/\partial x \, dt \to \operatorname{tr} \partial f(0, 0)/\partial x$$

as $\mu \to 0$. Consequently, the rate of attraction or repulsion of each p_μ will be exponential and uniform in μ if

$$(3.7) \qquad \sigma_0 \overset{\text{def}}{=} \operatorname{tr} \frac{\partial f(0, 0)}{\partial x} \neq 0$$

We can now prove

Theorem 3.3. *Let $\sigma_0 = \operatorname{tr} \partial f(0, 0)/\partial x$. The homoclinic orbit Γ is asymptotically stable if $\sigma_0 < 0$ and unstable if $\sigma_0 > 0$. There can be at most one periodic orbit bifurcating from Γ and it is asymptotically orbitally stable if $\sigma_0 < 0$ and unstable if $\sigma_0 > 0$.*

Proof. Let π_0 be the Poincaré map on L_0^+ defined before. In terms of local coordinates on L_0^+, π_0 may be expressed as follows. Let $p_0 \in \Gamma \cap L_0$ be fixed, $f_0(x) = f(x, 0)$, $f_0^\perp$ be a vector orthogonal to $f_0(p_0)$. Assume that L_0^+ is in the direction $f_0^\perp$. Thus,

$$L_0^+ = \{\alpha f_0^\perp + p_0 : \alpha > 0 \text{ small}\}.$$

If $\phi(t, p)$ is the solution of (2.2) through $p \in L_0^+$ and $\tau(p)$ is the first time to return to L_0^+ starting at p, then

$$\phi(\tau(p), p) = \pi_0(p) = \beta(\alpha) f_0^\perp + p_0$$

where $p = \alpha f_0^\perp + p_0$ and $\beta(\alpha)$ is smooth for $\alpha > 0$ small. The derivative of $\pi_0(p)$ is given by $\beta'(\alpha)$. Let $f_\delta = f(\delta f_0^\perp + p_0, 0)$ for any $\delta \in R$. By differentiating the above equality with respect to α, we obtain

$$\beta'(\alpha) f_0^\perp = \left[\frac{\partial}{\partial t} \phi(\tau(p), p) \right] \frac{\partial \tau(p)}{\partial p} f_0^\perp + \frac{\partial}{\partial p} \phi(\tau(p), p) f_0^\perp$$

$$= f_\beta \frac{\partial \tau}{\partial p} f_0^\perp + \frac{\partial \phi}{\partial p} f_0^\perp$$

where, for simplicity in notation, we have let $\beta = \beta(\alpha)$, suppressing the dependence on α. By taking the inner product with $f_\beta^\perp$,

$$\beta'(\alpha) = \frac{\left\langle f_\beta^\perp, \frac{\partial \phi}{\partial p} f_0^\perp \right\rangle}{\langle f_\beta^\perp, f_0^\perp \rangle}.$$

Note that $\partial \phi / \partial p$ is the fundamental matrix of the variational equation

$$\dot{u} = \frac{\partial f}{\partial x} (\phi(t, p), 0) u$$

for which $\phi(t, p)$ is a solution. By the definition of $\tau(p)$, this implies $(\partial \phi / \partial p) f_\alpha = f_\beta$. Let $(\partial \phi / \partial p) f_0^\perp = \xi f_\beta + \eta f_0^\perp$, where $\xi, \eta \in R$. By the continuity of f, $f_\beta = (1 + \varepsilon_1) f_\alpha + \varepsilon_2 f_0^\perp$, where $\varepsilon_1, \varepsilon_2 \to 0$ as $\alpha \to 0$. This implies that

$$\frac{\partial \phi}{\partial p} f_\beta = (1 + \varepsilon_1 + \varepsilon_2 \xi) f_\beta + \varepsilon_2 \eta f_0^\perp.$$

Hence, $\partial \phi / \partial p$ has the matrix representation in terms of bases f_β and $f_0^\perp$:

$$\begin{bmatrix} 1 + \varepsilon_1 + \varepsilon_2 \xi & \xi \\ \varepsilon_2 \eta & \eta \end{bmatrix}.$$

This says that $\det(\partial \phi / \partial p) = (1 + \varepsilon_1) \eta$. On the other hand,

$$\left\langle f_\beta^\perp, \frac{\partial \phi}{\partial p} f_0^\perp \right\rangle = \eta \langle f_\beta^\perp, f_0^\perp \rangle.$$

Hence, $\beta'(\alpha) = \eta = \det(\partial \phi / \partial p)/(1 + \varepsilon_1)$, or

$$\beta'(\alpha) = \frac{1}{1 + \varepsilon_1} \exp \int_0^{\tau(p)} \operatorname{tr} \frac{\partial}{\partial x} f(\phi(t, p), 0) \, dt.$$

Note that $\tau(p) \to \infty$ as $\alpha \to 0$. Since $\varepsilon_1 \to 0$ as $\alpha \to 0$, we have by continuity, either $\beta'(\alpha) \to 0$, as $\alpha \to 0$ if $\sigma_0 < 0$, or $\beta'(\alpha) \to \infty$, as $\alpha \to 0$ if $\sigma_0 > 0$. The stability properties of Γ are now easy to obtain.

Reversing the intuitive argument used in deriving the condition $\sigma_0 \neq 0$ before the statement of the theorem, one sees that any periodic orbit near Γ must have a nonzero characteristic multiplier which is approximately σ_0. Thus, each periodic orbit is hyperbolic, asymptotically orbitally stable if $\sigma_0 < 0$ and unstable if $\sigma_0 > 0$. The Poincaré–Bendixson Theorem implies therefore that there can be at most one periodic orbit. The details are omitted. $\qquad\square$

Remark 3.4. In the next chapter, it will be shown that the set of μ for which Equation (2.1) has a homoclinic orbit is a submanifold of codimension one (see Corollary 11.3.6). If $\sigma_0 \neq 0$, it is shown that there is no periodic orbit on one side of this submanifold and a unique hyperbolic periodic orbit on the other side (see Corollary 11.3.7).

EXAMPLE 3.5. If $\sigma_0 = 0$, one can have either periodic orbits in any neighborhood of Γ, or Γ can be asymptotically stable or unstable. To illustrate this, consider the system

$$(3.8) \qquad \dot{x} = 2y, \qquad \dot{y} = 12x - 3x^2$$

which has the first integral

$$(3.9) \qquad V(x, y) = x^3 - 6x^2 + y^2$$

and thus the solution curves are given in Fig. 3.6.

The origin 0 is a hyperbolic saddle and the point $A = (4, 0)$ is a center. There is a periodic orbit in any neighborhood of the homoclinic orbit Γ and $\sigma_0 = 0$. Note $\Gamma \subset \{(x, y): V(x, y) = 0\}$.

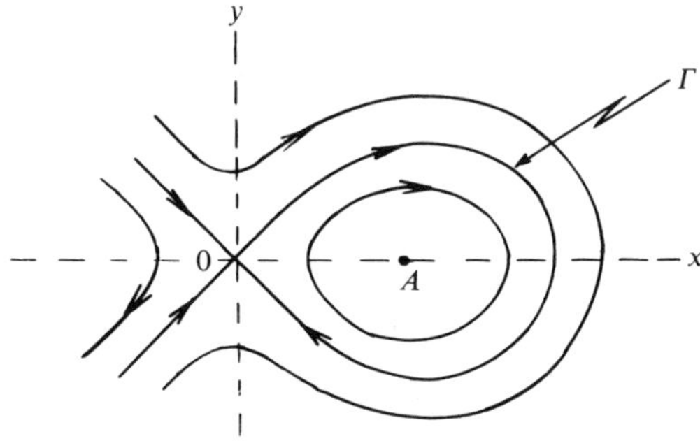

Figure 3.6

Now consider a perturbation of (3.8)

$$(3.10) \quad \begin{aligned} \dot{x} &= 2y - \mu V(x, y)(12x - 3x^2) \overset{\text{def}}{=} f_1(x, y, \mu) \\ \dot{y} &= 12x - 3x^2 + \mu V(x, y)2y \overset{\text{def}}{=} f_2(x, y, \mu) \end{aligned}$$

where V is the function in (3.9). It is easy to see that

$$\sigma_0(\mu) = \operatorname{tr} \partial(f_1(0, 0, \mu), f_2(0, 0, \mu))/\partial(x, y) = 0$$

for all μ. System (3.10) has only the equilibrium points 0 and A with 0 a hyperbolic saddle and A a hyperbolic focus for $\mu \neq 0$ sufficiently small. A is stable for $\mu > 0$ and unstable for $\mu < 0$. The curve in (x, y)-space defined by $V(x, y) = 0$ is invariant under (3.10). Thus, Γ is again a homoclinic orbit for (3.10) for every μ. By observing that (3.10) is obtained from (3.8) by a rotation through an angle $\tan^{-1} \mu V(x, y)$, it follows that no curve $V(x, y) = $ constant inside the curve Γ can be tangent to the vector field in (3.10). Thus, there can be no periodic orbits of (3.10) inside Γ for $\mu \neq 0$. Since $\mu > 0$ ($\mu < 0$) implies the focus A is stable (unstable), it follows that Γ is unstable (asymptotically stable) for $\mu > 0$ ($\mu < 0$).

10.4. Closed Curve with a Saddle-Node

In this section, we consider the two dimensional system (2.1) under the assumption that there is a smooth curve C_0 which is invariant under the flow defined by (2.2) and there is exactly one equilibrium point on C_0. The problem is to discuss the behavior of the solutions of (2.1) near C_0 for μ near zero.

To be specific, suppose C_0 is a closed curve in $\mathbb{R}^2$ defined by a C^k function, $k \geq 1$, and C_0 is invariant for system (2.2) with the origin 0 on C_0. Also, suppose the solution $x = 0$ is the only equilibrium point on C_0 and satisfies

$$(4.1) \quad \begin{aligned} \det \frac{\partial f}{\partial x}(0, 0) &= 0 \\ \sigma_0 &\overset{\text{def}}{=} \operatorname{tr} \partial f(0, 0)/\partial x \neq 0. \end{aligned}$$

Since zero is the only equilibrium point on C_0, the curve $C_0 \backslash \{0\}$ consists of a single orbit of (2.2) whose α- and ω-limit set is 0. The condition (4.1) implies the eigenvalues of $\partial f(0, 0)/\partial x$ are 0 and $\sigma_0 \neq 0$. Thus, we can define a center manifold M_0 in a neighborhood W of $x = 0$ which will be exponentially asymptotically stable as $t \to \infty$ ($t \to -\infty$) if $\sigma_0 < 0$ ($\sigma_0 > 0$) (see Section 9.2). From the fact that C_0 is invariant and the flow in the direction

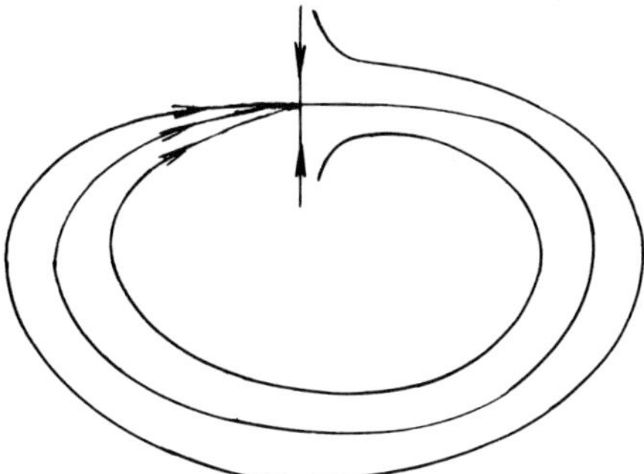

Figure 4.1

tangent to C_0 is not hyperbolic at zero, we may take M_0 as $C_0 \cap W$. In W, there is curve S, invariant for (2.2), tangent at 0 to the eigenvector of $\partial f(0,0)/\partial x$ associated with the eigenvalue σ_0, such that solutions on S approach zero exponentially as $t \to \infty$ $(t \to -\infty)$ if $\sigma_0 < 0$ $(\sigma_0 > 0)$. This implies that the flow near C_0 has the properties shown in Fig. 4.1 (drawn for $\sigma_0 < 0$).

Now consider the perturbed equation (2.1) for (x, μ) in a neighborhood of $C_0 \times \{0\}$. It is not difficult to show directly and is actually a consequence of the general theory of integral manifolds that there is a $\delta > 0$, a neighborhood V of C_0, a smooth closed curve $C_\mu \subset V$, $|\mu| < \delta$, invariant for Eq. (2.1) which is exponentially asymptotically stable as $t \to \infty$ $(t \to -\infty)$ if $\sigma_0 < 0$ $(\sigma_0 > 0)$ and $C_\mu \to C_0$ as $\mu \to 0$. Furthermore, there is a neighborhood W of $x = 0$ such that the only possible equilibrium points of (2.1) in the neighborhood V of C are in W.

The equilibrium points of (2.1) in V must lie on C_μ. Therefore, if C_μ contains no equilibrium points, then C_μ is a periodic orbit, unique in V, and is exponentially orbitally asymptotically stable as $t \to \infty$ $(t \to -\infty)$ if $\sigma_0 < 0$ $(\sigma_0 > 0)$. If C_μ contains equilibrium points, then the number and structure depend upon the nonlinearities of $f(x, 0)$ near $x = 0$. For example, suppose the equilibrium point on C_0 is a saddle-node; that is, the vector field $g(y)$ in the direction of the tangent to C_0 at zero satisfies $g(y) = \alpha y^2 + o(|y|^2)$ as $|y| \to 0$, $\alpha \neq 0$. Then an appropriate perturbation can lead to two equilibrium points, both hyperbolic, one a saddle and one a node (see Section 9.3).

We summarize these results in the following theorem.

Theorem 4.1. *Suppose C_0 is a smooth invariant closed curve for (2.2) containing only one equilibrium point $x = 0$. Then there are neighborhoods U of $\mu = 0$, V of C_0 and a neighborhood W of $x = 0$ such that, for $\mu \in U$, system (2.1) has a smooth invariant closed curve C_μ in V which is asymptotically stable as $t \to \infty$ $(t \to -\infty)$ if $\sigma_0 < 0$ $(\sigma_0 > 0)$, $C_\mu \to C_0$ as $\mu \to 0$. There is a periodic orbit of (2.1) in V (and then it is given by C_μ) if and only if there are no equilibrium points of (2.1) in W. If 0 is a saddle-node and $E = C^2(\mathbb{R}^2, \mathbb{R}^2)$, then there is a function*

$\gamma: U \to \mathbb{R}$ *such that* $U = \Gamma_+ \cup \Gamma_0 \cup \Gamma_-$, $\Gamma_+, \Gamma_0, \Gamma_-$ *are respectively the subsets of* U *where* $\gamma > 0$, $\gamma = 0$, $\gamma < 0$, Γ_0 *is a smooth submanifold of codimension one such that* C_μ *is a periodic orbit for* μ *in* Γ_+, C_μ *contains a saddle-node if* $\mu \in \Gamma_0$, C_μ *contains a saddle and a hyperbolic node if* $\mu \in \Gamma_-$.

10.5. Remarks on Structural Stability and Bifurcation

In the previous sections, we have discussed various types of dynamic bifurcation for autonomous systems—from a saddle-node in Section 9.3, a focus in 9.5, a periodic orbit in Section 2, a homoclinic orbit in Section 3 and a closed curve with a saddle-node in Section 4. In this section, we restrict the discussion to vector fields on the plane which are C^r, $r \geq 1$ and relate the above bifurcation phenomena to the concept of structural stability introduced by Andronov and Pontryagin [1] in 1937. No proofs will be given.

To avoid the difficulties that arise from the noncompactness of the plane, we restrict the discussion to the interior Ω of a closed curve Γ without contact to any of the vector fields to be considered. Let $\mathscr{X}_2^r$ be the set of all such C^r vector fields. Two vector fields X, Y in $\mathscr{X}_2^r$, $t \geq 1$, are *equivalent* if there is a homeomorphism on $\Omega \cup \Gamma$ which maps orbits of one onto orbits of the other and preserves the sense of direction in time. This is an equivalence relation among vector fields. X is *structurally stable* if every Y in a neighborhood of X is equivalent to X. The basic results on structurally stable vector fields were obtained by Andronov and Pontryagin [1] and Peixoto [1] and are stated in the following.

Theorem 5.1. *An* $f \in \mathscr{X}_2^r$ *is structurally stable if and only if every equilibrium point and every periodic orbit is hyperbolic and there are no connections between saddle points. Also, the set of structurally stable systems is open and dense in* $\mathscr{X}_2^r$.

A basic ingredient in the proof of this theorem is the Hartman–Grobman theorem in Section 3.6.

An $X \in \mathscr{X}_2^r$ is a *bifurcation point* (a vector field for which a perturbation could lead to a bifurcation) if X is *structurally unstable*; that is, not structurally stable. It is impossible to study the behavior of all vector fields in the neighborhood of an arbitrary bifurcation point; for example, $X = 0$ is a bifurcation point and any flow in the plane can be obtained by choosing an appropriate Y near zero. This is where the idea of genericity enters bifurcation theory. One must find those bifurcation points X which have the property that the simplest possible bifurcations occur near X. Andronov and Leontovich [1] made this precise by defining *structural instability of degree k* (or *bifurcation point of degree k*).

The vector field X is a *bifurcation point of degree* 0 if it is structurally stable. X is a *bifurcation point of degree* 1 if it is not of degree zero and there is a neighborhood of X which has only bifurcation points of degree 0, or, ones which are equivalent to X. It is a *bifurcation point of degree 2* if it is not of degree zero or one and there is a neighborhood containing only bifurcation points of degree 0 or 1, or, ones which are equivalent to X. Similarly, one defines degree k.

The fundamental results on bifurcation points of degree one are due to Andronov, Leontovich, Gordon and Maier [1], Sotomayor [1] and are contained in the following.

Theorem 5.2. *A vector field $f \in \mathscr{X}_2^r$, $r \geq 3$, is a bifurcation point of degree 1 if and only if there is a neighborhood W of f and a submanifold Γ of codimension one in W such that $W \backslash \Gamma = U_1 \cup U_2$ where each $g \in U_j$ is structurally stable but $g \not\sim h$ if $g \in U_1$, $h \in U_2$. For $g \in \Gamma$, only one of the following situations prevails:*

(i)   *$g \in \Gamma$ has an elementary saddle-node at x_0, there are no equilibrium points of g near x_0 if $g \in U_1$ and a saddle and node near x_0 if $g \in U_2$.*

(ii)   *$g \in \Gamma$ has an elementary focus at x_0, there is no periodic orbit of g near x_0 if $g \in U_1$ and a periodic orbit near x_0 if $g \in U_2$—the generic Hopf bifurcation.*

(iii)   *$g \in \Gamma$ has a periodic orbit γ which is stable from one side, unstable from the other, $g \in U_1$ has no periodic orbit near γ and $g \in U_2$ has two hyperbolic periodic orbits near γ.*

(iii)′   *γ is not both α- and ω-limit set of saddle separatrices.*

(iv)   *$\sigma_0 = \operatorname{tr} \partial f(0)/\partial x \neq 0$, $g \in \Gamma$ has a homoclinic orbit containing a saddle point x_0, $g \in U_1$ has a saddle near x_0 and no periodic orbit near γ, $g \in U_2$ has a saddle point and a unique hyperbolic periodic orbit near γ which coalesce as $g \to \Gamma$.*

(iv)′   *For $g \in \Gamma$, the homoclinic orbit is not the limit set of other saddle separatrices.*

(v)   *there is a connection between distinct saddle points.*

Each of the cases (i)–(v) is shown in Figure 5.1.

There is another interesting phenomena that can occur at a bifurcation point of degree 1 which satisfies (i) of Theorem 5.2. There is a smooth invariant curve containing a saddle and a node which coalesce, disappear and the invariant curve becomes a periodic orbit (see Section 4).

Although we do not give a proof of this theorem, we have discussed some of the important ingredients in previous sections and will discuss them in the next chapter. In particular, see Section 9.3 for (i), Section 9.6 for (ii), Section 10.2 for (iii) and Section 10.3 and Corollary 11.3.7 for (iv).

The fact that only two possibilities arise in a neighborhood of a bifurcation point of degree 1 suggests that this is the typical or generic situation that arises in the discussion of one parameter families of vector fields. Sotomayor [1] has proved, in fact, that this is the case.

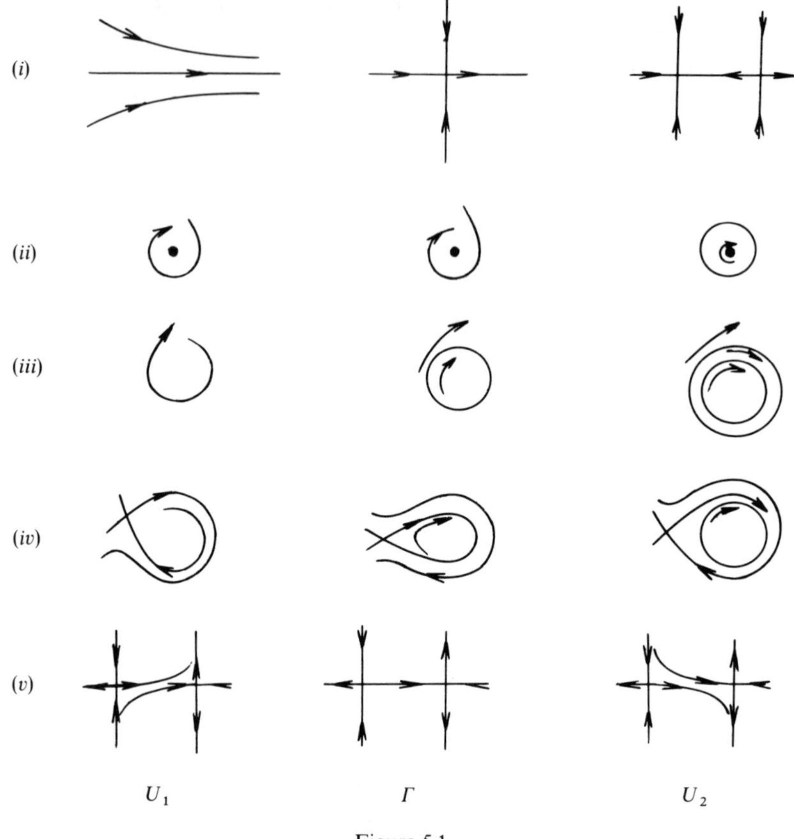

Figure 5.1

Theorem 5.3. *If $\Phi^k = \{\varphi : [0, 1] \to \mathscr{X}_2^k, \varphi \in C^k\}$, $k \geq 3$, and $\tilde{\Phi}^k$ is the set of $\varphi \in \Phi^k$ such that $\varphi(t)$ is structurally stable except at a countable number of points $\{t_j\}$ with finitely many accumulation points $\{T_k\}$ so that, at t_j, $\varphi(t_j)$ is a bifurcation point of codimension one and at T_k, $\varphi(T_k)$ violates either* (iii)′ *or* (iv)′ *of Theorem 5.2, then $\tilde{\Phi}^k$ is a residual set in Φ^k.*

Extensive applications of these results to the theory of nonlinear oscillations were made in the late 1930's (see Andronov, Vitt and Khaikin [1]). Most of the results in the literature on one parameter problems are a consequence of these results.

The characterization of bifurcation points of degree two has recently been completed, but the detailed results will not be given here (see Andronov, Leontovich, Gordon and Maier [1], Sotomayor [1], Takens [1–4], Carr [1]). We have discussed some things that can occur in Sections 9.3, 9.6 and Section 2 of this chapter. A few more examples will be given in Chapter 13. The theory of bifurcation points of degree two can be considered the typical or generic situation for two parameter families of vector fields.

There is also some discussion of bifurcation points of degree greater than two in Andronov, Leontovich, Gordon and Maier [1], Sotomayor [1, 2].

When the dimension of the system is >2, the class of structurally stable systems is small in the sense that it is not a residual set. Some of the types of strange phenomena that can occur will be exhibited in the next chapter when we discuss homoclinic points. Much of the modern theory of dynamical systems in an attempt to discover a residual set of "simple" dynamical systems—ones which can be classified and which will preserve their essential features when subjected to perturbations (see the survey articles and books of Smale [1], Nitecki [1], Peixoto [2, 3], Newhouse [1], Abraham and Marsden [1], Palis and Melo [1], Shub [1], Guckenheimer [3, 4]).

10.6. Remarks on Infinite Dimensional Systems and Turbulence

For infinite dimensional systems, the generic theory analogous to the one discussed previously is in its infancy. In this section, we outline an approach to the development of such a theory for a special class of semigroups of transformations. This class is general enough to include some types of parabolic and hyperbolic partial differential equations as well as retarded functional differential equations and some neutral functional differential equations.

Let X, Y, Z be Banach spaces and let $\mathcal{X}^r = C^r(Y, Z)$, $r \geq 1$, be the set of functions from Y to Z which are bounded and uniformly continuous together with their derivatives up through order r. We impose the usual topology on $\mathcal{X}^r$. For each $f \in \mathcal{X}^r$, let

$$T_f(t): X \to X, \qquad t \geq 0,$$

be a strongly continuous semigroup of transformations on X. For each $x \in X$, we suppose $T_f(t)x$ is defined for each $t \geq 0$ and is C^r in x.

In applications, one often is interested in open subsets of X, Y, but this presents mainly notational difficulties and therefore will not be discussed.

We say that a point x_0 has a *backward extension* (relative to $T_f(t)$) if there is a function $\phi:(-\infty, 0] \to X$ such that $\phi(0) = x_0$ and $T_f(t)\phi(\tau) = \phi(t + \tau)$ for all $0 \leq t \leq -\tau, \tau \in (-\infty, 0]$. If x_0 has a backward extension ϕ, we define $T_f(t)x_0$, $t \leq 0$, by the relation $T_f(t)x_0 = \phi(t)$ and say that $T_f(t)x_0$ is *defined for $t \leq 0$*. Let

$$(6.1) \qquad A_f = \{x \in X : T_f(t)x \text{ is defined and bounded for } t \leq 0\}.$$

This set contains all equilibrium points, periodic orbits and the ω-limit sets of any compact orbit. The set A_f is invariant. For any bounded open set

$N \subset X$, let

(6.2) $A_f(N) = \{x \in X : T_f(t)x \text{ is defined and belongs to } N \text{ for } t \geq 0\}$.

Definition 6.1. For $f, g \in \mathcal{X}^r$, we say f *is equivalent to* g, $f \sim g$, if there is a homeomorphism $h: A_f \to A_g$ such that h maps the orbits of $T_f(t)$ on A_f onto the orbits of $T_g(t)$ on A_g preserving the sense of direction in time. An $f \in \mathcal{X}^r$ is *structurally stable* if there is a neighborhood V of f such that $f \sim g$ for every $g \in V$. An $f \in \mathcal{X}^r$ is a *bifurcation point* if it is not structurally stable.

Definition 6.2. For $f, g \in \mathcal{X}^r$, we say f is *locally equivalent to* g in a neighborhood of x_0 if there are neighborhoods N, M of x_0 and a homeomorphism $h: A_f(N) \to A_g(M)$ such that h takes the orbits of $T_f(t)$ on $A_f(N)$ onto the orbits of $T_g(t)$ on $A_g(M)$ preserving the sense of direction in time. An $f \in \mathcal{X}^r$ is *locally structurally stable* if there is a neighborhood V of f such that f is locally equivalent to g at x_0 for every $g \in V$.

It is important to notice that the flows defined by $T_f(t)$, $T_g(t)$ are compared only on the invariant sets A_f, A_g. However, in order for there to be a homeomorphism between A_f and A_g for each g in a neighborhood of f, the set A_f must have some strong type of stability as an invariant set of $T_f(t)$. One can see several illustrations of this remark in infinite dimensional problems in Hale [13], but the idea is easily understood by comparing this definition with the one for ordinary differential equations in $\mathbb{R}^n$.

Consider the equation

(6.3) $\dot{x} = Bx + f(x)$

where $x \in \mathbb{R}^n$, $f(0) = 0$, $\partial f(0)/\partial x = 0$. If W_f^u is the unstable manifold of zero (it could contain only the point zero), suppose there is a neighborhood N of $x = 0$ such that if $x_0 \in N \backslash W_f^u$, then there is a $\tau < 0$ such that $T_f(\tau)x_0 \in \partial N$. It follows that $A_f(N) = W_f^u \cap N$. If f is locally structurally stable at zero according to Definition 6.2, then one can show that the equilibrium point $x = 0$ of (6.3) must be hyperbolic. Also, there is a neighborhood V of f such that, for each $g \in V$, there is a neighborhood M of zero, a unique equilibrium point x_g of $g \in M$ and $A_g(M) = W^u(x_g) \cap M$. In the Definition 9.4, the trajectories in a full neighborhood of zero are not compared and not even are they compared on the stable manifold. It frequently happens in applications in infinite dimensions that the stable manifold is infinite dimensional and the unstable manifold is finite dimensional. Comparison of trajectories on the infinite dimensional parts seems to be impossible. Thus, the definition of equivalence is chosen as above.

The following is an interesting result about the set A_f.

Theorem 6.3. *If A_f is compact and $D_x T_f(t)x$ for each $t > 0$, $x \in A_f$, is the sum of a contraction operator and a completely continuous operator, then A_f has finite Hausdorff dimension.*

For X a separable Hilbert space, this result was proved by Mallet–Paret [2]. The general case in Theorem 6.3 for X an arbitrary Banach space was proved by Mañé [1]. Ladyzenskaya [1] has obtained related results for the Navier–Stokes equation. Mañé [1] proved even more—that the limit capacity of A_f is finite. The limit capacity of A_f is defined as the exponential growth rate of the number of balls of radius $\exp(-t)$ that are required to cover A_f when $t \to \infty$. The limit capacity is always at least as large as the Hausdorff dimension.

Theorem 6.3 is also related to the transition to turbulence. It has been speculated that a k dimensional torus ($k \geq 1$) may bifurcate into a $(k+1)$ dimensional torus in the Navier–Stokes equations as the Reynolds number, or the viscosity, changes. In some experiments, as many as three independent frequencies have been observed. This suggests the existence of a three dimensional invariant torus in the phase space. One may even observe higher dimensional invariant tori. Theorem 6.3 indicates that if the transition to turbulence is by successive bifurcations of higher dimensional tori, then it has to occur after a finite number of such bifurcations. This theorem also shows that one cannot observe more than a finite number of independent frequencies which change continuously with respect to some parameters.

10.7. Bibliographical Notes

The discussion in this chapter follows closely the ideas of Andronov, Leontovich, Gordon and Maier [1] and Sotomayor [1, 2] even though some of the proofs are different. The definition of a periodic orbit of order k and example (2.10) are taken from Andronov, Leontovich, Gordon and Maier [1] as well as Example 3.5.

If $\sigma_0 = 0$, then it is possible that more than one periodic orbit can bifurcate from the homoclinic orbit Γ. Under this hypothesis, it can be shown that there is always a perturbation of (2.2) which will have at least two periodic orbits in a neighborhood of Γ (see Andronov, Leontovich, Gordon and Maier [1]). It is possible to give a more generic classification of Γ in the sense that one can introduce a concept of order k for a homoclinic orbit in the spirit of the one given for a periodic orbit in Section 2. This concept involves the Poincaré map used in Section 3 as well as specific properties of the vector field near the saddle point. Precise definitions have been given by Leontovich [1]. He has also discussed the precise number of limit cycles that can appear under perturbations of (2.2).

A few further references to structural stability and bifurcation are included in the text in Section 5. The work of Thom [1, 2] is also based on the ideas of structural stability and bifurcation.

For a further development of the infinite dimensional case and many references, see Hale [13].

Chapter 11

Bifurcation of Periodic Planar Equations

11.1. Introduction

In the previous chapter, we have given a rather complete discussion of the behavior of solutions near a periodic orbit or a homoclinic orbit of a two dimensional autonomous system subjected to an autonomous perturbation. In this case, new periodic orbits could occur or old periodic orbits could disappear at bifurcation points.

This chapter as well as parts of Chapters 12 and 13 are devoted to perturbations of two dimensional autonomous systems for which the perturbations are allowed to depend on the independent variable. For such equations, many new phenomena occur because the introduction of nonautonomous perturbations takes the problem out of the plane into higher dimensions. The discussion of periodic perturbations of planar autonomous systems corresponds to the discussion of autonomous systems on the plane cross a circle. Quasiperiodic perturbations correspond to autonomous equations in even higher dimensions.

In this chapter, the perturbations are assumed to be periodic of period one in the independent variable. Two specific problems are discussed. Suppose the unperturbed equation has a periodic orbit of period k, where k is an integer. The first problem in Section 11.2 is concerned with the existence of subharmonic solutions (that is, solutions of period k) in a neighborhood of this orbit for the perturbation small. A complete solution is given under generic hypotheses. Specific bifurcation diagrams are given for the perturbation corresponding to a small damping and small forcing.

The second problem in Section 11.3 deals with the case in which the unperturbed equation has a homoclinic orbit. For small perturbations, we give conditions for the existence of transverse homoclinic points near this orbit. For the case in which the unperturbed system is Hamiltonian, we show in Section 11.4 how the theory of subharmonic bifurcations mentioned above is related to the bifurcation to homoclinic points.

These problems have a character completely different from the ones considered in Chapter 9 where the solutions of the nonautonomous equations were restricted to lie in the neighborhood of an equilibrium point for the unperturbed equation. The above problems are more global. From the point of view of abstract bifurcation theory, the problems in Chapter 9 deal with bifurcation in a neighborhood of an isolated zero of a function. Abstractly, the problem of subharmonic bifurcation above is bifurcation in a neighborhood of a circle of zeros of a function—each point on the periodic orbit is a fixed point of the period k map. The problem of homoclinic bifurcation is abstractly the problem of bifurcation from a set whose closure is compact— the homoclinic orbit without the equilibrium point.

In Section 11.5, we show that the methods in Section 11.2 are easily adapted to the abstract problem of bifurcation from a circle.

11.2. Periodic Orbit-Subharmonics

Suppose $f:\mathbb{R}^2 \to \mathbb{R}^2$ is of class C^r, $r \geq 2$ and the two dimensional system

$$\dot{x} = f(x) \tag{2.1}$$

has a periodic orbit Γ of least period k, where k is a positive integer. Suppose $g:\mathbb{R} \times \mathbb{R}^2 \times \mathbb{R}^m \to \mathbb{R}^2$ is of class C^r, $g(t, x, \mu) = g(t + 1, x, \mu)$ for all t, x, μ, $g(t, x, 0) = 0$, for all t, x. We are interested in the k-periodic solutions of the perturbed equation

$$\dot{x} = f(x) + g(t, x, \mu) \tag{2.2}$$

for (x, μ) in a neighborhood of $\Gamma \times \{0\}$. These solutions are called *subharmonic solutions of order k* since the period of the solution is k times the period of the "forcing" function $g(t, x, \mu)$.

Equation (2.2) contains as a special case the equation

$$\ddot{y} + H(y) = -\mu_1 \dot{y} + \mu_2 F(t). \tag{2.3}$$

To obtain subharmonic solutions of (2.3), it is well known that the damping coefficient μ_1 cannot be too large with respect to the magnitude of μ_2. The interesting problem is to determine the precise region in (μ_1, μ_2)-space where subharmonic solutions can occur and, in particular, to determine the curves of bifurcation from the nonexistence of subharmonics to the existence of subharmonics. The stability properties of these subharmonics must also be discussed. For Eq. (2.2), this is our objective in the present section.

To obtain a complete picture of the bifurcation curves, we need some hypotheses. Let

$$\Gamma_k = \{p_k(t): t \in \mathbb{R}\} \tag{2.4}$$

where $p_k(t)$ is a k-periodic solution of (2.1). Throughout this section, we will have an annoying subscript k to denote the subharmonic of order k. This is retained in order to facilitate the discussion of subharmonic bifurcations and homoclinic points in Section 4. The function $\dot{p}_k$ is a nontrivial k-periodic solution of the linear variational equation around p_k,

$$(2.5) \qquad \dot{y} = A(t)y, \qquad A(t) = \partial f(p_k(t))/\partial x.$$

Thus, one is a characteristic multiplier of this system. Our first hypothesis is

$(\mathbf{H_1})$ *Every k-periodic solution of (2.5) is a multiple of $\dot{p}_k$*

Let us discuss the meaning of $(\mathbf{H_1})$ for (2.3). The second order equation (2.3) for $\mu_1 = \mu_2 = 0$ cannot have an isolated periodic orbit. Thus, the periodic orbit Γ_k must be a member of a smooth one parameter family of periodic orbits represented by a real parameter a. In this case, both characteristic multipliers of (2.5) are one. Let $w(a)$ be the period as a function of a and let a_0 correspond to Γ_k. Hypothesis $(\mathbf{H_1})$ is then equivalent to $w'(a_0) \neq 0$.

For the general equation (2.1), Hypothesis $(\mathbf{H_1})$ can be satisfied if Γ_k is a member of a smooth one parameter family of periodic orbits as above or Γ_k can have one characteristic multiplier $\neq 1$. In this latter case, the theory of integral manifolds implies there is a neighborhood V of Γ_k, $\delta > 0$, and a unique invariant cylinder $M_\mu \subset V \times \mathbb{R}$, $|\mu| < \delta$, which is either exponentially asymptotically stable as $t \to +\infty$ or $t \to -\infty$ depending on whether the nonzero characteristic exponent has modulus less than one or greater than one (Section 12.5). This invariant cylinder has a cross section of period one. Thus, the flow on the cylinder is the same as the flow on a torus T obtained by identifying the cross section of the cylinder at $t = 0$ and $t = 1$. Finding conditions on μ such that the rotation number of the flow on this torus is $1/k$ is equivalent to finding subharmonics of order k. Thus, our results below say something about this rotation number.

Let $q(t)$ be a nontrivial k-periodic solution of the equation

$$(2.6) \qquad \dot{y} = -yA(t)$$

adjoint to (2.5) and define

$$(2.7) \qquad h_k(\alpha) = \int_0^k q(t) \cdot [\partial g(t - \alpha, p_k(t), 0)/\partial \mu]\, dt$$

for $\alpha \in \mathbb{R}$. The function h_k is C^r, $r \geq 2$, and is 1-periodic.

A geodesic on the unit sphere S^{m-1} in $\mathbb{R}^m$ is the intersection of a hyperplane through the origin with S^{m-1}. Let

$$P_{h_k} = \{\beta \in \mathbb{R}^m : \beta \cdot h_k(\alpha) = 0 \text{ for some } \alpha \in [0, 1)\}$$

We now suppose that

(H₂)   *The boundary (in the relative topology of S^{m-1}) of $P_{h_k} \cap S^{m-1}$ is the union of two disjoint geodesies L_k^m, L_k^M.*

Hypothesis (H₂) implies there exists at least one pair of distinct real values α_m, $\alpha_M \in [0, 1)$ such that $h_k(\alpha_m)$ is orthogonal to L_k^m and $h_k(\alpha_M)$ is orthogonal to L_k^M.

Let $\gamma_{h_k} = \{h_k(\alpha) : \alpha \in [0, 1)\}$. Note that Hypothesis (H₂) excludes the possibility that $Ph_k \cap S^{m-1}$ is all of S^{m-1} so that, in particular, $0 \notin \gamma_{h_k}$. Also, γ_{h_k} cannot belong to a hyperplane through the origin. In Figure 2.1, we have indicated the situation for $m = 2$ and a particular curve γ_{h_k}. The symbol $[p]$ denotes the line through p. We now make the restriction that $\mu \in \mathbb{R}^2$ $(m = 2)$ and suppose

(H₃)
$$\begin{aligned}
\mu \cdot h_k(\alpha_m) = 0, \; \mu \cdot h_k'(\alpha_m) = 0, &\qquad |\mu| = 1 \Rightarrow \mu \cdot h_k''(\alpha_m) \neq 0 \\
\mu \cdot h_k(\alpha_M) = 0, \; \mu \cdot h_k'(\alpha_M) = 0, &\qquad |\mu| = 1 \Rightarrow \mu \cdot h_k''(\alpha_M) \neq 0
\end{aligned}$$

We can now prove the following result.

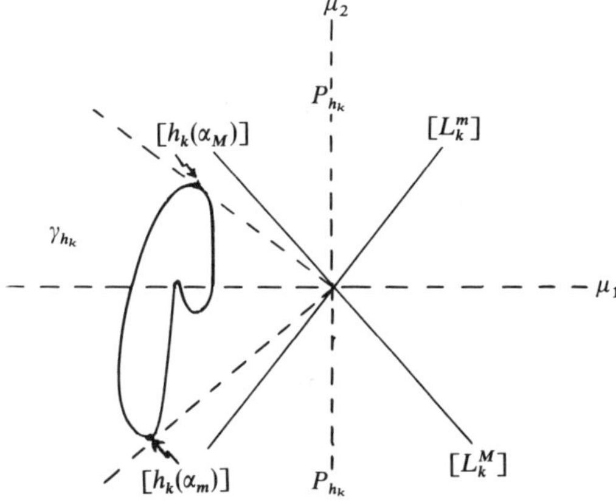

Figure 2.1

Theorem 2.1. *If Hypothesis (H₁)–(H₃) are satisfied, then there is a neighborhood U of Γ_k, V of $\mu = 0$ and two C^{r-1} manifolds C_m^k, $C_M^k \subseteq V$ of codimension one containing $\mu = 0$ whose tangent planes are orthogonal respectively to $h_k(\alpha_m)$, $h_k(\alpha_M)$ at $\mu = 0$, $C_m^k \cap C_M^k = \{0\}$, C_m^k, C_M^k divide $V \backslash \{0\}$ into two disjoint open sets S_1^k, S_2^k such that* Eq. (2.2) *has no subharmonic solutions in U for μ in S_1^k and at least $2k$ for μ in S_2^k (see Fig. 2.2).*

Remark 2.2. For $m \geq 3$, hypothesis (H₃) is vacuous. See the bibliographical notes for this case as well as $\mu \in E$, an arbitrary Banach space. In this way, one can consider arbitrary C^r perturbations of (2.1).

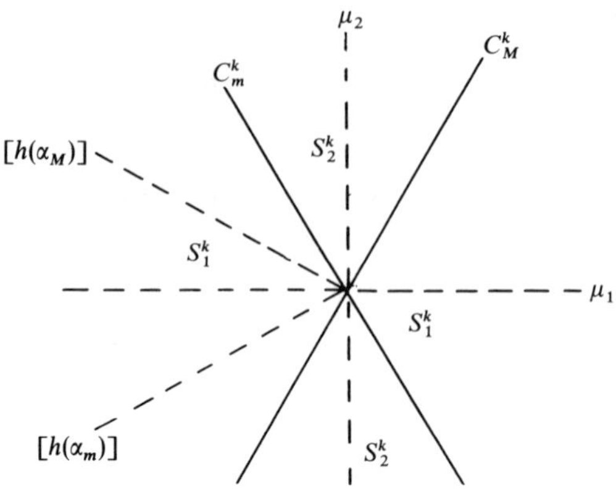

Figure 2.2

For the equation (2.3), Hypothesis (H_2) and (H_3) are much easier to state. In fact, writing (2.3) as a system

$$(2.8) \qquad \begin{aligned} \dot{x}_1 &= x_2 \\ \dot{x}_2 &= -H(x_1) - \mu_1 x_2 + \mu_2 F(t) \end{aligned}$$

one easily sees that $p_k(t) = (\varphi_k(t), \dot{\varphi}_k(t))$ where φ_k is k-periodic, $\ddot{\varphi}_k + H(\varphi_k) = 0$.

$$(2.9) \qquad h_k(\alpha) = \eta_k(-1, G_k(\alpha)), \qquad \eta_k = \int_0^k \dot{\varphi}_k^2(t)\,dt$$

$$G_k(\alpha) = \frac{1}{\eta_k} \int_0^k \dot{\varphi}_k(t) F(t - \alpha)\,dt.$$

Conditions (H_2) and (H_3) now become

(H_4)   *If α_m, α_M are the values in $[0, 1)$ for which the function $G_k(\alpha)$ in (2.9) attains respectively, its minimum and maximum, then $G_k''(\alpha_m) > 0$, $G_k''(\alpha_M) < 0$.*

We thus have the following result for Eq. (2.3).

Corollary 2.3. *Suppose (H_1), (H_4) are satisfied for Eq. (2.3). Then there is a neighborhood U of $\Gamma_k = \{(\varphi_k(t), \dot{\varphi}_k(t)) : t \in \mathbb{R}\}$ V of $\mu_1 = \mu_2 = 0$ and two C^{r-1} curves $C_m^k, C_M^k \subseteq V$ containing $\mu = 0$ which are tangent respectively to $\mu_1 = G_k(\alpha_m)\mu_2$, $\mu_1 = G_k(\alpha_M)\mu_2$ at $\mu = 0$, $C_m^k \cap C_M^k = \{0\}$, C_m^k, C_M^k divide $V \backslash \{0\}$ into two disjoint open sets S_1^k, S_2^k such that Eq. (2.3) has no subharmonic solutions in U for μ in S_1^k and at least $2k$ for μ in S_2^k.*

Let us now turn to a proof of Theorem 2.1.

Proof of Theorem 2.1. In this proof we drop all subscripts k to simplify notation. If $v(t) \in \mathbb{R}^2$ is orthogonal to $\dot{p}(t)$, then we have seen in Section 10.2 that the mapping

$$x = p(\alpha) + av(\alpha)$$

is a diffeomorphism of a neighborhood of Γ onto $[0, k) \times \{|a| < \delta\}$ for some $\delta > 0$. Thus, if $x(t)$ is a k-periodic solution of (2.2) which lies in a sufficiently small neighborhood of Γ, then the initial value $x(0)$ uniquely determines an (α, a) such that $x = p(\alpha) + av(\alpha)$. Thus, it is sufficient to consider only those k-periodic solutions of (2.2) of the form

$$(2.10) \qquad x(t) = p(t + \alpha) + z(t + \alpha)$$

where $z(\alpha) = av(\alpha)$ for some a; that is $z(\alpha)$ is orthogonal to $\dot{p}(\alpha)$.

If the transformation (2.10) is applied to (2.2) and $t + \alpha$ is replaced by t, then

$$\dot{z} = A(t)z + F(t, z, \mu, \alpha)$$
$$(2.11) \quad F(t, z, \mu, \alpha) = f(p(t) + z) - f(p(t)) - A(t)z + g(t - \alpha, p(t) + z, \mu)$$
$$A(t) = \partial f(p(t)/\partial x$$

We now apply the method of Liapunov–Schmidt. Let us normalize q so that $\int_0^k |q|^2 \, dt = 1$, where q is a k-periodic solution of Eq. (2.6). Let $\mathscr{P}_k = \{b: \mathbb{R} \to \mathbb{R}^2; b$ is continuous and k-periodic$\}$ and define $|b| = \sup_t |b(t)|$ for $b \in \mathscr{P}_k$. If we define

$$(2.12) \qquad P: \mathscr{P}_k \to \mathscr{P}_k \qquad Pb = q^{\mathrm{T}} \int_0^k q(t) \cdot b(t) \, dt$$

then P is a continuous projection and the equation

$$(2.13) \qquad \dot{z} = A(t)z + b(t)$$

for $b \in \mathscr{P}_k$ has a k-periodic solution if and only $Pb = 0$. If $Pb = 0$ and $\varphi(t, b)$ is a k-periodic solution of (2.13), then every k-periodic solution is given by

$$z(t) = \beta \dot{p}(t) + \varphi(t, b)$$

where $\beta \in \mathbb{R}$. To have $z(0)$ orthogonal to $\dot{p}(0)$, we choose

$$\beta |\dot{p}(0)|^2 = -\dot{p}(0) \cdot \varphi(0, b)$$

This uniquely defines $\beta = \beta(b)$. If we define

$$(Kb)(t) = \beta(b)\dot{p}(t) + \varphi(t, b), \qquad b \in (I - P)\mathscr{P}_k$$

then $K: (I - P)\mathscr{P}_k \to \mathscr{P}_k$ is a continuous linear operator.

With this definition of K, Eq. (2.11) will have a k-periodic solution with $z(0)$ orthogonal to $\dot{p}(0)$ if and only if

$$(2.14) \quad \begin{array}{ll} \text{(a)} & z = K(I - P)F(\cdot, z, \mu, \alpha) \\ \text{(b)} & PF(\cdot, z, \mu, \alpha) = 0 \end{array}$$

An application of the Implicit Function Theorem to (2.14a) gives a neighborhood $U \subseteq \mathscr{P}_k$ of zero and a neighborhood $V \subseteq \mathbb{R}^2$ of $\mu = 0$ and a unique C^r function $z^*(\alpha, \mu)$ in U satisfying (2.14a) for $\mu \in V$, $0 \leq \alpha \leq k$, $z^*(\alpha, 0) = 0$ for all α. Thus, the existence of k-periodic solutions of the original equation (2.2) is equivalent to finding (α, μ) which satisfy the bifurcation equation

$$(2.15) \quad B(\alpha, \mu) \stackrel{\text{def}}{=} \int_0^k q(t) \cdot F(t, z^*(\alpha, \mu)(t), \mu, \alpha)\, dt = 0.$$

With $h(\alpha) = h_k(\alpha)$ defined in (2.7), the function $B(\alpha, \mu)$ satisfies $B(\alpha, 0) = 0$,

$$(2.16) \quad B(\alpha, \mu) = \mu \cdot h(\alpha) + \tilde{B}(\alpha, \mu)$$

where $\tilde{B}(\alpha, \mu) = O(|\mu|^2)$ as $|\mu| \to 0$. For any $\mu \neq 0$, $\mu = \beta\zeta$, $\zeta \in \mathbb{R}$, $\beta \in \mathbb{R}^2$, $|\beta| = 1$ finding solutions of (2.15) is equivalent to finding solutions of the equation

$$(2.17) \quad C(\alpha, \beta, \zeta) \stackrel{\text{def}}{=} \beta \cdot h(\alpha) + C_0(\alpha, \beta, \zeta) = 0$$

where $C_0(\alpha, \beta, \zeta) = \tilde{B}(\alpha\beta, \zeta)/\zeta$ is a C^{r-1}-function of $\alpha \in \mathbb{R}$, $\beta \in \mathbb{R}^2$, $|\beta| = 1$, $\zeta \in \mathbb{R}$, satisfying $C_0(\alpha, \beta, 0) = 0$.

For $\zeta = 0$, the only possible solutions of (2.17) are those β_0, α_0 such that $\beta_0 \cdot h(\alpha_0) = 0$. For any $\delta > 0$, let $H_\delta = \{\beta : \beta h(\alpha_M) \geq \delta, \beta h(\alpha_m) \leq -\delta\}$. This set is nonempty for δ small. Then there is an $\varepsilon > 0$ such that, for any $\beta \in H_\delta$, the set $\{\beta \cdot h(\alpha), \alpha \in [0, 1]\}$ contains the interval $(-\varepsilon, \varepsilon)$. This implies there is an $\eta > 0$, such that, for any $\beta \in H_\delta$, $|\zeta| < \eta$, the set $\{C(\alpha, \beta, \zeta), \alpha \in [0, 1)\}$ contains the interval $(-\varepsilon/2, \varepsilon/2)$. Thus, for any $\beta \in H_\delta$, $|\zeta| < \eta$, there is an $\alpha = \alpha(\beta, \zeta)$ such that $C(\alpha, \beta, \zeta) = 0$. It may also happen that the set $\tilde{H}_\delta = \{\beta : \beta h(\alpha_M) \leq -\delta, \beta h(\alpha_m) \geq \delta\}$ is not empty (this is the case in Figure 2.1). The same argument as above shows there are $\varepsilon > 0$, $\eta > 0$ such that, for any $\beta \in \tilde{H}_\delta$, $|\zeta| < \eta$, there is an $\alpha = \alpha(\beta, \zeta)$ such that $C(\alpha, \beta, \zeta) = 0$. Therefore, it remains only to discuss a small neighborhood of the points α_m, α_M.

Let α_0 be either α_m or α_M, $\beta_0 \cdot h(\alpha_0) = 0$ and $\beta_0 \cdot h'(\alpha_0) = 0$. From Hypothesis (H$_3$), we have $\beta_0 \cdot h''(\alpha_0) \neq 0$. Thus, the Implicit Function Theorem implies there is a $\delta > 0$ and a unique C^{r-1}-function $\alpha^*(\beta, \zeta)$, defined for $|\beta - \beta_0| < \delta$, $|\zeta| < \delta$, satisfying $\partial C(\alpha^*(\beta, \zeta), \beta, \zeta)/\partial\alpha = 0$, $\alpha^*(\beta_0, 0) = \alpha_0$. Thus, $C(\alpha, \beta, \zeta)$ has a unique maximum or minimum $Q(\beta, \zeta)$. But, one can easily show that $Q(\beta_0, 0) = 0$, $\partial Q(\beta_0, 0)/\partial\beta = \gamma_0 \cdot h(\alpha_0)$ where $\gamma_0 \cdot \beta_0 = 0$. Since $h(\alpha_0) \neq 0$ and $\beta_0 \cdot h(\alpha_0) = 0$, it follows that $\partial Q(\beta_0, 0)/\partial\beta \neq 0$. The Implicit Function Theorem implies there is a $\delta(\beta_0) > 0$ and a function $\beta^*(\zeta)$, $|\beta^*(\zeta)| = 1$, $\beta^*(0) = \beta_0$ such that $Q(\beta^*(\zeta), \zeta) = 0$, $0 \leq \zeta < \delta(\beta_0)$. The set $\{\mu : \mu = \beta^*(\zeta)\zeta,$

$0 < \zeta < \delta(\beta_0)\}$ describes a manifold through the origin in $\mathbb{R}^m$ with the property that there are no solutions of (2.17) on one side of this manifold and two on the other. This is a bifurcation surface and clearly satisfies the properties stated in Theorem 2.1. This completes the proof. $\square$

Corollary 2.4. *Suppose the hypothesis of Theorem 2.1 are satisfied and $\mu_0 \neq 0$ belongs to one of the bifurcation surfaces C_m^k or C_M^k. Then there is a neighborhood U of Γ, a neighborhood $W = W(\mu_0)$ of zero in the space $\mathscr{F}^r$, $r \geq 2$, of C^k-vector fields $F(t, x)$ of period 1 in t and a function $\delta_k : W \to \mathbb{R}$ such that the system*

$$(2.18) \qquad \dot{x} = f(x) + g(t, x, \mu_0) + F(t, x)$$

satisfies one of the following three alternatives for $F \in W$:

(i) $\delta_k(F) > 0 \Rightarrow$ *no k-periodic solutions in Γ*
(ii) $\delta_k(F) = 0 \Rightarrow$ *exactly k k-periodic solutions in Γ*
(iii) $\delta_k(F) < 0 \Rightarrow$ *exactly $2k$ k-periodic solutions in Γ*

Proof. This is a consequence of the proof of Theorem 2.1. The function F plays the role of another parameter and the bifurcation function in (2.16) becomes

$$B(\alpha, \mu, F) = \mu \cdot h(\alpha) + \tilde{B}(\alpha, \mu, F)$$

Since $\mu_0 \neq 0$, the same proof as before can be applied to $B(\alpha, \mu, F)$ directly for $\mu - \mu_0$ and F small. $\square$

The next result is concerned with the limit points of a k-periodic solution of (2.2) as $\mu \to 0$ along some curve in the region S_2^k in Theorem 2.1 where k-periodic solutions exist. It is convenient to have some notation and a definition. If there is a continuous function $\mu(\beta)$, $0 \leq \beta \leq 1$, such that $\mu(\beta) \in S_2^k$, $0 < \beta \leq 1$, $\mu(0) = 0$ and $\gamma = \{\mu(\beta), 0 < \beta \leq 1\}$, then we say γ is a curve in S_2^k *approaching zero*. Now suppose the parameter space μ is $\mathbb{R}^2$. If $\bar{\gamma}$ is the closure of γ, then, for any $r > 0$, we can find a smallest cone $K_{\gamma, r}$ containing $\bar{\gamma} \cap \{\mu : |\mu| \leq r\}$. Define $K_{\gamma, 0} = \bigcap_{r > 0} K_{\gamma, r}$. Also, for any closed cone $K \subset \mathbb{R}^2$, let $K^{\perp} = \{h \in \mathbb{R}^2 : \text{there is a } \mu \in K \text{ with } \mu \cdot h = 0\}$.

With this notation, we can state

Theorem 2.5. *Suppose $m = 2$ in Theorem 2.1, $\gamma = \{\mu(\beta) : 0 < \beta \leq 1\}$ is a continuous curve in S_2^k approaching zero. If $x(t, \beta)$ is a k-periodic solution of (2.2), continuous in β, $0 < \beta \leq 1$, $x(t, \beta) \in V$, a bounded neighborhood of Γ_k, and*

$$\mathscr{S}(\gamma) = \{x(\cdot, \beta) \in \mathscr{P}_k : 0 < \beta \leq 1\}$$

then $\mathscr{S}(\gamma)$ is precompact in $\mathscr{P}_k$ and every limit point of $\mathscr{S}(\gamma)$ as $\beta \to 0$ is a k-periodic solution of (2.1). Furthermore, there is a closed interval $I(\gamma) \subset [0, 1]$

such that $K(h_k(I(\gamma)))^{\perp} = K_{\gamma,0}$, where $K(h_k(I(\gamma)))$ is the smallest cone containing the closed curve $\{h_k(\alpha): \alpha \in I(\gamma)\}$ and

$$\mathrm{cl}(\mathscr{S}(\gamma) \backslash \mathscr{S}(\gamma)) = \{p_k(\cdot + \alpha): \alpha \in I(\gamma)\}.$$

As a consequence of Theorem 2.5, for any $\alpha \in I(\gamma)$, there is a $\beta_j \to 0$ as $j \to \infty$ such that $x(t, \beta_j) - p(t + \alpha) \to 0$ as $j \to \infty$ uniformly in t. If the interval $I(\gamma)$ consists of more than one point, then $\mathscr{S}(\gamma)$ consists of more than one point and the solutions $x(t, \beta)$ are not continuous at $\beta = 0$. Since $I(\gamma)$ consisting of only one point is equivalent to $K_{\gamma,0}$ consisting of only one vector which is equivalent to either the ratio μ_1/μ_2 or μ_2/μ_1 approaching a limit as $\mu \to 0$ along γ, we have the following interesting corollary.

Corollary 2.6. *If the conditions of Theorem 2.5 are satisfied, then $x(\cdot, \beta)$ is continuous at $\beta = 0$ if and only if either μ_1/μ_2 or μ_2/μ_1 converge to some limit as $\mu \to 0$ along γ. In this case, there is an $\alpha_0 \in [0, 1)$ such that $x(\cdot, \beta) - p_k(\cdot + \alpha_0) \to 0$ as $\beta \to 0$.*

Proof of Theorem 2.5. We again drop the subscript k in the proof. The compactness of $\mathscr{S}(\gamma)$ is a consequence of Ascoli's theorem. The fact that the limit points of $\mathscr{S}(\gamma)$ as $\beta \to 0$ satisfy (2.1) is easily verified from the equivalent integral equation of (2.2). To prove the last part of the theorem, we note that the solution $x(t, \beta)$ can be obtained by the method of Liapunov–Schmidt given in the proof of Theorem 2.1. Thus, there are continuous functions $\alpha(\beta)$, $z(t, \beta)$ such that

$$x(t, \beta) = p(t + \alpha(\beta)) + z(t + \alpha(\beta), \beta)$$

where $z(t, \beta) = z^*(\alpha(\beta), \mu(\beta))(t)$ where $z^*(\alpha(\beta), \mu(\beta))$ satisfies (2.14)(a) with (α, μ) replaced by $(\alpha(\beta), \mu(\beta))$ and this latter vector must satisfy the bifurcation equation

$$B(\alpha(\beta), \mu(\beta)) = \mu(\beta) \cdot h(\alpha(\beta)) + \tilde{B}(\alpha(\beta), \mu(\beta)) = 0.$$

All of the assertions in the theorem are an easy consequence of this last equation. The theorem is proved. $\square$

It remains to discuss the stability of the k-periodic solutions for μ in S_2^k. The simplest case is when the periodic orbit Γ_k of (2.1) has one characteristic multiplier $\neq 1$.

Theorem 2.8. *Suppose S_2^k, U are defined in Theorem 2.1 and the periodic orbit Γ_k of (2.1) has a characteristic multiplier $\neq 1$. Assume there are only a finite number of points $\alpha_j \in [0, 1)$ for which there is a $\mu_{0j} \neq 0$ with $\mu_{0j} \cdot h_k(\alpha_j) = 0$, $\mu_{0j} \cdot h_k'(\alpha_j) = 0$. At each of these points, assume $\mu_{0j} \cdot h_k''(\alpha_j) \neq 0$. Then there* *are at least 2k subharmonics of order k of (2.2) in U for any $\mu \in S_2^k$ near a bifurcation curve, at least k are saddles and at least k are either nodes or foci for μ near a bifurcation curve.*

Proof. Suppose $\mu_0 \neq 0$ is on one of the bifurcation curves C_m, C_M in Theorem 2.1. In a sufficiently small neighborhood W of μ_0, there are exactly $2k$ subharmonics of (2.2) for $\mu \in W \cap S_2^k$, all obtained from two solutions $u_1(t)$, $u_2(t)$ by the translation $t \to t + j, j = 0, 1, 2, \ldots, k$. The solutions u_1, u_2 must be hyperbolic. Otherwise, we could perturb (2.2) by a small C^r-vector field $F(t, x)$ to obtain more than two solutions of Eq. (2.18). This contradicts the assertions in Corollary 2.4. These solutions u_1, u_2 bifurcate from a k-periodic solution v of (2.2) for $\mu = \mu_0$. Since we are assuming one characteristic multiplier of p for (2.1) is not one, it follows that one characteristic exponent of each of the functions v, u_1, u_2 is not zero. Since v does bifurcate to u_1 and u_2, the other characteristic exponent of v must be zero. In a neighborhood of μ_0, we have the situation described in Section 9.5. Applying the results there, the existence and stability properties of the periodic solutions are determined by a scalar autonomous equation

$$(2.19) \qquad\qquad \dot{a} = G(a, \mu).$$

For $\mu \in W \cap S_2^k$, this equation must have exactly two solutions which are hyperbolic since u_1, u_2 are. Thus one solution of (2.19) is asymptotically stable and one in unstable. This implies that we can take u_1 to be a saddle and u_2 to be a node. This proves the assertion for μ near a bifurcation curve. This completes the proof of the theorem. $\square$

For the case in which both multipliers of the orbit Γ_k of (2.1) are one, the stability question is more difficult and depends more specifically upon the way the parameters μ enter into the equation. To keep the discussion as simple as possible, we therefore consider only Eq. (2.3).

Theorem 2.9. *Suppose there are only a finite number of $\alpha_j \in [0, 1)$ such that $G_k(\alpha)$ in (2.9) satisfies $G_k'(\alpha_j) = 0$ and, at each of these points, suppose $G_k''(\alpha_j) \neq 0$. Then the conclusion of Theorem 2.8 holds for Eq. (2.3).*

Proof. Suppose $\mu_0 \neq 0$ is on one of the bifurcation curves C_m, C_M and $p(t, \mu_0)$ is a corresponding k-periodic solution. Then the special form of (2.3) implies that the characteristic multipliers λ_1^0, λ_2^0 of $p(t, \mu_0)$ satisfy $\lambda_1^0 \lambda_2^0 = \exp(-k\mu_{10})$, where $\mu_0 = (\mu_{10}, \mu_{20})$. Furthermore, $\mu_0 \neq 0$ implies $\mu_{10} \neq 0$. One multiplier is one since μ_0 is a bifurcation point. Thus, the other is $\neq 1$. The proof is completed as in the proof of Theorem 2.8. $\square$

Remark 2.10. The argument in the proof of Theorem 2.8 gives only stability properties near a bifurcation curve because the solutions can possibly bifurcate to other solutions; for example, to subharmonics of order $2k, 4k, \ldots$, in the interior of S_2^k.

Remark 2.11. The sectors S_1^k, S_2^k in Theorem 2.1 and Corollary 2.3 exist if it is required only that the functions $h_k(\alpha)$, $G_k(\alpha)$ are not the zero function. The Hypotheses (H_3), (H_4) are imposed to obtain a precise characterization of the boundaries of these sectors (see the proofs).

11.3. Homoclinic Orbit

In this section, we consider Eq. (2.1) under the assumption that

$$(3.1) \qquad\qquad f(0) = 0, \qquad \det[\partial f(0)/\partial x] < 0,$$

that is, zero is a saddle point of (2.1). We also assume the stable and unstable manifolds of 0 intersect at a point $p_0 \neq 0$; that is, there is a homoclinic orbit. If $0(p_0)$ is the orbit of (2.1) through p_0, we let

$$(3.2) \qquad\qquad \Gamma_\infty = 0(p_0) \cup \{0\}.$$

Along with Eq. (2.1), we consider the 1-periodic perturbed equation (2.2) with the objective being to study some of the properties of the solutions of (2.2) near $\Gamma_\infty \times \{0\}$. The case of autonomous perturbations was discussed in detail in Section 10.3 under the hypothesis that

$$(3.3) \qquad\qquad \sigma_0 \overset{\text{def}}{=} \operatorname{tr} \partial f(0)/\partial x \neq 0.$$

The conclusion was that one of three possibilities could occur for μ in a neighborhood of $\mu = 0$: either every orbit left a neighborhood for both increasing and decreasing time or there was a homoclinic orbit with strong stability properties or there was a unique limit cycle with strong stability properties.

For those μ for which a limit cycle occurs, the period $w(\mu) \to \infty$ as $\mu \to 0$. In particular, it must take on arbitrarily large integer values k as $\mu \to 0$. We have seen in the previous section that subharmonic bifurcation can occur near k-periodic orbits when 1-periodic perturbations are allowed in the vector field. If the auxiliary conditions in Theorem 2.1 for subharmonic bifurcation were to be satisfied uniformly as $\mu \to 0$, then a very complicated behavior is to be expected for (x, μ) near $\Gamma_\infty \times \{0\}$.

The complicated behavior is actually a consequence of a much more fundamental concept which we now describe. Since $x = 0$ is a saddle point of (2.1) and $g(t, x, 0) = 0$, there is a unique 1-periodic solution $\varphi(t, \mu)$ of (2.2) in a neighborhood of $x = 0$, $\mu = 0$, this solution is C^r in μ, $\varphi(t, 0) = 0$. Let $\gamma_\mu = \{(t, \varphi(t, \mu)), t \in \mathbb{R}\} \subseteq \mathbb{R} \times \mathbb{R}^2$ be the trajectory defined by this solution.

For this trajectory γ_μ of (2.2) and μ in a neighborhood of zero, there are a local stable manifold S_μ in $\mathbb{R} \times \mathbb{R}^2$ and an unstable manifold U_μ in $\mathbb{R} \times \mathbb{R}^2$ which are C^r in μ. These manifolds can be defined globally through the differential equation and have the property that the set

$$(3.4) \qquad \begin{aligned} S_\mu(t) &= \{x:(t,x) \in S_\mu\} \\ U_\mu(t) &= \{x:(t,x) \in U_\mu\} \end{aligned}$$

are periodic in t of period 1.

Definition 3.1. Let $p_\mu = \phi(0,\mu)$ be the initial value of the 1-periodic solution $\phi(t,\mu)$. A point $q \in \mathbb{R}^2$ is *homoclinic to* p_μ if $q \neq p_\mu$, $q \in S_\mu(0) \cap U_\mu(0)$. A point q is *transverse homoclinic to* p_μ if it is homoclinic to p_μ and $S_\mu(0)$, $U_\mu(0)$ intersect transversally at q.

If there is one point q homoclinic to p_μ, then there must be infinitely many since $x(k,q) \in S_\mu(0) \cap U_\mu(0)$ for $k = 1, 2, \ldots$ and $x(k,q,\mu) \to p_\mu$ as $k \to \pm\infty$, where $x(t,q,\mu)$ is the solution of (2.2) through q at $t = 0$. In fact, continuity of solutions of (2.2) with respect to initial data implies the sets $S_\mu(0)$ and $U_\mu(0)$ must have the complicated behavior shown in Fig 3.1. We have only drawn the types of intersections that occur with decreasing integers k. A similar phenomena occurs for increasing integers k.

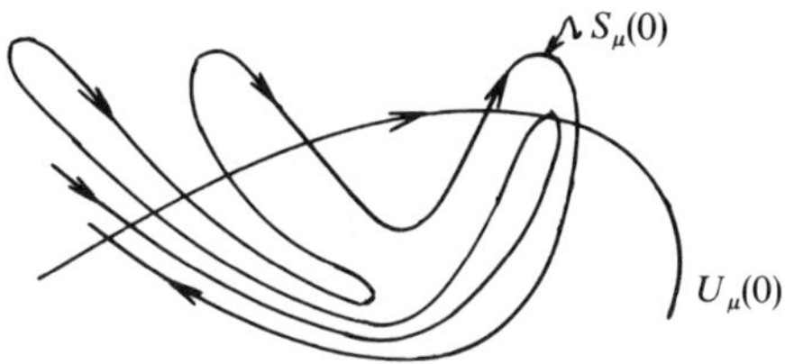

Figure 3.1

Our first objective is to give necessary and sufficient conditions on (2.2) in order that there be a homoclinic point in a neighborhood of $\{0\} \times \Gamma_\infty$. From the definition, we must determine a solution x of (2.2) which remains in a neighborhood of Γ_∞ for μ near zero with the property that $x(t) - \varphi(t,\mu) \to 0$ as $t \to \pm\infty$. The trajectory γ_μ being a saddle implies that any solution which remains in a small neighborhood of $\{0\} \times \mathbb{R}$ for $t \to \infty$ ($t \to -\infty$) must lie on the stable (unstable) manifold. Therefore, we need only look for solutions which remain in a neighborhood of $\mathbb{R} \times \Gamma_\infty$ for μ near zero.

Some preliminary results are needed. Let $\Gamma_\infty = \{p_\infty(t); t \in (-\infty, \infty)\} \cup \{0\}$ where $p_\infty(t)$ is a solution of (2.1). Since $p_\infty(t)$ lies on the stable and unstable manifolds of the saddle point 0 of (2.1), it follows that $p_\infty(t)$, $\dot{p}_\infty(t)$ approach zero exponentially at $t \to \pm\infty$. Also, $\dot{p}_\infty(t)$ is a vector tangent to these manifolds. If $-\gamma < 0, \eta > 0$ are the eigenvalues of $\partial f(0)/\partial x$, then known properties of solutions of a differential equation near a saddle point imply that there are nonzero vectors $c \in \mathbb{R}^2$, $d \in \mathbb{R}^2$ such that

$$(3.5) \qquad \lim_{t \to \infty} \dot{p}_\infty(t) e^{\gamma t} = c, \qquad \lim_{t \to -\infty} \dot{p}_\infty(t) e^{-\eta t} = d.$$

If

$$(3.6) \qquad \sigma_0 \overset{\text{def}}{=} \text{tr} \, \partial f(0)/\partial x = -\gamma + \eta$$

and

$$(3.7) \qquad \beta(t) = \exp\left(-\int_0^t \text{tr}[\partial f(p_\infty(s))/\partial x] \, ds\right)$$

then the following limits exist and are not zero

$$(3.8) \qquad \lim_{t \to \pm \infty} e^{\sigma_0 t} \beta(t) \neq 0.$$

Let $B(\mathbb{R}) = \{F: \mathbb{R} \to \mathbb{R}^2, \text{ bounded, continuous}\}$ with $|F| = \sup_{t \in \mathbb{R}} |F(t)|$ for $F \in B(\mathbb{R})$. For any $x = (x_1, x_2)$ in $\mathbb{R}^2$, let $x^\perp = (-x_2, x_1)$. From (3.5), (3.8), we have

$$|\beta(t)\dot{p}_\infty^\perp(t)| \leq (\text{constant}) \, e^{-\eta t}, \qquad t \geq 0$$
$$|\beta(t)\dot{p}_\infty^\perp(t)| \leq (\text{constant}) \, e^{\gamma t}, \qquad t \leq 0$$

Thus, $\beta \dot{p}_\infty^\perp \in B(\mathbb{R})$, approaches zero exponentially as $t \to \pm \infty$ and we can define the continuous projection P on $B(\mathbb{R})$ by

$$(3.9) \qquad PF = \frac{1}{b} \beta \dot{p}_\infty^\perp \int_{-\infty}^{\infty} \beta \dot{p}_\infty^\perp \cdot F, \qquad b = \int_{-\infty}^{\infty} \beta^2 |\dot{p}_\infty|^2$$

where $x \cdot y$ denotes the inner product of two vectors and $x \cdot x = |x|^2$.

Lemma 3.2. *With the notation above and $F \in B(\mathbb{R})$, the equation*

$$(3.10) \qquad \dot{x} = A(t)x + F(t), \qquad A(t) = \partial f(p_\infty(t))/\partial x$$

has a solution in $B(\mathbb{R})$ if and only if $PF = 0$. If $PF = 0$, there is a unique solution $KF \in B(\mathbb{R})$ with initial value orthogonal to $\dot{p}(0)$ and $K: (I - P)B(\mathbb{R}) \to B(\mathbb{R})$ is continuous and linear.

Proof. In the proof, we drop the subscript ∞. The function $\dot{p}(t)$ is a solution of

$$(3.11) \qquad \dot{y} = A(t)y$$

Let $q(t)$ be a solution of (3.11) such that

$$\det[\dot{p}(0), q(0)] = 1$$

Then the matrix $[\dot{p}(t), q(t)]$ is a fundamental matrix of solutions of (3.11) with determinant given by the function $[\beta(t)]^{-1}$. From (3.5) and (3.9), it follows that the following limits exist and are not zero,

$$(3.12) \qquad \lim_{t \to \infty} e^{-\eta t} q(t) \neq 0, \qquad \lim_{t \to -\infty} e^{\gamma t} q(t) \neq 0.$$

Also, there is a constant K such that, for every $x \in \mathbb{R}^2$,

$$(3.13) \qquad \begin{aligned} |\beta(s)q^{\perp}(s) \cdot x\dot{p}(t)| &\leq K e^{-\gamma(t-s)}|x|, & t \geq s \geq 0, \\ |\beta(s)q^{\perp}(s) \cdot x\dot{p}(t)| &\leq K e^{-\eta(s-t)}|x|, & t \leq s \leq 0. \end{aligned}$$

For the nonhomogeneous equation (3.10), the variation of constants formula is

$$(3.14) \qquad \begin{aligned} y(t) &= \left(A + \int_0^t \beta(s)\dot{p}^{\perp}(s) \cdot F(s)\,ds \right) q(t) \\ &+ \left(B - \int_0^t \beta(s)q^{\perp}(s) \cdot F(s)\,ds \right) \dot{p}(t) \end{aligned}$$

where A, B are real constants.

From (3.12), (3.13), (3.14), a solution of (3.10) is bounded on $(0, \infty)$ if and only if $A = -\int_0^{\infty} \beta\dot{p}^{\perp} \cdot F$, and a solution is bounded on $(-\infty, 0]$ if and only if $A = \int_{-\infty}^0 \beta\dot{p}^{\perp} \cdot F$. Thus, a solution is in $B(\mathbb{R})$ if and only if $\int_{-\infty}^{\infty} \beta\dot{p}^{\perp} \cdot F = 0$; that is $PF = 0$. When $PF = 0$, the solution is given by

$$(3.15) \qquad y(t) = \left(B - \int_0^t \beta q^{\perp} \cdot F \right) \dot{p}(t) + \left(\int_{\infty}^t \beta\dot{p}^{\perp}F \right) q(t).$$

To have $y(0)$ orthogonal to $\dot{p}(0)$, we must have

$$B|\dot{p}(0)|^2 = \left(\int_0^{\infty} \beta\dot{p}^{\perp} \cdot F \right) \dot{p}(0) \cdot q(0).$$

This defines B uniquely as a continuous linear function on $B(\mathbb{R})$. From (3.14) and our estimates, one observes that the corresponding solution KF is continuous and linear in F. This proves the lemma. $\square$

Let us now apply this lemma to determine those values of μ near $\mu = 0$ for which Eq. (2.2) has a homoclinic point in a neighborhood of the original homoclinic orbit Γ_∞ for $\mu = 0$. If

$$(3.16) \qquad x(t) = p_\infty(t + \alpha) + z(t + \alpha)$$

in (2.1) and $t + \alpha$ is then replaced by t, we obtain

$$(3.17) \qquad \begin{aligned} \dot{z} &= A(t)z + f(p_\infty + z) - f(p_\infty) - A(t)z + g(t - \alpha, p_\infty + z, \mu) \\ &\overset{\text{def}}{=} A(t)z + F(t, z, \mu, \alpha). \end{aligned}$$

Since $p_\infty(t) \to 0$ as $t \to \pm\infty$ and $p_\infty(0) \neq 0$, finding a homoclinic point for (2.2) for (x, μ) near $\Gamma_\infty \times \{0\}$ is equivalent to finding a solution $z(t)$ of (3.17) which remains in a small neighborhood of zero for all t and $z(t) - \varphi(t, \mu)$ approaches zero as $t \to \pm\infty$ for μ in a small neighborhood of zero. But, as remarked earlier, the fact that the trajectory γ_μ of $\varphi(t, \mu)$ is a saddle implies that any solution which remains in a small neighborhood of zero for $t \geq 0$ must lie on the stable manifold of γ_μ. Similarly, for $t \leq 0$ and the unstable manifold. Therefore, it is sufficient to consider small bounded solutions of (3.17). The parameter α designates the approximate points on Γ_∞ where $S_\mu(0) \cap U_\mu(0) \neq \varnothing$.

The method of Liapunov–Schmidt can now be applied in the usual way by using Lemma 3.2. In fact, in $B(\mathbb{R})$, Eq. (3.17) is equivalent to

$$(3.18) \quad \begin{aligned} &\text{(a)} & PF(\cdot, z, \mu, \alpha) &= 0 \\ &\text{(b)} & z &= K(I - P)F(\cdot, z, \mu, \alpha) \end{aligned}$$

An application of the Implicit Function Theorem implies there is a $\delta > 0$ and a unique solution $z^*(\alpha, \mu)$ of (3.18b) for $|z| < \delta$, $|\mu| < \delta$, $z^*(\alpha, \mu)$ has continuous derivatives up through order 2 and $z^*(\alpha, 0) = 0$. Using the definition of P, we have proved

Theorem 3.3. *Suppose $z^*(\alpha, \mu)$ is the solution of (3.18b). Then there is a $\delta > 0$ and a neighborhood V of the homoclinic orbit $\Gamma_\infty = \{p_\infty(t); t \in (-\infty, \infty)\} \cup \{0\}$ of (2.1) such that, for $|\mu| < \delta$, there is a homoclinic point of (2.2) in V if and only if there is a solution $x(t, \mu)$ of (2.2) in V, $t \in (-\infty, \infty)$, $x(t, \mu) - \varphi(t, \mu) \to 0$ as $t \to \pm\infty$ and this latter statement is true if and only if $x(t - \alpha, \mu) = p_\infty(t) + z^*(\alpha, \mu)(t)$ and μ, α satisfy*

$$(3.19) \qquad B(\alpha, \mu) \overset{\text{def}}{=} \int_{-\infty}^{\infty} \beta(t)\dot{p}_\infty^\perp(t) \cdot F(t, z^*(\alpha, \mu)(t), \mu)\, dt = 0$$

where F is defined in (3.17), $\beta(t) = \exp(-\int_0^t \mathrm{tr}[\partial f(p_\infty(s)/\partial x]\, ds)$, $\dot{p}^\perp = (-\dot{p}_2, \dot{p}_1)$.

Remark 3.4. Theorem 3.3 asserts that the points $q = p_\infty(\alpha_0) + z^*(\alpha_0, \mu_0)(0)$ are homoclinic to the point $\phi(0, \mu_0)$ if and only if $B(\alpha_0, \mu_0) = 0$. It is also true that q is transverse homoclinic to $\phi(0, \mu_0)$ if and only if $B(\alpha_0, \mu_0) = 0$

and $\partial B(\alpha_0, \mu_0)/\partial\alpha \neq 0$. In fact, if $\partial B(\alpha_0, \mu)/\partial\alpha \neq 0$ and q is not transverse homoclinic to $\phi(0, \mu_0)$, then one could perturb the vector field in (2.2) by a small function h to obtain a new function $\tilde\phi(t, \mu, h)$ and a new function $\tilde B(\mu, h, \alpha)$ whose zeros yield points homoclinic to $\phi(0, \mu, h)$. But $\tilde B(\mu, h, \alpha)$ will have a unique solution near the original α_0, μ_0; that is, there is always a point homoclinic to $\tilde\phi(0, \mu, h)$ for μ near μ_0 and h near zero. On the other hand, if the intersection of the stable and unstable manifolds were not transverse, we could choose h so that no homoclinic point exists, which is a contradiction. If the stable and unstable manifolds intersect at a point $q = p_\infty(\alpha) + z^*(\alpha, \mu)(0)$ which is transverse to $\phi(0, \mu_0)$, $B(\alpha_0, \mu_0) = 0$ and $\partial B(\alpha_0, \mu_0)/\partial\alpha = 0$, then one could perturb the vector field in (2.2) to eliminate the intersection which would be a contradiction.

Let $\mathscr{F}^r = \{f : R \times \mathbb{R}^2 \to \mathbb{R}^2 \text{ continuous}; f(t, x) \text{ is } C^r \text{ in } x, \text{ 1-periodic in } t\}$ and put the C^r topology on $\mathscr{F}^r$. We can now state the following

Corollary 3.5. *If Eq. (2.1) satisfies the condition above, then there are neighborhoods U of Γ, V of f in $\mathscr{F}^r$ and a C^r-function $\gamma_\infty : \mathbb{R} \times V \to \mathbb{R}$ such that, for $F \in V$, the equation*

$$(3.20) \qquad\qquad \dot x = F(t, x)$$

has a homoclinic point in U if and only if there is an $\alpha \in \mathbb{R}$ such that $\gamma_\infty(\alpha, F) = 0$. Furthermore, the set $\{F \in V : \gamma_\infty(\alpha, F) = 0\}$ is a submanifold of codimension 1 for each $\alpha \in \mathbb{R}$.

Proof. For any $F \in \mathscr{F}^r$, let $\mu = F - f$, $g(t, x, \mu) = \mu(t, x)$. A neighborhood of f in $\mathscr{F}^r$ is the same as a neighborhood of $\mu = 0$ in $\mathscr{F}^r$. Now apply Theorem 3.3 to obtain the function $B(\alpha, \mu)$ in (3.19). Theorem 3.3 was proved only for $\mu \in \mathbb{R}^m$, but the same proof is valid for μ in a Banach space. The function $\gamma(\alpha, F) = B(\alpha, F - f)$ satisfies the first property stated in the corollary. To prove the zeros of γ for fixed α form a submanifold of codimension one, we make a special perturbation of Eq. (2.1). Fix $T > 0$ and choose a tubular neighborhood V_T of that part of Γ described by $\{p_\infty(\theta), \theta \in [-(T+1), T+1]\}$ in the form $x = p_\infty(\theta) + \eta \dot p_\infty^\perp(\theta)/|\dot p_\infty^\perp(\theta)|$ where $\eta \in \mathbb{R}$, $0 \leq |\eta| \leq \delta$. For δ fixed and small, θ is uniquely determined by x. Suppose $h(\varepsilon)$ is C^1, $h(0) = 0$, $h'(0) = 1$, and define $g(x, \varepsilon) = 0$ outside V_T, $g(x, \varepsilon) = h(\varepsilon)\psi(\eta)\dot p^\perp(\theta)$, $\theta \in [-T, T]$, where ψ is any C^∞ function, positive on $(-\delta, \delta)$ and zero otherwise. For $\theta \in [-(T+1), -T] \cup [T, T+1]$ extend the definition of g so that it is continuously differentiable, $g(x, \varepsilon) = h(\varepsilon, \theta)\dot p^\perp(\theta)$, and $h(\varepsilon, \theta)$ decreasing (increasing) in θ for $\theta > 0$ $(\theta < 0)$. For this perturbation, one obtains the bifurcation function G as a function of ε with

$$\left.\frac{dG}{d\varepsilon}\right|_{\varepsilon=0} \geq \int_{-T}^{T} \eta^*(\alpha, \mu)(t)\beta(t)|\dot p_\infty^\perp(t)|^2 \, dt > 0,$$

where $\eta^*(\alpha, \mu)(t)$ is defined by $z^*(\alpha, \mu)(t) = \eta^*(\alpha, \mu)(t)\dot{p}_\infty^\perp(t)/|\dot{p}_\infty(t)|$. This proves the corollary. $\square$

Corollary 3.5 applies equally as well to the autonomous equation. In this case, we can always take $\alpha = 0$ in the theorem. We state this separately because of its importance in the applications. The proof is the same as before.

Corollary 3.6. *Suppose* $r \geq 2$, $f \in C^r(\mathbb{R}^2, \mathbb{R}^2)$ *and Equation* (2.1) *satisfies the conditions above. Then there are neighborhoods* U *of* Γ, V *of* f *in* $C^r(\mathbb{R}^2, \mathbb{R}^2)$ *and a* C^r-*function* $\gamma_\infty : V \to \mathbb{R}$ *such that, for* $F \in V$, *the equation*

$$(3.21) \qquad\qquad\qquad\qquad \dot{x} = F(x)$$

has a homoclinic orbit in U *if and only if* $\gamma_\infty(F) = 0$. *Furthermore, the set* $\{F \in V : \gamma_\infty(F) = 0\}$ *is a submanifold of codimension* 1.

If $V_0 = \{F \in V : \gamma_\infty(F) = 0\}$, $V_+ = \{F \in V : \gamma_\infty(F) > 0\}$, $V_- = \{F \in V : \gamma_\infty(F) < 0\}$, then $V = V_- \cup V_0 \cup V_+$. Without some further hypotheses on f in (2.1), it is possible to have $F_1 \in V_-$, $F_2 \in V_-$ and the flows defined by the corresponding differential equations (3.21) be topologically distinct in the neighborhood U of Γ. The number of components of topologically equivalent vector fields in V_- depends upon the stability properties of the orbit Γ of Equation (2.1); more precisely, the rate at which the Poincaré map induced by Γ converges or diverges to Γ. If the convergence or divergence is too slow, then the number of periodic orbits in the neighborhood U of Γ of a vector field F in V_- depends on F and not just on the fact that F belongs to V_-. See the discussion of the Example 10.3.5.

However, if $\sigma_0 = \mathrm{tr}\, \partial f(0)/\partial x \neq 0$, then we have seen in Theorems 10.3.2, 10.3.3 that only two possibilities arise if $F \in V \backslash V_0$; namely, there is a unique hyperbolic periodic orbit in U or every solution with initial value in U leaves U in both the forward and backward directions of time. By changing the sign of γ_∞^a if necessary, we can be sure that the first alternative occurs for $F \in V_-$ and the latter for $F \in V_+$. This result is summarized as follows.

Corollary 3.7. *Suppose the conditions of Corollary 3.5 are satisfied. If* $\sigma_0 = \mathrm{tr}\, \partial f(0)/\partial x \neq 0$, *then the following conclusions hold for Equation* (3.21) *for* $F \in V$:

(i) $\gamma_\infty^a(F) > 0$ *implies every solution leaves* U *in both forward and backward time.*

(ii) $\gamma_\infty^a(F) = 0$ *implies there is a homoclinic orbit in* U, *asymptotically stable for* $\sigma_0 < 0$ *and unstable for* $\sigma_0 > 0$.

(iii) $\gamma_\infty^a(F) < 0$ *implies there is a unique periodic orbit in* U, *asymptotically orbitally stable as* $t \to \infty$ $(t \to -\infty)$ *if* $\sigma_0 < 0$ $(\sigma_0 > 0)$.

To state the basic result on bifurcation to homoclinic points, we need some notation. For the 1-periodic function

$$h_\infty(\alpha) = \int_{-\infty}^{\infty} \beta(t)\dot{p}_\infty^\perp(t) \cdot \frac{\partial g}{\partial \mu}(t - \alpha, p_\infty(t), 0)\, dt$$

define the set

$$(3.22) \qquad P_{h_\infty} = \{\beta \in \mathbb{R}^m : \beta \cdot h_\infty(\alpha) = 0 \text{ for some } \alpha \in (0, 1)\}$$

and suppose that

(3.23)   The boundary (in the relative topology of S^{m-1}) of $P_{h_\infty} \cap S^{m-1}$
is the union of two disjoint geodesics L_∞^m, L_∞^M,

where S^{m-1} is the unit sphere in $\mathbb{R}^m$. If α_m, α_M are such that $h_\infty(\alpha_m), h_\infty(\alpha_M)$ are orthogonal respectively to L_∞^m, L_∞^M, suppose that

$$(3.24) \qquad \begin{aligned} \mu \cdot h_\infty(\alpha_m) &= 0, & \mu \cdot h_\infty'(\alpha_m) &= 0, & |\mu| = 1 &\Rightarrow \mu \cdot h_\infty''(\alpha_m) \neq 0 \\ \mu \cdot h_\infty(\alpha_M) &= 0, & \mu \cdot h_\infty'(\alpha_M) &= 0, & |\mu| = 1 &\Rightarrow \mu \cdot h_\infty''(\alpha_M) \neq 0 \end{aligned}$$

Theorem 3.8. *Suppose h_∞ satisfies (3.23), (3.24). Then there is a neighborhood U of Γ_∞, V of $\mu = 0$ and two C^{k-1} manifolds C_m^∞, $C_M^\infty \subseteq V$ of codimension one containing $\mu = 0$ whose tangent planes are orthogonal respectively to $h_\infty(\alpha_m), h_\infty(\alpha_M)$, at $\mu = 0$, $C_m^\infty \cap C_M^\infty = \{0\}$, C_m^∞, C_M^∞ divide $V \backslash \{0\}$ into two disjoint open sets S_1^∞, S_2^∞ such that Eq. (2.2) has no homoclinic points in U for μ in S_1^∞ and has a homoclinic point in U for μ in S_2^∞ (see Fig. 3.2).*

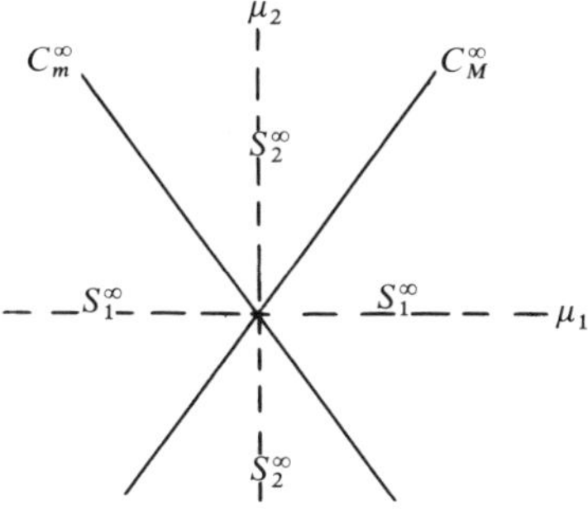

Figure 3.2

Proof. Using Theorem 3.3, the proof follows the same lines as the proof of Theorem 2.1 and will not be given.   □

Remark 3.9. If μ is infinite dimensional, the same comments as in Remark 2.2 apply.

As for subharmonic bifurcation, the results are easier to state for Eq. (2.3). Let

$$(3.25) \qquad G_\infty(\alpha) = \frac{1}{\eta_\infty} \int_{-\infty}^{\infty} \dot{p}_\infty(t) F(t - \alpha)\, dt$$

and suppose that the maximum (minimum) of $G_\infty(\alpha)$ occur in $[0, 1)$ at $\alpha_M(\alpha_m)$ and satisfy

$$(3.26) \qquad G''_\infty(\alpha_m) > 0, \qquad G''_\infty(\alpha_M) < 0.$$

Then a consequence of Theorem 3.8 is the following

Corollary 3.10. *Suppose G_∞ in (3.25) satisfies (3.26) and let $\Gamma_\infty = \{(\varphi_\infty(t), \dot{\varphi}_\infty(t)), t \in \mathbb{R}\} \cup \{(0,0)\}$ be the homoclinic orbit of Eq. (2.3) for $\mu = 0$, $\mu = (\mu_1, \mu_2)$. Then there is a neighborhood U of Γ_∞, V of $\mu = 0$ and two C^{k-1}-curves $C^\infty_m, C^\infty_M \subseteq V$ containing $\mu = 0$, tangent respectively to $\mu_1 = G_\infty(\alpha_m)\mu_2$, $\mu_1 = G_\infty(\alpha_M)\mu_2$, at $\mu = 0$, $C^\infty_m \cap C^\infty_M = \{0\}$, C^∞_m, C^∞_M divide $V\setminus\{0\}$ into two disjoint open sets S^∞_1, S^∞_2 such that Eq. (2.3) has no homoclinic points in U for μ in S^∞_1 and has one for μ in S^∞_2.*

The curves C^∞_m, C^∞_M are called the *curves of bifurcation* to homoclinic points. They have a very simple physical interpretation for Eq. (2.3). The parameter μ_1 represents the amplitude of the damping and μ_2 the amplitude of the forcing function. If $\mu_2 = 0$, the damping μ_1 breaks the intersection of the stable and unstable manifolds of zero and no homoclinic orbit exists. The forcing $\mu_2 F(t)$ has a tendency to restore this intersection. In fact, if (μ^0_1, μ^0_2) is in S^∞_2, then one can move along the line (μ^0_1, μ_2) increasing (or decreasing) μ_2 from μ^0_2 until one intersects either C^∞_m or C^∞_M. At this point, the stable and unstable manifolds of the saddle trajectory become tangent and intersect transversally for larger (or smaller) μ_2. The same remark holds by moving along (μ_1, μ^0_2).

In the next section, we discuss the nature of the solutions of (2.3) in a neighborhood of the bifurcation curves in more detail-especially the types of subharmonic bifurcations that occur.

We conclude this section with some further remarks on the general properties of mappings which possess transverse homoclinic points. No proofs will be given. For this discussion, it is convenient to use the *period map* for Eq. 2.2. If $x(t, p)$ is the solution of (2.2) with $x(0, p) = p$, let $\pi = \pi_\mu : \mathbb{R}^2 \to \mathbb{R}^2$ be the map $\pi p = x(1, p)$; that is, the solution of (2.2) at $t = 1$. The map π is a diffeomorphism and fixed points of π correspond to 1-periodic solutions of (2.2). For a given integer k, fixed points of π^k which are not fixed points of π^j for any $j < k$ are subharmonics of order k.

For Eq. (2.1), the homoclinic orbit $\Gamma_\infty = O(p_0) \cup \{0\}$ satisfies $\pi_0 \Gamma_\infty = \Gamma_\infty$. For μ small in Eq. (2.2) there is a 1-periodic solution through a point p_μ, $p_\mu \to 0$ as $\mu \to 0$; that is, a fixed point of π_μ. The stable manifold $W^s = W^s_\mu$ and unstable manifold $W^u = W^u_\mu$ of p_μ as a fixed point of π_μ are, respectively, the section of the stable and unstable manifolds of the 1-periodic solution through p_μ at $t = 0$.

With this notation, a point q is *homoclinic to p_μ* (relative to the map $\pi = \pi_\mu$) if $q \neq p_\mu$, $q \in W^s \cap W^u$. A point q is *transverse* homoclinic to p_μ if q is homoclinic to p_μ and W^s, W^u intersect transversally at q. In Fig. 3.1, we have indicated some of the complicated behavior that must occur at a transverse homoclinic point and we have given criteria for the existence of transverse homoclinic points. We now describe other implications of the existence of a transverse homoclinic point.

If A is a finite or countably infinite sequence of symbols, let S be the collection of doubly infinite sequences $s = \{s_k; k = 0, \pm 1, \ldots\}$ with each $s_k \in A$. The *shift automorphism* σ on S is defined by $\sigma s = \{\tilde{s}_k; k = 0, \pm 1, \ldots\}$, $\tilde{s}_k = s_{k-1}$ for all k.

Near a transverse homoclinic point q, we can construct a small quadrilateral Q, two of its sides consisting of parts of $W^u(p)$, $W^s(p)$ and the others parallel to the tangents of these sets at q (see Figure 3.3). For any point $r \in Q$, let $k = k(r)$ be the smallest positive integer such that $\pi^k(r) \in Q$, if it exists. Let $D(\tilde{\pi})$ be the set of $r \in Q$ for which such a k exists and define $\tilde{\pi}r = \pi^k(r)$ for $r \in D(\tilde{\pi})$. The map $\tilde{\pi}$ is called the *transversal map* of π for the quadrilateral Q.

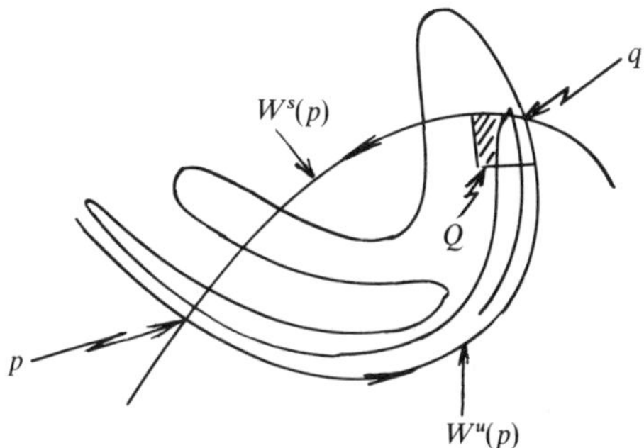

Figure 3.3

Theorem 3.11. *If π is a C^∞-diffeomorphism of the plane with a point q transverse homoclinic to a hyperbolic fixed point p, then in a neighborhood of q, the transversal map $\tilde{\pi}$ of a quadrilateral possesses an invariant set I homeomorphic to the sequence space S with an infinite number of symbols by a map $\tau: S \to I$ such that $\tilde{\pi}\tau = \tau\sigma$. Also, there is an integer k, an invariant set $\tilde{I}$ of π^k and a homeomorphism $\tilde{\tau}: S \to \tilde{I}$, where S is the sequence space of a finite number of symbols, such that $\pi^k\tilde{\tau} = \tilde{\tau}\sigma$.*

Note the difference in the two conclusions in the theorem. In the first part, the set I is invariant for $\tilde{\pi}$ and $\tilde{\pi}$ is equivalent on I to the shift automorphism on an infinite number of symbols. In the second part, the set $\tilde{I}$ is invariant under a fixed power k of π itself and π^k is equivalent on $\tilde{I}$ to the shift automorphism on a finite number of symbols.

It follows immediately from Theorem 3.11 that there are infinitely many periodic points in a neighborhood of the transverse homoclinic point and they are dense. Also, there is a random behavior to the orbits on the invariant set I (or $\tilde{I}$) since knowing the early terms of a sequence tells nothing about the later terms of a sequence.

Remark 3.12. The sectors S_1^∞, S_2^∞ in Theorem 3.8 and Corollary 3.10 exist if it is required only that the functions $h_\infty(\alpha)$, $G_\infty(\alpha)$ are not the zero functions. The Hypotheses (3.24), (3.26) are imposed to obtain a precise characterization of the boundaries of these sectors (see the proofs).

11.4. Subharmonics and Homoclinic Points

In this section, we consider Eq. (2.3); namely,

$$(4.1) \qquad \ddot{y} + H(y) = -\mu_1 \dot{y} + \mu_2 F(t)$$

where $F(t + 1) = F(t)$, $H(0) = 0$, $H'(0) < 0$ and there is a homoclinic orbit $\Gamma_\infty = \{(\varphi_\infty(t), \dot{\varphi}_\infty(t)) : t \in \mathbb{R}\} \cup \{(0,0)\} \subseteq \mathbb{R}^2$, where φ_∞ satisfies

$$(4.2) \qquad \ddot{z} + H(z) = 0$$

Corollary 3.10 gives a complete description of the existence or nonexistence of homoclinic points of (4.1) near Γ_∞ for μ in a neighborhood of zero. Since (4.2) is Hamiltonian, in any neighborhood of Γ_∞ and any $k \geq k_0$, sufficiently large, there is a periodic orbit $\Gamma_k = \{(\varphi_k(t), \dot{\varphi}_k(t)) : t \in \mathbb{R}\}$ of (4.2) where φ_k satisfies (4.2) and has least period k. Corollary 2.3 gives a complete description of the existence or nonexistence of subharmonic solutions of order k near Γ_k for μ in a neighborhood of zero. Furthermore, the conditions describing the bifurcation curves for subharmonics and homoclinic points are very similar. Therefore, one suspects that near the bifurcation curves C_m^∞, C_M^∞, there are infinitely many subharmonic bifurcations. We will prove this.

As in Sections 2, 3, define

$$(4.3) \qquad
\begin{aligned}
G_k(\alpha) &= \frac{1}{\eta_k} \int_{-k/2}^{k/2} \dot{\varphi}_k(t) F(t - \alpha)\, dt, & \eta_k &= \int_{-k/2}^{k/2} \dot{\varphi}_k^2 \\
G_\infty(\alpha) &= \frac{1}{\eta_\infty} \int_{-\infty}^{\infty} \dot{\varphi}_\infty(t) F(t - \alpha)\, dt, & \eta_\infty &= \int_{-\infty}^{\infty} \dot{\varphi}_\infty^2.
\end{aligned}$$

If α_m^k, α_M^k are the values of $\alpha \in [0, 1)$ for which $G_k(\alpha)$ attains respectively its minimum, maximum and $\alpha_m^\infty, \alpha_M^\infty$ are the corresponding values for $G_\infty(\alpha)$, suppose that

$$(4.4) \qquad G_\infty''(\alpha_m^\infty) > 0, \qquad G_\infty''(\alpha_M^\infty) < 0.$$

Conditions (4.4) are precisely the conditions imposed in Corollary 3.10 to ensure the existence of curves of bifurcation to homoclinic points. The corresponding conditions for G_k'' at α_M^k, α_m^k are the conditions imposed in Corollary 2.3 to ensure the existence of curves of bifurcation to subharmonics of order k. The next lemma shows that the conditions on G_∞ imply the ones on G_k if k is sufficiently large.

Lemma 4.1. *For any $F: \mathbb{R} \to \mathbb{R}$, continuous, $F(t + 1) = F(t)$ for all t and for any $\varepsilon > 0$, there is a $k_0 = k_0(\varepsilon, F) > 0$ such that, for $k \geq k_0$,*

$$\sup_\alpha |G_\infty^{(j)}(\alpha) - G_k^{(j)}(\alpha)| \leq \varepsilon \sup_t |F(t)|, \qquad j = 0, 1, 2.$$

Proof. In this proof, it is convenient for notational purposes to let $\Gamma_\infty = \{(p(t), \dot{p}(t)): t \in \mathbb{R}\} \cup \{(0, 0)\}$ and to normalize p so that $\dot{p}(0) = 0$. For any $0 \leq \delta \leq \delta_0, \delta_0 > 0$, let $q_\delta(t)$ be the solution of (4.2) with $q_\delta(0) = (1 - \delta)p(0)$, $\dot{q}_\delta(0) = 0$. Then for δ_0 sufficiently small, $0 < \delta \leq \delta_0$, $q_\delta(t)$ is periodic of least period $w(\delta)$, $w(\delta) \to \infty$ as $\delta \to 0$. Let

$$(4.5) \qquad \begin{aligned} h(\alpha, \delta, F) &= \int_{-w(\delta)/2}^0 \dot{q}_\delta(t) F(t - \alpha)\, dt \\ h(\alpha, 0, F) &= \int_{-\infty}^0 \dot{p}(t) F(t - \alpha)\, dt \end{aligned}$$

The operators $h(\alpha, \delta, \cdot)$, $h(\alpha, 0, \cdot)$ are continuous linear operators on the space of continuous 1-periodic functions. We now prove the assertion in the lemma for $h(\alpha, \delta, F) - h(\alpha, 0, F)$ and all δ sufficiently small. This will prove the lemma since the orbits of (4.2) are symmetric about the z-axis.

Let $T = w(\delta)/2$ and $\varepsilon > 0$ be given and let us first show that

$$|h(\alpha, \delta, F) - h(\alpha, 0, F)| < \varepsilon|F|, \qquad |F| = \sup_t |F(t)|$$

Since $T \to \infty$ as $\delta \to 0$, we can choose $\delta_1 > 0$ sufficiently small that

$$\int_{-\infty}^{-T} \dot{p}(t) F(t - \alpha)\, dt < \varepsilon|F|/3 \quad \text{for } 0 \leq \delta < \delta_1.$$

In each of the integrals in (4.5), we can change variables by integrating with respect to x along the curve $\Gamma_\delta = \{(q_\delta(t), \dot{q}_\delta(t)): t \in \mathbb{R}\}$, Γ_∞, respectively. Let $t(x), t_\delta(x)$ be the inverse functions, fix $0 < \eta < \varepsilon/6$ and choose δ_1 so that

$\eta > p(-T), q_\delta(-T)$ for $0 \leq \delta < \delta_1$. Then

$$\left| \int_{-T}^{0} \dot{p}(t)F(t - \alpha)\,dt - \int_{-T}^{0} \dot{q}_\delta(t)F(t - \alpha)\,dt \right|$$

$$= \left| \int_{p(-T)}^{p(0)} F(t(x) - \alpha)\,dx - \int_{q_\delta(-T)}^{q_\delta(0)} F(t_\delta(x) - \alpha)\,dx \right|$$

$$\leq \left| \int_{p(-T)}^{\eta} F(t(x) - \alpha)\,dx - \int_{q_\delta(-T)}^{\eta} F(t_\delta(x) - \alpha)\,dx \right|$$

$$+ \left| \int_{q_\delta(0)}^{p(0)} F(t(x) - \alpha)\,dx \right|$$

$$+ \left| \int_{\eta}^{q_\delta(0)} F(t(x) - \alpha)\,dx - \int_{\eta}^{q_\delta(0)} F(t_\delta(x) - \alpha)\,dx \right|$$

$$\leq 2\eta|F| + \delta|F| + \left| \int_{\eta}^{(1-\delta)p(0)} F(t(x) - \alpha)\,dx - \int_{\eta}^{(1-\delta)p(0)} F(t_\delta(x) - \alpha)\,dx \right|$$

Since η is fixed, $t_\delta(x) \to t(x)$ uniformly on $x \geq \eta$ as $\delta \to 0$. By further reduction in δ_1, if necessary, the right hand side of this inequality can be made less than $\varepsilon|F|/3$. This proves the assertion (4.6).

Since q_δ is at least C^3 and $\dot{q}_\delta(-T) = 0 = \dot{q}_\delta(0)$

$$\frac{\partial h}{\partial \alpha}(\alpha, \delta, F) = \int_{-T-\alpha}^{-\alpha} \ddot{q}_\delta(t + \alpha)F(t)\,dt$$

$$\frac{\partial^2 h}{\partial \alpha^2}(\alpha, \delta, F) = -\ddot{q}_\delta(0)F(-\alpha) + \ddot{q}_\delta(-T)F(-T - \alpha) + \int_{-T-\alpha}^{-\alpha} \dddot{q}_\delta(t + \alpha)F(t)\,dt$$

and $h(\alpha, \delta, F)$ is C^2 in α. Similarly, we can show $h(\alpha, 0, F)$ is C^2 in α.

We now show that, for any $\varepsilon > 0$, there is a $\delta_2 > 0$ such that

$$\left| \frac{\partial h(\alpha, \delta, F)}{\partial \alpha} - \frac{\partial h(\alpha, 0, F)}{\partial \alpha} \right| < \varepsilon|F| \quad \text{for } |\delta| < \delta_2.$$

If $G(t)$ is a 1-periodic function which is C^1, then

$$\frac{\partial}{\partial \alpha} h(\alpha, 0, G) - \frac{\partial}{\partial \alpha} h(\alpha, \delta, G) = -h(\alpha, 0, \dot{G}) + h(\alpha, \delta, \dot{G})$$

Thus, for any such G, we have

$$\left| \frac{\partial}{\partial \alpha} h(\alpha, 0, F) - \frac{\partial}{\partial \alpha} h(\alpha, \delta, F) \right| \leq \left| \frac{\partial}{\partial \alpha} h(\alpha, 0, F - G) \right| + \left| \frac{\partial}{\partial \alpha} h(\alpha, \delta, F - G) \right|$$

$$+ \left| h(\alpha, 0, \dot{G}) - h(\alpha, \delta, \dot{G}) \right|$$

$$< \varepsilon$$

if $|F - G|, \delta$ are sufficiently small. The estimate on the second derivatives are obtained in a similar way taking into account that $\ddot{q}_\delta(-T)$ approach zero as $\delta \to 0$. This completes the proof of the lemma. $\quad\square$

Theorem 4.2. *If condition* (4.4) *is satisfied, there is a neighborhood U of Γ in* $\mathbb{R}^2$, *a neighborhood V of $(\mu_1, \mu_2) = (0,0)$ in $\mathbb{R}^2$, an integer $k_0 > 0$ and curves* C_M^k, C_m^k *in V, $k = k_0, k_0 + 1, \ldots, \infty$ (including ∞) which have parametric representations with twice continuously differentiable functions, each C_M^k, C_m^k contains $(0,0)$, $C_M^k \to C_M^\infty$, $C_m^k \to C_m^\infty$ as $k \to \infty$ uniformly, $C_M^\infty[C_m^\infty]$ is tangent to the curve defined by $\mu_1 = G_\infty(\alpha_M)\mu_2[\mu_1 = G_\infty(\alpha_m)\mu_2]$ at $(\mu_1, \mu_2) = (0,0)$ and for each integer $k \in [k_0, \infty]$, the set $V \backslash (C_M^k \cup C_m^k) = S_1^k \cup S_2^k$, S_1^k, S_2^k open, disjoint, and the following conclusions hold for Eq. (4.1) and $\mu = (\mu_1, \mu_2)$:*

(i) $\mu \in S_1^\infty$ *implies no homoclinic point exists in U*
(ii) $\mu \in S_2^\infty$ *implies a transverse homoclinic point exists in U*
(iii) *If $k < \infty$, $\mu \in S_1^k$ implies no subharmonic solution in U*
(iv) *If $k < \infty$, $\mu \in S_2^k$ implies there are at least $2k$ subharmonics in U.*

Theorem 4.2 can be proved by using Corollary 2.3, Corollary 3.10, Lemma 4.1 and the following lemmas. We note that we will establish (Lemma 4.7) that the bifurcation equation (3.18) (b) is the limit of the bifurcation equations (2.14) (b) as $k \to \infty$.

Let $(\varphi_k(t), \dot{\varphi}_k(t))$ be a periodic solution of (4.2) with least period $k > 0$, $\dot{\varphi}_k(0) = 0$. The variational equation for this solution is given by

$$(4.6) \qquad \ddot{w} + H'(\varphi_k(t))w = 0.$$

Consider the nonhomogeneous equation:

$$(4.7) \qquad \ddot{w} + H'(\varphi_k(t))w = f(t)$$

where f is periodic with period 1.

  The following lemma is another way of expressing the solution of the auxiliary equation in the method of Liapunov-Schmidt.

Lemma 4.3. *For each continuous f, there exists a unique $\rho \in \mathbb{R}$ such that the equation*

$$(4.8) \qquad \ddot{w} + H'(\varphi_k(t))w = f(t) - \rho\dot{\varphi}_k(t)$$

has a unique periodic solution $w_k(t)$ with period k satisfying the relations:

$$\int_{-k/2}^{k/2} [f(t) - \rho\dot{\varphi}_k(t)]\dot{\varphi}_k(t)\, dt = 0,$$

$$\int_{-k/2}^{k/2} w_k(t)\dot{\varphi}_k(t)\, dt = 0.$$

In terms of the notations in Section 2, the solution $(w_k, \dot{w}_k)$ of (4.8) is given by $K(I - p)f(\cdot)$ (see (2.14) (a)). By using Lemma 4.3 and the variation of constants formula, we have the following:

Lemma 4.4. *The unique solution $w_k(t)$ in Lemma 4.3 can be expressed as follows:*

$$(4.9) \qquad w_k(t) = \int_{-k/2}^{k/2} J_k(t, s) f(s)\, ds, \qquad -\frac{k}{2} \le t \le \frac{k}{2}$$

where $J_k(t, s) = J_k(s, t)$ is the Green's function and is given by

$$J_k(t, s) = \frac{\dot{\varphi}_k(t) u_k(s)}{\|\dot{\varphi}_k\|_k^2} \left[\int_0^t \dot{\varphi}_k(\tau) u_k(\tau)\, d\tau + \int_0^s \dot{\varphi}_k(\tau) u_k(\tau)\, d\tau \right]$$

$$- \frac{2\dot{\varphi}_k(t) w_k(s)}{\|\dot{\varphi}_k\|_k^4} \left[\int_0^{k/2} \int_0^\sigma \dot{\varphi}_k^2(\sigma) \dot{\varphi}_k(\tau) u_k(\tau)\, d\tau\, d\sigma \right.$$

$$\left. + \int_0^{k/2} \int_\sigma^{k/2} \dot{\varphi}_k(\sigma) u_k(\tau) \dot{\varphi}_k^2(\tau)\, d\tau\, d\sigma \right]$$

$$(4.10) \qquad - \frac{c}{2\dot{u}_k\left(\dfrac{k}{2}\right)} u_k(t) u_k(s) + \tilde{J}_k(t, s),$$

$$\tilde{J}_k(t, s) = \begin{cases} -\dfrac{\dot{\varphi}_k(t) u_k(s)}{\|\dot{\varphi}_k\|_k^2} \displaystyle\int_{-k/2}^s \dot{\varphi}_k^2(\tau)\, d\tau + \dfrac{\dot{\varphi}_k(t) u_k(s)}{\|\dot{\varphi}_k\|_k^2} \displaystyle\int_t^{k/2} \dot{\varphi}_k^2(\tau)\, d\tau, & (s \le t) \\[3ex] \dfrac{\dot{\varphi}_k(t) u_k(s)}{\|\dot{\varphi}_k\|_k^2} \displaystyle\int_s^{k/2} \dot{\varphi}_k^2(\tau)\, d\tau - \dfrac{\dot{\varphi}_k(t) u_k(s)}{\|\dot{\varphi}_k\|_k^2} \displaystyle\int_{-k/2}^t \dot{\varphi}_k^2(\tau)\, d\tau, & (s \ge t) \end{cases}$$

where $|t|, |s| \le k/2$, $u_k(t)$ is the unique solution of (4.6) satisfying $u_k(0) = 1$, $\dot{u}_k(0) = 0$, $c = H(\varphi_k(k/2))$ and

$$\|\dot{\varphi}_k\|_k^2 = \int_{-k/2}^{k/2} |\dot{\varphi}_k(t)|^2\, dt.$$

Proof. The proof follows by simply differentiating (4.9) $\qquad\qquad\qquad$ $\square$

In other words, the operator $K(I - P)$ in (2.14)(a) is actually an integral operator with kernel $J_k(t, s)$. The bifurcation equation (2.14)(b) is given by

$$\rho = \int_{-k/2}^{k/2} \dot{\varphi}_k(t) F(t, w_k(t), \mu, \alpha)\, dt / \|\dot{\varphi}_k\|_k^2 = 0.$$

In order to show that these bifurcation equations are valid uniformly in the parameters μ, α for all large $k > 0$, we need to know that there is a uniform bound for all the integral operators.

Lemma 4.5. *There exists a constant $C > 0$ such that the norms of the integral operators (4.9) are uniformly bounded by C for all large $k > 0$; that is, for all large $k > 0$,*

$$\sup_{|t| \le k/2} \int_{-k/2}^{k/2} |J_k(t,s)| \, ds + \sup_{|t| \le k/2} \int_{-k/2}^{k/2} \left| \frac{\partial}{\partial t} J_k(t,s) \right| ds \le C.$$

Proof. By (4.10), it is sufficient to show that there exist constants $C_1 > 0$ and $\beta > 0$, independent of k, such that, for all large $k > 0$,

$$\text{(4.11)} \qquad
\begin{aligned}
|(\dot\varphi_k(t), \ddot\varphi_k(t))| &\le C_1 e^{-\beta|t|}, & |t| &\le k/2; \\
|(u_k(t), \dot u_k(t))| &\le C_1 e^{\beta|t|}, & |t| &\le k/2.
\end{aligned}$$

It is well known that u_k and φ_k are related as follows:

$$u_k(t) = \dot\varphi_k(t) \int_\sigma^t \frac{d\tau}{\dot\varphi_k^2(\tau)}$$

where $\sigma \ne 0$ is a fixed constant. Thus, (4.11) may be derived from the following inequalities for all large $k > 0$:

$$\text{(4.12)} \qquad C_2 e^{-\beta|t|} \le |(\dot\varphi_k(t), \ddot\varphi_k(t))| \le C_3 e^{-\beta|t|}, \qquad |t| \le k/2$$

where C_2 and $C_3 > 0$ are constants independent of k. Let the origin $(0,0)$ be the critical point which is on the homoclinic orbit Γ_∞ of (4.12). Hence, $H(0) = 0$ and $H'(0) < 0$. By using the stable and unstable manifolds as the new axes, (4.2) may be rewritten in the new coordinates (ξ, η) as follows:

$$\text{(4.13)} \qquad
\begin{aligned}
\dot\xi &= \xi H_1(\xi, \eta) \\
\dot\eta &= \eta H_2(\xi, \eta)
\end{aligned}$$

here $H_1(0,0) = \beta = \sqrt{-H'(0)} > 0$ and $H_2(0,0) = -\beta$. Thus,

$$\frac{d\xi}{\xi} = \beta + O(|\xi| + |\eta|), \qquad \frac{d\eta}{\eta} = -\beta + O(|\xi| + |\eta|).$$

If $T(\xi)$ is the time it takes for the solution of (4.13) to go from $(\xi, 1)$, with $\xi > 0$ sufficiently small, to a point on the line $\xi = 1$, then we have

$$C_4 T \le -\ln \xi \le C_5 T$$

where C_4 and C_5 are constants independent of $\xi > 0$. Inequalities (4.12) may now be obtained from the above inequalities. $\quad\square$

Lemma 4.6. *The solution $w_k(t)$ in (4.9) satisfies:*

$$(4.14) \qquad \left|(w_k(t), \dot{w}_k(t))\right| \le C_6 e^{-\beta|t|} \sup_{|t| \le k/2} |f(t)|, \qquad |t| \le k/2,$$

where $C_6 > 0$ is a constant independent of k.

Proof. Using (4.10) and the estimates (4.11) and (4.12), relation (4.14) follows immediately. $\square$

We may now compare the bifurcation functions (2.14)(b),

$$\|\dot{\varphi}_k\|_k^2 B_k(\alpha, \mu) = \int_{-k/2}^{k/2} \dot{\varphi}_k(t) F(t - \alpha, z_k^*(\alpha, \mu), \mu, \alpha) \, dt$$

for the subharmonics and the bifurcation function (3.18)(b),

$$\|\dot{\varphi}_\infty\|_\infty^2 B_\infty(\alpha, \mu) = \int_{-\infty}^{\infty} \dot{\varphi}_\infty(t) F(t - \alpha, z_\infty^*(\alpha, \mu), \mu, \alpha) \, dt$$

for the homoclinic orbits, where

$$\|\dot{\varphi}_\infty\|_\infty^2 = \int_{-\infty}^{\infty} \dot{\varphi}_\infty^2(t) \, dt.$$

Lemma 4.7. *For $|\mu|$ sufficiently small and all $|\alpha| \le 1$, $B_k(\alpha, \mu) \to B_\infty(\alpha, \mu)$ as $k \to \infty$.*

Proof. Because $\|\dot{\varphi}_k\|_k^2 \to \|\dot{\varphi}_\infty\|_\infty^2$ as $k \to \infty$, Lemma 4.1 implies that it is sufficient to show that

$$(4.15) \qquad \int_{-k/2}^{k/2} \dot{\varphi}_k(t) \left[F(t - \alpha, z_k^*(\alpha, \mu), \mu, \alpha) - F(t - \alpha, z_\infty^*(\alpha, \mu), \mu, \alpha) \right]$$

tends to zero as $k \to \infty$. By Lemmas 4.4 and 4.6, $z_k^* \to z_\infty^*$ pointwise in t as $k \to \infty$. By the Lebesque dominated convergence theorem, (4.15) converges to zero as $k \to \infty$. This proves the lemma. $\square$

The above theorem shows, in particular, the following. If (μ_1, μ_2) is a point of bifurcation to transverse homoclinic points; that is, $(\mu_1, \mu_2) \in C_M^\infty$ or C_m^∞, and W is any neighborhood of this point, then there are infinitely many subharmonic bifurcations in W.

As an example, consider the equation

$$(4.16) \qquad \ddot{x} - x + \tfrac{3}{2}x^2 = 0$$

which admits the first integral

$$I = \dot{x}^2 - x^2 + x^3.$$

Let $p(t)$ be the solution of (4.16) such that $p(0) = 1$, $\dot{p}(0) = 0$ so $I(p) = 0$. The function p is even, $\dot{p}$ is odd, $p > 0$ and p is decreasing on $(0, \infty)$.

The relation $I(p) = 0$ implies $dp/dt = -p(1 - p)^{1/2}$ for $t > 0$. An elementary integration implies

$$p(t) = 1 - \left(\frac{1 - e^t}{1 + e^t}\right)^2.$$

Let $F(t) = \cos t$ and define the function $G_\infty(\alpha)$ in (4.3),

$$G_\infty(\alpha) = \int_{-\infty}^{\infty} \dot{p}(t)\cos(t - \alpha)\,dt$$

$$= -\int_{-\infty}^{\infty} p(t)\sin(t - \alpha)\,dt$$

$$= -\sin \alpha \int_{-\infty}^{\infty} p(t)\cos t\,dt - \cos \alpha \int_{-\infty}^{\infty} p(t)\sin t\,dt$$

$$= -\varDelta \sin \alpha,$$

where $\varDelta = \int_{-\infty}^{\infty} p(t)\cos t\,dt$. Since $p(t) = \operatorname{sech}^2(t/2)$, we have

$$\varDelta = 2 \int_{-\infty}^{\infty} e^{2i\tau}/(\cosh^2 \tau)\,d\tau$$

$$= 4\pi i \quad \text{(sum of residues in the upper half plane)}$$

$$= 8\pi\omega \sum_{n \geq 0} e^{-2\pi(n + 1/2)}$$

$$= 4\pi\omega/\sinh \pi.$$

Thus, $\varDelta > 0$.

This shows that $G_\infty(\alpha)$ satisfies (4.4) and, thus, the conclusions of Theorem 4.2 apply to the equation

$$\ddot{x} - x + \tfrac{3}{2}x^2 = -\mu_1\dot{x} + \mu_2 \cos t$$

11.5. Abstract Bifurcation near a Closed Curve

In Chapters 6 and 7, we have given a procedure for analyzing the bifurcations near the zero solution of families of nonlinear equations

$$(5.1) \qquad\qquad M(x, \lambda) = 0, \qquad M : X \times \varLambda \to Z.$$

A complete analysis was possible under certain generic hypotheses on $M(x, \lambda)$. One of the hypotheses on the lowest order terms of $M(x, 0)$ implied in particular that $x = 0$ is an isolated solution of

$$(5.2) \qquad\qquad M(x, 0) = 0.$$

In this section, we suppose (5.1) has a family of solutions which is a smooth closed curve and give results on bifurcation near this family. The problem has independent interest because such families can occur when the equation (5.1) is invariant under certain groups of transformations, but the original motivation came from the problem of subharmonics in Section 10.2.

Suppose Equation (5.2) has a one parameter family of solutions $x = p(\alpha)$, $0 \le \alpha \le 1$, where p is continuous with derivatives up through order two and $p'(\alpha) = dp(\alpha)/d\alpha \ne 0$, $0 \le \alpha \le 1$. To avoid special treatment near $\alpha = 0$ and $\alpha = 1$, we assume $p(0) = p(1)$, $p'(0) = p'(1)$, $p''(0) = p''(1)$. Thus, we may assume p is 1-periodic and $p \in C^2(\mathbb{R}, X)$. If $\Gamma = \{p(\alpha): 0 \le \alpha \le 1\} \subseteq X$, the problem is to characterize the solutions (x, λ) of (5.1) in a neighborhood of $\Gamma \times \{0\} \subseteq X \times \Lambda$. Under certain hypotheses, we give such a characterization.

Since $M(p(\alpha), 0) = 0$ for $0 \le \alpha \le 1$, it follows that $p'(\alpha)$ is a nonzero element of the null space $\mathcal{N}(A(\alpha))$ of the linear operator

$$A(\alpha) = \partial M(p(\alpha), 0)/\partial x$$

for $0 \le \alpha \le 1$. Suppose that

(H$_1$) $\dim \mathcal{N}(A(\alpha)) = 1 = \operatorname{codim} \mathcal{R}(A(\alpha))$

for $0 \le \alpha \le 1$. For any element $x \in X$, let $[x]$ denote the span of x. Since $p'(\alpha) \ne 0$, Hypothesis (H$_1$) implies $[p'(\alpha)] = \mathcal{N}(A(\alpha))$. Also, suppose $q \in C^2(R, Z)$ is such that $[q(\alpha)] \oplus \mathcal{R}(A(\alpha)) = Z$. If $B(X)$ denotes the Banach space of bounded linear operators on X, let $U \in C^2(\mathbb{R}, B(X))$ be such that $U(\alpha)$ is a projection onto $\mathcal{N}(A(\alpha))$ and $E \in C^2(\mathbb{R}, B(Z))$, $E(\alpha)$ a projection onto $\mathcal{R}(A(\alpha))$, $I - E(\alpha)$ a projection onto $[q(\alpha)]$. We also suppose U, E are 1-periodic.

If $\Lambda = \mathbb{R}^m$, $\lambda = (\lambda_1, \dots \lambda_m) \in \Lambda$, define $h_j \in C^2(\mathbb{R}, \mathbb{R})$, $j = 1, \dots, m$, 1-periodic, by the relation

(5.3) $h_j(\alpha)q(\alpha) = (I - E(\alpha))\, \partial M(p(\alpha), 0)/\partial \lambda_j.$

If $h(\alpha) = (h_1(\alpha), \dots, h_m(\alpha))$, let

$$P_h = \{\beta \in \mathbb{R}^m : \beta \cdot h(\alpha) = 0 \text{ for some } \alpha \in [0, 1)\}$$

Our next hypothesis is

(H$_2$)   The boundary (in the relative topology of S^{m-1}) of $P_h \cap S^{m-1}$ is the union of two disjoint geodesics L^m, L^M, where S^{m-1} is the unit sphere in $\mathbb{R}^m$.

If α_m, α_M are such that $h(\alpha_m), h(\alpha_M)$ are orthogonal, respectively, to L^m, L^M, then suppose that $m = 2$ and

$$(\mathbf{H}_3) \qquad \begin{aligned} \mu \cdot h(\alpha_m) = 0, \qquad & \mu \cdot h'(\alpha_m) = 0, \qquad \mu \in \mathbb{R}^2, \qquad |\mu| = 1 \Rightarrow \mu h''(\alpha_m) \neq 0 \\ \mu \cdot h(\alpha_M) = 0, \qquad & \mu \cdot h'(\alpha_M) = 0, \qquad \mu \in \mathbb{R}^2, \qquad |\mu| = 1 \Rightarrow \mu h''(\alpha_M) \neq 0 \end{aligned}$$

Theorem 5.1. *Suppose $\lambda \in \mathbb{R}^2$ and $M \in C^r(X \times \Lambda, Z), r \geq 2$. If $(\mathrm{H}_1), (\mathrm{H}_2), (\mathrm{H}_3)$ are satisfied, then there is a neighborhood U of Γ, V of $\lambda = 0$ and two C^{r-1} manifolds $C_m, C_M \subseteq V$ of codimension one containing $\lambda = 0$ whose tangent vectors are orthogonal respectively to L^m, L^M at $\lambda = 0$, $C_m \cap C_M = \{0\}$, C_m, C_M divide $V \backslash \{0\}$ into two disjoint open sets S_1, S_2 such that Equation (5.1) has no solutions in U for $\lambda \in S_1$ and at least two solutions for λ in S_2 (see Figure 2.2).*

Proof. Since the proof is essentially the same as the proof of Theorem 2.1, we only indicate the idea. For $y \in (I - U(\alpha))X$, $|y| < \delta$ with δ sufficiently small, $\alpha \in [0, 1)$, one can introduce new coordinates (α, y) in a neighborhood of Γ by the relation $x = p(\alpha) + y$. Applying the method of Liapunov–Schmidt, one obtains a bifurcation function $f(\alpha, \lambda)$ with $f(\alpha, 0) = 0$, $\partial f(\alpha, 0)/\partial \lambda = h(\alpha)$ where $h(\alpha)$ is defined in (5.3). The remaining part of the proof is the same as the proof of Theorem 2.1. $\square$

11.6. Bibliographical Notes

All of the results in Section 2 except for the stability result in Theorem 2.8 as well as the proofs were essentially given in Hale and Táboas [1] for the perturbed Hamiltonian system (2.3). The stability results in Theorems 2.8, 2.9 were motivated by de Oliveira and Hale [1]. For Eq. (2.3), Loud [1] has also obtained results which are similar but less general than the ones in Theorem 2.1, Theorem 2.5 and Corollary 2.6. Results of a different nature on the existence of subharmonics are contained in Furimochi [1], Morris [1–2], Schmidt [1], Mazzanti and Schmitt [1].

It is interesting to study the problem in Section 2 when Hypothesis (H_1) is not satisfied; that is, the linear variational equation (2.5) around p has two linearly independent k-periodic solutions. Of course, this is much more complicated because the bifurcation equation will now be two dimensional. It is difficult to determine the analogues of Hypothesis $(\mathrm{H}_2), (\mathrm{H}_3)$ which will ensure that bifurcations are taking place in the simplest possible manner. For Eq. (2.3), Hale and Táboas [2] have given generic conditions for bifurcation. For this case, (H_1) not being satisfied implies the period $w(a)$ mentioned after (H_1) must satisfy $w'(a_0) = 0$. Rothe [2] has shown that the conditions in Hale and Táboas [2] are equivalent to $\omega''(a_0) \neq 0$ as one might expect.

The standard results from differential equations used in Section 3 can be found in most advanced texts on differential equations (for example, Hale [4]).

The results in Section 3 generalize the ones in Chow, Hale and Mallet–Paret [3] where only the Hamiltonian case (2.3) was considered. Corollary 3.6 is essentially due to Sotomayor [1] (see also Andronov, Leontovich, Gordon and Maier [1]) although the proof in the text is completely different. Corollary 3.10 is due to Chow, Hale and Mallet–Paret [3]. Poincaré [1] was the first to observe the importance of homoclinic points. Birkhoff [1] proved that every transverse homoclinic point is the limit of periodic points (that is, points p such that $\pi^k p = p$ for some integer k). Smale [1] carried the analysis even further. The presentation in the text follows Moser [2] (see also Conley [1]). For examples of transverse homoclinic points in Hamiltonian systems, see Moser [2], McGehee and Meyer [1], Churchill, Pecelli and Rod [1].

Mel'nikov [1], Sil'nikov [1, 2] and Morozov [1] have also investigated the existence of homoclinic points for second order periodic systems in an analytic way. The spirit is the same as in Section 3 and the analysis of the bifurcations is based on the function $G_\infty(\alpha)$ in Eq. (3.25). However, the overall analysis is different. Mel'nikov [1] bases his discussion on the analogue of the Poincaré map used in Section 10.3 to analyze the behavior near a homoclinic orbit for autonomous equations. The information obtained in this way applies directly to all solutions of the differential equation. The method in Section 3 using Liapunov-Schmidt is concerned only with the homoclinic points. Information about other solutions is obtained in an indirect way. If we recall the procedure for obtaining periodic solutions in Chapter 9, another analogy can be made. The procedure of Mel'nikov corresponds exactly to the period one map of the differential equation. The method in Section 3 corresponds precisely to the method of Liapunov–Schmidt for obtaining only the periodic solutions. We have seen in that chapter that the bifurcation function does carry the dynamics of the differential equation. As a result, the qualitative properties of the period one map are recoverable from the bifurcation function.

Various generalizations of the results in Section 3 are possible. For example, the equation

$$\ddot{x} - x(1 - x)(1 + ax) = 0, \qquad a > 0,$$

has a pair of homoclinic orbits Γ_1, Γ_2 containing zero generated by functions $p_1(t) < 0, p_2(t) > 0$ for $t \in \mathbb{R}, p_1(t) \to 0, p_2(t) \to 0$ as $t \to \pm\infty$. For a 1-periodic perturbation of this equation,

$$\ddot{x} - x(1 - x)(1 + ax) = \lambda_1 \dot{x} + \lambda_2 f(t)$$

one can look either for homoclinic points close to Γ_1 for which the bifurcation curves in (λ_1, λ_2)-space will be governed by the function $\int_{-\infty}^{\infty} \dot{p}_1(t)f(t - \alpha)\,dt$ or homoclinic points close to Γ_2 which will be governed by the function $\int_{-\infty}^{\infty} \dot{p}_2(t)f(t - \alpha)\,dt$. If transverse homoclinic points occur near both Γ_1 and Γ_2, then one obtains the complicated behavior shown in Figure 6.1. Holmes [1], [4] has discussed this situation in detail using the method of Mel'nikov [1]. Holmes and Marsden [1] have also extended the results to certain types of partial differential equations and have made applications to the equations of a beam.

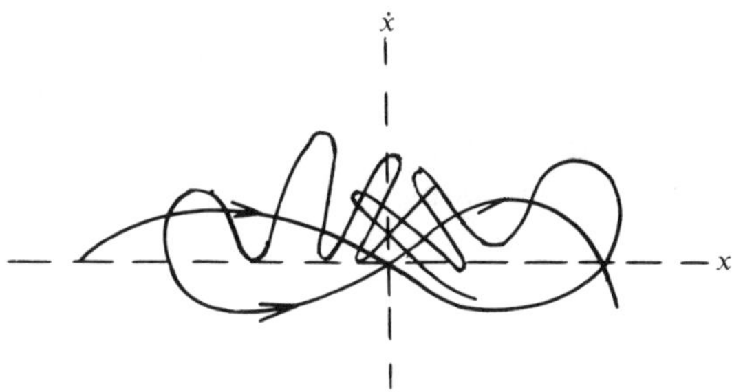

Figure 6.1

It is also possible to use the same ideas for situations where the unperturbed equation has an orbit Γ_∞ which is not a homoclinic orbit, but is the closure of an orbit containing two equilibrium points or the union of such orbits (a *heteroclinic orbit*). We give illustrations of this type of phenomena in Section 13.3.

For specific examples of systems without small parameters which possess or appear to possess transverse homoclinic points, see Levinson [1, 2], Cartwright [1], Cartwright and Littlewood [1], Lasota and Wazewska [1], Littlewood [1], Levi [1], Hayashi [1], Holmes [2], Kubicek, Marek and Raschman [1], Kubicek, Marek and Schreiber [1], Lorenz [1], Mackey and Glan [1], Moon and Holmes [1], Newhouse [2], Ray [1], Ray and Jensen [1], Rössler [1, 2], Ruelle and Takens [1, 2], Sethna [1], Takens [5].

Theorem 4.2 is due to Chow, Hale and Mallet–Paret [3]. The example (4.9) was a personal communication from K. Meyer and D. Henry. It is also possible to determine whether the bifurcation curves C_M^k, C_m^k lie to the right or to the left of the corresponding curves C_M, C_m. To do this, one needs to compute the manner in which the maximum (or minimum) of G_k varies with k. It is possible to express this in terms of the unperturbed equation (see Chow, Hale and Mallet-Paret [3]).

In Theorem 2.1, we have shown that the bifurcation to subharmonics of order k is of saddle-node type. Therefore, for μ in S_2^k and near a bifurcation curve, there are exactly k subharmonics which are saddles and k which are nodes. This is also true for each of the subharmonic bifurcation curves in Theorem 4.2 for $k \geq k_0$. These subharmonics of order k may bifurcate to subharmonics of order $2k$, $4k$, etc., a point which was overlooked in Chow, Hale, and Mallet-Paret [3]. Thus, the original nodes change their stability properties. However, for a generic class of perturbations, Newhouse [1, 2] shows there always are infinitely many subharmonics which are nodes for a residual set of μ in $\bigcap_{k \geq k_0} S_2^k$. Greenspan and Holmes [1] give a detailed discussion of this phenomenon for the Duffing equation.

The abstract bifurcation from a closed curve in Section 5 is based on Hale [11]. Chillingworth and Marsden [1] have encountered the same type of problem in their study of bifurcations in elastic bodies. The circle of solutions occurred there because of symmetry properties. Vanderbauwhede [7] has given further generalizations of the results in Section 5. Chillingworth [4] has given a characterization of the result in Section 5 in terms of singularity theory as well as many applications.

Chapter 12

Normal Forms and Invariant Manifolds

12.1. Introduction

In the previous chapter, we began the discussion of the bifurcations that occur in nonautonomous perturbations of autonomous equations. To understand more about these bifurcations, we need several general results and methods from the theory of differential equations—in particular, the theory of transformation to normal form and the method of integral manifolds. This material is also an important ingredient in the discussion in the next chapter on the behavior near an equilibrium point in dimension greater than two for which the linear variational equation has several eigenvalues on the imaginary axis.

In Section 12.2, we give a general theory of transformation to normal forms based on Lie transforms. This approach has been used extensively in Hamiltonian systems, but has not been employed systematically in general differential equations. Of course, it is equivalent to the usual theory of normal forms of Birkhoff and Poincaré, but it is more efficient to implement specific computations.

In Section 12.3, we give the implications of the general theory in Section 12.2 to autonomous equations and the normal form in the neighborhood of an equilibrium point. Normal forms for invariant manifolds near the equilibrium point are also given.

In Section 12.4, the theory is applied to systems of equations involving several dependent variables which play the role of angles. This gives the standard method of averaging except with the specific method for the computation being based on Lie transforms.

In Section 12.5, we state several classical results on the existence and stability of invariant tori for autonomous and nonautonomous equations. One result is proved in detail to illustrate the method. In a loose mathematical sense, the results can be stated as follows. Suppose T is an invariant torus of an autonomous equation which locally behaves as a saddle point. If the differential equation is perturbed by either an autonomous or nonautonomous perturbation, then this torus becomes an integral manifold for the new system and the stability properties are preserved.

In Section 12.6, these results are applied to obtain the bifurcation of a periodic orbit into a torus in autonomous equations. Applications are made to nonautonomous perturbations of an autonomous planar system with a degenerate focus. In Section 12.7, we consider the situation where the unperturbed autonomous planar system has a degenerate periodic orbit. Some partial results are given for the existence of two invariant tori when the system is subjected to a periodic nonautonomous perturbation. The difficulties encountered in obtaining a more complete theory are discussed.

12.2. Transformation Theory and Normal Forms

In the study of the local behavior of the solutions of nonlinear differential equations, the choice of coordinate system plays a major role in bringing out the essential features of the flow. It therefore becomes necessary to have efficient procedures to obtain the transformed vector field from the original one. The purpose of the present section is to present an approach to this subject using Lie transforms.

Throughout the section, ε is a scalar parameter, $\pi: \mathbb{C}^n \to \mathbb{C}^n$ is a continuous projection. For a function $f: \mathbb{C}^n \times \mathbb{C} \to \mathbb{C}^n$, consider the equation

$$(2.1) \qquad\qquad \dot{x} = f(x, \varepsilon).$$

For a given function $U: (\pi\mathbb{C}^n) \times \mathbb{C} \to \mathbb{C}^n$, let $u(y, \varepsilon)$, $y \in \mathbb{C}^n$, $\varepsilon \in \mathbb{C}$, be the solution (assumed to be unique) of the equation

$$(2.2) \qquad\qquad \frac{\partial u}{\partial \varepsilon} = U(\pi u, \varepsilon), \qquad u(y, 0) = y.$$

Let $y = v(x, \varepsilon)$ be the inverse of the transformation

$$(2.3) \qquad\qquad x = u(y, \varepsilon).$$

If the transformation (2.3) is applied to Eq. (2.1), then the differential equation for y is

$$(2.4) \qquad\qquad \dot{y} = g(y, \varepsilon) \stackrel{\text{def}}{=} \frac{\partial v(u(y, \varepsilon), \varepsilon)}{\partial x} f(u(y, \varepsilon), \varepsilon).$$

The problem is to give an efficient procedure for determining the partial derivatives of $g(y, \varepsilon)$ with respect to ε in terms of the partial derivatives of $f(y, \varepsilon)$ and $U(\pi y, \varepsilon)$. Once this procedure is given, one can turn to the problem of choosing the partial derivatives of $U(\pi y, \varepsilon)$ with respect to ε in order to make it easier to understand the flow given by (2.4). Even though we are generally interested only in the first few terms in the Taylor series of g, it is convenient to develop the method for formal infinite power series in ε. Thus, we assume that all functions of ε have formal infinite power series in ε. One could restrict the discussion to functions $f(x, \varepsilon)$, $U(\pi x, \varepsilon)$, which are analytic in ε in a neighborhood of $\varepsilon = 0$ since they are given a priori. Even when this is done, the function $g(y, \varepsilon)$ may not have a power series in ε which is convergent. This is the primary reason for considering formal power series. As remarked earlier, this will present no difficulty if we only use a finite number of terms of the series.

The projection π in the above discussion plays the following role. From the definition of $u(y, \varepsilon)$, it follows that

$$(2.5) \qquad u(y, \varepsilon) = y + \tilde{u}(\pi y, \varepsilon), \qquad \tilde{u}(\pi y, 0) = 0.$$

Thus, the essential part of the transformation of variables depends only on πy and not all of y; that is, in appropriate coordinates, the essential part of the transformation depends only on some of the coordinates.

Introducing the notation

$$(2.6) \qquad \begin{aligned} f \circ g &= \frac{\partial f}{\partial x} g \\[2mm] f \times g &= \frac{\partial f}{\partial x} g - \frac{\partial g}{\partial x} f \end{aligned}$$

we can prove

Theorem 2.1. *Suppose the notation as above and let*

$$U(s, \varepsilon) = \sum_{m=0}^{\infty} U_m(s)\varepsilon^m/m!, \qquad s \in \pi\mathbb{C}^n$$

$$(2.7) \qquad f(x, \varepsilon) = \sum_{m=0}^{\infty} f_m(x)\varepsilon^m/m!, \qquad x \in \mathbb{C}^n$$

$$g(y, \varepsilon) = (v \circ f)(u(y, \varepsilon), \varepsilon) = \sum_{m=0}^{\infty} g_m(y)\varepsilon^m/m!.$$

If we define the sequence $f_i^{(m)}(x)$, $i, m = 0, 1, 2, \ldots$ by the recursive relations

$$\begin{aligned} f_i^{(m)} &= f_{i+1}^{(m-1)} + \sum_{0 \le j \le i} C_j^i f_{i-j}^{(m-1)} \times U_j \pi \\ & \qquad i = 0, 1, 2, \ldots ; m = 1, 2, \ldots \\ f_i^{(0)} &= f_i, \qquad i = 0, 1, 2, \ldots, \end{aligned}$$

$$(2.8)$$

where $C_j^i = i!/j!(i-j)!$ is the binomial coefficient, then

$$(2.9) \qquad g_m = f_0^{(m)}, \qquad m = 0, 1, 2, \dots .$$

Before proving the theorem, we remark that the computations can be organized according to the following triangle.

$$(2.10) \qquad
\begin{array}{ccccc}
f_0^{(0)} & & & & \\
f_1^{(0)} & f_0^{(1)} & & & \\
f_2^{(0)} & f_1^{(1)} & f_0^{(2)} & & \\
f_3^{(0)} & f_2^{(1)} & f_1^{(2)} & f_0^{(3)} & \\
\vdots & \vdots & \vdots & \vdots & \ddots
\end{array}$$

The ith element of the mth column of this triangle can be computed by knowing only the first $i + 2$ elements of the $(m-1)^{\text{th}}$ column. The element $f_0^{(m)} = g_m$ on the diagonal of this triangle thus depends only on the triangle formed by $f_0^{(m)}$, $f_0^{(m-1)}$ and $f_1^{(m-1)}$.

Proof of Theorem 2.1. Let us first observe that, for any f,

$$(2.11) \qquad \frac{\partial}{\partial \varepsilon} \left[(v \circ f)(u(y, \varepsilon), \varepsilon) \right] = \left[v \circ \left(\frac{\partial f}{\partial \varepsilon} + f \times U \right) \right] (u(y, \varepsilon), \varepsilon)$$

In fact, since v is the inverse of u and u satisfies (2.2), we have $v(u(y, \varepsilon), \varepsilon) = y$ and

$$\frac{\partial v}{\partial \varepsilon} = -v \circ U$$

at $(x, \varepsilon) = (u(y, \varepsilon), \varepsilon)$. Thus, after a few computations,

$$\frac{\partial}{\partial \varepsilon} \left[(v \circ f)(u(y, \varepsilon), \varepsilon) \right] = \left[\frac{\partial}{\partial x} (v \circ f)^T U + \frac{\partial}{\partial \varepsilon} (v \circ f) \right] (u(y, \varepsilon), \varepsilon)$$

$$= \left[\frac{\partial}{\partial x} (v \circ f)^T U - \frac{\partial}{\partial x} (v \circ U)^T f + v \circ \frac{\partial f}{\partial \varepsilon} \right] (u(y, \varepsilon), \varepsilon)$$

$$= \left[v \circ \left(\frac{\partial f}{\partial \varepsilon} + f \times U \right) \right] (u(y, \varepsilon), \varepsilon).$$

This proves relation (2.11).

If $g(y, \varepsilon) = (v \circ f)(u(y, \varepsilon), \varepsilon)$, then a repeated application of (2.11) shows that

$$\frac{\partial^m g(y, \varepsilon)}{\partial \varepsilon^m} = \left[v \circ \left(\frac{\partial}{\partial \varepsilon} - U \times \right)^m f \right] (u(y, \varepsilon), \varepsilon).$$

Since at $\varepsilon = 0$, $\partial v/\partial x = I$, we have

$$\frac{\partial^m g(y,0)}{\partial \varepsilon^m} = \left[\left(\frac{\partial}{\partial \varepsilon} - U \times\right)^m f\right](y,0)$$

and

$$g(y,\varepsilon) = \sum_{m=0}^{\infty} \frac{\varepsilon^m}{m!} \left[\left(\frac{\partial}{\partial \varepsilon} - U \times\right)^m f\right](y,0).$$

In the triangle in relation (2.10), it is a simple matter to check that the mth column consists of the coefficients in the formal expansion of

$$\left[\left(\frac{\partial}{\partial \varepsilon} - U \times\right)^m f\right](x,\varepsilon).$$

This proves the theorem. $\square$

The next result is useful for obtaining the explicit formulas for the power series expansion of the transformation $x = u(y,\varepsilon)$. Recall that the transformed vector field was obtained without knowing these formulas. The next theorem is also useful for Hamiltonian systems since the vector field is not of particular interest. The transformations are made to simplify the Hamiltonian function.

Theorem 2.2. *Suppose* $u(y,\varepsilon)$ *is the solution of Eq. (2.2),* $p(x,\varepsilon)$ *is a given function of* x,ε,

$$p(x,\varepsilon) = \sum_{m=0}^{\infty} p_m(x)\varepsilon^m/m!$$

and

$$q(y,\varepsilon) = p(u(y,\varepsilon),\varepsilon) = \sum_{m=0}^{\infty} q_m(y)\varepsilon^m/m!.$$

If we define the sequence $p_i^{(m)}(y)$, $i,m = 0,1,2,\ldots$ *by the recursive relations*

$$p_i^{(m)} = p_{i+1}^{(m-1)} + \sum_{0 \le j \le i} C_j^i p_{i-j}^{(m-1)} \circ U_j \pi$$

(2.12)
$$m = 1,2,\ldots; i = 0,1,2,\ldots$$
$$p_i^{(0)} = p_i, \qquad i = 0,1,2,\ldots$$

then

(2.13)
$$q_m = p_0^{(m)}, \qquad m = 0,1,2,\ldots .$$

As remarked after Theorem 2.1, the computations can be arranged according to the triangle

$$
(2.14) \qquad
\begin{array}{ccccc}
p_0^{(0)} \\
p_1^{(0)} & p_0^{(1)} \\
p_2^{(0)} & p_1^{(1)} & p_0^{(2)} \\
p_3^{(0)} & p_2^{(1)} & p_1^{(2)} & p_0^{(3)} \\
\vdots & \vdots & \vdots & \vdots & \ddots
\end{array}
$$

Proof of Theorem 2.2. If $Dp(x,\varepsilon) = \partial p/\partial \varepsilon + p \circ U$, then

$$
p(u(y,\varepsilon), \varepsilon) = \sum_{m \geq 0} \frac{\varepsilon^m}{m!} (D^m p)(y, 0).
$$

The mth column of the triangle in Theorem 2.2 satisfies

$$
D^m p(x, \varepsilon) = \sum_{j \geq 0} \frac{\varepsilon^j}{j!} p_j^{(m)}(x).
$$

This proves the theorem. $\square$

Theorem 2.1 is directly applicable to the transformation of one differential equation to another. In fact, if $x = u(y,\varepsilon)$, $u(y,0) = y$ is the solution of (2.2), $v(x,\varepsilon) = u(x, -\varepsilon)$, then the transformation $x = u(y,\varepsilon)$ in (2.1) leads to the equation (2.4) where $g(y,\varepsilon)$ is given by (2.7), (2.9).

If one needs the transformation $x = u(y,\varepsilon)$

$$
u(y,\varepsilon) = y + \sum_{m=1}^{\infty} u_m(y)\varepsilon^m/m!
$$

then applying Theorem 2.2 with $p(x,\varepsilon) = U(x,\varepsilon)$ yields an expansion for $U(u(y,\varepsilon), \varepsilon) = \sum_{m=0}^{\infty} q_m \varepsilon^m/m!$ with $u_{m+1} = q_m$, $m = 0, 1, 2, \ldots$, given in (2.13).

For later purposes, we need the following elementary result.

Lemma 2.3. *If each $f_m(x)$ is a homogeneous polynomial in x of degree $m + 1$ and each $U_m(s)$ is a homogeneous polynomial in s of degree $m + 2$, then $g_m(y)$, $u_m(s)$ are homogeneous polynomials in y, s, respectively, of degree $m + 1$.*

Proof. The recursive relations (2.8) imply that $f_i^{(m)}(y)$ is a homogeneous polynomial in y of degree $m + i + 1$. Thus, $g_m(y)$ satisfies the stated properties. In the same way, the recursive relations (2.12) with $p(x,\varepsilon) = U(x,\varepsilon)$ imply $u_m(s)$ is homogeneous of degree $m + 1$. $\square$

As remarked earlier, the coefficients $U_m(s)$ of the power series expansion of the generator $U(\pi x, \varepsilon)$ of the transformation (2.3) may be used to simplify the transformed vector field in (2.4). The specific manner in which this is done depends upon the type of problem being discussed. A particularly important situation is when

$$(2.15) \qquad f(x, \varepsilon) = Ax + \sum_{m \geq 1} f_m(x)\varepsilon^m/m!$$

and A is an $n \times n$ constant matrix and $f_m(x)$ is a homogeneous polynomial in x of degree $m + 1$. Our objective is to give a prescription for determining the generator $U(s, \varepsilon)$ which will have the effect of specifying a priori those terms in the transformed vector field that can be taken to be zero regardless of the specific form of the terms $f_m(x)$. We consider the coefficients $U_m(s)$ of $U(s, \varepsilon)$ as homogeneous polynomials in s of degree $m + 2$.

As an illustration of the above remarks, let us attempt to change variables as in (2.2) in such a way as to transform the system

$$\dot{x}_1 = -x_1 + x_1 x_2, \qquad \dot{x}_2 = 2^{1/2} x_2$$

to the linear system $\dot{x}_1 = -x_1$, $\dot{x}_2 = 2^{1/2} x_2$. To write the equation in a form suitable for the application of the above results, scale the variables according to $x_1 \mapsto \varepsilon x_1$, $x_2 \mapsto \varepsilon x_2$, to obtain

$$\dot{x}_1 = -x_1 + \varepsilon x_1 x_2, \qquad \dot{x}_2 = 2^{1/2} x_2.$$

Theorem 2.1 can now be applied with $\pi = I$. Each $U_m(x)$ is a homogeneous polynomial in x of degree $m + 2$. A few computations show that $g_0(x) = f_0^{(0)}(x) = (-x_1, 2^{1/2} x_2)$, $g_1 = f_0^{(1)} \equiv 0$ if $U_0(x) = (2^{1/2} x_1 x_2, 0)$ and $g_m \equiv 0$ for $m \geq 2$ if $U_m \equiv 0$ for $m \geq 1$. Consequently, $U(x, \varepsilon) = (2^{1/2} x_1 x_2, 0)$ and the reduction to the linear system has been accomplished.

Applying Theorem 2.2, we obtain the specific transformation of variables

$$u(x, \varepsilon) = \left(y_1 \sum_{k=0}^{\infty} (2^{-1/2} y_2)^k \varepsilon^k/k!, \, y_2 \right) = (y_1 \exp(2^{-1/2} \varepsilon y_2), \, y_2).$$

This same transformation is also easily obtained by solving (2.2).

Let us now return to the general discussion of how to choose the U_m in a way that will simplify the transformed equations.

The coefficients $g_m(y)$ of the transformed vector field are the diagonal elements of the triangle (2.10), $g_m(y) = f_0^{(m)}$. From formula (2.8) and the special form of f in (2.15), it is easy to verify that

$$(2.16) \qquad g_{m+1} = \Gamma_{m-1} - M U_m,$$

$$(2.17) \qquad Mv(x) = \frac{\partial v(\pi x)}{\partial x} Ax - Av(\pi x),$$

where Γ_{m-1} depends on $U_0, U_1, \ldots, U_{m-1}$ and the elements $f_j^{(k)}, j + k \leq m$, $0 \leq k \leq m$, $0 \leq j \leq m + 1$, of the triangle (2.10). Thus, if $U_j, 0 \leq j \leq m - 1$, $f_j, 0 \leq j \leq m + 1$, are known, then Γ_{m-1} is known.

In the simplification of the g_m by an appropriate choice of the homogeneous polynomial U_m of degree $m + 2$, the operator M will play a fundamental role. Let us now make the assumption that $A\pi = \pi A$. Then

$$\frac{\partial v(\pi x)}{\partial x} Ax = \frac{\partial v(\pi x)}{\partial (\pi x)} \pi A \pi x$$

is a function of only πx. Let $m = \dim \pi \mathbb{C}^n$ and identify $\pi \mathbb{C}^n$ with $\mathbb{C}^m$ denoting by s the element of $\mathbb{C}^m$ corresponding to πx and by $\tilde{v}$ the map defined on $\mathbb{C}^n$ corresponding to $v \circ \pi$. Then there is an $m \times m$ matrix B such that

$$\frac{\partial v(\pi x)}{\partial x} Ax = \frac{\partial \tilde{v}(s)}{\partial s} Bs$$

and the operator M becomes

$$(2.18) \qquad (M\tilde{v})(s) = \frac{\partial \tilde{v}(s)}{\partial s} Bs - A\tilde{v}(s), \qquad s \in \pi \mathbb{C}^n$$

where

$$M : V(k, m, n) \to V(k, m, n)$$

where $V(k, m, n)$ is the linear space of all n-vector functions $v(x)$ of the m-vector x which are homogeneous polynomials in x of degree k. We identify $V(k, m, n)$ with $\mathbb{C}^N$ for some $N = N(k, m, n)$ by associating each $v \in V(k, m, n)$ with the coefficients in v. The coefficients we order according to the *canonical order* in the following way. If $q = (q_1, \ldots, q_m)$, $q_j \geq 0$ integers, let $|q| = q_1 + \cdots + q_m$. If $v = (v_1, \ldots, v_n)$, $x = (x_1, \ldots, x_m)$ and

$$v(x) = \sum_{|q| = k} v^{(q)} x_1^{q_1} \cdots x_m^{q_m}$$

then we say $v_k^{(q)}$ precedes $v_l^{(p)}$ if the first nonzero difference $l - k, p_1 - q_1, \ldots, p_m - q_m$ is positive.

If $\dim \pi \mathbb{C}^n = m$, then the operator M in (2.18) is a linear operator on $V(k, m, n)$. The space $V(k, m, n)$ has two decompositions

$$\begin{aligned} V(k, m, n) &= V_1^d(k, m, n) \oplus V_2^d(k, m, n) \\ &= V_1^r(k, m, n) \oplus V_2^r(k, m, n) \end{aligned}$$

where $V_2^d(k,m,n)$ is the null space of M and $V_1^r(k,m,n)$ is the range of M. Let P_1^d be the projection of $V(k,m,n)$ onto $V_1^d(k,m,n)$ along $V_2^d(k,m,n)$. Similarly, let P_1^r be the projection onto $V_1^r(k,m,n)$ along $V_2^r(k,m,n)$. Let M^{-1} be the right inverse of M taking $V_1^r(k,m,n)$ onto $V_1^d(k,m,n)$. The superscripts d, r are used to suggest respectively the decompositions in the domain and range of M.

We are now in a position to give a precise definition of a normal form for Eq. (2.1) relative to the matrix A and projection π and the decompositions of $V(k,m,n)$.

Definition 2.4. Suppose f satisfies (2.15) and $A\pi = \pi A$. Eq. (2.4) is a *normal form for* Eq. (2.1) *relative to the matrix A, projection π and projections P_1^d, P_1^r if the elements g_m, U_m satisfy*

$$(2.19) \qquad \begin{aligned} g_{m+1}(\pi y) &= (I - P_1^r)\Gamma_{m-1}(\pi y) \\ U_m(\pi y) &= M^{-1}P_1^d\Gamma_{m-1}(\pi y) \end{aligned}$$

where the Γ_{m-1} are defined in (2.16).

The meaning of this definition of normal form will become clear after the consideration of many special cases. To do examples, we need to know the spectrum $\sigma(M)$ of M.

Lemma 2.5. *For a given $n \times n$ matrix A and $m \times m$ matrix B define*

$$L(k,m,n): V(k,m,n) \to V(k,m,n)$$

$$(Lv)(x) = \frac{\partial v(x)}{\partial x}Bx - Av(x).$$

If $\sigma(A) = \{\lambda_1, \ldots, \lambda_n\}$, $\sigma(B) = \{\mu_1, \ldots, \mu_m\}$, then

$$\sigma(L(k,m,n)) = \{(q,\mu) - \lambda_j, \qquad j = 1, 2, \ldots, n; \; |q| = k\}$$

where $(q,\mu) = q_1\mu_1 + \cdots + q_m\mu_m$.

Proof. A complex number Λ is in $\sigma(L)$ if and only if there is a nonzero $h \in V(k,m,n)$ such that $Lh = \Lambda h$. If S is an $m \times m$ nonsingular matrix and T is an $n \times n$ nonsingular matrix and $g(y) = T^{-1}h(Sy)$ then $T^{-1}Lh(x) = \tilde{L}g(y)$ where $x = Sy$,

$$\tilde{L}g(y) = \frac{\partial g(y)}{\partial y}S^{-1}BSy - T^{-1}ATg(y).$$

Also, $Lh = \Lambda h$ is equivalent to $\tilde{L}g = \Lambda g$ and the eigenvalues of these two operators are the same.

Choose S, T so that $S^{-1}BS$, $T^{-1}AT$ are lower triangular Jordan canonical matrices with nondiagonal terms σ_k, $k = 2, 3, \ldots, m$, τ_k, $k = 2, 3, \ldots, n$, respectively. Put $\sigma_1 = \tau_1 = 0$, $x_0 = 0$, $g_0 = 0$. If $\tilde{L}g = f = (f_1, \ldots, f_n)$, then

$$f_j(x) = \sum_{i=1}^{m} \frac{\partial g_j}{\partial x_i} (\mu_i x_i + \sigma_i x_{i-1}) - \lambda_j g_j - \tau_j g_{j-1}.$$

Let $f_j^{(q)}$, $g_j^{(q)}$ be, respectively, the coefficients of $x_1^{q_1}$, $x_2^{q_2} \cdots x_m^{q_m}$ in f_j, g_j. Using the properties of differentiation of homogeneous polynomials, we have

$$(2.20) \qquad f_j^{(q)} = [(q, \mu) - \lambda_j] g_j^{(q)} + \sum_{i=2}^{m} q_i \sigma_i g_j^{(q-e_{i-1}+e_i)} - \tau_j g_{j-1}^{(q)}$$

where $e_i = (\delta_{1i}, \ldots, \delta_{mi})$, $\delta_{kj} = 1$ if $k = j$, $= 0$ if $k \neq j$. Since $g_j^{(q)}$ is preceded by $g_j^{(q-e_{i-1}+e_i)}$ and $g_{j-1}^{(q)}$, it follows that the equation $\tilde{L}g = \Lambda g$ has a nonzero solution if and only if $\Lambda = (q, \mu) - \lambda_j$ for some $q = (q_1, \ldots, q_m)$, λ_j. This proves the lemma. $\square$

Remark 2.6. In the application of the above transformation theory, it is useful to notice that (2.20) also gives the eigenvectors of the null space of L whenever it is representable in diagonal form. For the example $\dot{x}_1 = -x_1 + x_1 x_2$, $\dot{x}_2 = 2^{1/2} x_2$ considered above with $\pi = I$, the operator M is given by $Mv(x) = (\partial v/\partial x)Ax - Av(x)$ and it is easy to check that M has an inverse on $L(k, n, n)$ if $k \geq 2$. For $k = 1$, the null space of M is generated by the set $\{(x_1, 0), (0, x_2)\}$. The projection operators P_1^d, P_1^r can now be chosen so that the normal form is $\dot{x}_1 = -x_1$, $\dot{x}_2 = 2^{1/2} x_2$.

12.3. More on Normal Forms

If $z \in \mathbb{C}^n$, $z = (z_1, \ldots, z_n)$ and $q = (q_1, \ldots, q_n)$ where the q_j are nonnegative integers, let

$$z^q = z_1^{q_1} \cdots z_n^{q_n}$$
$$|q| = q_1 + \cdots + q_n.$$

For any C^∞-function $f : \mathbb{C}^n \to \mathbb{C}^n$, the formal power series of f is denoted by

$$f(z) = \sum_{|q| \geq 0} f^{(q)} z^q.$$

Suppose A is an $n \times n$ constant matrix, $Z(z)$ is a C^∞-function of z whose power series expansion begins with terms of degree ≥ 2 and consider the equation

$$\dot{z} = Az + Z(z) \tag{3.1}$$

in a neighborhood of $z = 0$. For a given projection $\pi: \mathbb{C}^n \to \mathbb{C}^n$, $A\pi = \pi A$, consider a transformation of variables

$$z = w + h(\pi w) = w + \sum_{|q| \geq 2} h^{(q)}(\pi w)^q \tag{3.2}$$

with the differential equation for w given by

$$\begin{aligned}
\dot{w} &= Aw + W(w) \\
W(w) &= \sum_{|q| \geq 2} W^{(q)} w^q.
\end{aligned} \tag{3.3}$$

The objective is to choose the coefficients $h^{(q)}$ in the transformation (3.2) in a way that will make the behavior of the solutions of (3.3) become more transparent. Rather than present the theory in the manner just described, we will use the results of the previous section.

If $z = \varepsilon x$, $\varepsilon \in \mathbb{C}$, $x \in \mathbb{C}^n$, then

$$\dot{x} = f(x, \varepsilon) = Ax + \sum_{m \geq 1} f_m(x)\varepsilon^m/m! \tag{3.4}$$

where

$$f_m(x) = m! \sum_{|q| = m+1} Z^{(q)} x^q, \qquad m \geq 1 \tag{3.5}$$

and $f_m(x)$ is a homogeneous polynomial in x of degree $m + 1$. Define $f_0(x) = Ax$.

Let $u(y, \varepsilon)$ be the solution (which we assume is unique) of the equation

$$\frac{\partial u}{\partial \varepsilon} = U(\pi u, \varepsilon), \qquad u(y, 0) = y \tag{3.6}$$

where $U(s, \varepsilon)$, $s \in \pi \mathbb{C}^n$, is given by

$$U(s, \varepsilon) = \sum_{m \geq 0} U_m(s)\varepsilon^m/m! \tag{3.7}$$

and $U_m(s)$ is a homogeneous polynomial of degree $m + 2$.

It follows that

$$u(y, \varepsilon) = y + \sum_{m \geq 1} u_m(\pi y)\varepsilon^m/m!. \tag{3.8}$$

If $x = u(y, \varepsilon)$ in (3.4), then the differential equation for y is

$$(3.9) \qquad \dot{y} = g(y, \varepsilon) = Ay + \sum_{m \geq 1} g_m(y)\varepsilon^m/m!$$

where the g_m are computed recursively by formulas (2.8) of Theorem 2.1.

Furthermore, Lemma 2.3 implies the $g_m(y)$ are homogeneous polynomials in y of degree $m + 1$ and the transformation u in (3.8) has $u_m(s)$ homogeneous in s of degree $m + 1$. Consequently, the change of variables

$$(3.10) \qquad \varepsilon x = z, \qquad \varepsilon y = w, \qquad z = w + \sum_{m=1}^{\infty} u_m(\pi w)/m!$$

yields the differential equation for w,

$$(3.11) \qquad \dot{w} = Aw + \sum_{m \geq 1} g_m(w)/m!.$$

Thus, the transformation of variables $x = u(y, \varepsilon)$ on (3.4) defined by (3.6) leads to a transformation of variables of the form (3.2) applied to (3.1).

Conversely, for any transformation (3.2) applied to (3.1), if we put $z = \varepsilon x$, $w = \varepsilon y$, and

$$u(y, \varepsilon) = y + \sum_{m \geq 1} \left(m! \sum_{|q| = m+1} h^{(q)}(\pi y)^q \right) \varepsilon^m/m!$$

then the transformation $x = u(y, \varepsilon)$ yields a differential equation for y of the form (3.9). Furthermore, it is easy to verify that $u(y, \varepsilon)$ satisfies a differential equation of the form (3.6).

Consequently, transforming Eq. (3.4) by transformations defined by (3.6) is equivalent to transforming (3.1) by transformations of the form (3.2). Even though these two processes are equivalent, they are computed much differently. The important thing to notice is that the coefficients in (3.9) are defined recursively from the coefficients $f_m(x)$ and $U_m(s)$. The transformation of variables is not needed to obtain the new equation as is the case for transformation (3.2) applied to (3.1). However, the transformation can be obtained by the recursive formulas mentioned above if desired.

Definition 3.1. The *normal form for* Eq. (3.1) relative to the projection π, $A\pi = \pi A$, is defined to be the differential equation (3.3) for which

$$(3.12) \qquad \sum_{|q|=m+1} W^{(q)}w^q = g_m(w)/m!$$

and the $g_m(w)$, $U_m(s)$ are determined by formulas (2.19), (2.16).

Before giving any consequences of this definition, let us first observe that the theory of normal forms is applicable to equations that depend on a small vector parameter. In fact, the equation

$$(3.13) \qquad \dot{\xi} = B(\mu)\xi + h(\xi, \mu)$$

is equivalent to the equation

$$(3.14) \qquad \begin{aligned} \dot{\xi} &= B(\mu)\xi + h(\xi, \mu) \\ \dot{\mu} &= 0 \end{aligned}$$

which has the form (3.1) with $x = (\xi, \mu)$, $A = \mathrm{diag}(B(0), 0)$. The normal form will be given by

$$(3.15) \qquad \begin{aligned} \dot{\eta} &= B(\mu)\eta + H(\eta, \mu) \\ \dot{\mu} &= 0 \end{aligned}$$

for an appropriate function $H(\eta, \mu)$. In the following, we will not write the equation $\dot{\mu} = 0$ when we speak of the normal form for Eq. (3.13).

We now state some consequences of the above concepts.

Theorem 3.2. *If the eigenvalues $(\lambda_1, \ldots, \lambda_n)$ of the $n \times n$ matrix A satisfy*

$$(3.16) \qquad (q, \lambda) - \lambda_k \neq 0 \quad \text{for } k = 1, 2, \ldots, n, |q| \geq 2$$

then the normal form for Eq. (3.1) relative to A and the identity projection is

$$(3.17) \qquad \dot{w} = Aw;$$

that is, there is a formal change of coordinates of the form

$$z = w + h(w), \qquad h(w) = \sum_{|q| \geq 2} h(w)w^q$$

taking (3.1) to (3.17).

Proof. For $\pi = I$, the identity, the operator M is given by

$$Mv(x) = \frac{\partial v(x)}{\partial x} Ax - Av(x)$$

for $v \in V(k, n, n)$. Hypothesis (3.16) implies M is nonsingular for every k. Thus, the range of M is $V(k, n, n)$ and the projection $P_1^r = I$, the identity. Thus, the definition of the normal form for (3.1) implies $g_m(y) = 0$ for all $m \geq 1$. This proves the theorem. $\square$

Let us consider some examples where (3.16) is not satisfied.

EXAMPLE 3.3. If $A = 0$, $\pi = I$, then $Mv = 0$ for all $v \in V(k, n, n)$ for every k. Thus, the range of M is $\{0\}$, the projection $P_1^r = 0$ and so any normal form for (3.1) has $U_m = 0$ for all m. From (2.19), (2.16) this implies $\Gamma_{m-1} = f_m$, and the normal form is the vector field itself.

EXAMPLE 3.4. Suppose $n = 2$, $\pi = I$, $A = \text{diag}(1, 3)$ and let M_k be the operator $M_k : V(k, 2, 2) \rightarrow V(k, 2, 2)$, $k \geq 2$. Then it is easy to verify that M_k is nonsingular if $k \neq 3$ and $\mathcal{N}(M_3)$ is spanned by the vector $(0, x_1^3)$. Thus the normal form for (3.1) is

$$\dot{y}_1 = y_1, \qquad \dot{y}_2 = 3y_2 + cy_1^3$$

where c is the coefficient of x_1^3 in the second equation of the original system.

EXAMPLE 3.5. Suppose $n = 2$, $\pi = I$, $A = \text{diag}(5, -2)$. With the notation of Example 3.4, M_k is nonsingular unless there is an integer $i \geq 1$ such that if $7i + 1 = k$, then

$$\mathcal{N}(M_k) = \text{span}\left\{ \begin{pmatrix} x_1^{2i+1}x_2^{5i} \\ 0 \end{pmatrix}, \begin{pmatrix} 0 \\ x_1^{2i}x_2^{5i-1} \end{pmatrix} \right\}$$

The norm form for (3.1) is given by

$$\dot{y}_1 = 5y_1 + \sum_{i \geq 1} a_i y_1^{2i+1} y_2^{5i}$$

$$\dot{y}_2 = -2y_2 + \sum_{i \geq 1} b_i y_1^{2i} y_2^{5i+1}$$

where the coefficients a_i, b_i are the coefficients of the corresponding monomials in the formal power series of the original system.

EXAMPLE 3.6. Suppose $n = 2$, $\pi = I$, $A = \text{diag}(\lambda, -\lambda)$, $\lambda \neq 0$. With the notation as in Example 3.4, M_k is nonsingular if k is even. If k is odd, let $V_2(k, 2, 2) \subset V(k, 2, 2)$ be the two dimensional space spanned by the functions $(x_1(x_1 x_2)^{(k-1)/2}, 0)$, $(0, x_2(x_1 x_2)^{(k-1)/2})$ and let

$$V_1(k, 2, 2) = \{v(x) = \sum_{|q| = k} v^{(q)} x_1^{q_1} x_2^{q_2},$$

$$v^{(q)} = (v_1^{(q)}, v_2^{(q)}) : v_i^{(q)} = 0 \quad \text{if } q_1 - q_2 = (-1)^{i+1}\}.$$

Then

$$V(k, 2, 2) = V_1(k, 2, 2) \oplus V_2(k, 2, 2)$$
$$MV_2(k, 2, 2) = \{0\}$$
$$MV_1(k, 2, 2) = V_1(k, 2, 2).$$

Thus, the normal form for Eq. (3.1) relative to this matrix A, the projection $\pi = I$ and the above decomposition of $V(k, 2, 2)$ has

$$g_m \in V_2(k, 2, 2), \qquad U_m \in V_1(k, 2, 2), \qquad m \geq 0.$$

Thus, the normal form is given by

(3.18)
$$\begin{aligned}
\dot{y}_1 &= \lambda y_1 + y_1 f_1(y_1 y_2) \\
\dot{y}_2 &= -\lambda y_2 + y_2 f_2(y_1 y_2)
\end{aligned}$$

where $f_j(y_1 y_2)$ is a formal power series in the product $y_1 y_2$.

If λ is real and the differential equation is real, then the linear part of this example corresponds to a saddle point with the normal form given by Eq. (3.18).

If $\lambda = i\omega$, ω real, then the equation could correspond to a real system

(3.19)
$$\begin{aligned}
\dot{z}_1 &= \omega z_2 + Z_1(z_1, z_2) \\
\dot{z}_2 &= -\omega z_1 + Z_2(z_1, z_2)
\end{aligned}$$

for which the linear part corresponds to a center. The normal form (3.18) in real coordinates has the form

(3.20)
$$\begin{aligned}
\dot{y}_1 &= \omega y_2 + y_1 f(r^2) + y_2 g(r^2) \\
\dot{y}_2 &= -\omega y_1 + y_2 f(r^2) - y_1 g(r^2)
\end{aligned}$$

where f, g are functions of $r^2 = y_1^2 + y_2^2$.

The reader should be aware of the fact that the normal form depends upon the representation of the linear part A of the equation. If A is a diagonal matrix, as in the above examples, then the normal form can never contain terms which are not in the expansions of the original vector field. However, this may not be the case if we take a nondiagonal representation of A.

This latter remark also means that new terms may appear in the normal form if it is not possible to diagonalize A.

The next example brings out the role of the operator π. Consider the system of differential equations

(3.21)
$$\begin{aligned}
\dot{\xi} &= B\xi + D(\xi, \eta) \\
\dot{\eta} &= C\eta + E(\xi, \eta)
\end{aligned}$$

where ξ, η are vectors of dimension n_1, n_2, respectively, $n_1 + n_2 = n$ and D, E have power series expansions in ξ, η beginning with terms of degree ≥ 2. If $\mu_1, \ldots, \mu_{n_1}$ are the eigenvalues of the $n_1 \times n_1$ matrix B and $v_1, \ldots, v_{n_2}$ are the eigenvalues of the $n_2 \times n_2$ matrix C, suppose

(3.22)
$$(q, \mu) - v_k \neq 0, \qquad k = 1, 2, \ldots, n_2, \qquad |q| \geq 2.$$

If $x = (\xi, \eta)$ and π is the projection, $\pi x = (\xi, 0)$, then the subspace $\pi \mathbb{R}^n$ is invariant under the solutions of the linear differential equation

$$(3.23) \qquad \dot{x} = Ax, \qquad A = \begin{bmatrix} B & 0 \\ 0 & C \end{bmatrix}$$

The natural question is the following: Is there an invariant surface of (3.21) near $(\xi, \eta) = (0, 0)$ which is tangent to $\pi \mathbb{R}^n$ at zero and what is a normal form for the vector field on this invariant surface? This question can be answered by the above concept of normal form using the projection π.

For A defined in (3.23) and $\pi x = (\xi, 0)$ we have $\pi Ax = A\pi x = (B\xi, 0)$. Thus, the operator M in (2.18) satisfies

$$(Mv)(\xi) = \frac{\partial v(\xi)}{\partial \xi} B\xi - Av(\xi)$$

where A is defined in (3.23). If $v = (v^1, v^2)$, $Mv = (M_1 v^1, M_2 v^2)$ where v^j, $M_j v^j$ have dimension n_j, then

$$(M_1 v^1)(\xi) = \frac{\partial v^1(\xi)}{\partial \xi} B\xi - Bv^1(\xi)$$

$$(M_2 v^2)(\xi) = \frac{\partial v^2(\xi)}{\partial \xi} B\xi - Cv^2(\xi)$$

By hypothesis (3.22), the operator M_2 on $V(k, n_1, n_2)$ is nonsingular for every $k \geq 2$. If $g_m = (g_m^1, g_m^2)$ and we choose g_m to be a normal form for (3.1) according to Def. 3.1, then

$$g_m^{(2)}(\pi y) = g_m^{(2)}(y^1, 0) = 0$$

where $y = (y^1, y^2)$ with y^j of dimension n_j. This means that the differential equation (3.21) can be transformed to a system in coordinates $(\tilde{\xi}, \tilde{\eta})$ which satisfy

$$(3.24) \qquad \begin{aligned} \dot{\tilde{\xi}} &= B\tilde{\xi} + \tilde{D}(\tilde{\xi}, \tilde{\eta}) \\ \dot{\tilde{\eta}} &= C\tilde{\eta} + \tilde{E}(\tilde{\xi}, \tilde{\eta}) \end{aligned}$$

with

$$(3.25) \qquad \tilde{E}(\tilde{\xi}, 0) = 0.$$

The set $\tilde{\eta} = 0$ is an invariant manifold of (3.24). Also the vector field $B\tilde{\xi} + \tilde{D}(\tilde{\xi}, 0)$ is in normal form. Finally, the terms of degree ≥ 2 of the transformation depend only on ξ.

The vector field in (3.24) is called a *normal form for an invariant manifold* of (3.21). These results are summarized in

Theorem 3.7. *If* (3.22) *is satisfied, then there is a formal change of variables* $\xi = \tilde{\xi} + h_1(\tilde{\xi})$, $\eta = \tilde{\eta} + h_2(\tilde{\xi})$ *which takes* (3.21) *to Eq.* (3.24) *having the invariant manifold* $\tilde{\eta} = 0$ *and the vector field on this invariant manifold in normal form.*

There are four particularly interesting cases where the hypothesis (3.22) is satisfied and thus where Theorem 3.7 is applicable:

(i)   Re $\lambda(B) < 0$, Re $\lambda(C) \geq 0$ (exponentially stable manifold)
(ii) Re $\lambda(B) > 0$, Re $\lambda(C) \leq 0$ (exponentially unstable manifold)
(iii) Re $\lambda(B) \geq 0$, Re $\lambda(C) < 0$ (center unstable manifold)
(iv) Re $\lambda(B) \leq 0$, Re $\lambda(C) > 0$ (center stable manifold).

If, in (i), Re $\lambda(C) > 0$, then the invariant manifold is the stable manifold. If, in (ii) Re $\lambda(C) < 0$, then the invariant manifold is the unstable manifold. If Re $\lambda(B) = 0$ in case (iii) or (iv), the invariant manifold is the center manifold.

Remark 3.8. If the functions D, E in (3.21) depend on a parameter α, then the normal form is obtained by adjoining the equation $\dot{\alpha} = 0$. This does not affect the condition (3.22) and so Theorem 3.7 is valid for this case.

EXAMPLE 3.9. Let us reconsider the example considered in Section 9.2.10; namely, the equation

$$\dot{x}_1 = \alpha x_1 - x_1^3 + x_1 x_2$$
$$\dot{x}_2 = -x_2 + x_2^2 - x_1^2$$

where α is a small parameter. Adjoining the equation $\dot{\alpha} = 0$, this becomes a special case of (3.1) with $A = \mathrm{diag}(0, -1, 0)$, $n = 3$. Let $\pi : \mathbb{R}^3 \to \mathbb{R}^3$, $\pi x = (x_1, 0, x_3)$. Finding a normal form for this equation is equivalent to finding a center manifold for the two dimensional system for α small.

We apply the previous procedure. The dimension of $\pi \mathbb{R}^3$ is 2 and the operator B in the definition of M is zero. Thus, $Mv = -Av$, $v = (v_1, v_2, -v_3)$. If $M_k = M | V(k, 2, 3)$, then $\mathscr{R}(M_2)$ is the span of the vectors $\{(0, x_1^2, 0),$ $(0, x_1 x_3, 0), (0, x_3^2, 0)\}$ and $\mathscr{N}(M) = \mathscr{R}(M_2)^{\perp}$. It is also easy to characterize the range and null space of M_k. Performing the computations according to the algorithm that has been given above, one obtains after several computations that the normal form up through the terms of order four is

$$\dot{\xi} = \alpha \xi + \xi \eta - 2\xi^3 + 2\alpha \xi^3, \qquad \dot{\eta} = -\eta + \eta^2.$$

The approximate center manifold is given by $\eta = 0$ and the flow on this manifold is approximately $\dot{\xi} = \alpha \xi - 2\xi^3 + 2\alpha \xi^3$.

The corresponding transformation that leads to this equation can be shown to be $x_1 = \xi$, $x_2 = \eta - \xi^2 + 2\alpha\xi^2 - 3\xi^4 - 4\alpha^2\xi^2$ up through terms of order four.

EXAMPLE 3.10. Suppose $\omega > 0$, α is a small real vector parameter and consider the equation

$$
\begin{aligned}
\dot{\xi}_1 &= \omega\xi_2 + D_1(\xi, \eta, \alpha) \\
\dot{\xi}_2 &= -\omega\xi_1 + D_2(\xi, \eta, \alpha) \\
\dot{\eta} &= C\eta + N(\xi, \eta, \alpha)
\end{aligned}
\tag{3.26}
$$

where $\xi_j \in \mathbb{R}$, $\xi = (\xi_1, \xi_2)$, $\eta \in \mathbb{R}^{n-2}$ and D_1, D_2, N are real valued functions of ξ, η, α vanishing for $(\xi, \eta) = (0, 0)$.

If the eigenvalues $v_1, \ldots, v_{n-2}$ of C satisfy

$$
v_k \neq im\omega, \qquad m = \text{integer}, \qquad k = 1, 2, \ldots, n - 2,
\tag{3.27}
$$

then Theorem 3.7 and Example 3.6 imply that a normal form for (3.26) can be chosen as

$$
\begin{aligned}
\dot{x}_1 &= (\omega + p(r^2, \alpha))x_2 + x_1 q(r^2, \alpha) + h_1(x, y, \alpha) \\
\dot{x}_2 &= -(\omega + p(r^2, \alpha))x_1 + x_2 q(r^2, \alpha) + h_2(x, y, \alpha) \\
\dot{y} &= Cy + Y(x, y, \alpha)
\end{aligned}
\tag{3.28}
$$

where $r^2 = x_1^2 + x_2^2$ and h_1, h_2, Y vanish for $y = 0$. On the invariant manifold $y = 0$, if $x_1 = r \cos \omega\theta$, $x_2 = -r \sin \omega\theta$, then

$$
\begin{aligned}
\dot{\theta} &= 1 + \omega^{-1} p(r^2, \alpha) \\
\dot{r} &= r q(r^2, \alpha)
\end{aligned}
$$

and the equation for r is independent of θ. Periodic orbits correspond to zeros of $r q(r^2, \alpha)$.

A case of particular interest here is when $\text{Re } v_k < 0$ and we are interested in the problem of bifurcation of a periodic orbit from an unstable focus. Condition (3.27) is satisfied. Suppose m is a fixed positive integer and we have obtained the approximate normal form up through terms of order m; that is, we know the expansions of $p(s, \alpha)$, $q(s, \alpha)$ in s up through terms of order $m - 1$; that is,

$$
\begin{aligned}
p(s, \alpha) &= p_{m-1}(s, \alpha) + (\text{terms of order } m \text{ in } s) \\
q(s, \alpha) &= q_{m-1}(s, \alpha) + (\text{terms of order } m \text{ in } s).
\end{aligned}
\tag{3.29}
$$

Then there is a change of variables (which can be computed explicitly if desired) in (3.26) taking (ξ, η) to (x, y) so that

$$
\begin{aligned}
\dot{x}_1 &= (\omega + p_{m-1}(r^2, \alpha))x_2 + x_1 q_{m-1}(r^2, \alpha) + h_1^m(x, y, \alpha) \\
\dot{x}_2 &= -(\omega + p_{m-1}(r^2, \alpha))x_1 + x_2 q_{m-1}(r^2, \alpha) + h_2^m(x, y, \alpha) \\
\dot{y} &= Cy + Y^m(x, y, \alpha)
\end{aligned}
$$
(3.30)

where the Taylor series expansions of h_1^m, h_2^m, Y^m in x_1, x_2 at $y = 0$ begin with terms of order $2m + 1$.

On an exact center manifold of (3.26), this means that the flow up through terms of order $2m + 1$ in x_1, x_2 is given by

$$
\begin{aligned}
\dot{x}_1 &= (\omega + p_{m-1}(r^2, \alpha))x_2 + x_1 q_{m-1}(r^2, \alpha) \\
\dot{x}_2 &= -(\omega + p_{m-1}(r^2, \alpha))x_1 + x_2 q_{m-1}(r^2, \alpha)
\end{aligned}
$$
(3.31)

If we introduce polar coordinates $x_1 = r \cos \omega\theta$, $x_2 = -r \sin \omega\theta$ in (3.31), then

$$
\dot{\theta} = 1 + \frac{1}{\omega} p_{m-1}(r^2, \alpha)
$$

(3.32)

$$
\dot{r} = r q_{m-1}(r^2, \alpha).
$$

Equations (3.32) represent the flow on the center manifold up through terms of order $m + 1$. Thus, we see that the determination of the approximate normal form has resulted in an equation for which the differential equation for r is independent of θ. This also means that the bifurcation function for periodic orbits of Eq. (3.30) is given approximately by $q_{m-1}(r^2, \alpha)$. The zeros of this function give rise to periodic orbits for the original equation as discussed in Section 9.5. The stability properties of these orbits are also discussed there.

We remark once more that the functions in (3.32) can be computed recursively by the formulas (2.19).

EXAMPLE 3.11. Consider the van der Pol equation

$$
\dot{x}_1 = x_2, \qquad \dot{x}_2 = -x_1 + \alpha(1 - x_1^2)x_2
$$

Adjoining the equation $\dot{\alpha} = 0$, we can determine a normal form. From Remark 2.6, it is more convenient to do this with the matrix A diagonal. Transforming coordinates $z_1 = x_1 - ix_2$, $z_2 = ix_1 - x_2$, one obtains a complex system $\dot{z} = Az + f(z, \alpha)$ where $A = \text{diag}(i, -i)$. Applying the procedure above and changing back to real variables, the normal form can be shown to be

$$\dot{\eta}_1 = \left(1 - \frac{\alpha^2}{4}\right)\eta_2 + \eta_1\left[\frac{\alpha}{2} - \frac{\alpha}{8}(\eta_1^2 + \eta_2^2)\right]$$

$$\dot{\eta}_2 = -\left(1 - \frac{\alpha^2}{4}\right)\eta_1 + \eta_2\left[\frac{\alpha}{2} - \frac{\alpha}{8}(\eta_1^2 + \eta_2^2)\right]$$

up through terms of order four in (α, η, η_2). As expected, this equation has the form (3.28).

12.4. The Method of Averaging

In applications, one often encounters systems of differential equations of the form

(4.1)
$$\dot{\theta} = \omega + \varepsilon\Theta(\theta, r, \varepsilon)$$
$$\dot{r} = \varepsilon R(\theta, r, \varepsilon)$$

where ε is a small parameter, θ, ω, Θ are vectors of dimension n_1, r, R of dimension n_2, $n_1 + n_2 = n$, and the functions Θ, R are 2π-periodic in each component of the vector θ. We assume Θ, R are C^∞-functions.

The method of averaging consists in determining a transformation of variables

(4.2)
$$\theta = u_1(\varphi, \rho, \varepsilon), \qquad\qquad r = u_2(\varphi, \rho, \varepsilon)$$
$$u_1(\varphi, \rho, 0) = \varphi, \qquad\qquad u_2(\varphi, \rho, 0) = \rho$$

which is 2π-periodic in each component of φ and chosen so that the transformed system

(4.3)
$$\dot{\varphi} = \omega + \varepsilon\Phi(\varphi, \rho, \varepsilon)$$
$$\dot{\rho} = \varepsilon\tilde{R}(\varphi, \rho, \varepsilon)$$

has the property that the functions $\Phi, \tilde{R}$ depend as little as possible upon φ. A precise definition will be given later.

The optimal situation would be to have $\Phi, \tilde{R}$ independent of φ. We will see that this is formally possible if the components of the vector ω are independent over the integers; that is, for every n_1-vector $k = (k_1, \ldots, k_{n_1})$ with integer components

(4.4)
$$(k, \omega) = k_1\omega_1 + \cdots + k_{n_1}\omega_{n_1} \neq 0 \quad \text{if } k \neq 0.$$

In general, the series representing the transformation (4.2) is divergent. In the applications, the coefficients of ε^m in the expansion of Θ, R are trigonometric polynomials in θ. In addition, we are only interested in the first few coefficients of ε^m in the transformed system. In this case, the averaging procedure given below will be exact up through terms of any finite order p provided the vector ω satisfies

$$(4.5) \qquad (k, \omega) \neq 0 \quad \text{for } k \neq 0, |k| \leq q$$

where q is a computable quantity depending on the number of terms p considered.

If $x = (\theta, r)$, then (4.1) has the form

$$(4.6) \qquad \dot{x} = f(x, \varepsilon)$$

where

$$(4.7) \qquad f(x, \varepsilon) = \begin{bmatrix} \omega + \varepsilon\Theta(x, \varepsilon) \\ \varepsilon R(x, \varepsilon) \end{bmatrix} = f_0 + \sum_{m \geq 1} f_m(x)\varepsilon^m/m!$$

If $U(x, \varepsilon) \in \mathbb{R}^n$ is a given function of x, ε and $u(y, \varepsilon)$ is the solution of

$$(4.8) \qquad \frac{\partial u}{\partial \varepsilon} = U(u, \varepsilon), \qquad u(y, 0) = y$$

Then we can transform Eq. (4.1) by the transformation $x = u(y, \varepsilon)$ as in Section 2 to obtain the new differential equation for $y = (\varphi, \rho)$ as

$$(4.9) \qquad \dot{y} = g(y, \varepsilon) = f_0 + \sum_{m \geq 1} g_m(y)\varepsilon^m/m!.$$

If

$$(4.10) \qquad U(y, \varepsilon) = \sum_{m \geq 0} U_m(y)\varepsilon^m/m!$$

then we attempt to determine the $U_m(y)$ so that the functions $g_m(y) = g_m(\varphi, \rho)$ depend as little as possible on φ.

From the definition of the g_m in (2.8), (2.9) of Theorem 2.1 in terms of f_m, U_m, it is easy to verify that

$$(4.11) \qquad g_{m+1} = \Gamma_{m-1} + f_0 \times U_m$$

where Γ_{m-1} is obtainable from the $U_j, 0 \leq j \leq m - 1, f_j, 1 \leq j \leq m + 1$.
From (4.7), $f_0 = (\omega, 0)$ and so

$$(4.12) \qquad f_0 \times U_m = -\frac{\partial U_m(\theta, r)}{\partial \theta}\, \omega.$$

The extent to which Eq. (4.1) can be simplified by the transformation generated by (4.8), (4.10) depends upon the properties that $\Gamma_{m-1} - g_{m+1}$ must satisfy in order for the equation

$$(4.13) \qquad -f_0 \times U_m = \frac{\partial U_m(\theta, r)}{\partial \theta} \, \omega = \Gamma_{m-1}(\theta, r) - g_{m+1}(\theta, r)$$

to have a solution $U_m(\theta, r)$ which is 2π-periodic in the components of θ. The function g_m will be chosen so that this equation always has a formal solution and U_m will be chosen as a solution.

For the vector ω in (4.1) and any n_1-vector function $h(\theta)$, 2π-periodic in the components of θ, define the *mean value of h with respect to* ω as

$$(4.14) \qquad (M_\omega h)(\theta) = \lim_{T \to \infty} \frac{1}{T} \int_0^T h(\theta + \omega t) \, dt.$$

This function exists since the function in the integrand is almost periodic in t. The computation of $M_\omega h$ is easily obtained from the Fourier series of h.

Lemma 4.1. *If* $h(\theta)$ *has the Fourier series*

$$h(\theta) \sim \sum_k h_k e^{i(k,\theta)}$$

then

$$(4.15) \qquad (M_\omega h)(\theta) = \sum_{(k,\omega)=0} h_k e^{i(k,\theta)}.$$

If the nonresonance condition (4.4) is satisfied, then $(M_\omega h)(\theta)$ *is independent of* θ *and*

$$(4.16) \qquad M_\omega h = \frac{1}{(2\pi)^{n_1}} \int_0^{2\pi} \cdots \int_0^{2\pi} h(\theta) \, d\theta_1 \cdots d\theta_{n_1}.$$

Proof. We have

$$h(\theta + \omega t) \sim \sum_k h_k e^{i(k,\theta) + i(k,\omega)t}.$$

Formulas (4.15) and (4.16) now follow immediately from the definition (4.14). $\square$

Lemma 4.2. *If* h *is an* n_1*-vector function, 2π-periodic in the components of* θ, $h(\theta) \sim \sum_k h_k \exp i(k, \theta)$, *then the equation*

$$(4.17) \qquad \frac{\partial v}{\partial \theta} \, \omega = h$$

has a formal solution which is 2π-periodic in the components of θ if and only if $M_\omega h = 0$ and then

$$(4.18) \qquad v(\theta) \sim -i \sum_k (k, \omega)^{-1} h_k e^{i(k,\theta)}.$$

Proof. If $v(\theta) \sim \sum_k v_k \exp i(k, \theta)$ and satisfies (4.17), then $i(k, \omega)v_k = h_k$. Using Lemma 4.1, the proof is completed. $\square$

We remark that the series (4.18) may not converge because of the term $(k, \omega)^{-1}$ which can introduce small divisors. On the other hand, if $h(\theta)$ is a trigonometric polynomial, this series will always converge.

Definition 4.3. The *method of averaging applied to* (4.1) consists of the application of the transformation of variables (4.8), (4.10), where the $U_m(\varphi, \rho)$, $g_m(\varphi, \rho)$ are chosen by the inductive formulas

$$(4.19) \qquad \begin{aligned} g_{m+1} &= M_\omega \Gamma_{m-1} \\ \frac{\partial U_m}{\partial \varphi} \omega &= \Gamma_{m-1} - g_{m+1} \end{aligned}$$

The transformed Eq. (4.9), (4.3) are called the *averaged equations*.

If the nonresonant condition (4.4) is satisfied, then the relation (4.19) becomes

$$(4.20) \qquad \begin{aligned} g_{m+1}(\rho) &= \frac{1}{(2\pi)^{n_1}} \int_0^{2\pi} \cdots \int_0^{2\pi} \Gamma_{m-1}(\theta, \rho)\, d\theta_1 \cdots d\theta_{n_1} \\ \frac{\partial U_m}{\partial \varphi} \omega &= \Gamma_{m-1} - g_{m+1} \end{aligned}$$

and the averaged equations are independent of φ.

An important generalization of the method of averaging is concerned with the equation

$$(4.21) \qquad \begin{aligned} \dot\theta &= \omega + \varepsilon \Theta(\theta, r, z, \varepsilon) \\ \dot r &= \varepsilon R(\theta, r, z, \varepsilon) \\ \dot z &= Cz + \varepsilon Z(\theta, r, z, \varepsilon) \end{aligned}$$

where θ, r, z are vectors of dimension n_1, n_2, n_3 respectively, the eigenvalues of the matrix C have nonzero real parts and the functions Θ, R, Z are 2π-periodic in the components of θ.

In this case, we make a transformation of variables

$$\theta = \varphi + \tilde{u}_1(\varphi, \rho, \varepsilon)$$
$$r = \rho + \tilde{u}_2(\varphi, \rho, \varepsilon) \qquad (4.22)$$
$$z = w + \tilde{u}_3(\varphi, \rho, \varepsilon)$$

where $\tilde{u}_j(\varphi, \rho, 0) = 0$ for $j = 1, 2, 3$. The essential part of the transformation depends only on φ, ρ and not on w. This type of transformation is a special case of Section 2 with $\pi(\varphi, \rho, w) = (\varphi, \rho, 0)$; that is, π is the projection onto the first $n_1 + n_2$ components.

Suppose $x = (\theta, r, z)$, $f(x, \varepsilon)$ is the vector field defined by (4.21), $y = (\varphi, \rho, w)$ and the generator of the transformation $u = (u_1, u_2, u_3)$ in (4.22) is a differential equation of the form

$$\frac{\partial u}{\partial \varepsilon} = U(\pi u, \varepsilon). \qquad (4.23)$$

The transformed equation has the form (4.9) with the g_m, U_m satisfying the recursive relations (4.11). In this more general case, the operator $f_0 \times U_m$ is given by

$$f_0 \times U_m = -\frac{\partial U_m}{\partial \theta} \omega + \tilde{C} U_m \qquad (4.24)$$

where $\tilde{C}$ is the $(n_1 + n_2 + n_3) \times (n_1 + n_2 + n_3)$ matrix given by $\tilde{C} = \operatorname{diag}(0, C)$.

Let $U_m = (U_m^1, U_m^2)$, $g_m = (g_m^1, g_m^2)$, $\Gamma_{m-1} = (\Gamma_{m-1}^1, \Gamma_{m-1}^2)$ where U_m^2, g_m^2, Γ_{m-1}^2 have dimension $n_3 \geq 0$. In analogy with Definition 4.3, we define the *method of averaging* for Eq. (4.21) as the choice of U_m, g_m by the formulas

$$g_{m+1}^1 = M_\omega \Gamma_{m-1}^1, \qquad g_m^2 = 0$$
$$\frac{\partial U_m^1}{\partial \varphi} \omega = \Gamma_{m-1}^1 - g_{m+1}^1 \qquad (4.25)$$
$$\frac{\partial U_m^2}{\partial \varphi} \omega - C U_m^2 = \Gamma_{m-1}^2$$

Remark 4.4. Under more severe restrictions on the matrix C, one can define an averaging procedure for which the transformation (4.23) has the form

$$\theta = \varphi + \tilde{u}_1(\varphi, \rho, z, \varepsilon)$$
$$r = \rho + \tilde{u}_2(\varphi, \rho, z, \varepsilon)$$
$$z = z$$

The vector z is unchanged, but the transformation of θ, r depends on z. For references, see Section 8.

As an example, consider the van der Pol equation

$$\ddot{x} - \varepsilon(1 - x^2)\dot{x} + x = 0.$$

If $x = r \cos \theta$, $y = -r \sin \theta$, then

$$\begin{aligned}
\dot{\theta} &= 1 + \varepsilon(1 - r^2 \cos^2 \theta)\sin \theta \cos \theta \\
\dot{r} &= \varepsilon r(1 - r^2 \cos^2 \theta)\sin^2 \theta
\end{aligned}$$

Since (4.4) is satisfied, there is a formal transformation taking (θ, r) to (φ, ρ) such that the vector field for the new differential equation for (φ, ρ) is independent of φ. Applying the formulas (4.19), one obtains

$$\begin{aligned}
\dot{\varphi} &= 1 + \varepsilon^2 G_1(\rho, \varepsilon) \\
\dot{\rho} &= (\varepsilon/2)\rho(1 - \rho^2/4) + \varepsilon^2 G_2(\rho, \varepsilon)
\end{aligned}$$

for the terms up through order ε.

12.5. Integral Manifolds and Invariant Tori

In this section, we give some of the fundamental results on integral manifolds for autonomous and nonautonomous differential equations.

Definition 5.1. For a differential equation $\dot{z} = Z(t, z)$, $z \in \mathbb{R}^n$, a set $S \subset \mathbb{R}^{n+1}$ is said to be a *local invariant set* if, for any $(t_0, z_0) \in S$, there is a $T > 0$ such that the solution $z(t, t_0, z_0)$ of the equation exists for $|t - t_0| < T$ and $(t, z(t, t_0, z_0)) \in S$ for $|t - t_0| < T$. If we can choose $T = \infty$, we say S is an *invariant set*. If S has a manifold structure, we use the terms *local invariant manifold* and *invariant manifold*.

Let us first consider the equation

(5.1)
$$\begin{aligned}
\dot{\theta} &= \omega + \Theta(\theta, \rho, \lambda) \\
\dot{\rho} &= A\rho + R(\theta, \rho, \lambda)
\end{aligned}$$

where λ is a parameter in a Banach space Λ, $\theta \in \mathbb{R}^k$, $\rho \in \mathbb{R}^n$, $\omega = (\omega_1, \ldots, \omega_k)$, $\omega_j > 0$, $j = 1, 2, \ldots, k$, the functions Θ, R are periodic of period 2π in each component of the vector θ, are continuous together with all derivatives up through order $r \geq 2$,

(5.2)
$$\begin{aligned}
\Theta(\theta, \rho, 0) &= O(|\rho|) \\
R(\theta, \rho, 0) &= O(|\rho|^2) \quad \text{as } \rho \to 0
\end{aligned}$$

and the eigenvalues of the matrix A have nonzero real parts.

For Equation (5.1), we are interested in integral manifolds which have a parametric representation of the form $\rho = f(\theta, \lambda)$ with f being 2π-periodic in each component of θ, $f(\theta, 0) = 0$. If

$$S_\lambda = \{(\theta, \rho) : \rho = f(\theta, \lambda),\ \theta \in \mathbb{R}^k\}$$

then $S_\lambda \times \mathbb{R}$ is an integral manifold of (5.1) in the sense of Definition 5.1. The set S_λ itself also is invariant with respect to the solution of (5.1) and represents a k-dimensional torus in $(n + k)$-dimensional space.

The idea for proving the existence of such an S_λ is the following. Construct a mapping which takes k-dimensional tori into k-dimensional tori with the property that the fixed points of this map are invariant with respect to the solutions of (5.1). Then apply the Implicit Function Theorem to this map. Let

$$F_1 = \{f \in C^1(\mathbb{R}^k, \mathbb{R}^n) : f(\theta)\ \text{is}\ 2\pi\text{-periodic in each component of}\ \theta\}.$$

If the topology on F is the C^1-topology, then F_1 is a Banach space.

For any $f \in F_1$, consider the equation

$$
\begin{aligned}
\dot\theta &= \omega + P(\theta, \lambda, f) \\
\dot\rho &= A\rho + Q(\theta, \lambda, f)
\end{aligned}
$$

(5.3)

where

$$
\begin{aligned}
P(\theta, \lambda, f) &= \Theta(\theta, f(\theta), \lambda) \\
Q(\theta, \lambda, f) &= R(\theta, f(\theta), \lambda)
\end{aligned}
$$

(5.4)

For a fixed $\varphi \in \mathbb{R}^k$, let $\theta(t, \varphi, \lambda, f)$ be the solution of the first equation in (5.3) with $\theta(0, \varphi, \lambda, f) = \varphi$. It is now obvious that Equation (5.1) has an integral manifold of the type S_λ if and only if, for each fixed λ, there is an $f = f_\lambda \in F_1$ such that the vector function $(\theta(t, \varphi, \lambda, f), f_\lambda(\theta(t, \varphi, \lambda, f)))$ satisfies (5.3) for all $\varphi \in \mathbb{R}^k$.

Without loss in generality, we may assume $A = \operatorname{diag}(A_1, A_2)$ with the eigenvalues of $A_1(A_2)$ having negative (positive) real parts. If

$$
\begin{aligned}
K(u) &= \operatorname{diag}(e^{-A_1 u}, 0) && \text{for } u < 0 \\
&= \operatorname{diag}(0, -e^{-A_2 u}) && \text{for } u \geq 0
\end{aligned}
$$

(5.5)

then there are positive constants β, α such that

(5.6) $$|K(u)| \leq \beta \exp(-\alpha|u|), \qquad u \in (-\infty, \infty).$$

The following lemma gives a characterization of the solutions of Equation (5.3) which belong to F_1.

Lemma 5.2. *The function*

$$(5.7) \qquad T(\lambda, f)(\varphi) = \int_{-\infty}^{\infty} K(u) Q(\theta(u, \varphi, \lambda, f), \lambda, f) \, du$$

is continuous for (λ, f, φ) in $\Lambda \times F_1 \times \mathbb{R}^k$ and is 2π-periodic in each component of φ. Furthermore, there is a neighborhood U of $\lambda = 0$, V of $f = 0$ such that $T(\lambda, f)(\varphi)$ has continuous first derivatives with respect to (λ, f, φ) if $(\lambda, f, \varphi) \in U \times V \times \mathbb{R}^k$. In particular, $T(\lambda, f) \in F_1$ if $(\lambda, f) \in U \times V$. Finally, Equation (5.3) for $(\lambda, f) \in U \times V$ has an integral manifold

$$S = \{ (\theta, \rho) : \rho = g(\theta) \} \quad \text{with } g \in F_1$$

if and only if $g = T(\lambda, f)$.

Proof. Let us first show that if S in the statement of the Lemma is our integral manifold, then g must be $T(\lambda, f)$. We then prove the continuity and differentiability properties of $T(\lambda, f)$ which will ensure that $T(\lambda, f)$ does define an integral manifold of the type S.

If S is to be an integral manifold, then $(\theta(t, \varphi, \lambda, f), g(\theta(t, \varphi, \lambda, f)))$ must satisfy Equation (5.3) for every $\varphi \in \mathbb{R}^k$. Using the variation of constants formula, this implies

$$g(\varphi) = e^{-At} g(\theta(t, \varphi, \lambda, f)) - \int_0^t e^{-As} Q(\theta(s, \varphi, \lambda, f), \lambda, f) \, ds$$

for all $\varphi \in \mathbb{R}^k$.

Suppose $A = \operatorname{diag}(A_1, A_2)$ with the eigenvalues of $A_1(A_2)$ having negative (positive) real parts. If A_1 is an $n_1 \times n_1$ matrix, $g = (g_1, g_2)$, $Q = (Q_1, Q_2)$ with g_1, Q_1 vectors of dimension n_1, then

$$g_1(\varphi) = e^{-A_1 t} g_1(\theta(t, \varphi, \lambda, f)) - \int_0^t e^{-A_1 s} Q_1(\theta(s, \varphi, \lambda, f), \lambda, f) \, ds$$

$$g_2(\varphi) = e^{-A_2 t} g_2(\theta(t, \varphi, \lambda, f)) - \int_0^t e^{-A_2 s} Q_2(\theta(s, \varphi, \lambda, f), \lambda, f) \, ds$$

Since g_1, g_2 are bounded and $e^{-A_1 t} \to 0$ as $t \to -\infty$, $e^{-A_2 t} \to 0$ as $t \to +\infty$, we have

$$g_1(\varphi) = \int_{-\infty}^{0} e^{-A_1 s} Q_1(\theta(s, \varphi, \lambda, f), \lambda, f) \, ds$$

$$g_2(\varphi) = - \int_0^{\infty} e^{-A_2 s} Q_2(\theta(s, \varphi, \lambda, f), \lambda, f) \, ds$$

for all $\varphi \in \mathbb{R}^k$. With $K(u)$ defined in (5.5), this implies $g = T(\lambda, f)$ where $T(\lambda, f)$ is defined in (5.7). The estimate (5.6) implies this function is well defined. If a component of φ is translated by 2π, then the corresponding component of the function $\theta(s, \varphi, \lambda, f)$ is translated by 2π. This shows that

$T(\lambda, f)(\varphi)$ has the periodicity properties stated. The function $\theta(s, \varphi, \lambda, f)$ is continuous together with its first derivatives in (s, φ, λ, f) from the basic theory of differential equations (see Section 3.1). This fact together with the estimate (5.6) proves that $T(\lambda, f)(\varphi)$ is continuous in (λ, f, φ).

Let $\theta^*(u) = \theta(u, \varphi, \lambda, f)$. We now prove that

$$D_\lambda T(\lambda, f)(\varphi) = \int_{-\infty}^{\infty} K(u) \left[\frac{\partial Q(\theta^*(u), \lambda, f)}{\partial \lambda} + \frac{\partial Q(\theta^*(u), \lambda, f)}{\partial \theta} \cdot \frac{\partial \theta^*(u)}{\partial \lambda} \right] du$$

$$(5.8) \quad D_f T(\lambda, f)(\varphi) = \int_{-\infty}^{\infty} K(u) \left[\frac{\partial Q(\theta^*(u), \lambda, f)}{\partial f} + \frac{\partial Q(\theta^*(u), \lambda, f)}{\partial \theta} \cdot \frac{\partial \theta^*(u)}{\partial f} \right] du$$

$$D_\varphi T(\lambda, f)(\varphi) = \int_{-\infty}^{\infty} K(u) \frac{\partial Q(\theta^*(u), \lambda, f)}{\partial \theta} \cdot \frac{\partial \theta^*(u)}{\partial \varphi} \, du$$

for (λ, f) in a neighborhood $U \times V$ of $(0, 0)$. We only give the details on the existence of the integrals in these expressions since the proof that they are the derivatives is fairly obvious.

For the last integral, we need estimates on the function $z(u) = \partial \theta^*(u) / \partial \varphi = \partial \theta(u, \varphi, \lambda, f) / \partial \varphi$. This function satisfies

$$\dot{z} = \frac{\partial P}{\partial \theta} z, \qquad z(0) = I.$$

Thus,

$$|z(u)| \le e^{\gamma(\lambda, f)|u|}, \qquad u \in \mathbb{R},$$

where $\gamma(\lambda, f)$ is an upper bound on the norm of $\partial P / \partial \theta$. Using the expression (5.4) for $P(\theta, \lambda, f)$, we see that

$$\frac{\partial P(\theta, \lambda, f)}{\partial \theta} = \frac{\partial \Theta(\theta, f(\theta), \lambda)}{\partial \theta} + \frac{\partial \Theta(\theta, f(\theta), \lambda)}{\partial \rho} \cdot \frac{\partial f(\theta)}{\partial \theta}$$

Consequently, if

$$(5.9) \qquad \Theta(\theta, \rho, \lambda) = p(\theta)\lambda + q(\theta)\rho + o(|\lambda| + |\rho|)$$

then, in a neighborhood of $(\lambda, \rho) = 0$, the function $\gamma(\lambda, f)$ can be taken to be

$$(5.10) \qquad \gamma(\lambda, f) = 2 \max_{\theta} \{|p'(\theta)|, |q'(\theta)|, |q(\theta)|\}(|\lambda| + |f|_1)$$

where $p'(\theta) = dp(\theta)/d\theta$, $q'(\theta) = dq(\theta)/d\theta$.

Consequently, if we choose the neighborhood $U \times V$ of $(0, 0) \in \Lambda \times F_1$ so that

$$(5.11) \qquad \gamma(\lambda, f) < \alpha, \qquad (\lambda, f) \in U \times V,$$

where α is the constant in (5.6), then the third integral in (5.8) is well defined. The other integrals are estimated in the same way to complete the proof of the lemma. $\square$

If $r \geq 1$ and

$$F_r = \{f \in C^r(\mathbb{R}^k, \mathbb{R}^n) : f(\theta) \text{ is } 2\pi\text{-periodic in each component of } \theta\}$$

then one can obtain an analogue of Lemma 5.2 for any fixed r. However, the proof shows that the region $U \times V$ will depend on r and, in general, becomes smaller as r becomes larger. This is easily seen, for example, by deriving the formula for the second derivative $D_{\varphi\varphi} T(\lambda, f)(\varphi)$. This involves a bilinear form in $\partial\theta(u, \varphi, \lambda, f)/\partial\varphi$ and, thus, relation (5.11) needs to be replaced by $2\gamma(\lambda, f) < \alpha$. In general, one can prove that

$$(5.12) \qquad\qquad r\gamma(\lambda, f) < \alpha, \qquad (\lambda, f) \in U \times V$$

implies $T(\lambda, f) : F_r \to F_r$ and is C^r in λ, f.

Let us now return to the existence of an integral manifold of (5.1) of the form S_λ. From Lemma 5.2, such a manifold exists if and only if f is a fixed point of $T(\lambda, f)$. If $G(\lambda, f) = f - T(\lambda, f)$, then $G(0, 0) = 0$. Furthermore, from formula (5.8), and the fact that $Q(\theta, \lambda, f) = R(\theta, f(\theta), \lambda)$ and R satisfies (5.2), we have $D_f G(0, 0) = I$. The Implicit Function Theorem implies there is a unique solution f_λ of $G(\lambda, f_\lambda) = 0$ in a neighborhood of zero, $f_0 = 0$; that is, a unique fixed point f_λ of $T(\lambda, f_\lambda)$ in a neighborhood of zero.

If we take into account the above remarks, we have proved the following result.

Theorem 5.3. *If Θ, R in (5.1) are C^r functions, then there is a neighborhood U of $\lambda = 0$ and a neighborhood V of $\rho = 0$ and a C^r-function $f_\lambda(\theta)$, $\lambda \in U$, $\theta \in \mathbb{R}^k$, 2π-periodic in each component of θ, $f_\lambda(\theta) \in V$, $f_0(\theta) = 0$, $\theta \in \mathbb{R}^k$, such that the set*

$$S_\lambda = \{(\theta, \rho) : \rho = f_\lambda(\theta), \theta \in \mathbb{R}^k\}$$

is an integral manifold of Equation (5.1). Furthermore, this is the only integral manifold of in U which can be expressed as a graph over the θ-axis.

Theorem 5.4. *If all the eigenvalues of A have negative real parts, then the integral manifold S_λ in Theorem 5.3 is exponentially stable, i.e., for $|\rho_0|$ and $|\lambda|$ sufficiently small, θ_0 arbitrary*

$$|\rho(t) - f_\lambda(\theta(t))| \leq Le^{-\beta t}, \qquad t \geq 0$$

where $L, \beta > 0$ are constants and $(\theta(t), \rho(t))$ is the unique solution of (5.1) with initial condition (θ_0, ρ_0) at $t = 0$.

Proof. The proof is similar to that of Lemma 9.2.12 and Theorem 9.2.13. We first construct a function $J(t, \psi, r)$ by the implicit function theorem as follows. Given $J(t, \psi, r)$, define

$$T(\lambda, J)(t, \psi, r) = e^{At}r + \int_0^t e^{A(t-s)} R(\theta(s), J(s, \theta(s), r), \lambda) \, ds$$

where $\theta(s)$ is the unique solution of the equation

$$\dot{\theta} = \omega + \Theta(\theta, J(t, \theta, r), \lambda), \qquad \theta(t) = \psi.$$

The implicit function theorem implies that there exists a unique function J_λ such that $J_\lambda - T(\lambda, J_\lambda) = 0$ and $J_0(t, \psi, r) = e^{At}r + o(|r|)$ as $r \to 0$. Moreover, for $|\rho_0| + |\lambda|$ sufficiently small and θ_0 arbitrary,

$$\rho(t) = J_\lambda(t, \theta(t), \rho_0)$$

where $(\theta(t), \rho(t))$ is the unique solution of (5.1) with initial condition (θ_0, ρ_0) at $t = 0$. Using the fact that $J_\lambda = T(\lambda, J_\lambda)$, it is not difficult to show that $\partial J_\lambda(t, \psi, r)/\partial \psi$ is bounded by a constant L in a neighborhood of $\lambda = 0, r = 0$ and there is a $\beta > 0$ such that $\partial J_\lambda(t, \psi, r)/\partial r$ is bounded by $L \exp(-\beta t)$, $t \geq 0$, in the same neighborhood. For a given t, the point $(\theta(t), \rho(t))$ on the trajectory through (t_0, ρ_0) corresponds to the point $(\theta(t), f_\lambda(\theta(t)))$ on the manifold S_λ. This point on the manifold can be reached by following the flow on the manifold at time t starting at a point (ψ_0, r_0) at time zero. In fact, if

$$\psi_0 = \theta(-t, \theta(t, \theta_0, \rho_0), f_\lambda(\theta(t, \theta_0, \rho_0))),$$
$$r_0 = \rho(-t, \theta(t, \theta_0, \rho_0), f_\lambda(\theta(t, \theta_0, \rho_0))),$$

then $\theta(t, \psi_0, r_0) = \theta(t, \theta_0, \rho_0)$, $\rho(t, \psi_0, r_0) = f_\lambda(\theta(t, \theta_0, \rho_0))$. Thus,

$$\rho(t, \psi_0, r_0) = J_\lambda(t, \theta(t, \psi_0, r_0), r_0) = J_\lambda(t, \theta(t, \theta_0, \rho_0), r_0)$$

and

$$|f_\lambda(\theta(t, \theta_0, \rho_0)) - J_\lambda(t, \theta(t, \theta_0, \rho_0), \rho_0)| \leq Le^{-\beta t}|r_0 - \rho_0|$$

for $t \geq 0$. This completes the proof. $\square$

Remark 5.5. It is possible to prove also that the integral manifold in Theorem 5.3 has the same stability properties as the manifold $\rho = 0$ of the equation $\dot{\theta} = \omega, \dot{\rho} = A\rho$; that is, each point on S_λ behaves locally as a saddle point. There is a stable manifold (unstable manifold) with dimension equal to the number of eigenvalues of A of negative (positive) real part. The proofs are similar to the ones in Section 9 for center manifolds.

Remark 5.6. The flow on the manifold S_λ in Theorem 5.3 is given by $\theta(t)$, $\rho(t) = f_\lambda(\theta(t))$ where $\theta(t)$ is the solution of the equation

$$(5.13) \qquad \dot{\theta} = \omega + \Theta(\theta(t), f_\lambda(\theta(t)), \lambda), \qquad \theta(0) = \phi.$$

Remark 5.7. The neighborhood U of $\lambda = 0$ (that is, the size of the perturbation) in Theorem 5.3 depends upon the smoothness properties (the integer r in the theorem) desired of the integral manifold defined by $f_\lambda(\theta)$. In fact, the proof requires that $r\gamma(\lambda, f_\lambda) < \alpha$. The estimate for $\gamma(\lambda, f)$ in (5.10) and the fact that the flow on the manifold is given by (5.13) shows that $\gamma(\lambda, f_\lambda)$ is a measure of the strength of the flow on the integral manifold. More precisely, it measures how rapidly the solutions on the manifold can converge toward each other. From (5.6), the number α measures the strength of the stability or instability of the manifold. Thus, the restrictions in Theorem 5.3 on the neighborhood of $\lambda = 0$ makes sure that the strength of the flow off the manifold is sufficiently great compared to the strength of the flow on the manifold.

Remark 5.8. The difficulties encountered in Remark 5.7 are not due to the method of proof of Theorem 5.3. In fact, one can give examples showing that the results are essentially the best possible. It is only necessary to consider an equation in $\mathbb{R}^2$ which has an invariant circle with two equilibrium points on it. The behavior at the node under perturbation is determined by the ratios of the eigenvalues.

Remark 5.9. In general, it is not possible to prove that the manifold $f_\lambda(\theta)$ is analytic if the vector field is analytic.

A special case of Equation (5.1) which frequently occurs in the applications is

$$\dot{\theta} = \omega + \varepsilon\Theta(\theta, \rho)$$
$$\dot{\rho} = A\rho + \varepsilon R(\theta, \rho)$$

where ε is a real small parameter.

The above proof can be used to give several variants of Theorem 5.3. For example, the same conclusions are valid for the equation

$$\dot{\theta} = \omega + \varepsilon\Theta(\theta, \rho)$$
$$\dot{\rho} = \varepsilon A\rho + \varepsilon^2 R(\theta, \rho)$$

where ε is a small real parameter. It is also possible for ω in (5.1) to depend on the parameter ε. For example, $\omega = v/\varepsilon$ or $\omega = \varepsilon v$. In the latter case, the coefficient of Θ should be ε^2.

We also remark that the proof of Theorem 5.3 can be adapted to the case where the functions Θ, R depend upon the independent variable t and are almost periodic in t uniformly in θ, ρ, λ. Of course, the integral manifold is considered in $\mathbb{R} \times \mathbb{R}^k \times \mathbb{R}^n$ and will have a parametric representation which is almost periodic in t.

12.6. Bifurcation from a Periodic Orbit to a Torus

In this section, we consider an unperturbed autonomous planar system satisfying the conditions for a generic Hopf bifurcation in Section 9.5 and then subject this system to nonautonomous periodic perturbations. We show that the bifurcation is then from a periodic solution to an invariant torus. We also discuss the problem of bifurcation from a periodic orbit to a torus for an n-dimensional autonomous system.

Suppose $f_0 \in C^r(\mathbb{R}^2, \mathbb{R}^2)$, $r \geq 3$, $f_0(0) = 0$, $\partial f_0(0)/\partial x = B$, a 2×2 matrix whose eigenvalues are purely imaginary. Without loss in generality, we may assume

$$(6.1) \qquad\qquad B = \begin{bmatrix} 0 & 1 \\ -1 & 0 \end{bmatrix}$$

Applying the theory of normal forms [see Example 3.4, formula (3.20)] to the equation

$$(6.2) \qquad\qquad \dot{x} = f_0(x)$$

we may assume that there are constants δ_0, β_0 such that $f_0(x)$ in a neighborhood of $x = (x_1, x_2) = 0$ has the form

$$(6.3) \quad f_0(x) = \begin{bmatrix} x_2 + \beta_0 x_1(x_1^2 + x_2^2) + \delta_0 x_2(x_1^2 + x_2^2) \\ -x_1 + \beta_0 x_2(x_1^2 + x_2^2) - \delta_0 x_1(x_1^2 + x_2^2) \end{bmatrix} + o((|x_1| + |x_2|)^3)$$

as $x_1, x_2 \to 0$. For Equation (6.2) to correspond to a generic Hopf bifurcation when subjected to perturbation (see Section 9.5), it is necessary and sufficient that

$$(6.4) \qquad\qquad \beta_0 \neq 0$$

we will assume this condition is satisfied.

Now suppose that λ is a parameter in a Banach space $\Lambda, f \in C^r(\mathbb{R} \times \mathbb{R}^2 \times \Lambda, \mathbb{R}^2)$, $f(t, x, \lambda) = f(t + T, x, \lambda)$ for some fixed $T > 0$, $f(t, x, 0) = f_0(x)$, and consider the equation

$$(6.5) \qquad\qquad \dot{x} = f(t, x, \lambda)$$

for (x, λ) in a neighborhood of $(0,0)$. For example, if $\lambda = (\mu, v) \in \mathbb{R}^2$, then a special case of (6.5) is

$$(6.6) \qquad \dot{x} = g(x, \mu) + vh(t, x)$$

where $h(t + T, x) = h(t, x)$, $g(x, 0) = f_0(x)$ and the matrix $\partial g(0, \mu)/\partial x$ is given by

$$(6.7) \qquad \frac{\partial g(0, \mu)}{\partial x} = \begin{bmatrix} \mu & 1 \\ -1 & \mu \end{bmatrix}$$

In this case, for $v = 0$, the equation $\dot{x} = g(x, \mu)$ experiences a generic Hopf bifurcation as μ passes through zero. Under some nonresonance conditions, we will see that, for each $\mu \neq 0$, there is a $v_0 = v_0(\mu)$ sufficiently small so that Equation (6.6) has a T-periodic solution and an invariant torus. We will actually prove more. We will determine a curve in (μ, v)-space in a neighborhood of zero such that one side of this curve, the above situation occurs and the other side, there is a unique T-periodic solution.

If

$$(6.8) \qquad \det(e^{BT} - I) \neq 0$$

then the Implicit Function Theorem implies there is a neighborhood of $(x, \lambda) = (0,0)$ such that Equation (6.5) has a unique T-periodic solution $\varphi(t, \lambda)$ in this neighborhood. If we replace x by $\varphi(t, \lambda) + x$, then we may assume without loss of generality that

$$(6.9) \qquad f(t, 0, \lambda) = 0.$$

The condition (6.9) permits the introduction of polar coordinates $x_1 = \rho \cos \theta$, $x_2 = -\rho \sin \theta$ to obtain the equations

$$(6.10) \qquad \begin{aligned} \dot{\theta} &= 1 + \Theta(t, \theta, \rho, \lambda) \\ \dot{\rho} &= R(t, \theta, \rho, \lambda) \end{aligned}$$

where Θ, R are periodic in both t, θ with periods $T, 2\pi$, respectively. Also, $\Theta(t, \theta, \rho, 0) = O(|\rho|^2)$, $R(t, \theta, \rho, 0) = O(|\rho|^3)$ as $\rho \to 0$.

If we impose the nonresonance condition that

$$(6.11) \qquad m + n2\pi/T \neq 0, \qquad |m| + |n| \leq 4, \qquad m, n \text{ integers}$$

then the method of averaging in Section 4 can be applied to Equation (6.10) to obtain the equivalent equations

$$(6.12) \qquad \begin{aligned} \dot{\theta} &= 1 + \gamma_\lambda + \delta_\lambda \rho^2 + \tilde{\Theta}(t, \theta, \rho, \lambda) \\ \dot{\rho} &= \alpha_\lambda \rho + \beta_\lambda \rho^3 + \tilde{R}(t, \theta, \rho, \lambda) \end{aligned}$$

where $\tilde{\Theta} = o(|\rho|^2)$, $\tilde{R} = o(|\rho|^3)$ as $\rho \to 0$. It is clear that $\gamma_0 = 0$, $\alpha_0 = 0$, β_0 is the same number as in (6.3) and, by our assumption (6.4), is $\neq 0$. The constants $\alpha_\lambda, \beta_\lambda, \gamma_\lambda, \delta_\lambda$ are C^r functions of λ in a neighborhood of $\lambda = 0$. Thus, there is a neighborhood U of $\lambda = 0$ such that $\beta_\lambda \neq 0$.

Let us further assume that

$$(6.13) \qquad\qquad D_\lambda \alpha_\lambda \neq 0 \quad \text{at } \lambda = 0$$

so that the set

$$(6.14) \qquad\qquad C = \{\lambda \in U : \alpha_\lambda = 0\}$$

is a smooth C^r submanifold of U of codimension one.

We can now prove the following result.

Lemma 6.1. *Suppose $\beta_0 \neq 0$, Relation (6.13) is satisfied and C is defined in (6.14). Then there is a neighborhood U of $\lambda = 0$, V of $\rho = 0$ such that $U \backslash C = U_1 \cup U_2$ where U_1, U_2 are disjoint open sets such that the following conclusions hold:*

(i) *If $\lambda \in U_1$, the integral manifold $\rho = 0$ of (6.12) is exponentially asymptotically stable (unstable) if $\beta_0 < 0$ (>0) and all solutions with $\rho_0 \in V$ approach this manifold as $t \to \infty$ ($-\infty$).*

(ii) *If $\lambda \in U_2$, there is a nontrivial invariant torus $S_\lambda = \{(t, \theta, \rho) : \rho = \psi(t, \theta, \lambda), t \in \mathbb{R}, \theta \in \mathbb{R}\}$ of (6.12) where $\psi(t, \theta, \lambda)$ is C^{r-1} in (t, θ, λ), which is exponentially asymptotically stable (unstable) if $\beta_0 < 0$ (>0) with the manifold $\rho = 0$ being unstable (exponentially asymptotically stable) if $\beta_0 < 0$ (>0). Furthermore, every solution of (6.12) with ρ_0 in U has its limit set in these two manifolds.*

Proof. We give the proof only for $\beta_0 < 0$. The case $\beta_0 > 0$ is treated in a similar way. Since $\beta_0 < 0$, it follows that the integral manifold $\rho = 0$ of (6.12) for $\lambda = 0$ is uniformly asymptotically stable. Thus, we can find a neighborhood V of $\rho = 0$ and a neighborhood U of $\lambda = 0$ such that the neighborhood V is positively invariant with respect to the solutions of (6.12) for $\lambda \in U$ and every solution approaches $\rho = 0$. Let $U \backslash C = U_1 \cup U_2$, where

$$U_1 = \{\lambda \in U : \alpha_\lambda < 0\}$$
$$U_2 = \{\lambda \in U : \alpha_\lambda > 0\}$$

If the neighborhood U of $\lambda = 0$ is small enough, it is clear that the integral manifold $\rho = 0$ is exponentially asymptotically stable for $\lambda \in U_1$ and every solution in V approaches $\rho = 0$. This proves part (i) of the lemma.

Now suppose $\lambda_0 \in U_2$. Since $\alpha_{\lambda_0} > 0$, the manifold $\rho = 0$ is unstable. Since $\alpha_{\lambda_0} \beta_{\lambda_0} < 0$, the theory of Section 5 implies there is a neighborhood of λ_0, say $|\lambda - \lambda_0| < \varepsilon$, such that Equation (6.12) for $|\lambda - \lambda_0| < \varepsilon$ has an invariant torus $S_\lambda = \{(t, \theta, \rho): \rho = \psi(t, \theta, \lambda),\ t \in \mathbb{R},\ \theta \in \mathbb{R}\}$ which is uniformly asymptotically stable and is given approximately by $\psi(t, \theta, \lambda) \sim (-\alpha_{\lambda_0}/\beta_{\lambda_0})^{1/2}$ for λ near λ_0. Using continuous dependence on initial data, one can show that every solution of the differential equation has its limit set in the manifold S_λ union the set $\rho = 0$. This proves the lemma. $\quad\square$

Remark 6.2. Lemma 6.1 remains valid if θ is a vector and the perturbations are almost periodic in t uniformly with respect to the other variable. The same proof applies.

Using this lemma, we can state a corresponding result for Equation (6.5).

Theorem 6.3. *Suppose f_0 satisfies (6.3) with $\beta_0 \neq 0$ and the nonresonance conditions (6.8), (6.11) are satisfied. Then there are neighborhoods U of $\lambda = 0$, W of $x = 0$ and a C^r submanifold C of U of codimension one such that $U\backslash C = U_1 \cup U_2$ are open disjoint sets and the following conclusions hold:*

(i)   *If $\lambda \in U_1$ there is a unique T-periodic solution $\varphi(\lambda)$ of (6.5) in W, $\varphi(\lambda) \to 0$ as $\lambda \to 0$ such that every solution with initial value in W approaches $\varphi(\lambda)$ as $t \to \infty$ $(t \to -\infty)$ if $\beta_0 < 0$ $(\beta_0 > 0)$.*

(ii)   *If $\lambda \in U_2$, there is a T-periodic solution $\varphi(\lambda)$ of (6.5) in W and an integral manifold*

$$S_\lambda = \{(t, x): x_1 = \psi(t, \theta, \lambda)\cos\theta,\ x_2 = -\psi(t, \theta, \lambda)\sin\theta,$$
$$t \in \mathbb{R},\ \theta \in \mathbb{R}\}$$

of (6.5) in $\mathbb{R} \times W$, where $\psi(t, \theta, \lambda)$ is T-periodic in t, 2π-periodic in θ, $\varphi(\lambda)$, $\psi(t, \theta, \lambda) \to 0$ as $\lambda \to 0$. Furthermore, the limit set of every solution which remains in W for either $t \geq 0$ or $t \leq 0$ approaches either S_λ or $\varphi(\lambda)$ as $t \to \infty$ or $-\infty$. Furthermore, $\varphi(\lambda)$ is asymptotically stable (unstable) and S_λ is unstable (asymptotically stable) if $\beta_0 > 0$ $(\beta_0 < 0)$.

We now use the theory of integral manifolds in Section 5 and Lemma 6.1 to discuss the problem of bifurcation from a periodic orbit to a torus. The situation arises in the following manner. Suppose the autonomous equation in $\mathbb{R}^n$,

$$\dot{u} = f_0(u), \tag{6.15}$$

has a periodic orbit $\Gamma = \{p_0(t), t \in \mathbb{R}\}$ where $p_0(t)$ is a periodic solution of (6.15) of least period $T > 0$. Suppose that the linear variational equation at Γ,

$$\dot{u} = \frac{\partial f_0(p(t))}{\partial u}\, u$$

has two complex multipliers on the unit circle and all other multipliers have moduli $\neq 1$ except the trivial multiplier 1 determined by $\dot{p}_0$. Now suppose λ

is a parameter in a Banach space Λ, $f : \mathbb{R}^n \times \Lambda \to \mathbb{R}^n$ is a C^r function, $r \geq 2$, $f(u, 0) = f_0(u)$.

$$(6.16) \qquad\qquad \dot{u} = f(u, \lambda)$$

in a neighborhood of $\lambda = 0$ and the periodic orbit Γ.

By the introduction of a moving orthonormal coordinate system of the form

$$u = p_0(\theta) + G(\theta)z$$

where $z \in \mathbb{R}^{n-1}$, $G(\theta)$ is an $n \times (n-1)$ matrix, T-periodic in θ, one obtains an equivalent system in a neighborhood of Γ,

$$(6.17) \qquad \begin{aligned} \dot{\theta} &= 1 + \Theta(\theta, z, \lambda) \\ \dot{z} &= Cz + Z(\theta, z, \lambda) \end{aligned}$$

where C has two simple eigenvalues on the imaginary axis and all other eigenvalues have nonzero real parts. The functions Θ, Z are T-periodic in θ and satisfy

$$\begin{aligned} \Theta(\theta, z, 0) &= o(|z|) \\ Z(\theta, z, 0) &= o(|z|) \end{aligned}$$

as $|z| \to 0$.

We are interested in the behavior of the solutions of Equation (6.17) in a neighborhood of $z = 0$. If $C = \mathrm{diag}(B, A)$ where B is a 2×2 matrix with pure imaginary eigenvalues and $z = (x, y)$, $x \in \mathbb{R}^2$, $y \in \mathbb{R}^{n-2}$, then Equation (6.17) may be written as

$$(6.18) \qquad \begin{aligned} \dot{\theta} &= 1 + \Theta(\theta, x, y, \lambda) \\ \dot{x} &= Bx + X(\theta, x, y, \lambda) \\ \dot{y} &= Ay + Y(\theta, x, y, \lambda) \end{aligned}$$

Since the eigenvalues of A have nonzero real parts, one can use the same reasoning as in Section 5 to obtain a C^r-local integral manifold of Equation (6.18).

$$S_\lambda = \{(\theta, x, y), \, y = \psi(\theta, x, \lambda), \, \theta \in \mathbb{R}, \, x \in U\}$$

for a neighborhood U of $x = 0$, V of $\lambda = 0$, W of $y = 0$, $\psi(\theta, x, 0) = 0$. Furthermore, the flow near each point on S_λ behaves like a saddle point.

Therefore, the discussion of the solutions of Equation (6.18) in a neighborhood of $x = 0$, $y = 0$, $\lambda = 0$ is reduced to the discussion of the solutions of the equation

$$\begin{aligned} \dot{\theta} &= 1 + \tilde{\Theta}(\theta, x, \lambda) \\ \dot{x} &= Bx + \tilde{X}(\theta, x, \lambda) \end{aligned}$$

near $x = 0$, $\lambda = 0$, where $\tilde{\Theta}, \tilde{X}$ are the functions Θ, X evaluated at $(\theta, x, \psi(\theta, x, \lambda), \lambda)$. Since $\Theta(\theta, 0, 0) = 0$, we may replace t by θ to obtain the second order nonhomogeneous equation

$$(6.19) \qquad \frac{dx}{d\theta} = Bx + f(\theta, x, \lambda)$$

where $f(\theta, x, \lambda)$ has the same smoothness properties as $\tilde{X}$, $f(\theta, x, 0) = o(|x|)$ as $|x| \to 0$.

We now make the nonresonant hypothesis

$$(6.20) \qquad \det(e^{BT} - I) \neq 0$$

Hypothesis (6.20) implies that the equation $dx/d\theta = Bx$ has no T-periodic solution except $x = 0$. The Implicit Function Theorem implies there is a unique T-periodic integral manifold $\varphi(\theta, \lambda)$ of (6.19) in a neighborhood of $x = 0$ for λ in a neighborhood of zero, $\varphi(\theta, 0) = 0$. If we make the transformation of variables $x \mapsto \varphi(\theta, \lambda) + x$, we may assume that

$$(6.21) \qquad f(\theta, 0, \lambda) = 0 \quad \text{for all } \lambda, \theta$$

The condition (6.21) makes it legitimate to introduce polar coordinates $x \mapsto (\rho, \varphi)$ to obtain equivalent equations

$$(6.22) \qquad \begin{aligned} \frac{d\varphi}{d\theta} &= \omega + \Phi(\theta, \varphi, \rho, \lambda) \\[1em] \frac{d\rho}{d\theta} &= R(\theta, \varphi, \rho, \lambda) \end{aligned}$$

where $\omega > 0$ is constant, Φ, R are T-periodic in θ, 2π periodic in φ, $\Phi(\theta, \varphi, \rho, 0) = O(|\rho|)$, $R(\theta, \varphi, \rho, \lambda) = O(|\rho|(|\rho| + |\lambda|))$ as $\rho \to 0$, $\lambda \to 0$.

Let us now suppose that the number of derivatives r of $f(u, \lambda)$ in (6.16) satisfies $r \geq 4$. Under certain nonresonance conditions on the frequency ω and the period T, Equation (6.22) can be assumed to have the form

$$(6.23) \qquad \begin{aligned} \frac{d\phi}{d\theta} &= \omega + \gamma_\lambda + \delta_\lambda \rho^2 + O(|\rho|^3) \\[1em] \frac{d\rho}{d\theta} &= \alpha_\lambda \rho + \beta_\lambda \rho^3 + O(|\rho|^4) \end{aligned}$$

as $\rho \to 0$, where $\alpha_0 = 0$. The nonresonant conditions can be explicitly determined if the periodic orbit is known. They certainly will be satisfied if

$$(6.24) \qquad T/2\pi \quad \text{is irrational.}$$

Equation (6.23) is the same as Equation (6.12). We will be able to apply Lemma 6.1 directly if we assume that

$$(6.25) \qquad\qquad D_\lambda \alpha_\lambda \neq 0 \quad \text{for } \lambda = 0, \ \beta_0 \neq 0.$$

Let us give an interpretation of the first condition in (6.25). If $\varphi(\theta, \lambda)$ is the T-periodic integral manifold of Equation (6.19), then the set

$$P_\lambda = \{(\theta, x, y) : x = \varphi(\theta, \lambda), \ y = \psi(\theta, \varphi(\theta, \lambda), \lambda), \ \theta \in \mathbb{R}\}$$

is a T periodic integral manifold of (6.18). We thus obtain a periodic orbit of (6.16) in the parametric form

$$u = P_0(\theta) + G(\theta)z(\theta, \lambda), \qquad \theta \in \mathbb{R},$$
$$z(\theta, \lambda) = (\varphi(\theta, \lambda), \ \psi(\theta, \varphi(\theta, \lambda), \lambda)).$$

The period τ_λ is determined by solving the equation

$$\dot\theta(t) = 1 + \Theta(\theta, z(\theta, \lambda), \lambda), \qquad \theta(0) = 0$$

and requiring that $\theta(\tau_\lambda) = T$. Let $p_\lambda(t)$ be this periodic solution of period τ_λ of (6.16). The linear variational equation with respect to $p_\lambda(t)$ is

$$(6.26) \qquad\qquad \dot u = \frac{\partial f(p_\lambda(t), \lambda)}{\partial u} u.$$

It is now relatively easy to show that the first condition in (6.25) is equivalent to saying that there is a $\lambda_0 \in \Lambda$, $\lambda_0 \neq 0$, such that, if $\lambda = \varepsilon\lambda_0$, then the two characteristic multipliers of (6.26) which are on the unit circle for $\varepsilon = 0$ have a nonzero derivative with respect to ε.

We now apply Lemma 6.1 and summarize the above results in the following

Theorem 6.4. *Suppose Equation* (6.15) *has a periodic orbit* Γ *with only the simple characteristic multipliers* 1, $\exp(\pm i\omega)$, *on the unit circle. Suppose also* (6.20), (6.24) *and* (6.25) *are satisfied. Then there is a neighborhood U of $\lambda = 0$, a neighborhood W of Γ and a C^r submanifold C of U of codimension one such that $U \backslash C = U_1 \cup U_2$ and the following conclusions hold for the solutions of* (6.23) *in* W:

(i) $\lambda \in U_1$ *implies there is a unique periodic orbit which is exponentially asymptotically stable (unstable) for* $\beta_0 < 0 \ (\beta_0 > 0)$.

(ii) $\lambda \in C$ *implies there is a unique periodic orbit which is asymptotically stable (unstable) if* $\beta_0 < 0 \ (\beta_0 > 0)$.

(iii) $\lambda \in U_2$ *implies there is a periodic orbit which is exponentially asymptotically stable (unstable) and an invariant torus which is unstable (exponentially asymptotically stable) if* $\beta_0 > 0 \ (\beta_0 < 0)$.

*Furthermore, the limit points of any solution in W belong to the invariant
sets in* (i), (ii), (iii).

12.7. Bifurcation of Tori

In this section, we discuss the possibility of having two invariant tori coalesce
and disappear as a parameter is varied. The complete solution of this type
of problem is not available. Since this is the case, we use a special type of
equation to illustrate the results that are obtained easily from the theory of
integral manifolds and then indicate why it is so difficult to solve the complete
problem.

Suppose λ, μ are real scalar parameters $f \in C^r(\mathbb{R}^2 \times \mathbb{R}, \mathbb{R}^2)$, $g \in C^r(\mathbb{R} \times \mathbb{R}^2 \times \mathbb{R}, \mathbb{R}^2)$, $r \geq 3$ and consider the equation

$$(7.1)_{\lambda,\mu} \qquad\qquad \dot{x} = f(x, \lambda) + g(t, x, \mu)$$

where $g(t + 1, x, \mu) = g(t, x, \mu)$, $g(t, x, 0) = 0$. We also suppose $f(\cdot, 0)$ is a
bifurcation point of degree one and, more specifically, that alternative (i)
of section 10.5 holds. That is, there is an open set $W \subset \mathbb{R}^2$, such that the
equation

$$(7.1)_{\lambda,0} \qquad\qquad \dot{x} = f(x, \lambda)$$

has no equilibrium points in W for any λ, has no periodic orbits in W for
$\lambda < 0$, has two hyperbolic periodic orbits in W for $\lambda > 0$ and a unique periodic
orbit for $\lambda = 0$ which is degenerate of order one.

The problem is to discuss what happens to the solutions of $(7.1)_{\lambda,\mu}$ in W
as (λ, μ) vary in a neighborhood of $(0,0)$. This problem is much more complicated than any we have considered before because there is no way to
obtain any solution whose qualitative properties are known for all λ, μ.
In the previous section, we used strongly the fact that a periodic orbit was
known to exist for all (λ, μ). Some results can be obtained, but they are not
complete.

For $\lambda = \mu = 0$, Equation $(7.1)_{0,0}$ has a unique periodic orbit γ_0 which is
asymptotically stable from one side and unstable from the other. Suppose
$\gamma_0 = \{x_0(\theta), \theta \in \mathbb{R}\}$ where $x_0(t)$ is a periodic solution of $(7.1)_{0,0}$ of least period
$2\pi/\omega_0$, $\omega_0 > 0$. If ρ represents distance from γ_0 along a normal, then one can
introduce a local coordinate system around γ_0 which takes $x \mapsto (\theta, \rho)$ to
obtain the equations

$$(7.2)_{\lambda,\mu} \qquad \begin{aligned} \dot{\theta} &= 1 + \Theta(t, \theta, \rho, \lambda, \mu) \\ \dot{\rho} &= R(t, \theta, \rho, \lambda, \mu) \end{aligned}$$

where Θ, R vanish for $(\rho, \lambda, \mu) = (0, 0, 0)$, are 1-periodic in t and $2\pi/\omega_0$ periodic
in θ.

Let us suppose that we can average to obtain an equivalent set of equations

$(7.3)_{\lambda,\mu}$
$$\dot{\theta} = 1 + a\lambda + b\mu + \tilde{\Theta}(t,\theta,\rho,\lambda,\mu)$$
$$\dot{\rho} = c\lambda + d\mu + e\rho + f\rho^2 + \tilde{R}(t,\theta,\rho,\lambda,\mu)$$

where a, b, c, d, e, f are constants, $\tilde{\Theta}, \tilde{R}$ for $\rho = 0$ are $O((|\lambda| + |\mu|)^2$ as $\lambda, \mu \to 0$ and $\tilde{R}$ for $\lambda = \mu = 0$ is $O(\rho^3)$. The hypothesis that $f(\cdot,0)$ is a bifurcation point of order one implies $cf < 0$, $e = 0$. Since $c \neq 0$, let us introduce new parameters (δ,ε) as $\delta = \mu$, $\varepsilon = c\lambda + d\mu$ to obtain the equations

$(7.4)_{\varepsilon,\delta}$
$$\dot{\theta} = 1 + a'\delta + b'\varepsilon + \Theta'(t,\theta,\rho,\varepsilon,\delta)$$
$$\dot{\rho} = \varepsilon + f\rho^2 + R'(t,\theta,\rho,\varepsilon,\delta)$$

with Θ', R' having the same type of order relations as before.

For any ε small enough such that $\varepsilon f > 0$, it is not difficult to show there is a $\delta_0(\varepsilon) > 0$, such that no solution of $(7.4)_{\varepsilon,\delta}$ remains in a neighborhood of $\rho = 0$ for all t if $0 \leq \delta < \delta_0(\varepsilon)'$. This is the analogue of $\lambda < 0$ for $(7.1)_{\lambda,0}$. If $\varepsilon f < 0$, then the theory of integral manifolds in Section 5 shows there is a $\delta_0(\varepsilon) > 0$ such that there are two invariant cylinders $T^1_{\varepsilon,\delta}$, $T^2_{\varepsilon,\delta}$ of $(7.4)_{\varepsilon,\delta}$, $0 \leq \delta < \delta_0(\varepsilon)$, with periodic cross section; that is, two invariant tori $T^1_{\varepsilon,\delta}$, $T^2_{\varepsilon,\delta}$. One of these tori is completely unstable and the other is uniformly asymptotically stable. Furthermore, any solution of $(7.4)_{\varepsilon,\delta}$ which remains in a sufficiently small neighborhood of $\rho = 0$ for $t \geq 0$ $(t \leq 0)$ must approach either $T^1_{\varepsilon,\delta}$ or $T^2_{\varepsilon,\delta}$ as $t \to \infty$ $(t \to -\infty)$. Thus, we have complete knowledge of the solutions in a subset of a neighborhood of $(\varepsilon,\delta) = (0,0)$. The subset is depicted in Figure 7.1.

As $\varepsilon \to 0$, the number $\delta_0(\varepsilon)$ may also approach zero. The reason for this fact is that the invariant tori $T^1_{\varepsilon,\delta}$, $T^2_{\varepsilon,\delta}$ for a fixed ε may fail to exist as smooth invariant sets for some small δ. This is a general fact in differential equations which was mentioned in Section 5. Given a smooth hyperbolic invariant

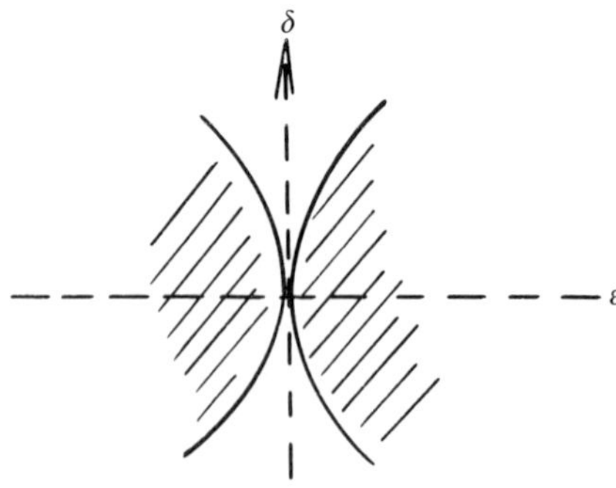

Figure 7.1

torus for a differential equation, the existence of a smooth invariant torus for a small perturbation of this equation generally requires that the strength of the hyperbolicity is greater than the rate of attraction of orbits on the torus [see the estimates (5.11), (5.12)]. If this is not the case, the invariant sets develop cusps at rational rotation numbers. For the sets $T^1_{\varepsilon,\delta}$, $T^2_{\varepsilon,\delta}$ above, the strength of the hyperbolicity approaches zero as $\varepsilon \to 0$ for any fixed δ. Thus, it is conceivable that the sets touch for some ε without coalescing in a uniform manner. This is why we cannot complete the picture in Figure 7.1. The behavior in the unshaded region will be extremely complicated. The classical theory of integral manifolds gives no information.

12.8. Bibliographical Notes

The transformation theory in Section 2 was motivated by de Prit [1], Gröbner [1], Gröbner and Knapp [1], Kamel [1, 2], Henrard [1], Kirchgraber and Stiefel [1]. The notation in Relation (2.6) comes from Kirchgraber and Stiefel [1]. The method is presented in this way since it provides a nice unification of various theories and also it lends itself to easy implementation on a computer (for example, see de Prit [1], Meyer and Schmidt [1]). Up to now, the primary applications using Lie transforms have been for Hamiltonian systems. This is probably due to the fact that the method is not well known to researchers in other areas.

The idea of using transformations of variables to bring a nonlinear differential equation into a form which exhibits more clearly the qualitative features of the flow is very old. We mention particularly the work of Poincaré [1, 2], in celestial mechanics, Liapunov [1] in his study of stability of an equilibrium when some eigenvalues are on the imaginary axis and the work of Dulac [1] in the study of the neighborhood of an equilibrium point in the plane. The theory of normal forms in the spirit of Section 3 was extensively developed by Birkhoff [1] and is still an important area of investigation today. For results and references, we mention the books of Siegel and Moser [1], Moser [1], Bibikov [1], Br'juno [1] and the papers of Br'juno [2]. In these references, one can also find specific results on the convergence of the formal power series. Several of the illustrations in Section 3 are based on Bibikov [1].

Some form of averaging has been used in differential equations for many years. Very important practical and theoretical contributions were made by Krylov and Bogoliubov [1] in the 1930's and Bogoliubov [1] in 1945. The importance of the average (4.14) was noted by Diliberto [1]. The principle of averaging has been extended in many directions for both finite and infinite dimensional problems (see, for example, Bogoliubov and Mitropolski [1], Kurzweil [2, 3], Mitropolski [1], Nayfeh [1], Volosov [1].) The details of the averaging procedure refered to in Remark 4.4 can be found in Chow and Mallet–Paret [2].

The theory of integral manifolds as presented in Section 5 began with the fundamental papers of Krylov and Bogoliubov [1] and successively developed by Bogoliubov and Mitropolski [1], the many colleagues of Mitropolski in Kiev (see Mitropolski [1] for references), Hale [3], Kurzweil [1], Pliss [1], Kelley [1, 2]. Levinson [3] also made important contributions on the existence of invariant tori of a periodically perturbed autonomous system. He used a different approach by showing that the period map had an invariant curve. This idea was extended in a significant way by Diliberto [1] and his colleagues (see Diliberto [1] or Hale [4] for references). The contributions of Sacker [1, 2] were made by using the partial differential equations that an integral manifold must satisfy. Many extensions have been made in recent years. Hirsch, Pugh and Shub [1]. [2] treat arbitrary hyperbolic invariant sets of diffeomorphisms. The theory is available for functional differential equations (see Hale [4] for results and references), for parabolic partial differential equations (see Henry [1] for results and references) and some types of hyperbolic equations (see Mitropolski [1] for results and references).

We remark that the proof of Theorem 5.3 can be adapted to the case where the functions Θ, R depend upon the independent variable t and are almost periodic in t uniformly in θ, ρ, λ. Of course, the integral manifold is considered in $\mathbb{R} \times \mathbb{R}^k \times \mathbb{R}^n$ and will have a parametric representation which is almost periodic in t (see, for example, Hale [4]). For a proof of Remark 5.5, see Pugh and Shub [1].

The dependence on λ as $\lambda \to 0$ of the rotation number of the flow on the invariant torus in Lemma 6.1 can be determined approximately. This gives information about the periods of periodic solutions near zero and they may approach ∞ as $\lambda \to 0$ (see Kozyakin [1]). Theorem 6.3 was first proved by Sacker [1] but the result was announced earlier by Neimark [1] with no proof. The proof in the text is different from Sacker. Another proof has been given by Lanford [1]. Using the methods of Section 5 and a general theory of characteristic exponents for invariant tori, Sell [1] has stated a generalization of Theorem 6.3 concerning the bifurcation from a torus of n-dimensions to one of $(n + 1)$-dimensions (see, also Iooss [4], Iooss and Langford [1]). In the study of traveling wave solutions of reaction diffusion equations on a circle, one encounters invariant tori bifurcating from a circle similar in spirit to the discussion in Section 6 (for references, see Auchmuty [1]). Wan [2] has also considered bifurcation into invariant tori. Chenciner [1] has obtained recently some results on the problem of bifurcation of tori mentioned in Section 7. He obtains the smooth coalesence of tori if the rotation number is badly approximated by rationals.

Chapter 13

Higher Order Bifurcation near Equilibrium

13.1. Introduction

Most of the discussion in the previous chapters on dynamics have centered around the case where the unperturbed autonomous equation is a bifurcation point of degree one. In particular, for any equilibrium point, this implies that the linear variational equation must either have only zero as a simple eigenvalue and no other eigenvalue on the imaginary axis or have only a pair of simple complex eigenvalues on the imaginary axis. Generically, the first alternative corresponds to a saddle-node bifurcation and the second alternative to the generic Hopf bifurcation. When the bifurcation point has degree greater than one, there can be more eigenvalues on the imaginary axis. In this chapter, we discuss situations where this occurs. As we shall see, the behavior of solutions of the perturbed equation near such a bifurcation point depends upon more global concepts; for example, homoclinic orbits can occur as well as all of the bifurcations encountered in the previous chapters. The method of investigation is to scale variables in such a way as to "blow up" the neighborhood of the equilibrium point in order to see the fine structure of the flow.

In Section 13.2, we consider a planar system with both eigenvalues zero near the equilibrium point, but this eigenvalue has nonsimple elementary divisors. Under generic conditions on the quadratic terms and with the bifurcation parameters being small linear perturbations, we give a complete description of the flow for these parameters varying in a neighborhood of zero. Furthermore, each bifurcation point for nonzero values of the perturbation parameters has degree one.

In Section 13.3, the same case as above is considered except the matrix has simple elementary divisors and the generic conditions are imposed on the quadratic and cubic terms. A complete description for which each bifurcation point has degree one for the parameters nonzero cannot be given by using only the quadratic terms. In Section 13.4, the quadratic terms in Section 13.3 all are assumed to be zero. Under generic conditions on the cubic and fifth degree terms, a complete bifurcation diagram is given. The fifth degree terms are necessary in the same way that the cubic terms were necessary in the previous example.

In Section 13.5, we discuss an example in three dimensions for which the linear variational equation has two pure imaginary and one zero root. Under a symmetry hypothesis, one can introduce polar coordinates and reduce the problem to a periodic system in two variables. The theory of normal forms or averaging can then be applied. The averaged equations are autonomous and of the form discussed in Sections 13.3 and 13.4 depending upon the hypotheses imposed. The perturbed equations are periodic. Using the results from Sections 13.3 and 13.4, the theory of integral manifolds in Chapter 12 and the ideas in Chapter 11 on homoclinic points, we give a discussion of the behavior of the solutions near the equilibrium point.

In Section 13.5, we also make some remarks on examples in four dimensions for which the linear variational equation has two pairs of imaginary roots. Under a symmetry hypothesis, new coordinates with two polar variables and two angle variables can be introduced. The averaged equations in this case involve no quadratic terms and are generically the ones considered in Section 13.4. The discussion for the perturbed equations is more complicated because two angle variables are involved. A partial discussion of the behavior near the equilibrium point is given. The complete behavior is not known.

13.2. Two Zero Roots I

In this section, we consider the effect of autonomous perturbations on the second order system

$$(2.1) \qquad \begin{aligned} \dot{x} &= y \\ \dot{y} &= \alpha x^2 + \beta xy \end{aligned}$$

under the assumption that

$$(2.2) \qquad \alpha < 0, \qquad \beta > 0.$$

The perturbed system will be taken to be of the form

$$(2.3) \qquad \begin{aligned} \dot{x} &= y \\ \dot{y} &= \varepsilon_1 x + \varepsilon_2 y + \alpha x^2 + \beta xy \end{aligned}$$

where $\varepsilon_1, \varepsilon_2$ are small parameters.

The behavior of the solutions of (2.3) only under the hypothesis that $\alpha \neq 0, \beta \neq 0$ is obtained from the discussion below by the change of variables $t \to -t, \varepsilon_2 \to -\varepsilon_2, x \to \pm x, y \to \mp y$.

The special form of the perturbation is chosen only for simplicity in notation. In fact, the same results apply to the system

$$\begin{aligned} \dot{x} &= y + X(x, y, \varepsilon) \\ \dot{y} &= \varepsilon_1 x + \varepsilon_2 y + Y(x, y, \varepsilon) \end{aligned}$$

where $X = O((|x| + |y|)^2 + |\varepsilon|(|x| + |y|))$, $Y = O((|x| + |y|)^2(1 + |\varepsilon|))$ as $|x|$, $|y|$, $|\varepsilon| \to 0$, $\varepsilon = (\varepsilon_1, \varepsilon_2)$, and $\partial^2 X(0,0,0)/\partial x^2 = 0$, $\partial^2 Y(0,0,0)/\partial x^2 \neq 0$, $\partial^2 Y(0,0,0)/\partial x\, \partial y \neq 0$.

First, we consider the case $\varepsilon_1 > 0$. The introduction of the scaled variables

$$(2.4) \qquad \begin{aligned} \varepsilon_1 &= \delta^2, \qquad \varepsilon_2 = \mu\delta^2, \qquad \delta > 0 \\ t &\mapsto \delta^{-1}t, \qquad x \mapsto \delta^2|\alpha|^{-1}x, \qquad y \mapsto \delta^3|\alpha|^{-1}y \end{aligned}$$

leads to new equations

$$(2.5) \qquad \begin{aligned} \dot{x} &= y \\ \dot{y} &= x - x^2 + \mu\delta y + \delta\gamma xy \end{aligned}$$

where $\gamma = \beta|\alpha|^{-1}$.

For $\delta = 0$, Equation (2.5) becomes the conservative system

$$(2.6) \qquad \begin{aligned} \dot{x} &= y \\ \dot{y} &= x - x^2 \end{aligned}$$

with first integral

$$(2.7) \qquad V(x, y) = \frac{y^2}{2} - \frac{x^2}{2} + \frac{x^3}{3}$$

The equilibrium point $(0,0)$ is a saddle with a homoclinic orbit through it while the equilibrium point $(1,0)$ is a center.

The problem is to discuss the behavior of the solutions of (2.5) for δ small and all μ. In planar systems, periodic orbits always play a fundamental role so we begin with the following lemma.

Lemma 2.1. *Every periodic orbit of Equation (2.5) must intersect the segment $(0,1) \times \{0\}$ in the (x, y)-plane. There is a continuous positive function δ_0: $(0,1) \to \mathbb{R}$ and a continuously differentiable function $\mu^*(b, \delta)$, $b \in (0,1)$, $|\delta| < \delta_0(b)$ such that there is a periodic orbit of Equation (2.5) through $(b,0)$ if and only if $\mu = \mu^*(b, \delta)$. Furthermore,*

$$\mu^*(b, 0) = -\gamma\beta(b)/\alpha(b)$$

$$(2.8) \qquad \alpha(b) = \int_b^{c(b)} y\, dx, \qquad \beta(b) = \int_b^{c(b)} xy\, dx > 0$$

$$y \geq 0, \qquad \frac{y^2}{2} = \frac{x^2 - b^2}{2} - \frac{x^3 - b^3}{3}$$

$$c(b) > 1, \qquad \frac{c^2(b) - b^2}{2} - \frac{c^3(b) - b^3}{3} = 0$$

and $d\mu^*(b,0)/db < 0$. *Also,* $\mu^*(b,0) \to -\gamma$ *as* $b \to 1$, $\mu^*(b,0) \to -\frac{6}{7}\gamma$ *as* $b \to 0$. *Finally, if* $\mu = \mu^*(b,\delta)$ *for a fixed* $b \in (0,1)$ *and* $|\delta| < \delta_0(b)$, *then the periodic orbit through* $(b,0)$ *is the only one corresponding to this* μ, δ.

Proof. Since the index of a periodic orbit is $+1$, every periodic orbit of (2.5) can encircle only the equilibrium point $(1,0)$. Let $0 < b < 1$ and suppose there is a periodic orbit $\Gamma(b,\delta,\mu)$ of (2.5) through $(b,0)$. Along the solutions of (2.5), the derivative $\dot{V}$ of the function V in (2.7) is given by

$$(2.9) \qquad\qquad \dot{V}(x,y) = \delta(\mu + \gamma x)y^2$$

For the periodic orbit $\Gamma(b,\delta,\mu)$ we must have $\int_\Gamma \dot{V}\,dt = 0$, or

$$(2.10) \qquad\begin{aligned} F(b,\delta,\mu) &= 0 \\ F(b,\delta,\mu) &= \int_{\Gamma(b,\delta,\mu)} (\mu + \gamma x)y^2\,dt \end{aligned}$$

The function $F(b,0,\mu)$ can be written explicitly. In fact, with the notation in (2.8)

$$F(b,0,\mu) = 2\mu\alpha(b) + 2\gamma\beta(b).$$

Thus $F(b,0,\mu) = 0$ implies $\mu = -\gamma\beta(b)/\alpha(b)$. Consequently, for any $b_0 \in (0,1)$, there is a $\delta_0 = \delta_0(b_0) > 0$ and a continuously differentiable function $\mu(b,\delta)$ defined for $|b - b_0| < \delta_0$, $|\delta| < \delta_0$ such that $\mu(b_0,0) = -\gamma\beta(b_0)/\alpha(b_0)$ and $F(b,\delta,\mu(\delta,b)) = 0$. This proves the first part of the lemma.

Since $y(b) = y(c(b)) = 0$, it is easy to verify that

$$\alpha'(b) = \int_b^{c(b)} (-b + b^2)\,\frac{dx}{y},$$

$$\beta'(b) = \int_b^{c(b)} (-b + b^2)\,\frac{x\,dx}{y}.$$

If $2T(b) > 0$ denotes the least period of the periodic solution through $(b,0)$, then

$$\alpha'(b) = (-b + b^2) \int_0^{T(b)} dt,$$

$$\beta'(b) = (-b + b^2) \int_0^{T(b)} x\,dt.$$

If $(x^*(b), y^*(b))$ denotes the periodic solution through $(b,0)$, then $x^*(b) = 1 + O(|b - 1|)$ as $b \to 1$. This implies

$$\lim_{b\to 1} \frac{\beta(b)}{\alpha(b)} = \lim_{b\to 1} \frac{\beta'(b)}{\alpha'(b)} = 1.$$

As $b \to 0$, $c(b) \to \frac{3}{2}$, a few computations yield

$$\lim_{b \to 0} \frac{\beta(b)}{\alpha(b)} = \frac{6}{7}.$$

Let $2a = -b^2 + 2b^3/3$. Since $a'(b) = -b(1 - b) < 0$ for $0 < b < 1$, we may consider β, α as functions of a, $-1 < 6a < 0$. If $\gamma = \beta/\alpha$ and "$'$" denotes differentiation with respect to a, then the proof of the theorem will be complete if we show that $\gamma' < 0$, $-1 < 6a < 0$. Suppose y, c are the functions defined in (2.8). The idea for the remainder of this proof is based on remarks by Carr[1, p. 84].

We first show that α, β can be expressed in terms of α', β'. Using the expression for yy_x and the fact that $y(b) = y(c) = 0$, one obtains

$$\beta' = \int_b^c x^2 y^{-1}\, dx.$$

Integrating α by parts to obtain

$$\alpha + \beta' = \int_b^c x^3 y^{-1}\, dx$$

and using the formula for y, one obtains

$$5\alpha = 6a\alpha' + \beta'$$

Integrating β by parts, using the expression for y and the above relations, one has

$$35\beta = 6a\alpha' + 6(1 + 5a)\beta'$$

These expressions also yield expressions for α'', β'' as

$$(1 + 6a)\beta'' = \beta' - \alpha'$$
$$6a(1 + 6a)\alpha'' = -6a\alpha' - \beta'.$$

We now claim that $\gamma' = 0$, $-1 < 6a < 0$, implies $\gamma'' < 0$. In fact, if $\gamma'(a) = 0$, then $\gamma = \beta/\alpha = \beta'/\alpha'$ and $\alpha\gamma'' = \beta'' - \gamma\alpha''$ at a and the expressions for α'', β'' imply that

$$-6a(1 + 6a)\alpha\gamma''/\alpha' = -(\gamma + 6a)^2 + 6a(1 + 6a) < 0$$

since $-1 < 6a < 0$. Since the coefficient of γ'' is positive, this proves the claim.

The next assertion is that $\gamma'(a) = 0$ implies $0 < \gamma(a) < 1$. Using the expressions for α, β in terms of α', β' one obtains, at this value of a,

$$7\gamma^2 + 6(2a - 1)\gamma - 6a = 0$$

If $\gamma(a) = 1$, then $6a = -1$. Since we know that $\gamma(0) = 6/7$, the assertion follows.

The final assertion is that $\gamma'(a) < 0$ for $-1 < 6a < 0$. In fact, if this is not the case, then there is an a_1, $-1 < 6a_1 < 0$, such that $\gamma'(a_1) = 0$, $\gamma''(a_1) < 0$. Since $\gamma(a_1) < 1$, $\gamma(-1/6) = 1$, $\gamma(0) = 6/7$, this is a contradiction. The proof of the theorem is complete. $\square$

Our next task is to determine the curves in the $(\varepsilon_1, \varepsilon_2)$-plane (that is, the values of δ, μ) at which the topological structure of the trajectories of Equation (2.5) changes. As we shall see, this structure can change only due to a homoclinic orbit or a change in stability of the equilibrium point $(1, 0)$.

Let $\Gamma_\infty = \{(q_\infty(t), \dot{q}_\infty(t)) : t \in (-\infty, \infty)\} \cup \{(0, 0)\}$ be the homoclinic orbit of Equation (2.6). From the special form of Equation (2.5), the function $\beta(t) = 1$ in Equation (11.3.19), the first component of the function B in (11.3.19) is identically zero and the second component vanishes when $\delta = 0$. If we write this second component of B as $\delta G(\mu, \delta)$, we see that

$$G(\mu, \delta) = \mu + \gamma v + \tilde{G}(\mu, \delta)$$

$$v = \int_{-\infty}^{\infty} q_\infty \dot{q}_\infty^2 \Big/ \int_{-\infty}^{\infty} \dot{q}_\infty^2$$

and $\tilde{G}(\mu, 0) = 0$. The constant v is actually the same as the limit of $\beta(b)/\alpha(b)$ as $b \to 0$ in Lemma 2.9. Thus, $v = \frac{6}{7}$. We thus see that there is a homoclinic orbit of (2.5) for δ in a neighborhood of zero if and only if μ, δ satisfy $G(\mu, \delta) = 0$. The Implicit Function Theorem implies that there is a unique solution $(\mu(\delta), \delta)$, $0 \le |\delta| \le \delta_0$, $\delta_0 > 0$, of this equation and it satisfies $\mu(0) = \mu_0 = -\gamma v$. Finally, the curve $\mathscr{C}_\infty$ in $(\varepsilon_1, \varepsilon_2)$ space (see Fig. 2.1) along which there is a homoclinic orbit is given by

$$\varepsilon_2 = \mu(\varepsilon_1^{1/2})\varepsilon_1, \qquad \mu(0) = -\gamma v.$$

On the curve $\mathscr{C}_\infty \backslash \{(0, 0)\}$, we have $\sigma_0 = \varepsilon_2 < 0$ in Corollary 11.3.7. The homoclinic orbit is asymptotically stable. For the vector field in (2.5), we can choose the function γ_∞^a of Corollary 11.3.7 to be $\gamma_\infty^a(\mu, \delta) = -\delta G(\mu, \delta)$ and $\mathscr{C}_\infty = \{\varepsilon_1 = \delta^2, \ \varepsilon_2 = \mu\delta^2 : \gamma_\infty^a(\mu, \delta) = 0\}$. Since $\partial G(\mu_0, 0)/\partial \mu = -1$, it follows that $\partial \gamma_\infty^a(\mu(\delta), \delta)/\partial \mu < 0$ if $\delta \ne 0$, which is the case in $\mathscr{C}_\infty \backslash \{(0, 0)\}$. Thus, both alternatives in Corollary 11.3.7 occur; that is, in a neighborhood U of the homoclinic orbit, on one side of $\mathscr{C}_\infty$, every solution leaves U in both the forward and backward time and on the other side there is a unique asymptotically orbitally stable periodic orbit in U. It is also easy to verify that the periodic orbit exists to the left of $\mathscr{C}_\infty$.

Note that our analysis shows that every periodic orbit of (2.5) in a neighborhood of Γ is obtained by the above process. Thus, we have considered all periodic orbits through $(b, 0)$ with $b \in (0, \varepsilon)$ for some $\varepsilon > 0$ for $|\delta| < \delta_0$

and some $\delta_0 > 0$. There is either a unique periodic orbit or no periodic orbit.

Next, we analyze the behavior of the solutions of (2.5) near $(1, 0)$. The point $(1, 0)$ is a stable focus if $\mu < -\gamma$, and an unstable focus if $\mu > -\gamma$, $\gamma = \beta|\alpha|^{-1}$, The curve $\mu = -\gamma$ is therefore a possible value for a Hopf bifurcation. One can apply the method of Section 9.5 for the periodic orbits near $(1, 0)$ and obtain a bifurcation function $G(a, \mu, \delta)$ for $|\delta| < \delta_0$, $|a| < a_0$, $|\mu + \gamma| < \eta$ for some constants $\delta_0 > 0$, $a_0 > 0$, $\eta > 0$. Since Equation (2.5) for $\delta = 0$ has a center at $(1, 0)$, it follows that $G(a, \mu, 0) = 0$ for all a, μ. Thus, the appropriate bifurcation function to consider is $H(a, \mu, \delta) = G(a, \mu, \delta)/\delta$. This function satisfies $H(0, \mu, \delta) = 0$, $\partial H(0, \mu, 0)/\partial a = (\mu + \gamma)/2$. Lemma 2.1 implies for each $\mu > -\gamma$ and sufficiently close to $-\gamma$, there is a unique periodic orbit of (2.5) through $(b, 0)$ with b near 1. Thus, if b is taken close enough to 1, this periodic orbit must correspond to a zero of the bifurcation function $H(a, \mu, \delta)$. This proves there is a Hopf bifurcation at $\mu = -\gamma$ and the periodic orbit is asymptotically stable. Also, we have this result for $|\delta| < \delta_0$ where δ_0 is independent of b in a neighborhood $1 - \varepsilon < b < 1$.

We are now able to complete the analysis of the behavior of the solutions of (2, 3) for any $\varepsilon_1 > 0$, ε_2 sufficiently small; that is, for any $\delta > 0$ sufficiently small and any $\mu \in \mathbb{R}$.

Near the curve $\mathscr{C}_\infty$ in the $(\varepsilon_1, \varepsilon_2)$-plane corresponding to the homoclinic orbit, we have a periodic orbit bifurcating from this orbit as we cross $\mathscr{C}_\infty$ by decreasing μ. Lemma 2.1 says this orbit continues to exist, through $(b, 0)$ and is unique for $\varepsilon_2 = \mu^*(b, \delta)\varepsilon_1$. Furthermore, if $b \in (0, 1 - \varepsilon])$, we can have this assertion valid for $0 \le \delta < \delta_0$ for some $\delta_0 = \delta_0(\varepsilon) > 0$. The analysis above near the critical point $(1, 0)$ furnished the same conclusion for a uniform δ neighborhood and $b \in (1 - \varepsilon, 1)$. Thus, we can assert that a neighborhood of $(\varepsilon_1, \varepsilon_2) = (0, 0)$ intersected with $\varepsilon_1 > 0$ can be divided into sectors as shown in Figure 2.1 for which the corresponding flows are shown in Figure 2.2.

The case $\varepsilon_1 < 0$ is much easier to analyze and is left as an exercise for the reader.

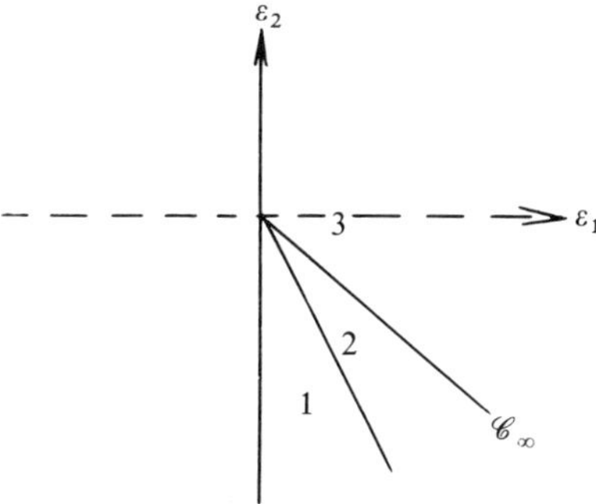

Figure 2.1

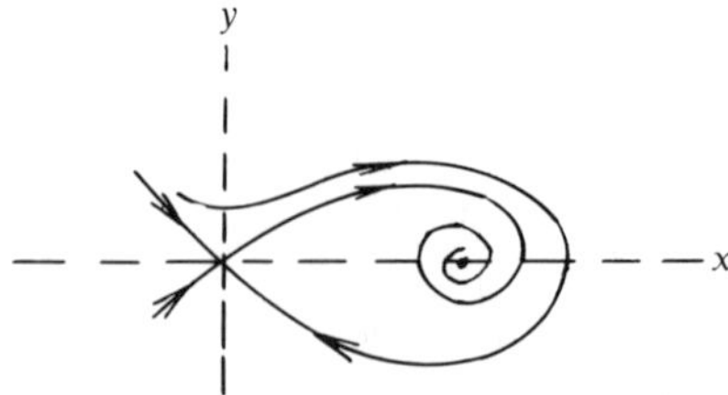

Sector 1

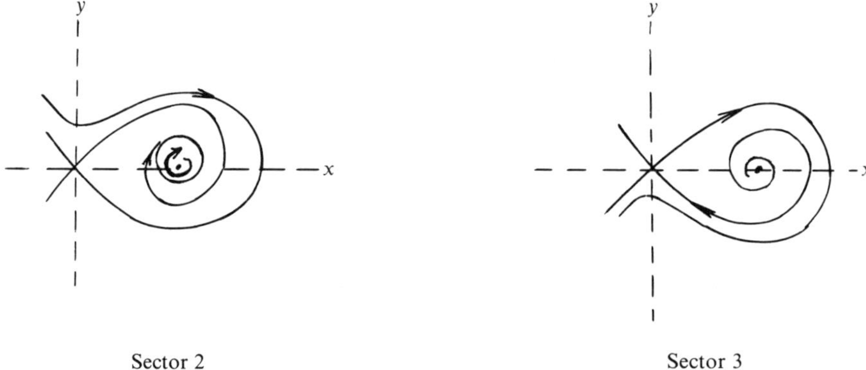

Sector 2   Sector 3

Figure 2.2

13.3. Two Zero Roots II

In this section, we consider the effects of autonomous perturbations on the behavior of the system

$$(3.1) \qquad \begin{aligned} \dot{x} &= axy + dx^3 \\ \dot{y} &= by^2 + cx^2 \end{aligned}$$

in a neighborhood of $(x, y) = (0,0)$. The perturbed system will be taken to be of the form

$$(3.2) \qquad \begin{aligned} \dot{x} &= \lambda x + axy + dx^3 \\ \dot{y} &= \beta y + by^2 + cx^2 \end{aligned}$$

where λ, β are small real parameters.

For fixed nonzero values of $a, b, c, d, a = -2b$, we determine the complete bifurcation diagram in the (λ, β)-plane. The nature of the bifurcation through a nonzero value (λ, β) on a bifurcation curve is either the coalescing of a saddle and node or a Hopf bifurcation or the creation of a periodic orbit near a homoclinic orbit. There are no bifurcation points corresponding to a degenerate periodic orbit. Finally, each bifurcation point is of the lowest possible degree subject to the symmetry in the equation.

If $bc < 0$, the term dx^3 is not needed to obtain the complete bifurcation diagram. It is essential when $bc > 0$.

When the system has only bifurcation points of the above types, perturbations of (3.2) by a vector field $(X(x, y, \lambda, \beta),\ Y(x, y, \lambda, \beta))$ with $X, Y = O((|x| + |y|)[(|x| + |y|)^3 + (|\lambda| + |\beta|)^2 + (|\lambda| + |\beta|)(|x| + |y|)])$ at $(x, y, \lambda, \beta) = (0, 0, 0, 0)$, $X(x, y, \lambda, \beta) = -X(-x, y, \lambda, \beta)$, $Y(x, y, \lambda, \beta) = Y(-x, y, \lambda, \beta)$ will have no qualitative effect on the bifurcation diagram or the solutions.

If a, b, c are nonzero, $a = -2b$, we may introduce the scaling in (3.2)

$$x \mapsto |b|^{1/2}x, \qquad y \mapsto |c|^{1/2}y, \qquad t \mapsto -(1/b|c|^{1/2})t,$$
$$\lambda \mapsto -(1/b|c|^{1/2})\lambda, \qquad \beta \mapsto -(1/b|c|^{1/2})\beta$$

to obtain the equivalent system

$$(3.3) \qquad \begin{aligned} \dot{x} &= \lambda x + 2xy + dx^3 \\ \dot{y} &= \beta y - y^2 - (\mathrm{sgn}\ bc)x^2 \end{aligned}$$

where d is another parameter.

It is convenient to introduce the change of variables $y \to \beta/2 + y$, $\lambda + \beta = \alpha$, to obtain the more symmetric form

$$(3.4) \qquad \begin{aligned} \dot{x} &= \alpha x + 2xy + dx^3 \\ \dot{y} &= \frac{\beta^2}{4} - y^2 - (\mathrm{sgn}\ bc)x^2. \end{aligned}$$

If we perform the scalings

$$(3.5) \qquad x \to \varepsilon x, \qquad y \to \varepsilon y, \qquad \beta \to \varepsilon, \qquad \alpha \to \varepsilon\alpha, \qquad t \to \varepsilon^{-1}t$$

the new equations become

$$(3.6) \qquad \begin{aligned} \dot{x} &= \alpha x + 2xy + d\varepsilon x^3 \\ \dot{y} &= \tfrac{1}{4} - y^2 - (\mathrm{sgn}\ bc)x^2. \end{aligned}$$

We are interested in the behavior of the solutions of Equation (3.6) for $\alpha \in \mathbb{R}$ and ε small.

For $\varepsilon = 0$ and $bc < 0$, the equations are

$$(3.7) \qquad \begin{aligned} \dot{x} &= \alpha x + 2xy \\ \dot{y} &= \tfrac{1}{4} - y^2 + x^2 \end{aligned}$$

The equilibrium points of this equation are $(0, \pm\tfrac{1}{2})$ and $(\pm(\alpha^2 - 1)^{1/2}/2, -\alpha/2)$ if $\alpha^2 > 1$. The equilibrium point $(0, \tfrac{1}{2})$ is a hyperbolic stable node if

$\alpha < -1$ and a saddle point if $\alpha > -1$. The point $(0, -\frac{1}{2})$ is a saddle point if $\alpha < 1$ and a hyperbolic unstable node if $\alpha > 1$. The other equilibrium points are always saddle points. At $\alpha = -1$, three hyperbolic equilibrium points coalesce at $(0, \frac{1}{2})$ and, at $\alpha = 1$, the same situation occurs at $(0, -\frac{1}{2})$. The symmetry in the problem for $\varepsilon = 0$ causes this type of degeneracy. There is also a degeneracy in the region $-1 < \alpha < 1$ because of the symmetry in the equation. There is an orbit connecting the saddle points $(0, \pm\frac{1}{2})$. This symmetry also shows that the term $\varepsilon\, dx^3$ will not change either the bifurcation curves or the qualitative properties of the solutions. In terms of the original coordinates, we can state the following result.

Theorem 3.1. *If $bc < 0$, there is a neighborhood U of $(x, y) = (0,0)$ and a neighborhood V of $(\lambda, \beta) = (0,0)$ such that the neighborhood V is divided into regions as shown in Figure 3.1 with the flow for Equation (3.3) in each region depicted in Figure 3.2 (drawn only for $\beta > 0$, $x \geq 0$ and symmetric with respect to the y axis).*

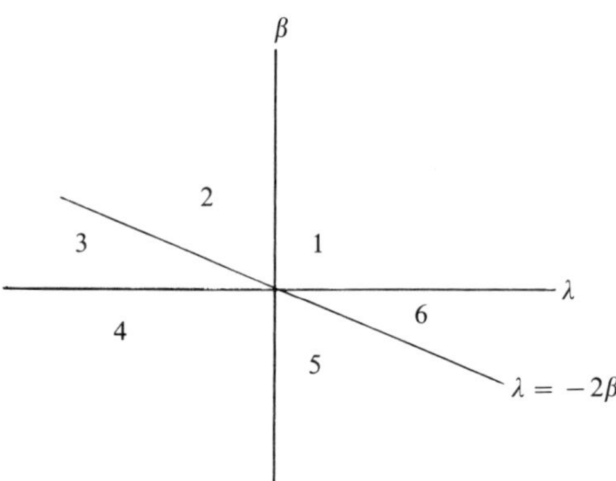

Figure 3.1. $(bc < 0)$.

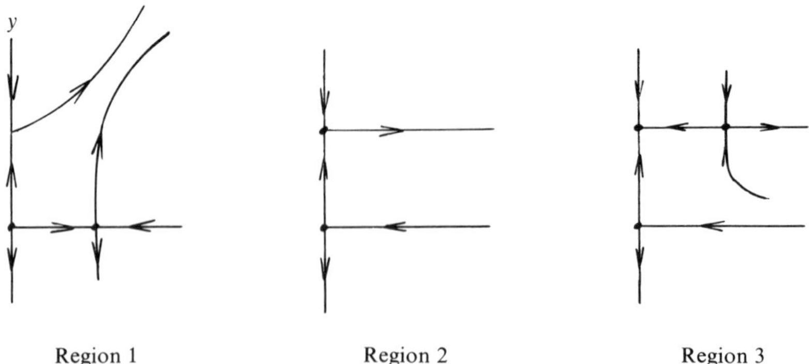

Region 1
Region 2
Region 3

Figure 3.2. $(bc < 0)$.

The more interesting case is when $bc > 0$. In this case, Equation (3.6) is

(3.8)
$$\dot{x} = \alpha x + 2xy + \varepsilon\, dx^3$$
$$\dot{y} = \tfrac{1}{4} - y^2 - x^2$$

These equations must be discussed for all $\alpha \in \mathbb{R}$ and ε small.

We note first that for $\alpha = 0$, $\varepsilon = 0$, the function

(3.9)
$$V(x, y) = \tfrac{1}{4}x - xy^2 - \frac{x^3}{3}$$

is a first integral of (3.8). Since the Jacobian of the vector field in (3.8) has trace α for $\varepsilon = 0$, it follows that there can be no periodic orbit of (3.8) unless $\alpha = 0$. This remark makes much of the discussion of (3.8) very simple for $\alpha \neq 0$, $\varepsilon = 0$. In fact, the topological structure of the flow is determined by the qualitative properties of the equilibrium points.

The equilibrium points of (3.8) for $\varepsilon = 0$ are $x = 0$, $y = \pm\tfrac{1}{2}$ for all α and the point $y = -\alpha/2$, $x^2 = (1 - \alpha^2)/4$ for $\alpha^2 \leq 1$. For $\alpha^2 \neq 1$, the points $x = 0$, $y = \pm\tfrac{1}{2}$ are hyperbolic saddles or nodes. In fact, for $\alpha > -1$ ($\alpha < -1$), the point $(0, \tfrac{1}{2})$ is a saddle (stable node). For $\alpha < 1$ (>1), the point $(0, -\tfrac{1}{2})$ is a saddle (unstable node). For $\alpha > 0$ (<0), the point $((1 - \alpha^2)^{1/2}/2, -\alpha/2)$ is a stable (unstable) focus. For $\alpha = 1$, the point $(0, -\tfrac{1}{2})$ is a pitchfork. For $\alpha = -1$, the point $(0, \tfrac{1}{2})$ is a pitchfork. These facts imply that the phase portrait for (3.8) for $\alpha \neq 0$, $\varepsilon = 0$ are the ones shown in Figure 3.3. The only bifurcation points are when $\alpha = \pm 1$ and these are of saddle-node type. For $\alpha = 0$, $\varepsilon = 0$, the function V in (3.9) is a first integral and the phase portrait is determined from the level curves of this function shown in Figure 3.3.

For each $\alpha \neq 0$, there is an $\varepsilon_0(\alpha)$ such that the same phase portraits are valid for the complete equations (3.8) for $0 \leq \varepsilon \leq \varepsilon_0(\alpha)$. This gives a uniform estimate in ε as long as α remains in a compact set. To obtain a uniform estimate for α large, we return to the original equation (3.4) and introduce the scaling

$$x \mapsto \varepsilon x, \qquad y \mapsto \varepsilon y, \qquad \alpha \mapsto \varepsilon, \qquad \beta \mapsto \varepsilon\beta, \qquad t \mapsto \varepsilon^{-1}t$$

to obtain the equations

$$\dot{x} = x + 2xy + \varepsilon^2\, dx^3$$
$$\dot{y} = \frac{\beta^2}{4} - y^2 - x^2.$$

In this equation, we need to discuss only the case where ε is near zero and β varies in a neighborhood of zero because the case for large β is the same as the previous one with the scaling (3.5). For $\varepsilon = 0$, $\beta = 0$, the equations have

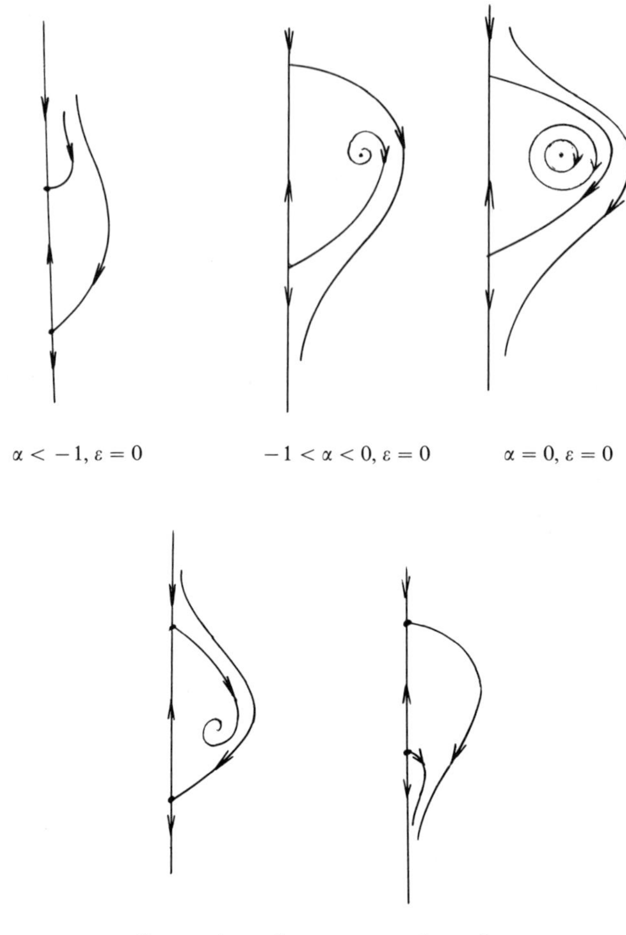

Figure 3.3

only one equilibrium point $x = 0$, $y = 0$ and this is a saddle-node. The analysis in the previous section applies to this case.

This shows that the only case remaining to discuss in Equation (3.8) is $\alpha = 0$, $\varepsilon = 0$. The manner in which the phase portrait in Figure 3.3 changes as we vary (α, ε) in a neighborhood of $(0,0)$ cannot be determined without specific knowledge of the term $\varepsilon\, dx^3$ in order ε.

For $\alpha = \varepsilon = 0$, the phase portrait is given in Figure 3.3. Let $(x_0(t), y_0(t))$, $x_0(t) > 0$, be the solution of (3.8) for $\alpha = \varepsilon = 0$ describing the heteroclinic orbit; that is, $x_0(t) \to 0$ as $t \to \pm\infty$, $y_0(t) \to -\frac{1}{2}$ as $t \to \infty$, $y_0(t) \to \frac{1}{2}$ as $t \to -\infty$. We may assume $y_0(0) = 0$. In a neighborhood of $\alpha = \varepsilon = 0$, there will be a curve with the property that Equation (3.8) has a heteroclinic orbit joining $(-\frac{1}{2}, 0)$, $(\frac{1}{2}, 0)$ for each (α, ε) belonging to this curve. The determination of this curve follows along the same ideas as in Section 11.3 as an application of the method of Liapunov–Schmidt. One constructs a scalar function $G(\alpha, \varepsilon)$, $G(0,0) = 0$, such that there is a heteroclinic orbit if and only if $G(\alpha, \varepsilon) = 0$. Furthermore,

$$G(\alpha, \varepsilon) = \alpha \int_{-\infty}^{\infty} x_0^2(t)\, dt + \varepsilon d \int_{-\infty}^{\infty} x_0^4(t)\, dt + O((|\alpha| + |\varepsilon|)^2)$$

as $\alpha, \varepsilon \to 0$. The Implicit Function Theorem implies there is a unique solution $\alpha = \alpha^*(\varepsilon)$ of $G(\alpha, \varepsilon) = 0$ for ε sufficiently small, $\alpha^*(0) = 0$ and

$$\delta \overset{\text{def}}{=} \frac{1}{d}\frac{d\alpha^*(0)}{d\varepsilon} = -\frac{\int_{-\infty}^{\infty} x_0^4(t)\, dt}{\int_{-\infty}^{\infty} x_0^2(t)\, dt} = -\frac{1}{2}.$$

The specific value of δ as $-\frac{1}{2}$ is obtained in the following way. Since

$$x_0^2\, dy_0 = (\tfrac{1}{4}x_0^2 - x_0^2 y_0^2 - x_0^4)\, dt - \int_{-\infty}^{\infty} x_0^4\, dt = \delta \int_{-\infty}^{\infty} x_0^2\, dt$$

$$0 = x_0 V(x_0, y_0) = \tfrac{1}{4}x_0^2 - x_0^2 y_0^2 - \tfrac{1}{3}x_0^4$$

we have

$$\int_{1/2}^{-1/2} x_0^2\, dy_0 = (\tfrac{1}{4} + \delta) \int_{-\infty}^{\infty} x_0^2\, dt - \int_{-\infty}^{\infty} x_0^2 y_0^2\, dt$$

$$= \frac{2\delta}{3} \int_{-\infty}^{\infty} x_0^2\, dt$$

$$= \frac{4\delta}{3} \int_{0}^{\infty} x_0^2\, dt$$

$$= \frac{2\delta}{3} \int_{0}^{x_0(0)} \frac{x_0}{y_0}\, dx_0$$

from $dx_0 = 2x_0 y_0\, dt$ and the symmetry in the equation. Since $V(x_0(t), y_0(t)) = 0$, we have $x_0^2(0) = \frac{3}{4}$, $x_0^2 = 3(\frac{1}{4} - y_0^2)$, $y_0 = (\frac{1}{4} - x_0^2/3)^{1/2}$. Using these relations we have, for $\theta = x_0^2$,

$$2 \int_{0}^{x_0(0)} \frac{x_0}{y_0}\, dx_0 = \int_{0}^{3/4} \left(\frac{1}{4} - \frac{\theta}{3}\right)^{-1/2} d\theta = 3$$

$$\int_{-1/2}^{1/2} x_0^2\, dy_0 = 3 \int_{-1/2}^{1/2} (\tfrac{1}{4} - y^2)\, dy = \tfrac{1}{2}.$$

Thus, $\delta = -\frac{1}{2}$.

The function $\alpha = \alpha^*(\varepsilon)$ satisfying $G(\alpha^*(\varepsilon), \varepsilon) = 0$ for $|\varepsilon| < \varepsilon_0$ has the property that there is a heteroclinic orbit of Equation (3.8) for $(\alpha, \varepsilon) = (\alpha^*(\varepsilon), \varepsilon)$.

Near the value $\alpha = 0$, $\varepsilon = 0$, there is also a curve where a Hopf bifurcation occurs. For $\alpha = 0$, $\varepsilon = 0$, Equation (3.8) has an equilibrium point $y = 0$, $x = \frac{1}{2}$. The Implicit Function Theorem implies there is an equilibrium close to this one for (α, ε) small and it is given approximately by $y_0 = -(\alpha + \varepsilon d/4)/2$, $x_0 = \frac{1}{2} + y_0$. Analyzing the stability properties of this solution, one sees that it has eigenvalues on the imaginary axis along a curve approximately given by $\alpha = -3\varepsilon d/4$.

We next analyze the periodic orbits of (3.8) for (α, ε) small. Any such orbit must be close to one of the periodic orbits for $(\alpha, \varepsilon) = (0,0)$. For $(\alpha, \varepsilon) \neq (0,0)$ $\alpha = \varepsilon\mu$, $b \in (0, \frac{1}{2})$, let $\Gamma(\alpha, \varepsilon, b)$ be a periodic orbit of (3.8) with $x(0) = b$, $y(0) = 0$. Following the procedure in Section 11.2, one can determine necessary and sufficient conditions on (α, ε) in order that (3.8) has a periodic orbit $\Gamma(\alpha, \varepsilon, b)$. Using the function V in (3.9), letting $\dot{V}$ denote the derivative of V along the solutions of (3.8) and requiring that $\int_{\Gamma(\alpha,\varepsilon,b)} \dot{V}\, dt = 0$, one obtains a bifurcation function $G(\mu, \varepsilon, b)$ which for $\varepsilon = 0$ is given by

$$G(\mu, 0, b) = \mu\alpha(b) + d\beta(b)$$

(3.10)
$$\alpha(b) = \int_b^{c(b)} x\, dy, \qquad \beta(b) = \int_b^{c(b)} x^3\, dy$$

and the function x is given in terms of y by the relation $V(x, y) = V(b, 0)$ and $c(b) > 1/2$ is defined by $V(c(b), 0) = V(b, 0)$.

It has been proved by Carr, Chow and Hale [1] that

$$d(\beta(b)/\alpha(b))/db \neq 0 \qquad \text{for } 0 < b < \tfrac{1}{2}.$$

This implies that the periodic orbit, if it exists, is unique. If we return to the original coordinate system in (3.3), we can state the following result.

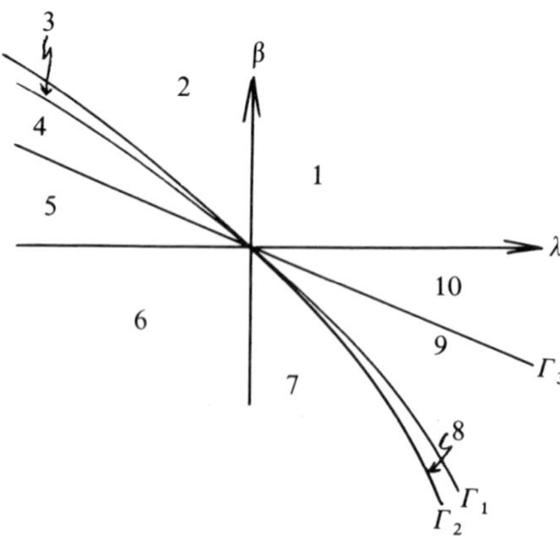

Figure 3.4. $(bc > 0)$.

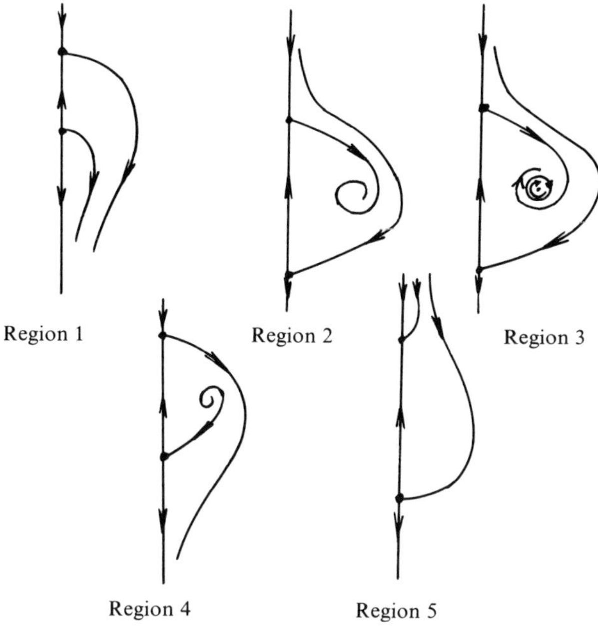

Figure 3.5. $(bc > 0)$.

Theorem 3.2. *If $bc > 0$ and $d \neq 0$, then there are neighborhoods U of $(x, y) = (0, 0)$ and V of $(\lambda, \beta) = 0$ such that V is divided into regions as shown in Figure 3.4 such that the flow for Equation (3.3) in each region is the one depicted in Figure 3.5 (drawn only from $\beta > 0$, $x \geq 0$ and symmetric with respect to the x-axis). The curves Γ_1, Γ_2, Γ_3 are given approximately by*

$$\Gamma_1 : \lambda \sim -\beta - d\beta^2/2$$
$$\Gamma_2 : \lambda \sim -\beta - 3d\beta^2/4$$
$$\Gamma_3 : \lambda \sim -2\beta$$

All curves in Figure 3.4 correspond to saddle-node bifurcations except Γ_1 and Γ_2. Each point on Γ_2 corresponds to a generic Hopf bifurcation and each point on Γ_1 corresponds to bifurcation through a heteroclinic orbit.

If $a \neq -2b$, one can obtain results similar to Theorem 3.2 by applying the same type of reasoning as above. The computations become more complicated because the first integral for the scaled equations with $\alpha = 0$, $\varepsilon = 0$ is not so simple. For more detail, see Carr, Chow and Hale [1].

It is possible also to consider other cubic terms in our original equation which preserve the same symmetry properties in x. More specifically, consider the equation

$$\dot{x} = \lambda x + axy + dx^3 + exy^2$$
$$\dot{y} = \beta y + by^2 + cx^2 + fx^2 y + gy^3$$

(3.11)

We introduce the following scaling

$$x \mapsto |b|^{1/2} x, \qquad y \mapsto |c|^{1/2} y, \qquad t \mapsto -b^{-1}|c|^{-1/2} t,$$
$$\lambda \mapsto -b|c|^{1/2}\lambda, \qquad \beta \mapsto -b|c|^{1/2}\beta, \qquad d \mapsto -|c|^{-1/2}(\operatorname{sgn} b) \cdot d,$$
$$e \mapsto -b^{-1}|c|^{1/2} e, \qquad f \mapsto -|c|^{-1/2}(\operatorname{sgn} b) f, \qquad g \mapsto -b^{-1}|c|^{1/2} g.$$

Hence, system (3.11) is transformed into

$$\begin{cases} \dot{x} = \lambda x + Bxy + dx^3 + exy^2, \\ \dot{y} = \beta y - y^2 + \eta x^2 + fx^2 y + gy^3, \end{cases}$$

(3.12)

where

$$B = -\frac{a}{b} \neq 0, \qquad \eta = -(\operatorname{sgn} bc).$$

Let

$$K = \eta\left(\frac{2}{B} + 2\right)d + \frac{2}{B}e + \eta f + 3g.$$

(3.13)

If $K \neq 0$, then the same conclusions as in Theorem 3.2 will be valid for Equation (3.11) in which the curves Γ_1 and Γ_2 will have different asymptotes. We may find detailed discussion in Carr, Chow and Hale [1], van Gils [1] Chow, Li and Wang [1] [2] and Zoladek [1].

13.4. Two Zero Roots III

In this section, we give another example of a bifurcation of degree greater than one. It is the situation where all of the quadratic terms vanish and there are some symmetries in the equation.

More specifically, we consider the equation

$$\dot{x} = x(\lambda - x^2 - by^2 + dx^4)$$
$$\dot{y} = y(\beta + cx^2 + y^2)$$

(4.1)

where $b > 0, c > 0, bc > 1, d \neq 0$ are fixed parameters and (λ, β) are bifurcation parameters varying in a neighborhood of $(0, 0)$. A complete description

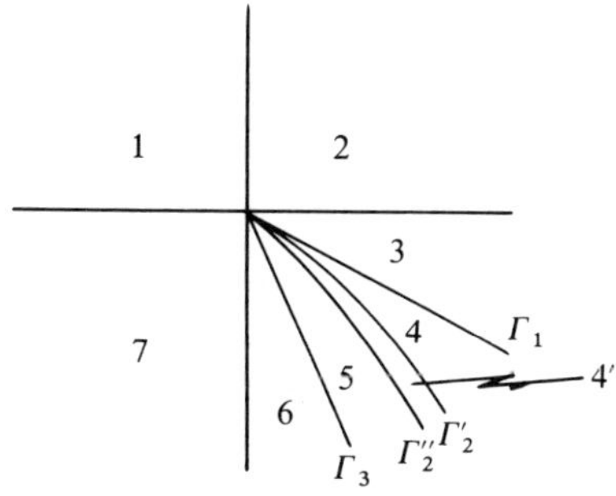

Figure 4.1

of the bifurcations that occur in this equation is summarized in the following result.

Theorem 4.1. *There is a neighborhood U of $(x, y) = (0,0)$ and a neighborhood V of $(\lambda, \beta) = (0,0)$ such that the neighborhood V is divided into regions as shown in Figure 4.1 such that the flow for Equation (4.1) for $x \geq 0$, $y \geq 0$, in each region is the one depicted in Figure 4.2 provided there is at most one periodic orbit of (4.1) for any (λ, β) between Γ_2' and Γ_2''. The curves Γ_1, Γ_2', Γ_2'', Γ_3*

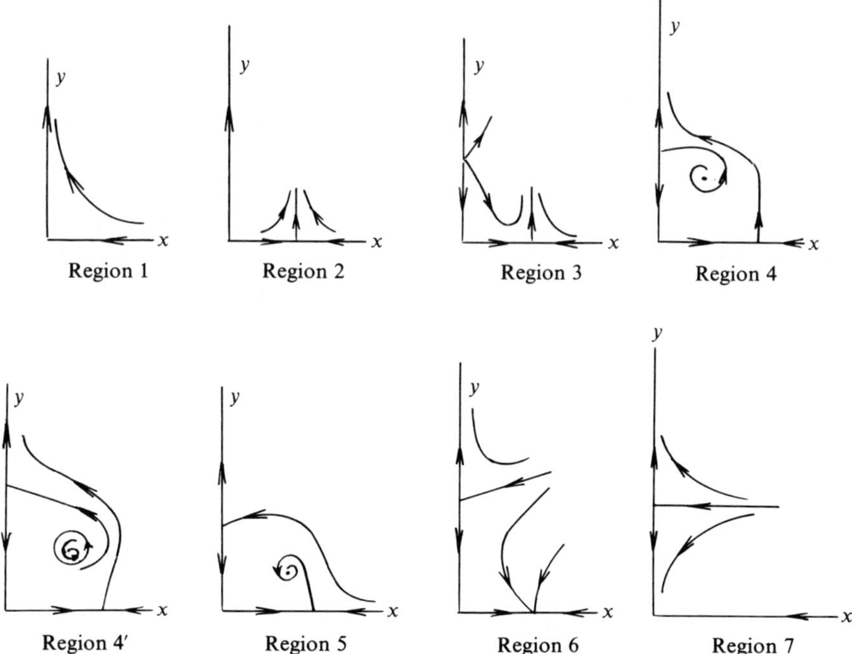

Figure 4.2

are given by

$$\Gamma_1 : \lambda + b\beta + o(|\beta|) = 0, \qquad \beta \le 0$$
$$\Gamma_2' : \lambda(1 + c) + \beta(1 + b) + d\delta'\beta^2 + o(\beta^2) = 0, \qquad \beta \le 0$$
$$\Gamma_2'' : \lambda(1 + c) + \beta(1 + b) + d\delta''\beta^2 + o(\beta^2) = 0, \qquad \beta \le 0$$
$$\Gamma_3 : \lambda c + \beta + o(|\beta|) = 0, \qquad \beta \le 0$$

as $\beta \to 0$, where δ', δ'' are constants computable from Equation (4.1) for $d = 0$, $\delta' < \delta'' < 0$. All of the bifurcations are of the saddle-node type except the ones on Γ_2', Γ_2'' where a Hopf bifurcation occurs at Γ_2'' and a heteroclinic orbit exists on Γ_2'. The flow in each region is depicted in Figure 4.2.

Remark 4.2. The term dx^5 in Equation (4.1) is needed only to resolve the bifurcation at the curves Γ_2', Γ_2''. If $d = 0$, we will see below that Equation (4.1) has a first integral with the flow given by the one shown in Figure 4.3.

Remark 4.3. The same qualitative behavior as depicted in Figures 4.1 and 4.2 will remain valid if the Equation (4.1) is subjected to higher order perturbations $(xX(x, y), yY(x, y))$ if the functions X, Y are even in x, y.

Remark 4.4. If fifth degree terms other than dx^5 occur in Equation (4.1), then the analysis will be similar to the one to be given below.

Remark 4.5. If $b > 0$, $c > 0$, $bc < 1$, then the equilibrium point (x_0, y_0) with $x_0 > 0$, $y_0 > 0$ is always a node, the curves Γ_2', Γ_2'' do not occur and the flow in the region between Γ_1 and Γ_2 is pictorially the same as shown in Regions 4 and 5 in Figure 4.2, except the interior point is a stable hyperbolic node. If $b > 0$, $c < 0$, then additional complications arise because there is the possibility of two more equilibrium points in the region $x > 0$, $y > 0$ and there can be a saddle-node bifurcation in this region. We do not discuss this case.

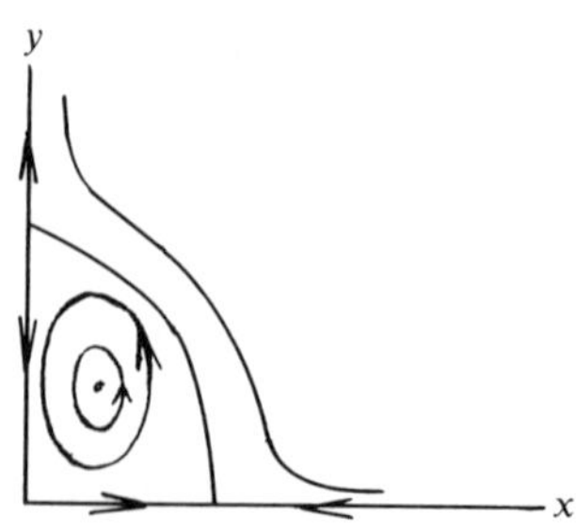

Figure 4.3

Outline of the proof. The lines $x = 0$ and $y = 0$ are invariant for all λ, β. In Regions 1, 2, 3, 6, 7, it is easy to verify that all equilibrium points lie on these lines. An analysis of the linear variational equation around these points gives the flows depicted in these regions. Furthermore, on the curves Γ_1, Γ_3 there is a saddle-node bifurcation which gives rise to an equilibrium point (x_0, y_0) with $x_0 > 0$, $y_0 > 0$. Thus, if there is to be a periodic orbit of (4.1), it must occur for values of λ, β in the complement of the regions 1, 2, 3, 6, 7. Also, for a periodic orbit to exist, there must be some value of α, β for which the eigenvalues of the linear variational equation about (x_0, y_0) are pure imaginary. If A is the matrix of this equation, then

$$
\begin{aligned}
\det A &= 4x_0^2 y_0^2 (bc - 1) \\
\operatorname{tr} A &= -2(x_0^2 - y_0^2)
\end{aligned}
$$
$$
x_0^2 = -(\lambda + \beta b)/(bc - 1), \qquad y_0^2 = (\beta + \lambda c)/(bc - 1)
$$

(4.2)

Since $bc > 1$, $x_0^2 > 0$, $y_0^2 > 0$, the eigenvalues are pure imaginary if and only if $\operatorname{tr} A = 0$; that is,

$$
\lambda(1 + c) + \beta(1 + b) = 0
$$

(4.3)

Equation (4.3) defines a curve Γ_2. The curves Γ_1, Γ_2 are obtained respectively when $x_0^2 = 0$ and $y_0^2 = 0$.

The local flow near Γ_1, Γ_2 is easily shown to be the one depicted in Regions 4 and 5. This picture remains global in both regions 4 and 5. In fact, any periodic orbit must enclose (x_0, y_0) since there are no other critical points. Such an orbit could only appear by the introduction of a saddle-node type bifurcation in region $x > 0$, $y > 0$. This is impossible.

Let us now show there is a first integral of Equation (4.1) for $d = 0$ on $\Gamma_2 = \{(\lambda, \rho): \lambda = -m_0 \beta, \beta \leq 0\}$, $m_0 = (1 + b)/(1 + c) > 0$. Let $\lambda = -m_0 \beta$,

$$
x \mapsto |\beta|^{1/2} x, \qquad y \mapsto |\beta|^{1/2} y, \qquad t \mapsto |\beta|^{-1} t.
$$

(4.4)

Since $\beta \leq 0$, we obtain the equivalent equations for $d = 0$,

$$
\begin{aligned}
\dot{x} &= +m_0 x - bxy^2 - x^3 \\
\dot{y} &= -y + cx^2 y + y^3
\end{aligned}
$$

(4.5)

If

$$
\mu = \frac{b + 1}{bc - 1}, \qquad v = \frac{c + 1}{bc - 1},
$$

(4.6)

then the function

$$
H(u, v) = u^{2v} v^{2\mu} \left[1 - \frac{v}{\mu} u^2 - v^2 \right]
$$

(4.7)

is a first integral of Equation (4.5). Using the first integral (4.7), it is easy to see that the flow for Equation (4.1) on Γ_2 for $d = 0$ is the one depicted in Figure 4.3.

We now analyze the solutions of Equation (4.1) for $d \neq 0$ and (λ, β) varying near Γ_2; that is, (λ, β) near the line $\lambda = -m_0\beta$, $m_0 = (1 + b)/(1 + c)$. Introduce the scaling

$$x \mapsto |\beta|^{1/2}x, \qquad y \mapsto |\beta|^{1/2}y, \qquad t \mapsto |\beta|^{-1}t, \qquad \lambda \mapsto -m_0\beta + \alpha\beta$$

to obtain

$$(4.8) \qquad \begin{aligned} \dot{x} &= x(-m_0 + \alpha - x^2 - by^2 + \beta \, dx^4) \\ \dot{y} &= y(-1 + cx^2 + y^2). \end{aligned}$$

One can proceed as in the proof of Theorem 3.2 to obtain the curve in the (λ, β)-plane along which Equation (4.8) has a heteroclinic orbit. In terms of the scaled variables, we observe that such a solution can exist only in a neighborhood of $\alpha = 0$, $\beta = 0$. We then obtain a bifurcation function $G(\alpha, \beta)$ for a heteroclinic orbit and observe that the equation $G(\alpha, \beta) = 0$ has a unique solution $\alpha = \alpha^*(\beta)$ in a neighborhood of $(\alpha, \beta) = (0, 0)$, $\alpha^*(0) = 0$. This gives a curve

$$\Gamma_2' = \{(\lambda, \beta): \lambda = -m_0\beta + \alpha^*(\beta)\beta\}$$

in the (λ, β)-plane such that Equation (4.1) for $(\lambda, \beta) \in \Gamma_2'$ has a heteroclinic orbit.

One can also obtain a curve Γ_2'' in the (λ, β)-plane

$$\Gamma_2'' = \{(\lambda, \beta): \lambda = -m_0\beta + \alpha^{**}(\beta)\beta\},$$

where $\alpha^{**}(0) = 0$, such that, for any $(\lambda, \beta) \in \Gamma_2''$, the linear variational equation for the solution (x_0, y_0), $x_0 > 0$, $y_0 > 0$, of (4.1) has both eigenvalues pure imaginary. Proceeding as before, one obtains the existence of a periodic orbit in the Region 4' between Γ_2' and Γ_2''.

The uniqueness of periodic orbits has been shown in Zoladek [2]. The flow in Region 4' is shown in Figure 4.2. This completes the proof of the theorem. $\square$

More generally, we consider the equation

$$(4.9) \qquad \begin{cases} \dot{x} = x(\lambda + ax^2 + by^2 + lx^4 + mx^2y^2 + ny^4), \\ \dot{y} = y(\beta + cx^2 + dy^2 + px^4 + qx^2y^2 + ry^4), \end{cases}$$

It is shown in Chow, Li and Wang [2] that if $abcd \neq 0$, $ad - bc \neq 0$ and

$$\tilde{K} := \frac{\alpha\xi^2}{(\alpha + 1)(\xi + 1)(\xi + 2)}\left(\frac{\alpha + 2}{\xi}l + p\right) - \frac{\alpha\xi\,\mathrm{sgn}(ab)}{(\alpha + 1)(\xi + 2)}\left(\frac{\alpha + 1}{\xi + 1}m + q\right)$$
$$+ \left(\frac{\alpha}{\xi + 2}\eta + r\right) \neq 0,$$

where

$$\alpha = \frac{d(c - a)}{ad - bc}, \qquad \xi = \frac{a(b - d)}{ad - bc},$$

then any perturbation of (4.9) by the terms of order higher than 5 will have no qualitative effect on the bifurcation diagrams or the flow. The total number of bifurcation diagrams for different regions in the $\alpha\xi$ plane is 13. For more details, see Zoladek [2] and Chow, Li and Wang [2].

If $\tilde{K} = 0$, then in order to describe the qualitative behavior of the flow completely, we need to add other higher order terms to system (4.9), and assume other non-degenerate conditions.

13.5. Several Pure Imaginary Eigenvalues

In this section, we consider a differential equation in $\mathbb{R}^3$ for which the linear variational equation near the equilibrium point zero has two pure imaginary eigenvalues and one zero eigenvalue. We also make a few remarks about an equation in $\mathbb{R}^4$ with all eigenvalues pure imaginary. To simplify the situation, it will be assumed also that a certain type of symmetry prevails. More specifically, consider the equation

$$\dot{x} = A(\lambda)x + f(x, y)$$
$$\dot{y} = \beta y + g(x, y)$$

(5.1)

where λ, β are small real parameters, $x \in \mathbb{R}^2$, $y \in \mathbb{R}$, f, g are C^4-functions,

$$A(\lambda) = \begin{pmatrix} \lambda & 1 \\ -1 & \lambda \end{pmatrix},$$

(5.2)

$$f(x, y) = O(|x|(|x| + |y|))$$
$$g(x, y) = O((|x| + |y|)^2)$$

as $|x|, |y| \to 0$. The hypothesis on f implies that the symmetry condition

(5.3)
$$f(0, y) = 0$$

is satisfied.

Since $f(0, y) = 0$, it is legitimate to introduce polar coordinates for $x = (x_1, x_2)$ as $x_1 = \rho \cos \theta$, $x_2 = -\rho \sin \theta$. If this is done and t is replaced by θ, one obtains the equations

$$
\begin{aligned}
\rho' &= \lambda\rho + R(\theta, \rho, y, \lambda, \beta) \\
y' &= \beta y + Y(\theta, \rho, y, \lambda, \beta)
\end{aligned}
$$
(5.4)

where "$'$" $= d/d\theta$, $R = O(|\rho|(|\rho| + |y|))$, $Y = O((|\rho| + |y|)^2)$ as $\rho, y \to 0$.

The problem is to determine the behavior of the solutions of (5.4) in a neighborhood of $(\rho, y) = (0, 0)$ for (λ, β) in a neighborhood of $(0, 0)$. To discuss (5.4), one can use the method of averaging to transform (5.4) by a transformation periodic in θ to the equivalent equation

$$
\begin{aligned}
\rho' &= \rho(\lambda + ay + d\rho^2 + ey^2) + R(\theta, \rho, y, \lambda, \beta) \\
y' &= \beta y + by^2 + c\rho^2 + f\rho^2 y + gy^3 + Y(\theta, \rho, y, \lambda, \beta)
\end{aligned}
$$
(5.5)

where R, Y are $O((|y| + |\rho|)^4 + (|\lambda| + |\beta|)(|y| + |\rho|)^2 + (|\lambda| + |\beta|)^2(|y| + |\rho|))$ in a neighborhood of zero. The function R is odd in ρ and Y is even in ρ.

For $R = Y = 0$ and nonzero a, b, c, we have indicated in Section 3 how to determine the complete bifurcation diagram in the (λ, β)-space. Also, under some additional conditions on the cubic terms, the bifurcations were of the simplest possible type; namely, saddle-node bifurcations, Hopf bifurcation and the creation of a periodic orbit by passing through a heteroclinic orbit. This information is contained in Theorem 3.1 and 3.2 and the remarks following these results.

It remains to discuss what happens to the solutions when R, Y are not zero. Since the functions R, Y are periodic in θ, the results in Section 9.5 give a complete description of the behavior of the complete equation (5.5) in the neighborhood of a saddle-node bifurcation. The hyperbolic saddle and node of the averaged equations become hyperbolic 2π-periodic solutions of (5.5) and they coalesce in a uniform way at the bifurcation curve.

When the curve of bifurcation of the averaged equations corresponds to a Hopf bifurcation, the situation is more difficult because a hyperbolic focus becomes a 2π-periodic solution for (5.5) and a hyperbolic periodic orbit for the averaged equations becomes an invariant torus. However, this situation was discussed in detail in Section 12.6 and is described fully in Lemma 12.6.1. The torus and periodic solution coalesce uniformly at the bifurcation curve. It is not necessary to impose any resonance between the frequency of the orbit that comes from the Hopf bifurcation and forcing frequency 1 in θ. This is due to the fact that the period of periodic orbit of the averaged equations approaches ∞ (and thus the frequency approaches zero) as $(\beta, \lambda) \to (0, 0)$.

This gives a complete description of the full equations (5.5) except in the neighborhood of a point (λ_0, β) which is on a curve for which the averaged equations have a heteroclinic orbit. The periodic perturbation for (λ, β) near such points can change the structure of the flow in a significant way. However, it is possible to give a description of the flow near such points by using the ideas in Section 11.3.

For simplicity, we assume e, f, g are zero, $a = -2b, d = 1, bc > 0$. Making the same transformations as in Section 3 including the scaling (3.5) (here (ρ, y) play the role of (x, y) in Section 3), one obtains the equations

$$
\begin{aligned}
\rho' &= 2\rho y + \alpha\rho + \varepsilon\rho^3 + O(|\varepsilon|^2) \\
y' &= \tfrac{1}{4} - y^2 - \rho^2 + O(|\varepsilon|^2)
\end{aligned}
\tag{5.6}
$$

where the $O(|\varepsilon|^2)$ terms are 2π-periodic in θ. Now suppose $(\alpha_0, \varepsilon_0)$ corresponds to a point (λ_0, β_0) on the curve Γ_1 in Figure 3.4; that is (λ_0, β_0) lies on the curve where the averaged equations have a heteroclinic orbit. Let φ represent a phase shift along the heteroclinic orbit, $(\rho_0(t), y_0(t))$ be the solution of (5.6) corresponding to $(\alpha_0, \varepsilon_0)$, $\rho_0(t) > 0$, $\to 0$ as $t \to \pm\infty$, $y_0(t) \to \pm\tfrac{1}{2}$ as $t \to \pm\infty$, and let $\alpha = \alpha_0 + v$, $\varepsilon = \varepsilon_0 + \mu$. Applying the same reasoning as in Section 11.3, one obtains a function $\tilde{G}(\varphi, v, \mu)$ for $\varphi \in \mathbb{R}$, μ, v close to zero, with the property that there is a heteroclinic point of (5.6) close to the point $(\rho_0(\alpha), y_0(\alpha))$ if and only if $\tilde{G}(\varphi, v, \mu) = 0$. Furthermore, the heteroclinic point is transverse if and only if $\partial\tilde{G}(\varphi, v, \mu)/\partial\varphi \neq 0$. The function $\tilde{G}$ has the form

$$
\tilde{G}(\phi, v, \mu) = v \int_{-\infty}^{\infty} \rho_0^2 + \mu \int_{-\infty}^{\infty} \rho_0^4 + \varepsilon_0\mu h_{\alpha_0, \varepsilon_0}(\phi) + O((|v| + |\mu|)^2)
$$

as $v, \mu \to 0$. If $h_{\alpha_0, \varepsilon_0}(\phi)$ is not the zero function (Remark 11.3.12), there are sectors in the (v, μ)-plane in which there are either no heteroclinic orbits or there are transverse heteroclinic orbits. This gives rise to some type of random behavior as described in Section 11.3 for transverse heteroclinic points.

The function $h_{\alpha_0, \varepsilon_0}(\phi)$, in principle, can be computed. One should be able to show that this function has nonzero second derivatives at its maximum and minimum. The results in Section 11.4 then imply that subharmonic bifurcations also occur near points in Γ_1. Due to the complexity of the computations, we do not dwell on this question.

If further symmetries occur in the problem, there may be no second order terms in the normal form for the vector field. In this case, the simplest averaged equations are

$$
\begin{aligned}
\dot{\rho} &= \rho(\lambda - a\rho^2 - by^2) \\
\dot{y} &= y(\beta + c\rho^2 + dy^2)
\end{aligned}
\tag{5.7}
$$

These equations were discussed in detail in Section 4. It was shown that a complete generic bifurcation diagram could not be obtained without the consideration of higher order terms. For $a = d = 1$, $bc > 1$, $b > 0$, $c > 0$, and the addition of a term $e\rho^5$ in the first equation in (5.7), Theorem 4.1 gives a description of the bifurcation pattern for the averaged equations.

When the higher order 2π-periodic terms are included, the same type of analysis that was given for (5.6) can be made in the present case.

Let us conclude this section with a few remarks about a more difficult problem. Consider the system of equations

$$\begin{aligned}
\dot{u} &= A(\lambda)u + U(u, v) \\
\dot{v} &= B(\beta)v + V(u, v)
\end{aligned} \tag{5.8}$$

where $u \in \mathbb{R}^2$, $v \in \mathbb{R}^2$,

$$A(\lambda) = \begin{bmatrix} \lambda & 1 \\ -\omega_1^2 & \lambda \end{bmatrix}, \qquad B(\beta) = \begin{bmatrix} \beta & 1 \\ -\omega_2^2 & \beta \end{bmatrix}$$

and $U, V = O((|u| + |v|)^2)$ as $u, v \to 0$, $U(0, v) = 0$, $V(u, 0) = 0$. The constants $\omega_1 > 0$, $\omega_2 > 0$, are fixed and λ, β are small bifurcation parameters.

If $u = (\rho \cos \theta_1, -\rho \sin \theta_1)$, $v = (y \cos \theta_2, -y \sin \theta_2)$, then Equations (5.8) are equivalent to the equations

$$\begin{aligned}
\dot{\theta}_j &= \omega_j + \Theta_j(\theta_1, \theta_2, \rho, y), \qquad j = 1, 2, \\
\dot{\rho} &= \lambda\rho + R(\theta_1, \theta_2, \rho, y) \\
\dot{y} &= By + Y(\theta_1, \theta_2, \rho, y)
\end{aligned} \tag{5.9}$$

where each function is 2π-periodic in θ_1 and θ_2.

Assuming certain nonresonance conditions between the frequencies ω_1, ω_2, one can apply averaging to obtain a set of equivalent equations

$$\begin{aligned}
\dot{\theta}_j &= \omega_j + \tilde{\Theta}_j(\theta_1, \theta_2, \rho, y, \lambda, \beta), \qquad j = 1, 2, \\
\dot{\rho} &= \rho(\lambda - a\rho^2 - by^2) + \tilde{R}(\theta_1, \theta_2, \rho, y, \lambda, \beta) \\
\dot{y} &= y(\beta + c\rho^2 + dy^2) + \tilde{Y}(\theta_1, \theta_2, \rho, y, \lambda, \beta)
\end{aligned} \tag{5.10}$$

where $\tilde{R}, \tilde{Y}$ are higher order terms. This is the simplest form that can be obtained for the averaged equations because of the nature of polar coordinates.

This problem looks very similar to the one before, but it is much more complicated because there are two angle variables. The theory of integral manifolds in Section 12.5 gives information away from the bifurcation curves of the averaged equations. In fact, any hyperbolic equilibrium point for the averaged equations becomes a hyperbolic two dimensional torus for the complete equations. Any hyperbolic periodic orbit for the averaged equations becomes a hyperbolic three-dimensional torus for the complete equations.

If the averaged equations have a saddle-node type of bifurcation or a Hopf bifurcation, then we can say very little about the complete equations for the reasons mentioned in Section 12.8.

If the averaged equations have a heteroclinic orbit, the theory of Section

11.3 does not apply since it allows only one angle variable. However, it seems plausible that an appropriate generalization of Section 11.3 to this case can be obtained. The manner in which the bifurcation function is obtained has nothing to do with periodicity.

13.6. Bibliographical Notes

The bifurcation diagram for Eq. (2.3) was considered by Howard and Kopell [1], Conley [1]. Arnold [2], [5] and Bogdanov [1] considered the same nonlinearities as in Eq. (2.3), but with different perturbation terms, namely, the equation

$$\dot{x} = y, \qquad \dot{y} = \varepsilon_2 + \varepsilon_1 x + \alpha x^2 + \beta xy$$

Bogdanov [1] has shown that every two parameter family of vector fields close to (2.1) in the C^3 topology is equivalent to the above equation. In the analysis of (2.3), the difficult part concerned the proof of uniqueness of the limit cycle in a certain region in parameter space. Yeh [1] also has a proof of uniqueness. The hypothesis on the derivative in Theorem 3.2 insures this uniqueness of the periodic orbit. Carr, Chow and Hale [1], van Giles [1] and Zholodek [1] have shown recently that this condition is always satisfied. Langford [1] has also considered Eqs. (3.1), (3.2).

Theorem 4.1 is related to results of Holmes [3]. He has also obtained partial results—not having analyzed the behavior near the homoclinic and Hopf bifurcation on the boundary of Region 4′ in Fig. 4.1. Carr [1] (see also Takens [1, 3] for partial results) has analyzed the behavior of the solutions of the equation

$$\dot{x} = y, \qquad \dot{y} = \varepsilon_1 x + \varepsilon_2 y + \alpha x^3 + \beta x^2 y$$

where $\varepsilon_1, \varepsilon_2$ are small real parameters and α, β are fixed, nonzero constants. He also has given very interesting applications of the analysis to a panel flutter problem. For other applications, see Sijbrand [1].

Guckenheimer [1] has discussed Eq. (5.5), Holmes [3] and Guckenheimer [2] Eq. (5.8) and Eq. (5.10).

The existence of almost periodic solutions for nonhomogeneous equations without lower terms is certainly related to the complete solution of the problems in Section 5 when there are two angle variables. The paper of Markus and Moore [1] deals with this question.

Chapter 14

Perturbation of Spectra of Linear Operators

14.1. Introduction

To describe the contents of this chapter, let us recall some previous concepts.

Suppose X, Z are Banach spaces with X continuously imbedded in Z by a continuous map $I: X \to Z$. We call I the identity map and let J be the inverse of I. If $A, B: X \to Z$ are bounded linear operators, the *resolvent set* $\rho(B, A)$ of the pair (B, A) is the set of $\lambda \in \mathbb{C}$ such that $B - \lambda A$ has a bounded inverse. The *spectrum* $\sigma(B, A)$ of the pair (B, A) is the complement of $\rho(B, A)$.

In Section 2, we give a basic result on the analytic dependence of the resolvent $(B - \lambda A)^{-1}$ on (B, A, λ). It is then a simple matter to prove that the spectrum $\sigma(B, A)$ is upper semicontinuous in B, A if it is bounded. It is also shown that $\sigma(B, A)$ is continuous in (B, A) at (B_0, A_0) if it is bounded and $\sigma(B_0, A_0)$ is totally disconnected. A simple application is made to numerical approximation of eigenvalues of differential operators. Sections 3–5 are independent of Section 2.

A point $\lambda \in \sigma(B, A)$ is an *eigenvalue* of (B, A) if zero is an eigenvalue of $B - \lambda A$. An eigenvalue of (B, A) is simple if $\dim \mathcal{N}(B - \lambda A) = 1 = \operatorname{codim} \mathcal{R}(B - \lambda A)$ and $A\mathcal{N}(B - \lambda A) \not\subset \mathcal{R}(B - \lambda A)$. Section 3 is devoted to a discussion of the dependence of simple eigenvalues on B, A and even more specifically on $\varepsilon \in E$, a Banach space, if $B(\varepsilon), A(\varepsilon)$ depend on ε. Basic continuity, differentiability and analyticity theorems are proved. Several examples from boundary value problems for differential equations are presented.

Assuming that $B(\varepsilon), A(\varepsilon)$ depend continuously on ε and that λ_0 is an isolated normal eigenvalue of $(B(0), A(0))$, it is proved that the eigenvalues near λ_0 are continuous in ε in Section 4. This is accomplished by using the method of Liapunov–Schmidt to reduce the problem to a finite dimensional problem. The method is constructive so that it is possible to obtain more structure about the eigenvalues by using the Newton polygon if the parameter is a scalar and the scaling techniques of Chapter 7 if the parameter is a vector.

In Section 5, for $B(\varepsilon)$ analytic in a scalar parameter, self adjoint and λ_0 an isolated normal eigenvalue of $B(0)$, we prove the eigenvalues near λ_0 are analytic in ε. This is again proved by the method of Liapunov–Schmidt using the fact that the reduced problem is also self-adjoint. Some remarks are made about the vector parameter case.

14.2. Continuity Properties of the Spectrum

To discuss the continuity properties of $\sigma(B, A)$ we use the notation

$$d(U, V) = \sup_{\lambda \in U} \text{dist}(\lambda, V)$$

for any sets $U, V \subseteq \mathbb{C}$. If $U = \phi$, we put $d(\phi, V) = 0$.

The continuity properties will depend on the type of convergence of (B, A) to (B_0, A_0). Any attempt to study the dependence of $\sigma(A, B)$ on A, B when (A, B) is considered close to (A_0, B_0) in a topology weaker than the norm topology seems to be very difficult. To illustrate the problems, let us consider some examples.

The simplest example is for the space $l_2 = \{x = (x_1, x_2, \ldots) : x_j \in \mathbb{C}, |x|^2 = \sum_j |x_j|^2 < \infty\}$. Let

$$B_n(x_1, x_2, \ldots, x_n, x_{n+1}, \ldots) = (x_1, x_2, \ldots, x_n, 0, 0, \ldots)$$

for $n = 1, 2, \ldots$. Then $B_n \to I$ strongly as $n \to \infty$, $\sigma(B_n) = \{0, 1\}$, $\sigma(I) = \{1\}$,. This can also be expressed by saying that $\sigma(B)$ is not *upper semicontinuous* at $B = I$ in the strong topology in the sense that there is a sequence $B_n \to I$ strongly such that $d(\sigma(B_n), \sigma(I)) \nrightarrow 0$ as $n \to \infty$. One may not have *lower semicontinuity* either in the sense that there are operators $A, A_n, A_n \to A$ strongly such that $d(\sigma(A), \sigma(A_n)) \nrightarrow 0$ as $n \to \infty$. In fact, if $A(x_1, x_2, x_3, \ldots) = (0, x_1, x_2, \ldots)$, then $|A^n x| = |x|$ for all x and so there is a point $\lambda \in \sigma(A^n)$ such that $|\lambda| = 1$. On the other hand, $AB_n \to A$ strongly, AB_n is nilpotent and so $\sigma(AB_n) = \{0\}$. Thus, $d(\sigma(A), \sigma(AB_n)) \nrightarrow 0$ as $n \to \infty$ and $\sigma(B)$ is not lower semicontinuous at $B = A$.

Other examples of this type occur naturally in delay equations. In fact, if an infinite delay equation is approximated by finite delay equations, one always has upper semicontinuity but does not always have lower semicontinuity of the spectrum.

In the study of the behavior of the solutions of difference equations as a function of the delays, the opposite situation occurs. One can prove lower semicontinuity of the spectrum but not upper semicontinuity.

In the following, we will consider convergence of operators in the operator topology. The first result concerns the dependence of $(B - \lambda A)^{-1}$ on (λ, B, A).

Theorem 2.1. *For any $\lambda_0 \in \rho(B_0, A_0)$, the operator $(B - \lambda A)^{-1}$ is analytic for (λ, B, A) in a neighborhood of (λ_0, B_0, A_0).*

Proof. This is an elementary application of the Implicit Function Theorem. In fact, if

$$F(\lambda, B, A, C) = (B - \lambda A)C - I : Z \to Z$$

then F is analytic in $\lambda, B, A, C, F(\lambda_0, B_0, A_0, (B - \lambda_0 A_0)^{-1}) = 0$, $\partial F(\lambda_0, B_0, A_0, (B_0 - \lambda_0 A_0)^{-1})/\partial C = B_0 - \lambda_0 A_0$. Since this latter operator has a bounded inverse, the Implicit Function theorem implies there is a $C^*(\lambda, B, A)$ for (λ, B, A) in a neighborhood of (λ_0, B_0, A_0), such that $(B - \lambda A)C^*(\lambda, B, A) = I$, $C^*(\lambda_0, B_0, A_0) = (B_0 - \lambda_0 A_0)^{-1}$. This operator $C^*(\lambda, B, A)$ is analytic in λ, B, A. Also, for all $x \in X$,

$$(B - \lambda A)x = (B - \lambda A)C^*(\lambda, B, A)(B - \lambda A)x.$$

Since $(B - \lambda A)$ is one-to-one, this implies

$$x = C^*(\lambda, B, A)(B - \lambda A)x$$

for all x. Thus, $C^*(\lambda, B, A) = (B - \lambda A)^{-1}$ and the theorem is proved. $\square$

Corollary 2.2. *If Γ is a compact set of $\rho(B_0, A_0)$, then there is a $\delta > 0$ such that $\Gamma \subset \rho(B, A)$ for $|B - B_0| < \delta, |A - A_0| < \delta$.*

Corollary 2.2 gives some information about the behavior of $\sigma(B, A)$ as $(B, A) \to (B_0, A_0)$. In fact, the following remark is valid.

Corollary 2.3. *For any compact set $K \subset \mathbb{C}$,*

$$d(K \cap \sigma(B, A), K \cap \sigma(B_0, A_0)) \to 0$$

as $(B, A) \to (B_0, A_0)$; that is, $\sigma(B, A)$ is upper-semicontinuous on compact sets of $\mathbb{C}$.

Proof. For any compact set $K \subset \mathbb{C}, \varepsilon > 0$, let $\Gamma = \{\lambda \in K : \operatorname{dist}(\lambda, \sigma(B_0, A_0)) \geq \varepsilon\}$. From Corollary 2.2, there is a $\delta > 0$ such that $\Gamma \subset \rho(B, A)$ if $|B - B_0| < \delta$, $|A - A_0| < \delta$. This proves the result. $\square$

A particular case of Corollary 2.3 is

Corollary 2.4. *If there is a $\delta > 0, r > 0$, such that*

$$\sigma(B, A) \subseteq \{\lambda \in \mathbb{C} : |\lambda| \leq r\} \quad for \ |B - B_0| < \delta, |A - A_0| < \delta,$$

then $\sigma(B, A)$ is upper semicontinuous in (B, A) for $|B - B_0| < \delta, |A - A_0| < \delta$.

Remark 2.5. If $X \neq Z$, then, in general, one cannot expect $\sigma(B_0, A_0)$ to be bounded. Even if $X = Z$, one can have $\sigma(B_0, A_0)$ unbounded. In fact, take

$$B_0 = 0, \qquad A_0 = \begin{bmatrix} 0 & 1 \\ 0 & 0 \end{bmatrix}$$

then $\mathbb{C} = \sigma(B, A)$. We also note that this example shows that $\sigma(B_0, A_0) = \mathbb{C}$ and $\sigma(B, A_0)$ may be empty for B close to B_0. In fact, take $B = \mu I, \mu \neq 0$.

If $A = I$, $X = Z$, then $\sigma(B) \overset{\text{def}}{=} \sigma(B, I) \subseteq \{\lambda : |\lambda| \leq |B|\}$ and Corollary 2.4 gives

Corollary 2.6. *If $X = Z$, $B : X \to X$ is a bounded linear operator then $\sigma(B)$ is upper semicontinuous in B.*

Corollary 2.6 states that $\sigma(B)$ cannot be significantly larger than $\sigma(B_0)$ if B is close to B_0. However, it can be much smaller. The following example illustrates this fact.

Let $X = l^2(-\infty, \infty)$, $x = (\ldots, x_{-1}, x_0, x_1, \ldots)$

$$T(\varepsilon)x = (\ldots, y_{-1}, y_0, y_1, \ldots)$$
$$y_{j-1} = x_j, \qquad j \neq 0, \qquad y_{-1} = \varepsilon x_0.$$

After some rather elementary computations, one obtains the following results: $\sigma(T(0)) = \{\lambda \in \mathbb{C} : |\lambda| < 1\}$, $\sigma(T(\varepsilon)) \subseteq \{\lambda \in \mathbb{C} : |\lambda| = 1\}$ for $\varepsilon \neq 0$. Thus, $d(\sigma(T(0)), \sigma(T(\varepsilon))) \nrightarrow 0$ as $\varepsilon \to 0$ and $\sigma(T(\varepsilon))$ is not lower semicontinuous at $\varepsilon = 0$.

A set $S_0 \subset \sigma(B, A)$ is called a *spectral set* for (B, A) if S_0 is open and closed in $\sigma(B, A)$.

Lemma 2.7. *Suppose $\sigma(B_0, A_0) = S_0 \cup S_1$ where S_0, S_1 are spectral sets for (B_0, A_0) with S_0 bounded $S_0 \neq \phi$, and $S_0 \cap S_1 = \phi$. Then there is a $\delta > 0$ such that $|B - B_0| < \delta$, $|A - A_0| < \delta$ implies $\sigma(B, A) = S_0(B, A) \cup S_1(B, A)$ with $S_0(B, A), S_1(B, A)$ spectral sets for (B, A), $S_0(B, A) \cap S_1(B, A) = \phi$, and $S_0(B, A)$ is upper semicontinuous at (B_0, A_0).*

Proof. Let U be any bounded open neighborhood of S_0, $\bar{U} \cap S_1 = \phi$ and let $\Gamma_\varepsilon = \{\lambda \in \bar{U} : \text{dist}(\lambda, S_0) \geq \varepsilon\}$ for $\varepsilon > 0$. From Corollary 2.2, there is a $\delta > 0$ such that $\Gamma_\varepsilon \subset \rho(B, A)$ for $|B - B_0| < \delta$, $|A - A_0| < \delta$. Thus, $\sigma(B, A) = S_0(B, A) \cup S_1(B, A)$ where $S_1(B, A) \subset \mathbb{C} \backslash \bar{U}$, $S_0(B, A) \subset \{\lambda \in \mathbb{C} : \text{dist}(\lambda, S_0) < \varepsilon\}$. Since $\sigma(B, A)$ is closed, the sets $S_0(B, A), S_1(B, A)$ are spectral sets. The fact that $S_0(B, A)$ is upper semicontinuous at (B_0, A_0) follows from Corollary 2.3. $\square$

From Remark 2.5, one cannot assert that $S_0(B, A)$ in Lemma 2.7 is nonempty without further restrictions on (B_0, A_0). To proceed further, we make the following hypothesis.

(H) *There is a $\delta > 0$ such that $S_0(B, A)$ in Lemma 2.7 is nonempty for each compact spectral set S_0 for (B_0, A_0).*

Hypothesis (H) is not very difficult to verify in the applications. In fact, if $A = I$, the hypothesis is verified by considering the integral

$$\frac{1}{2\pi i} \int_{\partial U} (B - \lambda I)^{-1}\, d\lambda$$

for U an open set containing S_0, $\bar{U} \cap (\sigma(B_0)\backslash S_0) = \phi$. This integral defines a projection on $S_0(B)$ and is zero if and only if $U \cap \sigma(B) = \phi$. Other cases in the applications where an operator A appears are as easy to verify, but a general result stating when (H) is valid is not known.

Theorem 2.8. *Suppose* (H) *is satisfied* $\sigma(B_0, A_0) = S_0 \cup S_1$; *where* S_0, S_1 *are spectral sets for* (B_0, A_0), S_0 *bounded*, $S_0 \neq \phi$, $S_0 \cap S_1 = \phi$. *If* S_0 *is totally disconnected, then* $S_0(B, A)$ *defined in Lemma* 2.7 *is continuous in* (B, A) *at* (B_0, A_0).

Proof. Suppose $\lambda \in S_0$ and U is any open neighborhood of λ, $U \cap S_1 = \phi$. Since S_0 is closed and totally disconnected, there is a spectral set $\tilde{S}_0(U) = \tilde{S}_0 \subset S_0 \cap U$, $\lambda \in \tilde{S}_0$. Lemma 2.7 implies the existence of a spectral set $\tilde{S}_0(B, A, U)$ for (B, A) near (B_0, A_0) and that $\tilde{S}_0(B, A, U)$ is upper semi continuous in (B_0, A_0). Thus, $\tilde{S}_0(B, A, U) \cap U \neq \phi$ since hypothesis (H) implies $\tilde{S}_0(B, A, U) \neq \phi$. Since this is true for every neighborhood U of $\lambda \in S_0$, for any sequence $(B_k, A_k) \to (B_0, A_0)$ as $k \to \infty$, we can find a sequence U_k such that $\bigcap_k U_k = \{\lambda\}$ and, thus, $\lambda \in \bigcup_{n=1}^{\infty} (\bigcap_{k=n}^{\infty} \tilde{S}_0(B_k, A_k, U_k))$. This clearly implies $d(S_0, S_0(B_k, A_k)) \to 0$ as $k \to \infty$, where $\tilde{S}_0(B_k, A_k)$ are the sets determined in Lemma 2.7. This is true for any sequence $(B_k, A_k) \to (B_0, A_0)$ and so $S_0(B, A)$ is lower semicontinuous at (B_0, A_0). Since it is upper semicontinuous by Lemma 2.7, the proof of the theorem is complete. $\square$

Corollary 2.9. *If there is a* $\delta > 0$, $r > 0$ *such that*

$$\sigma(B, A) \subseteq \{\lambda \in \mathbb{C} : |\lambda| \leq r\} \quad \text{for } |B - B_0| < \delta,\ |A - A_0| < \delta,$$

Hypothesis (H) *is satisfied and* $\sigma(B_0, A_0)$ *is totally disconnected, then* $\sigma(B, A)$ *is continuous at* (B_0, A_0).

Corollary 2.10. *If* $X = Z$ *and* $B_0 : X \to X$ *is a bounded linear operator, then* $\sigma(B_0)$ *is totally disconnected if* B_0 *is compact and* $\sigma(B)$ *is continuous at* B_0.

Proof. The fact that $\sigma(B_0)$ is totally disconnected is a classical result in the spectral theory of compact operators. The other statements follow from Corollary 2.9. $\square$

In the applications, it is often true that the operator B is a closed linear operator with domain $\mathscr{D}(B) \subset Z$ and A is a bounded linear operator on Z.

The space X is then taken to be $\mathscr{D}(B)$ with

$$|x|_X = |x|_Z + |Bx|_Z$$

so that both $B, A: X \to Z$ are bounded linear operators.

We end this section with an example illustrating an application of Corollary 2.10 to a numerical method for computation of eigenvalues of differential operators. The example is very simple but easily generalized.

EXAMPLE 2.11. Consider the boundary value problem

$$\ddot{x} + \lambda x = 0, \qquad 0 < x < 1$$

(2.1)

$$x(0) = x(1) = 0.$$

The eigenvalues of this problem are simple, given by $\lambda_k = k^2\pi^2$, $k = 1, 2, \ldots$. The corresponding eigenvectors are $\phi_k(t) = \sin k\pi t$. Using the Green's function for the differential operator d^2/dt^2, any solution x of (2.1) must satisfy

$$x(t) = -\lambda \int_0^1 k(t, s)x(s)\, ds$$

(2.2)

$$k(t, s) = \begin{cases} (1 - t)s, & 0 \le s \le t \\ t(1 - s), & 0 \le t \le s \end{cases}$$

and conversely. Equation (2.2) is equivalent to

$$x = \lambda Kx,$$

$$(Kx)(t) = \int_0^1 k(t, s)x(s)\, ds, \qquad 0 \le t \le 1.$$

Discretize the interval $[0, 1]$ into $N + 1$ equally spaced points and consider the eigenvalue problem for the matrix equation

$$y = \mu K_N y,$$

$$K_N = \frac{1}{N}\left(k\left(\frac{i}{N}, \frac{j}{N}\right), i, j = 0, 1, \ldots, N \right)$$

$$y = \mathrm{col}\left(x(0), x\left(\frac{1}{N}\right), \ldots, x\left(\frac{N}{N}\right) \right).$$

We claim $\sigma(K_N) \to \sigma(K)$ an $N \to \infty$ and, thus, for any eigenvalue λ^{-1} of K, there is a sequence λ_N^{-1} of eigenvalues of K_N such that $\mu_N \to \lambda$ as $N \to \infty$. To prove this, define $\hat{y} \in C([0, 1], \mathbb{R})$ by

$$\hat{y}(t) = \begin{cases} x\left(\dfrac{j}{N}\right) + N\left[x\left(\dfrac{j+1}{N}\right) - x\left(\dfrac{j}{N}\right) \right]\left(t - \dfrac{j}{N}\right), & \dfrac{j}{N} \le t \le \dfrac{j+1}{N}, \\ j = 0, 1, \ldots, N - 1 \end{cases}$$

and define

$$\hat{K}_N : C([0,1], \mathbb{R}) \to C([0,1], \mathbb{R})$$

$$\hat{K}_N z(t) = \begin{cases} \dfrac{1}{N} \displaystyle\sum_{j=0}^{N} k\left(\dfrac{i}{N}, \dfrac{j}{N}\right) z\left(\dfrac{j}{N}\right), & i = 0, 1, 2, \ldots, N \\ \text{linear between these points.} \end{cases}$$

Now suppose there exists a $z \neq 0$ and μ such that $z = -\mu K_N z$. Then z is a piecewise linear function y as above and $y = -\mu K_N y$. The converse is also true. Thus, $\sigma(K_N) = \sigma(\hat{K}_N)$. Also, $\hat{K}_N \to K$ as $N \to \infty$ in the operator topology. An application of Corollary 2.10 shows that $\sigma(\hat{K}_N) \to \sigma(K)$ as $N \to \infty$. This proves the assertion.

14.3. Simple Eigenvalues

Recall that λ is a *simple eigenvalue* of a pair of bounded linear operators (B, A) from a Banach space X to a Banach space Y if $\mathcal{N}(B - \lambda A)$ is one dimensional, codim $\mathcal{R}(B - \lambda A) = 1$ and $Ay \notin \mathcal{R}(B - \lambda A)$ for $y \in \mathcal{N}(B - \lambda A)$, $y \neq 0$.

Theorem 3.1. *Suppose E, X, Y are Banach spaces, $A(\varepsilon)$, $B(\varepsilon): X \to Y$ are bounded linear operators, continuous in ε in the operator topology for $\varepsilon \in E$. If λ_0 is a simple eigenvalue of $(B(0), A(0))$, then there is a constant $\delta > 0$ such that, for $\varepsilon \in E$, $|\varepsilon| < \delta$, there is a unique $\lambda(\varepsilon) \in \mathbb{C}$, $|\lambda(\varepsilon) - \lambda_0| < \delta$, continuous in ε, $\lambda(0) = \lambda_0$, such that $\lambda(\varepsilon)$ is a simple eigenvalue of $(B(\varepsilon), A(\varepsilon))$. Furthermore, if $A(\varepsilon)$, $B(\varepsilon)$ have continuous derivatives up through order k (or are analytic) in ε, then so is $\lambda(\varepsilon)$. The same remarks hold for a corresponding eigenvector of norm one.*

Proof. We must solve the equation

$$(3.1) \qquad\qquad (B(\varepsilon) - (\lambda_0 + v)A(\varepsilon))x = 0$$

for $v \in \mathbb{C}$, $x \in X$, $|x| = 1$, as functions of ε in a neighborhood of $\varepsilon = 0$. We apply the method of Liapunov–Schmidt to obtain the bifurcation equations. More specifically, if $X = X_0 \oplus X_1$, $Z = Z_0 \oplus Z_1$, $X_0 = \mathcal{N}(B(0) - \lambda_0 A(0))$, $Z_1 = \mathcal{R}(B(0) - \lambda_0 A(0))$, with fixed basis elements $y_0 \in X_0$, $w_0 \in Z_0$ and corresponding projections, P_0, P_1, Q_0, Q_1, then $A(0)y_0$ has a nonzero projection onto Z_0 and so we may assume

$$(3.2) \qquad\qquad w_0 = Q_0 A(0) y_0.$$

If $x = y + z$, $y \in P_0 X$, $z \in P_1 X$, then the Implicit Function Theorem implies there is a $\delta > 0$ such that the equation

$$Q_1(B(\varepsilon) - (\lambda_0 + v)A(\varepsilon))(y + z) = 0$$

has a unique solution $z = z^*(v, \varepsilon, y)$ for every $\varepsilon \in E$, $|\varepsilon| < \delta$, $|v| < \delta$ and all $y \in P_0 X$, $z^*(0, 0, y) = 0$ for all $y \in P_0 X$. The existence for all $y \in P_0 X$ rather than for y in some small ball around zero is a consequence of the fact that Equation (3.1) is linear in $x = y + z$ and the operators $A(\varepsilon)$, $B(\varepsilon)$ are continuous in the operator topology. Moreover, the function $z^*(v, \varepsilon, y)$ is linear in y, $z^*(v, \varepsilon, y) = c(v, \varepsilon)y$ where $c(v, \varepsilon): P_0 X \to P_1 X$ is continuous in v, ε, analytic in v in the operator topology uniformly with respect to ε, $\varepsilon \in E$, $|\varepsilon| < \delta$, $|v| < \delta$. Also, $c(0, 0) = 0$. If $A(\varepsilon)$, $B(\varepsilon)$ have continuous derivatives up through order k (or are analytic) for $\varepsilon \in E$, then $c(v, \varepsilon)$ has continuous derivatives up through order k (or are analytic) for $\varepsilon \in E$, $|\varepsilon| < \delta$.

If $c(v, \varepsilon)$ is defined as above and $y = y_0$, the basis for $\mathcal{N}(B(0) - \lambda_0 A(0))$, then the bifurcation equation for Equation (3.1) is

$$d(v, \varepsilon)w_0 \overset{\text{def}}{=} Q_0[B(\varepsilon) - (\lambda_0 + v)A(\varepsilon)](1 + c(v, \varepsilon))y_0 = 0$$

or, equivalently,

$$(3.3) \qquad d(v, \varepsilon) \overset{\text{def}}{=} d_0(\varepsilon) + d_1(\varepsilon)v + d_2(\varepsilon)v^2 + \cdots = 0.$$

From the properties of $c(v, \varepsilon)$ and the definitions of y_0, w_0, we have

$$(3.4) \qquad d_0(0) = 0, \qquad d_1(0) = 1.$$

The Implicit Function Theorem implies there is a unique continuous solution $v = v^*(\varepsilon)$, $v^*(0) = 0$, satisfying (3.3) for each $\varepsilon \in E$, $|\varepsilon| < \delta$, if δ is sufficiently small. Furthermore, $v^*(\varepsilon)$ is as smooth in ε as $A(\varepsilon)$, $B(\varepsilon)$. Since Equation (3.1) for $v = v^*(\varepsilon)$ has the solution $x(\varepsilon) \overset{\text{def}}{=} (1 + c(v^*(\varepsilon), \varepsilon))y_0$ for every $\varepsilon \in E$, $|\varepsilon| < \delta$, and $c(0, 0) = 0$, it follows that $x(\varepsilon) \neq 0$ and $\mu_0 + v^*(\varepsilon)$ is an eigenvalue of $(B(\varepsilon), A(\varepsilon))$. The fact that it is simple is a consequence of the continuity in ε and the first part of the theorem is proved. If $x^*(\varepsilon) = x(\varepsilon)/|x(\varepsilon)|$, then $x^*(\varepsilon)$ is an eigenvector of $\mu_0 + v^*(\varepsilon)$ of norm one and satisfies the smoothness properties stated in the theorem. This completes the proof. $\square$

Corollary 3.2. *If $B(\varepsilon)$ satisfies the conditions of Theorem 3.1, $A(\varepsilon) = I$ for all ε and λ_0 is a simple eigenvalue of $B(0)$, then there is a $\delta > 0$ such that, for each $\varepsilon \in E$, $|\varepsilon| < \delta$, there is a simple eigenvalue $\lambda(\varepsilon)$ of $B(\varepsilon)$ satisfying $|\lambda(\varepsilon) - \lambda_0| < \delta$, $\lambda(0) = \lambda_0$. Furthermore, $\lambda(\varepsilon)$ and the corresponding eigenvector of norm one satisfy the same smoothness properties as stated in the theorem.*

Let us now give some simple applications of this theorem. For $n \times n$ matrices, Corollary 3.2 states that to every simple eigenvalue λ_0 of the matrix

$B(0)$, there is a simple eigenvalue $\lambda(\varepsilon)$ of the matrix $B(\varepsilon)$ near λ_0 which is as smooth in ε as $B(\varepsilon)$. Theorem 3.1 states the same result is true for the generalized eigenvalue problem

$$\det[B(\varepsilon) - (\lambda_0 + \varepsilon)A(\varepsilon)] = 0$$

provided λ_0 is a simple eigenvalue of $(B(0), A(0))$. Of course, if $A(0)$ is a singular matrix, this puts restrictions on the relationship between $B(0)$ and $A(0)$. For example, if

$$B(0) = \begin{bmatrix} 0 & 1 \\ 0 & 0 \end{bmatrix}$$

then $A(0)$ must have a nonzero entry in the lower left-hand corner; that is, $A(0)$ may be assumed to be of the form

$$A(0) = \begin{bmatrix} - & - \\ 1 & - \end{bmatrix}$$

More interesting examples may be given in infinite dimensions; for example, for boundary value problems for differential equations.

EXAMPE 3.3.   Suppose $X = \{x \in C^2([0,1], \ \mathbb{R}); \ x(0) = x(1) = 0\}, \ Z = C([0,1], \mathbb{R})$ and consider the boundary value problem

$$(3.5) \qquad\qquad -\ddot{x} - q(t, \varepsilon)x = \lambda x$$

$$(3.6) \qquad\qquad x(0) = x(1) = 0$$

where $q(t, \varepsilon)$ is continuous in t, ε for $t \in [0,1]$, $\varepsilon \in E$, $q(t,0) = 0$, and λ is a real parameter. If

$$(3.7) \qquad\qquad \begin{aligned} B(\varepsilon) &: X \to Z \\ (B(\varepsilon)x)(t) &= -\ddot{x}(t) - q(t, \varepsilon)x(t) \end{aligned}$$

then finding a solution of Equations (3.5), (3.6) is equivalent to finding a solution of

$$(3.8) \qquad\qquad (B(\varepsilon) - \lambda I)x = 0, \qquad x \in X.$$

The eigenvalues of $B(0)$ are easily found to be $\lambda_n^0 = n^2\pi^2$, $n = 1, 2, \ldots$. These eigenvalues are simple with corresponding eigenvectors $\lambda_n(t) = \sin n\pi t$. According to Corollary 3.2, for each n there is a $\delta > 0$ such that for every $\varepsilon \in E, |\varepsilon| < \delta$, there is a simple eigenvalue $\lambda_n(\varepsilon)$ of $B(\varepsilon)$, continuous in ε, $\lambda_n(0) = \lambda_n^0$. The $\lambda_n(\varepsilon)$ as well as the corresponding eigenvector of norm one is as smooth in ε as $q(\cdot, \varepsilon)$.

If $q(t,0) = q_0(t)$, not necessarily identically zero, then it is known that every eigenvalue λ_0 of $B(0)$ is simple. Thus, Corollary 3.2 applies equally as well to this more general situation. The same argument shows that Corollary 3.2 applies to the more general equation

$$-\frac{d}{dt}\left(p(t,\varepsilon)\frac{dx}{dt}\right) - q(t,\varepsilon)x = \lambda x$$

with arbitrary two point boundary conditions.

EXAMPLE 3.4. The purpose of this example is to give an illustration of the applicability of Theorem 3.1 to a boundary value problem in which the boundary conditions contain the varying parameter. Consider the eigenvalue problem

$$(3.9) \qquad\qquad\qquad -\ddot{x} = \lambda x$$

with the boundary condition

$$(3.10) \qquad\qquad x(0) = 0, \qquad \dot{x}(1) + \varepsilon x(1) = 0, \qquad \varepsilon \in \mathbb{R}.$$

For $\varepsilon = 0$, the eigenvalues λ_n are easily seen to be $\lambda_n = (2n + 1)^2\pi^2/4$, $n = 0, 1, 2, \ldots$, each λ_n is simple and the corresponding eigenfunctions are multiples of $\lambda_n^0(t) = \sin \lambda_n^{1/2} t$. How do the eigenvalues of Equations (3.9), (3.10) depend on ε? One cannot incorporate the boundary conditions into the space X as in the previous example since the boundary conditions depend upon ε.

To overcome this difficulty, we define mappings on larger spaces in order to incorporate the boundary conditions. The idea is to let $x(t) = at + b + y(t)$, where $y(0) = \dot{y}(1) = 0$. More specifically, let

$$Z = C([0,1], \mathbb{R}) \times \mathbb{R}^2$$
$$X = \{y \in C^2([0,1], \mathbb{R}): y(0) = 0 = \dot{y}(1)\} \times \mathbb{R}^2$$
$$B(\varepsilon): X \to Z$$

$$(3.11a) \qquad B(\varepsilon)(y, a, b) = \begin{bmatrix} -\ddot{y} \\ b \\ a + \varepsilon(a + b + y(1)) \end{bmatrix}$$

$$A: X \to Z$$

$$(3.11b) \qquad A(y, a, b) = \begin{bmatrix} at + b + y(t), & 0 \le t \le 1 \\ 0 \\ 0 \end{bmatrix}$$

With these definitions, the function $x(t) = at + b + y(t)$ satisfies (3.9), (3.10) if and only if

$$(3.12) \qquad\qquad (B(\varepsilon) - \lambda A)(y, a, b) = 0.$$

Thus, λ is an eigenvalue of (3.9), (3.10) if and only if there is a nontrivial solution (y, a, b) of (3.12).

It remains to show that the hypotheses of Theorem 3.1 are satisfied. For $\varepsilon = 0$, the only solutions of (3.12) have been shown to be $\lambda_n = (2n + 1)^2\pi^2/4$, $n = 0, 1, 2, \ldots$, and (y, a, b) multiples of $(\phi_n^0, 0, 0)$, where $\phi_n^0(t) = \sin \lambda_n^{1/2}t$. We need to verify that λ_n is a simple eigenvalue of $(B(0), A)$. We have shown that $\dim \mathcal{N}(B(0) - \lambda_n A) = 1$. The triple $(z, c, d) \in Z$ is in $\mathcal{R}(B(0) - \lambda_n A)$ if and only if there exists $(y, a, b) \in X$ with $b = c$, $a = d$ and y satisfying the relations,

$$-\ddot{y} - \lambda_n y = z(t) + \lambda_n(dt + b), \qquad 0 \le t \le 1,$$
$$y(0) = 0 = \dot{y}(1).$$

The Fredholm alternative implies this problem has a solution if and only if

$$\int_0^1 [z(t) + \lambda_n(dt + b)]\sin \lambda_n^{1/2}t\, dt = 0$$

which shows that codim $\mathcal{R}(B(0) - \lambda_n A) = 1$. Since $(\phi_n^0, 0, 0)$ spans $\mathcal{N}(B(0) - \lambda_n A)$, $A(\phi_n^0, 0, 0) = (\phi_n^0, 0, 0)$, and $\int_0^1 \sin^2 \lambda_n^{1/2}t\, dt \ne 0$, we have $A(\phi_n^0, 0, 0) \notin \mathcal{R}(B(0) - \lambda_n A)$. Thus, λ_n is a simple eigenvalue of $(B(0), A)$ and the conclusions of Theorem 3.1 are valid. Since $B(\varepsilon)$ is analytic in ε, it follows that the eigenvalue $\lambda_n(\varepsilon)$ near λ_n and a corresponding eigenvector $\phi_n(\varepsilon)$ of unit norm are analytic in ε.

EXAMPLE 3.5. Consider the m dimensional differential system

$$(3.13) \qquad\qquad -\dot{x}(t) + G(t)x(t) = \lambda x(t), \qquad 0 \le t \le 1,$$

where $G(t)$ is an $m \times m$ matrix, continuous in t and λ is a complex number to be determined so that there is a nontrivial solution satisfying the boundary condition

$$(3.14) \qquad\qquad \Gamma(\varepsilon)x = 0$$

where $\Gamma(\varepsilon): C([0, 1], \mathbb{R}^m) \to \mathbb{R}^m$ is a continuous linear mapping for each ε in a Banach space E, continuous in ε in the operator topology. Under some hypotheses on $\Gamma(0)$, the same trick as in Example 3.4 can be used to transform this problem into one of the form considered in this section.

In fact, let

$$Z = C([0,1], \mathbb{R}^m) \times \mathbb{R}^m$$
$$X = \{ y \in C^1([0,1], \mathbb{R}^m) : \Gamma(0)y = 0 \} \times \mathbb{R}^m$$

and define $B(\varepsilon): X \to Z$, $A: X \to Z$ by the relations

$$B(\varepsilon)(y, a) = \begin{bmatrix} -\dot{x}(t) + G(t)x(t), & 0 \le t \le 1 \\ \Gamma(\varepsilon)(a + y) & \end{bmatrix}$$

$$A(y, a) = \begin{bmatrix} a + y(t), & 0 \le t \le 1 \\ 0 & \end{bmatrix}.$$

The operator $B(\varepsilon)$ has the same smoothness properties in ε as $\Gamma(\varepsilon)$.

By the Riesz representation theorem, there is an $n \times n$ matrix function η of bounded variation on $[0,1]$ such that

$$\Gamma(0)x = \int_0^1 [d\eta(t)]x(t)$$

for all $x \in C([0,1], \mathbb{R}^m)$. If we suppose $[\eta(1) - \eta(0)]^{-1}$ exists, then any $x \in C([0,1], \mathbb{R}^m)$ has a unique decomposition as $x = a + y, a \in \mathbb{R}^m, \Gamma(0)y = 0$. In fact, $a = [\eta(1) - \eta(0)]^{-1}\Gamma(0)x$. Consequently, the boundary value problem (3.13), (3.14) is equivalent to

$$(B(\varepsilon) - \lambda A)(y, a) = 0.$$

We can now prove the following result. The proof is omitted since it is so similar to the proof of the corresponding result in the previous example.

Lemma 3.6. *Suppose* $[\eta(1) - \eta(0)]^{-1}$ *exists. If* λ_0 *is a simple eigenvalue of the boundary value problem* (3.13), (3.14) *for* $\varepsilon = 0$, *then* λ_0 *is a simple eigenvalue of* $(B(0), A)$, *where these operators are defined above.*

For any simple eigenvalue of (3.13), (3.14) for $\varepsilon = 0$, Lemma 3.6 and Theorem 3.1 imply there is a simple eigenvalue $\lambda(\varepsilon)$ of (3.13), (3.14) close to λ_0 which has the same smoothness properties in ε as the boundary conditions $\Gamma(\varepsilon)$. The same holds for the corresponding eigenvector of norm one.

EXAMPLE 3.7. Consider the boundary value problem

$$-\ddot{x} = \lambda x$$
(3.15)
$$x(a(\varepsilon_1)) = x(b(\varepsilon_2)) = 0$$

where $\varepsilon_1 \in \mathbb{R}$, $\varepsilon_2 \in \mathbb{R}$, $a(\varepsilon_1)$, $b(\varepsilon_2)$ are smooth in $\varepsilon_1, \varepsilon_2$ in a neighborhood of $\varepsilon_1 = \varepsilon_2 = 0$, $a(0) = 0$, $a(1) = 1$.

For each $(\varepsilon_1, \varepsilon_2)$ in a neighborhood of $(0,0)$ we wish to map the interval $[a(\varepsilon_1), b(\varepsilon_2)]$ into the interval $[0, 1]$ by an analytic change of variable which is strictly increasing. Such a transformation is easily defined by the formula

$$\tau(t, \varepsilon_1, \varepsilon_2) = (-a(\varepsilon_1) + t)/[b(\varepsilon_2) - a(\varepsilon_1)]$$

If we let $x(t) = y(\tau)$ then

$$-\frac{d^2 y}{d\tau^2} = \mu(\varepsilon_1, \varepsilon_2)y,$$

(3.16)

$$y(0) = y(1) = 0$$

where $\mu(\varepsilon_1, \varepsilon_2) = [b(\varepsilon_2) - a(\varepsilon_1)]^2$.

Thus, the eigenvalues are multiplied by a scalar factor $[b(\varepsilon_2) - a(\varepsilon_1)]^2$. It was shown in Example 3.3 that the eigenvalues are simple for $\varepsilon = 0$. Therefore, the eigenvalues for Equation (3.15) are as smooth in $\varepsilon_1, \varepsilon_2$ as the functions $b(\varepsilon_2), a(\varepsilon_1)$. The next example shows that more complicated boundary conditions may lead to a more complicated looking equation when a similar reduction is made.

EXAMPLE 3.8. Consider the boundary value problem

(3.17)
$$-\ddot{y} = \dot{y}$$
$$y(0) = 0, \qquad \dot{y}(-1) = \dot{y}(1) = 0.$$

It is easy to verify that the eigenvalues are

$$\lambda_n = \left(n\frac{\pi}{2} \right)^2, \qquad n = 0, 1, 2, \ldots .$$

Furthermore, for $n = 2k + 1$, the eigenfunctions may be chosen to be

$$\phi_{2k+1}(t) = \sin(2k + 1)\frac{\pi}{2} t.$$

For $n = 2k$, the eigenfunctions may be chosen to be

$$\phi_{2k}(t) = -1 + \cos k\pi t$$

Each of these eigenvalues is simple in the sense that there is only one independent solution of Equation (3.17) for $\mu = \mu_n$ and there will be a solution of the nonhomogeneous equation with forcing function h if and only if

$\int_{-1}^{1} h\dot{\phi}_n = 0$. To put this equation in abstract form, let

$$X = C^3([-1,1], \mathbb{R}) \cap \{y : y(0) = \dot{y}(-1) = \dot{y}(1) = 0\}.$$
$$Z = C([-1,1], \mathbb{R})$$
$$B_0 y = -\ddot{y}, \qquad A_0 y = \dot{y}.$$

Equation (3.17) is now equivalent to the equation

$$(B_0 - \lambda A_0)y = 0$$
$$\mathcal{N}(B_0 - \lambda_n y_0) = [\phi_n]$$
$$\operatorname{coker}(B_0 - \lambda_n A_0) = [\phi_n].$$

Since $A_0\phi_n = \phi_n$, it follows that $A_0\phi_n \notin \mathcal{R}(B_0 - \lambda_n A_0)$ and μ_n is a simple eigenvalue of (B_0, A_0). Thus, we may subject the equation and boundary conditions to smooth perturbations and know that the eigenvalues have the same smoothness in the parameters. This type of perturbation is treated the same way as in Examples 3.3, 3.4, 3.5.

Let us illustrate in detail the procedure for a perturbation in the values of t at which the boundary conditions are imposed. Consider the problem

$$-\ddot{y} = \lambda\dot{y}$$
(3.18)
$$y(a_2) = 0, \qquad \dot{y}(a_1) = \dot{y}(a_3) = 0$$

where (a_1, a_2, a_3) are close to $(-1, 0, 1)$.

$$f(t, a) = \frac{a_1 + a_3 - 2a_2}{2} t^2 + \frac{a_3 - a_1}{2} t + a_2$$

$$\tau = f(t, a), \qquad \alpha(\tau, a) = \frac{\partial f(t, a)}{\partial t}$$

$$x(\tau) = y(t),$$

then $f(t, -1, 0, 1) = t$, $\alpha(\tau, -1, 0, 1) = 1$ and, for a_1, a_2, a_3 small, Equation (3.18) is equivalent to

$$-\alpha^2 x''' - 3\alpha\alpha' x'' = \lambda x'$$
(3.19)
$$x(0) = 0, \qquad x'(-1) = x'(1) = 0,$$

where $' = d/d\tau$. This reduction to a problem in which the boundary points are fixed has complicated the equation in the sense that the equation has become nonautonomous. However, this causes no difficulty in the general theory

since α is analytic in all variables and $\alpha(\tau, -1, 0, 1) = 1$. It is left to the reader
to show that all of the conditions of Theorem 3.1 are satisfied for an appropriate definition of $B(a)$ to obtain the result that the eigenvalues $\lambda(a_1, a_2, a_3)$
are analytic in a_1, a_2, a_3. Of course, if the $a_j = a_j(\varepsilon)$, then $\lambda(\varepsilon) = \lambda(a_1(\varepsilon),$
$a_2(\varepsilon), a_3(\varepsilon))$ is as smooth in ε as the $a_j(\varepsilon)$.

Remark 3.9. The examples above can be generalized to elliptic partial differential equations. We do not go into detail since a complete discussion of this
case involves considerable knowledge of Sobolev spaces and a priori estimates. The general form of perturbation for which the results are applicable
would be

$$-\Delta u - q(x, \varepsilon)u = \lambda x, \qquad x \in \Omega(\varepsilon) \subseteq \mathbb{R}^n$$

with boundary conditions

$$u + \beta(x, \varepsilon)u + \gamma(x, \varepsilon) = 0 \quad \text{on } \partial\Omega(\varepsilon)$$

or

$$\frac{\partial u}{\partial n} + \beta(x, \varepsilon)u = 0 \quad \text{on } \partial\Omega(\varepsilon)$$

or some mixture of these boundary conditions, where $q(x, 0) = 0$, $\beta(x, 0) =$
$\gamma(x, 0) = 0$ for all x, $\Omega(\varepsilon)$ is an open and bounded set with smooth boundary.

If $\Omega(\varepsilon)$ is independent of ε, the boundary operators can be incorporated
into operators on certain spaces in a manner similar to the previous examples.
If $\Omega(\varepsilon)$ varies with ε, but in such a way that it is possible to make a smooth
change of variables taking $\Omega(\varepsilon)$ onto $\Omega(0)$, then one can make this change of
coordinates to obtain a problem of the same type with $\Omega(\varepsilon)$ replaced by $\Omega(0)$.

14.4. Multiple Normal Eigenvalues

In the previous section, a rather complete theory was obtained for the dependence of a simple eigenvalue on a vector parameter. In this section, we
continue this type of investigation for the dependence of multiple proper
eigenvalues on parameters. We will see that continuity is easy to prove for
the case where the parameter is a vector. However, more detailed results
such as differentiability, some form of analyticity, etc. are not valid for vector
parameters.

For each $\varepsilon \in E$, a Banach space, suppose $A(\varepsilon), B(\varepsilon): X \to Z$ are bounded
linear operators, continuous in ε in the operator topology. For $\varepsilon = 0$, suppose
λ_0 is a normal isolated eigenvalue of $(B(0), A(0))$; that is, there is an integer

k such that

$$\dim \mathscr{N}(B(0) - \lambda_0 A(0))^k < \infty$$
$$X = \mathscr{N}(B(0) - \lambda_0 A(0))^k \oplus J\mathscr{R}(B(0) - \lambda_0 A(0))^k$$
$$Z = I\mathscr{N}(B(0) - \lambda_0 A(0))^k \oplus \mathscr{R}(B(0) - \lambda_0 A(0))^k.$$

This decomposition implies there are continuous projections

$$P_0, P_1 = I - P_0, \qquad Q_0, Q_1 = I - Q_0, \qquad P_0 = JQ_0$$

such that

$$P_0 : X \to \mathscr{N}(B(0) - \lambda_0 A(0))^k, \qquad P_1 : X \to J\mathscr{R}(B(0) - \lambda_0 A(0))^k$$
$$Q_0 : Z \to I\mathscr{N}(B(0) - \lambda_0 A(0))^k, \qquad Q_1 : Z \to \mathscr{R}(B(0) - \lambda_0 A(0))^k.$$

Determining eigenvalues $\lambda(\varepsilon)$ of $(B(\varepsilon), A(\varepsilon))$ near λ_0 for ε near zero is equivalent to finding those values of λ near λ_0 for which there is a nontrivial solution of the equation

$$(4.1) \qquad\qquad B(\varepsilon)x - \lambda A(\varepsilon)x = 0.$$

To solve Equation (4.1), we apply the method of Liapunov–Schmidt in the following way. If $x = y + z$, $y \in P_0 X$, $z \in P_1 X$, $v = \lambda - \lambda_0$, then there is a $\delta > 0$ such that the equation

$$(4.2) \qquad\qquad Q_1(B(\varepsilon)(y + z) - (\lambda_0 + v)A(\varepsilon)(y + z)) = 0$$

has a unique solution $z^*(v, \varepsilon, y)$ for every $|\varepsilon| < \delta$, $|v| < \delta$ and all $y \in P_0 X$, $z^*(0, 0, y) = 0$, for all $y \in P_0 X$. The existence for all $y \in P_0 X$ rather than y in some small ball around zero follows because Equation (4.1) is linear in x and the operators $A(\varepsilon), B(\varepsilon)$ are continuous in the operator topology. Moreover, the function $z^*(v, \varepsilon, y)$ is linear in y,

$$(4.3) \qquad\qquad \begin{aligned} z^*(v, \varepsilon, y) &= C(v, \varepsilon)y \\ C(v, \varepsilon) &: P_0 X \to P_1 X \end{aligned}$$

is a bounded linear operator, $C(v, \varepsilon)$ is continuous in v, ε, analytic in v uniformly with respect to ε in $|\varepsilon| < \delta$, $|v| < \delta$, in the operator topology. In addition, if $A(\varepsilon), B(\varepsilon)$ have continuous derivatives in ε up through order k in the operator topology, then $C(v, \varepsilon)$ has continuous derivatives up through order k in the operator topology. If $A(\varepsilon), B(\varepsilon)$ are analytic in ε in a neighborhood of zero in the operator topology, then $C(v, \varepsilon)$ is analytic in ε, v in the operator topology in a neighborhood of $(0, 0)$.

With the operator $C(v, \varepsilon)$ defined as in Relation (4.3), it follows that the eigenvalue Equation (4.1) is equivalent to the equation

$$(4.4) \qquad Q_0[B(\varepsilon) - (\lambda_0 + v)A(\varepsilon)][I + C(v, \varepsilon)]y = 0$$

Since $J: Q_0 Z \to P_0 X$ is an isomorphism, $JQ_0 = P_0$, it follows that if we define

$$(4.5) \qquad E(v, \varepsilon)y = P_0[B(\varepsilon) - (\lambda_0 + v)A(\varepsilon)][I + C(v, \varepsilon)]y$$

for all $y \in P_0 X$, then

$$(4.6) \qquad E(v, \varepsilon): P_0 X \to P_0 X$$

and Equation (4.4) in a neighborhood of $(v, \varepsilon) = (\lambda - \lambda_0, \varepsilon) = (0, 0)$ is equivalent to the equation

$$(4.7) \qquad E(v, \varepsilon)y = 0, \qquad y \neq 0.$$

Since $E(v, \varepsilon)$ is analytic in v uniformly with respect to ε for $|\varepsilon| < \delta$, $|v| < \delta$, the operator $E(v, \varepsilon)$ has a power series expansion

$$(4.8) \qquad E(v, \varepsilon) = E_0(\varepsilon) + vE_1(\varepsilon) + \frac{v^2}{2!}E_2(\varepsilon) + \cdots$$

In summary, the Liapunov–Schmidt procedure permits the reduction of the Eigenvalue Problem (4.1) in the space X to the Eigenvalue Problem (4.7) in the finite dimensional space $P_0 X$ where $E(v, \varepsilon)$ is given in Equation (4.5). Notice that the eigenvalue parameter v does not enter in a linear fashion in Equation (4.7) even though it did in the original problem. The reduction via Liapunov–Schmidt involved an application of the Implicit Function Theorem and resulted in the nonlinear dependence on v.

It is easy to see that

$$(4.9) \qquad \begin{aligned} E_0(\varepsilon) &= P_0[B(\varepsilon) - \lambda_0 A(\varepsilon)][I + C(0, \varepsilon)] \\ E_1(\varepsilon) &= -P_0 A(\varepsilon)[I + C(0, \varepsilon)] \end{aligned}$$

and so

$$(4.10) \qquad \begin{aligned} E_0(0) &= P_0[B(0) - \lambda_0 A(0)] \\ E_1(0) &= -P_0 A(0). \end{aligned}$$

Since $P_0 X$ is invariant under $B(0) - \mu_0 A(0)$, it follows that the only eigenvalue of $E_0(0)$ is zero.

If a system of basis vectors is chosen for the subspace $P_0 X$ of X, then the operator $E(v, \varepsilon)$ can be considered as a finite dimensional matrix. It is convenient to think of this operator as a matrix. We will do so without explicitly mentioning the basis and always write the matrix as $\tilde{E}(v, \varepsilon)$.

The remarks so far are summarized in the following result.

Lemma 4.1. *There is a neighborhood V of $(0,0) \in \mathbb{C} \times E$ such that Equation (4.1) has a nontrivial solution $x \in X$ for $\lambda = \lambda_0 + v$, $(v, \varepsilon) \in V$, if and only if there is a nontrivial solution $y \in P_0 X$ of Equation (4.7) for $(v, \varepsilon) \in V$. If $\tilde{E}(v, \varepsilon)$ is the matrix associated with $E(v, \varepsilon)$, this latter statement is equivalent to*

$$(4.11) \qquad\qquad \det \tilde{E}(v, \varepsilon) = 0.$$

The operator $E(v, \varepsilon)$ and matrix $\tilde{E}(v, \varepsilon)$ are analytic in v uniformly with respect to ε.

As an immediate consequence, we have

Corollary 4.2. *If λ_0 is a normal isolated eigenvalue of $(B(0), A(0))$, then there is a neighborhood V of $(\lambda_0, 0) \in \mathbb{C} \times E$ such that the eigenvalues $\lambda(\varepsilon)$ of $(B(\varepsilon), A(\varepsilon))$ with $(\lambda(\varepsilon), \varepsilon) \in V$ are continuous in ε. Furthermore, there are exactly n solutions with $(\lambda, \varepsilon) \in V$ where n is the least integer ≥ 1 such that $\det \tilde{E}(v, 0) = \beta_n v^n + \beta_{n+1} v^{n+1} + \cdots, \beta_n \neq 0$.*

Proof. From Lemma 4.1, one must find the solutions of Equation (4.11). If $f(v, \varepsilon) = \det \tilde{E}(v, \varepsilon)$, then $f(v, \varepsilon)$ is analytic in v uniformly with respect to ε. Furthermore, there is an integer $n \geq 1$ and number $\beta_n \neq 0$ such that

$$f(v, 0) = \beta_n v^n + \beta_{n+1} v^{n+1} + \cdots$$

In fact, if this were not the case, then the eigenvalue λ_0 would not be an isolated eigenvalue of $(B(0), A(0))$. The continuity and number of solutions in a neighborhood of zero is a consequence of Rouche's theorem. $\square$

If ε is a scalar and $A(\varepsilon), B(\varepsilon)$ are analytic in ε, then we can say more about the dependence of the eigenvalues upon ε by using the Weierstrass Preparation Theorem and a Newton's Polygon.

With the analyticity hypothesis on $A(\varepsilon), B(\varepsilon)$ and ε a scalar, we have

$$f(v, \varepsilon) = \det \tilde{E}(v, \varepsilon)$$
$$(4.12)$$
$$= \sum_{j=0}^{n} b_j(\varepsilon) v^{n-j} + O(|v|^{n+1}) \qquad b_0(0) \neq 0.$$

From the Weierstrass Preparation Theorem, we know that

$$(4.13) \qquad\qquad f(v, \varepsilon) = [v^n + a_1(\varepsilon)^{n-1} + \cdots + a_n(\varepsilon)] g(v, \varepsilon)$$

where $g(0,0) \neq 0$. Furthermore, Lemma 2.6.7 implies the following result: If

$$b_j(\varepsilon) = c_{jl_j}\varepsilon^{l_j} + O(|\varepsilon|^{l_j+1}) \quad \text{as } |z| \to 0$$
$$c_{jl_j} \neq 0, \qquad j = 1, 2, \ldots, n,$$

(4.14)

then

(4.15)
$$a_j(\varepsilon) = \frac{c_{jl_j}}{c_n}\varepsilon^{l_j} + O(|\varepsilon|^{l_j+1}) \quad \text{as } |\varepsilon| \to 0.$$

These coefficients are computable from the original operators $A(\varepsilon), B(\varepsilon)$.

For the function $f(v, \varepsilon)$ in Relation (4.12), one can now associate the Newton polygon for the polynomial in the decomposition in Relation (4.13). We shall refer to this polygon as the *Newton polygon for* $(B(\varepsilon), A(\varepsilon))$. This polygon allows one to determine all of the solutions of Equation (4.1) as analytic functions in certain powers of $\varepsilon^{1/n}$. Thus, we have proved the following result.

Theorem 4.3. *If ε is a scalar parameter, $A(\varepsilon), B(\varepsilon)$ are analytic in ε in a neighborhood of zero and λ_0 is a normal eigenvalue of $(B(0), A(0))$, then there is a $\delta > 0$ and a neighborhood U of $\{\lambda_0\}$ such that $(B(\varepsilon), A(\varepsilon))$ has exactly n eigenvalues $\{\lambda_1(\varepsilon), \ldots, \lambda_n(\varepsilon)\}$ in U for $|\varepsilon| < \delta$ and these eigenvalues are analytic in $\varepsilon^{1/n}$. The precise form of the expansions in powers of $\varepsilon^{1/n}$ are determined from the Newton polygon for $(B(\varepsilon), A(\varepsilon))$.*

If ε is a vector parameter, then we know that $f(v, \varepsilon) = \det \tilde{E}(v, \varepsilon)$ has the property that there is an integer $n \geq 1$, $\beta_n \neq 0$, such that

$$f(v, \varepsilon) = [v^n + a_1(\varepsilon)v^{n-1} + \cdots + a_n(\varepsilon)]g(\varepsilon, v)$$

with $g(0,0) = \beta_n \neq 0$.

In specific problems, one can still discover the behavior of the solutions of $f(v, \varepsilon) = 0$ by using the scaling techniques in Chapter 7. We also saw some simple examples in the Introduction where this could be done. The scaling techniques were actually used in proving the validity of the Newton polygon in Section 2.8.

14.5. Self-adjoint Operators

In this section, we consider families of self-adjoint operators depending on a scalar parameter and the dependence of the eigenvalues on this parameter.

Suppose $\varepsilon \in \mathbb{C}$, X is a Hilbert space, $B(\varepsilon): X \to X$ is a family of bounded self-adjoint linear operators depending continuously on ε and λ_0 is an isolated normal eigenvalue of $B(0)$; that is, λ_0 is an isolated point of the spectrum of

$B(0)$. Since the ascent of the self-adjoint operator $B(0) - \lambda_0 I$ is one, this implies that

$$\dim \mathcal{N}(B(0) - \lambda_0 I) < \infty$$
$$X = \mathcal{N}(B(0) - \lambda_0 I) \oplus \mathcal{R}(B(0) - \lambda_0 I).$$

Let

$$P_0 : X \to \mathcal{N}(B(0) - \lambda_0 I),$$
$$P_1 = I - P_0 : X \to \mathcal{R}(B(0) - \lambda_0 I)$$

be continuous orthogonal projections.

From the previous section, the method of Liapunov–Schmidt implies the following results. There is a neighborhood V of $(0,0) \in \mathbb{C} \times \mathbb{C}$ such that the equation

$$(5.1) \qquad\qquad [B(\varepsilon) - (\lambda_0 + v)I]x = 0$$

has a nontrivial solution x for $(v, \varepsilon) \in V$ if and only if

$$(5.2) \qquad\qquad x = (I + C(v, \varepsilon))y, \qquad y \in P_0 X$$

where $C(v, \varepsilon) : P_0 X \to P_1 X$ is a bounded linear operator analytic in v uniformly with respect to ε, $C(v, \varepsilon)y$ is the unique solution of the equation

$$(5.3) \qquad\qquad [P_1 B(\varepsilon) - (\lambda_0 + v)I]z + P_1 B(\varepsilon)y = 0$$

for $(v, \varepsilon) \in V$, $y \in P_0 X$, and (v, ε, y), $y \neq 0$, satisfy the equation

$$(5.4) \qquad \begin{aligned} &E(v, \varepsilon)y = 0 \\ &E(v, \varepsilon) = P_0[B(\varepsilon) - (\lambda_0 + v)I][I + C(v, \varepsilon)]. \end{aligned}$$

One can now prove the following result.

Lemma 5.1. *The operator $E(v, \varepsilon) : P_0 X \to P_0 X$ is self-adjoint. If $\tilde{E}(v, \varepsilon)$ is the matrix associated with $E(v, \varepsilon)$, then $\tilde{E}(v, \varepsilon)$ is a self-adjoint matrix.*

Proof. Since $B(\varepsilon)$ is self-adjoint, the eigenvalue problem is variational, i.e.,

$$df(x, \varepsilon) = B(\varepsilon)x = \lambda x$$

where $f(x, \varepsilon) = \langle x, B(\varepsilon)x \rangle / 2$. Thus, $E(v, \varepsilon)$ is self-adjoint because of Corollary 4.11.2. $\square$

The main result of this section is contained in the following theorem.

Theorem 5.2. *If ε is a scalar, X is a Hilbert space, $B(\varepsilon): X \to X$ is analytic in a neighborhood of $\varepsilon = 0$, self-adjoint for each ε and λ_0 is an isolated normal eigenvalue of $B(0)$ of multiplicity n, then there is a neighborhood U of $\{\lambda_0\}$, $\delta > 0$, such that $B(\varepsilon)$ has exactly n eigenvalues $\{\lambda_1(\varepsilon), \ldots, \lambda_n(\varepsilon)\}$ in U for $|\varepsilon| < \delta$, $\lambda_j(0) = \lambda_0$, $j = 1, 2, \ldots, n$. These eigenvalues as well as corresponding eigenvectors may be chosen to be analytic in ε.*

Proof. From the reduction principle expressed in Lemma 4.1, we need only consider the n-dimensional equation

$$\tilde{E}(v, \varepsilon)b = 0$$

for the self-adjoint matrix $\tilde{E}(v, \varepsilon)$ in Lemma 5.1, The matrix $\tilde{E}(v, \varepsilon)$ is analytic in v, ε in a neighborhood of zero, $\tilde{E}(0, 0)$ has only the eigenvalue zero,

$$\tilde{E}(v, \varepsilon) = \tilde{F}_0(\varepsilon) - v\tilde{F}_1(\varepsilon) - v^2\tilde{F}_2(\varepsilon) \cdots$$

where $\tilde{F}_0(0)$ has only the eigenvalue zero, $\tilde{F}_1(0) = I$.

To simplify the notation in the proof, let us assume that

$$\tilde{E}(v, \varepsilon) = F(\varepsilon) - vI,$$

where $F(\varepsilon)$ is self-adjoint. It will be clear from the proof that the higher order terms can be handled in the same way. If

$$F(\varepsilon) = F_0 + \varepsilon F_1 + \varepsilon^2 F_2 + \cdots$$

then we must find the v such that there are nonzero solutions y of the equation

$$(F_0 + \varepsilon F_1 + \varepsilon^2 F_2 + \cdots - vI)y = 0.$$

Since F_0 is self-adjoint with only zero as an eigenvalue we can make a unitary transformation of variables to make $F_0 = 0$ and retain the self-adjointness of the other terms. Now let $v \to v\varepsilon$, divide by ε and consider the equation

$$(5.5) \qquad\qquad (F_1 + \varepsilon F_2 + \cdots - vI)y = 0$$

To any simple eigenvalue λ_1 of F_1, there is a simple eigenvalue $v_1(\varepsilon)$ of $F_1 + \varepsilon F_2 + \cdots$, analytic in ε in a neighborhood of $\varepsilon = 0$, $v_1(0) = \lambda_1$ by Theorem 3.1. Also, the same theorem implies there is a corresponding eigenvector which is analytic in ε. For each multiple eigenvalue of F_1 we may use the reduction principle in Lemma (4.1) to obtain a problem of the same type as Equation (5.5) except now the only eigenvalue of F_1 is zero. Since it is self-adjoint, we may assume $F_1 = 0$. Let $v \to \varepsilon v$ and repeat the process. It is clear that all eigenvalues will be obtained in this way and the theorem is proved. □

The dependence of the eigenvalues and eigenvectors upon a parameter can be extremely complicated if the perturbation terms are not analytic. The next examples illustrate some peculiarities.

EXAMPLE 5.3. We give an example in two dimensions of a self-adjoint matrix subject to C^∞ perturbations in ε for which the eigenvalues are C^∞ and the eigenfunctions are not even continuous at $\varepsilon = 0$. Let $B(0) = 0$ and for $\varepsilon \neq 0$, let

$$B(\varepsilon) = e^{-\varepsilon^{-2}} \begin{bmatrix} \cos 2\varepsilon^{-1} & \sin 2\varepsilon^{-1} \\ \sin 2\varepsilon^{-1} & -\cos 2\varepsilon^{-1} \end{bmatrix}$$

The eigenvalues and eigenvectors of $B(\varepsilon)$ are

$$\lambda_1(\varepsilon) = -e^{-\varepsilon^{-2}}, \qquad \lambda_2(\varepsilon) = e^{-\varepsilon^{-2}}$$
$$y_1(\varepsilon) = (\cos \varepsilon^{-1}, \sin \varepsilon^{-1}), \qquad y_2(\varepsilon) = (\sin \varepsilon^{-1}, -\cos \varepsilon^{-1}).$$

For $\varepsilon \neq 0$, this is the only possible choice for the eigenvectors (except for a scalar multiple) since $\lambda_1(\varepsilon) \neq \lambda_2(\varepsilon)$, $\varepsilon \neq 0$. It is clear that the eigenvectors are not continuous at $\varepsilon = 0$ and the eigenvalues are C^∞ at $\varepsilon = 0$.

EXAMPLE 5.4. The following example is a 2×2 matrix which is C^1 in a parameter ε, is diagonalizable for every ε and yet the eigenvalues are not C^1 in ε.

Let $B(0) = 0$ and, for $\varepsilon \neq 0$, let

$$B(\varepsilon) = \begin{bmatrix} |\varepsilon|^\alpha & |\varepsilon|^\alpha - |\varepsilon|^\beta \left(2 + \sin \dfrac{1}{|\varepsilon|} \right) \\ -|\varepsilon|^\alpha & |\varepsilon|^\alpha \end{bmatrix}$$

where $\alpha > 1$, $\beta > 2$, $\alpha + \beta < 4$, are given constants. The matrix $B(\varepsilon)$ is continuous and continuously differentiable in ε in a neighborhood of $\varepsilon = 0$. The two eigenvalues are

$$\mu_\pm(\varepsilon) = \pm |\varepsilon|^{\alpha + \beta/2} \left(2 + \sin \frac{1}{|\varepsilon|} \right)^{1/2}, \qquad \varepsilon \neq 0,$$

$$\mu_\pm(0) = 0.$$

Since these eigenvalues are distinct, the matrix $B(\varepsilon)$ is diagonalizable for each ε. The eigenvalues are differentiable at each point but the derivatives are not continuous at zero since $\alpha + \beta \leq 4$.

It was very easy to prove continuity of the eigenvalues in a parameter. Example 5.4 shows that the eigenvalues may not be differentiable even though they remain distinct. For the self-adjoint case, we state without proof

Theorem 5.5. *If ε is a scalar, X is a Hilbert space, $B(\varepsilon): X \to X$ is C^1 in ε in a neighborhood of zero, self-adjoint and λ_0 is a normal eigenvalue of $B(0)$ of multiplicity n, then there is a $\delta > 0$ and a neighborhood U of $\{\lambda_0\}$ such that $B(\varepsilon)$ has exactly n eigenvalues $\{\lambda_1(\varepsilon), \ldots, \lambda_n(\varepsilon)\}$ in U for $|\varepsilon| < \delta$ and these eigenvalues can be selected as C^1-functions in ε, $\lambda_j(0) = \lambda_0$, $j = 1, 2, \ldots, n$.*

It is natural to enquire if there are any results on analyticity of the eigenvalues for the case in which the self-adjoint operator $B(\varepsilon): X \to X$ depends analytically on two parameters $\varepsilon = (\varepsilon_1, \varepsilon_2)$. In general, the eigenvalues are not analytic as the following example shows.

EXAMPLE 5.6. Let

$$B(\varepsilon_1, \varepsilon_2) = \begin{bmatrix} \varepsilon_1 & \varepsilon_2 \\ \varepsilon_2 & -\varepsilon_1 \end{bmatrix}.$$

$B(\varepsilon_1, \varepsilon_2)$ is self-adjoint for $\varepsilon_1, \varepsilon_2$ real and has eigenvalues $\pm(\varepsilon_1^2 + \varepsilon_2^2)^{1/2}$ which are not differentiable at $(0, 0)$.

In general if $B(\varepsilon)$ is self-adjoint matrix analytic in a vector parameter ε and $B(0)$ has an eigenvalue μ_0 of multiplicity > 1, then the eigenvalues near μ_0 will be analytic in ε only if $B(\varepsilon)$ satisfies some very special conditions. To be more precise, let $L(\mathbb{C}^n)$ be the set of linear transformations on $\mathbb{C}^n$, $S(\mathbb{C}^n)$ be the real subspace of self-adjoint transformations in $L(\mathbb{C}^n)$. A subspace $\mathscr{A} \subset S(\mathbb{C}^n)$ is *analytic* for A if, for every pair B, C in $\mathscr{A}$, the eigenvalues and eigenvectors of $A + \varepsilon_1 B + \varepsilon_2 C$ are represented as convergent power series in ε_1, ε_2 in some neighborhood of zero. Theorem 5.2 shows that any $\mathscr{A}$ of dimension one is analytic. If A has simple eigenvalues, Theorem 3.1 shows that the eigenvalues of $A + \varepsilon_1 B + \varepsilon_2 C$ are analytic in $\varepsilon_1, \varepsilon_2$ for every B, C in $L(\mathbb{C}^n)$. The following theorem is stated without proof.

Theorem 5.7. *Suppose that $n \leq 3$. If $\mathscr{A} \subset S(\mathbb{C}^n)$ is analytic for A and $\dim \mathscr{A} = 2$ modulo the span of A and I, then A has simple eigenvalues.*

It is not known if this result is true for arbitrary n.

Even though it is not possible to discuss the precise dependence of the eigenvalues on vector parameters, the scaling techniques in Chapter 7 can be used very effectively to determine this behavior in specific applications.

14.6. Bibliographical Notes

The first examples in Section 2 are due to Newburgh [1]. For the discussion of the remarks in Section 2 on infinite delay equations, see Naito [1, 2] and for the ones on difference equations see Avellar and Hale [1]. The example in (2.1) is due to Kato [1, p. 210]. Theorem 2.8 was motivated by Newburg [1].

Paine and De Hoog [1] have obtained interesting uniform estimates of the eigenvalues of a second order ordinary differential equation.

In the applications, it often happens that $D(B)$ changes with B; for example, in the study of differential equations where the boundary data depends on parameters or where the region of definition of the equation depends on parameters. We have seen in some examples that these difficulties often can be overcome either by simple changes of variables or by incorporating boundary terms in the operator. In this way, the results in Section 2 become applicable.

It is possible to introduce another measure of distance between closed operators which takes into account the fact that the domains are different. This is accomplished by considering the Hausdorff metric on the graphs of the operators. Results similar to those in Section 2 can be proved (see Newburgh [1], Kato [1]).

The smoothness properties of simple eigenvalues stated in Theorem 3.1 and Corollary 3.2 have been known for a long time (see, for example, Kato [1], Rellich [1]). For some interesting engineering applications, see Haug and Rousellet [1].

For a different proof of Theorem 5.2, see Kato [1], Rellich [1]. Example 5.3 is due to Rellich [1, p. 52] (see, also Kato [1, p. 111]). Example 5.4 is due to Kato [1, p. 115]. For a proof of Theorem 5.5, see Kato [1, p. 123]. It is very difficult and a simpler proof would certainly be desirable. Example 5.6 is due to Rellich [1, p. 47] (see also Kato [1, p. 116]). Theorem 5.7 is due to Turner [2]. For $(\cos \theta)A_1 + (\sin \theta)A_2$ where A_1, A_2 are symmetric, Yen and Bazer [1] have shown the eigenvalues are algebraic functions of $(\sin \theta, \cos \theta)$. Porsching [1] has obtained some interesting results on the analyticity of the eigenvalues of a matrix function if the eigenvalues are always real. Chapter 2 of Wilkinson [1] also has interesting material on matrices containing parameters.

Bibliography

R. Abraham and J. Marsden [1], *Foundations of Mechanics*, Benjamin/Cummings, Reading. Mass., 1978.

R. Abraham and J. Robbin [1], *Transversal Mapping and Flows*, Benjamin, New York. 1967.

R. A. Adams [1], *Sobolev Spaces*, Academic Press, New York, 1975.

S. Agmon [1], "Lectures on elliptic boundary value problems," *Van Nostrand Math. Studies, No. 2*, Van Nostrand, Princeton. N.J., 1965.

S. Agmon, A. Douglis and L. Nirenberg [1], "Estimates near the boundary for solutions of elliptic partial differential equations satisfying general boundary conditions," I. *Comm. Pure Appl. Math.*, **12** (1959), 623–727; II. *Comm. Pure Appl. Math.*, **17** (1964), 35–92.

S. Ahmad, A. C. Lazer and J. L. Paul [1], "Elementary critical point theory and perturbations of elliptic boundary value problems at resonance," *Ind. Univ. Math. J.*, **25** (1976), 933–944.

P. G. Aizengendler, "Some problems in the theory of the ramification of solutions of nonlinear equations," *Uspehi Mat. Nauk* **21** (1966) 182–184 (Russian).

J. C. Alexander [1], "Bifurcation of zeros of parametrized functions," *J. Functional Analysis* **29** (1978), 37–53.

J. C. Alexander and J. F. G. Auchmuty [1], "Global bifurcation of waves," *Manuscripta Math.* **27** (1979), 159–166.

J. C. Alexander and J. Yorke [1], "Global bifurcation of periodic orbits," *Amer. J. Math.*, **100** (1978), 263–292.

   [2], "The implicit function theorem and the global methods of cohomology," *J. Funct. Analy.*, **21** (1976). 330–339, MR 53, 1630.

   [3], "Homotopy continuation methods: numerically implementable topological procedures," *Trans. Am. Math. Soc.*, **242** (1978), 271–284.

R. K. Alexander and B. Fleishman [1], "Perturbation and bifurcation in a free boundary value problem," *J. Differential Eqns.*, **45** (1982), 34–52.

E. Allgower and K. Georg [1], "Simplicial and continuation methods for approximating fixed points and solutions to systems of equations," *SIAM Review*, **22** (1980), 28–85.

H. Amann [1]. "Ljusternik—Schnirelman theory and nonlinear eigenvalue problems," *Math. Ann.*, **199** (1972), 55–72.

A. Ambrosetti [1], "A note on the Ljusternik—Schnirelman theory for functionals which are not even," *Ricerche Mat.*, **25** (1976), 179–186, MR 56, 13264.

   [2], "Some remarks on the buckling problem for a thin clamped shell," *Ricerche di Matematica*, **23** (1974), 161–170, MR 53, 4680.

   [3], "On the existence of multiple solutions for a class of nonlinear boundary value problems," *Rend. Sem. Mat. Univ. Padova*, **49** (1973), 195–204.

A. Ambrosetti and G. Mancini [1], "Sharp nonuniqueness results for some nonlinear problems," Nonlinear Analy. TMA, **3** (1979), 635–645.

A. Ambrosetti and G. Prodi [1], "On the inversion of some differentiable mappings with singularities between Banach spaces," *Annali Mat. Pura Appl.*, **93** (1972), 231–246, MR 47, 9377.

A. Ambrosetti and P. H. Rabinowitz [1], "Dual variational methods in critical point theory and applications," *J. Funct. Anal.*, **14** (1973), 349–381.

A. A. Andronov and C. E. Chaiken [1], *Theory of Oscillations*, 1st ed., English transl., Princeton Univ. Press, Princeton, N.J., 1949. *Also*: A. A. Andronov, A. A. Vitt and S. E. Khaiken, *Theory of Oscillators*, 2nd ed., English transl., Pergamon Press, London, 1966.

A. Andronov and E. A. Leontovich [1], "Sur la théorie de la variation de la structure qualitative de la division du plan en trajectoires," *Dokl. Akad. Nauk.*, **21** (1938), 427–430.

A. A. Andronov, E. A. Leontovich, I. I. Gordon and A. G. Maier [1], *Theory of Bifurcations of Dynamical Systems on a Plane*, Wiley, 1973.

A. Andronov and L. Pontrjagin [1], "Systemes grossiers," *Dokl. Akad Nauk.*, SSSR, **14** (1937), 247–251.

A. Andronov, A. A. Vitt and S. E. Khaiken [1], *Theory of Oscillators*, Pergamon, 1966.

S. S. Antman, [1], "Buckled states of nonlinearly elastic plates," *Arch. Rational Mech. Anal.*, **67** (1978), 111–149.

[2], "Bifurcation problems for nonlinear elastic structures," p. 73–125 in *Applications of Bifurcation Theory* (ed. Rabinowitz), Academic Press, 1977.

H. A. Antosiewisz [1], "Un analogue du principe du point fixé de Banach," *Annali di Mat. Pura Appl.*, (4) **74**, 61–64.

R. Aris and C. R. Kennedy [1], "Bifurcations of a model diffusion reaction system," p. 211–234 in *New Approaches to Nonlinear Problems in Dynamics* (Ed. Holmes), SIAM, Philadelphia, 1980.

V. I. Arnol'd [1], "On matrices depending upon a parameter," *Russian Math. Surveys*, **26** (1971), No. 2, 29–43.

[2], "Lectures on bifurcation in versal families," *Russian Math. Surveys*, **27** (1972), 54–123.

[3], "Singularities of smooth mappings," *Uspekhi Mat. Nauk*, **23**:1 (1968), 3–44, *Russian Math. Surveys*, **23**:1 (1968), 1–43.

[4], "Integrals of rapidly oscillating functions and singularitis of projections of Lagrangian submanifolds," *Funk. Ana. Prilozhen.*, **6** (1972), 61–62. Eng. Transl., *Func. Ana. Appl.*, **6** (1972), 222–224.

[5], *Additional Chapters on the Theory of Ordinary Differential Equations*, (Russian) Izd. Nauk, Moscow, 1978.

[6], "Proof of A. N. Kolmogorov's theorem on the preservation of quasi-periodic motions under small perturbations of the Hamiltonian," *Usp. Mat. Nauk SSSR*, **18** (1963), 13–40.

M. Ashkenazi and S. N. Chow [1], "A Hopf bifurcation theorem for non-simple eigenvalues," *Advances in Appl. Math.*, **1** (1980), 360–372.

[2], "Existence of periodic solutions at resonance for functional differential equations," *Tôhoku Math. J.*, **32** (1980), 235–254.

J. F. G. Auchmuty [1], "Bifurcating waves," p. 263–278 in *Bifurcation Theory and Applications in Scientific Disciplines*, N.Y. Acad Sci., **316**, 1979.

J. F. G. Auchmuty and G. Nicolis [1], "Bifurcation analysis of non-linear reaction-diffusion equations I," *Bull. Math. Biology*, **37** (1975), 323–365.

C. Avellar and J. K. Hale [1], "On the zeros of exponential polynomials," *J. Math. Ana. Appl.*, **73** (1980), 434–452.

S. Bancroft, J. K. Hale and D. Sweet [1], "Alternative problems for nonlinear functional equations," *J. Differential Equations*, **4** (1968), 40–56, MR 36, 3184.

V. Barbu [1], *Nonlinear Semigroups and Differential Equations in Banach Spaces*, Noordhoff, Leyden, 1976.

R. G. Bartle [1], "Singular points of functional equations," *Trans. Am. Math. Soc.*, **75** (1953), 366–384.

R. W. Bass [1], "Equivalent linearization, nonlinear circuit analysis and the stabilization and optimization of control systems," Proc. Symp. Nonlin. Circuit Anal., 6 (1956).

L. Bauer, H. B. Keller and E. L. Reiss [1], "Axisymmetric buckling of hollow spheres and hemispheres," *Comm. Pure Appl. Math.*, **23** (1970), 529–568.

[2], "Multiple eigenvalues lead to secondary bifurcation," *SIAM J. Appl. Math.*, **17** (1975), 101–122.

L. Bauer and E. L. Reiss [1], "Nonlinear buckling of rectangular plates," *SIAM J.*, **13** (1965), 603–626.

[2], "Numerical bifurcation and secondary bifurcation—case history," p. 443–467 of *Numerical Solutions of Partial Differential Equations*, III. (1976), Academic Press.

N. W. Bazley and J. B. McLeod [1], "Bifurcation from infinitely and singular eigenvalue problems," *Proc. London Math. Soc.*, (3), **34** (1977), 231–244; MR 55, 11117.

N. W. Bazley and G. H. Pimbley [1], "A region of no secondary bifurcation for nonlinear Hammerstein operators," *J. Appl. Math. Phys.* (ZAMP), **25** (1974), 743–751.

N. W. Bazley and B. Zwalen [1], "Estimation of the bifurcation coefficient for nonlinear eigenvalue problems," *J. Appl. Math. Phys.* (ZAMP), **20** (1969), 281–288.

[2], "A branch of positive solutions of nonlinear eigenvalue problems," *Manuscripta Math.*, **2** (1970), 365–374. MR 42, 3628.

V. Benci [1], "Some critical point theorems and applications," *Comm. Pure Appl. Math.*, **33** (1980), 147–172.

M. S. Berger [1], *Nonlinearity and Functional Analysis*, Acad. Press, New York, 1977.

[2], "Applications of global analysis to specific nonlinear eigenvalue problems," *Rocky Mtn. Math. J.*, **3** (1973), 319–354.

[3], "On von Kármán's equations and the buckling of a thin elastic plate, I. The clamped piate," *Comm. Pure Appl. Math.*, **20** (1967), 687–719, MR 36, 4860.

[4], "On nonlinear perturbations of the eigenvalues of a compact self-adjoint operator," *Bull. Amer. Math. Soc.*, **73** (1967), 704–708. MR 35, 3494.

M. S. Berger and M. S. Berger [1], *Perspectives in Nonlinearity*, Benjamin, New York, 1968.

M. S. Berger and P. T. Church [1], "Compiete integrability and perturbation of a nonlinear Dirichlet problem, I," *Indiana Univ. Math. J.*, **28** (1979), 935–952; II, **29** (1980), 715–735.

M. S. Berger and P. C. Fife [1], "On von Kármán's equations and the buckling of a thin elastic plate," *Bull. A.M.S.*, **72** (1966), 1006–1011.

[2], "Von Kármán's equations and the buckling of a thin elastic plate, II. Plate with general edge conditions," *Comm. Pure Appl. Math.*, **21** (1968), 227–241.

M. S. Berger and L. E. Fraenkel [1], "Nonlinear desingularization in certain free-boundary value problems," *Comm. Math. Phy.*, **77** (1980), 149–172.

M. S. Berger and M. Schechter [1], "On the solvability of semilinear gradient operator equations," *Adv. Math.*, **25** (1977), 97–132.

S. R. Bernfeld, P. Negrini and L. Salvadori [1], "Stability and generalized Hopf bifurcation through a reduction principle," *Nonlinear Differential Equations* (Proc. Internat. Conf. Trento, 1980), pp. 29–40, Acad. Press, New York, 1981.

Yu. N. Bibikov [1], *Local Theory of Nonlinear Analytic Ordinary Differential Equations*, Lecture Notes in Math., **702**, Springer-Verlag, 1979.

G. D. Birkhoff [1], *Dynamical Systems*, Am. Math. Soc. Colloq. Publ., **9**, 1927.

J. A. Boa and D. S. Cohen [1], "Bifurcation of localized disturbances in a model biochemical reaction," *SIAM J. Appl. Math.*, **30** (1976), 123–135.

R. I. Bogdanov [1], "Versal deformations of a singular point of a vector field on the plane in the case of zero eigenvalues," *Func. Ana. Appl.*, **9** (1975), 144–145; Ibid., **10** (1976), 61–62.

N. N. Bogoliubov [1], "On some statistical methods in mathematical physics," *Izv. vo Akad. Nauk Ukr. SSR*, Kiev, 1945.

N. Bogoliubov and Yu. Mitropolskii [1], *Asymptotic methods in the Theory of Nonlinear Mechanics*, Gordon and Breach, New York, 1961.

R. Böhme [1], "Die Lösung der Verweigungsgleichungen für nichtlineare Eigenwertprobleme," *Math. Z.*, **127** (1972), 105–126; MR 47, 910.

M. Bottkol [1], Bifurcation of periodic orbits on manifolds and Hamiltonian systems, Thesis, New York University, 1977.

A. Boucherif [1], "Nonlinear three point boundary value problems," *J. Math. Ana. Appl.*, **77** (1980), 577–600.

H. Brézis [1], *Operateurs Maximaux Monotones et Semi-groupes*, North-Holland, 1973.

H. Brézis, J. M. Coron and L. Nirenberg [1], Free vibrations for a nonlinear wave equation and a theorem of P. Rabinowitz. *Comm. Pure Appl. Math.*, **33** (1980), 667–684.

H. Brézis and L. Nirenberg [1], "Forced vibrations of a nonlinear wave equation," *Comm. Pure Appl. Math.*, **31** (1978), 1–31.

A. D. Br'juno, [1], *Local Methods in Nonlinear Analysis in Differential Equations* (Russian), Izd. Nauk. Moscow, 1979.

[2], "Analytic form of differential equations, I," *Trans. Moscow Math. Soc.*, **25** (1971), 131–288. II, Ibid., **26** (1972), 199–239.

Th. Brocker and L. Lander [1], "Differentiable Germs and Catastrophes," *London Math. Soc. Lecture Notes 17*, Cambridge Univ. Press, Cambridge, 1975.

F. E. Browder [1], "Nonlinear eigenvalue problems and group invariance," appearing in *Functional Analysis and Related Fields*, F. E. Browder, editor, Springer-Verlag (1970), 1–58.

[2], "Existence theorems for nonlinear partial differential equations," *Proc. Sym. Pure Math.*, **16**, A.M.S. Providence, (1970), 1–60.

[3], "Nonlinear operators and nonlinear equations of evolution in Banach spaces," *Proc. Symp. Pure Math.*, **18**, part 2, Amer. Math. Soc., Providence, 1976.

P. Brunovsky [1], "On one parameter families of diffeomorphisms," *Comment. Math. Univ. Carolinae*, **11** (1970), 559–582.

N. N. Bruslinskaya [1], "Qualitative integration of a system of n differential equations in a region containing a singular point and a limit cycle," *Dokl. Akad. Nauk SSSR*, **139** (1961), 9–12.

S. Caprino, C. Maffei, and P. Negrini [1], Hopf furcation at $1:1$ resonance, *Nonlinear Anal. TMA*, **8** (1984), 1011–1032.

J. Carr [1], "Applications of Centre Manifold Theory," *Applied Mathematical Sciences* Vol. 35, 1981, Springer-Verlag, New York.

J. Carr, S.-N. Chow, and J. K. Hale [1], Abelian integrals and bifurcation theory, *J. Diff. Eq.*, **59** (1985), 413–463.

M. Cartwright [1], "Forced oscillations in nonlinear systems," *Contr. Theory Nonlin. Osc.*, **1** (1950), 149–241. *Annals Math. Studies*, No. 20, Princeton.

M. L. Cartwright and J. E. Littlewood [1], "On nonlinear differential equations of the second order, I," The equation $\ddot{y} - k(1 - y^2)\dot{y} + y = b\lambda k \cos(\lambda t + \alpha)$, k large," *J. London Math. Soc.*, **20** (1945), 180–189.

R. G. Casten and C. J. Holland [1], "Instability results for a reaction diffusion equation with Neumann boundary conditions," *J. Differential Eqns.*, **27** (1978), 266–273.

A. Castro and A. Lazer [1], "Applications of a max-min principle," *Rev. Columbia Mat.*, **10** (1976), 141–149.

[2], "Critical point theory and the number of solutions of a nonlinear Dirichlet Problem," *Annali Mat. Pura Appl.* **120** (1979), 113–137.

L. Cesari [1], "Functional analysis and periodic solutions of nonlinear differential equations," In: *Contributions to Differential Equations*, **1** (1963), 149–187.

[2], "Functional analysis and Galerkin's Method," *Mich. Math. J.*, **11** (1964), 385–418.

[3], "Functional analysis, nonlinear differential equations, and the alternative method," in *Nonlinear Functional Analysis and its Applications* (Eds. Cesari, Kannan and Schur), Dekker, New York, 1976, 1–197.

[4], "Alternative methods in nonlinear analysis," Int. Conf. Diff. Eqns., Los Angeles, 1974, p. 95–198. Academic Press, New York.

[5], "Nonlinear analysis," *Lecture Notes C.I.M.E.* (1972), Ed. Cremonese, Roma, 1973.

[6], "Sulla stabilita delle soluzioni dei sistemi di equazioni differenziali lineari a coefficienti periodici," *Atti Accad. Italia Mem. Cl. Sci. Fis. Mat. Nat.* (6), **11** (1940), 633–695.

[7], "Asymptotic behavior and stability problems in ordinary differential equations," 2nd ed., *Ergebnisse der Mathematik und ihrer Grenzgebiete*, N. F., **16**, Academic Press, New York; Springer-Verlag, Berlin, 1963. MR 27, 1661.

L. Cesari and R. Kannan [1], *Nonlinear Analysis and Differential Equations*. Proc. of Michigan State Univ. Conf., edited by Cesari, Kanner and Schuur. Lecture Notes in Pure and Applied Mathematics, Vol. 19., Marcel Dekker, New York, 1976.

L. Cesari and P. J. McKenna [1], "On Grothendieck approximation properties". *Bull. Inst. Mat. Acad Sin.* **6** (1978), 569–583.

N. Chafee [1], "The bifurcation of one or more closed orbits from an equilibrium point of an autonomous differential equation," *J. Diff. Eqns.*, **4** (1968), 661–679.

[2], "A bifurcation problem for a functional differential equation of finitely retarded type," *J. Math. Anal. Appl.*, **35** (1971), 312–348.

[3], "Generalized Hopf bifurcation and perturbation in a full neighborhood of a given vector field," *Indiana Univ. Math. J.*, **27** (1978), 173–194.

[4], "Asymptotic behavior for solutions of a one-dimensional wave equation with homogeneous Neumann boundary conditions," *J. Differential Eqns.*, **18** (1975), 111–134.

N. Chafee and E. F. Infante [1], "A bifurcation problem for a nonlinear partial differential equation of parabolic type," *Applicable Ana.*, **4** (1974), 17–37.

A. Chenciner [1], Courbes fermées invariantes non normalement hyperboliques au voisinage d'une bifurcation de Hopf dégénerée diffeomorphismes de $(\mathbb{R}^2, 0)$. C.R. Acad. Sci. Paris Ser. I Math., **292** (1981), No. 10, 507–510.

D. R. J. Chillingworth [1], *Differential Topology with a View to Applications*, Research Notes in Math., **9**, Pitman, London, 1976.

[2], A Global Genericity Theorem for Bifurcations in Variational Problems. J. Funct. Anal. **35** (1980), No. 2, 251–278.

[3], "The catastrophe of a buckling beam," In *Dynamical Systems*, Warwick 1974, 88–91. Lecture Notes in Math., **488**, Springer-Verlag (1975).

D. R. J. Chillingworth and J. Marsden [1], personal communication.

S. N. Chow [1], "Periodic solutions of nonlinear autonomous hyperbolic equations," *Functional Differential Equations and Bifurcations*, Springer-Verlag, Lecture Notes in Math., **799** (1980), 126–139.

S. N. Chow, J. K. Hale and J. Mallet-Paret [1], "Application of generic bifurcations, I." *Arch. Rat. Mech. Anal.*, **59** (1975), 159–188.

[2], "Application of generic bifurcation, II." *Arch. Rat. Mech. Anal.*, **62** (1976), 209–236.

[3], "An example of bifurcation to homoclinic orbits," *J. Differential Eqns.*, **37** (1980), 351–373.

S. N. Chow and J. Kurtz [1], "Singular nonlinear equations on unbounded interval," *J. Math. Anal. Appl.*, **77** (1980), 642–653.

S. N. Chow and J. Mallet-Paret [1], "The Fuller index and global Hopf bifurcation," *J. Diff. Eqns.*, **29** (1978), 66–85.

[2], "Integral averaging and bifurcation," *J. Diff. Eqns.*, **26** (1977), 112–159.

[3], Periodic solutions near an equilibrium of a nonpositive definite Hamiltonian system. Preprint.

S.-N. Chow, C. Li, and D. Wang [1], Uniqueness of periodic orbits of some vector fields with codimension two singularities, *J. Diff. Eq.*, **77** (1989), 231–253.

[2], *Normal Forms and Bifurcation of Planar Vector Fields*, Cambridge University Press, 1994.

S. N. Chow, J. Mallet-Paret and J. Yorke [1], "Global Hopf bifurcation from a multiple eigenvalue," *Nonlin. Anal., Th., Meth., Appl.*, **2** (1978), 753–763.

[2], "Finding zeros of maps: homotopic methods that are constructive with probability one," *Math. Comp.*, **32** (1978), 887–899.

[3], "A homotopy method for locating zeros of a system of polynomials," Proc. FDE and Approx. Fixed Points, *Lect. Notes in Math.* **730**, Springer-Verlag 1979, 77–88.

[4], "A periodic orbit index which is a bifurcation invariant," LCDS 81-15, Division of Applied Math., Brown Univ., Providence, 1981.

R. C. Churchill, G. Pecelli and D. L. Rod [1], "Stability transitions for periodic orbits in Hamiltonian systems," *Arch. Rat. Mech. Ana.*, **73** (1980), 313–347.

P. G. Ciarlet [1], "A justification of the von Kármán equations," *Arch. Rat. Mech. Ana.* **73** (1980), 349–389.

D. C. Clark [1], "A variant of the Ljusternik—Schnirelman Theory," *Ind. Univ. Math. J.*, **22** (1972), 65–74.

[2], "On periodic solutions of autonomous Hamiltonian systems of ordinary differential equations," *Proc. A.M.S.*, **39** (1973), 579–584.

[3], "Eigenvalue bifurcation for odd gradient operators," *Rocky Mountain J. Math.*, **5** (1975), 317–336.

[4], "Periodic solutions of variational differential equations," *J. Diff. Eqns.*, **28** (1978), 354–368.

F. C. Clarke and I. Ekeland [1], "Hamiltonian trajectories having prescribed minimal period," *Comm. Pure Appl. Math.*, **33** (1980), 103–116.

E. A. Coddington and N. Levinson [1], *Theory of Ordinary Differential Equations*, McGraw-Hill, New York, 1955.

C. V. Coffman [1], "A minimum-minimax principle for a class of nonlinear integral equations," *J. Analyse Math.*, **22** (1969), 391–419.

D. S. Cohen [1], "Multiple solutions of nonlinear partial differential equations," *Lecture Notes in Math.*, **322**, Springer-Verlag, Berlin, (1973), 15–77.

H. Cohen [1], "Nonlinear diffusion problems," p. 27–63 in *Studies in Math.*, **7**, (Ed. Taub) MAA Studies in Applied Math., Prentice-Hall, 1971.

C. Conley [1], "Isolated invariant sets and the Morse index," N.S.F., *C.B.M.S. Lecture Notes 38*, Amer. Math. Soc., Providence, R. I., 1978.

K. Cooke and J. A. Yorke, [1], "Some equations modelling growth processes and gonorrhea epidemics," *Math. Biosci.*, **16** (1973), 75–101.

G. Cooperman [1], A note on the alternative method. Unpubushed preprint, June, 1977.

C. Corduneanu [1], *Integral Equations and Stability of Feedback Systems*, Academic Press, 1973.

R. Courant and D. Hilbert [1], *Methods of Mathematical Physics*, Vol. 1, Interscience, New York, 1953.

[2], *Methods of Mathematical Physics*, Vol. II, Interscience, New York, 1962.

M. G. Crandall and P. H. Rabinowitz [1], "Bifurcation from simple eigenvalue," *J. Funct. Anal.*, **8** (1971), 321–340.

[2], "Bifurcation, perturbation of simple eigenvalues and linearized stability," *Arch. Rat. Mech. Anal.*, **52** (1973), 161–180.

[3], "The Hopf bifurcation theorem in infinite dimensions," *Arch. Rat. Mech. Anal.*, **67** (1978), 53–72.

[4], "The principle of exchange of stability," *Proc. Int. Symp. Dyn. Syst.*, Gainesville, Fla., 1976, Eds. Bedernek and Cesari.

[5], "Nonlinear Sturm—Liouville eigenvalue problems and topological degree," *J. Math. Mech.*, **19** (1970), 1083–1102.

J. Cronin [1], "Bifurcation of periodic solutions," *J. Math. Anal. Appl.*, **68** (1979), 130–151.

[2] "Branch points of solutions of equations in Banach space," *Trans. Amer. Math. Soc.*, **69** (1950), 208–231. MR 12, 716.

[3], "Branch points of solutions of equations in Banach space, II," *Trans. Amer. Math. Soc.*, **76** (1954), 207–222. MR 16, 47.

[4], "Analytic functional mappings," *Ann. of Math.* (2) **58** (1953), 175–181. MR 15, 234.

[5], "Fixed points and topological degree in nonlinear analysis," *Math. Surveys*, No. 11, Amer. Math. Soc., Providence, R. I., 1964. MR 29, 1400.

[6], "One-side bifurcation points," *J. Diff. Eqns.*, **9** (1971), 1–12.

J. M. Cushing [1], "Integrodifferential equations and delay models in population dynamics," *Lecture Notes in Biomath.*, **20**, Springer-Verlag, 1977.

[2], "Nontrivial periodic solutions of some Volterra integral equations," to appear in: *Proc. of the Helsinki Conf. on Int. Eqns.*, Lecture Notes in Math., Springer-Verlag, Berlin.

E. N. Dancer [1], "Bifurcation theory in real Banach spaces," *J. London Math. Soc.*, **23** (1971), 699–734.

[2], "Bifurcation theory for analytic operators," *Proc. London Math. Soc.*, **26** (1973), 359–384.

[3], "Global structure of the solutions of nonlinear real analytic eigenvalue problems," *J. London Math. Soc.*, **27** (1973), 747–765.

[4], "Global solution branches for positive mappings," *Arch. Rat. Mech. Ana.*, **54** (1973), 181–192.

[5], On the existence of bifurcating solutions in the presence of symmetry. *Proc. Roy. Soc. Edinburgh*, *A*, **85** (1980), 321–336.

Y.-M. J. Demoulin and Y. U. Chen [1], "An iteration method for solving nonlinear eigenvalue problems," *SIAM J. Appl. Math.*, **28** (1975), 588–595. MR 51, 1506.

J. C. de Oliveira [1], "Hopf bifurcation for functional differential equations," *J. Nonlinear Anal.*, **4** (1980), 217–230.

J. C. de Oliveira and J. K. Hale [1], "Dynamic behavior from bifurcation equations," *Tôhoku Math. J.*, **32** (1980), 189–199.

A. DePrit [1], "Canonical transformations depending on a small parameter," *Cel. Mech.*, **1** (1969), 12–30. MR 40, 6825.

L. de Simon and G. Torelli [1], "Soluzioni periodiche di equazioni a derivate parziali di tipo iperbolico non lineari," *Rend. Sem. Mat. Padova*, **40** (1968), 380–401.

R. W. Dickey [1], "Bifurcation problems in nonlinear elasticity," *Research Notes in Math.*, **3**, Pitman, London, 1976.

J. Dieudonné [1], *Foundations of Modern Analysis*, Academic Press, New York, 1960.

S. P. Diliberto [1], "Perturbation theorems for periodic surfaces, I, II," *Rend. Circ. Mat. Palermo*, **9** (2) (1960), 265–299; **10** (2) (1961), 111–161.

R. C. DiPrima and J. T. Stuart [1], "Flow between eccentric rotating cylinders," *Lubr. Tech. Trans.*, *A.S.M.E.*, **94** (1972), 266–274.

Yu. A. Dubinskii [1], "Nonlinear elliptic and parabolic equations," *Jour. Sov. Math.*, **12** (1979), 475–554.

J. J. Duistermaat [1], "Oscillatory integrals, Lagrange immersions, and unfoldings of singularities," *Comm. Pure Appl. Math.*, **27** (1974), 207–281.

H. Dulac [1], "Solutions d'un système d'équations differentielles dans le voisinage de valeurs singulières," *Bullertin Soc. Math. de France*, **40** (1912), 324–392.

N. Dunford and J. J. Schwartz [1], *Linear Operators, Part I*, Interscience, New York, 1958.

[2], *Linear Operators, Part II*, Interscience, New York, 1963.

G. Duvaut and J. L. Lions [1], *Inequalities in Mechanics and Physics*, Springer-Verlag, Berlin, 1976.

I. Ekeland and R. Temam [1], *Convex Analysis and Variational Problems*, North Holland, 1976.

E. Fadell and P. H. Rabinowitz [1], "Generalized cohomological index theories for Lie group actions with an application to bifurcation questions for Hamiltonian systems," *Invent. Math.*, **45** (1978), 139–174.

N. Fenichel [1], "Persistence and smoothness of invariant manifolds for flows," *Indiana Univ. Math. J.*, **21** (1971), 193–226.

[2], "Asymptotic stability with rate conditions," *Indiana Univ. Math. J.*, **23** (1974), 1109–1137.

[3], "Asymptotic stability and rate condition," *Indiana Univ. Math. Jour.*, **26** (1977), 81–93.

[4], "Geometric singular perturbation theory for ordinary differential equations," *J. Differential Eqns.*, **31** (1979), 53–98.

P. Fife [1], *Mathematical Aspects of Reaction Diffusion Systems*, Lecture Notes in Biomathematics, **28**, Springer-Verlag, 1979.

[2], "Stationary patterns for reaction diffusion equations," in *Nonlinear Diffusion, Res. Notes in Math.*, **14**, Pitman, 81–121.

[3], "Nonlinear deflection of thin elastic plates under tension," *Comm. Pure Appl. Math.*, **14** (1961), 81–112.

[4], "The Bénard problem for general fluid dynamical equations and remarks on the Boussinesq approximation," *Indiana Univ. Math. J.*, **20** (1970/71), 303–326.

P. Fife and D. D. Joseph [1], "Existence of convective solutions of the generalized Bénard problem which are analytic in their norm," *Arch. Rat. Mech. Anal.*, **33** (1969), 116–138.

D. Flockerzi [1], "Bifurcation formulas for O.D.E.'s in $\mathbb{R}^n$," *Nonlinear Analysis*, **5** (1981), 249–263.

A. Friedman [1], *Partial Differential Equations*, Holt-Rinehart & Winston, 1969.

K. Friedriches [1], *Special Topics in Analysis*, Lecture Notes, New York Univ., 1953–54.

[2], *Advanced Ordinary Differential Equations*, Gordon and Breach, New York, 1965.

K. O. Friedrichs and J. J. Stoker [1], "The nonlinear boundary value problem of the buckled plate," *Amer. J. Math.*, **63** (1941), 839–888.

S. Fucik, J. Necas, J. Soucek, and V. Soucek [1], *Spectral Analysis of Nonlinear Operators*, Lecture Notes in Mathematics **343**, Springer-Verlag, 1973.

F. B. Fuller [1], "An index of fixed point type for periodic orbits," *Amer. J. Math.*, **89** (1967), 133–148.

T. Furimochi [1], "Existence of periodic solutions of two dimensional differential-delay equations," *Applicable Math.*, **9** (1979), 279–289.

R. E. Gaines and J. Mawhin [1], *Coincidence Degree and Nonlinear Differential Equations*, Lecture Notes in Math., **568**, Springer-Verlag, 1977.

G. R Gavalas [1], *Nonlinear Differential Equations of Chemically Reacting Systems*, Springer-Verlag, 1968.

K. Georg [1], "On tracing an implicitly defined curve by quasiNewton steps and calculating bifurcation by local perturbations," *SIAM J. Sci. Statist. Comp.*, **2** (1981), 35–50.

D. Gilbarg and N. S. Trudinger [1], *Elliptic Partial Differential Equations of Second Order*, Springer-Verlag, New York, 1977.

M. Golubitsky [1], "An introduction to catastrophe theory and its applications," *SIAM Review*, **20** (1978), 353–387.

M. Golubitsky and V. Guillemin [1], *Stable Mappings and Their Singularities*, Springer-Verlag, Berlin, 1973.

M. Golubitsky and B. L. Keyfitz [1], "A qualitative study of the steady state solutions for a continuous flow stirred tank chemical reactor," *SIAM J. Math. Ana.* **11** (1980), 316–339.

M. Golubitsky, B. L. Keyfitz and D. Schaeffer [1], "A singularity theory analysis of the thermal chainbranching model for the explosion peninsula in the H_2–O_2 reaction, *Comm. Pure Appl. Math.*, **34** (1981), No. 4, 433–463.

M. Golubitsky and W. F. Langford [1], "Classification and unfoldings of degenerate Hopf bifurcations," *J. Differential Eqns.* **41** (1981), 375–415.

M. Golubitsky and D. Schaeffer [1], "A theory for imperfect bifurcation via singularity theory," *Comm. Pure Appl. Math.*, **32** (1979), 21–98.

[2], "Imperfect bifurcation in the presence of symmetry," *Comm. Math. Phys.*, **67** (1979), 205–232.

[3], "Boundary conditions and mode jumping in the buckling of a rectangular plate," *Comm. Math. Phys.*, **69** (1979), 209–230.

[4], "Bifurcation analysis near a double eigenvalue of a model chemical reactor," *Arch. Rat. Mech. Anal.* **75** (1981), 315–348.

[5], *Singularities and Groups in Bifurcation Theory*, Vol. 1, Springer-Verlag, 1985. M. Golubitsky, I. Stewart and D. Schaeffer, *Singularities and Groups in Bifurcation Theory*, Vol. 2, Springer-Verlag, 1988.

W. B. Gordon [1], "A theorem on the existence of periodic solutions to Hamiltonian systems with convex potential," *J. Differential Eqns.*, **10** (1971), 324–335.

M. I. Granero, A. Poreti and D. Zanaeca [1], "A bifurcation analysis of pattern formation in a diffusion governed morphogenetic field," *J. Math. Bio.*, **4** (1977), 21–27.

L. M. Graves [1], "Remarks on singular points of functional equations," *Trans. Am. Math. Soc.*, **79** (1955), 150–157.

W. M. Greenlee [1], "Remarks on branching from multiple eigenvalues," *Lecture Notes in Math.*, **322**, 101–112, Springer-Verlag, New York 1973.

B. Greenspan and P. Holmes [1], "Homoclinic orbits, subharmonics and global bifurcations in forced oscillations." *Nonlinear dynamics and turbulence*, 172–214, Interaction Mech. Math. Ser., Pitman, Boston, Mass-London, 1983.

D. M. Grobman [1], "Topologicsal classification of a neighborhood of a singular point in n-dimensional space," *Mat. Sb.*, **56** (1962), 77–94.

[2], "Homeomorphisms of systems of differential equations," *Dokl. Akad. Nauk*, **128** (1959), 880–881. MR 22, 12282.

W. Gröbner and H. Knapp [1], "Contributions to the method of Lie series," Hochschulskripten, Bibliographisches Institut., 1967.

W. Gröbner [1], *Die Lie-Reihen und ihre Anwendimgen*, VEB Deutscher Verlag der Wissenschaften, 2nd Edition, Berlin, 1967.

D. Gromoll and W. Meyer [1], "On differentiable functions with isolated critical points," *Topology*, **8** (1969), 361–369.

J. Guckenheimer [1], "On a codimensional two bifurcation," *Lect. Notes Math.* 898, 994–142 Springer-Verlag, 1982.

[2], "On quasiperiodic flow with three independent frequencies," preprint.

[3], "A brief introduction to dynamical systems," *Nonlinear Oscillations in Biology*, 187–253, Lectures in Appl. Math., **17**, Am. Math. Soc., Providence, R. I., 1979.

[4], "Bifurcation of dynamical systems," pp. 115–233 in *Dynamical Systems*, Progress in Math. **8**, Birkhäuser, Boston, Basel, Stuttgart 1980.

J. Guckenheimer and P. Holmes, "Dynamical Systems and Bifurcations of Vector Fields," *Nonlinear Oscillations*, Springer-Verlag, 1983.

V. Guillemin and S. Sternberg [1], *Geometric Asymptotics*, Math. Surveys, **14**, Am. Math. Soc., Providence, R. I., 1977.

O. Gurel and O. E. Rössler, eds. [1], *Bifurcation Theory and Applications in Scientific Disciplines*, New York Acad. Sci., New York, 1979.

J. K. Hale [1], "Periodic solutions of non-linear systems of differential equations," *Riv. Mat. Univ. Parma*, **5** (1954), 281–311. MR 17, 1088.

[2], *Oscillations in Nonlinear Systems*, McGraw-Hill, New York, 1963. MR 27, 401.

[3], "Integral manifolds of perturbed differential systems," *Ann. Math.*, **73** (1961), 496–531.

[4], *Ordinary Differential Equations*, Wiley-Interscience, New York, 1969. 2nd Edition, Kreiger Publ. Co., 1980.

[5], *Applications of Alternative Problems*, Lecture Notes, Brown University, 1971.

[6], "Behavior near constant solutions of functional differential equations," *J. Differential Equations*, **15** (1974), 278–294.

[7], Bifurcation with Several Parameters. VII. Internat. Knof. Nichtlin. Schwing., Berlin (1975), **I**, 1, 279–288, Abh. Akad. Wiss. D.D.R., Akademie-Verlag, Berlin, 1977.

[8], "Bifurcation near families of solutions," *Proc. Int. Conf. on Diff. Eqns.*, Uppsala 1977, 91–100.

[9], *Theory of Functional Differential Equations*, Springer-Verlag, Berlin, 1977.

[10], "Retarded equations with infinite delays," p. 157–193 in *Functional Differential Equations and Approximation of Fixed Points* (Eds. Peitgen and Walther), Lecture Notes in Math., **730**, Springer-Verlag, 1979.

[11], "Bifurcation from simple eigenvalues for several parameter values," *Nonlin. Anal.*, **2** (1978), 491–497.

[12], "Nonlinear oscillations in equations with delays," p. 157–185 in *Nonlinear Oscillations in Biology* (Ed. F. Hoppenstadt), Lectures in Applied Mathematics, **17**, Am. Math. Soc., Providence, R. I., 1979.

[13], *Topics in Dynamic Bifurcation Theory*, NSF-CBMS Lectures, **47**, Am. Math. Soc., Providence, R. I., 1981.

[14], "Generic properties of an integro-differential equation," *Proc. Conf. on Diff. Eqns. in Analysis and Geometry*, 1980, Baltimore, MD, 131–137.

[15], "Generic bifurcation with applications," pp. 59–157 in *Nonlinear Analysis and Mechanics—Heriot-Watt Symposium* (Ed. R. Knopps), **1**, Pitman, 1977.

J. K. Hale and J. C. de Oliveira [1], "Hopf bifurcation for functional equations," *J. Math. Ana. Appl.*, **74** (1980), 41–59.

J. K. Hale and J. Kato [1], "Phase space for retarded equations with infinite delays," *Funkcial. Ekvac.*, **21** (1978), 297–315.

J. K. Hale and H. M. Rodrigues [1], "Bifurcation in the Duffing equation with independent parameters, I," *Proc. Roy. Soc. Edinburgh*, **77A** (1977), 57–65; II. *Proc. Roy. Soc. Edinburgh*, **79A** (1978), 317–326.

J. K. Hale and P. Z. Taboas [1], "Interaction of damping and forcing in a second order equation," *Nonlin. Anal. Th. Meth. Appl.*, **2** (1978), 77–84.

[2], "Bifurcation near degenerate families," *Applicable Anal.*, **11** (1980), 21–37.

J. K. Hale and S. M. Verduyn Lunel, *Introduction to Functional Differential Equations*, Springer-Verlag, 1993.

W. S. Hall [1], "On the existence of periodic solutions for the equations $D_u u + (-1)^p D_x^2 p_u = \varepsilon f(\cdot, \cdot, u)$, *J. Differential Eqns.*, **7** (1970), 509–526.

[2], "The bifurcation of solutions in Banach spaces," *Trans. Am. Math. Soc.*, **161** (1971), 207–218.

A. Haraux [1], *Special topics in Differential Equations*, Lect. Notes in Math., **841**, Springer-Verlag, 1981.

G. H. Hardy, J. E. Littlewood and G. Polya [1], *Inequalities*, University Press, Cambridge, 1952.

P. Hartman [1], "A lemma in the theory of structural stability of differential equations," *Proc. Amer. Math. Soc.*, **11** (1960), 610–620.

[2], "On the local linearization of differential equations," *Proc. Am. Math. Soc.*, **14** (1963), 568–573.

[3], *Ordinary Differential Equations*, Wiley, New York, 1964.

B. D. Hassard, N. D. Kazarinoff and Y. H. Wan [1], *Theory and Applications of the Hopf Bifurcation*, Cambridge University Press, 1981.

E. J. Haug [1], "Optimization of distributed parameter structures with repeated eigenvalues," Proc. Conf. on Problems in Nonlinear Dynamics, Asilomar, Dec. 9–14, 1979.

E. J. Haug and B. Rousellet [1], "Design sensitivity analysis in structural mechanics I: Static response variations," *J. Structural Mech.*, **8** (1980), 17–41; "II: Eigenvalue variations," *J. Struct. Mech.*, **8** (1980), 161–186. "III: Effects of shape variation," *J. Struct. Mech.*, **10** (1982/83), 273–310.

A. Hausrath [1], "Stability in the critical case of purely imaginary roots for neutral functional differential equations," *J. Differential Equas.*, **13** (1973), 329–357.

C. Hayashi [1], *Selected Papers on Nonlinear Oscillations*, Nippon, Japan, 1975.

J. A. Hempel [1], "Multiple solutions of a class of boundary value problems," *Indiana Univ. Math. J.*, **20** (1971), 983–996. MR 43, 5145.

J. Henrard [1], "On a perturbation theory using Lie transforms," *Cel. Mech.*, **3** (1970), 107–120. MR 43, 4427.

D. Henry [1], *Geometric Theory of Semilinear Parabolic Equations*, Lecture Notes in Math., Springer-Verlag, **840** (1981).

[2], Implicit functions in finite dimensions. What to do when the Jacobian vanishes, Univ. Ky., 1971. Personal communication.

G. Hetzler [1], "Bifurcation theorems of Rabinowitz type," *J. Nonlinear Anal.*, **1** (1977), 471–479.

M. W. Hirsch, C. C. Pugh and M. Shub [1], "Invariant manifolds," *Bull. Amer. Math. Soc.*, **76** (1970), 1015–1019.

[2], *Invariant Manifolds*, Lecture Notes in Mathematics 583, Springer-Verlag, Berlin, 1977.

T. H. Hildebrandt and L. M. Graves [1], "Implicit functions and their differentials in general analysis," *Trans. Amer. Math. Soc.*, **29** (1927), 127–153.

P. J. Holmes [1], "Averaging and chaotic motion in forced oscillations," *SIAM J. Appl. Math.*, **38** (1980), 65–80.

[2], "Domains of stability in a wind-induced oscillation problem," *J. Appl. Mech.*, **46** (1979), 672–676.

[3], "Unfolding a degenerate nonlinear oscillator: a codimensional two bifurcation," p. 473–488 of *Nonlinear Dynamics* (Ed. Helleman) N.Y. Acad. Sci., **357**, 1980.

[4], "A nonlinear oscillator with a strange attractor," *Phil. Trans. Roy. Soc., London*, Ser. A, **292** (1979), 419–448.

[5], Ed., *New Approaches to Nonlinear Problems in Dynamics*, SIAM, 1980.

P. J. Holmes and J. Marsden [1], "A partial differential equation with infinitely many periodic orbits: chaotic oscillations of a forced beam," *Arch. Rat. Mech. Ana.*, **76** (1981), 135–165.

P. J. Holmes and D. A. Rand [1], "Bifurcation of the forced van der Pol oscillator," *Quart. Appl. Math.*, **35** (1978), 495–509.

E. Hopf [1], "Abzweigung einer periodischer Losung von einer stationären Losung eines Differential systems," *Ber. Math. Phys.*, Sachsische Akad. der Wissenschaften, Leipzig, **94** (1942), 1–22. Translated in Marsden and McCracken [1].

[2], "Über die Anfangswertaufgabe für die hydrodynamischen Grundgleichungen," *Math. Nachr.*, **4** (1950/51), pp. 213–231.

L. N. Howard [1], "Nonlinear oscillations," p. 1–68 in *Nonlinear Oscillations in Biology* (Ed. F. Hoppenstadt), Vol. 17, Lectures in Appl. Math., Am. Math. Soc., 1979.

[2], "Bifurcation in reaction-diffusion problems," *Advances in Math.*, **16** (1975), 246–258. MR 51, 6057.

L. N. Howard and N. Kopell [1], "Bifurcations and trajectories joining critical points," *Adv. Math.*, **18** (1976), 306–358.

[2], "Bifurcations under nongeneric situations," *Adv. Math.*, **13** (1974), 274–283.

S. Invernizzi and F. Zanolin [1], Existence and Uniqueness Theorems for Periodic Solutions in Delay Differential Systems. *Math. Z.*, **163** (1978), 25–37.

G. Iooss [1], "Théorie nonlinéarie de la stabilité des écoulements laminaires dans le cas de l'échange des stabilités," *Arch. Rational Mech. Anal.*, **40** (1971), pp. 166–208.

[2], "Existence et stabilité de la solution périodique secondaire intervenant dans les problèms d'évolution du type Navier-Stokes," *Arch. Rat. Mech. Anal.*, **47** (1972), 301–329.

[3], Bifurcation et stabilité, Lecture Notes, Université de Paris Xl, 1973.

[4], *Bifurcation of Maps and Applications*. Elsevier, North-Holland, New York, 1979.

G. Iooss and D. D. Joseph [1], "Bifurcation and stability of nT-periodic solutions branching from T-periodic solutions at points of resonance," **66** (1977), 135–172.

[2], *Elementary Stability and Bifurcation Theory*, Springer-Verlag, New York, 1980.

G. Iooss and W. F. Langford [1], Conjectures on the Routes to Turbulence via Bifurcation. Preprint.

T. Itoh [1], "Bifurcation and stability of stationary solutions of nonlinear evolution equations." *J. Fac. Sci.* Univ. Tokyo, IA MATH., **29** (1982), 443–472.

V. I. Iudovich, [1], "On the stability of stationary flows of a viscious incompressible fluid," *Dokl. Akad. Nauk SSSR*, **161** (1965), pp. 1037–1045 = Soviet Physics. Dokl. **10** (1965), 293–295.

[2], "Secondary flows and fluid instability between rotating cylinders," *Prikl. Mat. Mech.*, **30** (1966), 688–698 = *J. Appl. Math. Mech.*, **39** (1966), 822–833.

[3], "On the origin of convection," *Prikl. Mat. Mech.*, **30** (1966), 1000–1005 = *J. Appl. Math. Mech.*, **39** (1966), 1193–1199.

[4], "Free convection and bifurcation," *Prikl. Mat. Mech.*, **31** (1967), 101–111 = *J. Appl. Math. Mech.*, **31** (1967), 103–114.

[5], "Stability of convection flows," *Prikl. Mat. Mech.*, **31** (1967), 272–281 = *J. Appl. Math. Mech.*, **31** (1967), 294–303.

[6], "The onset of autooscillations in a fluid," *Prikl. Mat. Mech.*, **35** (1971), 587–603.

J. Ize [1], Bifurcacion global de orbitas periodicas. Preprint.

[2], "Bifurcation theory for Fredholm operators," *Memoirs of the A.M.S.*, **174**, 1976.

H. Jacobowitz [1], "Periodic solutions of $x'' + f(x, t) = 0$ via the Poincaré—Birkhoff theorem," *J. Diff. Eq.*, **20** (1976), 37–52.

D. D. Joseph [1], "Nonlinear stability of the Boussinesq equations by the method of energy," *Arch. Rat. Mech. Anal.*, **22** (1966), 163–184.

[2], *Stability of Fluid Motions Vol. I, II*, Springer-Verlag, New York, 1976.

D. D. Joseph and D. A. Nield [1], "Stability of bifurcating time solutions of arbitrary amplitude," *Arch. Rat. Mech. Ana.*, **58** (1975), 369–380.

D. D. Joseph and D. H. Sattinger [1], "Bifurcating time periodic solutions and their stability," *Arch. Rat. Mech. Anal.*, **45** (1972), 79–109.

R. Jost and E. Zehnder [1], "A generalization of the Hopf bifurcation theorem," *Helv. Phys. Acta*, **45** (1972), 258–276.

A. A. Kamel [1], "Perturbation method in the theory of nonlinear oscillations," *Cel. Mech.*, **3** (1970), 90–106, MR 43, 4426.

[2], "Lie transforms and the Hamiltonization of nonlinear Hamiltonian systems," *Cel. Mech.*, **4** (1971), 397–405. MR 47, 1343.

R. Kannan and J. Locker [1], "On a class of nonlinear boundary value problems," *J. Diff. Eqns.*, **26** (1977), 1–8.

T. Kato [1], *Perturbation Theory for Linear Operators*, Springer-Verlag, Berlin, 1966.

Th. von Karman [1], "Festigkeitsproblem in Machinenbau," *Enz. d Math. Wiss.*, **IV-4** (1910), 348–352.

J. P. Keener [1], "Perturbed bifurcation theory at multiple eigenvalues," *Arch. Rat. Mech. Anal.*, **56** (1974), 348–366. MR 50, 8184.

[2], "Buckling imperfection sensitivity of columns and spherical caps," *Quart. Appl. Math.*, **32** (1974), 173–188.

[3], Secondary Bifurcation and multiple eigenvalues. *SIAM J. Appl. Math.*, **37** (1979), 330–349.

J. P. Keener and H. B. Keller [1], "Perturbed bifurcation theory," *Arch. Rat. Mech. Anal.*, **50** (1973), 159–175.

[2], Perturbed bifurcation and buckling of circular plates, *Lecture Notes in Math.*, **280**, Springer-Verlag, Berlin, 1972, 286–293.

H. B. Keller [1], *Bifurcation Theory and Nonlinear Eigenvalue Problems*, Unpublished Lecture Notes, Cal. Tech. (1976).

[2], "Numerical solution of bifurcation and nonlinear eigenvalue problems," p. 359–384 in *Applications of Bifurcation Theory* (Ed. Raboniwitz) Academic Press, 1977.

[3], "Nonlinear bifurcation," *J. Diff. Eqns.*, **7** (1970), 417–434.

H. B. Keller, J. B. Keller and E. L. Reiss [1], "Buckled states of circular plates," *Quart. Appl. Math.*, **20** (1962), 55–65.

J. B. Keller and S. Antman [1], *Bifurcation Theory and Nonlinear Eigenvalue Problems*, Benjamin, New York, 1969.

A. Kelley [1], "The stable, center-stable, center, center-unstable and unstable manifolds," *J. Diff. Eqns.*, **3** (1967), 546–570.

[2], "Stability of the center stable manifold," *J. Math. Anal. Appl.*, **18** (1967), 336–344.

H. Kielhöfer [1], "Stability and semilinear evolution equations in Hilbert space," *Arch. Rat. Mech. Anal.*, **57** (1974), 150–165.

[2], "Generalized Hopf bifurcation in Hilbert space," *Math. Meth. Appl. Sci.*, **1** (1979), 498–513.

[3], "Hopf bifurcation at multiple eigenvalues," *Arch. Rat. Mech. Anal.*, **69** (1979), 53–83.

K. Kirchgässner [1], "Die instabilität der Strömung zwischen zwei rotierenden Zylindern gegenüber Taylor-Wirbeln für beliebige Spaltbreiten," *Z. Angew. Math. Phys.*, **12** (1961), 14–30.

[2], "Bifurcation in nonlinear hydrodynamic stability," *SIAM Review*, **17** (1975), 652–683.

[3], "Preference in pattern and cellular bifurcation in fluid dynamics," p. 149–174 in *Applications of Bifurcation Theory* (Ed. Rabinowitz), Academic Press, 1977.

[4], "Multiple eigenvalue bifurcation for holomorphic mappings" (1971), p. 69–99, in *Contributions to Nonlinear Functional Analysis*, (Ed. E. H. Zarantello), Academic Press.

K. Kirchgässner and H. Kielhöfer [1], "Stability and bifurcation in fluid dynamics," *Rocky Mountain J. Math.*, **3** (1972), 275–318.

[2], "Stability and bifurcation in fluid mechanics," *Rocky Mountain J. Math.*, **3** (1973), 258–318.

K. Kirchgässner and P. Sorger [1], "Stability analysis of branching of the Navier-Stokes equations," Proc. 12th Int. Congr. Appl. Mech., Stanford, 1968, pp. 257–268.

[2], "Branching analysis for the Taylor Problem," *Quart. J. Mech. Appl. Math.*, **32** (1969), 183–209.

U. Kirchgraber and E. Stiefel [1], *Methoden der analytischen Störungsrechnung und ihre Anwendungen.* Teubner, Stuttgart, 1978.

G. H. Knightly [1], "An existence theorem for the von Kármán equations," *Arch. Rat. Mech. Anal.*, **27** (1967), 233–242.

[2], "Some mathematical problems from plate and shell theory," pp. 245–268 in *Nonlinear Functional Analysis and Differential Equations*, Dekker, New York, 1976.

G. H. Knightly and D. Sather [1], "On nonuniqueness of solutions of the von Karman equations," *Arch. Rat. Mech. Anal.*, **36** (1970), 65–78.

[2], "Nonlinear buckled states of rectangular plates," *Arch. Rat. Mech. Anal.*, **54** (1974), 356–372.

[3], "Nonlinear axisymmetric buckled states of shallow spherical caps," *SIAM J. Math. Anal.*, **6** (1975), 913–924.

[4], "Existence and stability of axisymmetric buckled states of spherical shells," *Arch. Rat. Mech. Anal.*, **63** (1977), 305–319.

[5], "Nonlinear buckling of cylindrical panels," *SIAM J. Math. Anal.*, **10** (1979), 389–403.

[6], "Buckled states of a spherical shell under uniform external pressure," *Arch. Rat. Mech. Anal.*, **72** (1980), 315–380.

W. T. Koiter [1], Over de stabiliteit van het elastisch evenwicht, Thesis, Delft: H. J. Paris, Amsterdam, 1945. English Translation issued as Technical Report AFFDL-TR-70-25, Air Force Flight Dynamics Lab., Wright-Patterson Air Force Base, Ohio, 1970.

[2], "Elastic stability and post-buckling behavior," in Nonlinear Problems, ed. R. Langer, University of Wisconsin Press, Madison, 1963.

[3], "On the nonlinear theory of thin elastic shells," *Proc. Kon. Nederl. Akad., Wet.* Amsterdam **B69** (1966), 1–54.

A. Kolmogorov [1], Théorie générale des systèmes dynamique et mechanique classique, Proc. Int. Congr. Math., Amsterdam, 1954.

I. I. Kolodner [1], "Heavy rotating string—A nonlinear eigenvalue problem," *Comm. Pure Appl. Math.*, **8** (1955), 395–408.

V. S. Kozyakin [1], "Subfurcation in loss of stability of equilibrium position for a system of differential equations with lag," *Siberian Math. J.* **18** (1977), 414–424; *Sibirski Mat. Z.* **18** (1977), 580–594.

M. A. Krasnosel'skii [1], *Topological Methods in the Theory of Nonlinear Integral Equations*, Pergamon, 1964.

[2], "Small solutions of a class of nonlinear operators," *Sov. Math. Dokl.*, **9** (1968), 579–581.

[3], "Investigation of the bifurcation of small eigenfunctions in the case of multiple dimensional degeneration," *Sov. Math. Dokl.*, **11** (1970), 1609–1613.

[4], "The projection method for the study of bifurcation of the zero solution of a nonlinear operator equation with multidimensional degeneracy," *Sov. Math. Dokl.*, **12** (1971), 967–971.

N. Krylov and N. N. Bogoliubov [1], The Application of Methods of Nonlinear Mechanics to the Theory of Stationary Oscillations, Publ. 8 of the Ukrainian Acad. Sci., Kiev, 1934.

M. Kubicek, M. Marek and P. Raschman [1], Waves in distributed chemical systems—experiments and computations, Proc. Symp. on Problems in Nonlinear Dynamics, Asilomar, Dec. 9–14, 1979.

M. Kubicek, M. Marek and I. Schreiber [1], On coupled cells, Proc. Conf. on Problems in Nonlinear Dynamics, Asilomar, Dec. 9–14, 1979.

J. Kurzweil [1], "Invariant manifolds for flows," *Proc. Sym. Diff. Eqns. and Dynamical Systems*, Academic Press, 1967.

[2], "The averaging principle in certain special cases of boundary value problems for partial differential equations," *Casopia Pest. Mat.*, **88** (1963), 444–456.

[3], "van der Pol perturbation of the equation for a vibrating string," *Czechoslovak Math. J.*, **17** (92) (1967), 558–608.

O. A. Ladyzenskaya [1], "A dynamical system generated by the Navier-Stokes equation," Zap. Nauch. Steklov Math. Inst. Leningrad **27** (1972), 91–115; also, *J. Sov. Math.*, **3** (1975), 458–479.

E. M. Landesman and A. C. Lazer [1], "Linear eigenvalues and a nonlinear boundary value problem," *Pac. J. Math.*, **33** (1970), 311–328.

[2], "Nonlinear perturbation of linear elliptic boundary value problems at resonance," *Indiana Univ. Math. J.*, **19** (1970), 609–623.

S. Lang [1], *Differential Manifolds*, Addison-Wesley, Reading, Mass., 1972.

[2], *Real Analysis*, Addison-Wesley, Reading, Mass., 1969.

C. G. Lange and A. C. Newell [1], "Spherical shells like hexagons: cylinders prefer diamonds," *J. Appl. Mech.*, June, 1973, 575–581.

[2], "The post buckling problem for thin elastic shells," *SIAM J. Appl. Math.*, **21** (1971), 605–629.

O. E. Lanford [1], "Bifurcation of Periodic Solutions into Invariant Tori: The Work of Ruelle and Takens," Lecture Notes **322**, pp. 159–192, Springer-Verlag, Berlin, 1972.

W. F. Langford, [1], "Periodic and Steady mode interactions lead to tori," *SIAM Appl. Math.*, **37** (1979), 22–48.

[2], "Numerical solutions of bifurcation problems for ordinary differential equations," *Numer. Math.*, **28** (1977), 171–190.

A. Lasota and M. Wazewska [1], "Mathematical models of the red cell system," *Mat. Stosewana,* **6** (1976), 25–40.

A. C. Lazer [1], "Small periodic perturbations of a class of conservative systems," *J. Diff. Eqns.,* **13** (1973), 438–456. MR 50, 13732.

A. C. Lazer and D. E. Leach [1], "Bounded perturbations of forced harmonic oscillators at resonance," *Ann. Mat. Pura Appl.* (4) **82** (1969), 49–68.

A. C. Lazer and D. Sanchez [1], "On a periodically perturbed conservative system," *Mich. Math. J.,* **16** (1969), 193–200.

V. F. Lazutkin and T. F. Pankratova [1], "Normal forms and versal deformations for Hill's equation," *Func. Ana. Appl.,* **9** (1975), 306–311.

S. Lefschetz [1], "The critical case in differential equations," *Bol. Soc. Mat. Mexicanna,* (2) **6** (1961), 5–18; MR 24, A3354.

[2], *Differential Equations: Geometric Theory,* Interscience, New York, 1963.

[3], *Algebraic Geometry,* Princeton Univ. Press, Princeton, 1953.

F. A. Leontovich [1], "On the creation of limit cycles from a separatrix," *Dokl. Akad. Nauk SSSR,* **78** (1951), 641–644.

M. Levi [1], Qualitative Analysis of the Periodically Forced Relaxation Oscillations, Thesis, New York Univ., 1979.

N. Levinson [1], "Transformation theory of nonlinear differential equations of the second order," *Ann. Math.,* **45** (1944), 723–737.

[2], "A second order differential equation with singular solutions," *Ann. of Math.* (2) **50** (1949), 127–153.

[3], "Small periodic perturbations of an autonomous equation with a stable orbit," *Ann. Math.,* **52** (1950), 727–738.

D. C. Lewis, [1], "On the role of first integrals in the perturbation of periodic solutions," *Ann. Math.,* **63** (1956), 535–548.

[2], "Autosynartetic solutions of differential equations," *Amer. J. Math.,* **83** (1961), 1–32.

A. Liapunov [1], *Problème Général de la Stabilité du Mouvement,* Princeton Univ. Press, 1947.

[2], "Sur les figures d'équilibre peu différentes des ellipsoids d'une masse liquide homogène doudée d'un mouvement de rotation," *Zap. Akad. Nauk.,* Petersburg, 1 (1960), 1–225.

P. Lima [1], Hopf Bifurcation in Equations with Infinite Delays, Ph.D. Thesis, Brown University, 1977.

J. L. Lions [1], *Quelques méthodes de resolution des problèmes aux limites non linéaires,* Dunod, Paris, 1969.

[2], *Optimal Control of Systems Governed by Partial Differential Equations,* Springer-Verlag, 1971.

J. L. Lions and E. Magenes [1], *Problèmes aux limites non homogènes et applications, Vol. 1 & 2,* Dunod, Paris, 1968.

S. List [1], Generic Bifurcation with Applications to the Von Kármàn Equations, Ph.D. Thesis, Brown University, 1976. *J. Diff. Eqns.,* **30** (1978), 89–118.

L. M. Likhtarinkov [1], "Spectra of a class of linear integral equations with two parameters," *Diff. Eqns.,* **11** (1975), 833–840.

J. E. Littlewood [1], "On nonlinear differential equations of the second order: IV. The general theory," *Acta Math.,* **98** (1957), 1–110.

L. A. Ljusternik [1], "Topologische Grundlagen der allgemainen Eigenwerttheorie," *Monatsch. Math. Phys.,* **37** (1930), 125–130.

L. A. Ljusternik and L. G. Schnirelmann [1], *Topological Methods in the Calculus of Variations,* Hermann, Paris, 1934.

N. G. Lloyd [1], *Degree Theory,* Cambridge Univ. Press, 1978.

B. V. Loginov and V. A. Trenogin [1], "Application of continuous groups in branching theory," *Sibirisk Mat. Z.,* **12** (1971), 353–366. English translation, *Siberian Math. J.,* **12** (1971), 251–260. MR 44, 866.

[2], "Employment of group invariance in branching theory," *Diff. Eqns.,* **11** (1975), 1137–1140.

E. Lorenz [1], "Deterministic nonperiodic flow," *J. Atmospheric Sci.,* **20** (1963), 130–141.

W. S. Loud [1], "Branching of solutions of a two-parameter boundary value problem for second order differential equations," *Ingenieur-Archiv.,* **45** (1976), 347–359.

H. Lovicarova [1], "Periodic solutions of a weakly nonlinear wave equation in one dimension," *Czech. Math. J.,* **19** (1969), 324–342.

Y.-C. Lu [1], *Singularity Theory and an Introduction to Catastrophe-Theory,* Springer-Verlag, 1976.

J. MacBain [1], "Global bifurcation theorems for noncompact operators," *Bull. Am. Math. Soc.*, **80** (1974), 1005–1009, MR 50, 14403.

M. Mackey and L. Glass [1], "Oscillation and chaos in physiological control systems," *Science*, **197** (1977), 287–289.

J. W. Macki and P. Waltman [1], "A boundary value problem for a two-dimensional system with a parameter," *SIAM J. Math. Anal.*, **6** (1975), 91–95. MR 51, 988.

W. MacMillan [1], "A reduction of a system of power series to an equivalent system of polynomials," *Math. Ann.*, **72** (1912), 157–179.

[2], "A method for determining the solutions of a system of analytic functions in the neighborhood of a branch point," *Math. Ann.*, **72** (1912), 180–202.

R. J. Magnus [1], "On the local structure of the zero set of a Banach space valued mapping," *J. Funct. Anal.*, **22** (1976), 58–72.

[2], "Topological stability of the solution set of a nonlinear equation," Battelle, Geneva, Math. Report No. 85 (1974).

[3], "The reduction of a vector valued function near a critical point," Battelle, Geneva, Math. Report No. 93 (1975).

[4], "On universal unfoldings of certain real functions on a Banach space," *Math. Proc. Cambridge Philos. Soc.*, **81** (1977), 91–95; MR 54, 13986.

[5], "A generalization of multiplicity and the problem of bifurcation," *Proc. London Math. Soc.*, (3), **32** (1976), 251–278. MR 53, 6380.

J. G. Malkin [1], *Theory of Stability of Motion*, Atomic Energy Comm. Translation AEC-TR-3352.

J. Mallet-Paret [1], "Buckling of cylindrical shells with small curvature," *Quart. Appl. Math.*, **35** (1977), 383–400.

[2], "Negatively invariant sets of compact maps and extension of a theorem of Cartwright," *J. Diff. Eqns.*, **22** (1976), 331–348.

R. Mañé [1], On the Dimension of the Compact Invariant Set of Certain Nonlinear Maps. *Dynamical Systems and Turbulence*, Warwick 1980 (Conventry, 1979/1980), 230–242, Lecture notes in Math., 898, Springer-Verlag, Berlin-New York, 1981.

A. Marino [1], "La biforcazione nel caso variazionale." Conf. Sem. Mat. Univ. Bari, No. 132 (1973), 14pp. MR 50, 1068.

A. Marino and G. Prodi [1], "La teoria di Morse per gli spazi di Hilbert," *Rend. Sem. Mat. Univ. Padova*, **41** (1968), 43–68. MR 41, 2715.

L. Markus and R. A. Moore [1], "Oscillation and disconjugacy for linear differential equations with almost periodic coefficients," *Acta Mat.*, **96** (1956), 99–123. MR 18, 306.

J. Marsden [1], "The Hopf bifurcation for nonlinear semigroups," *Bull. A.M.S.*, **79** (1973), 537–541.

[2], "Qualitative methods in bifurcation theory," *Bull. A.M.S.*, **84** (1978), 1125–1148.

[3], *Geometric Methods in Mathematical Physics*, NSF-CBMS Conference, Lowell, Mass., Mar. 19–23, 1979.

J. Marsden and M. F. McCracken [1], *The Hopf Bifurcation and its Applications*, Springer-Verlag, New York, 1976.

R. H. Martin [1], *Nonlinear Operators and Differential Equations in Banach Spaces*, Wiley-Interscience, 1976.

J. L. Massera and J. J. Schaffer [1], *Linear Differential Equations and Function Spaces*, Academic Press, 1966.

H. Matano [1], "Asymptotic behavior and stability of solutions of semilinear diffusion equations," *Publ. Res. Inst. Math. Sci., Kyoto Univ.*, **15** (1979), 401–454.

B. J. Matkowsky and L. J. Putnick [1], "Multiple buckled states of rectangular plates," *Int. J. Nonlin. Mech.*, **9** (1974), 89–103.

B. J. Matkowsky and E. L. Reiss [1], "Singular perturbations of bifurcations," *SIAM J. Appl. Math.*, **33** (1977), 230–255.

J. Mawhin [1], "Equivalence theorems for nonlinear operator equations and coincidence degree theory in for some mappings in locally convex topological spaces," *J. Diff. Eqns.*, **12** (1972), 610–636.

[2], "Contractive mappings and periodically perturbed conservative systems," *Arch. Math.* (Brno), **12** (1976), 67–63. MR 55, 10779.

[3], "Solutions périodiques d'equations aux dérivées partielles hyperboliques nonlinéaires," in "Mélanger Th. Vogel," Rybak, Jenssenr, Jessel eds., Presses Univ. Bruxelles, 1978, 301–319.

[4], "Topological Degree methods in nonlinear boundary value problems," CBMS 40, Am. Math. Soc., Providence, R.I., 1979.

R. May [1], *Stability and Complexity in Model Ecosystems*, Princeton Univ. Press, Princeton, N.J., 1973.

S. Mazzanti and B. V. Schmitt [1], "Solutions periodiques symetriques de l'equation de Duffing sans dissipation," *J. Diff. Eqns.*, **42** (1981), 199–214.

R. McGehee and K. R. Meyer [1], "Homoclinic points of area preserving diffeomorphisms," *Am. J. Math.*, **96** (1974), 409–421. MR 50, 11321.

P. J. McKenna and H. Shaw [1], "On the structure of the set of solutions to some nonlinear boundary value problems," *J. Diff. Eqns.*, **35** (1980), 183–199.

J. McLeod and D. Sattinger [1], "Loss of stability and bifurcation at a double eigenvalue," *Journ. Funct. Analy.*, **14** (1973), 62–84. MR 50, 5565.

J. B. McLeod and R. E. L. Turner [1], "Bifurcation for nondifferentiable operators with an application to elasticity," *Arch. Rat. Mech. Anal.*, **63** (1976/77), 1–45. MR 57, 13611.

M. Medved [1], "General properties of parametrized vector fields, I," *Czech. Math. J.*, **25** (1975), 376–388.

[2], "Generic bifurcations of second order ordinary differential equations on differentiable manifolds," *Mathematica Siovaca*, **27** (1977), 9–24.

V. K. Mel'nikov [1], "On the stability of a center for time periodic solutions," *Trans. Moscow Math. Soc.* (Trudy), **12** (1963), 3–52. MR 27, 5981.

K. Meyer [1], "Generic bifurcation in Hamiltonian systems," p. 62–70 in *Dynamical Systems— Warwick 1974*, Lecture Notes in Math., Vol. 468, Springer-Verlag, 1975.

[2], "Generic bifurcation of periodic points," *Trans. Am. Math. Soc.*, **149** (1970), 95–107.

K. Meyer and D. S. Schmidt [1], "Entrainment domains," *Funkcial. Ekvac.*, **20** (1977), 171–192. MR 57, 1557.

[2], "Periodic orbits near $\mathscr{L}_4$ for mass ratio near the critical mass ratio of Routh," *Cel. Mech.*, **4** (1971), 99–109. MR 47, 6164.

P. Michor [1], The Division Theorem on Banach Spaces. Preprint. Math. Inst. Univ. Wein, Austria.

J. W. Milnor [1], *Topology from the Differentiable Viewpoint*, Univ. Press of Virginia, Charlottesville, 1965.

[2], *Morse Theory*, Princeton Univ. Press, Princeton, N.J., 1963.

N. Minorsky [1], *Introduction to Nonlinear Mechanics*, J. W. Edwards, PUbl., Ann Arbor, Mich., 1947.

[2], *Nonlinear Oscillations*, Van Nostrand, 1962.

Y. A. Mitropolski [1], *The Method of Averaging in Nonlinear Mechanics* (Russian) Izd. Nauk. Dymka, Kiev, 1971.

F. C. Moon and P. J. Holmes [1], "A magnetoelastic strange attractor," *J. Sound and Vibration*, **65** (1979), 275–296.

A. D. Morozov [1], "On the complete qualitative investigation of the equation of Duffing," *Differentialniye Uravneniya*, **12** (1976), 241–255.

G. Morris [1], "A case of boundedness in Littlewood's problem on oscillatory differential equations," *Bull. Austral. Math. Soc.*, **14** (1976), 71–93.

[2], "A differential equation for undamped forced nonlinear oscillations, I, II." *Proc. Camb. Philos. Soc.*, **5** (1955), 297–312, **54** (1958), 426–438.

J. Moser [1], *Stable and Random Motions in Dynamical Systems*, Princeton Univ. Press, Princeton, N.J., 1973.

[2], "Periodic orbits near an equilibrium and a theorem by Alan Weinstein," *Comm. Pure Appl. Math.*, **29** (1976), 727–747. MR 54, 13998.

[3], "A rapidly convergent iteration method and nonlinear partial differential equations, I," *Ann. Scuola Norm. Sup.*, Pisa, **20** (1966), 226–315. II, Ibid., 449–535.

M. Nagumo [1], "A theory of degree of mapping based on infinitesimal analysis," *Amer. J. Math.*, **73** (1951), 485–496.

T. Naito [1], "Adjoint equations of autonomous linear functional differential equations with infinite retardations," *Tohoku Math. J.*, (2), **28** (1976), 135–143; MR 57, 6719.

[2], "On linear autonomous retarded equations with an abstract phase space for infinite delay," *J. Diff. Eqns.*, **33** (1979), 74–91.

J. Naumann [1], Uber Bifurkation bei flachen Kugelschalen. Tagungsbericht sur ersten Tagen der WK Analysis (1970). *Beiträge Anal. Heft* **4** (1972), 75–79. MR 51 #3988.

[2], "On bifurcation buckling of thin elastic shells, *J. Mécanique*, **13** (1974), 715–741.

[3], "Variationsmethoden für Existenz und Bifurkation von Lösungen nichlinearer Eigenwert-probleme I & II," *Math. Nacht.*, **54** (1972), 285–296, **55** (1973), 325–344.

A. Nayfeh [1], *Perturbation Methods*, Wiley-Interscience, John Wiley & Sons, New York, 1973.

P. Negrini and L. Salvadori [1], "Attractivity and Hopf bifurcation," *Nonlinear Anal.*, **3** (1979), 87–99.

J. Neimark [1], "On some cases of periodic motions depending on parameters," *Dokl. Akad. Nauk. SSSR*, **129** (1959), 736–739.

J. D. Newburgh [1], "The variation of spectra," *Duke Math. J.*, **18** (1951), 165–176.

S. Newhouse [1], Lectures on Dynamical Systems, C.I.M.E. Bressanove, Italy, June 19–27, 1978.

[2], "Diffeomorphisms with infinitely many sinks," *Topology*, **13** (1974), 9–18.

L. Nirenberg [1], "A proof of the Malgrange preparation theorem," *Liverpool Singularities*, **1**, 97–105.

[2], *Topics in Nonlinear Functional Analysis*, N.Y.U. Lecture Notes, 1973–74.

Z. Nitecki [1], *Differentiable Dynamics*, M.I.T. Press, Cambridge, Mass., 1971.

R. D. Nussbaum [1], "A global bifurcation theorem with applications to functional differential equations," *J. Func. Anal.*, **19** (1975), 319–338.

[2], "Periodic solutions of nonlinear autonomous functional differential equations," p. 283–325 in *Functional Differential Equations and Approximation of Fixed Points*, (Eds. Peitgen and Walther), Lecture Notes in Math., **730**, Springer-Verlag, 1979.

[3], "The range of periods of periodic solutions of $x'(t) = -\alpha f(x(t-1))$," *J. Math. Anal. Appl.*, **58** (1977), 280–292.

[4], "The radius of the essential spectrum," *Duke Math. J.*, **37** (1970), 473–488.

[5], "Degree theory for local condensing maps," *J. Math. Ana. Appl.*, **37** (1972) 741–766.

R. D. Nussbaum and C. A. Stuart [1], "A singular bifurcation problem," *J. London Math. Soc.* (2), **14** (1976), 31–38. MR 55, 755.

N. Olhoff and S. H. Rasmussen [1], "On single and bimodel optimum buckling loads of clamped columns," *Int. J. Solids and Structures*, **13** (1977), 605–614.

Z. Opial [1], *Nonexpensive and Monotone Mappings in Banach Spaces*, Lect. Notes. 67–1, Lefschetz CDS, Brown Univ., Providence, R.I., 1967.

J. Paine and F. DeHoog [1], "Uniform estimation of the eigenvalues of Sturm-Liouville problems," *J. Australian Math. Soc., Ser. B, Appl. Math.*, **23** (1980), 365–383.

R. Palals [1], "Ljusternik—Schnirelman theory on Banach manifolds," *Topology*, **5** (1966), 115–132.

[2], "Critical point theory and the minimax principle," *Proc. Symp. Pure Math.*, **15**, A.M.S., Providence, R.I. (1970), 185–212.

[3], "Morse theory on Hilbert manifolds," *Topology*, **2** (1963), 299–340.

R. Palais and S. Smale [1], "A generalized Morse Theory," *Bull. Amer. Math. Soc.*, **70** (1964), 165–171.

J. Palis [1], "On the local structure of hyperbolic points in a Banach space," Am Acad. Brasil. Ci. **40** (1968), 263–266.

J. Palis and W. deMelo [1], *Introducao aos Sistemas Dinamicos*, Inst. Math. Pura Aplicada (IMPA), 1978.

K. J. Palmer [1], "Topological equivalence and the Hopf bifurcation," *J. Math. Anal. Appl.*, **66** (1978), 586–598.

[2], "Linearization near integral manifolds," *Jour. Math. Anal. Appl.*, **51** (1975), 243–255.

[3], Qualitative behaviour of a system of ODE near an equilibrium point—a generalization of the Hartman–Grobman theorem. Instit. of Appl. Math., University of Bonn, Preprint, 372.

G. C. Papanicolaou [1], Stochastically perturbed bifurcation. *Computing methods in applied sciences and engineering* (Proc. Fourth Internat. Sympos., Versailles, 1979), 659–673. North-Holland, Amsterdam, 1980.

H. O. Peitgen and M. Prüfer [1], "The Leray—Schauder continuation method is a constructive element in the numerical study of nonlinear eigenvalue and bifurcation problems," pp. 326–409, *Lecture Notes in Math.*, **730**, Springer-Verlag, 1979.

M. Peixoto [1], "Structural stability on two dimensional manifolds," *Topology*, **1** (1962), 101–120.

[2], "Generic properties of ordinary differential equations," in *Studies in Differential Equations* (Ed. J. K. Hale), 52–92; vol. 14 of Studies in Math., Math. Assoc. of Amer., 1977.

[3], *Dynamical Systems*, Academic Press, 1973.

G. H. Pimbley [1], "Eigenfunction branches of nonlinear operators and their bifurcation," Lecture Notes in Math., **104**, Springer-Verlag, Berlin, 1969.

R. H. Plaut [1], "Postbuckling analysis of continuous, elastic systems under multiple loads. Part 1: theory, Part 2, Applications," Trans. ASME, Ser. E, *J. Appl. Mech.*, **46** (1979), 393–403.

V. A. Pliss [1], "A reduction principle in the theory of stability of motion," *Izv. Akad. Nauk SSSR Ser. Mat.*, **28** (1964), 1297–1324; MR 32, 7861.

V. Poenaru [1], *Singularities C^∞ en présence de symétrie.* Lecture Notes in Math., **510**, Springer-Verlag, Berlin, 1976.

H. Poincaré [1], *Les Méthodes Nouvelles de la Mécanique Céleste*, Vol. 3, Gauthier-Villars, Paris, 1892.

[2], "Sur l'equilibre d'une masses fluide animés d'un mouvement de rotation," *Acta Math.*, **7** (1885), 259–380.

A. B. Poore [1], "On theory and application of the Hopf–Friedrichs bifurcation theory," *Arch. Rat. Mech. Anal.*, **60** (1976), 371–393.

T. A Porsching [1], "Analytic eigenvalues and eigenvectors," *Duke Math. J.*, **35** (1968), 363–367.

T. Poston and I. Stewart [1], *Catastrophe Theory and Its Applications*, Pitman, London, 1978.

M. Potier-Ferry [1], "Bifurcation et stabilité pour des systèmes dérivant d'un potential," *J. Mecanique*, **17** (1978), 579–608.

[2], "Perturbed bifurcation theory," *J. Diff. Eqns.*, **33** (1979), 112–146.

G. Prodi and A. Ambrosetti [1], *Analisi Non Lineare I Quaderno*, Scuola Normale Superiore, Pisa, 1973.

C. C. Pugh [1], "On a theorem of Hartman," *Amer. J. Math.*, **91** (1969), 363–367.

C. C. Pugh and M. Shub [1], "Linearization of normally hyperbolic diffeomorphisms and flows," *Invent. Math.*, **10** (1970), 187–198.

A. S. Pyartli [1], "Birth of complex invariant manifolds close to a singular point of a parametrically dependent vector field," *Functional Anal. Appl.*, **6** (1972), 339–340.

P. Rabinowitz [1], "Periodic solutions of a nonlinear wave equation," *Comm. Pure Appl. Math.*, **20** (1967), 145–205.

[2], "Nonlinear Sturm–Liouville problems for second order ordinary differential equations," *Comm. Pure Appl. Math.*, **23** (1970), 939–961.

[3], "Some global results for nonlinear eigenvalue problems," *J. Func. Anal.*, **7** (1971), 487–513.

[4], "Existence and nonuniqueness of rectangular solutions of the Bénard Problem," *Arch. Rat. Mech. Anal.*, **29** (1968), 32–57.

[5], "Some aspects of nonlinear eigenvalue problems," *Rocky Mtn. J. Math.*, **3** (1973), 161–202.

[6], "Variational methods for nonlinear eigenvalue problems," in *Eigenvalues of Nonlinear Problems* (Ed. G. Prodi) pp. 141–195, Roma: Edizion Cremonese, 1974.

[7], "A bifurcation theorem for potential operators," *J. Funct. Analy.*, **26** (1977), 48–67.

[8], "Some critical point theorems and applications to semilinear elliptic partial differential equations," *Ann. Scuol. Norm. Sup. Pisa*, Ser IV, Vol. II, 215–223 (1978).

[9], "Some minimax theorems and applications to nonlinear partial differential equations," in *Nonlinear Analysis* (Cesari, Kannan, and Weinberger, ed.), pp. 161–177, New York, Academic Press, 1978.

[10], "A variational method for finding periodic solutions of differential equations," in *Nonlinear Evolution Equations* (M. G. Crandall, ed.), pp. 225–251, New York, Academic Press, 1978.

[11], "Free vibrations for a semilinear wave equation," *Comm. Pure Appl. Math.*, **31** (1978), 31–68.

[12], "Periodic solutions of Hamiltonian systems," *Comm. Pure Appl. Math.*, **31** (1978), 157–184.

[13], "Periodic solutions of a Hamiltonian system on a prescribed energy surface," *J. Diff. Eqns.*, **33** (1979), 336–352.

[14], *Applications of Bifurcation Theory*, Proc. Symp. Wisc., Oct. 1976, Academic Press, 1977.

[15], "On bifurcation from infinity," *J. Diff. Eqns.*, **3** (1973), 462–475.

H. J. Raveché and C. A. Stuart [1], "Bifurcation of solutions with crystalline symmetry," *J. Math. Phys.*, **17** (1976), 1949–1953; MR 54, 6812.

[2], Bifurcation and Stability of Solutions with Lattice Symmetries, Preprint.

W. H. Ray [1], "Bifurcation phenomena in chemically reacting systems," p. 285–315 in *Applications of Bifurcation Theory* (Ed. Rabinowitz), Academic Press, 1977.

W. H. Ray and K. F. Jensen [1], "Bifurcation phenomena in stirred tank and catalytic reactors," New Approaches to Nonlinear Problems in Dynamics (P. Holmes, Ed.), SIAM, Philadelphia, 1980, 235–256.

M. Reeken [1], "A general theorem on bifurcation and its application to the Hartree equation of the helium atom," *J. Math. Phys.*, **11** (1970), 2505–2512.

E. L. Reiss [1], "Bifurcation buckling of spherical caps," *Comm. Pure Appl. Math.*, **18** (1965), 65–82.

F. Rellich [1], *Perturbation Theory of Eigenvalue Problems*, Lecture Notes, New York Univ., 1953; Gordon and Breach, 1969.

J. W. Robbin [1], "On the existence theorem for differential equations," *Proc. Am. Math. Soc.*, **19** (1968), 1005–1006.

R. Rockafellar [1], *Convex Analysis*, Princeton Univ. Press, 1970.

H. M. Rodrigues and A. L. Vanderbauwhede [1], "Symmetric perturbations of nonlinear equations: symmetry of small solutions," *Nonlin. Anal. Th. Meth. and Appl.*, **2** (1978), 27–46.

S. Rosenblatt [1], "Asymmetric bifurcation," *J. Australian Math. Soc.*, Ser. B, *Appl. Math.*, **23** (1980), Part 3, 297–304.

O. E. Rössler [1], "An equation for continuous chaos," *Phys. Lett.* **A57** (1976), 397–398.

[2], "Different types of chaos in two simple differential equations," *Z. Naturforsch.*, **31a** (1976), 1664–1670.

E. H. Rothe [1], "Completely continuous scalars and variational methods," *Ann. of Math.*, **49** (1948), 265–278.

[2], The energy-period function and perturbation of Hamiltonian systems in the plane, *Cambridge Math. Soc. Conf. Proc.*, **8** (1987), 621–635.

D. Ruelle [1], "Bifurcations in the presence of a symmetry group," *Arch. Rat. Mech. Anal.*, **51** (1973), 136–152.

D. Ruelle and F. Takens [1], "On the nature of turbulence," *Comm. Math. Phys.*, **20** (1971), 167–192.

[2], "Note concerning our paper 'On the nature of turbulence'", *Comm. Math. Phys.*, **23** (1971), 343–344.

J. Ruiz-Claeyssen [1], "Effect of delays on functional differential equations," *J. Diff. Eqns.*, **20** (1976), 404–440.

R. J. Sacker [1], "A new approach to the perturbation theory of invariant surfaces," *Comm. Pure Appl. Math.*, **18** (1965), 717–732.

[2], "A Perturbation theorem for invariant manifolds and Hölder continuity," *J. Math. Mech.*, **18** (1969), 705–762.

B. N. Sadovskii [1], "Limit compact and condensing operators," *Uspehi Math. Nauk*, **27** (163) (1972), 81–146. Russian Math. Surveys, **27** (1972), 85–146.

D. Sather [1], "Nonlinear gradient operators and tbe method of Liapunov—Schmidt," *Arch. Rat. Mech. Anal.*, **43** (1971), 222–244.

[2], "Branching of solutions of nonlinear equations," *Rocky Mtn. J. Math.*, **3** (1973), 203–250.

[3], *Branching and stability for Nonlinear Shells*, Lecture Notes in Math., **503**, Springer-Verlag, Berlin, 1976, pp. 462–473.

[4], "Bifurcation and stability for a class of shells," *Arch. Rat. Mech. Anal.*, **63** (1977), 295–304.

D. H. Sattinger, [1], "The mathematical problem of hydrodynamic stability," *J. Math. Mech.*, **19** (1969/70), 797–817.

[2], "Bifurcation of periodic solutions of the Navier-Stokes equations," *Arch Rat. Mech. Anal.*, **41** (1971), 66–80.

[3], "Stability of bifurcating solutions by Leray–Schauder degree," *Arch. Rat. Mech. Analy.*, **43** (1971), 154–166.

[4], *Topics in Stability and Bifurcation Theory*. Springer-Verlag, Lecture Notes **309**, 1973.

[5], "Pattern formation in convective phenomena. Turbulence and Navier-Stokes Equations," (Proc. Conf., Univ. Paris-Sud, Orsay, 1975), pp. 159–173, Lecture Notes in Math., **565**, 1976; MR 57, 7663.

[6], "Transformation groups and bifurcation at multiple eigenvalues," *Bull. A.M.S.*, **79** (1973), 709–711.

[7], "Group representation theory and branch points of nonlinear functional equations," *SIAM J. Math. Anal.*, **8** (1977), 179–201. MR 55, 11297.

[8], "Group representation theory, bifurcation theory and pattern formation," *J. Functional Anal.*, **28** (1978), 58–101.

[9], "Bifurcation from rotationally invariant states," *J. Math. Phys.*, **19** (1978), 1720–1732.

[10], *Group theoretic methods in bifurcation theory*, Lecture Notes in Math., **762**, Springer-Verlag, Berlin, 1979.

J. C. Saut and B. Scheurer [1], "Remarks on a nonlinear equation arising in population genetics," *Comm. Partial Differential Eqns.*, **3** (1978), 907–931.

D. G. Schaeffer [1], "Nonuniqueness in the equilibrium shape of a confned plasma," *Comm. Partial Differential Eqns.*, **2** (1977), 587–600.

E. Schmidt [1], "Zur Theorie der linearen und nichtlinearen Integralgleichungen," *Math. Ann.*, **65** (1980), 370–379.

D. S. Schmidt [1], "Hopf's bifurcation theorem and the center theorem of Liapunov with resonance cases," *J. Math. Anal. Appl.*, **63** (1978), 354–370.

D. S. Schmidt and D. Sweet [1], "A unifying theory in determining periodic families for Hamiltonian systems at resonance," *J. Differential Eqns.*, **14** (1973), 597–609.

K. Schmidt and H. L. Smith [1], "On eigenvalue problems for nondifferentiable mappings," *J. Diff. Eqns.*, **33** (1979), 294–319.

B. V. Schmitt [1], "Determination, à l'aide d'index, des solutions périodiques de l'equation de Duffing $x'' + 2x^3 = \lambda \cos t$." *Lect. Notes. Math.*, **280**, Springer-Verlag, 1972, 330–334.

P. K. Schuster, K. Sigmund and R. Wolff [1], "Dynamical systems under constant organization. I. Topological analysis of a family of nonlinear differential equations—A model for catalytic hypercycles," *Bull. Math. Biol.*, **40** (1978), 743–770.

J. T Schwartz [1], *Nonlinear Functional Analysis*, Lecture Notes, Courant Inst. of Math., Sc., New York Univ., 1965.

L. A. Segel and J. T. Stuart [1], "On the question of the preferred mode in cellular thermal convection," *J. Fluid Mech.*, **13** (1962), 289–306. MR 25, 3659.

H. Seifert [1], "Periodische Bewegungen mechanischer Systeme," *Math. Z.*, **51** (1948), 197–216.

G. Sell [1], "Bifurcation of higher dimensional tori," *Arch. Rat. Mech. Anal.*, **69** (1979), 199–230.

F. Sergeraert [1], "Une Généralisation du théoreme des fonctions implicites de Nash," C.R., **270** (1970), Série A, 861–863.

P. R. Sethna [1], Bifurcation theory and averaging in mechanical systems. Proc. Symp. on Problems in Nonlinear Dynamics. Asilomar, Dec. 9–14, 1979.

M. J. Sewell [1], "Some mechanical examples of catastrophe theory," *Bull. Inst. Math. and Appl.*, **12** (1976), 163–172.

M. Shearer [1], "Secondary bifurcation near a double eigenvalue," *SIAM J. Math. Ana.*, **11** (1980), 365–389.

[2], "Small solutions of a non-linear equation in Banach space for a dgenerate case," *Proc. Roy. Soc. Edinburgh* **79A** (1977), 35–49. MR 57, 17428.

[3], Coincident Bifurcation of Equilibrium and Periodic Solutions of Evolution Equations. MRC Tech. Rpt., Oct. 15, 1979.

A. N. Shoshitaishvili [1], "On the bifurcation of topological type of a vector field near a singular point," *Funk. Anal. i Prilozhen.* **6** (1972), 97–98.

M. Shub [1], "Stabilité globale des systèmes dynamiques," *Asterisque* **56** (1978).

M. Shub and D. Sullivan [1], "A remark on the Lefschetz fixed point formula for differentiable maps," *Topology*, **13** (1974), 189–191.

C. L. Siegel and J. Moser [1], *Lectures on Celestial Mechanics*, Grundlehren der math. Wiss., Springer-Verlag, 1971.

J. Sijbrand [1], "Studies in nonlinear stability and bifurcation theory," Doctoral Thesis, Univ. of Utrecht, June, 1981.

L. P. Sil'nikov [1], "On the generation of a periodic motion from trajectories doubly asymptotic to an equilibrium state of saddle type," *Math. USSR-Sb.*, **6** (1968), 428–438.

[2], "A contribution to the problem of the structure of an extended neighborhood of a rough equilibrium state of a saddle-focus type," *Math. Sb.*, USSR, **10** (1970), 91–102.

G. I. Silova [1], "The decomposition of multiple limit cycles of the second kind," *Volz. Mat. Sb. Vyp.*, **16** (1973), 314–322. MR 51, #1001.

C. Simon [1], "A bound for fixed point index of an area-preserving map with applications to mechanics," *Inv. Math.*, **26** (1976), 187–200; **32** (1976), 101.

I. V. Skrypnik [1], "Solvability and properties of solutions of nonlinear elliptic equations," *Jour. Sov. Math.*, **12** (1979), 555–629.

S. Smale [1], "Differentiable dynamical systems," *Bull. Amer. Math. Soc.*, **73** (1967), 747–817.

[2], "Diffeomorphisms with many periodic points," *Differential and Combinatorial Topology*, Princeton Univ. Press, Princeton, N.J., 1965, pp. 63–80.

K. T. Smith [1], *Primer of Modern Analysis*, Bogden and Quigley, Inc., Publishers, 1971.
J. Smoller and A. Wasserman [1], "Global bifurcation of steady state solutions," *J. Diff. Eqns.*, **39** (1981), 269–290.
S. L. Sobolev [1], *Applications of Functional Analysis in Mathematical Physics*, Am. Math. Soc., Providence, R.I., 1963.
J. Sotomayor [1], "Generic one-parameter families of vector fields on two-dimensional manifolds," *Publ. Math. I.H.E.S.*, **43** (1974), 5–46.
    [2], "Structural stability and bifurcation theory," pp. 549–560 in *Dynamical Systems* (Ed. M. Peixoto), Academic Press, 1973.
M. Sova [1], "Abstract semilinear equations with small nonlinearities," *Comment. Math. Univ. Carolinae*, **12** (1971), 785–805.
I. Stakgold [1], "Branching of solutions of nonlinear equations," *SIAM Rev.*, **13** (1971), 289–332.
I. Stakgold, D. Joseph and D. Sattinger [1], Eds., *Nonlinear Problems in the Physical Sciences and Biology*, Lecture Notes in Math., **322**, Springer-Verlag, 1973.
S. Sternberg [1], *Lectures on Differential Geometry*, Prentice-Hall, Englewood Cliffs, N.J., 1964.
C. A. Stuart [1], "Some bifurcation theory for k-set contractions," *Proc. London Math. Soc.* (3), **27** (1973), 531–550; MR 48, 12178.
    [2], "The measure of non-compactness of some linear integral operators," *Proc. Roy. Soc. Edinburgh Sect.*, A, **71** (1973), 167–179; MR 48, 4845.
    [3], "Solutions of large norm of nonlinear eigenvalue problems," *Quart. J. Math., Oxford Ser.* (2) **24** (1973), 129–139; MR 47, 7112.
    [4], "Existence theory for the Hartree equation," *Arch. Rat. Mech. Anal.*, **51**, No. 1 (1973), 60–69.
    [5], "An example in nonlinear functional analysis: the Hartree equation," *J. Math. Anal. Appl.*, **49** (1975), 725–733; MR 52, 11685.
    [6], "Spectral theory of rotating chains," *Proc. Roy. Soc. Edinburgh*, A, **73** (1975), 199–214. MR 52, 9736.
    [7], "The number of solutions of boundary value problems with discontinuous nonlinearities," *Arch. Rat. Mech. Anal.*, **66** (1977), 225–235. MR 57, 6593.
    [8], "An introduction to bifurcation theory based on calculus." p. 76–135 in *Nonlinear Analysis and Mechanics*: Heriot-Watt Symp., Vol. 4 (Ed. Knops), Pitman, 1979.
    [9], "Differential equations with discontinuous nonlinearities," *Arch. Rat. Mech. Anal.*, **63** (1976/77), 59–75.
    [10], "The number of solutions of equations with discontinuous nonlinearities," *Arch. Rat. Mech. Anal.*, **66** (1977), 225–235.
J. T. Stuart [1], "Bifurcation theory in nonlinear hydrodynamic stability," p. 127–147 in *Applications of Bifurcation Theory* (Ed. Rabinowitz) Academic Press, 1977.
F. Takens [1], "Singularities of functions and vector fields," *Nieuw Arch. Wisk.*, **20** (1972), 107–130.
    [2], "Unfolding of certain singularities of vector fields: generalized Hopf bifurcations," *J. Diff. Eqns.*, **14** (1973), 476–493.
    [3], "Singularities of vector fields," *Publ. Math. I.H.E.S.*, **43** (1974), 47–100.
    [4], "Forced oscillations and bifurcations," in *Applications of Global Analysis*, Comm. 3 of the Math. Inst. Rikjsuniversitat Utrecht (1974).
    [5], "Global phenomena in bifurcations of dynamical systems with simple recurrence," *Jber. d. Dt. Math.-Verein.*, **81** (1979), 87–96.
E. Tarafdar and S. K. Teo [1], "On the existence of solutions of the equation $Lx \in Nx$ and a coincidence degree theory," *J. Austral. Math. Soc.* (Series A), **28** (1979), 139–173.
N. M. Temme (ed.) [1], *Nonlinear Analysis, I and II*, Mathematisch Centrum, Amsterdam, 1976.
R. Thom [1], "Topological methods in biology," *Topology*, **8** (1968), 313–335.
    [2], *Structural Stability and Morphogenesis*, Benjamin, Reading, Mass., 1975.
R. Thom and E. C. Zeeman [1], Bibliography on catastrophe theory, Dynamical Systems— Warwick (1974), 390–401. Lecture Notes in Math., **468**, Springer-Verlag.
J. W. Thomas and D. W. Zachmann [1], "A nonlinear three point boundary value problem," *J. Math. Anal. Appl.*, **58** (1977), 647–652, MR 55, 8455.
J. M. T. Thompson [1], "The rotationally-symmetric branching behavior of a complete spherical shell," *Proc. Kon. Nederl. Akad. Wet. Amsterdam*, **B67** (1964), 295–311.
J. M. T. Thompson and G. W. Hunt [1], *A General Theory of Elastic Stability*, Wiley, London, 1973.

G. Torelli [1], "Soluzioni periodiche dell'equazioni non lineari $u_{tt} - u_{xx} + \varepsilon F(x,t,u) = 0$." *Rend. Ist. Mat. Univ. Trieste*, **1** (1969), 123–137.

W. C. Troy [1], "Bifurcation phenomena in Fitzhugh's nerve conduction equations," *J. Math. Ana. Appl.*, **54** (1976), 678–690.

R. E. L. Turner [1], "Transversality and cone maps," *Arch. Rat. Mech. Ana.*, **58** (1975), 151–179.

[2], "Perturbation with two parameters," *Linear Alg. and Appl.*, **2** (1969), 1–11.

A. Uppal, W. H. Ray and A. B. Poore [1], "On the dynamic behavior of continuous stirred tank reactors," *Chem. Eng. Soc.*, **29** (1974), 967–985.

[2], "On the dynamic behavior of catalytic wires," *Chem. Eng. Soc.*, **29** (1974), 1330–1334.

M. M. Vainberg [1], *Variational Method for the Study of Nonlinear Operators*, Holden-Day, Inc., San Francisco, 1964.

M. M. Vainberg and V. A. Tregonin [1], "The methods of Liapunov and Schmidt in the theory of nonlinear equations and their further development," *Russian Math. Surveys*, **17** (1962), 1–60.

[2], *Theory of Branching of Solutions of Nonlinear Equations*, Noordhoff, 1974.

A. Vanderbauwhede [1], *Local Bifurcation Theory and Symmetry*, Habilitation, Ghent, 1980.

[2], "Alternative problems and symmetry," *J. Math. Anal. Appl.*, **62** (1978), 483–494.

[3], "Alternative problems and invariant subspaces," *J. Math. Anal. Appl.*, **63** (1978), 1–8.

[4], "Some remarks on symmetry and bifurcation," *Simon Stevin*, **51** (1978), 173–194.

[5], "Generic and nongeneric bifurcation for the von Karman equations," *J. Math. Anal. Appl.*, **66** (1978), 550–573.

[6], "Generic bifurcation and symmetry, with an application to the von Karman equations," *Proc. Roy. Soc. Edinburgh*, **81A** (1978), 211–235.

[7], Bifurcation near multi-parameter families of solutions, Proc. Equadiff 78 Conference, Firenze, 1978, p. 9–23.

[8], "An abstract setting for the study of the Hopf bifurcation," *Nonlin. Anal., Th. Meth. Appl.*, **4** (1980), 547–566.

[9], Hopf furcation at non-semisimple eigenvalues, *Univ. Gent Inst. Theor. Mech.*, Preprint (1986).

S. A. van Gils [1], A note on abelian integrals and bifurcation theory, *J. Diff. Eq.*, **59** (1985), 437–439.

J. Vegas [2], Linear stability of bifurcating branches of solutions, *Equadiff 82*, Lectures Notes in Math. **1017** (1983), (Eds. Knobloch and Schmitt) 625–631.

V. M. Volosov [1], "Averaging in systems of ordinary differential equations," *Russian Math. Surveys*, **17** (1962), 1–126.

N. T. Waller [1], Bifurcation Theory with Applications to Chemical Reaction Equations, Dissertation, Mich. State Univ., 1976.

Y. H. Wan [1], "Computation of stability condition for the Hopf bifurcation of diffeomorphisms on R^2," *SIAM J. Appl. Math.*, **34** (1978), 167–175.

[2], "Bifurcation into invariant tori at points of resonance," *Arch. Rat. Mech. Anal.*, **68** (1978), 343–357.

L. T. Watson, T. Y. Li and C. Y. Wang [1], "The elliptic porous slider—a homotopy method," *J. Appl. Mech.*, **45** (1978), 435–436.

G. Wasserman [1], *Stability of Unfoldings*, Lecture Notes in Math., **393**, Springer-Verlag, Berlin, 1974.

D. Weistrich [1], "Bifurcation at eigenvalues of odd multiplicity," *Proc. Am. Math. Soc.*, **41** (1973), 609–614.

[2], "Banach space bifurcation theory," *Trans. Amer. Math. Soc.*, **171** (1972), 135–156.

[3], "Bifurcation of operator equations with unbounded linearized parts," *Pac. J. Math.*, **57** (1975), 611–618.

[4], "Periodic solutions of second order Lagrangian systems," *Duke Math. J.*, **41** (1974), 405–411.

H. F. Weinberger [1], "On the stability of bifurcating solutions," p. 219–233 in *Nonlinear Analysis*, (eds. Cesari, Kannan, Weinberger), Academic Press, 1978.

[2], "The stability of solutions bifurcating from steady or penodic solutions," Univ. Fla. Symp. Dynamical Systems, Academic Press, 1977.

[3], "Variational methods for eigenvalue approximation," Reg. Conf. Ser. in Appl. Math., **15**, SIAM, Philadelphia, 1974.

A. Weinstein [1], "Normal modes for non-linear Hamiltonian systems," *Inv. Math.*, **20** (1973), 47–57.

[2], "Lagrangian submanifolds and Hamiltonian systems," *Ann. Math.*, **98** (1973), 377–410.

G. T. Whyburn [1], *Topological Analysis*, Princeton Univ. Press, Princeton, N.J., 1968.

J. H. Wilkinson [1], *The Algebraic Eigenvalue Problem*, Oxford, 1965.

S. Williams [1], "A connection between the Cesari and Leray—Schauder methods," *Mich. Math. J.*, **15** (1968), 441–448.

J. H. Wolkowisky [1], "Existence of buckled states of circular plates," *Comm. Pure Appl. Math.*, **20** (1967), 549–560. MR 35, 3952.

A. E. R. Woodcock and T. Poston [1], *A Geometric Study of the Elementary Castastrophes*, Lecture Notes in Math., **373**, 1974.

D. Yen and J. Bazer [1], "The Riemann matrix of $(2 + 1)$-dimensional symmetric-hyperbolic systems," *Comm. Pure Appl. Math.*, **20** (1967), 329–363.

K. Yoshida [1], *Functional Analysis*, 2nd Ed., Springer-Verlag, New York, 1968.

D. W. Zachmann [1], "Existence of secondary solutions to a generalized Taylor problem," *Applicable Anal.*, **4** (1974), 145–160.

[2], "Branching solutions of equations containing several parameters," *SIAM J. Math. Anal.*, **5** (1974), 898–907; MR 51, 1521.

E. Zeidler [1], "The Ljusternik—Schnirelman theory for indefinite and not necessary odd nonlinear operators and its applications," *Nonlinear Analy.*, **4** (1980), 451–489.

[2], "Bifurcation theory and permanent waves," p. 203–223 in *Applications of Bifurcation Theory* (ed. Rabinowitz), Academic Press, 1977.

E. J. Zehnder [1], "Generalized implicit function theorems with applications to some small divisor problems, I, II," *Comm. Pure Appl. Math.*, **28** (1975), 91–140; **29** (1976), 49–111.

H. Zoladek [1], On the versality of certain family of vector fields on the plane, *Math. USSR Sb.*, **48** (1984), 463–492 (in Russian).

[2], Bifurcations of certain family of planar vector fields tangent to axes, *J. Diff. Eq.*, **67** (1987), 1–55.

V. I. Zubov [1], *Methods of A. M. Liapunov and Their Applications*, Noordhoff, 1964.

Index

Grundlehren der mathematischen Wissenschaften

A Series of Comprehensive Studies in Mathematics
A Selection